QUANTUM
TRANSPORT THEORY

Jørgen Rammer

CRC Press
Taylor & Francis Group
Boca Raton London New York

CRC Press is an imprint of the
Taylor & Francis Group, an **informa** business

To the memory of my father

Hans Christian Rammer

Hardcover edition first published in 1998 by Perseus Books.
Paperback edition first published in 2004 by Westview Press

Published 2018 by CRC Press
Taylor & Francis Group
6000 Broken Sound Parkway NW, Suite 300
Boca Raton, FL 33487-2742

CRC Press is an imprint of the Taylor & Francis Group, an informa business

Visit the Taylor & Francis Web site at
http://www.taylorandfrancis.com

and the CRC Press Web site at
http://www.crcpress.com

Cover design by Lynne Reed
This book was typeset by the author using LaTex

A Cataloging-in-Publication data record for this book is available from the Library of Congress.

ISBN 13: 978-0-8133-4284-9 (pbk)

Frontiers in Physics

David Pines, Editor

Volumes of the Series published from 1961 to 1973 are not officially numbered. The parenthetical numbers shown are designed to aid librarians and bibliographers to check the completeness of their holdings.

Titles published in this series prior to 1987 appear under either the W. A. Benjamin or the Benjamin/Cummings Imprint; titles published since 1986 appear under the Westview Press imprint.

15. E. J. Squires Complex Angular Momenta and Particle Physics: A Lecture Note and Reprint Volume, 1963

16. H. L. Frisch The Equilibrium Theory of Classical Fluids: A Lecture J. L. Lebowitz Note and Reprint Volume, 1964

17. M. Gell-Mann The Eightfold Way (A Review—with a Collection of Y. Ne'eman Reprints), 1964

18. M. Jacob Strong-Interaction Physics: A Lecture Note Volume, 1964 G. F. Chew

19. P. Nozières Theory of Interacting Fermi Systems, 1964

20. J. R. Schrieffer Theory of Superconductivity, 1964 (revised 3rd printing, 1983)

21. N. Bloembergen Nonlinear Optics: A Lecture Note and Reprint Volume, 1965

22. R. Brout Phase Transitions, 1965

23. I. M. Khalatnikov An Introduction to the Theory of Superfluidity, 1965

24. P. G. deGennes Superconductivity of Metals and Alloys, 1966

25. W. A. Harrison Pseudopotentials in the Theory of Metals, 1966

26. V. Barger Phenomenological Theories of High Energy Scattering: D. Cline An Experimental Evaluation, 1967

27. P. Choquàrd The Anharmonic Crystal, 1967

28. T. Loucks Augmented Plane Wave Method: A Guide to Performing. Electronic Structure Calculations—A Lecture Note and Reprint Volume, 1967

29. Y. Ne'eman Algebraic Theory of Particle Physics: Hadron Dynamics in Terms of Unitary Spin Currents, 1967

30. S. L.Adler Current Algebras and Applications to Particle Physics, R. F. Dashen 1968

31. A. B. Migdal Nuclear Theory: The Quasiparticle Method, 1968

32. J. J. J. Kokkede The Quark Model, 1969

33. A. B. Migdal Approximation Methods in Quantum Mechanics, 1969 V. Krainov

34. R. Z. Sagdeev Nonlinear Plasma Theory, 1969 A. A. Galeev

35. J. Schwinger Quantum Kinematics and Dynamics, 1970

36. R. P. Feynman Statistical Mechanics: A Set of Lectures, 1972

37. R. P. Feynman Photon-Hadron Interactions, 1972

38. E. R. Caianiello Combinatorics and Renormalization in Quantum Field Theory, 1973

39. G. B. Field The Redshift Controversy, 1973 H. Arp J. N. Bahcall

40. D. Horn Hadron Physics at Very High Energies, 1973 F. Zachariasen

41. S. Ichimaru Basic Principles of Plasma Physics: A Statistical Approach, 1973 (2nd printing, with revisions, 1980)

42. G. E. Pake The Physical Principles of Electron Paramagnetic T. L. Estle Resonance, 2nd Edition, completely revised, enlarged, and reset, 1973 [cf. (9)—1st edition]

Volumes published from 1974 onward are being numbered as an integral part of the bibliography.

43. R. C. Davidson Theory of Nonneutral Plasmas, 1974
44. S. Doniach Green's Functions for Solid State Physicists, 1974 E. H. Sondheimer
45. P. H. Frampton Dual Resonance Models, 1974
46. S. K. Ma Modern Theory of Critical Phenomena, 1976
47. D. Forster Hydrodynamic Fluctuations, Broken Symmetry, and Correlation Functions, 1975
48. A. B. Migdal Qualitative Methods in Quantum Theory, 1977
49. S. W. Lovesey Condensed Matter Physics: Dynamic Correlations, 1980
50. L. D. Faddeev Gauge Fields: Introduction to Quantum Theory, 1980 A. A. Slavnov
51. P. Ramond Field Theory: A Modern Primer, 1981 [cf. 74—2nd ed.]
52. R. A. Broglia Heavy Ion Reactions: Lecture Notes Vol. I: Elastic and A. Winther Inelastic Reactions, 1981
53. R. A. Broglia Heavy Ion Reactions: Lecture Notes Vol. II, 1990 A. Winther
54. H. Georgi Lie Algebras in Particle Physics: From Isospin to Unified Theories, 1982
55. P. W. Anderson Basic Notions of Condensed Matter Physics, 1983
56. C. Quigg Gauge Theories of the Strong, Weak, and Electromagnetic Interactions, 1983
57. S. I. Pekar Crystal Optics and Additional Light Waves, 1983
58. S. J. Gates Superspace *or* One Thousand and One Lessons in M. T. Grisaru Supersymmetry, 1983 M. Rocek W. Siegel
59. R. N. Cahn Semi-Simple Lie Algebras and Their Representations, 1984
60. G. G. Ross Grand Unified Theories, 1984
61. S. W. Lovesey Condensed Matter Physics: Dynamic Correlations, 2nd Edition, 1986
62. P. H. Frampton Gauge Field Theories, 1986
63. J. I. Katz High Energy Astrophysics, 1987
64. T. J. Ferbel Experimental Techniques in High Energy Physics, 1987
65. T. Appelquist Modern Kaluza-Klein Theories, 1987 A. Chodos P. G. O. Freund
66. G. Parisi Statistical Field Theory, 1988
67. R. C. Richardson Techniques in Low-Temperature Condensed Matter E. N. Smith Physics, 1988
68. J. W. Negele Quantum Many-Particle Systems, 1987 H. Orland
69. E. W. Kolb The Early Universe, 1990 M. S. Turner
70. E. W. Kolb The Early Universe: Reprints, 1988 M. S. Turner
71. V. Barger Collider Physics, 1987 R. J. N. Phillips
72. T. Tajima Computational Plasma Physics, 1989 (updated 2004)

73. W. Kruer The Physics of Laser Plasma Interactions, 1988 (updated 2003)

74. P. Ramond Field Theory: A Modern Primer, 2nd edition, 1989 [cf. 51—1st edition]

75. B. F. Hatfield Quantum Field Theory of Point Particles and Strings, 1989

76. P. Sokolsky Introduction to Ultrahigh Energy Cosmic Ray Physics, 1989 (updated 2004)

77. R. Field Applications of Perturbative QCD, 198978. G. Baym Heavy Ion Collisions, 1991 L. McLerran79. H. Frannfelder Physucs if Biomolecules, 1991 P. Debrunner

80. J. F. Gunion The Higgs Hunter's Guide, 1990 H. E. Haber G. Kane S. Dawson

81. R. C. Davidson Physics of Nonneutral Plasmas, 1990

82. E. Fradkin Field Theories of Condensed Matter Systems, 1991

83. L. D. Faddeev Gauge Fields, 1990 A. A. Slavnov

84. R. Broglia Heavy Ion Reactions, Parts I and II, 1990 A. Winther

85. N. Goldenfeld Lectures on Phase Transitions and the Renormalization Group, 1992

86. R. D. Hazeltine Plasma Confinement, 1992 J. D. Meiss

87. S. Ichimaru Statistical Plasma Physics, Volume I: Basic Principles, 1992 (updated 2004)

88. S. Ichimaru Statistical Plasma Physics, Volume II: Condensed Plasmas, 1994 (updated 2004)

89. G. Grüner Density Waves in Solids, 1994

90. S. Safran Statistical Thermodynamics of Surfaces, Interfaces, and Membranes, 1994 (updated 2003)

91. B. d'Espagnat Veiled Reality: An Analysis of Present Day Quantum Mechanical Concepts, 1994 (updated 2003)

92. J. Bahcall Solar Neutrinos: The First Thirty Years, 1994 (updated 2002) R. Davis, Jr. P. Parker A. Smirnov R. Ulrich

93. R. Feynman Feynman Lectures on Gravitation, 1995 (reprinted 2003) F. Morinigo W. Wagner

94. M. Peskin An Introduction to Quantum Field Theory, 1995 D. Schroeder

95. R. Feynman Feynman Lectures on Computation, 1996 (reprinted 1999)

96. M. Brack Semiclassical Physics, 1997 (updated 2003) R. Bhaduri

97. D. Cline Weak Neutral Currents, 1997 (reprinted 2004)

98. T. Tajima Plasma Astrophysics, 1997 (updated 2002) K. Shibata

99. J. Rammer Quantum Transport Theory, 1998

100. R. Hazeltine The Frameworkof Plasma Physics, 1998 (updated 2004) F. Waelbroeck

101. P. Ramond Journeys Beyond the Standard Model, 1999 (updated 2004)

102. Y. Nutku Conformal Field Theory: New Non-PerturbativeC. Saclioglu Methods in String and Field Theory, 2000 (reprinted 2004) T. Turgut

103. P. Philips Advanced Solid State Physics, 2003

Editor's Foreword

The problem of communicating in a coherent fashion recent developments in the most exciting and active fields of physics continues to be with us. The enormous growth in the number of physicists has tended to make the familiar channels of communication considerably less effective. It has become increasingly difficult for experts in a given field to keep up with the current literature; the novice can only be confused. What is needed is both a consistent account of a field and the presentation of a definite "point of view" concerning it. Formal monographs cannot meet such a need in a rapidly developing field, while the review article seems to have fallen into disfavor. Indeed, it would seem that the people who are most actively engaged in developing a given field are the people least likely to write at length about it.

Frontiers in Physics was conceived in 1961 in an effort to improve the situation in several ways. Leading physicists frequently give a series of lectures, a graduate seminar, or a graduate course in their special fields of interest. Such lectures serve to summarize the present status of a rapidly developing field and may well constitute the only coherent account available at the time. One of the principal purposes of the *Frontiers in Physics* series is to make notes on such lectures available to the wider physics community.

As *Frontiers in Physics* has evolved, a second category of book, the informal text/monograph, an intermediate step between lecture notes and formal text or monographs, has played an increasingly important role in the series. In an informal text or monograph an author has reworked his or her lecture notes to the point at which the manuscript represents a coherent summation of a newly developed field, complete with references and problems, suitable for either classroom teaching or individual study.

Quantum Transport Theory is just such a volume. The author, who has made significant contributions to the scientific literature on this topic, provides for the non-specialist a self-contained account of the developments in quantum transport theory which have led to our present understanding of transport in semiconductors and normal metals. His careful pedagogical presentation makes his book a useful text or reference volume for both undergraduate and graduate courses, while his chapters on localization and weak localization provide graduate students and experienced researchers alike with an excellent introduction to these frontier topics in condensed matter. It gives me great pleasure to welcome Jørgen Rammer to *Frontiers in Physics*.

David Pines
Urbana, Illinois
July, 1998

Contents

Preface

This book is an introduction to transport theory, the kinetic equation approach as well as linear response theory. The main physical applications are to electronic transport in semiconductors and metals in the normal state. However, the techniques presented are general, and the book should therefore be of interest to a wider audience than students of condensed matter physics and physicists in general, as for example electrical engineers.

A purpose of the book is also to show the utility of Feynman diagrams in nonequilibrium quantum statistical mechanics. The presentation attempts to emphasize the aspects of quantum transport theory in the simplest and most illustrative fashion. Since emphasis has been put on an approach appealing to physical intuition, the real-time description of nonequilibrium quantum statistical mechanics has been adopted, and the diagrammatic technique for systems out of equilibrium is developed systematically.

The physical systems studied are complicated many-body systems. However, for the phenomena of interest a mean-field description can be adopted, and an effective single-particle description applies. Although the description of fermions in terms of quantum fields is not introduced, the presented single-particle approach can, from a methodological point of view, be considered an application of field theoretic methods in a quantum mechanical context.

The understanding of transport in disordered systems has matured to the extent that it constitutes an important part of condensed matter physics, and a comprehensive presentation of transport in disordered systems is given. A complete allocation of the credit for the recent progress in the understanding of transport in disordered systems has not been attempted. However, the references in particular the review articles, should make it possible for the interested reader to trace this information.

Chapter 1 starts with an introduction to quantum mechanics based on Feynman's space-time approach in terms of path integrals. In section 1.2 quantum mechanics is then reformulated in the equivalent operator calculus, and elementary representation and transformation theory is presented following Dirac. A reader with a basic knowledge of quantum mechanics, as presented in sections 1.1 and 1.2, should be able to understand the content in the rest of this book. The reader only interested in the main topics of the book, though, can in fact skip the rest of chapter 1 (except perhaps for consulting section 1.8, where the notion of the density matrix is introduced). For the reader not satisfied with the intuitive

and deductive introduction to quantum mechanics presented in sections 1.1 and
1.2, the rest of chapter 1 shows, following Schwinger, how the kinematic structure
of quantum mechanics follows from a few experimental facts. Chapter 1 ends with
a presentation of the interpretation of quantum mechanics in terms of consistent
histories.

In chapter 2 propagators and their role in perturbation theory are presented,
and Feynman diagrams are introduced. The analytic properties of propagators,
and the consequences of the discrete symmetries, space inversion and time reversal, are discussed. In chapter 3 propagation in a random potential is considered.
The diagrammatic impurity-average technique is presented, and the self-energy
and topological notions of skeleton diagrams are introduced. Order-of-magnitude
estimation of a diagram based on its topological appearance is presented, thereby
paving the way for identifying the small parameter of the problem, and a systematic treatment of transport in disordered conductors. In chapter 4 the general
quantum kinetic equation for a particle in a random potential is analyzed in terms
of diagrams. The Wigner function is introduced, and the criterion for the validity
of the Boltzmann transport theory is established from the quantum mechanical
description. The physical system of main interest is the conduction electrons in a
metal, and in chapter 5 an account of the implications of the fermionic nature of
the electrons is given and the Boltzmann theory for an electron gas in a random
potential presented. In chapter 6 the nonequilibrium quantum statistical mechanics of a particle interacting with an environment of oscillators is considered and
the Feynman diagrammatics established.

In chapter 7 linear response theory is presented. The consequences of causality and time-reversal symmetry for the response functions are explored, and the
stability of the thermal equilibrium state is established. The important fluctuation-
dissipation theorem is derived, and used to study current and electric field fluctuations. The measurability of correlation functions is demonstrated by considering
neutron scattering off a substance. In chapter 8 the linear response of a disordered
conductor is studied using the diagrammatic technique. The Boltzmann theory
and Brownian motion are identified in terms of diagrams. The quantum kinetic
equation that allows a treatment of transport beyond the weak-disorder limit is
constructed. The implications of particle conservation and time-reversal symmetry
are discussed in terms of diagrams.

In chapter 9 the phenomenon of localization due to impurity scattering is considered. The scaling theory of localization is presented. The quantum interference process responsible for localization of a particle is identified, and the metal-
insulator transition discussed using the self-consistent theory of localization.

In chapter 10 the interaction between the electrons and ions in a metal is considered. The collective excitations are identified, and electron-phonon and electron-
electron interaction discussed.

In the final chapter quantum transport in weakly disordered systems is presented. The small but important corrections to the Boltzmann results, weak-
localization effects, are discussed in detail. The destruction of the phase coherence
of the electronic wave function due to electron-phonon and electron-electron inter-

action is considered. The anomalous magnetoresistance and the Aharonov-Bohm effect are explored in detail. The weak antilocalization effect due to spin-orbit scattering is investigated as well as the effect of spin-flip scattering. Finally mesoscopic fluctuations are considered, and conductance fluctuations discussed in detail.

The book is intended to be sufficiently broad to serve as a text for a one- or two-semester graduate course on transport theory. Moreover, in view of the simple single-particle approach, the book should be useful in undergraduate teaching as well. It is also hoped that the book can serve as a useful reference book for courses on nonequilibrium statistical mechanics, physics of disordered systems, and quantum mechanics courses in general. The book is self-contained to the extent that it should be useful for students with only elementary knowledge of quantum and statistical mechanics to read it on their own. A number of exercises with solutions (indeed always the case when the result is later used in the main text) has been provided in order to aid self-instruction.

I would like to thank Dr. Dierk Bormann, Professor Ulrich Eckern, and Professor Paul Muzikar for correcting misprints and for helpful suggestions. I am grateful to Tekn. Lic. Staffan Grundberg for providing figures and Latex help. I am especially indebted to Dr. Andrei L. Shelankov for critical reading of the manuscript, and for numerous helpful suggestions.

Umeå, Sweden   Jørgen Rammer
June 1998

The paperback edition has given the opportunity to correct misprints in the original edition and make a few additions.

Umeå, Sweden   Jørgen Rammer
May 2004

Chapter 1

Quantum Mechanics

In this chapter an introduction to quantum mechanics is given. After a brief description of quantum mechanics, we follow Feynman in section 1.1, and deduce the content of quantum mechanics from the few basic principles using the path integral formulation. Having thus established the notions of quantum mechanics, we introduce in section 1.2 the formulation in terms of the operator calculus. The reader with this basic knowledge of quantum mechanics has the necessary tools for understanding the content of the rest of the book. The rest of this chapter is thus not necessary reading (except perhaps for consulting section 1.8 where the statistical operator and the density matrix are introduced). The main part of chapter 1 is thus for the reader not satisfied with the intuitive and deductive introduction to quantum mechanics presented in sections 1.1 and 1.2. In section 1.3 and 1.4 we follow Schwinger, and induce from a few experimental facts the kinematic structure of quantum mechanics, i.e., the notions needed for describing physical systems. In section 1.5 we show how nonrelativistic quantum kinematics emerges from the general theory. In section 1.6 we discuss how physical quantities are related to symmetries through unitary transformations, and consider translations in detail. The presentation in section 1.7 of quantum dynamics, i.e., how the properties of a physical system change in time, stresses the unitary equivalence aspect, and in section 1.8 we introduce the statistical operator and the density matrix. The last section gives an introduction to the consistent history interpretation of quantum mechanics.

Experimental knowledge of the behavior of physical systems shows that they can be described as consisting of point particles. In the quantum mechanical description, the particles are at a given moment in time t completely described by a probability amplitude function, a complex function $\psi(\mathbf{x}_1, \mathbf{x}_2, ..; t)$ of the particle positions (and possible internal degrees of freedom). The probability amplitude, which is also referred to as the wave function, has the significance that its absolute square $|\psi(\mathbf{x}_1, \mathbf{x}_2, ..; t)|^2$ determines the probability of finding the particles at the specified positions at the time in question. The dynamics of the particles is determined by the Schrödinger equation, which specifies the rate of change in time of the probability amplitude function in terms of a linear operator H

$$i\hbar \frac{\partial \psi(\mathbf{x}_1, \mathbf{x}_2, ..; t)}{\partial t} \; = \; H\psi(\mathbf{x}_1, \mathbf{x}_2, ..; t) \, . \tag{1.1}$$

The operator H is called the Hamiltonian and is ultimately determined by experimental knowledge. The symbol $\hbar$ is a constant of nature, and as such also must be empirically determined.

The use of position coordinates to describe the state of the system might be said to have a preferred practical status since experiments are immediately described in space and time. However, to specify the laws of nature, i.e., the Hamiltonian, we need to invoke other physical properties of a system such as momentum. By employing the dynamics of a system, as given by the Hamiltonian, we can obtain an amplitude description in terms of the momentum degree of freedom. The specification of the momentum amplitudes in terms of the position amplitudes can be obtained by considering the method of detection of the momentum. For example, employing a time-of-flight technique to measure the momentum of a free particle demonstrates that the momentum probability amplitude function is given by a Fourier transformation of the position probability amplitude function.[1]

After this brief description we turn to a detailed discussion of quantum mechanics. In order to elucidate the conceptual structure of quantum mechanics we begin with a presentation of Feynman's space-time approach since it offers the most intuitive formulation of quantum mechanics.

1.1   The Principles of Quantum Mechanics

Instead of formulating quantum dynamics in terms of a differential equation, i.e., the Schrödinger equation, we can obtain an integral characterization following Feynman [1]. For clarity of presentation, we first postulate the three basic principles of quantum mechanics. The content and consequences of the basic principles are then illustrated by deriving the Schrödinger equation for the example of a single particle.

The quantitative description of a physical system is provided by the Born rule or

Kinematic Principle:
To any possible event is attributed a complex number ψ called the probability amplitude. The probability P of the event taking place is given by the absolute square of the amplitude

$$P \; = \; |\psi|^2. \tag{1.2}$$

In quantum mechanics probability is thus attributed as a fundamental feature of the physical world.

In the description of a single particle, the probability amplitude associated with the event of the particle being at position $\mathbf{x}$ at time t (space-time point $(\mathbf{x}, t)$ for

---

[1]As demonstrated in section 1.1.2.

short) is denoted $\psi(\mathbf{x}, t)$, and by taking the absolute square of the amplitude we get the probability density $P(\mathbf{x}, t) = |\psi(\mathbf{x}, t)|^2 = \psi(\mathbf{x}, t)\, \psi^*(\mathbf{x}, t)$ for the particle to be at position $\mathbf{x}$ at time t ($*$ denotes complex conjugation).[2]

The probability of two events at different times, events 1 and 2, to take place is specified by the joint probability $P(2, 1)$. The joint probability can be expressed in terms of the conditional probability $P(2\,;1)$, relating the probability P_1 for event 1 to take place and the joint probability, $P(2, 1) = P(2\,;1)P_1$. We can then introduce the conditional probability amplitude, $K(2\,;1)$, *the amplitude for event 2 to take place given that event 1 has taken place*, according to $|K(2\,;1)|^2 \equiv P(2\,;1)$. The amplitude for the sequence of events 1 and 2 is then $K(2\,;1)\,\psi_1$, up to an irrelevant phase factor.

So far only the *probability amplitude* concept has been introduced and made operational by its physical interpretation. To get a handle on the use of the concept we now specify the two basic rules for ascribing (conditional) amplitudes to events.

The temporal ordering of events allows us to consider an event as the termination of a sequence of events. The assignment of the amplitude to an event is in this case specified by the

Multiplication Principle:

For a sequence of successive individual events, the amplitude for this sequence of events is the product of (conditional) amplitudes for the consecutive events

$$\psi = \prod_i \psi_i \ , \tag{1.3}$$

and the associated probability for the sequence of events is therefore

$$P = |\psi|^2 = \prod_i |\psi_i|^2. \tag{1.4}$$

Holding two pieces of polarizer between a light source and your eye, you can by rotating the two pieces relative to each other gradually block out the light. Each particle, a photon, in the stream of light, is assigned an amplitude for passing the first polarizer and an amplitude for then passing the second. The amplitude for a photon to reach your eye is proportional to the product of amplitudes for passing the two polarizers (and the amplitude for the photon, after passing the second polarizer, to reach your eye). The multiplication rule reflects the fact that time is just a parameter in the theory, and the rule is motivated by the preceding example, or equivalently, by the following analogous consideration. Consider for example the situation where a particle in order to be detected (to arrive at a point in space where detection takes place) has to pass a screen which might or might not have holes for the particle to pass through. The amplitude for the event of arrival at the detector depends on whether the particle can actually pass the screen or

---

[2]The principle of special relativity requires that particles can be created and annihilated. Including this feature in the theory would lead to a reformulation of the dynamics in terms of quantum fields, and to a generalization of the probability amplitude to a set of amplitudes: one probability amplitude for each possible configuration of particle species. However, it would not change the present discussion in principle.

not (the screen having holes or not!). A non-or completely transparent screen has probabilities zero or one for the events of passing, and the probability for reaching a detector must be zero or unaffected. Multiplication of amplitudes for consecutive events is the only general way to implement such features.

The true core of quantum mechanics is the rule for assigning the amplitude to an event that can be effected in indistinguishable ways. More precisely, if an event is realized under conditions without distinction to the alternative ways it can be effected, we must for asserting its probability use the[3]

Superposition Principle:

If an event can be effected in indistinguishable ways, the amplitude for the event is the sum of the individual amplitudes ψ_i for the alternative ways the event can be effected

$$\psi = \sum_i \psi_i \ , \tag{1.5}$$

and the associated probability for the event is therefore

$$P = |\psi|^2 = \left| \sum_i \psi_i \right|^2 = \sum_i |\psi_i|^2 + \sum_{i \neq j} \psi_i \psi_j^*. \tag{1.6}$$

The superposition principle is enforced by the characteristic interference phenomena exhibited in nature. In order to illustrate the physical implication of the superposition principle we consider the double slit, shown in figure 1.1, where a particle after being emitted from a source can arrive at space points (the *detector* screen) in only two ways, viz. through two holes in an otherwise impenetrable wall.

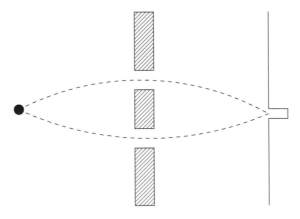

Figure 1.1 A particle passing a double slit has two alternatives for reaching a point on the screen.

---

[3]If the amplitudes in the two rules were interpreted as probabilities, the stated rules are just the rules of probability calculus; however, the rules are for probability *amplitudes*!

The two ways of passing the double slit are indistinguishable if there is no signature that in principle distinguishes which one of the holes the particle passed through. Arrival through either hole is thus an indistinguishable alternative, and the individual amplitudes for each of the two ways of possible arrival should be added before squaring to get the probability for the event of arrival, thereby producing the characteristic quantum interference arrival pattern in accordance with the last term in eq.(1.6), which in this case equals $2\Re e\psi_1\,\psi_2^* = 2|\psi_1\,\psi_2|\cos(\phi_1 - \phi_2)$. Since the phases depend on the alternative paths in question, this interference term oscillates as a function of the position on the detector screen. Opening up the second hole through which the particle can reach the detector screen thus does not for all detector points increase the number of arriving particles, it can in fact decrease the number, even to zero!

If on the other hand, say, spin-polarized electrons impinge on the double slit, and a magnetic atom in each trial is prepared in a proper metastable state at one of the holes such that the passage of the electron through that hole will flip the atom spin, the interference pattern will not appear (we assume total efficiency of the flip-interaction; otherwise there will be only partial smearing of the interference pattern). The interference feature in the arrival pattern is now absent because we have distinguishable conditions, since each time a definite alternative is realized. Either the atom spin is flipped or not, and the electron has definitely passed through one or the other of the two holes. Each event of detection is the succession of first the event of passing through one of the holes, and then the event of arrival at a detector.[4] The possible event of the particle reaching hole 1 has ascribed, say, the probability amplitude ψ_1, and according to the multiplication principle the arrival events occurring with passage through hole 1 happens with a probability P_1 proportional to $|\psi_1|^2$, whereas the events where the particle passes through hole 2 occur with a frequency P_2 proportional to $|\psi_2|^2$. The arrival pattern is obtained by the addition of the two frequencies of arrival, $P = P_1 + P_2$, and the state of affairs is in accordance with probability calculus where each event sequence, arriving through hole 1 or hole 2, is assigned a probability.[5]

Since time is just a parameter in quantum mechanics, i.e., not a property of the system, and in practice the one we read off our watches, we readily introduce the time dependence of the amplitudes.[6] For example, at times t' and t the particle is somewhere in space and according to the kinematic principle has associated the amplitude functions $\psi(\mathbf{x}', t')$ and $\psi(\mathbf{x}, t)$. The application of the two rules specifies the relationship between the conditional probability amplitude $K(\mathbf{x}, t; \mathbf{x}', t')$,[7] the

---

[4]Of course, initially the particle is emitted from a source with an associated amplitude ψ_S. So the amplitude for the alternative to arrive at a detector position D passing through hole 1 and emitted at S is $K(D; 1)\,K(1; S)\,\psi_S$.

[5]The state of affairs is of course independent of whether the spin state of the magnetic atom is actually inspected or not, i.e., whether *we* actually have knowledge of the distinguishability of the alternatives or not.

[6]Theories of quantum gravity consider quantized spacetime metric.

[7]In order for this conditional probability amplitude to be meaningful, it is of course vital that the position of the particle alone constitute a complete description. That indeed a description in terms solely of the position, i.e., the wave function, is complete is here taken for granted. The

probability amplitude for finding the particle at position $\mathbf{x}$ at time t *given* it was at position $\mathbf{x}'$ at time t', and the amplitude functions at different times

$$\psi(\mathbf{x}, t) \;=\; \int\! d\mathbf{x}'\, K(\mathbf{x}, t; \mathbf{x}', t')\, \psi(\mathbf{x}', t') \,. \tag{1.7}$$

The amplitude to arrive at the position $\mathbf{x}$ at time t via the space-time point $(\mathbf{x}', t')$ equals, according to the multiplication principle, the product of amplitudes $K(\mathbf{x}, t; \mathbf{x}', t')\, \psi(\mathbf{x}', t')$ for the two events in sequence: arriving at $(\mathbf{x}', t')$ followed by the event of arriving at $(\mathbf{x}, t)$ via $(\mathbf{x}', t')$. Since the particle can arrive at $(\mathbf{x}, t)$ coming via any alternative point $\mathbf{x}'$ at time t' we have to sum over all the alternatives according to the superposition principle (for a continuum we have to sum over all the alternative volumes $\Delta\mathbf{x}'$ the particle can arrive via; i.e., we must integrate). The dynamics of a particle is thus specified in terms of the kernel of the integral equation, eq.(1.7), the conditional probability amplitude $K(\mathbf{x}, t; \mathbf{x}', t')$, which is also referred to as the propagator, since it propagates the wave function. Once the propagator is known and the wave function at a given moment in time specified, the probabilities for the whereabouts of the particle in the future and the past can according to eq.(1.7) be predicted.[8]

We can obtain an integral expression for the propagator by the following consideration. At any intermediate moment in time the particle has associated events to be at any position. We illustrate each such possible sequence of events pictorially by dots in a space-time plot, as shown in figure 1.2.

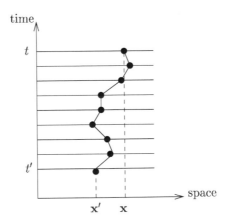

Figure 1.2 A set of consecutive space-time points representing possible events and their associated straight-line path.

---

[8]In the above discussion we assumed $t > t'$, but *retrodiction*, the *whereabouts* of the particle in the past, i.e., the wave function for an isolated particle in the past, can be obtained from the result of exercise 1.3.

Increasing the number of intermediate time slices between t' and t, each such alternative space-time event sequence corresponds to a path in space, $\mathbf{x}_t$, and has an associated probability (density) amplitude $A_{\mathbf{x},t;\mathbf{x}',t'}[\mathbf{x}_t]$ (we suppress the parametric dependence on the fixed starting and end points in the following). According to the superposition principle the conditional amplitude for the particle is the sum of the amplitudes for all the alternative paths connecting the space-time points $(\mathbf{x}',t')$ and $(\mathbf{x},t)$[9]

$$K(\mathbf{x},t;\mathbf{x}',t') \;=\; \sum_{\mathbf{x}_t} A[\mathbf{x}_t] \,. \tag{1.8}$$

A path $\mathbf{x}_t$ can, for example, be represented by its successive straight-line paths, and each such subpath $\mathbf{x}_t^{(i)}$ has associated a conditional amplitude $A[\mathbf{x}_t^{(i)}]$. According to its successive straight-line construction, a path $\mathbf{x}_t$ is the sequence or *sum* of its subpaths $\mathbf{x}_t^{(i)}$, " $\mathbf{x}_t \equiv \sum_i \mathbf{x}_t^{(i)}$ ", and the amplitude for the path $\mathbf{x}_t$ is according to the multiplication principle the product of the sequential amplitudes of its constituent subpaths

$$A[\mathbf{x}_t] \;\equiv\; A\left[\sum_i \mathbf{x}_t^{(i)}\right] \;-\; \prod_i A[\mathbf{x}_t^{(i)}] \,. \tag{1.9}$$

The amplitude for a path is therefore of the exponential form

$$A[\mathbf{x}_t] \;=\; e^{\frac{i}{[S]}S_{\mathbf{x}t\mathbf{x}'t'}[\mathbf{x}_t]} \tag{1.10}$$

where $S_{\mathbf{x}t\mathbf{x}'t'}[\mathbf{x}_t]$ is a functional of the path, and equals according to eq.(1.9) the sum of the contributions from its constituent subpaths

$$S_{\mathbf{x}t\mathbf{x}'t'}[\mathbf{x}_t] \;=\; \sum_i S_{\mathbf{x}_{i+1}t_{i+1}\mathbf{x}_it_i}[\mathbf{x}_t^{(i)}] \,. \tag{1.11}$$

We call this functional the action. The presence of the imaginary unit is on purpose, as the conservation of the total probability – the particle must be somewhere in space at all times – then requires the action to be a real functional. The quantity $[S]$ is inserted in order to make the exponent dimensionless, i.e., to account for an eventual dimension of the action. We immediately introduce the notation $\hbar \equiv [S]$, and refer to this quantity as the quantum of action.

Since the actions of sub-paths are additive, eq.(1.11), the action for an infinitesimal straight-line segment is proportional to its time step Δt and is further uniquely characterized by the segment's position $\mathbf{x}_t$ and direction $\dot{\mathbf{x}}_t$. The action for an isolated particle can therefore be written as an integral over a function $L(\mathbf{x}_t, \dot{\mathbf{x}}_t)$ we call the Lagrangian[10]

$$S[\mathbf{x}_t] \;=\; \sum_i S_i \;\equiv\; \sum_i \Delta t_i \, L(\mathbf{x}_{t_i}, \dot{\mathbf{x}}_{t_i}) \;=\; \int_{t'}^{t} d\bar{t}\, L(\mathbf{x}_{\bar{t}}, \dot{\mathbf{x}}_{\bar{t}}) \,. \tag{1.12}$$

---

[9]Summing over paths is an infinite-dimensional continuum summation, called a path integral, and should be approached with care. How to *sum over paths* quantitatively will be addressed in section 1.1.3.

[10]Any explicit time dependence due to the system (here the single particle) not being isolated is handled with equal ease, leaving the Lagrangian explicitly depending on time.

We have thus arrived at the following path integral expression for the conditional probability amplitude

$$K(\mathbf{x}, t; \mathbf{x}', t') \;=\; \sum_{\mathbf{x}_t} e^{\frac{i}{\hbar} S_{\mathbf{x}t\mathbf{x}'t'}[\mathbf{x}_t]} \;=\; \sum_{\mathbf{x}_{\bar{t}}} e^{\frac{i}{\hbar} \int_{t'}^{t} d\bar{t}\, L(\mathbf{x}_{\bar{t}}, \dot{\mathbf{x}}_{\bar{t}})} \;. \qquad (1.13)$$

Each alternative path contributes to the propagator through a phase factor determined by the action of the path, a so far unknown functional of the path however.

This is how far the two principles for ascribing probability amplitudes can take us. The form of the action can in general only be obtained by comparison of theoretical ideas and empirical data. However, for the discussion of the low-energy behavior of a single massive particle, the path integral formulation offers an easy assessment of the action. We can namely appeal to the empirical fact that under certain conditions, such as the low-velocity motion of a particle in homogeneous fields, classical mechanics correctly describes the motion. Or evidently, the motion of the center of mass of a large collection of particles follows the trajectory dictated by Newton's equation.[11] This state of affairs is achieved from the path integral expression, eq.(1.13), if in this case only the classical path makes a substantial contribution to the path integral. This, in turn, is achieved if we choose the action for any path as simply given by the expression offered by classical mechanics, i.e., the Lagrangian is identified as Lagrange's function, and it so happens that the quantum of action is small compared to the action of the classical path. Since the classical path is determined by stationarity of the classical action, Hamilton's principle,

$$\left. \frac{\delta S}{\delta \mathbf{x}_t} \right|_{\mathbf{x}_t = \mathbf{x}_t^{cl}} = \mathbf{0} \qquad (1.14)$$

only paths close to the stationary one will in that case contribute to the sum over paths in eq.(1.13). Contributions from paths deviating from the classical path cancel each other on account of the rapidly oscillating phase factor, $S_{\mathbf{x}t\mathbf{x}'t'}[\mathbf{x}_t] \gg \hbar$, (assuming here that the classical path minimizes the action). In this way it is ascertained that probabilities for the particle to be found at places other than those dictated by classical mechanics are vanishingly small. The conditional probability amplitude is then given with sufficient accuracy by the expression obtained in the stationary phase approximation (of the expression in eq.(1.13) or more specificly eq.(1.40)), and we obtain for the quasi-classical propagator[12]

$$K_{cl}(\mathbf{x}, t; \mathbf{x}', t') \;=\; A(t, t')\, e^{\frac{i}{\hbar} S_{cl}(\mathbf{x}, t; \mathbf{x}', t')} \qquad (1.15)$$

where S_{cl} denotes the action for the classical path, $S_{cl}(\mathbf{x}, t; \mathbf{x}', t') \equiv S_{\mathbf{x}t\mathbf{x}'t'}[\mathbf{x}_t^{cl}]$. The prefactor is captured with sufficient accuracy by the Gaussian fluctuations around

---

[11]In this case the classical Lagrange function is an effective Lagrangian, and the emergence of the classical path as the only one of importance is the result of decoherence due to the multitude of degrees of freedom left unobserved (as discussed further at the end of section 1.9.).

[12]In most situations there is more than one classical path connecting $\mathbf{x}'$ and $\mathbf{x}$ in the time span in question, and in that case their contributions should be added in accordance with the superposition principle.

the classical path, and is specified by the Van Vleck determinant [2] (up to a phase factor [3], [4])[13]

$$A(t, t') = (2\pi i\hbar)^{-d/2} \left| \det_{\alpha\beta} \left(\frac{\partial^2 S_{cl}(\mathbf{x}, t; \mathbf{x}', t')}{\partial x'_\alpha \partial x_\beta} \right) \right|^{1/2} \tag{1.16}$$

representing the density of classical paths arriving in a small volume around $\mathbf{x}$, all originating from $\mathbf{x}'$ and taking the time span in question, but starting out with different momentum values, or more precisely, the Jacobian for the involved transformation, i.e., the aforementioned initial small momentum volume divided by the corresponding final position volume, since the classical action specifies the initial momentum according to $\mathbf{p}'_\alpha = -\partial S/\partial \mathbf{x}'_\alpha$.[14]

From correspondence with the classical limit it follows that the action and thereby $\hbar$, the quantum of action, has the dimension of energy *times* time. The theory does not fix the value of the quantum of action; it has to be taken from comparison of theoretical predictions with experimental results (say, calculate for the hydrogen atom the energy levels and compare with spectroscopic data using measured values of the mass and charge of the electron). Experiments of highest accuracy (employing the Josephson and quantum Hall effects) set the present value at $\hbar = 1.05457266(63) \cdot 10^{-34}$ Js, indeed a small value compared to common experienced actions (moving a mass of 1 gram the distance 1 centimeter in 1 second requires the action $5 \cdot 10^{-8}$ Js).

The recipe for calculating the fundamental dynamic quantity, the propagator, is thus stated in simple terms in Feynman's space-time approach: calculate the classical action for any path connecting the space-time points in question; the conditional probability amplitude is then conceived as the sum of amplitudes for propagation along the different alternative paths. However, the conceptual simplicity of the path integral formulation has been obtained at the price of having to perform quite a horrendous summation!

By appealing to correspondence with classical mechanics we have thus revealed the form of the propagator and thereby the fundamental dynamical law of quantum mechanics.[15] Before figuring out *how to sum over paths* quantitatively let us first show that this knowledge is not needed since it can be circumvented by deriving the differential characterization of the dynamics, the Schrödinger equation.

---

[13] The prefactor is calculated in appendix A.

[14] The path integral formulation is consequently a useful calculational tool in the quasi-classical limit, as we exploit in chapter 11. Furthermore, for a particle moving in spatially homogeneous fields the integrals occurring are Gaussian, and as shown in appendix A the path integral formalism allows for an easy assessment of the propagator. However, as a general calculational tool the path integral is often clumsy in comparison to the expediency of the operator calculus where physical quantities are represented by operators. We shall discuss operator calculus shortly.

[15] It is in fact possible to avoid the embarrassing appeal to correspondence and instead appeal to symmetries. The free propegator can be inferred from Galilean invariance and the translational and rotational invariance of space. In the relativistic quantum theory where degrees of freedom without classical analog proliferate, particle dynamics is described in terms of the interaction of quantum fields. The propagators and the structure of the interactions is determined by symmetry principles, such as the requirement of Lorentz invariance, but ultimately the strength of particle interactions is at present determined empirically.

1.1.1   The Schrödinger Equation

From the path integral formulation we can easily obtain the differential characterization of the dynamics, the Schrödinger equation, even without explicit knowledge of *how to sum over paths*. We simply need to know the propagator for an infinitesimal time step. Let us consider the case of a particle in a potential. We note that for an infinitesimal time step only the straight-line path contributes to the path integral since no intermediate events are considered, and we have for the propagator

$$K(\mathbf{x}, t + \Delta t; \mathbf{x}', t) \;\propto\; e^{\frac{i}{\hbar}\Delta t \, L\left(\frac{\mathbf{x}+\mathbf{x}'}{2}, \frac{\mathbf{x}-\mathbf{x}'}{\Delta t}, t\right)} \tag{1.17}$$

where the proportionality sign reflects our present ignorance of how to assign the measure for the density of paths involved in the path integral. Appealing to correspondence and considering for definiteness the case of a particle of mass m moving in a potential V for which the Lagrangian is

$$L(\mathbf{x}_t, \dot{\mathbf{x}}_t, t) \;=\; \frac{1}{2} m \dot{\mathbf{x}}_t^2 \;-\; V(\mathbf{x}_t, t) \tag{1.18}$$

we get

$$K(\mathbf{x}, t + \Delta t; \mathbf{x}', t) \;\propto\; e^{\frac{i}{\hbar}\frac{m(\mathbf{x}-\mathbf{x}')^2}{2\Delta t} - \frac{i}{\hbar}\Delta t \, V\left(\frac{\mathbf{x}+\mathbf{x}'}{2}, t\right)} \;. \tag{1.19}$$

At equal times, the propagator, according to eq.(1.7), is Dirac's delta function[16]

$$K(\mathbf{x}, t'; \mathbf{x}', t') \;=\; \delta(\mathbf{x} - \mathbf{x}') \tag{1.20}$$

defined according to

$$\int d\mathbf{x}' \, \delta(\mathbf{x} - \mathbf{x}') \, \psi(\mathbf{x}') \;=\; \psi(\mathbf{x}) \tag{1.21}$$

for arbitrary wave functions ψ. The prefactor in eq.(1.17) is now immediately asserted from eq.(1.19) (see the representation of the delta function eq.(B.17) of appendix B), and we have for the propagator for an infinitesimal time step (in three spatial dimensions)

$$K(\mathbf{x}, t + \Delta t; \mathbf{x}', t) \;=\; \frac{1}{\left(\frac{2\pi\hbar i \Delta t}{m}\right)^{3/2}} \, e^{\frac{i}{\hbar}\frac{m(\mathbf{x}-\mathbf{x}')^2}{2\Delta t} - \frac{i}{\hbar}\Delta t \, V\left(\frac{\mathbf{x}+\mathbf{x}'}{2}, t\right)} \;. \tag{1.22}$$

The wave function is propagated an infinitesimal time step by inserting into eq.(1.7) the propagator for an infinitesimal time step, eq.(1.22), whereby we obtain

$$\psi(\mathbf{x}, t + \Delta t) \;=\; \int \frac{d\mathbf{x}'}{\left(\frac{2\pi\hbar i \Delta t}{m}\right)^{d/2}} \, e^{\frac{i}{\hbar}\frac{m(\mathbf{x}-\mathbf{x}')^2}{2\Delta t} - \frac{i}{\hbar}\Delta t \, V\left(\frac{\mathbf{x}+\mathbf{x}'}{2}, t\right)} \, \psi(\mathbf{x}', t) \;. \tag{1.23}$$

---

[16]Dirac's delta function is discussed further in appendix B.

Introducing the variable $\bar{\mathbf{x}} = \mathbf{x} - \mathbf{x}'$, and Taylor-expanding on both sides we get (with only terms up to linear order in Δt displayed)

$$\psi(\mathbf{x}, t) + \Delta t \frac{\partial \psi(\mathbf{x}, t)}{\partial t} + \dots = \left(\frac{m}{2\pi i \hbar \Delta t}\right)^{d/2} \int d\bar{\mathbf{x}} \; e^{\frac{i}{\hbar} \frac{m\bar{\mathbf{x}}^2}{2\Delta t}} \left[1 - \frac{i}{\hbar} \Delta t V(\mathbf{x}, t) + \dots\right]$$

$$\left[\psi(\mathbf{x}, t) - \bar{\mathbf{x}} \cdot \frac{\partial \psi(\mathbf{x}, t)}{\partial \mathbf{x}} + \frac{1}{2} \bar{x}_\alpha \frac{\partial^2 \psi(\mathbf{x}, t)}{\partial x_\alpha \partial x_\beta} \bar{x}_\beta + \dots\right]. \quad (1.24)$$

Performing the Gaussian integrals we obtain the Schrödinger equation for a particle in a potential

$$i\hbar \frac{\partial \psi(\mathbf{x}, t)}{\partial t} = \left(-\frac{\hbar^2}{2m} \frac{\partial^2}{\partial \mathbf{x}^2} + V(\mathbf{x}, t)\right) \psi(\mathbf{x}, t) \quad (1.25)$$

whereby we identify the Hamiltonian for a particle in a potential V as[17]

$$H = -\frac{\hbar^2}{2m} \frac{\partial^2}{\partial \mathbf{x}^2} + V(\mathbf{x}, t). \quad (1.26)$$

Exercise 1.1 *Deduce from the Schrödinger equation, eq.(1.25), the conservation of probability; i.e., the integral over space of $|\psi(\mathbf{x}, t)|^2$ is independent of time.*

Exercise 1.2 *Show that the probability density $n(\mathbf{x}, t) = |\psi(\mathbf{x}, t)|^2$ satisfies the continuity equation*

$$\frac{\partial n(\mathbf{x}, t)}{\partial t} + \nabla \cdot \mathbf{j}(\mathbf{x}, t) = 0 \quad (1.27)$$

where

$$\mathbf{j}(\mathbf{x}, t) = \frac{\hbar}{2im} \left(\psi^*(\mathbf{x}, t) \frac{\partial \psi(\mathbf{x}, t)}{\partial \mathbf{x}} - \psi(\mathbf{x}, t) \frac{\partial \psi^*(\mathbf{x}, t)}{\partial \mathbf{x}}\right) \quad (1.28)$$

and consequently is the probability current density.

From eq.(1.25), and eq.(1.7), we obtain that the kernel K is the solution of the Schrödinger equation

$$i\hbar \frac{\partial K(\mathbf{x}, t; \mathbf{x}', t')}{\partial t} = H K(\mathbf{x}, t; \mathbf{x}', t') \quad (1.29)$$

satisfying the initial condition eq.(1.20).[18] The path integral expression for K is a way of expressing the solution of this differential equation in integral form.

---

[17]We note that the Hamiltonian can be obtained from Hamilton's function, $H(\mathbf{x}, \mathbf{p}, t) = \frac{\mathbf{p}^2}{2m} + V(\mathbf{x}, t)$, by substituting for the momentum the differential operator, $\mathbf{p} \to \frac{\hbar}{i}\nabla_\mathbf{x}$, i.e., $H = H\left(\mathbf{x}, \frac{\hbar}{i}\frac{\partial}{\partial \mathbf{x}}, t\right)$. This is a very useful form of the correpondence principle and is therefore referred to as the canonical quantization rule! A more profound understanding of this is provided by the next section.

[18]We shall discuss such propagator equations in detail in chapter 2.

Exercise 1.3 *Deduce from the conservation of probability (see exercise 1.1 on page 11) that the propagator satisfies*

$$\int d\mathbf{x}_2 \, K^*(\mathbf{x}_2, t_2; \mathbf{x}_1, t_1) \, K(\mathbf{x}_2, t_2; \mathbf{x}_1', t_1) \;=\; \delta(\mathbf{x}_1 - \mathbf{x}_1') \tag{1.30}$$

and interpret the result.

The free particle propagator is immediately obtained from the free particle Schrödinger equation by Fourier transformation giving the expression

$$K_0(\mathbf{x}, t; \mathbf{x}', t') \;=\; \left(\frac{m}{2\pi\hbar i(t - t')} \right)^{d/2} e^{\frac{im}{2\hbar} \frac{(\mathbf{x} - \mathbf{x}')^2}{t - t'}} \tag{1.31}$$

in d spatial dimensions.

We now show that Fourier transformation has a profound *physical* meaning in quantum mechanics.

1.1.2   Momentum Measurement

Let us consider a time-of-flight experiment. Suppose that we at a given time $(t = 0)$ ascertain that a particle is in a definite region of space (say, by administering fast opening and closing of a shutter to a beam of approaching particles). We can then ascribe to a particle making it through the shutter the wave function

$$\psi(\mathbf{x}, t = 0) \;=\; \psi_i(\mathbf{x}) \tag{1.32}$$

which is nonvanishing only in the region near the shutter (a so-called wave packet); i.e., for $\mathbf{x} \simeq \mathbf{0}$ as we choose our reference frame to have its origin in that region. We assume free motion of the particle at times subsequent to $t = 0$. Suppose the particle after a time span t is detected to be in a volume element $\Delta\mathbf{x}$ situated at position $\mathbf{x}$. We shall then *say* that the particle at time $t = 0$ had a velocity vector in the volume $\Delta\mathbf{v} \equiv \Delta\mathbf{x}/t^3$ around the velocity vector $\mathbf{v} \equiv \mathbf{x}/t$. We shall also *say* that the particle at time $t = 0$ had a momentum in the volume $\Delta\mathbf{p} \equiv (m/t)^3 \Delta\mathbf{x}$ around the momentum vector $\mathbf{p} \equiv m\mathbf{x}/t$. Upon repeating the measurement (ascertaining the positions of an identical particle according to the foregoing procedure), we will find a distribution in the final position outcomes, and accordingly a distribution in ascribed initial momentum values. In accordance with the preceding definition the two probability distributions are related according to

$$
\begin{aligned}
P_p(\mathbf{p}, t = 0) \, \Delta\mathbf{p} \;&\equiv\; P_x(\mathbf{x}, t) \, \Delta\mathbf{x} \\[2mm]
&=\; \left| \int d\mathbf{x}' \, K_0(\mathbf{x}, t; \mathbf{x}', 0) \psi_i(\mathbf{x}') \right|^2 \Delta\mathbf{x} \\[2mm]
&=\; \left(\frac{m}{2\pi\hbar t} \right)^3 \left| \int d\mathbf{x}' \, e^{\frac{im}{2\hbar} \frac{(\mathbf{x} - \mathbf{x}')^2}{t}} \, \psi_i(\mathbf{x}') \right|^2 \Delta\mathbf{x} \\[2mm]
&=\; \left(\frac{m}{2\pi\hbar t} \right)^3 \left| \int d\mathbf{x}' \, e^{-\frac{i}{\hbar} \mathbf{x}' \cdot \left(\frac{m\mathbf{x}}{t} - \frac{m\mathbf{x}'}{2t} \right)} \, \psi_i(\mathbf{x}') \right|^2 \Delta\mathbf{x}
\end{aligned} \tag{1.33}
$$

and the momentum probability distribution density at time $t = 0$ is given by

$$P_p(\mathbf{p}, t=0) \;=\; \left(\frac{t}{m}\right)^3 P_x\left(\mathbf{x} = \frac{\mathbf{p}t}{m}, t\right) \;=\; \frac{1}{(2\pi\hbar)^3}\left|\int d\mathbf{x}'\, e^{-\frac{i}{\hbar}\mathbf{x}'\cdot\left(\mathbf{p}-\frac{m\mathbf{x}'}{2t}\right)}\psi_i(\mathbf{x}')\right|^2. \quad (1.34)$$

Since the initial state is localized near the shutter with a linear extension of size L_s (determined by the functioning of the shutter), we choose the time of measurement so that for the region of momentum values of interest we have $mL_s/t \ll \min\{|p_x|, |p_y|, |p_z|\}$. We can then neglect the quadratic term in $\mathbf{x}'$ in the exponent in eq.(1.34), and we get for the momentum probability density at time $t = 0$

$$P_p(\mathbf{p}, t = 0) \;=\; \frac{1}{(2\pi\hbar)^3}\left|\int d\mathbf{x}'\, e^{-\frac{i}{\hbar}\mathbf{x}'\cdot\mathbf{p}}\,\psi_i(\mathbf{x}')\right|^2 \quad (1.35)$$

and we have, up to the usual arbitrary overall phase factor of an amplitude, that the position and momentum probability amplitudes are related through Fourier transformation[19]

$$\psi(\mathbf{p}, t=0) \;=\; \frac{1}{(2\pi\hbar)^{3/2}}\int d\mathbf{x}\, e^{-\frac{i}{\hbar}\mathbf{x}\cdot\mathbf{p}}\,\psi(\mathbf{x}, t = 0)\ . \quad (1.36)$$

Since the duration of the experiment t is a choice of the experimenter, it can always be chosen large enough (for a given range of momenta of interest) so that the initial confinement due to the shutter arrangement is irrelevant, and we have in general that the position and momentum probability amplitudes are related through Fourier transformation

$$\psi(\mathbf{p}, t) \;=\; \frac{1}{(2\pi\hbar)^{3/2}}\int d\mathbf{x}\, e^{-\frac{i}{\hbar}\mathbf{x}\cdot\mathbf{p}}\,\psi(\mathbf{x}, t)\ . \quad (1.37)$$

We infer from eq.(1.37) and the multiplication and superposition principles, that $(2\pi\hbar)^{-3/2}e^{-\frac{i}{\hbar}\mathbf{x}\cdot\mathbf{p}}$ is the conditional probability amplitude for the particle to have momentum $\mathbf{p}$ *given* it has position $\mathbf{x}$, i.e., the amplitude that if the particle has position $\mathbf{x}$ then it has momentum $\mathbf{p}$. For the state of affairs where the particle is at a definite position, corresponding to the probability amplitude function $\psi_\mathbf{x}(\mathbf{x}') = \delta(\mathbf{x}' - \mathbf{x})$, the corresponding momentum probability distribution has equal probability for any value of the momentum. We have struck on the essential feature of quantum mechanics, complementarity, that if the particle is at a definite position, complete ignorance of the momentum probability distribution reigns.

1.1.3   Path Integrals

To get a quantitative handle on how to *sum over all paths*, we note that repeated use of eq.(1.7) gives for the propagator the equation

$$K(\mathbf{x}, t; \mathbf{x}', t') \;=\; \int d\mathbf{x}_N .. \int d\mathbf{x}_2 \int d\mathbf{x}_1\, K(\mathbf{x}, t; \mathbf{x}_N, t_N)\, K(\mathbf{x}_N, t_N; \mathbf{x}_{N-1}, t_{N-1})$$

$$K(\mathbf{x}_{N-1}, t_{N-1}; \mathbf{x}_{N-2}, t_{N-2}) \;\cdots\; K(\mathbf{x}_1, t_1; \mathbf{x}', t')\ . \quad (1.38)$$

---

[19]In appendix A we establish the same result for a time-of-flight measurement in a magnetic field.

In order to evaluate the conditional probability amplitude for any time interval we thus only need to know the infinitesimal time step propagator.

By introducing sufficiently many intermediate times, N, we can in eq.(1.38) employ the infinitesimal propagator, eq.(1.22) and encounter $(t - t' = (N + 1)\Delta t)$

$$K(\mathbf{x}, t; \mathbf{x}', t') = \frac{1}{\left(\frac{2\pi\hbar i \Delta t}{m}\right)^{3/2}} \int \prod_{n=1}^{N} \frac{d\mathbf{x}_n}{\left(\frac{2\pi\hbar i \Delta t}{m}\right)^{3/2}} e^{\frac{i}{\hbar}\Delta t \sum_{n=1}^{N+1} \left[\frac{m}{2} \frac{(\mathbf{x}_n - \mathbf{x}_{n-1})^2}{\Delta t^2} - V(\mathbf{x}_n, t_n)\right]} \qquad (1.39)$$

because there are $N + 1$ infinitesimal time step propagators. We have introduced the notation $\mathbf{x}_0 \equiv \mathbf{x}'$ and $\mathbf{x}_{N+1} \equiv \mathbf{x}$. Increasing the number of intermediate time slices N, we obtain the limiting expression for the path integral

$$K(\mathbf{x}, t; \mathbf{x}', t') = \lim_{N \to \infty} \frac{1}{\left(\frac{2\pi\hbar i \Delta t}{m}\right)^{3/2}} \int \prod_{n=1}^{N} \frac{d\mathbf{x}_n}{\left(\frac{2\pi\hbar i \Delta t}{m}\right)^{3/2}} e^{\frac{i}{\hbar}\Delta t \sum_{n=1}^{N+1} \left[\frac{m}{2} \frac{(\mathbf{x}_n - \mathbf{x}_{n-1})^2}{\Delta t^2} - V(\mathbf{x}_n, t_n)\right]}$$

$$\equiv \int_{\mathbf{x}_{t'}=\mathbf{x}'}^{\mathbf{x}_t=\mathbf{x}} \mathcal{D}\mathbf{x}_{\bar{t}} \; e^{\frac{i}{\hbar} \int_{t'}^{t} d\bar{t}\, L(\mathbf{x}_{\bar{t}}, \dot{\mathbf{x}}_{\bar{t}}, \bar{t})} \qquad (1.40)$$

where L in the continuum limit is seen to be Lagrange's function.

For a free particle, the integrations over the intermediate positions are Gaussian, and they can be performed.[20] After performing first the Gaussian integration over $\mathbf{x}_1$ in eq.(1.40) (in the absence of the potential) an expression identical to the starting expression is obtained, except that there is one integration less and the first time step is $2\Delta t$. Upon integrating over $\mathbf{x}_2$ this feature repeats and the first time step becomes $3\Delta t$. The form of the free propagator for a finite time step will thus become identical to the free propagator for an infinitisimal time step, and for its evaluation it is irrelevant how many intermediate time steps we choose to include, giving for the free particle propagator the expression in eq.(1.31).

In eq.(1.17) we used the *midpoint* rule, $V(?) = V(\frac{\mathbf{x}+\mathbf{x}'}{2})$, instead of, for example, $V(?) = V(\mathbf{x})$ as in eq.(1.39). This, however, turned out to be irrelevant as the additional terms resulting upon Taylor-expanding the potential term in this case are of higher than linear order in Δt. On the contrary, for the case of a particle with charge e in the presence of a classical electromagnetic field described by the vector potential $\mathbf{A}(\mathbf{x}, t)$, we have by correspondence with classical mechanics the additional term in the Lagrangian describing the interaction with the vector potential

$$L_{\mathbf{A}}(\mathbf{x}_t, \dot{\mathbf{x}}_t, t) = e\, \dot{\mathbf{x}}_t \cdot \mathbf{A}(\mathbf{x}_t, t) \qquad (1.41)$$

---

[20]Recall the integral $\int_{-\infty}^{\infty} dx\, e^{-ax^2+bx} = \sqrt{\frac{\pi}{a}}\, e^{\frac{b^2}{4a}}$, $\Im ma \neq 0$ or $\Re ea > 0$. The quickest way of obtaining the free particle propagator is to solve the Fourier transformed free particle Schrödinger equation (confer section 1.1.1) and then employ eq.(1.7) thereby obtaining the free propagator by a *single* Gaussian integration. The free propagator is also immediately obtained by appealing to the result of appendix A. Since we are dealing with a quadratic action, the free propagator attains the form $K_0(\mathbf{x}, t; \mathbf{x}', t') = A(t - t') e^{\frac{i}{\hbar} S_{\mathbf{x}t\mathbf{x}'t'}[\mathbf{x}_{\bar{t}}^{cl}]}$, and $A(t - t')$ is determined by the initial condition at $t = t'$, eq.(1.20).

and to get the Schrödinger equation one should use the midpoint rule.[21] The integral over time appearing in eq.(1.12) is therefore not of the Riemann kind, as we have to invoke extra rules for its evaluation (such as the midpoint rule). The reason for this lack of uniqueness can be attributed to the wild zigzag character of the paths entering in the evaluation of the path integral. In the continuum limit these paths become nondifferentiable, the mean square of the velocity does not exist at any point on the path (complementarity in the language of the path integral; for a discussion of this feature we refer to reference [5]).[22]

Exercise 1.4 *Derive from the Lagrange function*

$$L = \frac{1}{2}m\dot{\mathbf{x}}_t^2 + e\,\dot{\mathbf{x}}_t \cdot \mathbf{A}(\mathbf{x}_t, t) \tag{1.42}$$

and with the use of the midpoint rule, i.e., using $\mathbf{A}(\frac{\mathbf{x}+\mathbf{x}'}{2}, t)$ in the expression for the infinitesimal propagator, the Schrödinger equation

$$i\hbar \frac{\partial \psi(\mathbf{x}, t)}{\partial t} = \frac{1}{2m}\left(\frac{\hbar}{i}\frac{\partial}{\partial \mathbf{x}} - e\mathbf{A}(\mathbf{x}, t)\right)^2 \psi(\mathbf{x}, t) \tag{1.43}$$

for a particle with charge e coupled to a vector potential. Show that if instead of the midpoint rule, for example $\mathbf{A}(\mathbf{x}, t)$ or $\mathbf{A}(\mathbf{x}', t)$ is used in the expression for the infinitesimal propagator, an equation violating particle conservation is produced. Show that in the presence of a vector potential, the probability current density is given by

$$\mathbf{j}(\mathbf{x}, t) = \frac{\hbar}{2im}\left(\psi^*(\mathbf{x}, t)\frac{\partial \psi(\mathbf{x}, t)}{\partial \mathbf{x}} - \psi(\mathbf{x}, t)\frac{\partial \psi^*(\mathbf{x}, t)}{\partial \mathbf{x}}\right) - \frac{e^2}{m}\mathbf{A}(\mathbf{x}, t)\,|\psi(\mathbf{x}, t)|^2 \,. \tag{1.44}$$

For numerous applications of path integrals we refer to the classic reference [5]. We note that the perturbation theory expressions of chapter 2 can be immediately derived from the path integral expression for the propagator (see the exercise in section 2.6), and the entire diagrammatic perturbation theory of chapter 2 can be obtained without use of the operator formalism.

---

[21] Otherwise an equation is obtained which does not respect the conservation of probability, i.e., that the particle is always somewhere in space. Equivalently we can say that the form of the Schrödinger equation follows from gauge invariance; i.e., the gauge transformation of the electromagnetic field, $\mathbf{A}(\mathbf{x}, t) \rightarrow \mathbf{A}(\mathbf{x}, t) + \nabla\Lambda(\mathbf{x}, t)$, $\phi(\mathbf{x}, t) \rightarrow \phi(\mathbf{x}, t) - \dot{\Lambda}(\mathbf{x}, t)$, and the transformation of the wave function $\psi(\mathbf{x}, t) \rightarrow \psi(\mathbf{x}, t)\,e^{\frac{ie}{\hbar}\Lambda(\mathbf{x},t)}$, leaves all physical quantities invariant. The gauge invariance of quantum mechanics is a consequence of the wave function obtained by the above phase transformation equally well represents the probability distribution of the particle.

[22] This feature has its analog in the path integral treatment of stochastic processes, as is well known from the diffusion process, where diffusive trajectories are nondifferentiable as $<\Delta x^2>$ $\propto \Delta t$. This shared feature is not surprising in view of the relationship between the Schrödinger equation for a free particle and the diffusion equation, a topic we shall return to in sections 5.6 and 8.7.

1.2   Representation Theory

For calculational purposes it is useful to represent physical quantities by linear operators on the vector space of wave functions. To see how this comes about, let us first consider the geometry of the vector space of wave functions.

1.2.1   State Space

Let us assume that at most with given equipment we can measure the position of a particle with linear precision a. Therefore we might as well imagine space partitioned into boxes of linear size a. In each box i we label a chosen point $\mathbf{x}_i$. A probability amplitude $\psi_a(\mathbf{x}_i)$ is ascribed the event of the particle to be in box i, or as we say at lattice point $\mathbf{x}_i$, and the probability for the event of the particle to be at lattice point $\mathbf{x}_i$ is

$$P_i^{(a)} \;=\; |\psi_a(\mathbf{x}_i)|^2 \;=\; \psi_a^*(\mathbf{x}_i)\,\psi_a(\mathbf{x}_i) \,. \tag{1.45}$$

When the particle is definitely at lattice point $\mathbf{x}_i$ it is described (with proper phase choice) by the wave function, the Kronecker function,

$$\psi_{\mathbf{x}_i}(\mathbf{x}_j) \;=\; \delta_{\mathbf{x}_i,\mathbf{x}_j} \;\equiv\; \begin{cases} 1 & \mathbf{x}_j = \mathbf{x}_i \\ 0 & \mathbf{x}_j \neq \mathbf{x}_i \,. \end{cases} \tag{1.46}$$

Any possible wave function is specified by giving the amplitudes at each lattice point, $\psi_a(\mathbf{x}_i) = a_i$, and can be expressed as the superposition of wave functions of definite position

$$\psi_a(\mathbf{x}_i) \;=\; \sum_j a_j\,\psi_{\mathbf{x}_j}(\mathbf{x}_i) \,. \tag{1.47}$$

The absolute square of the coefficients in the superposition, $|a_i|^2$, represents the probability for the event of the particle to be at site i as

$$|\psi_a(\mathbf{x}_i)|^2 \;=\; \sum_{j,j'} a_j^*\,\psi_{\mathbf{x}_j}^*(\mathbf{x}_i)\,a_{j'}\,\psi_{\mathbf{x}_{j'}}(\mathbf{x}_i) \;=\; |a_i|^2 \,. \tag{1.48}$$

The set of wave functions constitutes a vector space over the complex numbers, spanned by the basis of wave functions of definite position, and can be equipped with a scalar product defined for arbitrary wave functions ψ and ϕ by (we suppress here and often in the following the reference to the lattice, $\psi(\mathbf{x}_i) \equiv \psi_a(\mathbf{x}_i)$):

$$(\phi,\psi) \;\equiv\; \sum_i \phi^*(\mathbf{x}_i)\,\psi(\mathbf{x}_i) \,. \tag{1.49}$$

In particular the scalar product has the property

$$(\phi,\psi) \;=\; (\psi,\phi)^* \,. \tag{1.50}$$

The wave functions of definite position form an orthonormal basis for the vector space of wave functions

$$(\psi_{\mathbf{x}_j},\psi_{\mathbf{x}_k}) \;=\; \sum_i \psi_{\mathbf{x}_j}^*(\mathbf{x}_i)\,\psi_{\mathbf{x}_k}(\mathbf{x}_i) \;=\; \sum_i \delta_{\mathbf{x}_j,\mathbf{x}_i}\,\delta_{\mathbf{x}_k,\mathbf{x}_i} \;=\; \delta_{\mathbf{x}_j,\mathbf{x}_k} \,. \tag{1.51}$$

The probability amplitude, $\psi(\mathbf{x}_i) = a_i$, for the event of the particle to be at lattice point $\mathbf{x}_i$, can be viewed as the projection coefficient of the wave function onto the basis wave function of definite position $\mathbf{x}_i$ as expressed by the scalar product

$$\psi(\mathbf{x}_i) = \sum_j \delta_{\mathbf{x}_i, \mathbf{x}_j} \psi(\mathbf{x}_j) = \sum_j \psi^*_{\mathbf{x}_i}(\mathbf{x}_j) \psi(\mathbf{x}_j) = (\psi_{\mathbf{x}_i}, \psi). \tag{1.52}$$

This is analogous to the coordinate representation $(x_1, x_2, ..) \leftrightarrow \mathbf{x}$ of a spatial vector $\mathbf{x}$ in a frame of reference, the only difference being that now the coordinates are *complex* numbers. The spatial vector, an arrow, can be represented in terms of the sum of its projections onto a set of (orthonormal) basis vectors $\mathbf{x} = (\mathbf{e}_1 \cdot \mathbf{x}) \mathbf{e}_1 + (\mathbf{e}_2 \cdot \mathbf{x}) \mathbf{e}_2 + ..$, and its projection coefficients, given by the scalar products, are the coordinates, $x_i = \mathbf{x} \cdot \mathbf{e}_i$. Representing the lattice wave function by a column vector with the entries $\psi(\mathbf{x}_i) = (\psi_{\mathbf{x}_i}, \psi)$ is analogous to the coordinate representation of a spatial vector, and to a wave function we can associate a geometric object, $|\psi>$, whose *coordinates* are the values of the wave function at the corresponding lattice point. The vector, $|\psi>$, is referred to as the state vector. A wave function of definite position, zero everywhere in the column except at one place where the entry is one, has as any wave function according to the above prescription associated a state vector, $|\psi_{\mathbf{x}_i}>_a$, , and equivalently for the whole orthonormal basis set of definite position wave functions. Just as for a spatial vector the coordinates are the projections of the vector onto the basis vectors, we can also introduce a scalar product between *state vectors* to produce its *coordinates*, the wave function. The scalar product between two arbitrary state vectors $|\phi>$ and $|\psi>$ is defined to be equal to the scalar product between their corresponding wave functions

$$<\phi|\psi> = (\phi, \psi) = \sum_i \phi^*_a(\mathbf{x}_i) \psi_a(\mathbf{x}_i) \tag{1.53}$$

and we note for the scalar product of state vectors the property

$$<\phi|\psi> = <\psi|\phi>^* \tag{1.54}$$

inherited from eq.(1.50).

In particular the scalar product of an arbitrary state vector, $|\psi>$, and the basis state vector of definite position, $|\psi_{\mathbf{x}_i}>_a$, is according to eq.(1.52)

$$_a<\psi_{\mathbf{x}_i}|\psi> = (\psi_{\mathbf{x}_i}, \psi) = \psi(\mathbf{x}_i) \tag{1.55}$$

the value of the corresponding wave function at the position in question. We have hereby obtained the geometrical interpretation of the values of the wave function ψ at different lattice points as the projections of the state vector $|\psi>$ onto the orthonormal basis set of state vectors $|\psi_{\mathbf{x}_i}>_a$, and the projection coefficients specifies the state vector as a superposition of position basis vectors

$$|\psi> = \sum_j {}_a<\psi_{\mathbf{x}_j}|\psi> |\psi_{\mathbf{x}_j}>_a \tag{1.56}$$

as we then obtain the wave function, eq.(1.47), by taking the scalar product with $|\psi_{\mathbf{x}_i}>_a$. The state vector, $|\psi>$, carries by construction equivalent information of the state of the system as the wave function.

Introducing the notation for the position basis state vectors, $|\mathbf{x}_i>_a \equiv |\psi_{\mathbf{x}_i}>_a$, we can rewrite eq.(1.56)

$$|\psi> = \sum_i {}_a<\mathbf{x}_i|\psi> \; |\mathbf{x}_i>_a \; . \tag{1.57}$$

The vector $|\mathbf{x}_i>_a$, the state vector of definite position $\mathbf{x}_i$, represents the state of the particle where it definitely is in box i, or as we shall say in state $\mathbf{x}_i$. The scalar product between an arbitrary state vector and the state vectors of definite position was defined to reproduce the wave function, i.e., rewriting eq.(1.55)

$$\psi_a(\mathbf{x}_i) = {}_a<\mathbf{x}_i|\psi> \tag{1.58}$$

and is referred to as the *position* representation of the state vector $|\psi>$, or as we shall say, of the particle in state ψ. We recall that the wave function is a probability amplitude, and infer that the scalar product, eq.(1.58), can be interpreted as the *conditional* probability amplitude that the particle is in state $\mathbf{x}_i$ *given* it is in state ψ.

The vector space of state vectors is referred to as the state space. The vector space of linear functionals on the state space, the dual space, is isomorphic to the state space, and the scalar product between state vectors can also be viewed as the value of the dual vector, denoted $<\phi|$, on the vector $|\psi>$. Using that the dual vector is a linear functional we obtain using eq.(1.57) and eq.(1.54)

$$<\phi|\psi> = \sum_i <\phi|\mathbf{x}_i>_a {}_a<\mathbf{x}_i|\psi> = <\phi| \left(\sum_i |\mathbf{x}_i>_a {}_a<\mathbf{x}_i| \right) |\psi> \tag{1.59}$$

where in the last equality we have introduced the operator (a mapping of the state space onto itself) in the parenthesis, which resolve the vector $|\psi>$ on the position basis according to $(\sum_i |\mathbf{x}_i>_a {}_a<\mathbf{x}_i|)|\psi> \equiv \sum_i {}_a<\mathbf{x}_i|\psi> \; |\mathbf{x}_i>_a$. We can read off that this operator is the identity operator, expressing the completeness of the position basis vectors

$$\sum_i |\mathbf{x}_i>_a {}_a<\mathbf{x}_i| = \hat{I} \tag{1.60}$$

the resolution of the identity operator,[23] or interpreted as the sum of the projection operators

$$\hat{P}(\mathbf{x}_i) = |\mathbf{x}_i>_a {}_a<\mathbf{x}_i| \; . \tag{1.61}$$

---

[23]It is common practice to write 1 instead of $\hat{I}$, since the effect of the identity operator is identical to multiplycation by 1, $\hat{I}|\psi> = |\psi>$.

1.2.2 Position and Momentum Operators

A wave function of definite position is seen to be characterized by the equation

$$\mathbf{x}_j \, \psi_{\mathbf{x}_i}(\mathbf{x}_j) \; = \; \mathbf{x}_i \, \psi_{\mathbf{x}_i}(\mathbf{x}_j) \, . \tag{1.62}$$

If we define the position operator as the linear operator $\hat{\mathbf{x}}$, which has the following effect on the wave function for a state of definite lattice position:[24]

$$\hat{\mathbf{x}} \, \psi_{\mathbf{x}_i}(\mathbf{x}_j) \; \equiv \; \mathbf{x}_j \, \psi_{\mathbf{x}_i}(\mathbf{x}_j) \tag{1.63}$$

we can express the previous equation as an eigenvalue equation

$$\hat{\mathbf{x}} \, \psi_{\mathbf{x}_i}(\mathbf{x}_j) \; = \; \mathbf{x}_i \, \psi_{\mathbf{x}_i}(\mathbf{x}_j) \tag{1.64}$$

for a position eigenfunction, i.e., the wave function for a state for which the particle is at a definite lattice point. We can then easily find the effect of the position operator on an arbitrary wave function. The wave function that the position operator turns the superposition in eq.(1.47) into, is by linearity of the operator

$$\hat{\mathbf{x}} \, \psi(\mathbf{x}_i) \; = \; \sum_j \, a_j \, \hat{\mathbf{x}} \, \psi_{\mathbf{x}_j}(\mathbf{x}_i) = \sum_j \, a_j \, \mathbf{x}_i \, \psi_{\mathbf{x}_j}(\mathbf{x}_i) = a_i \, \mathbf{x}_i = \mathbf{x}_i \, \psi(\mathbf{x}_i) \, . \tag{1.65}$$

The position operator thus acts as the so-called multiplication operator on an arbitrary wave function.

We can rewrite the eigenvalue equation, eq.(1.64), on the form

$$_a{<}\mathbf{x}_j|\hat{\mathbf{x}}|\mathbf{x}_i{>}_a \; = \; \mathbf{x}_i \, {}_a{<}\mathbf{x}_j|\mathbf{x}_i{>}_a \tag{1.66}$$

and obtain, using the linear property of the scalar product, the eigenvalue equation in the state space

$$\hat{\mathbf{x}} \, |\mathbf{x}_i{>}_a \; = \; \mathbf{x}_i \, |\mathbf{x}_i{>}_a \, . \tag{1.67}$$

We have achieved the goal of expressing any possible position outcome $\mathbf{x}_i$ for the particle as an eigenvalue with the eigenvector being the definite position state vector corresponding to the position in question.[25]

We observe by operating on a position basis vector that the position operator on the state space can be expressed as

$$\hat{\mathbf{x}} \; = \; \sum_i \, \mathbf{x}_i \, |\mathbf{x}_i{>}_a \, {}_a{<}\mathbf{x}_i| \, . \tag{1.68}$$

---

[24]We use the notation $\hat{\mathbf{x}} \, \psi(\mathbf{x}) \equiv (\hat{\mathbf{x}}\psi)(\mathbf{x}) = {<}\mathbf{x}|\hat{\mathbf{x}}|\psi{>}$ for the function obtained by the operation of an operator, here $\hat{\mathbf{x}}$, on the function ψ, and the last equality specifies the operator on the state space (in terms of the expansion coefficients on the position basis vectors). The last expression is referred to as the matrix element of the operator between the states in question.

[25]We could equally well have started out defining the position operator in state space according to eq.(1.67) and obtained that in the position representation it acts as the multiplication operator (which stictly speaking should be distinguished from the position operator on state space, for example by a subscript $\hat{\mathbf{x}}_{(x)}$).

In the continuum limit where the lattice constant approaches zero, $a \to 0$, we have for the probability to find the particle in the volume element $\Delta \mathbf{x} = a^d$ around position $\mathbf{x} \equiv \mathbf{x}_i$ (d is the spatial dimension)

$$P(\mathbf{x})\, \Delta \mathbf{x} = \lim_{a \to 0} |\psi_a(\mathbf{x})|^2 \frac{\Delta \mathbf{x}}{a^d} = \lim_{a \to 0} |{}_a\!<\mathbf{x}|\psi_a>|^2 \frac{\Delta \mathbf{x}}{a^d} \tag{1.69}$$

and thereby for the wave function in the continuum limit, $|\psi(\mathbf{x})|^2 \equiv P(\mathbf{x})$, in terms of the lattice wave function (up to a phase factor)

$$\psi(\mathbf{x}) \equiv \lim_{a \to 0} a^{-d/2}\, \psi_a(\mathbf{x}) \tag{1.70}$$

or introducing continuum position state vectors, $|\mathbf{x}> \equiv \lim_{a \to 0} a^{-d/2} |\mathbf{x}_i>_a$:

$$\psi(\mathbf{x}) = \lim_{a \to 0} a^{-d/2}\, {}_a\!<\mathbf{x}|\psi_a> = <\mathbf{x}|\psi> . \tag{1.71}$$

For the resolution of the identity, eq.(1.60), we get in terms of continuum projection operators (suppressing that the continuum limit, $a \to 0$, is taken):

$$1 = \sum_i |\mathbf{x}_i>_a\,{}_a\!<\mathbf{x}_i| = \sum_i a^d |\mathbf{x}><\mathbf{x}| = \int\! d\mathbf{x}\, |\mathbf{x}><\mathbf{x}| . \tag{1.72}$$

Using the resolution of the identity, we obtain

$$\psi(\mathbf{x}) = <\mathbf{x}|\psi> = \int\! d\mathbf{x}' <\mathbf{x}|\mathbf{x}'> \psi(\mathbf{x}') \tag{1.73}$$

and we read off that the scalar product between continuum position basis vectors is Dirac's delta function

$$<\mathbf{x}|\mathbf{x}'> = \delta(\mathbf{x} - \mathbf{x}') . \tag{1.74}$$

The wave function for a position eigenstate $|\mathbf{x}>$ is thus in the continuum limit the delta function

$$\psi_{\mathbf{x}}(\mathbf{x}') = <\mathbf{x}'|\psi_{\mathbf{x}}> = <\mathbf{x}'|\mathbf{x}> = \delta(\mathbf{x} - \mathbf{x}') . \tag{1.75}$$

We note that a continuum position eigenfunction is not normalizable.

In the continuum limit eq.(1.65) becomes

$$\hat{\mathbf{x}}_{(x)}\, \psi(\mathbf{x}) = \mathbf{x}\, \psi(\mathbf{x}) \tag{1.76}$$

i.e., the continuum position operator is the multiplication operator. According to eq.(1.67) we have in the continuum limit the corresponding eigenvalue equation in the state space

$$\hat{\mathbf{x}}\,|\mathbf{x}> = \mathbf{x}\,|\mathbf{x}> . \tag{1.77}$$

We have achieved the goal of representing the continuum of possible values of positions for the particle as eigenvalues of the position operator with the eigenvectors being the definite position state vectors corresponding to the position values in question.

We will now, similarly to above, construct a linear operator representing the momentum of a particle. The inverse to the Fourier relationship, eq.(1.37), between position and momentum probability amplitudes is

$$\psi(\mathbf{x}, t) \; = \; \int \frac{d\mathbf{p}}{(2\pi\hbar)^{d/2}} \; e^{\frac{i}{\hbar}\mathbf{p}\cdot\mathbf{x}} \; \psi(\mathbf{p}, t) \; . \tag{1.78}$$

Inserting eq.(1.78) into eq.(1.37), we obtain the representation of the delta function

$$\delta(\mathbf{x} - \mathbf{x}') \; = \; \int \frac{d\mathbf{p}}{(2\pi\hbar)^d} \; e^{\frac{i}{\hbar}\mathbf{p}\cdot(\mathbf{x}-\mathbf{x}')} \tag{1.79}$$

or vice versa

$$\delta(\mathbf{p} - \mathbf{p}') \; = \; \int \frac{d\mathbf{x}}{(2\pi\hbar)^d} \; e^{-\frac{i}{\hbar}\mathbf{x}\cdot(\mathbf{p}-\mathbf{p}')} \; . \tag{1.80}$$

A particle that definitely has the momentum $\mathbf{p}$ is in the momentum representation specified by the momentum probability amplitude function, up to a phase factor,[26]

$$\psi_\mathbf{p}(\mathbf{p}') \; = \; \delta(\mathbf{p} - \mathbf{p}') \tag{1.81}$$

and is in the position representation, according to eq.(1.78), described by the wave function, the plane wave,

$$\psi_\mathbf{p}(\mathbf{x}) \; = \; \frac{1}{(2\pi\hbar)^{d/2}} \; e^{\frac{i}{\hbar}\mathbf{p}\cdot\mathbf{x}} \tag{1.82}$$

corresponding to equal (relative) probability for finding the particle anywhere in space.[27] We have again encountered the essential feature of quantum mechanics, complementarity, that all the physical attributes of a system, here position and momentum of a particle, can not simultaneously be ascribed with arbitrary accuracy. If the particle with definiteness has momentum $\mathbf{p}$, all position outcomes have equal probability (and vice versa).

Introducing the notation for the differential operator[28]

$$\hat{\mathbf{p}} = \frac{\hbar}{i}\nabla_\mathbf{x} \tag{1.83}$$

a state of definite momentum $\mathbf{p}$ satisfies the eigenvalue equation[29]

$$\hat{\mathbf{p}} \, \psi_\mathbf{p}(\mathbf{x}) \; = \; \mathbf{p} \, \psi_\mathbf{p}(\mathbf{x}) \; . \tag{1.84}$$

[26]We consider immediately the continuum case, as the transition from discrete to continuum description is analogous to the one performed above for the position. The discrete case is discussed in section 1.6. Eq.(1.81) is also immediately obtained by considering the Fourier transformed of eq.(1.7).

[27]A state of definite momentum $\mathbf{p}$ is described by a wave function which oscillates in space with the de Broglie wavelength $\lambda = 2\pi\hbar/|\mathbf{p}|$.

[28]A more general representation of the momentum operator is discussed in section 1.6.

[29]We use the notation $\hat{\mathbf{p}} \, \psi(\mathbf{x}) \equiv (\hat{\mathbf{p}}\psi)(\mathbf{x}) = <\mathbf{x}|\hat{\mathbf{p}}|\psi>$ for the function obtained by the operation of the operator $\hat{\mathbf{p}}$ on the function ψ, and the last equality specifies the operator on the state space (in terms of the expansion coefficients on the position basis vectors).

We shall call $\hat{\mathbf{p}}$ the momentum operator, and the coefficient on the right-hand side is the eigenvalue of the state of definite momentum, the momentum eigenstate. Since the momentum operator is linear, it acts according to eq.(1.78) also on an arbitrary wave function as the differential operator, eq.(1.83).[30]

Like any other wave function, the wave function for the state of definite momentum $\psi_{\mathbf{p}}(\mathbf{x})$ has a corresponding state vector denoted $|\psi_{\mathbf{p}}>$ or $|\mathbf{p}>$ for short. In terms of the scalar product of state vectors of definite position and momentum, we can rewrite eq.(1.82)

$$<\mathbf{x}|\mathbf{p}> \equiv <\mathbf{x}|\psi_{\mathbf{p}}> = \psi_{\mathbf{p}}(\mathbf{x}) = \frac{1}{(2\pi\hbar)^{d/2}} e^{\frac{i}{\hbar}\mathbf{p}\cdot\mathbf{x}} \qquad (1.85)$$

the amplitude for the event of the particle at position $\mathbf{x}$ *given* it has momentum $\mathbf{p}$.

In terms of the scalar product of state vectors, we can rewrite the position representation of the eigenvalue equation, eq.(1.84), as

$$<\mathbf{x}|\hat{\mathbf{p}}|\mathbf{p}> = \mathbf{p} <\mathbf{x}|\mathbf{p}> \qquad (1.86)$$

and thereby the momentum operator eigenvalue equation in the state space [31]

$$\hat{\mathbf{p}}|\mathbf{p}> = \mathbf{p}|\mathbf{p}> . \qquad (1.87)$$

We have again achieved the goal of representing the values of a physical quantity, here momentum, as eigenvalues of an operator which thereby represents the physical quantity in question.

For the scalar product, using eq.(1.78) and eq.(1.79), we have in terms of the momentum probability amplitudes

$$<\phi|\psi> \equiv \int d\mathbf{x}\ \phi^*(\mathbf{x})\,\psi(\mathbf{x}) = \int d\mathbf{p}\ \phi^*(\mathbf{p})\,\psi(\mathbf{p}) = \int d\mathbf{p}\ <\phi|\mathbf{p}><\mathbf{p}|\psi>$$
$$(1.88)$$

and we have the following expression for the resolution of the identity in terms of the momentum projection operators

$$\int d\mathbf{p}\ |\mathbf{p}><\mathbf{p}| = 1 \qquad (1.89)$$

expressing that the momentum eigenstates constitute a complete basis in the state space. Orthonormality

$$<\mathbf{p}|\mathbf{p}'> = \delta(\mathbf{p}-\mathbf{p}') \qquad (1.90)$$

---

[30]The identification of the momentum operator, makes it possible to rephrase the canonical quantization rule: the Hamiltonian is obtained from Hamilton's function by substituting for the momentum the momentum *operator* !

[31]We could equally well have defined the momentum operator in the state space according to eq.(1.87), or equivalently $\hat{\mathbf{p}} \equiv \int d\mathbf{p}\ \mathbf{p}\,|\mathbf{p}><\mathbf{p}|$, and obtained that in the position representation it acts as the differential operator (which stictly speaking should be distinguished from the momentum operator on state space, for example by a superscript, for example for the momentum operator in the y-direction $\hat{p}_y^{(x)}$).

follows, according to eq.(1.80), by inserting the resolution of the identity in terms of position eigenstates on the left side of the equation.

The Fourier transformations between the position and momentum representations, eq.(1.37) and eq.(1.78), can now be written as simply expressing the various resolutions of the identity

$$<\mathbf{x}|\psi> = \int d\mathbf{p} <\mathbf{x}|\mathbf{p}><\mathbf{p}|\psi> \tag{1.91}$$

and

$$<\mathbf{p}|\psi> = \int d\mathbf{x} <\mathbf{p}|\mathbf{x}><\mathbf{x}|\psi> . \tag{1.92}$$

We refer to the position and momentum probability amplitude functions as the position and momentum representations of a particle in the state described by the state vector $|\psi>$, respectively.

It follows $((\hat{x}\hat{p}_x - \hat{p}_x\hat{x})\psi(\mathbf{x}) = -i\hbar(x\partial_x - \partial_x x)\psi(\mathbf{x}) = i\hbar\psi(\mathbf{x}))$ that equal Cartesian components of position and momentum operators satisfy the canonical commutation relation

$$[\hat{x}, \hat{p}_x] = i\hbar \tag{1.93}$$

where we have introduced the commutator of two operators, here position and momentum operators,

$$[\hat{x}, \hat{p}_x] \equiv \hat{x}\,\hat{p}_x - \hat{p}_x\,\hat{x} . \tag{1.94}$$

For different Cartesian components, the position and momentum operators clearly commute; i.e., their commutator vanishes, and we have generally

$$[\hat{x}_\alpha, \hat{x}_\beta] = 0 \quad , \quad [\hat{p}_\alpha, \hat{p}_\beta] = 0 \quad , \quad [\hat{x}_\alpha, \hat{p}_\beta] = i\hbar\,\delta_{\alpha,\beta} \quad , \quad \alpha, \beta = x, y, z. \tag{1.95}$$

Exercise 1.5 *Show that the momentum operator in the position representation has the matrix elements*

$$<\mathbf{x}|\hat{\mathbf{p}}|\mathbf{x}'> = \frac{\hbar}{i}\frac{\partial}{\partial \mathbf{x}}\,\delta(\mathbf{x} - \mathbf{x}') . \tag{1.96}$$

Exercise 1.6 *Show that in the momentum representation the momentum operator is the multiplication operator and the position operator the differential operator*

$$\hat{\mathbf{x}}^{(p)} = -\frac{\hbar}{i}\frac{\partial}{\partial \mathbf{p}} . \tag{1.97}$$

Exercise 1.7 *Show that the average value of the momentum in an arbitrary state is expressed by the momentum operator according to*

$$<\mathbf{p}(t)> \equiv \int d\mathbf{p}\, \mathbf{p}\, |\psi(\mathbf{p}, t)|^2 = \int d\mathbf{x}\, \psi^*(\mathbf{x}, t)\frac{\hbar}{i}\nabla_\mathbf{x}\,\psi(\mathbf{x}, t) \equiv <\hat{\mathbf{p}}> . \tag{1.98}$$

1.2.3   Conservation Laws and Energy

Conservation laws are the basis for a rational understanding of the world. *Chaos* does not reign because an account has to be balanced. If a subsystem does not carry the total amount of a conserved quantity another one must. Clearly, at a very practical level conservation laws are used to interpret experimental data.

A physical quantity is said to be conserved if it for any state of the system stays constant in time. In quantum mechanics this therefore means that its probability distribution in any state is independent of time. If a quantity is conserved then in particular its average value must be independent of time. Taking the time derivative of the average value and using the Schrödinger equation, conservation of a quantity A is equivalent to the Hamiltonian commuting with the operator representing the quantity in question, i.e., $[\hat{A}, \hat{H}] = 0$.[32] It follows from eq.(1.110) that if at some moment in time a system is in a state of a definite value of a conserved quantity it will always stay an eigenstate of this quantity with the value in question.

We have already encountered a situation where a quantity is conserved, viz. that of the momentum of a *free* particle. The position of a particle is never conserved due to the kinetic energy term in the Hamiltonian.

In general only the Hamiltonian commutes with itself, and we shall therefore also call it the energy operator, and the physical quantity it represents the *energy*. This conclusion could also be arrived at much earlier, directly from the Schrödinger equation. This we shall now do by considering a special set of states.

1.2.4   Stationary States

For an isolated system, the Hamiltonian is time independent, and for a time-independent situation we can find solutions of the Schrödinger equation for which the probability distribution is time independent, $|\psi(\mathbf{x}, t)|^2 = P(\mathbf{x})$. Such a state $\psi(\mathbf{x}, t)$, called a stationary state, is seen to have the form (realizing that the average value of the position in a stationary state is time independent)

$$\psi_E(\mathbf{x}, t) = \psi_E(\mathbf{x})\, e^{-\frac{i}{\hbar} E t} \tag{1.99}$$

where $\psi_E(\mathbf{x})$ satisfies the time independent Schrödinger equation

$$H\left(\mathbf{x}, \frac{\hbar}{i} \nabla_{\mathbf{x}}\right) \psi_E(\mathbf{x}) = E\,\psi_E(\mathbf{x}) \tag{1.100}$$

the eigenvalue equation for the Hamiltonian, and E is a real number completely characterizing the wave function.

At any time the stationary state corresponding to eigenvalue E remains the eigenstate with eigenvalue E as

$$H\left(\mathbf{x}, \frac{\hbar}{i} \nabla_{\mathbf{x}}\right) \psi_E(\mathbf{x}, t) = H\left(\mathbf{x}, \frac{\hbar}{i} \nabla_{\mathbf{x}}\right) e^{-\frac{i}{\hbar} E t} \psi_E(\mathbf{x}) = E\,\psi_E(\mathbf{x}, t)\,. \tag{1.101}$$

---

[32]We note that commuting operators have common eigenfunctions.

In general the Hamiltonian is the only operator for which its eigenfunctions stay eigenfunctions at all times with the same eigenvalue (for an isolated system of course). We therefore call the conserved eigenvalue for the isolated system its energy, and the Hamiltonian the energy operator.

We have stated that the position is never a conserved quantity as the position operator does not commute with the Hamiltonian. However, we have identified the special circumstances under which its probability distribution is time independent, viz. for the case of energy eigenstates.

1.2.5 Quantum Dynamics

We have previously considered the Schrödinger equation in the position representation, we now wish to consider the time dependence of a state vector. For the state vector which at time t is $|\psi(t)>$ we have the wave function $\psi(\mathbf{x}, t) = <\mathbf{x}|\psi(t)>$, and the Schrödinger equation, eq.(1.25), can be rewritten

$$i\hbar \frac{\partial <\mathbf{x}|\psi(t)>}{\partial t} = <\mathbf{x}| \left(\frac{\hat{\mathbf{p}}^2}{2m} + V(\hat{\mathbf{x}}, t) \right) |\psi(t)> \tag{1.102}$$

since (using the resolution of the identity eq.(1.72), i.e., expanding $|\psi(t)>$ on the complete set of position eigenstates)

$$<\mathbf{x}|V(\hat{\mathbf{x}}, t)|\psi(t)> = \int d\mathbf{x}' <\mathbf{x}|V(\hat{\mathbf{x}}, t)|\mathbf{x}'> <\mathbf{x}'|\psi(t)>$$

$$= V(\mathbf{x}, t)\, \psi(\mathbf{x}, t) \tag{1.103}$$

and (using the resolution of the identity eq.(1.89), i.e., expanding $|\psi(t)>$ on the complete set of momentum eigenstates)

$$<\mathbf{x}|\hat{\mathbf{p}}^2|\psi(t)> = <\mathbf{x}|\hat{\mathbf{p}}^2 \int d\mathbf{p} <\mathbf{p}|\psi(t)> |\mathbf{p}> = \int d\mathbf{p}\, \mathbf{p}^2 <\mathbf{p}|\psi(t)> <\mathbf{x}|\mathbf{p}>$$

$$= \left(\frac{\hbar}{i} \nabla_\mathbf{x} \right)^2 \int d\mathbf{p} <\mathbf{x}|\mathbf{p}> <\mathbf{p}|\psi(t)> = \left(\frac{\hbar}{i} \nabla_\mathbf{x} \right)^2 \psi(\mathbf{x}, t) \ . \tag{1.104}$$

For the time evolution of the state vector we then have the Schrödinger equation in the state space

$$i\hbar \frac{d|\psi(t)>}{dt} = \hat{H}(t)\, |\psi(t)> \tag{1.105}$$

where

$$\hat{H}(t) = H(\hat{\mathbf{x}}, \hat{\mathbf{p}}, t) \tag{1.106}$$

is the Hamiltonian on the state space.

Let us obtain the formal solution of the Schrödinger equation. For a small time step $\Delta t = t - t'$ we have from the Schrödinger equation (we assume the Hamiltonian is time independent; the time-dependent case is considered in section 2.4)

$$|\psi(t)> \; = \; |\psi(t')> + \frac{\Delta t}{i\hbar}\hat{H}|\psi(t')> + \mathcal{O}(\Delta t^2) = \left(\hat{I} + \frac{\Delta t}{i\hbar}\hat{H}\right)|\psi(t')> + \mathcal{O}(\Delta t^2)$$

(1.107)

where $\mathcal{O}(\Delta t^2)$ signifies all terms beyond linear order. To generate a finite displacement in time, we use the tactic of small steps. Stepping repeatedly back in time

we obtain successively using the Schrödinger equation, $\Delta t = (t - t')/N$,

$$|\psi(t)> \; = \; \left(\hat{I} + \frac{\Delta t}{i\hbar}\hat{H}\right)|\psi(t_{N-1})> \; = \; \left(\hat{I} + \frac{\Delta t}{i\hbar}\hat{H}\right)^N |\psi(t')> \; \equiv \; \hat{U}_N(t,t')\,|\psi(t')>$$

(1.108)

where higher order terms in Δt have been dropped as we eventually take the limit of large N. Introducing still shorter time steps, we obtain for the time-evolution operator

$$|\psi(t)> \; = \; \hat{U}(t,t')\,|\psi(t')>$$

(1.109)

the expression

$$
\begin{aligned}
\hat{U}(t,t') \; &= \; \lim_{N\to\infty} \hat{U}_N(t,t') \; = \; \lim_{N\to\infty}\left(\hat{I} + \frac{\Delta t}{i\hbar}\hat{H}\right)^N \\[2mm]
&= \; \lim_{N\to\infty}\left(\hat{I} + \frac{\Delta t}{i\hbar}N\hat{H} + \left(\frac{\Delta t}{i\hbar}\right)^2 \frac{N(N-1)}{2!}\hat{H}^2\right. \\[2mm]
&\quad + \; \left.\left(\frac{\Delta t}{i\hbar}\right)^3 \frac{N(N-1)(N-2)}{3!}\hat{H}^3 + .. + \left(\frac{\Delta t}{i\hbar}\right)^N \hat{H}^N\right) \\[2mm]
&= \; \hat{I} + \frac{t-t'}{i\hbar}\hat{H} + \frac{1}{2!}\left(\frac{t-t'}{i\hbar}\right)^2 \hat{H}^2 + \frac{1}{3!}\left(\frac{t-t'}{i\hbar}\right)^3 \hat{H}^3 + ... \\[2mm]
&= \; e^{-\frac{i}{\hbar}\hat{H}(t-t')}
\end{aligned}
$$

(1.110)

where we have used the binomial formula and have in the last line defined the exponential of an operator in terms of its series expansion.

Exercise 1.8 *Show that the Hamiltonian for a free particle in the position representation has the matrix elements*

$$<\mathbf{x}|\frac{\hat{\mathbf{p}}^2}{2m}|\mathbf{x}'> \; = \; \frac{1}{2m}\left(\frac{\hbar}{i}\right)^2 \frac{\partial^2}{\partial\mathbf{x}^2}\,\delta(\mathbf{x}-\mathbf{x}')\,.$$

(1.111)

Finally we wish to express the propagator in terms of the Hamiltonian (again we consider an isolated system). If a particle starts out at the definite position $\mathbf{x}'$ at time t', its state vector has evolved at time t, according to eq.(1.109), to the state vector

$$|\psi_{\mathbf{x}'t'}(t)> \ = \ e^{-\frac{i}{\hbar}\hat{H}(t-t')}|\mathbf{x}'> \qquad (1.112)$$

and the amplitude to find the particle at position $\mathbf{x}$ at time t is $\psi_{\mathbf{x}'t'}(\mathbf{x}, t) = <\mathbf{x}|\psi_{\mathbf{x}'t'}(t)>$, which by definition equals the conditional probability amplitude $K(\mathbf{x}, t; \mathbf{x}', t')$, for which we then have the operator expression[33]

$$K(\mathbf{x}, t; \mathbf{x}', t') \ = \ <\mathbf{x}|e^{-\frac{i}{\hbar}\hat{H}(t-t')}|\mathbf{x}'> \ . \qquad (1.113)$$

We can also identify the operator expression for the propagator by noting that for an arbitrary state at time t', $|\psi(t')>$, we obtain, by using eq.(1.109) and inserting the resolution of the identity

$$<\mathbf{x}|\psi(t)> = \int d\mathbf{x}' \ <\mathbf{x}|e^{-\frac{i}{\hbar}\hat{H}(t-t')}|\mathbf{x}'><\mathbf{x}'|\psi(t')> \qquad (1.114)$$

and identify the propagator according to eq.(1.7).

In appendix A we come full circle between the operator and path integral formulations of quantum mechanics, since we there start from the above operator expression for the propagator, and obtain the path integral form eq.(1.40).

1.2.6 Summary of Quantum Mechanics

Instead of stating the principles of quantum mechanics as in section 1.1, we can equivalently state them as axioms in terms of the operator calculus.

Axiom I. (Quantum Kinematics) The description of a physical system is provided by a space of state vectors, and its physical quantities by an algebra of operators.

Axiom II. (Quantum Dynamics) The change in time of a state vector is determined by the Schrödinger equation

$$i\hbar \frac{d|\psi(t)>}{dt} \ = \ \hat{H}|\psi(t)> \ . \qquad (1.115)$$

where the Hamiltonian $\hat{H}$ is a hermitian operator, and $\hbar$ is a constant of nature determined by experiment to be $\hbar \simeq 1.055 \cdot 10^{-34}$ Js.[34]

A reader with the above basic knowledge of quantum mechanics should be able to understand the rest of this book. The reader only interested in the content of the following chapters can in fact skip the rest of this chapter (except perhaps for consulting section 1.8, where the notion of the density matrix is introduced).

---

[33] Alternatively, one immediately shows that the expression on the right-hand side of eq.(1.113) satisfies the Schrödinger equation and the same initial condition as the propagator K.

[34] How to obtain the Hamiltonian? Ask your local experimentalist and use the canonical quantization rule!

However, for the reader not satisfied with the preceding intuitive and deductive introduction to quantum mechanics, and interested in how the kinematic structure of quantum mechanics follows from a few experimental facts without referring to the dynamics of the system, we start from scratch and develop quantum mechanics following Schwinger [7].

1.3 Quantum Kinematics

In the preceding introduction to quantum mechanics we used the quantum dynamics of a system, i.e., How do properties change in time?, in order to derive statements about quantum kinematics, i.e., How are states of a physical system described? This was the case when we obtained the momentum amplitude as the Fourier transform of the position amplitude. Furthermore, we deduced the content of the theory from a few basic principles. In the following we shall instead follow Schwinger [7] and develop the kinematic structure of quantum mechanics from a minimal input of experimental facts whereby the obtained physical theory is the generalization of present experimental knowledge.

1.3.1 State Symbols

Based on the practical experience of the ability to select physical systems to have definite properties we state the

Kinematic Principle:

A system selected to have a definite value a of the property A is described by a symbol $\hat{P}(a)$, the property symbol.

The simplest type of selection is where we, by administering impenetrable walls with holes (and shutters), select a particle to have a definite position in space (at a definite time).[35] For the ensuing discussion, however, it is often useful to have the following example of a selection in mind. Consider repeating sending identical systems (say silver atoms) through an inhomogeneous magnetic field (the Stern-Gerlach experiment). The experiment will reveal that upon emerging from the magnetic field region, the systems can only be found in definite regions of space, as witnessed by the finite number of arrival dots made on a screen detecting arrivals of the systems.[36] Punching a hole in the screen at a definite arrival dot allows one to select only systems that *pass through hole so and so*, and from this datum we can infer that we have selected the system to have the definite property *magnetic*

---

[35]We are clearly not contemplating attempts to localize the particle with such accuracy that special relativity effects become important, i.e., with an accuracy the size of the Compton wavelength. For an electron this length scale is $\lambda_e = 2\pi\hbar/m_e c \simeq 2.4 \cdot 10^{-12}$ m $= 2.4 \cdot 10^{-2}$Å.

[36]In the original 1922 experiment of Stern and Gerlach, actually two lines of silver accumulated on a glass plate forming a pair of lips, and not simply two dots. However, in our discussion we abstract from such fringe effects. A selection is thus an idealization of a measurement.

moment component so and so along the common direction of the magnetic field and its inhomogeneity.[37]

We shall say that a system which is selected to have the value a of property A, is in state a or has been prepared in state a. In the following we shall also use the terminology that a system in state a is represented by the state symbol $\hat{P}(a)$. We shall also refer to property values as quantum numbers.[38]

In a selection the system experiences interaction (in the Stern-Gerlach experiment with the inhomogeneous magnetic field), which makes it possible to reveal a property value of the system. The final act of the selection is to select only the systems which have a definite property value (in the example, the datum being the passage of the system through a chosen space region).

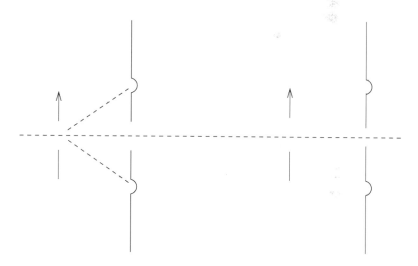

Figure 1.3 Schematic representation of a repeated selection for the Stern-Gerlach experiment. The broken arrows represent the inhomogeneous magnetic field.

From a selection only systems with the selected property emerge, and as a matter of experimental fact, a repetition of the selection is without effect; exactly

---

[37]In the example of silver atoms the number of dots is 2 and attributed to the two internal spin states of the outermost electron. The datum, i.e., a dot is not significant; it is what it means. The *value* of a property is the number inferred from the datum. For details of how the interaction between the magnetic moment of the silver atom and the magnetic field through the dynamics leads to the correlation between the spin and the spatial part of the wave function, and the role of the nucleus of the atom we refer to standard discussions, for example reference [6] (see also exercise 1.17 on page 58).

[38]In the Stern-Gerlach type of experiment the property values is a subset of the natural numbers, and in general a property value is represented by a real number.

the same selection results.[39] In the Stern-Gerlach example all systems passing the first selection will pass the subsequent identical selection, as illustrated in figure 1.3.[40] Using the multiplication symbol for *subsequent,* and the equality sign for *identical to,* we can algebraically symbolize this property of the state of the system after a sequence of identical selections as

$$\hat{P}(a) \cdot \hat{P}(a) = \hat{P}(a) \qquad (1.116)$$

the state symbol thus being idempotent.

The effect on the state of a system subjected to the selection process which does not discriminate properties; i.e., where no selection is taking place, we symbolize by $\hat{I}$. Since such a nondiscriminating selection is equivalent to the absence of selection, it does not influence any prior or subsequent selection. The state of a prior or subsequent selected system is insensitive to the nondiscriminating selection

$$\hat{P}(a) \cdot \hat{I} = \hat{P}(a) = \hat{I} \cdot \hat{P}(a) \ . \qquad (1.117)$$

The nondiscriminating selection has the algebraic property of the identity; it leaves the state of the system unchanged.

The effect on state of a system resulting from a selection process which selects systems with property values in a subset, $a_1, a_2, ..$, without distinguishing any of the possible property values, is represented by the addition of the state symbols for the property values in question. We speak of the system as either in state a_1 or in state a_2 or ... Since no ordering among nondistinguished property values exists, the addition of state symbols is commutative[41]

$$\hat{P}(a_1) + \hat{P}(a_2) + .. \ = \ \hat{P}(a_2) + \hat{P}(a_1) + .. \ = \ .. \ = \ \sum_{i=1,2,..} \hat{P}(a_i) \ . \qquad (1.118)$$

In the case of the Stern-Gerlach experiment the incoming silver atom interacts with the magnetic field, and only the holes corresponding to the magnetic moment property values not included in the subset are blocked.

The nondiscriminating selection, $\hat{I}$, therefore produces a state of the system which is represented by the sum of state symbols over the set of all possible property

---

[39]It might not always be possible to arrange for conditions for repeating a selection, as for instance is the case for position, except for the trivial case where the repetition takes place at almost the same instant. However, in the case where the property in question is the magnetic moment of a particle, we just arrange for no additional magnetic fields besides the ones used in the Stern-Gerlach separators, and a repeated selection is easily arranged for as illustrated in figure 1.3, as the magnetic moment is a constant of the motion. A selection is thus an idealization of a measurement, and in the literature a repeatable measurement is referred to as a measurement of type I. On the contrary, in typical high-energy experiments the measured system ceases to exist.

[40]When repeating the selection for one of the other magnetic moment components, the system can without changing its state of magnetic moment be brought to approach the second magnet.

[41]As shown in the discussion of quantum theory and logic in section 1.9, this state represents, in the usual logical sense of *or,* a system which is either in state a_1 *or* state a_2 *or* ...

values, and thus for a total number N of property values, $a_1, a_2, .., a_N$, we have[42]

$$\sum_{i=1}^{N} \hat{P}(a_i) = \hat{I} \, . \tag{1.119}$$

Adding all state symbols gives a resolution of the identity for the property in question.

The state of a system resulting from a selection process which do not select any system irrespective of property value, the total rejection, is symbolized by $\hat{0}$, the selected state of the system being the absence of the system. A subsequent selection has no system to select, and the state of the system for the combined selection is therefore a selection of no system. Hence we have for the state of a system resulting from selections involving rejection

$$\hat{P}(a) \cdot \hat{0} \, = \, \hat{0} \, = \, \hat{0} \cdot \hat{P}(a) \tag{1.120}$$

because also ending a sequence of selections with total rejection makes the combined selection no selection at all, the system is absent.

A subsequent selection of only systems with a different property value (of the same property as in the first selection) results in the empty selection, absence of the system,

$$\hat{P}(a_i) \cdot \hat{P}(a_j) = \hat{0} = \hat{P}(a_j) \cdot \hat{P}(a_i) \qquad a_i \neq a_j \tag{1.121}$$

as a system accepted in the first selection is subsequently rejected. The different properties are mutually exclusive.

The total rejection, being an empty selection, adds no extra property value, as absence of the system is not a property of the system. Addition of the rejection to a selection does therefore not change the state of the selected system since no property value is added to the set of selected property values

$$\hat{P}(a) + \hat{0} = \hat{P}(a) = \hat{0} + \hat{P}(a) \, . \tag{1.122}$$

The selection that selects systems with either property a_i or a_j (i.e., selects without distinction systems with property value a_i or a_j) followed by the selection that only selects a_i-systems is identical to the a_i-selection, the state of the selected system is a_i,

$$\hat{P}(a_i) \cdot [\hat{P}(a_i) + \hat{P}(a_j)] = \hat{P}(a_i) \, . \tag{1.123}$$

If instead the a_i-a_i and a_i-a_j subsequent selections are added (i.e., no distinction between the two types of selections takes place), we have according to the previously established algebra for the state symbols

$$\hat{P}(a_i) \cdot \hat{P}(a_i) + \hat{P}(a_i) \cdot \hat{P}(a_j) = \hat{P}(a_i) + \hat{0} = \hat{P}(a_i) \, . \tag{1.124}$$

Multiplication of state symbols is thus distributive with respect to addition.

---

[42]The limit of a continuous degree of freedom is treated in section 1.5.

Finally, the usual conventions for multiplying the numbers zero and one with symbols are introduced

$$\hat{P}(a)\ 1 = \hat{P}(a) = 1\ \hat{P}(a)\quad ,\quad \hat{P}(a)\ 0 = \hat{0} = 0\ \hat{P}(a) \tag{1.125}$$

and we can then combine the results for repeated selections of arbitrary values of the same property in the formula

$$\hat{P}(a_i) \cdot \hat{P}(a_j)\ =\ \delta_{a_i,a_j}\ \hat{P}(a_i) \tag{1.126}$$

where we have introduced the Kronecker function

$$\delta_{a_i,a_j}\ =\ \begin{cases} 1 & a_i = a_j \\ 0 & a_i \neq a_j \end{cases} . \tag{1.127}$$

1.3.2   Incompatible Properties

Two properties, $A^{(1)}$ and $A^{(2)}$, are called compatible if their order of selection is irrelevant, resulting in either case in the selected system to have the property values in question of the two properties

$$\hat{P}(a_{i_1}^{(1)}) \cdot \hat{P}(a_{i_2}^{(2)})\ =\ \hat{P}(a_{i_2}^{(2)}) \cdot \hat{P}(a_{i_1}^{(1)})\ \equiv\ \hat{P}(a_{i_1}^{(1)}, a_{i_2}^{(2)}) \tag{1.128}$$

i.e., the subsequent selection does not influence the value of the property selected previously. The two properties can thus be specified simultaneously. We have already encountered a compatible set of properties, viz. position and magnetic moment component of a particle, as the way of selecting a system with a definite magnetic moment component simultaneously could be arranged as a selection of the system to have a definite position.

A complete description of a system is a maximal possible simultaneous specification of properties, and a complete property coordinate is thus a collective coordinate $a = (a^{(1)}, a^{(2)}, ..)$ specifying a maximal number of compatible properties. The corresponding complete state symbol is the product of the compatible state symbols

$$\hat{P}(a) = \prod_\alpha \hat{P}(a^{(\alpha)}) \tag{1.129}$$

providing a complete description of the system.

The arguments establishing all the previous formulas, where a single property was considered, are due to the compatibility property eq.(1.128), equally valid for a complete description as well. For the Kronecker function for a complete property we introduce the abbreviated notation

$$\delta_{a_k,a_j}\ \equiv\ \delta_{a_{k_1}^{(1)},a_{j_1}^{(1)}}\ \delta_{a_{k_2}^{(2)},a_{j_2}^{(2)}}\ \cdots\ =\ \prod_{\alpha=1,2,..} \delta_{a_{k_\alpha}^{(\alpha)},a_{j_\alpha}^{(\alpha)}} . \tag{1.130}$$

In the following we shall always assume we are dealing with a complete description.

We now come to the essential feature of the quantum theory, viz. the experimentally verified fact that not all properties of a physical system are compatible: the order of selection of properties, say A and B, can matter

$$\hat{P}(a) \cdot \hat{P}(b) \neq \hat{P}(b) \cdot \hat{P}(a) \qquad (1.131)$$

they are different selections, resulting in different states of the selected system. We call such properties incompatible, and we can therefore have different mutually incompatible complete descriptions $A, B, C, D, ..$ of a system. Interchanging the selections of two incompatible properties can lead to the emergence of systems in different states, as reflected algebraically by the corresponding state symbols being noncommuting.[43]

1.3.3   Measurement Algebra

Experience with phenomena at the atomic level teach us that identically prepared initial states of a system in general lead to different subsequent observed outcomes. For example, consider the experiment where in each trial we first select the position of the particle to be in a definite space region at a given point in time, and measure its position again after a chosen time span. In each trial of this experiment we will find that the outcomes for the position of the particle measured the second time are erratic. We interpret this fact as reflecting a noncausal aspect of transitions between states that is inherent in nature. Similarly, a system selected to be in a definite property state, say a, will lead to erratic outcomes for an incompatible property upon immediate selection, as demonstrated by a Stern-Gerlach type of experiment by choosing the direction of the magnetic field differently in the second magnetic component analyzer.[44] In the theory there must therefore be an element symbolizing this aspect of unpredictable transition of the system being in some state a to the system being in state b. We represent this transition by the symbol $\hat{P}(b\,;a)$ and refer to it as the transition symbol.

A transition can also be monitored to happen in a controlled way. For example, after a selection where the system is selected to have a definite magnetic moment component, we let it enter a region of space with a constant magnetic field perpendicular to the inhomogeneous direction of the Stern-Gerlach selector, thereby precessing the magnetic moment.[45] Adjusting the time of precession, a subsequent selection can be made to encounter only systems with a definite magnetic moment component different from that initially selected. This type of selection, illustrated

---

[43]We have thus reached the true core of the quantum mechanical description, that we can not simultaneously attribute all the properties of an object with arbitrary precision, Heisenberg's uncertainty principle. We shall more fully discuss this matter in the discussion of complementarity in section 1.4.

[44]The sequence of outcomes seems unpredictable, and the statements of the theory will consequently be of a probabilistic character. Local hidden-variables theories, invoking a deeper level of description where the erratic behavior of outcomes simply reflects our ignorance of the exact values attained by these hidden variables, has been ruled out experimentally [8].

[45]In the previous discussion we have assumed that the system has no dynamics.

in figure 1.4, where a system selected to be in state a_i only can emerge in state a_j is symbolized by $\hat{P}(a_j \, ; a_i)$, and represents the transition of the system between states of different property values of the same property: a transition from state a_i to state a_j. Transitions describe changes in the state of a system and thus reflect interaction processes.[46]

A transition $\hat{P}(b \, ; a)$ can be viewed as a generalized type of selection: accepting solely systems in state a to emerge only in state b. A transition between identical states, $\hat{P}(a \, ; a)$, is therefore identical to two subsequent identical a-selections and thereby to the corresponding state symbol

$$\hat{P}(a \, ; a) = \hat{P}(a) \cdot \hat{P}(a) = \hat{P}(a) \, . \tag{1.132}$$

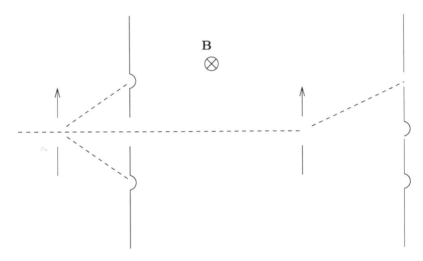

Figure 1.4 A particle first selected to be in the state *passing through the middle hole* is monitored to be transformed into the state *passing through the upper hole.*

Above we have established the rules governing state symbols as simple consequences entailed by experimental knowledge. Similarly we immediately establish the rule for sequential transitions between states of the same property

$$\hat{P}(a_l \, ; a_k) \cdot \hat{P}(a_j \, ; a_i) = \delta_{a_k, a_j} \, \hat{P}(a_l \, ; a_i) = \begin{cases} \hat{P}(a_l \, ; a_i) & a_k = a_j \\ \\ \hat{0} & a_k \neq a_j \end{cases} \tag{1.133}$$

---

[46]The ultimate content of a transition, i.e., where we pursue the finest resolution possible at present, is the creation and annihilation of elementary particles, objects whose properties are not further reducible. The reduction to elementary particles of matter and the forces between them seems to have been successfully carried through for the four forces of nature, except for gravitation, i.e., for the weak, electromagnetic, and strong interactions.

as it is the statement that after the first transition process, where the emerging system has property a_j, we can get a nonempty total selection only if the subsequent transition accepts the system emerging with property a_j. In particular we have

$$\hat{P}(a_i\,;a_j) \cdot \hat{P}(a_j\,;a_i) = \hat{P}(a_i\,;a_i) = \hat{P}(a_i) \tag{1.134}$$

and

$$\hat{P}(a_j\,;a_i) \cdot \hat{P}(a_i\,;a_j) = \hat{P}(a_j\,;a_j) = \hat{P}(a_j) \ . \tag{1.135}$$

For identical property values, $a_i = a_j$, we recover the property of repeated selection, but for different values, $a_i \neq a_j$, we see that the order of transitions is important, the transition symbols are noncommuting objects. Such objects are called operators.

Practical experience shows that a transition allowing only a system in state a to emerge in state b, followed by a transition of only systems in state c into a system in state d, does not in general result in the empty selection, but in a transition of a-systems into d-systems. A succession of transitions of possibly incompatible properties A, B, C, D is thus related to a single transition according to the

Multiplication Law:

$$\hat{P}(d\,;c) \cdot \hat{P}(b\,;a) \ = \ <c|b>\hat{P}(d\,;a) \tag{1.136}$$

where $<c|b>$ denotes a number specified jointly by the property values c and b, and are called Dirac's transformation function. The transformation function describes a statistical relationship between states b and c of the system, describing, in a yet to be determined fashion, the fraction of b-systems that are selected as c-systems. The multiplication law specifies the multiplication of transition symbols, thus establishing the noncommutative algebra for transition operators, the measurement algebra.

Rewriting eq.(1.126) in view of the multiplication law we have for the transformation function

$$<a_k|a_j> \ = \ \delta_{a_k,a_j} \ . \tag{1.137}$$

Applying the multiplication law to an a-c selection, and using the resolution of the identity for the property B, and the fact that multiplication with respect to addition is distributive, we obtain

$$\begin{aligned} <c|a> \hat{P}(c\,;a) \ &= \ \hat{P}(c) \cdot \hat{P}(a) = \hat{P}(c) \cdot \left(\sum_b \hat{P}(b) \right) \cdot \hat{P}(a) \\ &= \ \sum_b \left(\hat{P}(c) \cdot \hat{P}(b) \cdot \hat{P}(a) \right) \\ &= \ \sum_b <c|b><b|a> \hat{P}(c\,;a) \end{aligned} \tag{1.138}$$

and we can read off the relationship between transformation functions

$$\sum_b <c|b><b|a> \ = \ <c|a> \ . \tag{1.139}$$

In the event that we choose the C and A descriptions as identical, we get the
relation

$$\sum_b <a|b><b|a'> \ = \ \delta_{a,a'} \tag{1.140}$$

and upon interchanging the role of a and b

$$\sum_a <b|a><a|b'> \ = \ \delta_{b,b'} . \tag{1.141}$$

The number of property values in any complete set of compatible properties is
therefore the same since

$$\sum_b 1 \ = \ \sum_b \sum_a <b|a><a|b> \ = \ \sum_a \sum_b <a|b><b|a> \ = \ \sum_a 1 \tag{1.142}$$

as the freedom of choice of any complete description should demand.

Preceding a transition by the identity selection has no effect on the state of a
system, nor has a subsequent identity selection, and we have as a consequence of
the multiplication law the linear relationship

$$\hat{P}(d\,;c) \ = \ \hat{I} \cdot \hat{P}(d\,;c) \cdot \hat{I} \ = \ \sum_{a,b} \hat{P}(b) \cdot \hat{P}(d\,;c) \cdot \hat{P}(a)$$

$$= \ \sum_{a,b} <c|a><b|d> \hat{P}(b\,;a) \ \equiv \ \sum_{a,b} <ba|dc> \hat{P}(b\,;a) . \tag{1.143}$$

The underlying vector space of the measurement algebra of transition operators is
of dimensionality N^2, and eq.(1.143) describes how the basis vectors of the mixed
d-c description are linearly related to the basis vectors of the b-a description. In
accordance with eq.(1.122) the rejector is the zero vector, and (using $0\,\hat{0} = \hat{0}$) we
have $c\,\hat{0} = \hat{0}$ for an arbitrary complex number.

1.3.4   State Space

In the transition symbolized by the operator $\hat{P}(b\,;a)$ the system with property a
disappears and the system with property b appears, and there is no reference to
any causal relationship between the appearance and disappearance of the systems
with the two property values in question. Indeed, according to experience there is
in general no causal relationship between a system in state a being revealed in state
b, and the inherently probabilistic character of quantum phenomena is expressed
by the multiplication law.[47] The transition process from state a to state b is
therefore equivalent to two independent processes: the annihilation of the system
with property a, and the creation of the system with property b. We symbolize
the noncausal aspect of a transition by introducing for the transition operator the

---

[47] Quantum mechanics thus does not provide any understanding as to why a certain event takes
place; it will only allow us to calculate the probabilities that events will happen. Probability is
thus a fundamental feature of the physical world.

asymmetric notation $\hat{P}(b;a) \equiv |b><a|$, where the difference between creation and annihilation is manifest. With this split, the disappearance and appearance of systems as symbolized in the transition are represented as separate acts.

Rewriting eq.(1.133) in the new notation and using the multiplication law, we have

$$\hat{P}(a_l\,;a_k) \cdot \hat{P}(a_j\,;a_i) \;=\; (|a_l><a_k|) \cdot (|a_j><a_i|) \;=\; <a_k|a_j> \,|a_l><a_i|$$

$$= \left\{ \begin{array}{ll} |a_l><a_i| & a_k = a_j \\[2ex] \hat{0} & a_k \neq a_j \,. \end{array} \right. \tag{1.144}$$

In terms of the separate acts of annihilation and creation we can interpret the sequence of transitions in eq.(1.144) as follows: If $a_k = a_j$ the symbol $<a_k|$ in $\hat{P}(a_l\,;a_k)$ symbolizes the annihilation of the system emerging in state $a_k = a_j$, and the rest of the symbol $|a_l>$ the creation of the system in state a_l. If the two property values are different, $a_k \neq a_j$, the two subsequent transitions equal the rejector, the system is absent, expressing that a system can not be annihilated unless it has previously been created. In general for sequential transitions we do not have total but only partial restoration, as described by the multiplication law.

As noted previously, the state symbol is the special case of a transition, $\hat{P}(a) = \hat{P}(a\,;a) = |a><a|$. If we rewrite the formula for the resolution of the identity in the introduced notation, and recall the property eq.(1.137) of the transformation function

$$\sum_a |a><a| \;=\; \hat{I} \,, \qquad\qquad <a|a'> \;=\; \delta_{a,a'} \tag{1.145}$$

we recognize these formulas as the statement of $\{|a>\}_a$ being a complete orthonormal basis in a vector space.[48] With each (complete) property value a of a physical system we have thus associated a vector $|a>$, called the state vector, or equivalently the dual vector $<a|$, the linear functional on the vector space having the value $<a|\psi>$ on a vector $|\psi>$. The vector space of state vectors[49] is referred to as the state space (in the present case of dimension N), and the states corresponding to an orthonormal basis are said to constitute a complete orthonormal set of states.

The representation of creation and annihilation of a system as separate acts in the transition symbol $|b><a|$ identify the transition operator as the linear operator on the state space[50]

$$\hat{P}(c\,;b)|a> \;=\; (|c><b|)|a> \;=\; <b|a>\,|c> \tag{1.146}$$

---

[48]Expressed here in Dirac's *ket* and *bra* notation for vectors and dual vectors [9]. Needless to say, reference [9] should be consulted for its monumental exposition of the principles of quantum mechanics.

[49]Just as the vector space of transition operators, as we realize in section 1.3.5, the state space is a vector space over the field of complex numbers.

[50]We note conversely that granted the specified interpretation of the transition operator, the multiplication law is a simple consequence.

mapping a state vector into the state vector for the emerging property multiplied by the transformation function for the initial and accepted property values. In particular, the state symbol $\hat{P}(a)$ is the projection operator onto the basis vector $|a>$.

In order to ease the language, we shall adhere to the fiction that any transition operator of the form $\hat{P}(\psi;\psi) \equiv |\psi><\psi| \equiv \hat{P}(\psi)$ corresponds to a physical property, thus allowing us to say that a system in state ψ, corresponding to state vector $|\psi>$, has the property ψ. According to eq.(1.145) the identity state symbol is the linear operator leaving all basis vectors unchanged

$$\hat{I}\,|a> \,=\, |a> \ . \tag{1.147}$$

A general state vector is characterized by its expansion coefficients on a basis

$$|\psi> \,=\, \sum_a <a|\psi> |a> \,\equiv\, \sum_a \psi(a)|a> \ . \tag{1.148}$$

The unique expansion coefficients $<a|\psi>$ are the transformation functions between property ψ and the properties of the basis. The set of numbers $\psi(a) \equiv <a|\psi>$ is called the wave function in the a-representation, or the a-representation of the state ψ. The operator $\hat{I}$, being linear, leaves any state vector unchanged

$$\hat{I}\,|\psi> \,=\, |\psi> \ . \tag{1.149}$$

We note that we can rewrite eq.(1.139)

$$<c|a> \,=\, <c|\left(\sum_b |b><b|\right)|a> \,=\, \sum_b <c|b><b|a> \ . \tag{1.150}$$

This identity, the insertion of a complete set of states, will be used repeatedly to unfold transformation functions.

1.3.5   Physical Interpretation

The multiplication law is invariant under the joint transformations

$$\hat{P}(b;a) \,\rightarrow\, [\lambda(a)]^{-1}\hat{P}(b;a)\,\lambda(b) \ , \qquad <a|b> \,\rightarrow\, \lambda(a)<a|b>\,[\lambda(b)]^{-1} \tag{1.151}$$

where λ is an arbitrary function, and the transformation function can therefore not be attributed a direct physical interpretation. Using the multiplication law on the sequence $a - b - a$ of selections we have

$$\hat{P}(a) \cdot \hat{P}(b) \cdot \hat{P}(a) \,=\, p(b;a)\,\hat{P}(a) \tag{1.152}$$

where we have introduced the notation

$$p(b;a) \,\equiv\, <a|b><b|a> \,=\, p(a;b) \ . \tag{1.153}$$

If for the $a-b-a$ selection we sum over all b-values (i.e., no intermediate B-selection takes place) we obtain from eq.(1.152)

$$\sum_b p(b;a) = 1 . \tag{1.154}$$

From this normalization condition we are provided with the quantity having the interpretation of a probability. In view of the multiplication law, we interpret the number $p(b;a)$ as the probability to find the system in state b given the system is known to be in state a, or vice versa as $p(b;a)$ is invariant with respect to interchange of a and b. As a test of the validity of such an interpretation we note that the certainty of a system in state a to exhibit the value a is ascertained as $p(a;a) = <a|a>^2 = \delta_{a,a}^2 = 1$. For incompatible properties $p(b;a)$ is a number between zero and one, and precise knowledge of a property value thus excludes precise knowledge of an incompatible property.

The probability assertion demands positivity, $p(b;a) \geq 0$. Since a dual vector belongs to a different vector space than the state vectors (though the dual space is isomorphic to the state space), a transformation function need not be invariant under interchange of its property values. The property that the probability is real therefore does not guarantee the transformation function to be real. Positivity, however, is guaranteed by imposing that the transformation function is complex and obeys the involution property[51]

$$<b|a> = <a|b>^* \tag{1.155}$$

where $*$ denotes complex conjugation.

In order for the second of the scale transformations in eq.(1.151) to respect the involution property, eq.(1.155), we must restrict the allowed scale transformations to functions satisfying $[\lambda(a)]^* = [\lambda(a)]^{-1}$, leaving freedom only for a change by a simple phase factor, $\lambda(a) = \exp\{i\varphi(a)\}$.[52] The representation of the state of a system by a vector is thus not unique since the theory is invariant with respect to changes by an overall phase factor, a so-called phase transformation (or pure gauge transformation). Instead the state of a physical system in state ψ is properly represented by a ray, the equivalence class of vectors $e^{i\varphi}|\psi>$ differing only by an overall phase factor.

1.3.6   Physical Quantities and Operators

The vector space of linear operators on the state space is spanned by N^2 independent transition operators, and an arbitrary operator has expansions on operator

---

[51]In view of the involution property, eq.(1.155), a transformation function can be viewed as the scalar product of the two state vectors involved. The involution property equips the state space with a metric, a measure of distance between vectors $dist(|\phi>, |\psi>) \equiv ||(|\phi> - |\psi>)||$, where the length or norm of a vector is well defined due to the scalar product being nonnegative, $|| |\psi> || \equiv (<\psi|\psi>)^{1/2} \geq 0$.

[52]Instead of expressing the invariance in terms of the transformation function we can express the second transformation in eq.(1.151) as the transformation property of the state vector $|a> \rightarrow e^{i\phi(a)}|a>$.

bases

$$\hat{X} = \sum_{a,a'} <a|\hat{X}|a'> \hat{P}(a\,;a')$$

$$= \sum_{a,b} <b|\hat{X}|a> \hat{P}(b\,;a) \tag{1.156}$$

expressed, for example, either in the A-basis or in a mixed AB-basis. The expansion coefficients are called the matrix elements of the operator, and they constitute a matrix representation of the operator. In particular we have for the unit and zero element of the measurement algebra $<b|\hat{I}|a> = <b|a>$, and $<b|\hat{0}|a> = 0$.

A physical property A is characterized by the total set of possible values $\{a\}_a$ it can exhibit, and we can represent the same information in the measurement algebra, or operator vector space, through the definition of the linear operator[53]

$$\hat{A} \equiv \sum_a a\,\hat{P}(a) = \sum_a |a> a <a| \tag{1.157}$$

where the property value a, being a number, is allowed to roam freely in between vectors and dual vectors.

Operating on the state vector $|a>$ describing a system in state a with the operator $\hat{A}$ corresponding to the quantity A does not change the state vector but gives us according to eq.(1.157) the eigenvalue equation

$$\hat{A}|a> = a|a> \tag{1.158}$$

or equivalently on bra form

$$<a|\hat{A} = <a|a\,. \tag{1.159}$$

The set of property values is in this context referred to as the spectrum of the operator $\hat{A}$, and the state vector $|a>$ as the eigenvector corresponding to the eigenvalue a.

The rejector $\hat{O}$ operating on an arbitrary state results in the zero vector in the state space, which we denote

$$|\mathcal{O}> \equiv \hat{O}|\psi> = 0|\psi>\,. \tag{1.160}$$

Since $\hat{A}\cdot\hat{O} = \hat{O}\cdot\hat{A}$ any physical property has zero eigenvalue in this state. The dual zero vector, $<\mathcal{O}|$, annihilates any vector $<\mathcal{O}|\psi> = 0$.

Subjecting a system in state a' to a selection of the system only with property value a gives an empty selection or recovers the state depending on whether a equals a' or not. The state of the system is represented by $\hat{P}(a)\cdot\hat{P}(a')$ since we are simply dealing with this sequence of selections. We can represent this state of affairs in

---

[53]This is the mathematical expression of the statement of completeness, the spectral theorem, the one being made functional in the theory of spectral representations of linear essentially self-adjoint operators on infinite-dimensional Hilbert spaces [10].

the space of operators as $\hat{P}(a) \cdot \hat{P}(a') = |a><a|a'><a'| = \delta_{a,a'} |a><a'|$, or equivalently as an equation in the state space, the vector identity

$$\hat{P}(a)|a'> = |a><a|a'> = \delta_{a,a'} |a> . \qquad (1.161)$$

If the two property values are different, $a \neq a'$, we have the empty selection: the state is turned into the vacuum state, the absence of the system. In terms of the previously introduced terminology we say that state vector $|a'>$ is annihilated by the dual vector $<a|$. If the two property values are equal, $a = a'$, the selection leaves the state unchanged, or equivalently, destroyed and recreated. The last statement is equivalent to the statement that a repetition of a selection does not change the selected state. The eigenvalue equation, eq.(1.158), thus has the interpretation, that when the property A is measured for a system in state $|a>$ the measured value is the eigenvalue (and ideally the state of the system is unchanged).

The operator corresponding to the property $f(A)$ we can represent by

$$f(\hat{A}) = \sum_a f(a) \hat{P}(a) = \sum_a |a> f(a) <a| \qquad (1.162)$$

as

$$f(\hat{A}) |a> = f(a) |a> . \qquad (1.163)$$

We note that the identity operator corresponds to the function for which $f(a) = 1$ for all a. Furthermore, an operator $\hat{X}$ which commutes with an operator $\hat{A}$ representing a physical quantity, is seen to be a function of $\hat{A}$, $\hat{X} = f(\hat{A})$.[54] The inverse operator to $f(\hat{A})$, $[f(\hat{A})]^{-1} \cdot f(\hat{A}) = \hat{I}$, is , assuming no value $f(a)$ equals zero,

$$[f(\hat{A})]^{-1} = \sum_a |a> \frac{1}{f(a)} <a| \equiv \frac{1}{f(\hat{A})} . \qquad (1.164)$$

In particular for the inverse of an operator we write

$$\hat{A}^{-1} = \frac{1}{\hat{A}} . \qquad (1.165)$$

The state symbol $\hat{P}(a)$ represents the quantity which equals the value unity if the property A exhibits the value a and is zero otherwise, as expressed by the Kronecker function

$$\hat{P}(a) = \sum_{a'} |a'> \delta_{a,a'} <a'| = \delta_{\hat{A},a} \qquad (1.166)$$

i.e., the property *the system has the property value a*, in agreement with the original definition of the state symbol.

The introduced notation allows us to express an arbitrary state vector on the form

$$|\psi> = \sum_a \psi(a)|a> = \sum_a \psi(\hat{A})|a> . \qquad (1.167)$$

---

[54]The proof is elementary as commutation of two operators implies that common eigenvectors can be chosen, and a linear operator is specified by how it transforms a basis.

The characteristic function ϕ of an operator equals the rejector

$$\phi(\hat{A}) \ \equiv \ \prod_a (\hat{A} - a\hat{I}) \ = \ \hat{0} \tag{1.168}$$

showing that the values of a property A are determined as solutions of the N'th order polynomial equation, the characteristic equation,

$$\prod_a (x - a) = 0 \quad . \tag{1.169}$$

If we delete the a'-term from the characteristic product in a normalized fashion, we obtain the state symbol

$$\prod_{a \neq a'} \frac{\hat{A} - a\hat{I}}{a' - a} \ = \ \hat{P}(a') \ . \tag{1.170}$$

For later use we note that if an operator $\hat{X}$ satisfies the equation

$$\phi(\hat{A}) \ = \ (\hat{A} - a\hat{I}) \cdot \hat{X} \tag{1.171}$$

the operator is proportional to the projector corresponding to the property value a, $\hat{X} = c\,\hat{P}(a)$.

1.3.7   Adjoint

The reading of transition symbols from right to left is conventional. The other possible convention where transition symbols are read from left to right will be referred to as the adjoint reading, and the transition symbols in this reading are called adjoint transition symbols. Labeling the adjoint way of reading the transition operators by †, we have the relationship between the two conventions

$$\hat{P}(b\,;a)^\dagger \ = \ (|b{>}{<}a|)^\dagger \ \equiv \ \hat{P}(a\,;b) \ = \ |a{>}{<}b| \tag{1.172}$$

in either case representing the transition of a system from state b to state a.

Inverting between left and right reading we have

$$[\hat{P}(d\,;c) \cdot \hat{P}(b\,;a)]^\dagger \ = \ \hat{P}(a\,;b) \cdot \hat{P}(c\,;d) \ = \ \hat{P}(b\,;a)^\dagger \cdot \hat{P}(d\,;c)^\dagger \tag{1.173}$$

and applying the multiplication law on the above equation we obtain that

$$[{<}c|b{>} \, \hat{P}(d\,;a)]^\dagger \ = \ {<}b|c{>} \, \hat{P}(a\,;d) \ = \ {<}c|b{>}^* \, \hat{P}(d\,;a)^\dagger \ . \tag{1.174}$$

The adjoint operation is thus an antilinear isomorphic mapping of the operator space onto itself, and in general in the space of operators we define for any complex number c

$$(c\hat{X})^\dagger \ = \ c^*\hat{X}^\dagger \quad . \tag{1.175}$$

From the antilinearity of the adjoint operation we get for arbitrary operators

$$(\hat{X} + \hat{Y})^\dagger = \hat{X}^\dagger + \hat{Y}^\dagger \quad . \tag{1.176}$$

The adjoint reading induces an interchange of vectors and dual vectors, i.e., a mapping between the state space and its dual space, the vector space of linear functionals on the state space. By virtue of the involution property, eq.(1.155), this mapping is antilinear and isomorphic, and we use the same symbol as for the mapping of the transition symbols, $|\psi>^\dagger = <\psi|$ and $<\psi|^\dagger = |\psi>$.

By expanding on a basis, we recognize the adjoint operator or hermitian conjugate as the one that has the matrix elements

$$<b|\hat{X}^\dagger|a> \ = \ <a|\hat{X}|b>^* \tag{1.177}$$

or equivalently expressed in terms of the adjoint mapping between the state and dual space

$$(\hat{X}|\psi>)^\dagger \ = \ <\psi|\hat{X}^\dagger \ . \tag{1.178}$$

In particular we have for an eigenstate of an arbitrary operator

$$<a|\,\hat{A}^\dagger \ = \ <a|\,a^* \ . \tag{1.179}$$

Combining the adjoint rules for multiplication and addition we get

$$(\hat{X} \cdot \hat{Y})^\dagger \ = \ \left(\sum_{a,a'} <a|\hat{X}|a'> \hat{P}(a\,;a') \cdot \sum_{a'',a'''} <a''|\hat{Y}|a'''> \hat{P}(a''\,;a''') \right)^\dagger$$

$$= \ \hat{Y}^\dagger \cdot \hat{X}^\dagger \tag{1.180}$$

i.e., the adjoint transforms to the opposite sequence of adjoint operators.

An operator $\hat{X}$ satisfying the relation

$$\hat{X}^\dagger = \hat{X} \tag{1.181}$$

is said to be hermitian, and consequently its matrix representation is a hermitian matrix $<b|\hat{X}|a> \ = \ <a|\hat{X}|b>^*$. A hermitian operator is a normal operator, i.e., $\hat{X} \cdot \hat{X}^\dagger = \hat{X}^\dagger \cdot \hat{X}$, and has accordingly an orthogonal set of eigenvectors spanning the state space, and being hermitian its eigenvalues are real.

The state symbols are identical in the left and right conventions

$$\hat{P}(a)^\dagger = \hat{P}(a) \tag{1.182}$$

and are thus hermitian operators, and as a consequence so is any operator representing a physical quantity.

Reversing the reading direction twice is identical to the original reading and we have for any operator

$$(\hat{X}^\dagger)^\dagger \ = \ \hat{X} \ . \tag{1.183}$$

1.3.8   Unitary Equivalence

Two complete orthonormal sets of basis vectors in the state space, $\{|a_n>\}_n$ and $\{|b_n>\}_n$, $n = 1, .., N$; can be put in one-to-one correspondence through the linear operators

$$\hat{U}_{ab} = \sum_n |a_n><b_n| \,, \qquad \hat{U}_{ba} = \sum_n |b_n><a_n| \qquad (1.184)$$

because

$$<a_n|\hat{U}_{ab} = <b_n| \quad, \quad \hat{U}_{ab}|b_n> = |a_n> \quad, \quad \hat{U}_{ba}|a_n> = |b_n> \quad, \quad <b_n|\hat{U}_{ba} = <a_n| \,.$$
$$(1.185)$$

An operator connecting two orthonormal basis sets is said to be a unitary operator.

From orthonormality of the bases follows

$$\hat{U}_{ab} \cdot \hat{U}_{ba} = \hat{I} = \hat{U}_{ba} \cdot \hat{U}_{ab} \qquad (1.186)$$

i.e., the two operators are each other's inverse:

$$\hat{U}_{ab}^{-1} = \hat{U}_{ba} \quad . \qquad (1.187)$$

An equivalent statement for an operator $\hat{U}$ to be unitary is, according to eq.(1.177) and eq.(1.187), that the inverse equals the adjoint

$$\hat{U}^{-1} = \hat{U}^{\dagger} \quad . \qquad (1.188)$$

The existence of incompatible properties implies the existence of different but unitary equivalent descriptions of a system. Two unitary equivalent descriptions describe a physical system in terms of two different sets of basis vectors. State vectors in the two descriptions $|\underline{\psi}>$ and $|\psi>$ are thus unitarily related $|\underline{\psi}> = \hat{U}^{-1}|\psi>$. A physical quantity is represented by different operators in the two descriptions, $\hat{X}$ and $\hat{\underline{X}}$, respectively. However, the set of property values is one and the same in the two descriptions. The operator $\hat{\underline{X}}$ measured in the state $|\underline{x}>$ we therefore demand to give the same value x as the equivalent operator $\hat{X}$ measured in the equivalent state $|x>$, as expressed by the eigenvalue equations

$$\hat{X} |x> = x |x> \,, \qquad \hat{\underline{X}} |\underline{x}> = x |\underline{x}> \quad . \qquad (1.189)$$

From this requirement we obtain

$$\hat{\underline{X}} = \hat{U}^{-1}\hat{X}\hat{U} \,. \qquad (1.190)$$

as

$$\hat{X} = \sum_x |x> x <x| \,, \qquad \hat{\underline{X}} = \sum_{\underline{x}} |\underline{x}> x <\underline{x}| \,. \qquad (1.191)$$

For the adjoint we have $\hat{\underline{X}}^{\dagger} = \hat{U}^{-1}\hat{X}^{\dagger}\hat{U}$, and all algebraic relations, such as

$$<\underline{x}|\underline{\psi}> = <x|\psi> \,, \qquad <\underline{\psi}| \hat{\underline{X}} |\underline{\psi}'> = <\psi| \hat{X} |\psi'> \qquad (1.192)$$

are preserved by the transformation.

Writing a unitary operator on the form

$$\hat{U} = \hat{I} + i\hat{G} \tag{1.193}$$

and inserting into the unitarity condition, $\hat{U} \cdot \hat{U}^\dagger = \hat{I}$, gives

$$\hat{G} \cdot \hat{G}^\dagger + i\hat{G} - i\hat{G}^\dagger = \hat{0} \tag{1.194}$$

which shows that a unitary transformation close to the identity operator is specified by a hermitian operator, the so-called generator. The eigenvalues of a unitary operator have the absolute value 1, $|u|^2 = 1$, and a unitary operator can be expressed as the exponential of the generator $\hat{G}$, $\hat{U} = e^{i\hat{G}}$, and $\hat{U}^\dagger = e^{-i\hat{G}} = \hat{U}^{-1}$.

Since the unit operator is unitary, and the product of two unitary operators is a unitary operator, the set of unitary operators on the state space forms a group with $\hat{I}$ as the identity element.

1.3.9   Trace

The linear functional on the measurement algebra mapping the transition operator onto the corresponding transformation function

$$Tr\hat{P}(b;a) \;=\; <a|b> \tag{1.195}$$

is called the trace. The trace of an arbitrary operator

$$Tr\hat{X} \;=\; \sum_{a,a'} <a|\hat{X}|a'> Tr\hat{P}(a;a') \;=\; \sum_{a} <a|\hat{X}|a> \tag{1.196}$$

is seen to be the sum of its diagonal elements in an arbitrary orthonormal basis.

Using the multiplication law we have (we shall from now on feel free to drop the *subsequent* symbol $\cdot$ for the operator product)

$$Tr(\hat{P}(d;c)\,\hat{P}(b;a)) = Tr(\hat{P}(b;a)\,\hat{P}(d;c)) \tag{1.197}$$

and it follows that any pair of operators can be commuted under the trace

$$Tr(\hat{A}\hat{B}) = Tr(\hat{B}\hat{A}) \tag{1.198}$$

or, in general, a sequence of operators can under the trace be permuted cyclically; for example,

$$Tr(\hat{A}\hat{B}\hat{C}\hat{D}) \;=\; Tr(\hat{B}\hat{C}\hat{D}\hat{A}) \;=\; .. \;=\; Tr(\hat{D}\hat{A}\hat{B}\hat{C})\,. \tag{1.199}$$

We note that any matrix element of an operator can be expressed through the trace

$$<a|\hat{X}|b> \;=\; Tr(\hat{X}\hat{P}(b;a)) \tag{1.200}$$

and in particular the left-right reading invariance of the formalism corresponds to the various trace formulas for the conditional probability

$$p(b\,;a) = Tr(\hat{P}(b)\hat{P}(a)) = <a|\hat{P}(b)|a> = <b|\hat{P}(a)|b> = Tr(\hat{P}(a)\hat{P}(b)) = p(a\,;b).$$
$$(1.201)$$

In general, given the system is in state ψ, we can express the probability of finding the system in state a in terms of the trace of the state symbols

$$p(a\,;\psi) = Tr(\hat{P}(\psi)\hat{P}(a)) \ . \tag{1.202}$$

Let us say we are interested in a physical quantity A of a system prepared in the state ψ. Since our statements about the possible results of a subsequent A-selection, or A-measurement, are probabilistic, we must in general perform a series of measurements in order to verify the predictions of the quantum theory. The result of such a series of $\mathcal{N}$ repeated measurements on identically prepared states will give the outcome a, say, an $r_a^{(\mathcal{N})}$ number of times. In the limit of many experiments the relative frequency r_a^{ψ} (the number of measurements with outcome a divided by the total number of measurements $\mathcal{N}$) approaches the value predicted by the theory

$$r_a^{\psi} \ \equiv \ \lim_{\mathcal{N}\to\infty} \frac{r_a^{\mathcal{N}}}{\mathcal{N}} \ = \ p(a\,;\psi) \ = \ \frac{|<a|\psi>|^2}{<\psi|\psi>} \ . \tag{1.203}$$

The last expression is invariant with respect to the magnitude of the initial state vector, and allows for a nonnormalized choice of state vector.[55] For a normalized initial state vector

$$1 \ = \ <\psi|\psi> \ = \ \sum_a \ <\psi|a><a|\psi> \ = \ \sum_a \ |\psi(a)|^2 \ = \ \sum_a \ p(a\,;\psi) \tag{1.204}$$

the probability frequencies for the series of outcomes of the measurements are specified by eq.(1.202), or alternatively stated on the various forms

$$r_a^{\psi} = <\psi|\hat{P}(a)|\psi> = Tr(\hat{P}(\psi)\,\hat{P}(a)) = Tr(\hat{P}(\psi)\,\delta_{\hat{A},a}) \ . \tag{1.205}$$

The average value for the measurements of the quantity A on a system in state ψ (the relative frequency of the outcome a times the value a, summed over all outcomes a) is, in statistics, referred to as the mean or expectation value of the quantity A

$$\bar{A}_{\psi}^{Av} \ \equiv \ \sum_a a\,r_a^{\psi} \ = \ \sum_a a\,p(a;\psi) = Tr(\hat{P}(\psi)\,\hat{A}) = <\psi|\hat{A}|\psi> \ \equiv \ <\hat{A}>_{\psi} \tag{1.206}$$

and is seen to be the matrix element of the operator and state in question.

---

[55]We note that addition of a state vector with itself does thus not result in a state vector representing a different state, the marked difference of the quantum mechanical superposition principle as compared to superposition principles in classical physics (stretching the pattern on a drumhead to twice its height gives again a solution to the linear wave equation, but corresponding to a different physical state). However, unlike in classical physics the principle of superposition is obeyed exactly.

Exercise 1.9 *On the vector space of operators*

$$<\hat{X}|\hat{Y}> \equiv Tr(\hat{X}^\dagger\hat{Y}) \qquad (1.207)$$

defines an inner product, and the transition operators constitute an orthonormal basis

$$<\hat{P}(b\,;a)|\hat{P}(b'\,;a')> \ = \ \delta_{b,b'}\,\delta_{a,a'} \ . \qquad (1.208)$$

Show that the completeness of an orthonormal set $\{\hat{X}_\alpha\}_{\alpha=1,2,..,N^2}$ is equivalent to the validity of

$$\sum_{\alpha=1}^{N^2} \hat{X}_\alpha\,\hat{X}\,\hat{X}_\alpha^\dagger \ = \ \hat{I}\,Tr\hat{X} \qquad (1.209)$$

for an arbitrary operator $\hat{X}$.

  Show that it is always possible to choose an orthonormal basis of hermitian operators and that the matrix $<\alpha|\beta> \ \equiv \ <\hat{X}_\alpha|\hat{Y}_\beta>$ connecting two hermitian orthonormal bases $\{\hat{Y}_\beta\}_{\beta=1,2,..,N^2}$ and $\{\hat{X}_\alpha\}_{\alpha=1,2,..,N^2}$

$$\hat{Y}_\beta \ = \ \sum_{\alpha=1}^{N^2} <\alpha|\beta> \ \hat{X}_\alpha \qquad (1.210)$$

is real and orthogonal. The subspace of hermitian operators is thus a Euclidean space.

  Show that for a hermitian orthonormal basis $\{\hat{Y}_\alpha\}$, $Tr(\hat{Y}_\alpha\hat{Y}_\beta) \propto \delta_{\alpha,\beta}$, where one of the basis vectors is proportional to the identity operator, $\hat{Y}_\gamma \propto \hat{I}$, we have

$$Tr\hat{Y}_\alpha = 0 \ , \quad \alpha \neq \gamma. \qquad (1.211)$$

Solution

  Expanding $\hat{P}(b\,;a)$ and $\hat{P}(b'\,;a')$ on the basis results in

$$\delta_{b,b'}\,\delta_{a,a'} = Tr(\hat{P}(b\,;a)^\dagger\,\hat{P}(b'\,;a')) = \sum_{\alpha=1}^{N^2} <b|\hat{X}_\alpha|a><a'|\hat{X}_\alpha^\dagger|b'> \ . \qquad (1.212)$$

Multiplying this equation with $<a|\hat{X}|a'>$, and summing over a and a' results in eq.(1.209). A possible choice for a hermitian orthonormal operator basis is the set consisting of $\hat{P}(a\,;a)$, and for $a \neq a'$, $2^{-1/2}[\hat{P}(a\,;a')+\hat{P}(a'\,;a)]$ and $i2^{-1/2}[\hat{P}(a\,;a')-\hat{P}(a'\,;a)]$. The matrix $r_{\alpha\beta} = <\alpha|\beta>$ is clearly real since the operators involved are hermitian, and using that the two operator bases are orthonormal, for example $Tr(\hat{Y}_\alpha\hat{Y}_\beta) = \delta_{\alpha,\beta}$, it follows that the matrix $r_{\alpha\beta}$ is orthogonal

$$\delta_{\alpha,\beta} = Tr\left(\hat{Y}_\alpha\,\hat{Y}_\beta\right) = Tr\left(\sum_{\alpha',\beta'} r_{\alpha\alpha'}\hat{X}_{\alpha'}r_{\beta\beta'}\hat{X}_{\beta'}\right) = \sum_{\alpha',\beta'} r_{\alpha\alpha'}\,r_{\beta\beta'}\,\delta_{\alpha',\beta'} = \sum_{\alpha'} r_{\alpha\alpha'}\,r_{\beta\alpha'} \ . \qquad (1.213)$$

The $N^2 - 1$ operators differing from the identity operator are traceless as $\alpha \neq \gamma$, $Tr(\hat{Y}_\alpha) = Tr(\hat{Y}_\alpha\hat{I}) \propto Tr(\hat{Y}_\alpha\hat{Y}_\gamma) = 0$.

Exercise 1.10 *Show that if* $\{\hat{X}_\alpha\}_{\alpha=1,2,..,N^2}$ *is an orthonormal basis on the vector space of operators, so is the set of vectors*

$$\hat{Y}_\alpha \equiv \hat{U}^{-1}\,\hat{X}_\alpha\,\hat{U} \tag{1.214}$$

for any unitary operator $\hat{U}$, *and*

$$<\hat{X}_\alpha|\hat{X}_{\alpha''}\hat{X}_{\alpha'}> \;=\; <\hat{Y}_\alpha|\hat{Y}_{\alpha''}\hat{Y}_{\alpha'}> \;\equiv\; <\alpha|\alpha''\alpha'> \;. \tag{1.215}$$

Show, conversely, that if two orthonormal bases obey the above multiplication property; i.e.,

$$\hat{X}_{\alpha''}\hat{X}_{\alpha'} \;=\; \sum_{\alpha=1}^{N^2} <\alpha|\alpha''\alpha'>\,\hat{X}_\alpha \tag{1.216}$$

and

$$\hat{Y}_{\alpha''}\hat{Y}_{\alpha'} \;=\; \sum_{\alpha=1}^{N^2} <\alpha|\alpha''\alpha'>\,\hat{Y}_\alpha \tag{1.217}$$

they are related according to eq.(1.214), with the unitary operator given (uniquely up to a phase factor) by

$$\hat{U} \;=\; \frac{1}{\left(\sum\limits_{\alpha=1}^{N^2} <\hat{X}_\alpha|\hat{Y}_\alpha>\right)^{1/2}} \sum_{\alpha=1}^{N^2} \hat{X}_\alpha \hat{Y}_\alpha^\dagger \;. \tag{1.218}$$

Solution

Orthonormality of the basis $\{\hat{Y}_\alpha\}_{\alpha=1,2,...,N^2}$ *follows from the cyclic property of the trace, for example,* $<\hat{Y}_\alpha|\hat{Y}_{\alpha'}> = Tr(\hat{U}^{-1}\hat{X}_\alpha\hat{U}\hat{U}^{-1}\hat{X}_{\alpha'}\hat{U}) = <\hat{X}_\alpha|\hat{X}_{\alpha'}> = \delta_{\alpha,\alpha'}$, *and completeness simply because there are* N^2 *operators. The converse statement follows from the easily proven identity*

$$\left(\sum_{\alpha'=1}^{N^2} \hat{X}_{\alpha'}\,\hat{Y}_{\alpha'}^\dagger\right) \hat{Y}_\alpha \;=\; \hat{X}_\alpha \left(\sum_{\alpha'=1}^{N^2} \hat{X}_{\alpha'}\,\hat{Y}_{\alpha'}^\dagger\right) . \tag{1.219}$$

1.4 Complementarity

We have already in section 1.3.5 stressed the all important aspect that all the attributes of a physical system are not compatible. The incompatibility of the properties of a system can be elucidated by representing the measurement algebra in a special basis. This basis is expressible in terms of products of simply two operators which represent maximally incompatible properties, i.e., properties for which if either one of the quantities has a definite value, the probabilities for the other property are equally distributed. We shall also refer to maximally incompatible properties as complementary properties of the system. Definite knowledge of a property value thus excludes completely any knowledge of the value of the complementary one.

The construction of the complementary operator basis is achieved by considering the simplest of unitary transformations. Given an orthonormal basis in the state space we can generate a unitary transformation $\hat{\mathcal{U}}$ through a permutation of the basis[56]

$$\hat{\mathcal{U}} \, |a_n> = \, |a_{n+1}> \, , \quad n = 1, 2, .., N - 1. \qquad \hat{\mathcal{U}} \, |a_N> = \, |a_1> \, . \tag{1.220}$$

Explicitly we have

$$\hat{\mathcal{U}} \, = \, \sum_{n=1}^{N} |a_{n+1}><a_n| \tag{1.221}$$

where we define $|a_{N+1}> \equiv |a_1>$.[57]

Repeating the transformation through the cycle gives

$$\hat{\mathcal{U}}^N - \hat{I} = \hat{0} \, . \tag{1.222}$$

Hence the eigenvalues u_n of the unitary operator $\hat{\mathcal{U}}$

$$\hat{\mathcal{U}} \, |u_n> = \, u_n \, |u_n> \, , \quad n = 1, 2, .., N; \tag{1.223}$$

satisfy the equation $(u_n)^N = 1$, the roots of the characteristic equation thus being

$$u_n = e^{\frac{2\pi i n}{N}} \, , \quad n = 1, 2, .., N; \tag{1.224}$$

the eigenvalues of the orthonormal set $\{|u_n>\}_{n=1,2,..,N}$ of eigenvectors of $\hat{\mathcal{U}}$. Making the phase choice for the $|u_n>$'s

$$<a_N|u_n> \, \equiv \, \frac{1}{\sqrt{N}} \, , \quad n = 1, 2, .., N; \tag{1.225}$$

we readily verify that the orthonormal basis of eigenvectors of the unitary transformation $\hat{\mathcal{U}}$ has the expansion on the orthonormal a-basis

$$|u_n> \, = \, \frac{1}{\sqrt{N}} \sum_{m=1}^{N} |a_m> e^{-\frac{2\pi i n m}{N}} \, , \quad n = 1, 2, .., N. \tag{1.226}$$

---

[56]The labeling is of course arbitrary; we could equally well have used another labeling, say, $n = 0, 1, 2, .., N - 1$, or $n = 0, \pm 1, \pm 2, .., \pm(N - 1)/2$ for the case of an odd number of property values. The different choices of labeling can be compensated by different phase choices for basis vectors.

[57]For the adjoint we have $\hat{\mathcal{U}}^\dagger = \hat{\mathcal{U}}^{-1} = \sum_{n=1}^{N} |a_n><a_{n+1}|$, or equivalently, the inverse operator corresponds to the inverse permutation, $\hat{\mathcal{U}}^\dagger|a_n> = |a_{n-1}>, n = 2, .., N. \; \hat{\mathcal{U}}^\dagger \, |a_1> = |a_N>$.

Accordingly we have for the dual basis

$$< u_n| \ = \ \frac{1}{\sqrt{N}} \sum_{m=1}^{N} < a_m| \, e^{\frac{2\pi i n m}{N}} \ , \qquad n = 1, 2, .., N. \qquad (1.227)$$

Rewriting eq.(1.222) we have

$$\hat{0} \ = \ u_n \left(\hat{\mathcal{U}}^N - \hat{I} \right) \ = \ u_n \left(\left(\frac{\hat{\mathcal{U}}}{u_n} \right)^N - \hat{I} \right) \ = \ (\hat{\mathcal{U}} - u_n \hat{I}) \sum_{m=0}^{N-1} \left(\frac{\hat{\mathcal{U}}}{u_n} \right)^m . \qquad (1.228)$$

In view of eq.(1.222) we have

$$c \left(\hat{\mathcal{U}}^N - \hat{I} \right) \ = \ \prod_{n=1}^{N} (\hat{\mathcal{U}} - u_n \hat{I}) \ \equiv \ \phi(\hat{\mathcal{U}}) . \qquad (1.229)$$

and recalling eq.(1.171), we identify the operator sum in eq.(1.228) as the projector corresponding to state $|u_n>$[58]

$$\hat{P}(u_n) \ = \ |u_n><u_n| \ = \ \frac{1}{N} \sum_{m=0}^{N-1} \hat{\mathcal{U}}^m \, e^{-\frac{2\pi i n m}{N}} \qquad (1.230)$$

where the prefactor is fixed by normalization, $\hat{P}(u_n) \cdot \hat{P}(u_n) = \hat{P}(u_n)$.

The orthonormal u-basis can also be subdued to a cyclic permutation defining yet another unitary transformation $\hat{\mathcal{V}}$

$$< u_n| \, \hat{\mathcal{V}} \ = \ < u_{n+1}| \ , \ n = 1, 2, .., N-1. \qquad < u_N| \, \hat{\mathcal{V}} \ = \ < u_1| \qquad (1.231)$$

and similarly to the previous analysis we obtain the projectors and eigenvalues

$$\hat{P}(v_n) \ = \ |v_n><v_n| \ = \ \frac{1}{N} \sum_{m=0}^{N-1} \hat{\mathcal{V}}^m \, e^{-\frac{2\pi i n m}{N}} \ , \ v_n = e^{\frac{2\pi i n}{N}} \ , \ n = 1, 2, .., N. \qquad (1.232)$$

Because of the phase choice for the $|v_m>$'s

$$< u_N|v_m> = \frac{1}{\sqrt{N}} \ , \qquad m = 1, 2, .., N. \qquad (1.233)$$

we find that the orthonormal basis of dual eigenvectors of the unitary transformation $\hat{\mathcal{V}}$ has the expansion on the orthonormal u-basis

$$< v_m| \ = \ \frac{1}{\sqrt{N}} \sum_{n=1}^{N} < u_n| \, e^{-\frac{2\pi i m n}{N}} \ , \ m = 1, 2, .., N. \qquad (1.234)$$

Observing the identity of the matrix elements

$$< u_n|v_m> \ = \ \frac{1}{\sqrt{N}} \, e^{\frac{2\pi i n m}{N}} \ = \ < u_n|a_m> \qquad (1.235)$$

---

[58] The expression for the projector is also immediately verified by use of the identity $\sum_{m=0}^{N-1} e^{-\frac{2\pi i m (n - n')}{N}} = N \, \delta_{n,n'}$.

we realize that the orthonormal basis of eigenvectors for the last unitary transformation is identical to the original basis

$$|v_n> \ = \ |a_n> \quad , \quad n = 1, 2, .., N; \tag{1.236}$$

and we have

$$\hat{\mathcal{U}} |v_n> \ = \ |v_{n+1}> \quad , \quad <u_n| \hat{\mathcal{V}} \ = \ <u_{n+1}| \ . \tag{1.237}$$

The set of operators $\hat{\mathcal{U}}^n \hat{\mathcal{V}}^m, n, m = 1, 2, .., N$, spans the vector space of transition operators since we, for example, can express the completeness through explicit construction of any basis vector in the complete operator basis:

$$|u_n> <u_{n+m}| \ = \ \frac{1}{N} \sum_{n'=1}^{N} e^{-\frac{2\pi i n n'}{N}} \hat{\mathcal{U}}^{n'} \hat{\mathcal{V}}^m \ . \tag{1.238}$$

Exercise 1.11 *Show that an operator $\hat{X}$ which commutes with both $\hat{\mathcal{U}}$ and $\hat{\mathcal{V}}$ is proportional to the identity operator.*

Solution

The quantity corresponding to $\hat{\mathcal{U}}$ is a complete description, and since $\hat{X}$ and $\hat{\mathcal{U}}$ commute we have $\hat{X} = f(\hat{\mathcal{U}})$. From

$$0 \ = \ <u_n|[f(\hat{\mathcal{U}}), \hat{\mathcal{V}}]|u_{n+1}> \ = \ f(u_n) - f(u_{n+1}) \tag{1.239}$$

it follows that f takes on the same value for all property values, and $f(\hat{\mathcal{U}})$ is therefore proportional to the identity.

The properties represented by the operators $\hat{\mathcal{U}}$ and $\hat{\mathcal{V}}$ each provide a complete description. Equivalently, the property represented by the generator of $\hat{\mathcal{U}}$ constitutes a complete description

$$\hat{\mathcal{U}} = e^{i\frac{2\pi}{N}\hat{x}} \tag{1.240}$$

and equivalently for $\hat{\mathcal{V}}$

$$\hat{\mathcal{V}} = e^{i\frac{2\pi}{N}\hat{p}} \ . \tag{1.241}$$

The generators are seen to be defined according to

$$\hat{x} |u_n> \ = \ n|u_n> \quad , \quad n = 1, 2, .., N; \tag{1.242}$$

and

$$\hat{p} |v_n> \ = \ n|v_n> \quad , \quad n = 1, 2, .., N. \tag{1.243}$$

The generators are usually chosen as representing the physical quantities as they are the ones with real eigenvalues.[59]

The constructed $\mathcal{U}$ and $\mathcal{V}$ descriptions of a physical system (or equivalently the x and p descriptions) have a unique relationship, they are maximally incompatible, or as we say, complementary. The probability of finding the system in state u_n given that it is in state v_m is

$$p(u_n; v_m) \;=\; |<u_n|v_m>|^2 \;=\; \frac{1}{N} \; . \tag{1.244}$$

Precise knowledge of the quantity $\mathcal{V}$ thus excludes any knowledge of the quantity $\mathcal{U}$, and vice versa. The physical quantities represented by x and p are similarly complementary properties, $p(x_n; p_m) = |<u_n|v_m>|^2 = p(u_n; v_m)$.

We have thus constructed two complementary operators in terms of which we can represent any physical property F of the system

$$\hat{F} = \sum_{n,m=1}^{N} f_{nm} \, \hat{\mathcal{U}}^n \, \hat{\mathcal{V}}^m \;=\; \sum_{n,m=1}^{N} f_{nm} \, e^{i\frac{2\pi n}{N}\hat{x}} \, e^{i\frac{2\pi m}{N}\hat{p}} \; . \tag{1.245}$$

The quantum kinematics of a system is thus expressed through a set of operators which form a group [11].

Noting for the commutator of $\hat{\mathcal{U}}$ and $\hat{\mathcal{V}}$, $[\hat{\mathcal{U}}, \hat{\mathcal{V}}] \equiv \hat{\mathcal{U}}\hat{\mathcal{V}} - \mathcal{V}\hat{\mathcal{U}}$,

$$<u_n| \, [\hat{\mathcal{U}}, \hat{\mathcal{V}}] \;=\; (u_n - u_{n+1}) <u_{n+1}| \;=\; (1 - e^{\frac{2\pi i}{N}}) <u_n|\hat{\mathcal{U}}\hat{\mathcal{V}} \tag{1.246}$$

we obtain

$$\hat{\mathcal{V}}\hat{\mathcal{U}} \;=\; e^{\frac{2\pi i}{N}} \, \hat{\mathcal{U}}\hat{\mathcal{V}} \tag{1.247}$$

and by repeated use we get the fundamental identity

$$\hat{\mathcal{V}}^n \hat{\mathcal{U}}^m \;=\; e^{\frac{2\pi i n m}{N}} \, \hat{\mathcal{U}}^m \, \hat{\mathcal{V}}^n \; . \tag{1.248}$$

Since

$$<\hat{\mathcal{U}}^n \, \hat{\mathcal{V}}^m | \hat{\mathcal{U}}^{n'} \, \hat{\mathcal{V}}^{m'}> \;=\; N \, \delta_{n,n'} \, \delta_{m,m'} \tag{1.249}$$

we can choose as orthonormal basis in the operator space the set of vectors $\{\hat{B}(n,m)\}_{n,m=1,2,..,N}$,

$$\hat{B}(n, m) \;\equiv\; \frac{1}{\sqrt{N}} e^{\frac{i\pi n m}{N}} \, \hat{\mathcal{U}}^n \, \hat{\mathcal{V}}^m \; . \tag{1.250}$$

We note the equivalent expressions for these basis vectors

$$\frac{1}{\sqrt{N}} e^{\frac{i\pi n m}{N}} \, \hat{\mathcal{U}}^n \, \hat{\mathcal{V}}^m \;=\; \frac{1}{\sqrt{N}} e^{\frac{-i\pi n m}{N}} \, \hat{\mathcal{V}}^m \, \hat{\mathcal{U}}^n \; . \tag{1.251}$$

---

[59]We could equally well have defined the unitary transformations inversely, i.e., by the inverse permutation, $\hat{\mathcal{U}}|a_n> = |a_{n-1}>$, corresponding to the interchange $\hat{\mathcal{U}} \to \hat{\mathcal{V}}, \hat{\mathcal{V}} \to \hat{\mathcal{U}}^{-1}$, or equivalently, $\hat{x} \to \hat{p}, \hat{p} \to -\hat{x}$, reflecting an invariance in the kinematic description.

Exercise 1.12 *Show that the operator $\sum_{n,m=1}^{N} \hat{B}(n,m) \, \hat{X} \, \hat{B}(n,m)^{\dagger}$, for an arbitrary operator $\hat{X}$, commutes with both $\hat{U}$ and $\hat{V}$, and conclude that the operator is given by $\sum_{n,m=1}^{N} \hat{B}(n,m) \, \hat{X} \, \hat{B}(n,m)^{\dagger} = \hat{I} \, Tr\hat{X}$.*

Exercise 1.13 *Consider a system that can take on only two property values, say ± 1, with the corresponding eigenstates $|\pm>$. Making the phase choice eq.(1.233), show that the complementary operators have the matrix representations*

$$<\pm|\hat{U}|\pm> \;=\; \begin{pmatrix} 0 & 1 \\ 1 & 0 \end{pmatrix} \equiv \tau^{(1)} \tag{1.252}$$

and

$$<\pm|\hat{V}|\pm> \;=\; \begin{pmatrix} 1 & 0 \\ 0 & -1 \end{pmatrix} \equiv \tau^{(3)}. \tag{1.253}$$

The matrices $\tau^{(1)}$ and $\tau^{(3)}$ are thus the matrix representations of the two complementary operators. The operator representing the physical quantity in question (for the chosen phase choice) is $\hat{V}$ as

$$\hat{V} \;=\; \sum_{v=\pm 1} |v> v <v| = |+><+| \;-\; |-><-|. \tag{1.254}$$

Solution

The matrix representation of $\hat{U}$ follows directly from eq.(1.221), as we have $\hat{U} = |+><-| \;+\; |-><+|$. From eq.(1.232) we obtain, $|v_2> = |+>$,

$$|+><+| \;=\; \frac{1}{2}(\hat{I} + e^{-2\pi i}\hat{V}) \tag{1.255}$$

and since $\hat{I} = |+><+| \;+\; |-><-|$, we obtain $\hat{V} = |+><+| \;-\; |-><-|$, and thereby the stated matrix representation.

Exercise 1.14 *Introduce for a two-state system the operator $\hat{\tau}_2$ with the matrix representation[60]*

$$<\pm|\hat{\tau}_2|\pm> \;=\; \begin{pmatrix} 0 & -i \\ i & 0 \end{pmatrix} \equiv \tau^{(2)} \tag{1.256}$$

---

[60]The matrices $\tau^{(1)}$, $\tau^{(2)}$, and $\tau^{(3)}$ are called the Pauli matrices, and are in the context of the spin of the electron denoted σ_x, σ_y, and σ_z, respectively. The two states are referred to as the spin up and down states.

and the notation $\hat{\tau}_0 \equiv \hat{I}$, $\hat{\tau}_1 \equiv \hat{U}$, *and* $\hat{\tau}_3 \equiv \hat{V}$. *Express the* τ-*operators in terms of the orthonormal basis* $\hat{P}(\pm\,;\pm)$ *of the four-dimensional operator space. Show that the four operators* $\hat{\tau}_0/\sqrt{2}, \hat{\tau}_1/\sqrt{2}, \hat{\tau}_2/\sqrt{2}, \hat{\tau}_3/\sqrt{2}$ *constitute a hermitian orthonormal basis for which*

$$\hat{\tau}_k\hat{\tau}_l = \delta_{k,l}\,\hat{I} + i\sum_{m=1}^{3}\epsilon_{klm}\hat{\tau}_m \quad , \quad k,l = 1, 2, 3; \tag{1.257}$$

and thereby

$$Tr(\hat{\tau}_k\hat{\tau}_l\hat{\tau}_m) = 2i\epsilon_{klm}\,. \tag{1.258}$$

For the two-state system the complementary operators thus anticommute, and have unit square.

   Introduce the vector notation $\hat{\vec{\tau}} \equiv (\hat{\tau}_1, \hat{\tau}_2, \hat{\tau}_3)$, *and show that,* $\hat{s} \equiv \hat{\vec{\tau}}/2$,

$$\hat{s} \times \hat{s} = i\,\hat{s} \quad \text{and that} \quad \hat{s}^2 = \frac{1}{2}\left(\frac{1}{2} + 1\right)\hat{I}\,. \tag{1.259}$$

Solution

   For $\hat{\tau}_2$ *we have*

$$\hat{\tau}_2 = -i\,|+><-| \quad + \quad i\,|-><+| \tag{1.260}$$

and the expressions for the other operators were obtained in the previous exercise. The operator $\hat{\tau}_2$ *is clearly hermitian, as are the other three. By straightforward calculation we obtain eq.(1.257), for example,*

$$\hat{\tau}_1\hat{\tau}_2 = (|+><-| \quad + \quad |-><+|)\,(-i\,|+><-| \quad + \quad i\,|-><+|)$$

$$= i\,|+><+| \quad - \quad i\,|-><-|) = i\hat{\tau}_3 \tag{1.261}$$

and orthonormality of the operators follows from eq.(1.257), since according to exercise 1.9 on page 47 we have $Tr\,\hat{\tau}_k = 0, k = 1, 2, 3.$

Exercise 1.15 *Let* $\{\hat{\tau}_\alpha\}_{\alpha=0,1,2,3.}$ *and* $\{\hat{\vec{\tau}}\}_{\alpha=0,1,2,3.}$ *both be orthogonal bases for which* $\hat{\tau}_0 \propto \hat{I}$ *and* $\hat{\vec{\tau}}_0 \propto \hat{I}$, *so that*

$$\hat{\vec{\tau}}_k = \sum_{l=1}^{3} r_{kl}\,\hat{\tau}_l \quad , \quad k = 1, 2, 3; \tag{1.262}$$

where r_{kl}, *according to exercise 1.9 on page 47, is a real orthogonal matrix. We can thus conceive of* $\hat{\vec{\tau}}$ *(or* $\hat{s}$) *as representing a vector quantity in a (internal quantum) space with orthogonal axes labeled* $1, 2, 3$. *Choosing* $\hat{\tau}_3$ *as representing the physical property (as we have done with our phase choice in exercise 1.13) any* $\hat{\tau}$-*operator in the perpendicular* 1, 2-*plane,* $\hat{\tau} = c_1 \hat{\tau}_1 + c_2 \hat{\tau}_2$, *represents a complementary property. Only for a proper orthogonal matrix, det* $\underline{r} = +1$, *does a unitary transformation exist*

$$\hat{\hat{\tau}}_k = \hat{U}^{-1} \hat{\tau}_k \hat{U} \ , \ \ k = 1, 2, 3. \tag{1.263}$$

since otherwise the multiplication properties, eq.(1.216) and eq.(1.217), are not preserved, but involve a minus sign. Show that the above unitary operator is given by

$$2(1 + tr\,\underline{r})^{1/2} \hat{U} = \sum_{\alpha=0}^{3} \hat{\tau}_\alpha \hat{\hat{\tau}}_\alpha = \sum_{\alpha,\beta=0}^{3} \hat{\tau}_\alpha < \alpha|\beta > \hat{\tau}_\beta = (1 + tr\,\underline{r})\hat{I} + i \sum_{k,l,m=1}^{3} r_{kl}\, \epsilon_{klm}\, \hat{\tau}_m \tag{1.264}$$

where

$$tr\,\underline{r} \equiv r_{11} + r_{22} + r_{33} \tag{1.265}$$

is the trace of the real orthogonal matrix $\underline{r} = \{r_{kl}\}_{k,l=1,2,3}$.

Introducing the parametrization of the orthogonal matrices in terms of the Euler angles, an arbitrary rotation in three dimensions

$$\underline{r} = \begin{pmatrix} \cos\psi\cos\phi - \cos\theta\sin\phi\sin\psi & \cos\psi\sin\phi + \cos\theta\cos\phi\sin\psi & \sin\psi\sin\theta \\ -\sin\psi\cos\phi - \cos\theta\sin\phi\cos\psi & -\sin\psi\sin\phi + \cos\theta\cos\phi\cos\psi & \cos\psi\sin\theta \\ \sin\theta\sin\phi & -\sin\theta\cos\phi & \cos\theta \end{pmatrix}$$

$$= \begin{pmatrix} \cos\psi & \sin\psi & 0 \\ -\sin\psi & \cos\psi & 0 \\ 0 & 0 & 1 \end{pmatrix} \begin{pmatrix} 1 & 0 & 0 \\ 0 & \cos\theta & \sin\theta \\ 0 & -\sin\theta & \cos\theta \end{pmatrix} \begin{pmatrix} \cos\phi & \sin\phi & 0 \\ -\sin\phi & \cos\phi & 0 \\ 0 & 0 & 1 \end{pmatrix}$$

$$= \underline{r}^{(3)}(\psi)\, \underline{r}^{(1)}(\theta)\, \underline{r}^{(3)}(\phi) \equiv \underline{r}(\phi, \theta, \psi) \tag{1.266}$$

is described in terms of three successive rotations $(0 \le \phi, \psi < 2\pi; 0 \le \theta < \pi)$. *For the corresponding unitary transformation, eq.(1.263), we have* $\hat{U}(\phi, \theta, \psi) = \hat{U}_3(\psi)\, \hat{U}_1(\theta)\, \hat{U}_3(\phi)$ *and for the matrix representation* $\underline{U}(\phi, \theta, \psi) = \underline{U}^{(3)}(\psi)\, \underline{U}^{(1)}(\theta)\, \underline{U}^{(3)}(\phi)$. *Show that the corresponding unitary operator, eq.(1.264), has the matrix representation*[61]

$$\underline{U}(\phi, \theta, \psi) \equiv \ <\pm|\hat{U}|\pm> = \ \begin{pmatrix} e^{\frac{i}{2}(\phi+\psi)} \cos\frac{\theta}{2} & i\sin\frac{\theta}{2} e^{\frac{i}{2}(\psi-\phi)} \\ i\sin\frac{\theta}{2} e^{-\frac{i}{2}(\psi-\phi)} & e^{-\frac{i}{2}(\phi+\psi)} \cos\frac{\theta}{2} \end{pmatrix} \equiv \underline{U} \ . \tag{1.267}$$

Solution

---

[61]The matrices $\underline{r}(\phi, \theta, \psi)$ form a continuously connected $(2^2 - 1)$-parameter compact group, the special (we are not including inversion) three-dimensional orthogonal group $SO(3)$ of 3×3 real orthogonal matrices $\underline{r}$ with determinant 1. The corresponding proper unitary matrices, i.e., $det\,\underline{U}(\phi, \theta, \psi) = +1$, form a continuously connected, $(2^2 - 1)$-parameter compact Lie group $SU(2)$, a subgroup of $U(2)$, the group of unitary 2×2 matrices.

Using the result of exercise 1.10 on page 48, we obtain

$$\sum_{\alpha=0}^{3} \left\langle \frac{1}{\sqrt{2}}\hat{\tau}_\alpha \left| \frac{1}{\sqrt{2}}\hat{\tau}_\alpha \right\rangle \right. = 1 + tr\,\underline{r} \qquad (1.268)$$

and according to eq.(1.218) and eq.(1.257) we have

$$\sum_{\alpha=0}^{3} \hat{\tau}_\alpha \hat{\tau}_\alpha = \hat{I} + i\sum_{k,l=1}^{3} \hat{\tau}_k\, r_{kl}\, \hat{\tau}_l = (1 + tr\,\underline{r})\hat{I} + i\sum_{k,l,m=1}^{3} r_{kl}\, \epsilon_{klm}\, \hat{\tau}_m \qquad (1.269)$$

and thereby the stated result, eq.(1.264). Using eq.(1.264) we get

$$\underline{U} \equiv\ <\pm|\hat{U}|\pm> \ = \frac{1}{2}(1 + tr\,\underline{r})^{-1/2}((1 + tr\,\underline{r})\tau^{(0)}$$

$$+\ i(r_{23} - r_{32})\tau^{(1)} + i(r_{31} - r_{13})\tau^{(2)} + i(r_{12} - r_{21})\tau^{(3)}) \qquad (1.270)$$

and using $(1 + tr\,\underline{r}) = (1 + \cos\theta)(1 + \cos(\phi + \psi))$, $r_{23} - r_{32} = \sin\theta(\cos\psi + \cos\phi)$, $r_{31} - r_{13} = \sin\theta(\sin\phi - \sin\psi)$, $r_{12} - r_{21} = \sin(\psi + \phi)(1 + \cos\theta)$, *and noting the identity* $(1 + \exp\{\pm i(\phi + \psi)\})/\cos((\phi + \psi)/2) = 2\exp\{\pm i(\phi + \psi)/2\}$ *we obtain the stated result.*

Exercise 1.16 *Consider a rotation through the angle* ϕ *around the 3-axis, then*

$$\underline{r} = \begin{pmatrix} \cos\phi & \sin\phi & 0 \\ -\sin\phi & \cos\phi & 0 \\ 0 & 0 & 1 \end{pmatrix} \qquad (1.271)$$

with the corresponding unitary matrix $(\theta = 0 = \psi)$

$$\underline{U}^{(3)}(\phi) \equiv \underline{U}(\phi, 0, 0) = \begin{pmatrix} e^{\frac{i}{2}\phi} & 0 \\ 0 & e^{-\frac{i}{2}\phi} \end{pmatrix} = \cos\frac{\phi}{2}\underline{1} + i\sin\frac{\phi}{2}\underline{\tau}^{(3)}. \qquad (1.272)$$

Use the identity $(\vec{\alpha}\cdot\hat{\vec{\tau}})^2 = \alpha^2\hat{I}$, *which follows from eq.(1.257), to show that*

$$e^{i\vec{\alpha}\cdot\hat{s}} = \cos\frac{\alpha}{2}\hat{I} + 2i\sin\frac{\alpha}{2}\frac{\vec{\alpha}}{|\vec{\alpha}|}\cdot\hat{s} \qquad (1.273)$$

from which it follows that

$$\underline{U}^{(3)}(\phi) = \ <\pm|\hat{U}_3(\phi)|\pm> \ = \begin{pmatrix} e^{\frac{i}{2}\phi} & 0 \\ 0 & e^{-\frac{i}{2}\phi} \end{pmatrix} \qquad (1.274)$$

or $\hat{U}_3(\phi) \equiv \hat{U}_3(\vec{\phi}) = e^{i\vec{\phi}\cdot\hat{s}} = e^{i\phi\hat{s}_3}, \vec{\phi} \equiv \phi\,\mathbf{e}_3$; *i.e.,* $\hat{s}_3$ *is the generator of rotations around the 3-axis, specified by the unit vector* $\mathbf{e}_3$. *Similarly* $\hat{s}_1$ *and* $\hat{s}_2$ *are the generators of rotations around the 1- and 2-axis. The three generators of the group form a closed Lie algebra.*[62]

---

[62]We note that a 2π-rotation, which is no rotation at all, does not leave the operators invariant, $\underline{U}(,\phi = 2\pi, \theta = 0, \psi = 0) = -\underline{1}$, the *spin* representation of rotations is double connected.

Solution

Since

$$e^{i\vec{\alpha}\cdot\hat{\vec{\tau}}} = e^{\frac{i}{2}\vec{\alpha}\cdot\hat{\mathbf{s}}} = \hat{I} + \frac{i}{2}\vec{\alpha}\cdot\hat{\vec{\tau}} + \frac{1}{2!}\left(\frac{i}{2}\right)^2(\vec{\alpha}\cdot\hat{\vec{\tau}})^2 + \frac{1}{3!}\left(\frac{i}{2}\right)^3(\vec{\alpha}\cdot\hat{\vec{\tau}})^3 + \ldots$$

$$= \hat{I}\left(1 - \frac{1}{2!}\left(\frac{\alpha}{2}\right)^2 + \frac{1}{4!}\left(\frac{\alpha}{2}\right)^4 - \frac{1}{6!}\left(\frac{\alpha}{2}\right)^6 + \ldots\right)$$

$$+ i\frac{\vec{\alpha}}{|\vec{\alpha}|}\cdot\hat{\vec{\tau}}\left(\frac{\alpha}{2} - \frac{1}{3!}\left(\frac{\alpha}{2}\right)^3 + \frac{1}{5!}\left(\frac{\alpha}{2}\right)^5 - \frac{1}{7!}\left(\frac{\alpha}{2}\right)^7 + \ldots\right) \quad (1.275)$$

we obtain eq.(1.273).

Analogously we have for rotations around the 1-axis

$$\underline{U}^{(1)}(\theta) = \underline{U}(0,\theta,0) = \begin{pmatrix} \cos\frac{\theta}{2} & i\sin\frac{\theta}{2} \\ i\sin\frac{\theta}{2} & \cos\frac{\theta}{2} \end{pmatrix} = <\pm|\hat{U}_1(\theta)|\pm> \quad (1.276)$$

where (introducing standard units) $\hat{U}_1(\theta) \equiv \hat{U}_1(\vec{\theta}) = e^{\frac{i}{\hbar}\vec{\theta}\cdot\hat{\mathbf{s}}} = e^{\frac{i}{\hbar}\theta\hat{s}_1}, \vec{\theta} \equiv \theta\,\mathbf{e}_1$.

The generator (and any function thereof) is the only quantity which is invariant under the symmetry operation. In the present case the generator of rotations around the k-axis is $\hat{s}_k$, *and we choose in accordance with custom a component of* **s** *(and not* $\vec{\tau}$*) to represent the physical property. With our phase choice we have arbitrarily chosen the generator of rotations around the 3-axis to represent the physical property, and we have the eigenvalue equation (in standard units)*

$$\hat{s}_3|\pm> = \pm\frac{1}{2}\hbar\,|\pm> . \quad (1.277)$$

The spin-1/2 degree of freedom is prolific, and describes the internal angular momentum of, for example, electrons, muons, tauons, quarks.

The matrices

$$\underline{U}(\phi,\theta,\psi) = <\pm|e^{\frac{i}{\hbar}\psi\hat{s}_3}e^{\frac{i}{\hbar}\theta\hat{s}_1}e^{\frac{i}{\hbar}\phi\hat{s}_3}|\pm> \equiv \mathcal{D}_{m,m'}^{1/2}(\phi,\theta,\psi) , \quad m,m' = \pm \quad (1.278)$$

constitute the $s = 1/2$ *irreducible representation of the group of rotations in a three-dimensional space. We note that the group of rotations is nonabelian; for example,*

$$\hat{U}_3(\psi)\,\hat{U}_1(\theta) \neq \hat{U}_1(\theta)\,\hat{U}_3(\psi) . \quad (1.279)$$

For a spin-1/2 particle, say in the nonrelativistic treatment of an electron, it is customary to introduce the spinor representation of the wave function (choosing the z-direction along the quantization axis)

$$\psi(\mathbf{x}, s_z = \pm) \rightarrow \begin{pmatrix} \psi_+(\mathbf{x}) \\ \psi_-(\mathbf{x}) \end{pmatrix} \quad (1.280)$$

for the individual amplitudes for the particle to be in the two spin states. The spinor transforms under rotations in spin-space according to the rotation matrix eq.(1.267). Associated with the spin is a magnetic moment represented by the operator $\hat{\mathbf{m}} = g\,\hat{\mathbf{s}}\,e\hbar/2m$, where g for an electron is close to the value 2. In the presence of a magnetic field the dynamics of the spin is governed by the Hamiltonian

$$\hat{H} = -\,\hat{\mathbf{m}} \cdot \mathbf{B}(\hat{\mathbf{x}}, t) \qquad (1.281)$$

Exercise 1.17 *Consider a particle in the superposition of the two spin states,* $|\psi> = |+><+|\psi> \;+\; |-><-|\psi>$, *or in the position and spinor representation*

$$\psi(\mathbf{x}, s_z = \pm) \;=\; \begin{pmatrix} \psi_+(\mathbf{x}) \\ \psi_-(\mathbf{x}) \end{pmatrix} \;=\; \psi_+(\mathbf{x}) \begin{pmatrix} 1 \\ 0 \end{pmatrix} + \psi_-(\mathbf{x}) \begin{pmatrix} 0 \\ 1 \end{pmatrix}. \qquad (1.282)$$

Show that when the particle traverses a region of space with an inhomogeneous magnetic field, the Hamiltonian eq.(1.281) will cause a deflection of the different spin components of the wave packet into the different regions of space where the energy eq.(1.281), of the spin up and down states respectively, is minimized. Show that after leaving the region of space with the inhomogeneous magnetic field, assuming that the time span is long enough, the two components have no spatial overlap (assume the particle has no electric charge, otherwise the coupling of the vector potential to the velocity, "the Lorentz force," must also be taken into account). Show that the probability for finding the particle in the spin up (or down) state after leaving the field region is equal to the probability for finding the particle in the spin up (or down) state when the particle enters the field region.
Hint: For a smoothly varying magnetic field the center of a wave packet follows the classical trajectory.[63]

Any quantity with N property values can have its one-dimensional labeling broken down into its unique many-dimensional decomposition in view of its unique product of primes $N = N_1 \cdot N_2 \cdot \cdot N_M$, and within each subset of basis vectors $\{|a_{n_j}>\}_{n_j=1,..,N_j}$, the complementary operators can be constructed satisfying

$$\hat{\mathcal{U}}_j{}^{N_j} = \hat{I} = \hat{\mathcal{V}}_j{}^{N_j} \;, \;\; \hat{\mathcal{V}}_j\,\hat{\mathcal{U}}_j = e^{\frac{2\pi i}{N_j}}\hat{\mathcal{U}}_j\,\hat{\mathcal{V}}_j \;, \;\; j = 1,..,M. \qquad (1.283)$$

With each prime, $N_j = 2, 3, 5, ...$, a distinct quantum degree of freedom is associated. We shall in particular be concerned with the two extreme cases $N = 2$ and $N = \infty$,[64] characterizing the two spin values, and a component of the position (or momentum) of a (low-energy) electron, respectively. In the preceding exercises, we discussed the former case, and we now turn to the discussion of the latter.

---

[63]For a detailed account of the Stern-Gerlach experiment we refer to the references [6] and [12].
[64]The $N = \infty$ case also provides the quantum field description of bosons.

1.5 Nonrelativistic Quantum Kinematics

We shall be concerned with low-energy phenomena where special relativity effects can be neglected. In that case a particle can not disappear and its probability to be found somewhere in space is constantly 1. In this limit we can treat the position of the particle as a degree of freedom as we are always able to select the particle to be somewhere in space. The position variable is continuous, and we shall now show how the continuum limit emerges from the general quantum theory presented above as the case $N = \infty$.

Consider two complementary physical degrees of freedom X and P.[65] Instead of expressing the equidistant property values in terms of unity as before, $n = 0, \pm1, \pm2, .., \pm(N-1)/2$ (the number of property values being odd except for the spin-1/2 case), we scale the spacing between the physical property values to $d \equiv \sqrt{2\pi/N}$, thereby labeling the property values by

$$x_n = nd = p_n = 0, \pm d, \pm 2d, .., \pm \left(\frac{\pi}{d} - \frac{d}{2} \right). \tag{1.284}$$

The corresponding unitary operators $\hat{\mathcal{U}}$ and $\hat{\mathcal{V}}$ are complex functions of the hermitian operators $\hat{x}$ and $\hat{p}$ representing the physical quantities in question

$$\hat{\mathcal{U}} = e^{id\hat{x}} \ , \ \hat{\mathcal{V}} = e^{id\hat{p}} \tag{1.285}$$

where

$$\hat{x} |x_n> \ = \ x_n |x_n> \ , \quad n = 0, \pm1, \pm2, .., \pm(N-1)/2; \tag{1.286}$$

and

$$\hat{p} |p_n> \ = \ p_n |p_n> \ , \quad n = 0, \pm1, \pm2, .., \pm(N-1)/2. \tag{1.287}$$

We have for eq.(1.220) and eq.(1.223)

$$e^{id\hat{x}} |p_n> \ = \ |p_n + d> \ , \ e^{id\hat{x}} |x_n> \ = \ e^{idx_n} |x_n> \tag{1.288}$$

and for eq.(1.231) and the eigenvalue equation for $\hat{\mathcal{V}}$[66]

$$<x_n| \, e^{id\hat{p}} \ = \ <x_n + d| \ , \ e^{id\hat{p}} |p_n> \ = \ e^{idp_n} |p_n> \tag{1.289}$$

The fundamental relation, eq.(1.248), expressed in terms of the operators $\hat{x}$ and $\hat{p}$ reads

$$e^{ix\hat{p}} \, e^{ip\hat{x}} = e^{ixp} \, e^{ip\hat{x}} \, e^{ix\hat{p}} \tag{1.290}$$

where we have introduced the notation (appropriate for the eventual continuum limit)

$$x \equiv x_n = nd \ , \ p \equiv p_m = md \ . \tag{1.291}$$

---

[65]Eventually we shall take the continuum limit, and let X denote (one Cartesian coordinate of) the position of a particle.

[66]We observe that the complementary property to position describe spatial translations, a point we shall elaborate in the next section.

The equation can be rewritten on the form

$$e^{ip(\hat{x}-x)} = e^{-ix\hat{p}}e^{ip\hat{x}}e^{ix\hat{p}} = \exp\{ipe^{-ix\hat{p}}\hat{x}e^{ix\hat{p}}\} \tag{1.292}$$

where the last equality follows from the observation that expansion of the exponentials generates the same power series on the two sides.

The choice of labeling numerically the property values is arbitrary. We can choose another p-interval where one property value is new, say we shift the numerical labeling of the state with the lowest p-value, $p \to p + 2\pi/d$. The factor $\exp\{-ipx\}$ on the left-hand side of eq.(1.292) is unchanged by this shift, and at this point it is therefore illegitimate to identify powers in p in the above equation. However, if we restrict the acceptable physical states to have nonzero values of their transformation functions only for finite p-values (and thereby finite x-values), the periodicity requirement is ineffective in the continuum limit $d \to 0$, $N \to \infty$, $Nd^2 = 2\pi$, and we can identify each term in the expansion in p. Inspecting the first-order term we get the relation

$$e^{-ix\hat{p}}\,\hat{x}\,e^{ix\hat{p}} = \hat{x} - x \ . \tag{1.293}$$

Identifying terms in powers of the continuous real variable x in the above relation gives for the linear term the canonical commutation relation[67] (introducing standard units[68])

$$[\hat{x}, \hat{p}] = i\hbar\hat{I} \ . \tag{1.294}$$

We should keep in mind that its validity is restricted to the space of physically acceptable states. The commutator of the position operator and its complementary operator is said to be a c-number (since its operation just amounts to multiplication by a complex number).

Before taking the continuum limit, the property values are discrete, and the state symbol $\hat{P}_\Delta(x_n) = |x_n>_\Delta {}_\Delta<x_n|$ can be viewed as referring to the property value being within a region of the size of the resolution $\Delta x \equiv d$; i.e., $|x_n>_\Delta$ represents the state where the property value is in the interval of size Δx around value x_n. The probability $P_\Delta(x)$ to find the property value of X in the range Δx around the value $x \equiv x_n$ given that the system is in state ψ is

$$P_\Delta(x\,;\psi) \ = \ Tr(\hat{P}(\psi)\,\hat{P}_\Delta(x)) \ = \ |{}_\Delta<x|\psi>|^2 \ . \tag{1.295}$$

In the idealized case of an arbitrary fine resolution, the continuum limit, the probability becomes specified in terms of the probability density, $x \equiv x_n$,

$$P_\Delta(x\,;\psi) \ = \ p(x\,;\psi)\,\Delta x \tag{1.296}$$

---

[67]We note, that the canonical commutation relation is of course invariant under the previously noted invariance of quantum kinematics, $\hat{x} \to \hat{p}, \hat{p} \to -\hat{x}$.

[68]Comparison of the quantum theory with experiment renders the physical quantities with dimensions – an aspect that the formalism is not aware of. We have introduced standard units in which position and momentum have dimensions, position is specified by length etc., and we have used the experimental fact that the amount of action dividing the product of position and momentum is the quantum of action.

and we shall therefore introduce continuum states $|x>$ according to the substitution $|x_n>_\Delta \to (\Delta x)^{1/2}|x_n>$. The resolution of the identity

$$\hat{I} = \sum_n |x_n>_\Delta {}_\Delta<x_n| = \sum_n \hat{P}_\Delta(x_n) \qquad (1.297)$$

then becomes the integral expression

$$\hat{I} = \lim_{\Delta x_n \to 0} \sum_n \Delta x_n |x_n><x_n| = \int_{-\infty}^{\infty} dx \, |x><x| = \int_{-\infty}^{\infty} dx \, \hat{P}(x) \qquad (1.298)$$

where $\hat{P}(x) \equiv |x><x|$ is the probability density projector representing the probability density for the particle to be at position x. Equivalently we have $\hat{P}_\Delta(x) = \hat{P}(x)\Delta x$. For a particle in state ψ we have for the probability density to find it at position x

$$p(x\,;\psi) = Tr(\hat{P}(\psi)\,\hat{P}(x)) = |<x|\psi>|^2 = |\psi(x)|^2\,. \qquad (1.299)$$

In the continuum limit we have for the wave function by use of the resolution of the identity

$$\psi(x) \equiv <x|\psi> = \int_{-\infty}^{\infty} dx' \, <x|x'><x'|\psi> \qquad (1.300)$$

revealing that in the continuum limit, the transformation function is Dirac's delta function[69]

$$<x|x'> = \delta(x - x')\,. \qquad (1.301)$$

Dirac's delta function is a caricature of a function that allows us to use the language of eigenvalue equations also for the case of continuous property values. For example, we write for the position operator the eigenvalue equation

$$\hat{x}\,|x> = x\,|x> \qquad (1.302)$$

and have in the position representation

$$x'\,<x'|x> = x\,<x'|x>\,. \qquad (1.303)$$

1.6 Symmetries and Physical Quantities

In the following we shall refer to the x-representation as the possible positions of a particle, and recognize that the complementary degree of freedom describes the momentum of the particle.

If we physically displace a system the distance a, then, under the assumption of homogeneity of space, the properties of the system at the original and translated

---

[69]We observe that the continuum limit of the $<x|x'>$ transformation function cannot be a function in the usual sense, because no function can be singular enough to satisfy the above equation for arbitrary wave functions, and the integral expression is just a suggestive way of representing a functional equation $\delta_x[\psi] = \psi(x)$, as discussed in appendix B.

positions are the same, and the descriptions of the states at the two places are therefore unitarily related. For example, a particle in state $|x>$ will upon displacement a distance a be in the state which (up to a phase factor) is specified by the state vector $|x+a>$. Since $|x>$ and $|x+a>$ both are basis vectors, they are related by a unitary transformation, the one generated by the complementary operator to the position operator according to eq.(1.289). Alternatively, we could describe the same system in two reference frames, differing only by location and orientation, the passive view. The homogeneity of space is then ascertained by requiring that the two descriptions are equivalent. For the case of translation, a position in the new reference frame is related to the old one through $\underline{x} = x + a$, and the two position descriptions are unitarily related $|\underline{x}> = [\hat{U}(a)]^{-1}|x>$ by homogeneity of space. The active and passive view of translations are just each other's inverse $[\hat{U}_P(a)]^{-1} = \hat{U}_A(a) = \hat{U}_P(-a)$.

Since the resulting displacement of two translations of a frame is independent of their order and is simply specified by the sum of the vectors defining the translations, we have the abelian group property

$$\hat{U}(a_1) \cdot \hat{U}(a_2) = \hat{U}(a_2) \cdot \hat{U}(a_1) = \hat{U}(a_1 + a_2) \ . \tag{1.304}$$

For an infinitesimal translation a distance Δa, the unitary operator has the form $\hat{U}(\Delta a) = \hat{I} + i\hat{G}\Delta a$, as clearly $\hat{U}(0) = \hat{I}$, and the group property implies the deviation from the identity operator to be linear in Δa. From eq.(1.289) we know that the generator of translations is identical to the complementary property of position $\hat{G} = \hat{p}$. The generator (and functions thereof) is the only operator which in general is invariant under the transformation in question, giving way to defining physical properties through symmetry invariances. The generator of translations is called the momentum operator, momentum being defined as the quantity invariant under translations. The physical attribute of momentum thus reflects a symmetry property of space, its translational invariance.[70] We hereby recognize that position and momentum are the complementary degrees of freedom describing the (nonrelativistic) motion of a particle.

Operating on a position eigenstate $|x>$ with the commutator $[\hat{x}, \hat{U}(\Delta a)]$ we obtain to lowest order in the distance of translation Δa[71]

$$[\hat{x}, i\Delta a\hat{p}]|x> \ = \ [\hat{x}, \hat{U}(\Delta a)]|x> \ = \ (x + \Delta a)|x + \Delta a> \ -x|x + \Delta a>$$

$$= \ \Delta a|x + \Delta a> \ = \ \Delta a|x> \ . \tag{1.305}$$

We thus again obtain the canonical commutation relation, eq.(1.93), for position and momentum

$$[\hat{x}, \hat{p}] = i\hbar\hat{I} \ . \tag{1.306}$$

_____

[70]The homogeneity of time implies that for an isolated system the generator of displacements in time of a state vector is a constant of the motion, and since it is in general the only constant of motion, we call it the energy operator or Hamiltonian. In general we only know that it is a function of the complementary quantities $\hat{H} = H(\hat{x}, \hat{p})$.

[71]The last estimate refers to the distance in the norm on the state space.

In order to generate a finite translation of a system we use the tactic of small steps (just as we did for displacements in time in section 1.2). A finite translation the distance a

can be divided into $N = a/\Delta a$ infinitesimal repeated translations

$$\hat{U}(a) = \lim_{N \to \infty} (\hat{U}(\Delta a))^N = \lim_{N \to \infty} (\hat{I} + i\Delta a \hat{G})^N$$

$$= \lim_{N \to \infty} \left(\hat{I} + i\Delta a N \hat{G} + (i\Delta a)^2 \frac{N(N-1)}{2!} \hat{G}^2 \right.$$

$$+ \left. (i\Delta a)^3 \frac{N(N-1)(N-2)}{3!} \hat{G}^3 + .. + (i\Delta a)^N \hat{G}^N \right)$$

$$= \hat{I} + ia\hat{G} + \frac{1}{2!}(ia)^2 \hat{G}^2 + \frac{1}{3!}(ia)^3 \hat{G}^3 + ...$$

$$\equiv e^{ia\hat{G}} . \tag{1.307}$$

We have here encountered a general feature: that repeating an infinitesimal unitary transformation generates a finite unitary transformation expressed through the generator.

We have thus recovered the result, eq.(1.289), that for a finite translation we then have for the unitary operator of translations

$$\hat{U}(a) = e^{\frac{i}{\hbar} a \hat{p}} \tag{1.308}$$

and translation of a state is generated by the momentum operator according to

$$|x + a> = e^{-\frac{i}{\hbar} a \hat{p}} |x> . \tag{1.309}$$

The corresponding position operator in the translated reference frame is according to eq.(1.293) the expected result

$$\hat{\underline{x}} = [\hat{U}(-a)]^{-1} \hat{x} \, \hat{U}(-a) = e^{\frac{i}{\hbar} a \hat{p}} \hat{x} \, e^{-\frac{i}{\hbar} a \hat{p}} = \hat{x} + a\hat{I} . \tag{1.310}$$

The preceding relations are immediately generalized to the three-dimensional case by interpreting the parameters as three-dimensional vectors. The three spatial positions can be assigned simultaneously, and because of their compatibility their state symbols and thereby their associated operators commute:

$$[\hat{x}_i, \hat{x}_j] = \hat{0} . \tag{1.311}$$

Repeating the preceding calculation for a translation in the y-direction (now using boldface notation for spatial vectors and their associated generators)

$$\hat{U}(\Delta a\,\mathbf{e}_y) = \hat{I} + \frac{i}{\hbar}\Delta \mathbf{a}\cdot\hat{\mathbf{p}} = \hat{I} + \frac{i}{\hbar}\Delta a\,\hat{p}_y \qquad (1.312)$$

we get

$$\frac{i}{\hbar}\Delta a[\hat{x},\hat{p}_y]\,|\mathbf{x}> = [\hat{x},\hat{U}(\Delta a\mathbf{e}_y)]\,|\mathbf{x}> = x\,|x,y+\Delta a,z> -x\,|x,y+\Delta a,z>$$

$$= |\mathcal{O}> \qquad (1.313)$$

and the commutation relation

$$[\hat{x},\hat{p}_y] = \hat{0}\,. \qquad (1.314)$$

The equation states that a translation in the y-direction does not influence the position labeling in the x-direction, leaving the position operator in the x-direction unchanged: $\hat{\underline{x}} = \hat{U}(y\mathbf{e}_y)^\dagger\,\hat{x}\,\hat{U}(y\mathbf{e}_y) = \hat{x}$. The position in the x-direction, and the momentum component in the perpendicular y-direction are compatible degrees of freedom. Similarly we have for the other Cartesian coordinates

$$[\hat{x}_i,\hat{p}_j] = i\hbar\,\delta_{ij}\,\hat{I}\,, \quad i,j = x,y,z. \qquad (1.315)$$

Considering addition of translations in different spatial directions and using the homogeneity of space to infer

$$\hat{U}(x\mathbf{e}_x + y\mathbf{e}_y) = \hat{U}(x\mathbf{e}_x)\cdot\hat{U}(y\mathbf{e}_y) = \hat{U}(y\mathbf{e}_y)\cdot\hat{U}(x\mathbf{e}_x) \qquad (1.316)$$

we get upon expanding the exponentials

$$\hat{0} = [\hat{U}(x\mathbf{e}_x),\hat{U}(y\mathbf{e}_y)] = [e^{\frac{i}{\hbar}x\hat{p}_x}, e^{\frac{i}{\hbar}y\hat{p}_y}]$$

$$= -\frac{xy}{\hbar^2}\,[\hat{p}_x,\hat{p}_y] + (\text{higher polynomial terms in } x \text{ and } y)\,. \qquad (1.317)$$

In particular we have from the uniqueness of the expansion coefficients:

$$[\hat{p}_x,\hat{p}_y] = \hat{0}\,. \qquad (1.318)$$

Similarly for the other Cartesian components, and we have specified the commutation properties of position and momentum completely.

The translation operators constitute an abelian subgroup of the unitary operators on the state space

$$\hat{U}(\mathbf{a}_1)\cdot\hat{U}(\mathbf{a}_2) = \hat{U}(\mathbf{a}_1+\mathbf{a}_2)\,, \quad \hat{U}(\mathbf{a})\cdot\hat{U}(-\mathbf{a}) = \hat{I} = \hat{U}(-\mathbf{a})\cdot\hat{U}(\mathbf{a})\,, \quad \hat{U}(-\mathbf{a}) = \hat{U}^{-1}(\mathbf{a}) = \hat{U}^\dagger(\mathbf{a})\,. \qquad (1.319)$$

In general a system possesses a symmetry if its properties are invariant with respect to different ways of describing the system. If a system possesses a symmetry

S described by the operator $\hat{S}$, equivalent matrix elements must satisfy the probability constraint $(|S\psi> \equiv \hat{S}|\psi>)$, $|<S\psi|S\phi>| = |<\psi|\phi>|$. Wigner's theorem states that we can always choose phases so that a symmetry is either represented by a unitary or an antiunitary operator.[72] The continuous symmetries of translations and rotations, and the discrete symmetries of space inversion through a point or a plane, are described by unitary operators, whereas the case of time-reversal symmetry by an antiunitary operator (to be discussed in section 2.9).

Exercise 1.18 *Show that for an operator of the form $\hat{f} \equiv f(\hat{x}, \hat{p})$ we have*

$$\frac{\partial \hat{f}}{\partial \hat{x}} = \frac{1}{i\hbar}[f(\hat{x}, \hat{p}), \hat{p}] , \qquad \frac{\partial \hat{f}}{\partial \hat{p}} = -\frac{1}{i\hbar}[f(\hat{x}, \hat{p}), \hat{x}] . \qquad (1.320)$$

Solution

Using eq.(1.293) we obtain

$$\Delta \hat{f} \equiv f(\hat{x} + \Delta x \hat{I}, \hat{p}) - f(\hat{x}, \hat{p}) = f(e^{\frac{i}{\hbar}\Delta x \hat{p}} \hat{x} e^{-\frac{i}{\hbar}\Delta x \hat{p}}, \hat{p})$$

$$= e^{\frac{i}{\hbar}\Delta x \hat{p}} f(\hat{x}, \hat{p}) e^{-\frac{i}{\hbar}\Delta x \hat{p}} = \frac{\Delta x}{i\hbar}[f(\hat{x}, \hat{p}), \hat{p}] + \mathcal{O}(\Delta x^2) \qquad (1.321)$$

and as

$$\frac{\partial \hat{f}}{\partial \hat{x}} \equiv \frac{\Delta \hat{f}}{\Delta x} \qquad (1.322)$$

the stated result. The second equality follows from the invariance $\hat{x} \to \hat{p}, \hat{p} \to -\hat{x}$ of quantum kinematics. We note, that for the case of f being a polynomial, the relation follows from the identity $[\hat{x}^n, \hat{p}] = in\hat{x}^{n-1}$, obtained trivially by repeated use of the canonical commutation relation.

Exercise 1.19 *Show that the generator of rotations $\hat{\mathbf{J}}$ satisfies the commutation relations*

$$\hat{\mathbf{J}} \times \hat{\mathbf{J}} = i\hbar \hat{\mathbf{J}} . \qquad (1.323)$$

For the transformation function we have according to eq.(1.235)

$$<x|p> = \frac{1}{d} \lesssim x|p \gtrsim = \frac{1}{\sqrt{Nd^2}} e^{\frac{2\pi i}{Nd^2}xp} . \qquad (1.324)$$

In the continuum limit we thus have (in three spatial dimensions and in standard units)

$$<\mathbf{x}|\mathbf{p}> = \frac{1}{(2\pi\hbar)^{3/2}} e^{\frac{i}{\hbar}\mathbf{x}\cdot\mathbf{p}} \qquad (1.325)$$

---

[72]We shall not need to refer to Wigner's theorem since we shall explicitly construct the needed symmetry operators, but for a proof of Wigner's theorem we refer to reference [12].

expressing the complementarity between position and momentum. For a state of definite momentum, the position of the particle is completely undetermined, and vice versa.

From eq.(1.289) in standard units, we obtain in the continuum limit for the matrix element of the momentum operator for a physical state ψ (expand to lowest order in the distance d)[73]

$$<\mathbf{x}|\hat{\mathbf{p}}|\psi> \; = \; \frac{\hbar}{i}\frac{\partial\psi(\mathbf{x})}{\partial\mathbf{x}} \; . \tag{1.326}$$

We have thus established that in the continuum limit the operator $\hat{\mathbf{p}}$ in its complementary representation act as the differential operator

$$\hat{\mathbf{p}}^{(x)} \; \equiv \; \frac{\hbar}{i}\frac{\partial}{\partial\mathbf{x}} \tag{1.327}$$

where the superscript indicates that the operator acts on a wave function in the position representation. Referring back to eq.(1.26) we observe that we can obtain the Hamiltonian for a particle in a potential by the so-called canonical quantization procedure: replace position and momentum in the corresponding Hamilton function with the position and momentum operators!

If we in eq.(1.326) insert $|\psi> \; = \; |\mathbf{x}'>$ we obtain

$$<\mathbf{x}|\hat{\mathbf{p}}|\mathbf{x}'> \; = \; \frac{\hbar}{i}\frac{\partial}{\partial\mathbf{x}}\,\delta(\mathbf{x}-\mathbf{x}') \; . \tag{1.328}$$

The position representation of the momentum operator, eq.(1.326), is of course not unique; we are always free to make a phase transformation $|\mathbf{x}> \to e^{-i\alpha(\mathbf{x})}|\mathbf{x}>$ resulting in the transformation of the matrix element

$$<\mathbf{x}|\hat{\mathbf{p}}|\mathbf{x}'> \; \to \; e^{i\alpha(\mathbf{x})}<\mathbf{x}|\hat{\mathbf{p}}|\mathbf{x}'>e^{-i\alpha(\mathbf{x}')} \; = \; <\mathbf{x}|\hat{\mathbf{p}}|\mathbf{x}'> - \hbar\frac{\partial\alpha(\mathbf{x})}{\partial\mathbf{x}}\,\delta(\mathbf{x}-\mathbf{x}') \; . \tag{1.329}$$

Instead of representing the momentum operator in the position representation as in eq.(1.328) we could thus equally well use the representation

$$\hat{\mathbf{p}}^{(x)} \; = \; \frac{\hbar}{i}\frac{\partial}{\partial\mathbf{x}} + \hbar\frac{\partial\alpha(\mathbf{x})}{\partial\mathbf{x}} \tag{1.330}$$

where $\alpha(\mathbf{x})$ is an arbitrary function.

Expressing the wave function in its Taylor series

$$<\mathbf{x}+\mathbf{a}|\psi> \; = \; \psi(\mathbf{x}+\mathbf{a}) \; = \; e^{\mathbf{a}\cdot\nabla_{\mathbf{x}}}\psi(\mathbf{x}) \; = \; <\mathbf{x}|e^{\frac{i}{\hbar}\mathbf{a}\cdot\hat{\mathbf{p}}}|\psi> \tag{1.331}$$

we recover that the momentum operator translates states according to

$$<\mathbf{x}+\mathbf{a}| \; = \; <\mathbf{x}|\,e^{\frac{i}{\hbar}\mathbf{a}\cdot\hat{\mathbf{p}}} \tag{1.332}$$

or equivalently, relates the unitary equivalent descriptions of two reference frames shifted by the distance $\mathbf{a}$.

---

[73]In the limit $N \to \infty$, the space of physical states is thus characterized by the existence of such limits.

Exercise 1.20 *Show that for two complementary components of the position and momentum operators we have*

$$e^{\frac{i}{\hbar}p\hat{x}+\frac{i}{\hbar}x\hat{p}} = e^{\frac{i}{\hbar}x\hat{p}}\, e^{+\frac{i}{\hbar}p\hat{x}}e^{-\frac{i}{2\hbar}xp} \, . \tag{1.333}$$

Solution

Differentiating the operator $\hat{A}(s) \equiv e^{\frac{i}{\hbar}\hat{p}xs}\, e^{\frac{i}{\hbar}\hat{x}ps}$, *and using eq.(1.293) we obtain*

$$\frac{\hbar}{i}\frac{d\hat{A}(s)}{ds} = (x\hat{p} + p\hat{x} + pxs)\,\hat{A}(s) \, . \tag{1.334}$$

Since $\hat{x}$ *and* $\hat{p}$ *commute with* $[\hat{x}, \hat{p}]$,

$$\hat{A}(s) = e^{\frac{i}{\hbar}(p\hat{x}+x\hat{p})s}\, e^{\frac{px}{2\hbar^2}[\hat{x},\hat{p}]s^2} \tag{1.335}$$

satisfies the same first-order differential equation and initial condition $\hat{A}(s=0) = \hat{I}$, *and the two operators are therefore identical. Choosing the value* $s = 1$, *we obtain the desired formula.*

For a particle constrained to a box of volume V we have by normalization for the transformation function (up to the ubiquitous overall phase factor)

$$<\mathbf{x}|\mathbf{p}> = \frac{1}{\sqrt{V}}\, e^{\frac{i}{\hbar}\mathbf{x}\cdot\mathbf{p}} \, . \tag{1.336}$$

The allowed momentum values are discrete as the wave function is zero outside the volume V. For a free particle (except constrained to the box) the basis set of momentum eigenfunctions is trigonometric functions and there is one momentum state per momentum volume $(2\pi\hbar)^3/V$ (in three dimensions).[74] Using the resolution of the identity eq.(1.298) for a particle in a finite volume, and the corresponding resolution of the identity in terms of the momentum eigenstates $\hat{I} = \sum_{\mathbf{p}}|\mathbf{p}><\mathbf{p}|$, we obtain

$$\psi(\mathbf{x}) = \frac{1}{V}\sum_{\mathbf{p}} e^{\frac{i}{\hbar}\mathbf{p}\cdot\mathbf{x}} \int_V d\mathbf{x}'\, e^{-\frac{i}{\hbar}\mathbf{x}'\cdot\mathbf{p}}\, \psi(\mathbf{x}') \, . \tag{1.337}$$

Cavalierly interchanging the summation and integration before letting the volume approach infinity gives according to eq.(1.300) the following useful representation of Dirac's delta function:

$$\delta(\mathbf{x} - \mathbf{x}') = <\mathbf{x}|\mathbf{x}'> = \lim_{V\to\infty}\frac{1}{V}\sum_{\mathbf{p}} e^{\frac{i}{\hbar}\mathbf{p}\cdot(\mathbf{x}-\mathbf{x}')} = \int\frac{d\mathbf{p}}{(2\pi\hbar)^3}\, e^{\frac{i}{\hbar}\mathbf{p}\cdot(\mathbf{x}-\mathbf{x}')} \tag{1.338}$$

in accordance with eq.(1.325).

---

[74]A free particle in a definite momentum state is also in a definite energy state as the Hamiltonian is independent of position, and the position and energy are complementary. For the case where a particle is experiencing a potential, the position and energy are still incompatible, but no longer complementary. For a particle in an attractive potential, such as is the case for an electron in a hydrogen atom, position and energy are therefore not complementary. For a state of definite energy we are not completely ignorant about the position of the electron, the possible position property values are distributed over the size of the atom, of the order of one Ångström $(10^{-10}m)$.

1.7   Quantum Dynamics

The kinematic structure of quantum mechanics was obtained without any direct reference to time, except for the necessary temporal order in which selections have to be done. All states introduced previously thus depend on the time the system is in the state ψ in question. Following the common notation, the state vector corresponding to the property $\psi(t)$ at time t will be denoted by $|\psi(t)>$. At each instant of time a complete description of an isolated system is provided by a state vector, $|\psi(t)>$, thereby defining an operator, the time-evolution operator, connecting state vectors at different times

$$|\psi(t)> \; = \; \hat{U}(t,t')\,|\psi(t')> \; . \tag{1.339}$$

At equal times the operator is the identity operator

$$\hat{U}(t,t) = \hat{I} \tag{1.340}$$

and the linear structure of the operator space determines the evolution operator to be linear. By repeated use of the defining relation we note the semigroup property

$$\hat{U}(t,t'')\,\hat{U}(t'',t') = \hat{U}(t,t') \; . \tag{1.341}$$

The completeness of the description determines the equation of motion to be first order in time

$$\frac{d|\psi(t)>}{dt} \; = \; \lim_{\Delta t \to 0} \frac{|\psi(t+\Delta t)> \; - |\psi(t)>}{\Delta t}$$

$$= \; \left(\lim_{\Delta t \to 0} \frac{\hat{U}(t+\Delta t, t) - \hat{I}}{\Delta t} \right) |\psi(t)>$$

$$\equiv \; -i\hat{K}_t \,|\psi(t)> \; . \tag{1.342}$$

Since the state vector is arbitrary we obtain that the time-evolution operator satisfies the equation[75]

$$\frac{\partial \hat{U}(t,t')}{\partial t} \; = \; -i\hat{K}_t \,\hat{U}(t,t') \; . \tag{1.343}$$

The probabilistic interpretation of the state vector requires the evolution operator to evolve a state vector in such a way that its length is kept independent of time; i.e., we have the normalization condition

$$<\psi(t')|\psi(t')> \; = \; <\psi(t)|\psi(t)> \; = \; <\psi(t')|\hat{U}^{\dagger}(t,t')\,\hat{U}(t,t')|\psi(t')> \tag{1.344}$$

and thereby

$$\hat{U}^{\dagger}(t,t')\,\hat{U}(t,t') \; = \; \hat{I} \tag{1.345}$$

---

[75]From a mathematical point of view, convergence properties of limiting processes for operator sequences $\hat{A}_t$ are inherited from the topology of the vector space; i.e., convergence is defined by convergence of an arbitrary vector $\hat{A}_t|\psi>$.

i.e., the evolution operator is unitary, $\hat{U}^\dagger(t,t') = \hat{U}^{-1}(t,t')$ (from which follows $U(t,t')\,\hat{U}^\dagger(t,t') = \hat{I}$). Interchanging t and t' in eq.(1.339), we get for the inverse evolution operator, evolving a system backward in time,

$$\hat{U}^{-1}(t,t') = \hat{U}(t',t) \tag{1.346}$$

and thereby the relationship

$$\hat{U}^\dagger(t,t') = \hat{U}(t',t) \ . \tag{1.347}$$

The adjoint operation interchanges the time arguments in the evolution operator, the equivalent of changing the direction of time propagation.

In the above so-called Schrödinger picture, the state vector is the dynamic variable, and the operators representing the physical quantities are time independent. Since the time-evolution operator is unitary, the state vectors at different times provide equivalent descriptions

$$\hat{U}^{-1}(t,t_r)\,|\psi(t)> \ = \ |\psi(t_r)> \ = \ |\psi_H \gg \tag{1.348}$$

the two descriptions being identical at the arbitrary reference time t_r. In the unitary equivalent Heisenberg picture, the role of the time dependence is interchanged so that the dynamic law is expressed in terms of the dynamics of the operators representing the physical quantities (reminiscent of classical mechanics) and the state vector of the system $|\psi_H>$ is time independent. Operators in the two equivalent descriptions are related according to eq.(1.190)

$$\hat{A}_H(t) \ \equiv \ \hat{U}^\dagger(t,t_r)\,\hat{A}\,\hat{U}(t,t_r) \ = \ \hat{U}^\dagger(t,t')\,\hat{A}_H(t')\,\hat{U}(t,t') \tag{1.349}$$

and an operator in the unitary equivalent Heisenberg description becomes time dependent as the unitary transformation contains the time as parameter.

For an isolated system, the time-evolution operator, as a consequence of the homogeneity of time, only depends on the time difference $\hat{U}(t,t') = \hat{U}(t-t')$. For short time differences we have according to eq.(1.343)

$$\hat{U}(t,t_r) \ = \ \hat{I} \ - \ i(t-t_r)\hat{K} \ . \tag{1.350}$$

The generator of time displacements is the only operator in the Heisenberg picture which in general is independent of time, and the quantity it represents we call the energy of the system. In standard units the energy operator is denoted $\hat{H}$, $\hat{H} = \hbar\hat{K}$, and is also referred to as the Hamiltonian. The conversion factor is empirically determined to be the quantum of action.

We obtain, from eq.(1.343) and its adjoint, the Heisenberg equation of motion

$$\frac{d\hat{A}_H(t)}{dt} \ = \ \frac{i}{\hbar}[\hat{H}_H(t), \hat{A}_H(t)] \ + \ \hat{U}^\dagger(t,t_r)\frac{\partial\hat{A}}{\partial t}\,\hat{U}(t,t_r) \tag{1.351}$$

where

$$\hat{H}_H(t) \ = \ \hat{U}^\dagger(t,t_r)\,\hat{H}_t\,\hat{U}(t,t_r) \tag{1.352}$$

is the Hamiltonian in the Heisenberg picture, and for the case $\hat{A} = \hat{H}_t$ we have anticipated that the Hamiltonian in the Schrödinger picture $\hat{H}_t$ can have an explicit time dependence due to external fields.

Establishing the form of the energy operator (as a function of the complementary operators), or equivalently the dynamic law, is ultimately to be determined by experimental data and theoretical ideas.[76] However, in the event that we can appeal to correspondence with classical mechanics, such as is the case for a massive particle at low energies, the energy operator is specified by Hamilton's function

$$\hat{H} = H(\hat{x}, \hat{p}) \tag{1.353}$$

in accordance with the result, eq.(1.26), of section 1.1.

The formal solution of the Schrödinger equation for an isolated system

$$i\hbar \frac{d|\psi(t)>}{dt} = \hat{H} |\psi(t)> \tag{1.354}$$

or equivalently, the solution for the evolution operator

$$i\hbar \frac{\partial \hat{U}(t, t')}{\partial t} = \hat{H} \hat{U}(t, t') \tag{1.355}$$

is obtained by propagating in small steps; see eq.(1.110),

$$\hat{U}(t, t') = e^{-\frac{i}{\hbar}\hat{H}(t-t')} . \tag{1.356}$$

State vectors and operators in two equivalent descriptions at different times are related so that the spectrum is the same. For the position operator, say, we have (recall eq.(1.190))

$$\hat{\mathbf{x}}_H(t) = \hat{U}^{-1}(t, t_r) \, \hat{\mathbf{x}} \, \hat{U}(t, t_r) \tag{1.357}$$

and the corresponding eigenstates for $\hat{\mathbf{x}}_H(t)$ are related to the Schrödinger ones of eq.(1.302) by the relation

$$|\mathbf{x}, t> = \hat{U}^{-1}(t, t_r)|\mathbf{x}> \tag{1.358}$$

so that

$$\hat{\mathbf{x}}_H(t) |\mathbf{x}, t> = \mathbf{x} |\mathbf{x}, t> . \tag{1.359}$$

The state symbol representing the state where the particle has position $\mathbf{x}$ is in the Schrödinger picture (recall eq.(1.299))

$$\hat{P}(\mathbf{x}) = |\mathbf{x}><\mathbf{x}| = \delta(\hat{\mathbf{x}} - \mathbf{x}) \equiv \hat{n}(\mathbf{x}) \tag{1.360}$$

and in the last identity we have introduced the standard notation for the probability density operator. The corresponding Heisenberg operator can be expressed in various ways:

$$\hat{P}(\mathbf{x}, t) = \hat{U}^{\dagger}(t, t_r) \, \hat{P}(\mathbf{x}) \, \hat{U}(t, t_r) = |\mathbf{x}, t><\mathbf{x}, t| = \delta(\hat{\mathbf{x}}_H(t) - \mathbf{x}) \equiv \hat{n}(\mathbf{x}, t) . \tag{1.361}$$

---

[76]It is presently believed that all the laws of nature are known except those that govern strong gravitational forces; i.e., there is presently no satisfactory quantum theory of gravitation.

In view of eq.(1.298) and eq.(1.300) we have for the resolution of the identity

$$\int d\mathbf{x} \; |\mathbf{x}, t> <\mathbf{x}, t| \; = \; \int d\mathbf{x} \; \hat{P}(\mathbf{x}, t) \; = \; \hat{I} \tag{1.362}$$

and for the wave function at time t

$$\psi(\mathbf{x}, t) \; \equiv \; <\mathbf{x}|\psi(t)> \; = \; <\mathbf{x}, t|\psi_H> \; . \tag{1.363}$$

Furthermore, we can express the probability density in terms of the trace

$$|\psi(\mathbf{x}, t)|^2 \; = \; Tr(\hat{P}(\psi) \, \hat{P}(\mathbf{x}, t)) \; . \tag{1.364}$$

For the Schrödinger equation in the position representation we obtain

$$i\hbar \frac{\partial \psi(\mathbf{x}, t)}{\partial t} \; = \; \int d\mathbf{x}' \; <\mathbf{x}|\hat{H}|\mathbf{x}'> \; \psi(\mathbf{x}', t) \; . \tag{1.365}$$

Appealing to correspondence with classical mechanics, we have for a particle in a potential V the Hamiltonian

$$\hat{H} \; = \; \frac{\hat{\mathbf{p}}^2}{2m} \; + \; V(\hat{\mathbf{x}}) \tag{1.366}$$

and using eq.(1.328), we obtain the Schrödinger equation, eq.(1.25).

Using eq.(1.339) we get the time evolution of the wave function on integral form

$$\psi(\mathbf{x}, t) \; = \; \int d\mathbf{x}' \; <\mathbf{x}|e^{-\frac{i}{\hbar}\hat{H}(t-t')}|\mathbf{x}'> \; \psi(\mathbf{x}', t) \tag{1.367}$$

and we have identified the propagator[77]

$$K(\mathbf{x}, t; \mathbf{x}', t') \; = \; <\mathbf{x}|e^{-\frac{i}{\hbar}\hat{H}(t-t')}|\mathbf{x}'> \; = \; <\mathbf{x}, t|\mathbf{x}', t'> \; . \tag{1.368}$$

Exercise 1.21 *Show that the density operator in the Heisenberg picture satisfies the continuity equation*

$$\frac{\partial \hat{n}(\mathbf{x}, t)}{\partial t} \; + \; \nabla_{\mathbf{x}} \cdot \hat{\mathbf{j}}(\mathbf{x}, t) \; = \; 0 \tag{1.369}$$

where

$$\hat{\mathbf{j}}(\mathbf{x}, t) \; = \; \frac{1}{2}\{\hat{n}(\mathbf{x}, t), \hat{\mathbf{v}}(t)\} \tag{1.370}$$

is expressed in terms of the anticommutator $\{\hat{A}, \hat{B}\} \equiv \hat{A}\hat{B} + \hat{B}\hat{A}$, *and* $\hat{\mathbf{v}}(t) = \dot{\hat{\mathbf{x}}}(t)$ *is the velocity operator.*

---

[77]In appendix A we start from the matrix element expression for the propagator, and obtain the path integral form.

Solution

From eq.(1.351) we obtain

$$\frac{\partial \hat{n}(\mathbf{x},t)}{\partial t} = -\frac{i}{\hbar} [\hat{n}(\mathbf{x},t), \hat{H}] = -\frac{i}{\hbar} \hat{U}^\dagger(t,t_r) [\hat{n}(\mathbf{x}), \hat{H}] \hat{U}(t,t_r) \qquad (1.371)$$

and using the identity $[\hat{A}, \hat{B}\hat{C}] = \{\hat{A}, \hat{B}\}\hat{C} - \hat{B}\{\hat{C}, \hat{A}\}$, and the result of exercise 1.18 on page 65, we obtain, assuming $\hat{H} = \hat{\mathbf{p}}^2/2m + V(\hat{\mathbf{x}})$,

$$[\hat{n}(\mathbf{x}), \hat{H}] = \left[\hat{n}(\mathbf{x}), \frac{\hat{\mathbf{p}}^2}{2m}\right] = \frac{1}{2m}\left(-\hat{\mathbf{p}} \cdot \{\hat{n}(\mathbf{x}), \hat{\mathbf{p}}\} + \{\hat{n}(\mathbf{x}), \hat{\mathbf{p}}\} \cdot \hat{\mathbf{p}}\right)$$

$$= \frac{1}{2m}[\{\hat{n}(\mathbf{x}), \hat{\mathbf{p}}\}; \hat{\mathbf{p}}] = \frac{i\hbar}{2} \frac{\partial}{\partial \hat{\mathbf{x}}}\{\hat{n}(\mathbf{x}), \hat{\mathbf{v}}\}$$

$$= -\frac{i\hbar}{2} \frac{\partial}{\partial \mathbf{x}}\{\hat{n}(\mathbf{x}), \hat{\mathbf{v}}\} \qquad (1.372)$$

and upon unitary transforming with the evolution operator, we obtain the stated result. We have hereby identified $\hat{\mathbf{j}}$ as the probability current density operator, and for the expectation value in state ψ, $\mathbf{j}(\mathbf{x},t) = <\psi|\hat{\mathbf{j}}(\mathbf{x},t)|\psi>$, we obtain the expression eq.(1.28).

Exercise 1.22 *Verify the Baker-Hausdorff formula*

$$e^{\hat{B}} \hat{A} e^{-\hat{B}} = \hat{A} + [\hat{B}, \hat{A}] + \frac{1}{2!}[\hat{B}, [\hat{B}, \hat{A}]] + \frac{1}{3!}[\hat{B}, [\hat{B}, [\hat{B}, \hat{A}]]] + \dots . \quad (1.373)$$

Solution

By differentiating

$$\hat{A}(s) = e^{s\hat{B}} \hat{A} e^{-s\hat{B}} \qquad (1.374)$$

with respect to s, we successively obtain

$$\frac{d\hat{A}(s)}{ds} = [\hat{B}, \hat{A}(s)] \ , \quad \frac{d^2\hat{A}(s)}{ds^2} = [\hat{B}, [\hat{B}, \hat{A}(s)]] \ , \quad \frac{d^3\hat{A}(s)}{ds^3} = [\hat{B}, [\hat{B}, [\hat{B}, \hat{A}(s)]]] \ , \quad \dots$$

$$(1.375)$$

and since $\hat{A}(s=0) = \hat{A}$, the result is obtained by comparing with the Taylor series expansion around $s = 0$

$$\hat{A}(1) = \hat{A}(0) + \frac{d\hat{A}(0)}{ds} + \frac{1}{2!}\frac{d^2\hat{A}(0)}{ds^2} + \frac{1}{3!}\frac{d^3\hat{A}(0)}{ds^3} + \dots . \qquad (1.376)$$

For a unitary transformation described by $\hat{S}$, we have that $\hat{S}^{-1}|\psi(t)>$ is a solution of the Schrödinger equation (given $|\psi(t)>$ is), provided that $\hat{S}^{-1}\hat{H}\hat{S} = \hat{H}$. This equality is, by the Baker-Hausdorff formula, equivalent to the statement that the generator of the transformation commutes with the Hamiltonian $[\hat{G}, \hat{H}] = 0$. According to the equation of motion the physical property defined by the symmetry is then a constant of the motion

$$\frac{d\hat{G}_H(t)}{dt} = 0 . \tag{1.377}$$

1.8 The Statistical Operator

The description of the states of a physical system must be generalized to the case where the state is prepared in a random way for purely practical reasons, as under the repetition of measurements needed for comparison with the probabilistic statements of the theory, it is in general impossible to keep preparation conditions completely identical. If, for example, atoms possessing a magnetic moment are let out through a small hole in a cold oven kept in a huge magnetic field, the atoms will have their magnetic moment component prepared predominantly along the magnetic field direction. However, out of a hot oven will come atoms with random values of their magnetic moment component.

For a system prepared in a random way, we can for a complete property A only attribute a set of probabilities $\{p_a\}_a$ for the system to be prepared in the complete set of states labeled by a. In this case the probability p_b to find the value b of the property B is

$$p_b = \sum_a p(b\,;a)\,p_a = \sum_a |<b|a>|^2\,p_a = <b|\sum_a |a>\,p_a\,<a|b>$$

$$= <b|\hat{\rho}|b> = Tr(\hat{\rho}\,\hat{P}(b)) \tag{1.378}$$

where we have introduced the statistical operator[78]

$$\hat{\rho} \equiv \sum_a p_a\,|a><a| = p(\hat{A}) \tag{1.379}$$

describing the situation where the system is prepared in a random way according to the probability distribution p_a.[79] We shall say that the state is a mixture in contrast to a pure state a_0 for which $p_a = \delta_{a,a_0}$. We shall use the convenient phrase that a system in a mixture is in the state ρ.

---

[78]In general a statistical operator is specified by a set of (normalized) state vectors, $|\psi_1>$, $|\psi_2>$, .., $|\psi_n>$, not necessarily orthogonal, and a set of non-negative numbers adding up to one, $\sum_{i=1}^n p_i = 1$, according to $\hat{\rho} = \sum_{i=1}^n p_i\,|\psi_i><\psi_i|$. Since the statistical operator is hermitian and non-negative, it is always possible to find an orthogonal set of states $|\phi_1>, |\phi_2>, .., |\phi_N>$, so that $\hat{\rho} = \sum_{n=1}^N \pi_n\,|\phi_n><\phi_n|$, where $\pi_n \geq 0$ and $\sum_{n=1}^N \pi_n = 1$.

[79]The completely random case, $p_a = 1/N$, corresponds to a selection without discriminating any property value, in accordance with eq.(1.118) and the normalization condition imposed on the statistical operator, $\hat{\rho} = 1/N \sum_a |a><a| = \hat{I}/N$.

Future property values exhibited by a system in a mixture of states are not only probabilistic due to the inherently probabilistic nature of quantum mechanics, but, since the initial value of the property A is a random variable (fluctuating according its distribution) the outcome is also weighted according to the initial distribution.

The statistical operator is seen to be hermitian, positive ($<\psi|\,\hat{\rho}\,|\psi> \geq 0$), and have unit trace

$$Tr\hat{\rho} \; = \; \sum_a p_a \; = \; 1 \tag{1.380}$$

for a normalized probability distribution. The statistical operator is only idempotent for the case of a pure state. For a mixture we have $\hat{\rho}^2 \neq \hat{\rho}$, and $Tr\,\hat{\rho}^2 < 1$.

A system in a mixture of states is an effective description of a system interacting with another system of whose state we have only statistical knowledge, and the statistical operator incorporates this extrinsically induced statistical uncertainty. In chapter 6 we shall discuss the generic situation of a system interacting with an environment in detail.

The most complete information we can have about the state of a system is when we can assert it to be in a pure state. For example, after a selection or measurement[80] in which we have ascertained the property values in a complete description A, the system is in the state a described by the state symbol $\hat{P}(a)$

$$\hat{P}(a) \; \equiv \; \hat{P}(a_{i_1}^{(1)}, a_{i_2}^{(2)}, ..) \; = \; \delta_{\hat{A}_1, a_1} \delta_{\hat{A}_2, a_2} \cdots \; \equiv \; \delta_{\hat{A}, a} \tag{1.381}$$

or, equivalently, up to a phase factor described by the state vector in question

$$|a> \; = \; |a_{i_1}^{(1)}, a_{i_2}^{(2)}, ..> \; . \tag{1.382}$$

An example of such a state of affairs is the situation where an energy measurement has better resolution than the level spacing, and the system can be determined to be in the energy-eigenstate $|\psi> \; = \; |E_n >$, corresponding to the statistical operator being the projector $\hat{\rho} = \delta_{E_n, \hat{H}} = |E_n><E_n|$. Clearly, when dealing with macroscopic systems consisting of a huge number of particles, this is not feasible, since the density of levels depends exponentially on the number of particles.

Statistical physics is concerned with systems with many degrees of freedom, and the knowledge we have of a macroscopic system is through the macroparameters such as (local) temperature, pressure, etc. A measurement of a macroparameter will not provide the myriad of information encompassed in a complete projector (tantamount to have gained knowledge of, say, the position of 10^{23} particles). From a typical measurement we will only be justified in ascribing a statistical operator where the probability distribution will be constrained by the physical situation in question, specified in terms of the macroparameters. For example, for a system allowed to exchange energy with a system with many degrees of freedom the temperature specifies the energy distribution for the system.

---

[80]By measurement is here thus meant a measurement which when repeated after having been done results in the same outcome; i.e., the state of the system is the same before and after the second measurement.

The Schrödinger equation governing the temporal evolution of the state vector gives for the temporal evolution of the statistical operator

$$\hat{\rho}(t) = \hat{U}(t, t') \, \hat{\rho}(t') \, \hat{U}^\dagger(t, t') . \qquad (1.383)$$

Equivalently, the statistical operator satisfies the von Neumann equation

$$i\hbar \frac{d\hat{\rho}(t)}{dt} = [\hat{H}_t, \hat{\rho}(t)] \qquad (1.384)$$

where we anticipate a system coupled to external fields as described by a time-dependent Hamiltonian.

The matrix elements of the statistical operator

$$\rho(a, a', t) \equiv <a|\hat{\rho}(t)|a'> \qquad (1.385)$$

is called the density matrix in the a-description. The statistical operator specified by eq.(1.383), which is diagonal in the a-representation at some initial time t'

$$\hat{\rho}(t') = \sum_a p_a \, |a><a| \qquad (1.386)$$

will at subsequent times develop off-diagonal elements, unless A is a constant of the motion, and never again be diagonal in any representation (that would violate the inequality $Tr\hat{\rho}^2 < 1$). Equivalently stated, no unitary transformation exists which can change a mixture into a pure state, $\hat{U}^\dagger \hat{\rho} \hat{U} \neq \hat{P}(\psi)$. However, the diagonal elements in any representation is the distribution function since

$$\rho(b, b, t) = <b|\hat{\rho}(t)|b> \qquad (1.387)$$

is the probability p_b for the outcome b for the quantity B at time t as

$$p_b(t) \equiv \sum_a p_a \, |<b|\hat{U}(t, t')|a>|^2 = <b|\hat{\rho}(t)|b> . \qquad (1.388)$$

For the expectation value of property B at time t we have

$$B(t) \equiv \sum_b b \, p_b(t) = \sum_b <b|\rho(t)\hat{B}|b>$$

$$= Tr(\hat{\rho}(t)\hat{B}) = Tr(\hat{\rho}(t_r)\hat{B}_H(t)) \qquad (1.389)$$

where t_r is the arbitrary reference time where the Schrödinger and Heisenberg pictures coincide.

A stationary state is characterized by having time-independent properties, and is therefore described by a time-independent statistical operator

$$\frac{d\hat{\rho}}{dt} = 0 \qquad (1.390)$$

which according to eq.(1.384) implies that the statistical operator commutes with the Hamiltonian, $[\hat{H}, \hat{\rho}] = 0$. The statistical operator is then a real (since $\hat{\rho}$ is hermitian) function, ρ, of the Hamiltonian[81]

$$\hat{\rho} = \rho(\hat{H}) . \tag{1.391}$$

In the following we shall need the statistical operator for a large system characterized by the macroscopic parameter the temperature T, or equivalently the statistical operator for a system allowed to exchange energy with a thermal reservoir. Applying the zeroth law of thermodynamics, that two systems in equilibrium at temperature T will upon being brought in thermal contact be in equilibrium at the same temperature, we obtain the canonical or thermal equilibrium statistical operator

$$\hat{\rho}_T = \frac{1}{Z} e^{-\hat{H}/kT} \tag{1.392}$$

where the partition function Z is determined by the normalization of the canonical statistical operator

$$Z(T, V, N) = Tr \, e^{-\hat{H}/kT} = e^{-F(T,V,N)/kT} \tag{1.393}$$

and specified by the free energy F, a function of the temperature T, volume V, and number N of particles in the system. We recall that all thermodynamic information is contained in the partition function or equivalently the free energy.

Exercise 1.23 *A particle with mass m and charge e in a homogeneous magnetic field parallel to the z-axis, $\hat{\mathbf{B}} = B\hat{\mathbf{z}}$, has the energy spectrum (consult reference [13] or section 11.4)*

$$\epsilon_n = \hbar\omega_c(n + 1/2) + \frac{p_z^2}{2m} , \qquad n = 0, 1, 2, ... \tag{1.394}$$

where $\omega_c = |eB|/m$ is the Larmor frequency, and each so-called Landau level is macroscopically degenerate with the degeneracy factor $|eB|A/2\pi\hbar$, where A is the the cross-sectional area of the system perpendicular to the magnetic field direction. Show that the partition function Z_B for a particle confined to the volume $V = AL$ is given by

$$Z_B = AL \left(\frac{mkT}{2\pi\hbar^2}\right)^{3/2} \frac{\frac{\hbar\omega_c}{2kT}}{\sinh\frac{\hbar\omega_c}{2kT}} . \tag{1.395}$$

Solution

---

[81]We assume that the energy is the only conserved quantity. In general, the ensemble must take into account all conserved quantities.

The partition function, evaluated in the complete basis of eigenstates of the Hamiltonian, is the sum

$$Z_B = \sum_\lambda e^{-\epsilon_\lambda/kT} = \frac{|eB|A}{2\pi\hbar} \sum_{n,p_z} e^{-(\hbar\omega_c(n+1/2)+\frac{p_z^2}{2m})/kT}$$

$$= \frac{|eB|}{2\pi\hbar} AL \int_{-\infty}^{\infty} \frac{dp_z}{2\pi\hbar} \, e^{-\frac{p_z^2}{2mkT}} \sum_{n=0}^{\infty} e^{-\hbar\omega_c(n+1/2)/kT} . \tag{1.396}$$

The summation over the allowed p_z-values for confined motion in the z-direction, $p_z = 2\pi\hbar n_z/L, n_z = 0, \pm1, \pm2, ...$, becomes in the limit of large system size in the z-direction, $L \gg \hbar/\sqrt{mkT}$, a Gaussian integral, and the summation over the orbital index n is a simple geometric series. The integration and summation are therefore immediately performed, and we obtain the stated result. The partition function for a free particle is seen to be recovered in the limit of vanishing field.

1.9 Consistency of Quantum Mechanics

When illustrating the content of the formalism of quantum mechanics we have referred to measurements. However, when formulating the principles of quantum mechanics we have avoided any reference to acts of measurements [14], and in doing so attributed probability as a fundamental feature of the physical world. In recent years proposals for a realistic interpretation of the quantum mechanics of a closed system has progressed. The basic ingredients in such interpretations, which avoids reference to anything beyond the symbols of quantum mechanics (such as measurements and the realm of classical physics), we describe in the following.

If a system in state a is subjected to a B-selection and a subsequent C-selection, we can ask for the probability $p(c, b, a)$ that the system passes these discriminations, or equivalently, the probability that the system exhibits the sequence of property values a, b, and c. This probability is given by (the last three equalities simply being equivalent ways of stating the original identity)

$$p(c,b,a) = p(c\,;b)\,p(b\,;a) = |{<}c|b{>}|^2\,|{<}b|a{>}|^2$$

$$= |{<}c|\hat{P}(b)|a{>}|^2 = {<}a|\hat{P}(b)\hat{P}(c)\hat{P}(b)|a{>} . \tag{1.397}$$

We note that we can express the probability for this sequence of selections, or events, to occur in terms of the trace in various forms

$$p(c,b,a) = Tr(\hat{P}(c)\,\hat{P}(b))\ Tr(\hat{P}(b)\,\hat{P}(a))$$

$$= Tr((\hat{P}(a)\,\hat{P}(b)\,\hat{P}(c))^\dagger\,(\hat{P}(a)\,\hat{P}(b)\,\hat{P}(c)))$$

$$= Tr(\hat{P}(c)\,\hat{P}(b)\,\hat{P}(a)\,\hat{P}(b)\,\hat{P}(c))$$

$$= Tr(\hat{P}(b)\,\hat{P}(a)\,\hat{P}(b)\,\hat{P}(c)) . \tag{1.398}$$

In general, the probability that a system, initially in state a, will successively exhibit the property values $b, .., f$ is

$$p(f, .., b, a) \ = \ p(f \, ; \, .) \, .. \, p(. \, ; b) \, p(b \, ; a)$$

$$= \ Tr((\hat{P}(f) \, .. \, \hat{P}(b)) \hat{P}(a) (\hat{P}(b) \, .. \, \hat{P}(f))) \, . \qquad (1.399)$$

When the initial state is not a pure state, described by a projector $\hat{P}(a)$, but a mixture described by a statistical operator, $\hat{\rho} = \sum_a p_a \, |a><a|$, the probability $p_\rho(f, .., b)$ the system will successively exhibit the property values $b, .., f$ is

$$p_\rho(f, \ .. \, , b) \ = \ \sum_a \ p(f \, ; \, .) \, .. \, p(. \, ; b) \, p(b \, ; a) \, p_a = Tr((\hat{P}(f) \, .. \, \hat{P}(b)) \hat{\rho} (\hat{P}(b) \, .. \, \hat{P}(f))) \, .$$

$$(1.400)$$

Quantum mechanics thus establishes statistical correlations between subsequent observations. To exonerate any reference to measurement, this number is now ascribed as the probability that in an isolated system, initially in state ρ, the events $b, .., f$ occur.

The possible evolution of an isolated system is described in terms of a sequence of possible exhibited property values. A sequence of possible exhibited property values $a, .., f$ of an isolated system, initially in state ρ, occurring at certain times is referred to as a history, or elementary history,[82]

$$h : (a, t_1) \to (b, t_2) \to .. \to (f, t_n) \qquad (1.401)$$

and the specified possible history is ascribed the probability[83]

$$p_\rho(h) = Tr((\hat{P}(f, t_n) \, .. \, \hat{P}(b, t_2) \hat{P}(a, t_1)) \, \hat{\rho} \, (\hat{P}(a, t_1) \hat{P}(b, t_2) \, .. \, \hat{P}(f, t_n))) \, . \quad (1.402)$$

We have amended the previous notation, where we suppressed the moments in time where the property values in question are exhibited. The projectors are therefore in the Heisenberg picture, and we have for convenience chosen as reference time the instant of the exhibited initial state, $\hat{\rho} \equiv \hat{\rho}(t_r)$. A history is thus characterized by an initial state ρ, a sequence of times, and a set of property values $a, b, .., f$. The probability of an elementary history, a product of nonnegative numbers, is of course guaranteed positive, $p_\rho(h) \geq 0$. A history is a narrative of a possible sequence of events (at time t_1 the system exhibited property a, at time t_2 the system exhibited property b, etc.) occurring in an isolated system. The most primitive history corresponds to a single event

$$h : (a, t) \qquad (1.403)$$

---

[82]In the path integral formalism for a single particle, a history is simply a path $\mathbf{x}_t$ for the particle at a certain level of discretization, and in accordance with the multiplication principle it can be assigned the probability $|A[\mathbf{x}_t]|^2 = |\prod_i A[\mathbf{x}_t^{(i)}]|^2$, eq.(1.9); i.e., $\Delta\mathbf{x} \, |A_{\mathbf{x},t;\mathbf{x}',t'}[\mathbf{x}_t]|^2 \Delta\mathbf{x}'$ is the probability for the particle to arrive at time t in the volume $\Delta\mathbf{x}$ around position $\mathbf{x}$ *following the path* $\mathbf{x}_t$, and starting in the volume $\Delta\mathbf{x}'$ around position $\mathbf{x}'$ at time t'. Coarser histories corresponds to summing over all paths crossing certain volumes at intermediate times, as discussed at the end of appendix A.

[83]One of the projectors $\hat{P}(f, t_n)$ can be omitted due to the cyclic invariance of the trace.

and has, according to eq.(1.402), associated the probability for the occurrence of the event a at time t

$$p_\rho(h) = Tr(\hat{P}(a,t)\,\hat{\rho}\,\hat{P}(a,t)) = Tr(\hat{\rho}\,\hat{P}(a,t)) \tag{1.404}$$

the Born rule.

We introduce the notion of a family of histories, the set of histories sharing the same initial state and sequence of times, and the same properties $A, B, .., F$ exhibited at these times. The only difference between histories in a family being that their set of exhibited property values differ. The histories can be represented as track records, where the exhibited property values at the moments in time in question for the histories are followed, giving a *motion picture* representation of histories as shown in figure 1.5.

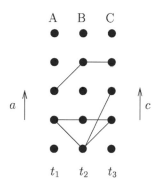

Figure 1.5 Four possible histories depicted for a three-time family.

The moments in time t_i and t_{i+1} in a history can be identical, and so can the property values at different times (at equal times, due to $\hat{P}(a,t)\,\hat{P}(a,t) = \hat{P}(a,t)$, the sequence $(a,t) \rightarrow (a,t)$ is redundant). Furthermore, in a history we do not encounter the sequence $(a,t) \rightarrow (a',t)$, since for the exhibition of mutually exclusive property values at the same time we have $\hat{P}(a,t)\,\hat{P}(a',t) = \hat{0}$, and such a nonsensical history has zero probability. A history where the same property occurs at the next time, taking into account the time evolution governed by the Schrödinger equation, i.e., a sequence $\hat{P}(a,t_{n+1})\,\hat{P}(a,t_n)$, is also not of interest for a conserved quantity A, as it just asserts at the subsequent time what was already asserted about the system at the immediate prior time.

We can also represent the set of elementary histories by all the possible sequences of property values $(a, b, .., f)$, i.e., a lattice of points, the history lattice of all $(a, b, .., f)$-tuples. As an example we consider the family of histories with only two properties A and B exhibited, and only two moments in time appearing in the history. In this case the set of all possible elementary histories is represented by all possible (a, b)-tuples as in figure 1.6.

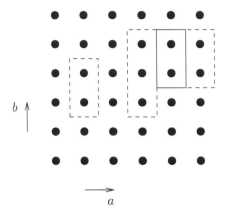

Figure 1.6 History lattice for a family with two times and properties A and B (here displayed for six possible property values).

An elementary history can in this case be represented in various ways

$$h : (a, t_1) \rightarrow (b, t_2) \quad , \quad h : \hat{P}(a, t_1)\hat{P}(b, t_2) \quad , \quad h : (a, b) . \tag{1.405}$$

This history has the associated probability (we suppress from now on the reference to the initial state ρ on the history probability)

$$p(h) = Tr((\hat{P}(b, t_2)\, \hat{P}(a, t_1))\, \hat{\rho}\, (\hat{P}(a, t_1)\, \hat{P}(b, t_2))) . \tag{1.406}$$

We can introduce compound, or coarse-grained, histories by the notion of addition of different histories according to the definition: the sum, $h = h' + h''$, of two histories corresponds to the compound narrative of the two histories. Adding in our example two different elementary histories $h' \equiv (a', b')$ and $h'' \equiv (a'', b'')$ (at least one of the property values must be different in order for the two histories to be different), the sum of the two histories corresponds in this case to the narrative "at time t_1 the system exhibited the property value a' *or* a'' , and at time t_2 the property value b' *or* b''." The sum of these two elementary histories is thus the compound history (assuming $a'' \neq a'$ and $b'' \neq b'$)

$$h = h' + h'' : (\hat{P}(a', t_1) + \hat{P}(a'', t_1))(\hat{P}(b', t_2) + \hat{P}(b'', t_2)) \tag{1.407}$$

represented by the sum of the respective projectors in accordance with eq.(1.118). The compound history has associated the union of the lattice points associated with the two histories in the history lattice as illustrated in figure 1.6. For the compound history in question we have the associated probability

$$\begin{aligned} p(h) &= Tr((\hat{P}(a', t_1) + \hat{P}(a'', t_1))\hat{\rho}(\hat{P}(a', t_1) + \hat{P}(a'', t_1))(\hat{P}(b', t_2) + \hat{P}(b'', t_2))) \\ &= Tr((\hat{P}(b', t_2) + \hat{P}(b'', t_2))(\hat{P}(a', t_1) + \hat{P}(a'', t_1))\, \hat{\rho} \\ &\quad (\hat{P}(a', t_1) + \hat{P}(a'', t_1))(\hat{P}(b', t_2) + \hat{P}(b'', t_2))) . \end{aligned} \tag{1.408}$$

Adding histories gives us a complete family of histories, reflecting all possible narratives for the given family specified by initial state, moments in time, and properties. The empty narrative where at each instant any property values can have been exhibited is described by the history I, which is the sum over all the elementary histories

$$I : \sum_a \hat{P}(a,t_1) \sum_b \hat{P}(b,t_2) = \hat{I} \cdot \hat{I} = \hat{I} \qquad I : \sum_{i,j} (a_i, b_j) \ . \qquad (1.409)$$

This maximally compounded history, which has no information on exhibited values, has unit probability

$$p(I) = Tr((\hat{I}\hat{I}) \, \hat{\rho} \, (\hat{I}\hat{I})) = 1 \qquad\qquad (1.410)$$

by normalization of the initial state. This normalization simply reflects that we are dealing with complete properties. The probability that some history or narrative takes place is 1.

We can thus envisage applying probability calculus to a complete family. However, in order for p, as function on a set of histories, to be a probability measure it should satisfy (besides the already proven properties of positivity and normalization) the property of additivity; i.e., given two different histories h' and h'', the function p must satisfy that its value on the compound history $h = h' + h''$ obeys the relation

$$p(h) \; = \; p(h') \; + \; p(h'') \ . \qquad\qquad (1.411)$$

This is a nontrivial requirement due to the incompatibility of properties, and as noted by Griffiths [15], this leads to consistency conditions ensuring that there is no quantum interference between different histories.[84]

Let us examine the implications of the additivity requirement for the chosen example. For the case $a'' = a'$ we have for the sum of the two histories (both in the same a-column in the history lattice)

$$h = h' + h'' : \hat{P}(a',t_1) \, (\hat{P}(b',t_2) + \hat{P}(b'',t_2)) \qquad\qquad (1.412)$$

and additivity is trivially satisfied as the properties b' and b'' are mutually exclusive in order to have two different histories in the first place, and consequently $\hat{P}(b'',t_2)\hat{P}(b',t_2) = \hat{0}$, giving

$$\begin{aligned} p(h) \; &= \; Tr((\hat{P}(b',t_2) + \hat{P}(b'',t_2)) \, \hat{P}(a',t_1) \, \hat{\rho} \, \hat{P}(a',t_1) \, (\hat{P}(b',t_2)) + \hat{P}(b'',t_2))) \\ &= \; Tr(\hat{P}(a',t_1) \, \hat{\rho} \, \hat{P}(a',t_1) \, (\hat{P}(b',t_2) + \hat{P}(b'',t_2))) \\ &= \; p(h') + p(h'') \ . \end{aligned} \qquad (1.413)$$

For the case $b'' = b'$ we have for the sum of the histories (now property value a'' must be different from a' in order to have two different histories)

$$h = h' + h'' : (\hat{P}(a',t_1) + \hat{P}(a'',t_1)) \, \hat{P}(b',t_2) \qquad\qquad (1.414)$$

---

[84]At the end of appendix A this is also illustrated by describing the histories in terms of path integrals.

and the associated probability is given by

$$
\begin{aligned}
p(h) &= Tr((\hat{P}(a',t_1) + \hat{P}(a'',t_1))\,\hat{\rho}\,(\hat{P}(a',t_1) + \hat{P}(a'',t_1))\hat{P}(b',t_2))) \\
&= p(h') + p(h'') + Tr(\hat{P}(b',t_2)\hat{P}(a',t_1)\,\hat{\rho}\,\hat{P}(a'',t_1)) \\
&+ Tr(\hat{P}(b',t_2)\hat{P}(a'',t_1)\hat{\rho}\hat{P}(a',t_1)) \\
&= p(h') + p(h'') + 2\,\Re e\Big(Tr(\hat{P}(b',t_2)\hat{P}(a',t_1)\,\hat{\rho}\,\hat{P}(a'',t_1))\Big)\ . \qquad (1.415)
\end{aligned}
$$

In order for the function p to be additive for the two histories in question, thus requires the consistency condition

$$
\Re e\left(Tr(\hat{P}(a',t_1)\,\hat{\rho}\,\hat{P}(a'',t_1)\,\hat{P}(b',t_2))\right) = 0\ . \qquad (1.416)
$$

The need for consistency conditions thus reflects the existence of incompatible properties, the noncommutativity of property symbols. The fulfillment of a consistency condition necessitates proper choices of histories.

For the case $a'' \neq a'$ and $b'' \neq b'$ (neither histories is in the same a-column or b-row in the history lattice), we have according to eq.(1.408) for the probability of the compound history

$$
\begin{aligned}
p(h) &= p(h') + p(h'') + 2\,\Re e\,Tr(\hat{P}(a'',t_1)\,\hat{\rho}\,\hat{P}(a',t_1)\hat{P}(b',t_1)) \\
&+ 2\,\Re e\,Tr(\hat{P}(a',t_1)\,\hat{\rho}\,\hat{P}(a'',t_1)\hat{P}(b'',t_2)) + Tr(\hat{P}(a'',t_1)\,\hat{\rho}\,(\hat{P}(a'',t_1)\hat{P}(b',t_2)) \\
&+ Tr(\hat{P}(a',t_1)\,\hat{\rho}\,(\hat{P}(a',t_1)\hat{P}(b'',t_2)) \qquad (1.417)
\end{aligned}
$$

and we get the following consistency conditions (the last two terms in the above equation are clearly real):

$$
\begin{aligned}
\Re e\,Tr(\hat{P}(a'',t_1)\,\hat{\rho}\,\hat{P}(a',t_1)\hat{P}(b',t_2)) &= 0 & (1.418) \\
\Re e\,Tr(\hat{P}(a',t_1)\,\hat{\rho}\,\hat{P}(a'',t_1)\hat{P}(b'',t_2)) &= 0 & (1.419) \\
\Re e\,Tr(\hat{P}(a'',t_1)\,\hat{\rho}\,\hat{P}(a'',t_1)\hat{P}(b',t_2)) &= 0 & (1.420) \\
\Re e\,Tr(\hat{P}(a',t_1)\,\hat{\rho}\,\hat{P}(a',t_1)\hat{P}(b'',t_2)) &= 0\ . & (1.421)
\end{aligned}
$$

The additivity requirement for having a probability measure on a family of histories splits the complete family of histories into consistent families of histories on each of which we have a probability measure. Within the context of a consistent family of histories, a probability has thus been assigned to a possible history of an isolated system in the usual statistical sense. We note that checking additivity (consistency) of histories is a matter of computation. The consistent history approach codifies in a quantitative way the complementary aspect of quantum mechanics.[85]

---

[85]The so-called Copenhagen interpretation of quantum mechanics disposes this fruition as *speaking of the unspeakable*, and gives preference to classical mechanics in the realm of macro-

For the primitive family of histories, eq.(1.403), the complete family of histories, corresponding to all possible a-values, is of course consistent, because the different a-values are mutually exclusive properties thereby securing additivity, the Born rule is universal.[86]

Histories are equivalent to propositions. An elementary history is equivalent to the proposition: "at time t_1 the system exhibited the property a, at time t_2 the system exhibited the property b, ..." Such an elementary history is represented by a single point in the history lattice. A history is a proposition about the properties of an isolated system. The proposition associated to a history does not affirm the exhibition of the property values in question but has only ascribed a probability for the events to occur at the times in question: the probability for the corresponding history. Propositions being equivalent to histories thus appear in the corresponding family context. Adding elementary histories, we can construct all possible histories in the complete family. Such compound histories correspond to less restrictive propositions "at time t_1 the system exhibited the property a' or a'', at time t_2 the system exhibited the property b' or b'', ..," and expresses the logical operation or for propositions, as we soon introduce. A general proposition is associated with a set of points in the history lattice. A predicate attributes a history to a system. For example, the proposition "at time t_2 the system exhibited the property b" corresponds to the sum of histories "at time t_2 the system exhibited the property b, and at the other times in question any property value could have been exhibited" (i.e., no property value except the one at time t_2 is asserted). This proposition is represented by the whole row of history lattice points where the property value b occurs. Having a representation of propositions in terms of sets in the history lattice at hand, we can then introduce the fundamental logical operations and , or and $negation$, $\wedge, \vee, \neg$, for the propositions as the usual definitions in terms of set operations, thereby respecting the rules of formal logic. We denote propositions by Greek letters α, β, .., and the proposition "α or β" ($\alpha \vee \beta$) is associated with the union of the two corresponding proposition sets, and "α and β" ($\alpha \wedge \beta$) corresponds to the intersection of the two proposition sets, and negation, "not α" ($\neg\alpha$) is associated with the complement of the proposition set for the proposition α. In figure 1.6 compound histories are illustrated (dashed line boxes) as well as the intersection or union ($\alpha \wedge \beta$) of two proposition sets (bold line box).

Since the compound propositions are associated with sets in the history lattice, we can ascribe to them a probability according to the history they are equivalent to. The probability for the proposition $\alpha \vee \beta$, $p(\alpha \vee \beta)$, is defined as the probability

---

scopic bodies such as the case of a measurement apparatus. However, such dogmatism clearly has its own problems in coping with the grand task of describing the universe as a closed system. The consistent history interpretation can be said to have problems of its own as we touch upon later, but it allows for approaching the real mystery in quantum mechanics, viz. the status of reality, how does things happen, by relegating it to the status of a stochastic feature of how Nature works.

[86]Similarly, histories differing only in their property value at the latest time are consistent.

of the corresponding compound history

$$h_{\alpha \vee \beta} : \left(\sum_{(a,b)\in\alpha} (a,b) \right) \cup \left(\sum_{(a,b)\in\beta} (a,b) \right) \tag{1.422}$$

giving for the probability of the compound proposition

$$p(\alpha \vee \beta) \equiv p(h_{\alpha\vee\beta}) \, . \tag{1.423}$$

Similarly the probability for the proposition $\alpha \wedge \beta$, $p(\alpha \wedge \beta)$, is the probability for the history corresponding to the intersection of their sets in the history lattice

$$h_{\alpha \wedge \beta} : \left(\sum_{(a,b)\in\alpha} (a,b) \right) \cap \left(\sum_{(a,b)\in\beta} (a,b) \right) \tag{1.424}$$

giving for the probability of the compound proposition

$$p(\alpha \wedge \beta) \equiv p(h_{\alpha\wedge\beta}) \, . \tag{1.425}$$

Having related proposition probabilities to sets, all probability definitions within a consistent family are as in usual probability calculus. The conditional probability, $p(\beta; \alpha)$, the probability for the events described in proposition β to occur given the events described in proposition α have occurred, is defined as

$$p(\beta; \alpha) \equiv \frac{p(\alpha \wedge \beta)}{p(\alpha)} \tag{1.426}$$

which of course presupposes that the proposition α is a sensible one, $p(\alpha) \neq 0$.

The all important logic operation for reasoning, the implication, $\alpha \Rightarrow \beta$, is defined in the usual way: proposition α is said to imply proposition β when the conditional probability equals unity (which of course again presupposes that the proposition α is a sensible one)

$$p(\beta; \alpha) = 1 \tag{1.427}$$

i.e., $\alpha \Rightarrow \beta$ means per definition

$$\frac{p(\alpha \wedge \beta)}{p(\alpha)} = 1 \quad , \quad p(\alpha) \neq 0 \, . \tag{1.428}$$

Two propositions are said to be logically equivalent, $\alpha = \beta$, if proposition α implies proposition β *and* proposition β implies proposition α, or equivalently

$$p(\alpha; \beta) = 1 \quad \text{and} \quad p(\beta; \alpha) = 1 \tag{1.429}$$

provided both propositions are sensible.

A quantum logic describing the properties of an isolated system is now defined as a set of propositions associated with a consistent family of histories, and quantum reasoning about the fate of an isolated system is then according to commonsense logic.

A quantum logic is free of inconsistencies as the consistency requirements simply reflect the set operations, $\cup$, $\cap$, $\neg$, in the history lattice. However, the definition entails that we can have many quantum logics. Two logics, L' and L'', are said to be mutually consistent if a consistent logic L exists which contains both, i.e., $\exists L : L', L'' \in L$. However, two quantum logics need not be mutually consistent and are in that case said to be complementary logics. The existence of complementary logics, however, does not lead to any logical contradictions for quantum reasoning if the universal rule of interpretation is adapted:[87] Any description of an isolated physical system must be done in terms of properties belonging to a common consistent logic, and a valid reasoning relating these properties of the system should consist of implications holding in that logic. The absence of contradiction is demonstrated by the following consideration. A proposition α corresponds to a history h_α and is represented by a set of points in the history lattice

$$\alpha : \sum_{(a,b)\in\alpha} (a, b) \qquad (1.430)$$

a sum of elementary histories $h_{(a,b)}$. The probability $p(h_\alpha)$ for history h_α is the sum of probabilities for the elementary histories it consists of, and the value is therefore determined by the α-set of points in the history lattice, and independent of the logic, L or L', it is considered belonging to, $p(h_\alpha) = p'(h_\alpha)$. Similarly the proposition "$\alpha \wedge \beta$" is the proposition corresponding to the intersection of the α- and β-sets

$$h_{\alpha\wedge\beta} : \left(\sum_{(a,b)\in\alpha} (a, b) \right) \cap \left(\sum_{(a,b)\in\beta} (a, b) \right) \qquad (1.431)$$

which has the same probability in all logics of which it is a member, $p(h_{\alpha\wedge\beta}) = p'(h_{\alpha\wedge\beta})$, and the condition for implication is therefore the same in all these logics. We can therefore never end in the contradictory situation, where the proposition α implies proposition β in one logic, and proposition α implies proposition $\neg\beta$ in another. We can have the implication $\alpha \Rightarrow \beta$ in one logic and not in another, say because the proposition α is simply not a proposition in the other logic. This, however, is not a contradiction, but reflects complementarity, with its consequence of the existence of many different possible quantum logics describing the same system though in mutually exclusive ways.

As an illustration of reasoning in a quantum logic, let us consider Mott's analysis [17] of cloud chamber tracks within the consistent history interpretation of quantum mechanics.[88] Let us assume that a particle of mass m is created in a nuclear decay near position $\mathbf{x} = \mathbf{0}$, and subsequently detected at position $\mathbf{x}$. We can then address the question whether we logically can conclude that the particle

---

[87]The logical framework of the consistent history approach has been stressed by Omnés; see reference [16].

[88]We follow the presentation of reference [16].

in its (free) motion followed a straight line. We are therefore interested in the question whether the particle was in a certain volume V_1 at an intermediate time. In the consistent history approach we are therefore dealing with the two-time family characterized by the time of detection t_2, and an intermediate time t_1, $0 < t_1 < t_2$. The events of interest correspond to the following propositions

α: the particle is in detection volume V_d at time t_2.

β: the particle is in volume V_1 at the intermediate time t_1.

The two events or propositions of interest correspond to the histories

$$h_\alpha : \hat{P}_{V_d}(t_2) , \qquad \hat{P}_{V_d}(t_2) = \int_{V_d} dx \, |\mathbf{x}, t_2><\mathbf{x}, t_2| \qquad (1.432)$$

and

$$h_\beta : \hat{P}_{V_1}(t_1) , \qquad \hat{P}_{V_1}(t_1) = \int_{V_1} dx \, |\mathbf{x}, t_1><\mathbf{x}, t_1| . \qquad (1.433)$$

The history h_α and thereby the corresponding proposition α has associated the probability

$$p(\alpha) = Tr(\hat{\rho}\hat{P}_{V_d}(t_2)) = \int_{V_d} dx \, \rho(\mathbf{x}, \mathbf{x}, t_2) = \int_{V_d} dx \, |\psi(\mathbf{x}, t_2)|^2$$

$$= \int_{V_d} dx \left| \int dx' \, K_0(\mathbf{x}, t_2; \mathbf{x}', 0) \, \psi_i(\mathbf{x}') \right|^2 \qquad (1.434)$$

where ψ_i is the initial wave function for the particle. As the initial state we assume a spherical symmetric outgoing wave with average radial momentum p

$$\psi_i(\mathbf{x}) = \left(\frac{1}{2\pi \Delta x^2} \right)^{3/4} e^{-\frac{|\mathbf{x}|^2}{4\Delta x^2} + \frac{i}{\hbar} p |\mathbf{x}|} . \qquad (1.435)$$

In order for $p(\alpha)$ to be nonzero it follows from eq.(1.434) that t_2 is essentially equal to $m|\mathbf{x}_d|/p$, where $\mathbf{x}_d$ is a position in the detection volume.

We shall only be interested in whether the particle is in the specified volumes or not, and we thus have the following possible evolutions of the particle motion depicted in figure 1.7, where the initial statistical operator is a pure state $\hat{\rho} = |\psi_i><\psi_i|$.

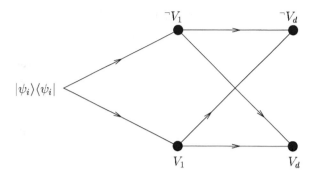

Figure 1.7 The possible four histories for Mott's cloud chamber analysis.

The history lattice therefore consists of the four elementary histories as shown in figure 1.8.

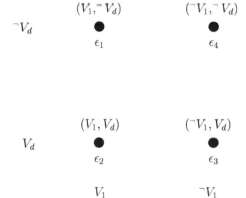

Figure 1.8 History lattice of interest.

Only addition of histories adjacent to each other is of interest for this history lattice since the propositions $\epsilon_1 \vee \epsilon_3$ and $\epsilon_2 \vee \epsilon_4$ simply state that the particle at times t_1 and t_2 are somewhere in space and has associated the probability 1. Elementary histories on top of each other in the history lattice, ϵ_1 and ϵ_2, and ϵ_3 and ϵ_4, are clearly consistent as they only differ in property value at the latest time. We note that $\alpha = \epsilon_2 \vee \epsilon_3$, and in view of the above $p(\alpha) = p(h_{\epsilon_2 \vee \epsilon_3}) = p(\epsilon_2) + p(\epsilon_3)$.

The consistency of ϵ_2 and ϵ_3 requires (the case $b' = b''$ treated above)

$$\Re e \, Tr(\hat{P}_{V_1}(t_1)\, \hat{\rho}\, \hat{P}_{\neg V_1}(t_1)\, \hat{P}_{V_d}(t_2)) = 0 \,. \tag{1.436}$$

Noting that the consistency condition can be rewritten

$$Tr(\hat{P}_{V_1}(t_1)\, \hat{\rho}\, \hat{P}_{\neg V_1}(t_1)\, \hat{P}_{V_d}(t_2)) = -Tr(\hat{P}_{\neg V_1}(t_1)\, \hat{\rho}\, \hat{P}_{V_1}(t_1)\, \hat{P}_{V_d}(t_2)) \tag{1.437}$$

we find that the consistency of ϵ_1 and ϵ_4 results in the same consistency condition. The consistency of the logic consisting of the four elementary propositions thus only requires one consistency condition.

Corresponding to the proposition $\alpha \wedge \beta$, which is identical to ϵ_2, we have the history

$$h_{\alpha \wedge \beta} : \hat{P}_{V_1}(t_1) \cdot \hat{P}_{V_d}(t_2) \tag{1.438}$$

with the associated probability

$$
\begin{aligned}
p(\alpha \wedge \beta) \;=\;& Tr(\hat{P}_{V_1}(t_1) \, |\psi_i><\psi_i| \, \hat{P}_{V_1}(t_1) \, \hat{P}_{V_d}(t_2)) \\[2mm]
=\;& \int_{V_d} d\mathbf{x}_2 \left| \int_{V_1} d\mathbf{x}_1 \!\int\! d\tilde{\mathbf{x}} \; K_0(\mathbf{x}_2, t_2; \mathbf{x}_1, t_1) \, K_0(\mathbf{x}_1, t_1; \tilde{\mathbf{x}}, 0) \, \psi_i(\tilde{\mathbf{x}}) \right|^2 \\[2mm]
=\;& \int_{V_d} d\mathbf{x}_2 \left| \int_{V_1} d\mathbf{x}_1 \; K_0(\mathbf{x}_2, t_2; \mathbf{x}_1, t_1) \, \psi(\mathbf{x}_1, t_1) \right|^2 \tag{1.439}
\end{aligned}
$$

In order to show that α logically implies β we must according to eq.(1.428) show that

$$p(\alpha \wedge \beta) \;=\; p(\alpha) \,. \tag{1.440}$$

The calculations needed for verifying the consistency condition and calculating $p(\alpha \wedge \beta)$ are similar to Mott's where an atom acting as a detector was contemplated placed in the volume V_1.

By properly choosing the volume V_1 the consistency condition can be exactly fulfilled, $\Re e\, C = 0$, where

$$
\begin{aligned}
C \;\equiv\;& Tr(\hat{P}_{V_1}(t_1) \, \hat{\rho} \, \hat{P}_{\neg V_1}(t_1) \, \hat{P}_{V_d}(t_2)) \\[2mm]
=\;& \int_{V_d} d\mathbf{x}_2 \!\int_{V_1} d\mathbf{x}_1 K_0(\mathbf{x}_2, t_2; \mathbf{x}_1, t_1) \, \psi(\mathbf{x}_1, t_1) \!\int_{\neg V_1} d\mathbf{x}_1' \, K_0^*(\mathbf{x}_2, t_2; \mathbf{x}_1', t_1) \psi^*(\mathbf{x}_1', t_1). \tag{1.441}
\end{aligned}
$$

But the integrals over the finite volumes can not be done exactly.[89] However, we shall take the pragmatic point of view and be satisfied if consistence conditions are fulfilled to a certain accuracy. We therefore introduce the notion of a relaxed quantum logic. We shall say that proposition α "almost-ϵ" implies proposition β ($\alpha \overset{\epsilon}{\Rightarrow} \beta$) if $p(\beta; \alpha) \geq 1 - \epsilon$, ϵ being a number between zero and one (preferably close to one).

To get an estimate of the integrals involved we recall the path integral expression of the propagator. The consistency condition involves one factor where only paths passing through the volume V_1 at time t_1 contributes, and a factor where

---

[89]We can perform the $\tilde{\mathbf{x}}$-integration in eq.(1.439), and obtain a closed expression for the wave function at time t_1 in terms of the probability integral.

only paths passing through the complement volume $\bar{\ }V_1$ at time t_1 contribute. If we choose the position of volume V_1 on the line joining $\mathbf{0}$ and $\mathbf{x}$ at the distance $|\mathbf{x}_c| = pt_1/m$ from the origin, the classical path traverses the volume V_1 at time t_1. If the main contribution to the path integral is from the stationary path then $\Re e\, C$ is indeed a very small number, as the path integral involving paths that passes through $\bar{\ }V_1$ at time t_1 does not contain the stationary path, and the wildly oscillating contributions average almost to zero. In fact, the integrals can be evaluated in the stationary phase approximation, giving

$$\Re e\, C \;\simeq\; p(\alpha)\, e^{-R_1^2/\Delta x^2}\, \cos\left(\frac{pR_1}{\hbar} + \varphi\right) \tag{1.442}$$

where R_1 is the linear size of the volume V_1, and φ an irrelevant phase. By choosing $R_1 \gg \Delta x$, the probability for violating the consistency condition is insignificant.

Similarly, when estimating the probability $p(\alpha \wedge \beta)$, which only involves contributions from paths passing the volume V_1 at time t_1, this is really no restriction as the stationary path is included, and we can integrate $\mathbf{x}_1$ over all space to get

$$p(\alpha \wedge \beta) \;\simeq\; \int_{V_d} d\mathbf{x}_2\, |\psi(\mathbf{x}_2, t_2)|^2 \;=\; p(\alpha) \tag{1.443}$$

i.e., we can conclude that in the chosen logic (specified by fixing the parameters according to the requirements above), the proposition that the particle at an intermediate time is in the chosen volume is a logical consequence of the detection of the particle.[90]

Quantum phenomena are indeed not intuitive, and in fact are often counterintuitive to our perception of phenomena, which is exclusively expressed by ordinary language and common sense, or formally in terms of classical mechanics. Since we have taken the point of view that the quantum theory is fundamental, the perceived classical objective reality is considered to emerge from quantum mechanics. This state of affairs comes about as follows. Our perception relates to macroscopic objects which are described in a very coarse-grained fashion. The myriad of undetected microscopic degrees of freedom constituting a macroscopic object (and its environment with which it interacts) has the effect of averaging the quantum interference terms to zero to an exceedingly good approximation. The different coarse-grained histories of the macroscopic object are said to decohere, and we have true probabilities for the possible series of events, histories, emerging from the quantum description. In order to reconcile the probabilistic predictions of quantum mechanics with almost certainty to the deterministic predictions of classical mechanics, it is convenient to use a relaxed quantum logic.[91] In this way the

---

[90]In connection with the above conclusion of straight-line motion, one should keep in mind that we could introduce another consistent logic, and in that conclude as a logical consequence of the initial state that the state of the particle at time t_1 is still isotropic. This raises the important question of choosing the proper logic. In the program of Gell-Mann and Hartle, which is concerned with understanding the quasiclassical realm that includes familiar experience, the proper choice of finding consistent families is produced by the physical mechanism of decoherence [18].

[91]For details on recovering classical physics we refer to the references [16] and [18].

certainty of implication becomes a quantitative question of probability: proposition α implies proposition β with a probability at most in error with the magnitude ϵ if $p(\beta; \alpha) \geq 1 - \epsilon$. Having implication or to certainty limiting implication in our possession, we can reason about the probable course of events exhibited by an isolated system akin to how an experimenter goes about discussing an experiment or designing its equipment.[92] Sound reasoning about a quantum system has become a quantitative matter, a matter of calculation. The smaller the value of ϵ, the better will the reasoning and conclusions be in concordance with reality.

---

[92]Extending the above analysis, we can conclude that particles keep on circulating in a storage ring in accordance with the classical equations of motion, because the probability for not doing so is by proper design insignificant.

Chapter 2

Diagrammatic Perturbation Theory

This chapter is concerned with propagators. After introducing the retarded and advanced propagator we study their perturbation theoretic structure in a potential in terms of diagrams. The scattering cross section is introduced, and the implications for the propagators of the discrete symmetries of space inversion and time reversal are established. The analytical properties of the propagators are discussed, and the spectral function introduced.

At present, the only general method available for gaining knowledge from the fundamental principles about the dynamics of a system is the perturbative study. This consists in dividing the Hamiltonian into one part representing a simpler well-understood problem and a nontrivial part, the effect of which is studied order by order. The expressions resulting from perturbation theory quickly become unwieldy. A convenient method of representing perturbative expressions by diagrams was invented by Feynman. Besides the appealing aspect of representing perturbative expressions by drawings, the diagrammatic method can also be used directly for reasoning and problem solving. The easily recognizable topology of diagrams makes the diagrammatic method a powerful tool for constructing approximation schemes as well as exact equations that may hold true beyond perturbation theory. Furthermore, by elevating the diagrams to be a representation of possible alternative physical processes, the diagrammatic representation becomes a suggestive tool providing physical intuition into quantum dynamics. We now embark on the construction of the diagrammatic representation starting from the canonical formalism presented in the first chapter.

2.1 Green's Functions and Propagators

The Schrödinger equation describing the dynamics of a single particle in the position representation is

$$i\hbar \frac{\partial \psi(\mathbf{x}, t)}{\partial t} = H\psi(\mathbf{x}, t) .$$

(2.1)

In order to describe a physical problem we need to specify particulars, typically in the form of an initial condition. Such general initial condition problems can be solved through the introduction of the Green's function. The Green's function $G(\mathbf{x}, t; \mathbf{x}', t')$ represents the solution to the Schrödinger equation for the particular initial condition where the particle is definitely at position $\mathbf{x}'$ at time t'

$$\lim_{t \searrow t'} \psi(\mathbf{x}, t) = \delta(\mathbf{x} - \mathbf{x}') = <\mathbf{x}, t'|\mathbf{x}', t'> . \tag{2.2}$$

The solution of the Schrödinger equation corresponding to this initial condition therefore depends parametrically on $\mathbf{x}'$ and t', and is by definition the conditional probability density amplitude for the dynamics in question[1]

$$\psi_{\mathbf{x}', t'}(\mathbf{x}, t) = K(\mathbf{x}, t; \mathbf{x}', t') \equiv G(\mathbf{x}, t; \mathbf{x}', t') \tag{2.3}$$

connecting the two incompatible complete descriptions defined by the operators $\hat{\mathbf{x}}_H(t)$ and $\hat{\mathbf{x}}_H(t')$.

The Green's function, being the kernel of the Schrödinger equation on integral form, eq.(1.7), specifies the dynamics of the system. We shall therefore refer to the Green's function as the propagator.[2] The propagator, being a transformation function, is the trace of the transition operator $\hat{P}(\mathbf{x}, t; \mathbf{x}', t')$[3]

$$G(\mathbf{x}, t; \mathbf{x}', t') = Tr(\hat{P}(\mathbf{x}, t; \mathbf{x}', t')) = Tr(|\mathbf{x}', t'><\mathbf{x}, t|) = <\mathbf{x}, t|\mathbf{x}', t'> . \tag{2.4}$$

Since the Green's function is defined to be a solution of the Schrödinger equation, we have (as also verified by differentiating eq.(2.4))

$$\{i\hbar \frac{\partial}{\partial t} - H\} G(\mathbf{x}, t; \mathbf{x}', t') = 0 . \tag{2.5}$$

We note that the partition function and the trace of the evolution operator are related by analytical continuation:

$$Z = Tr\, e^{-\hat{H}/kT} = \int d\mathbf{x} \ <\mathbf{x}|e^{-\hat{H}/kT}|\mathbf{x}> = Tr\, \hat{U}(-i\hbar/kT, 0)$$

$$= \int d\mathbf{x} \ G(\mathbf{x}, -i\hbar/kT; \mathbf{x}, 0) \tag{2.6}$$

showing that the partition function is obtained from the propagator at the imaginary time $\tau = -i\hbar/kT$.

---

[1] In the continuum limit the Green's function is not a normalizable solution of the Schrödinger equation as is clear from eq.(2.2).

[2] In appendix A the path integral expression for the propagator is derived starting from the transformation function.

[3] The absolute square of the propagator, the conditional probability density $P(\mathbf{x}, t; \mathbf{x}', t')$, can be viewed as the probability density for a (one-time) history since $P(\mathbf{x}, t; \mathbf{x}', t') = |G(\mathbf{x}, t; \mathbf{x}', t')|^2 = Tr(\hat{P}(\mathbf{x}, t)\hat{P}(\mathbf{x}', t'))$ is the probability for the history where the particle is at position $\mathbf{x}$ at time t given it was at position $\mathbf{x}'$ at time t'.

Exercise 2.1 *Derive for a particle in a potential the path integral expression for the imaginary-time propagator (consider the one-dimensional case for simplicity)*

$$\mathcal{G}(x, x', \hbar/kT) \equiv G(x, -i\hbar/kT; x', 0) = <x|e^{-\hat{H}/kT}|x'> = \int\limits_{x(0)=x'}^{x(\hbar/kT)=x} \mathcal{D}x_\tau \; e^{-S_\mathcal{E}[x_\tau]/\hbar}$$

(2.7)

where the Euclidean action

$$S_\mathcal{E}[x_\tau] = \int_0^{\hbar/kT} d\tau \; L_\mathcal{E}(x_\tau, \dot{x}_\tau)$$

(2.8)

is specified in terms of the Euclidean Lagrange function

$$L_\mathcal{E}(x_\tau, \dot{x}_\tau) = \frac{1}{2}m\dot{x}_\tau^2 + V(x_\tau)$$

(2.9)

where the potential energy is "added" to the kinetic energy.

Solution

According to eq.(2.6) we obtain by writing $e^{-\hat{H}/kT}$ as the product of $N+1$ identical operators $e^{-\hat{H}/kT} = e^{-\hat{H}/(N+1)kT} \, .. \, e^{-\hat{H}/(N+1)kT}$, and inserting N complete sets of states

$$<x|e^{-\hat{H}/kT}|x'> = \int dx_1 \int dx_2 .. \int dx_N \; <x|e^{-\hat{H}/(N+1)kT}|x_N><x_N|e^{-\hat{H}/(N+1)kT}|x_{N-1}>$$

$$\cdot \; <x_{N-1}|e^{-\hat{H}/(N+1)kT}|x_{N-2}> .. <x_1|e^{-\hat{H}/(N+1)kT}|x'> \; .$$

(2.10)

We have introduced N time slices in the so-called imaginary time interval $[0, \hbar/kT]$, each separated by the amount $\Delta\tau = \hbar/(N+1)kT$. The calculation is now analogous to the one of appendix A, eq.(A.3), except for the substitution $i\Delta t \to \Delta\tau$, and we obtain

$$<x_n|e^{-\hat{H}/(N+1)kT}|x_{n-1}> = <x_n|e^{-\frac{\Delta\tau}{\hbar}\hat{H}}|x_{n-1}>$$

$$= \delta(x_n - x_{n-1}) - \frac{\Delta\tau}{\hbar}<x_n|\hat{H}|x_{n-1}> +\mathcal{O}(\Delta\tau^2).$$

$$= \int \frac{dp_n}{2\pi\hbar} e^{\frac{i}{\hbar}p_n(x_n-x_{n-1}) - \frac{\Delta\tau}{\hbar}H(x_n,p_n)} + \mathcal{O}(\Delta\tau^2) \quad (2.11)$$

where $H(x_n, p_n)$ is Hamilton's function, eq.(A.6), and we get the path integral expression for the imaginary-time propagator

$$<x|e^{-\hat{H}/kT}|x'> = \lim_{N\to\infty} \int \prod_{n=1}^{N} dx_n \prod_{n=1}^{N+1} \frac{dp_n}{2\pi\hbar} e^{\frac{i}{\hbar}p_n(x_n-x_{n-1}) - \frac{\Delta\tau}{\hbar}H(x_n,p_n)}$$

$$\equiv \int \frac{\mathcal{D}x_\tau \mathcal{D}p_\tau}{2\pi\hbar} e^{\frac{i}{\hbar}\int_0^{\hbar/kT} d\tau \, [p_\tau \dot{x}_\tau + iH(x_\tau, p_\tau)]}$$

(2.12)

and upon performing the Gaussian momentum integrations the stated result (which is the expression in eq.(1.40) after a so-called Wick rotation, it $\to \tau = \hbar/kT$). Interpreting τ as a length, we note that the Euclidean Lagrange function $L_{\mathcal{E}}$ equals the potential energy of a string of "length" $L \equiv \hbar/kT$ and tension m, placed in the external potential V. The classical partition function for the string with ends fixed at x' and x is

$$Z_{cl}(x, x') \;=\; \int_{x(0)=x'}^{x(\hbar/kT)=x} \mathcal{D}x_\tau \; e^{-S_{\mathcal{E}}[x_\tau]/\hbar} \;=\; \mathcal{G}(x, \hbar/kT; x', 0) \qquad (2.13)$$

and we have established that the imaginary-time propagator is specified in terms of the classical partition function for the string. The propagator evaluated at imaginary time $-i\hbar/kT$, $G(x, -i\hbar/kT; x', 0)$, equals the classical partition function $Z_{cl}(x, x')$ for a string of "length" $\hbar/kT$ evaluated at the "temperature" $1/\hbar$.

2.2 Retarded and Advanced Propagators

For later use we introduce the retarded Green's function or propagator (the choice of phase is for later convenience)

$$G^R(\mathbf{x}, t; \mathbf{x}', t') \;\equiv\; \begin{cases} -iG(\mathbf{x}, t; \mathbf{x}', t') & \text{for } t \geq t' \\ 0 & \text{for } t < t'. \end{cases} \qquad (2.14)$$

The retarded propagator satisfies the equation

$$\{i\hbar\frac{\partial}{\partial t} - H\}G^R(\mathbf{x}, t; \mathbf{x}', t') = \hbar\, \delta(\mathbf{x} - \mathbf{x}')\, \delta(t - t') \qquad (2.15)$$

which in conjunction with the condition

$$G^R(\mathbf{x}, t; \mathbf{x}, t') \;=\; 0 \qquad \text{for} \quad t < t' \qquad (2.16)$$

specifies the retarded propagator. The source term on the right-hand side of eq.(2.15) represents the discontinuity in the retarded propagator at time $t = t'$, and is recognized by integrating eq.(2.15) over an infinitesimal time interval around t'. The retarded Green's function propagates the wave function forward in time.[4]

According to eq.(2.4), the retarded propagator is given by

$$G^R(\mathbf{x}, t; \mathbf{x}', t') \;=\; -i\theta(t - t') <\mathbf{x}|\hat{U}(t, t')|\mathbf{x}'> \qquad (2.17)$$

---

[4]The retarded propagator also has the following interpretation: prior to time t' the particle is absent, and at time $t = t'$ the particle is created at point $\mathbf{x}'$, and is subsequently propagated according to the Schrödinger equation. In contrast to the relativistic quantum theory, this point of view of propagation is not mandatory in nonrelativistic quantum mechanics where the quantum numbers describing the particle species are conserved.

and is immediately seen to satisfy the initial condition

$$G^R(\mathbf{x}, t' + 0; \mathbf{x}', t') = -i\,\delta(\mathbf{x} - \mathbf{x}')$$

(2.18)

and due to the step function the condition eq.(2.16). By direct differentiation of eq.(2.17) with respect to time it also immediately follows that the retarded propagator satisfies eq.(2.15). The source term can thus be viewed as incorporating the equal time condition, eq.(1.340), for the evolution operator.

We note the path integral expression for the retarded propagator (see also appendix A)

$$G^R(\mathbf{x}, t; \mathbf{x}', t') = -i\theta(t - t')G(\mathbf{x}, t; \mathbf{x}', t') = -i\theta(t - t')K(\mathbf{x}, t; \mathbf{x}', t')$$

$$= -i\,\theta(t - t') \int_{\mathbf{x}_{t'}=\mathbf{x}'}^{\mathbf{x}_t=\mathbf{x}} \mathcal{D}\mathbf{x}_{\bar{t}}\ e^{\frac{i}{\hbar}\int_{t'}^{t} d\bar{t}\, L(\mathbf{x}_{\bar{t}}, \dot{\mathbf{x}}_{\bar{t}})}\ .$$

(2.19)

We shall also need the advanced propagator

$$G^A(\mathbf{x}, t; \mathbf{x}', t') \equiv \begin{cases} 0 & \text{for } t > t' \\ iG(\mathbf{x}, t; \mathbf{x}', t') & \text{for } t \leq t' \end{cases}$$

(2.20)

which propagates the wave function backwards in time, as we have for $t < t'$ for the wave function at time t

$$\psi(\mathbf{x}, t) = -i \int d\mathbf{x}'\, G^A(\mathbf{x}, t; \mathbf{x}', t')\,\psi(\mathbf{x}', t')$$

(2.21)

in terms of the wave function at the later time t'.

The retarded and advanced propagators are related according to

$$G^A(\mathbf{x}, t; \mathbf{x}', t') = [G^R(\mathbf{x}', t'; \mathbf{x}, t)]^* \ .$$

(2.22)

The advanced propagator is also a solution of eq.(2.15), but zero in the opposite time region as compared to the retarded propagator.

We note, that in the spatial representation we have

$$G(\mathbf{x}, t; \mathbf{x}', t') = <\mathbf{x}|\hat{U}(t, t')|\mathbf{x}'> = i[G^R(\mathbf{x}, t; \mathbf{x}', t') - G^A(\mathbf{x}, t; \mathbf{x}', t')]$$

$$\equiv A(\mathbf{x}, t; \mathbf{x}', t')$$

(2.23)

where we have introduced the notation A for the Green's function G, and also refer to it as the spectral function.

Introducing the retarded and advanced Green's operators

$$\hat{G}^R(t, t') \equiv -i\theta(t - t')\,\hat{U}(t, t')\ , \quad \hat{G}^A(t, t') \equiv i\theta(t' - t)\,\hat{U}(t, t')$$

(2.24)

we have for the evolution operator

$$\hat{U}(t, t') = i(\hat{G}^R(t, t') - \hat{G}^A(t, t')) \equiv \hat{G}(t, t') \equiv \hat{A}(t, t')$$

(2.25)

and the unitarity of the evolution operator is reflected in the hermitian relationship of the Green's operators

$$\hat{G}^A(t, t') = [\hat{G}^R(t', t)]^\dagger \ . \tag{2.26}$$

The retarded and advanced Green's operators are characterized as solutions to the same differential equation

$$\left(i\hbar \frac{\partial}{\partial t} - \hat{H}\right) \hat{G}^{R(A)}(t, t') = \hbar\, \delta(t - t')\, \hat{I} \tag{2.27}$$

but are zero for different time relationship.

The various representations of the Green's operators are obtained by taking matrix elements. For example, in the momentum representation we have for the retarded propagator

$$G^R(\mathbf{p}, t; \mathbf{p}', t') = -i\theta(t - t') <\mathbf{p}, t|\mathbf{p}', t'> = <\mathbf{p}|\hat{G}^R(t, t')|\mathbf{p}'> \ . \tag{2.28}$$

Exercise 2.2 *Defining in general the imaginary-time propagator*

$$\mathcal{G}(\mathbf{x}, \tau; \mathbf{x}', \tau') \equiv \theta(\tau - \tau') <\mathbf{x}|e^{-\frac{\hat{H}(\tau - \tau')}{\hbar}}|\mathbf{x}'> \tag{2.29}$$

show that for the Hamiltonian for a particle in a magnetic field

$$\hat{H} = \frac{1}{2m}\left(\hat{\mathbf{p}} - e\mathbf{A}(\hat{\mathbf{x}})\right)^2 \tag{2.30}$$

the imaginary-time propagator satisfies the equation

$$\left(\hbar\frac{\partial}{\partial \tau} + \frac{1}{2m}\left(\frac{\hbar}{i}\nabla_\mathbf{x} - e\mathbf{A}(\mathbf{x})\right)^2\right) \mathcal{G}(\mathbf{x}, \tau; \mathbf{x}', \tau') \equiv \hbar\,\delta(\mathbf{x} - \mathbf{x}')\,\delta(\tau - \tau') \tag{2.31}$$

and write down the path integral representation of the solution.

2.3   Free Particle Propagator

In the previous chapter we established, by appealing to correspondence, that the Hamiltonian for a (low-energy) free particle of mass m is Hamilton's function of the momentum operator

$$\hat{H}_0 = H_0(\hat{\mathbf{p}}) = \frac{\hat{\mathbf{p}}^2}{2m} \ . \tag{2.32}$$

The free particle propagator in the momentum representation

$$G_0^R(\mathbf{p}, t; \mathbf{p}', t') = -i\theta(t - t') < \mathbf{p}|e^{-\frac{i}{\hbar}\hat{H}_0(t - t')}|\mathbf{p}' > \tag{2.33}$$

is therefore given by

$$G_0^R(\mathbf{p}, t; \mathbf{p}', t') = G_0^R(\mathbf{p}, t, t') <\mathbf{p}|\mathbf{p}'> = G_0^R(\mathbf{p}, t - t') \begin{cases} \delta(\mathbf{p} - \mathbf{p}') \\ \delta_{\mathbf{p}, \mathbf{p}'} \end{cases} \quad (2.34)$$

where the Kronecker or delta function (depending on whether the particle is confined to a box or not) reflects the spatial translation invariance of free propagation. The compatibility of the energy and momentum of a free particle is reflected in the definite temporal oscillations of the propagator

$$G_0^R(\mathbf{p}, t, t') = -i\theta(t - t') e^{-\frac{i}{\hbar}\epsilon_{\mathbf{p}}(t-t')} \quad (2.35)$$

determined by the energy of the state in question

$$\epsilon_{\mathbf{p}} = \frac{\mathbf{p}^2}{2m} \quad (2.36)$$

the dispersion relation for a free particle.

Fourier transforming we obtain for the free particle propagator in the spatial representation

$$G_0^R(\mathbf{x}, t; \mathbf{x}', t') = -i\theta(t - t') <\mathbf{x}|e^{-\frac{i}{\hbar}\hat{H}_0(t-t')}|\mathbf{x}'>$$

$$= -i\theta(t - t') \left(\frac{m}{2\pi\hbar i(t - t')}\right)^{d/2} e^{\frac{im}{2\hbar}\frac{(\mathbf{x}-\mathbf{x}')^2}{t-t'}} \quad (2.37)$$

in accordance with eq.(1.31).

Exercise 2.3 *Show that the free retarded propagator in the momentum representation satisfies the equation*

$$\{i\hbar\frac{\partial}{\partial t} - \epsilon_{\mathbf{p}}\} G_0^R(\mathbf{p}, t; \mathbf{p}', t') = \hbar\delta(\mathbf{p} - \mathbf{p}')\delta(t - t') . \quad (2.38)$$

2.4 Perturbation Theory

Situations are ubiquitous where an interaction with a system is adequately described in terms of a time-dependent classical field. Furthermore, in perturbation theory we shall for calculational reasons encounter time-dependent Hamiltonians (though the Hamiltonian for a closed system is time independent). We therefore in the following consider the time evolution of a system for the case of a time-dependent Hamiltonian, $\hat{H}(t)$, for which we have the Schrödinger equation[5]

$$i\hbar \frac{d|\psi(t)>}{dt} = \hat{H}(t) |\psi(t)> \ . \qquad (2.39)$$

We can use the same method for generating a finite time translation as in the time-independent case, i.e., generate a finite time step by repeating taking small steps. Using the Schrödinger equation on infinitesimal form, eq.(1.107), repeatedly, we obtain in the second step, and after N steps

$$
\begin{aligned}
|\psi(t)> &= |\psi(t_{N-2})> +\frac{\Delta t}{i\hbar}\hat{H}(t_{N-2})|\psi(t_{N-2})> +\frac{\Delta t}{i\hbar}\hat{H}(t_{N-1})|\psi(t_{N-1})> \\
&= |\psi(t')> + \sum_{n=1}^{N-1} \frac{\Delta t}{i\hbar}\hat{H}(t_n)|\psi(t_n)> \\
&\equiv |\psi(t')> +\frac{1}{i\hbar} \int_{t'}^{t} d\bar{t}\, \hat{H}(\bar{t})|\psi(\bar{t})> \qquad (2.40)
\end{aligned}
$$

the last equality defining the integral of a state vector.

Since the Hamiltonian at different times may fail to commute, the equation can not be straightforwardly integrated. Iterating the equation, i.e., we substitute the right-hand side into the integrand, we get a power series solution in $\hat{H}$. To zeroth order in $\hat{H}$ we have $|\psi(t)>^{(0)}= |\psi(t')>$, the state vector being unchanged in time since the evolution operator is the identity operator. Inserting the zeroth-order term in the second term on the right-hand side of eq.(2.40) gives for the first iteration

$$|\psi(t)>^{(1)} = |\psi(t')> +\frac{1}{i\hbar} \int_{t'}^{t} d\bar{t}\, \hat{H}(\bar{t})|\psi(t')> \qquad (2.41)$$

the solution correct to first order in $\hat{H}$, which thereupon inserted in the last term on the right-hand side gives for the second iteration

$$
\begin{aligned}
|\psi(t)>^{(2)} &= |\psi(t')> +\frac{1}{i\hbar} \int_{t'}^{t} dt_1\, \hat{H}(t_1)|\psi(t_1)>^{(1)} \\
&= \left(1 + \frac{1}{i\hbar}\int_{t'}^{t} dt_1\, \hat{H}(t_1) + \frac{1}{(i\hbar)^2}\int_{t'}^{t} dt_2\, \hat{H}(t_2)\int_{t'}^{t_2} dt_1\, \hat{H}(t_1)\right)|\psi(t')> \qquad (2.42)
\end{aligned}
$$

and so on ad infinitum producing the formula

$$|\psi(t)> = \sum_{n=0}^{\infty} \left(\frac{-i}{\hbar}\right)^n \int_{t'}^{t} dt_n \int_{t'}^{t_n} dt_{n-1} .. \int_{t'}^{t_3} dt_2 \int_{t'}^{t_2} dt_1\, \hat{H}(t_n)\hat{H}(t_{n-1}) .. \hat{H}(t_2)\hat{H}(t_1)|\psi(t')> \ . \qquad (2.43)$$

---

[5]We could equally well discuss perturbation theory in the path integral formalism, as done in exercise 2.6 on page 106.

The $n = 0$ term denoting the zeroth-order term. We can then read off the perturbative solution for the evolution operator[6]

$$\hat{U}(t,t') = \sum_{n=0}^{\infty} \left(\frac{-i}{\hbar}\right)^n \int_{t'}^{t} dt_n \int_{t'}^{t_n} dt_{n-1} .. \int_{t'}^{t_3} dt_2 \int_{t'}^{t_2} dt_1 \, \hat{H}(t_n)\hat{H}(t_{n-1})..\hat{H}(t_2)\hat{H}(t_1)$$
(2.44)

where the $n = 0$ term denotes the zeroth-order term, the identity operator. This is the power series solution in $\hat{H}$ for the evolution operator we shall need for practical purposes, but for reasons of compactness we introduce a notation that takes care of the specific time ordering appearing in the sequence of Hamiltonians.

Introducing the time-ordering operation which orders a product of time-dependent operators into its time-descending sequence (displayed here for the case of three operators)

$$T(\hat{A}(t_1)\hat{B}(t_2)\hat{C}(t_3)) = \begin{cases} \hat{A}(t_1)\hat{B}(t_2)\hat{C}(t_3) & \text{for} \quad t_1 > t_2 > t_3 \\ \hat{B}(t_2)\hat{A}(t_1)\hat{C}(t_3) & \text{for} \quad t_2 > t_1 > t_3 \\ \text{etc.} & \quad \text{etc.} \end{cases}$$
(2.45)

we can lift the restrictions on the time integrations compensating by a factor of $n!$

$$\hat{U}(t,t') = \sum_{n=0}^{\infty} \left(\frac{-i}{\hbar}\right)^n \frac{1}{n!} \int_{t'}^{t} dt_n \int_{t'}^{t} dt_{n-1} .. \int_{t'}^{t} dt_2 \int_{t'}^{t} dt_1 \, T(\hat{H}(t_n)\hat{H}(t_{n-1})..\hat{H}(t_2)\hat{H}(t_1))$$
(2.46)

since the $n!$ differently ordered time regions give identical expressions (modulo a dummy integration relabeling). The evolution operator can therefore be rewritten

$$\hat{U}(t,t') = \sum_{n=0}^{\infty} \left(\frac{-i}{\hbar}\right)^n \frac{1}{n!} T\left(\left(\int_{t'}^{t} d\bar{t}\, \hat{H}(\bar{t})\right)^n\right) = T e^{-\frac{i}{\hbar}\int_{t'}^{t} d\bar{t}\, \hat{H}(\bar{t})}$$
(2.47)

where the second equality is realized by expanding the exponential and using the definition of the time-ordering operation.

To be remembered for the following is the simple rule: differentiating the expression for the time-ordered exponential, eq.(2.47), with respect to the upper integration limit, brings down the Hamiltonian to the left with the time label given by the upper limit of the integral according to eq.(1.343)

$$i\hbar\frac{\partial}{\partial t} T e^{-\frac{i}{\hbar}\int_{t'}^{t} d\bar{t}\, \hat{H}(\bar{t})} = \hat{H}(t)\, T e^{-\frac{i}{\hbar}\int_{t'}^{t} d\bar{t}\, \hat{H}(\bar{t})} .$$
(2.48)

One verifies readily that

$$\hat{U}^{\dagger}(t,t') \equiv [\hat{U}(t,t')]^{\dagger} = \left(T e^{-\frac{i}{\hbar}\int_{t'}^{t} d\bar{t}\, \hat{H}(\bar{t})}\right)^{\dagger} = \sum_{n=0}^{\infty} (\frac{i}{\hbar})^n \frac{1}{n!} \tilde{T}\left(\left(\int_{t'}^{t} d\bar{t}\, \hat{H}(\bar{t})\right)^n\right)$$

$$= \tilde{T} e^{\frac{i}{\hbar}\int_{t'}^{t} d\bar{t}\, \hat{H}(\bar{t})}$$
(2.49)

---

[6]This result could of course also be arrived at by noting that from eq.(2.40) we read off for the evolution operator

$$\hat{U}(t,t') = \hat{I} - \frac{i}{\hbar}\int_{t'}^{t} d\bar{t}\, \hat{H}(\bar{t})\, \hat{U}(\bar{t},t') .$$

Iterating this equation, we generate eq.(2.44).

where the antitime-ordering symbol, $\tilde{T}$, orders the time sequence oppositely as compared to T, as the adjoint inverts the order of a sequence of operators.

Exercise 2.4 *Verify for an arbitrary operator $\hat{X}(t)$ the following property for time-ordered exponentials for the time relationship $t' < t'' < t$*

$$Te^{\int_{t'}^{t} d\bar{t}\,\hat{X}(\bar{t})} = Te^{\int_{t''}^{t} d\bar{t}\,\hat{X}(\bar{t})}\, Te^{\int_{t'}^{t''} d\bar{t}\,\hat{X}(\bar{t})}\,. \tag{2.50}$$

From the unitarity of the evolution operator, $\hat{I} = \hat{U}^\dagger(t,t')\,U(t,t')$, and eq.(2.48), one verifies readily (as also obtained by taking the adjoint of eq.(2.48)) that

$$-i\hbar\,\frac{\partial \hat{U}(t,t')^\dagger}{\partial t} = \hat{U}^\dagger(t,t')\,\hat{H}(t) \tag{2.51}$$

and using eq.(1.347) we get

$$-i\hbar\frac{\partial \hat{U}(t,t')}{\partial t'} = \hat{U}(t,t')\,\hat{H}(t') \tag{2.52}$$

thereby establishing that differentiating the time-ordered exponential, eq.(2.47), with respect to the lower integration limit brings down the Hamiltonian to the right with the time label given by the lower limit of the integral.

Exercise 2.5 *Consider a particle in the potential V (vanishing in the far past) for which we have the Hamiltonian $\hat{H} = \hat{H}_0 + \hat{V}$. Show that*

$$|\psi(t)> = |\phi(t)> + \frac{1}{i\hbar}\int_{-\infty}^{t} dt'\, e^{-\frac{i}{\hbar}\hat{H}_0(t-t')}\,\hat{V}(t')\,|\psi(t')> \tag{2.53}$$

is a solution of the Schrödinger equation provided $|\phi(t)>$ is a solution of the Schrödinger equation in the absence of the potential.

Solution

Upon Taylor-expanding the exponential to lowest or in Δt we obtain from eq.(2.53)

$$\frac{|\psi(t+\Delta t)> - |\psi(t)>}{\Delta t} - \frac{|\phi(t+\Delta t)> - |\phi(t)>}{\Delta t}$$

$$= \frac{1}{i\hbar\Delta t} \left(\int_{-\infty}^{t+\Delta t} dt' \; e^{-\frac{i}{\hbar}\hat{H}_0(t+\Delta t-t')} \; \hat{V}(t') \; |\psi(t')\rangle - \int_{-\infty}^{t} dt' \; e^{-\frac{i}{\hbar}\hat{H}_0(t-t')} \; \hat{V}(t') \; |\psi(t')\rangle \right)$$

$$= \frac{1}{\Delta t} \left(\frac{\Delta t}{i\hbar}\hat{H}_0 \int_{-\infty}^{t} dt' \; e^{-\frac{i}{\hbar}\hat{H}_0(t-t')} \; \hat{V}(t') \; |\psi(t')\rangle + \Delta t \; \hat{V}(t) \; |\psi(t)\rangle \right) \qquad (2.54)$$

and thereby the sought result.

2.5   Interaction Picture

Let us consider a Hamiltonian consisting of two parts:

$$\hat{H}_t = \hat{H}_0 + \hat{H}'_t . \qquad (2.55)$$

We can rewrite the evolution operator:

$$\hat{U}(t,t') = Te^{-\frac{i}{\hbar}\int_{t'}^{t} d\bar{t}\, \hat{H}_{\bar{t}}} = \hat{U}_0(t,t_r)\, \hat{U}_I(t,t')\, \hat{U}_0^\dagger(t',t_r) \qquad (2.56)$$

in terms of the evolution operator in the so-called interaction picture, the Heisenberg picture with respect to $\hat{H}_0$,

$$\hat{U}_I(t,t') = Te^{-\frac{i}{\hbar}\int_{t'}^{t} d\bar{t}\, \hat{H}'_I(\bar{t})} . \qquad (2.57)$$

An operator in the interaction picture is specified by the operator in the Schrödinger picture according to[7]

$$\hat{H}'_I(t) \equiv \hat{U}_0^\dagger(t,t_r)\, \hat{H}'_t\, \hat{U}_0(t,t_r) \qquad (2.58)$$

and assuming that $\hat{H}_0$ is time independent, we have for the evolution operator in the absence of $\hat{H}'_t$

$$\hat{U}_0^\dagger(t,t_r) = e^{\frac{i}{\hbar}\hat{H}_0(t-t_r)} . \qquad (2.59)$$

The arbitrary reference time where the interaction and Schrödinger pictures coincide we denote by t_r.

We can derive the construction, eq.(2.56), explicitly, but let us here use our above derived differentiation rules, thereby noticing that the operators on the two sides of eq.(2.56) satisfy the same first-order differential equation, and are therefore identical as they satisfy the same initial condition.

Often it is convenient to take the reference time as zero, $t_r = 0$, and we have

$$Te^{-\frac{i}{\hbar}\int_{t'}^{t} d\bar{t}\, \hat{H}_{\bar{t}}} = e^{-\frac{i}{\hbar}\hat{H}_0 t}\left(Te^{-\frac{i}{\hbar}\int_{t'}^{t} d\bar{t}\, \hat{H}'_I(\bar{t})}\right)e^{\frac{i}{\hbar}\hat{H}_0 t'} \qquad (2.60)$$

where

$$\hat{H}'_I(t) = e^{\frac{i}{\hbar}\hat{H}_0 t}\hat{H}'_t\, e^{-\frac{i}{\hbar}\hat{H}_0 t} . \qquad (2.61)$$

---

[7] We note, that had we studied the case where $\hat{H}'$ is time independent we would still encounter a time-dependent operator in the interaction picture, $\hat{H}'_I(t)$.

2.6   Propagation in a Potential

The simplest example of diagrammatic perturbation theory is the case of a particle in a scalar potential $V(\mathbf{x}, t)$; i.e., we consider the particle to be in an environment whose influence on the particle can be described in terms of a classical potential. We then have the Hamiltonian

$$\hat{H}_t = \hat{H}_p + \hat{V}_t \tag{2.62}$$

where the effect of the potential is represented by the operator

$$\hat{V}_t = V(\hat{\mathbf{x}}, t) . \tag{2.63}$$

The Hamiltonian in the absence of the potential, $\hat{H}_p$, we assume to be time independent.

The retarded propagator in the external potential is specified by the matrix element of the evolution operator

$$
\begin{aligned}
G^R(\mathbf{x}, t; \mathbf{x}', t') &= -i\theta(t - t') < \mathbf{x}|\hat{U}(t, t')|\mathbf{x}' > \\
&= -i\theta(t - t') <\mathbf{x}|Te^{-\frac{i}{\hbar}\int_{t'}^t d\bar{t}\,\hat{H}_{\bar{t}}}|\mathbf{x}' > .
\end{aligned}
\tag{2.64}
$$

The perturbative expansion of the propagator is obtained by introducing the time-ordered exponential expressed in the interaction picture

$$Te^{-\frac{i}{\hbar}\int_{t'}^t d\bar{t}\,\hat{H}_t} = \hat{U}_0(t, t_r)\, Te^{-\frac{i}{\hbar}\int_{t'}^t d\bar{t}\, V(\hat{\mathbf{x}}(\bar{t}), \bar{t})}\, \hat{U}_0^\dagger(t', t_r) \tag{2.65}$$

where we have used that the potential operator in the interaction picture is the potential function of the position operator in the interaction picture

$$\hat{V}_I(t) = \hat{U}_0^\dagger(t, t_r)\, V(\hat{\mathbf{x}}, t)\, \hat{U}_0(t, t_r) = V(\hat{\mathbf{x}}(t), t) \tag{2.66}$$

now dropping the index indicating the interaction picture as no confusion should arise

$$\hat{\mathbf{x}}(t) \equiv \hat{\mathbf{x}}_I(t) = \hat{U}_0^\dagger(t, t_r)\, \hat{\mathbf{x}}\, \hat{U}_0(t, t_r) . \tag{2.67}$$

Expanding the time-ordered exponential, we get the perturbative expansion of the propagator

$$
\begin{aligned}
G^R(\mathbf{x}, t; \mathbf{x}', t')) &= -i\theta(t - t') <\mathbf{x}, t|T \exp\{-\frac{i}{\hbar}\int_{t'}^t d\bar{t}\hat{V}_I(\bar{t})\}|\mathbf{x}', t' > \\
&\equiv \sum_{n=0}^\infty G_n^R(\mathbf{x}, t; \mathbf{x}', t')
\end{aligned}
\tag{2.68}
$$

where the n'th order term is equal to

$$G_n^R(\mathbf{x}, t; \mathbf{x}', t')) = -i\theta(t - t')\left(\frac{-i}{\hbar}\right)^n \frac{1}{n!}\int_{t'}^t \prod_{m=1}^n dt_m <\mathbf{x}, t|T(V(\hat{\mathbf{x}}(t_n), t_n)$$

$$V(\hat{\mathbf{x}}(t_{n-1}), t_{n-1}) .. V(\hat{\mathbf{x}}(t_2), t_2)\, V(\hat{\mathbf{x}}(t_1), t_1))\, |\mathbf{x}', t' > \tag{2.69}$$

and the time-labeled states now denotes the eigenstates of the position operator in the interaction picture

$$\hat{x}_I(t)\,|\mathbf{x},t> \;=\; \mathbf{x}\,|\mathbf{x},t> , \qquad\qquad |\mathbf{x},t> \;=\; \hat{U}_0^{-1}(t,t_r)|\mathbf{x}> . \qquad (2.70)$$

By inserting a complete set of such states[8]

$$\hat{I} \;=\; \int\!d\mathbf{x}_i\,|\mathbf{x}_i,t_i><\mathbf{x}_i,t_i| \qquad\qquad (2.71)$$

in front of each operator $\hat{V}_I(t_i) = V(\hat{\mathbf{x}}(t_i),t_i)$ in the perturbative expression, the operation of the potential operator is turned into multiplication by the value of the potential at the space-time point in question as

$$\hat{V}_I(t_i)|\mathbf{x}_i,t_i> \;=\; V(\hat{\mathbf{x}}(t_i),t_i)|\mathbf{x}_i,t_i> \;=\; \left(\int\!d\mathbf{x}\,|\mathbf{x},t_i> V(\mathbf{x},t_i)\,<\mathbf{x},t_i|\right)|\mathbf{x}_i,t_i>$$

$$=\; V(\mathbf{x}_i,t_i)\,|\mathbf{x}_i,t_i> . \qquad\qquad (2.72)$$

The zeroth-order propagator, the propagator in the absence of the potential V, is given by

$$G_0^R(\mathbf{x},t;\mathbf{x}',t') = -i\theta(t-t')\,<\mathbf{x}|e^{-\frac{i}{\hbar}\hat{H}_p(t-t')}|\mathbf{x}'> . \qquad (2.73)$$

The first-order correction to the propagator is given by

$$G_1^R(\mathbf{x},t;\mathbf{x}',t') \;=\; -\frac{1}{\hbar}\theta(t-t')\int_{t'}^{t}\!dt_1\ <\mathbf{x},t|V(\hat{\mathbf{x}}(t_1),t_1)|\mathbf{x}',t'> . \qquad (2.74)$$

For the time relations $t' < t_1 < t$ we have for the step function

$$\theta(t-t') \;=\; \theta(t-t_1)\,\theta(t_1-t') \qquad\qquad (2.75)$$

and for the first-order term we therefore have the expression

$$G_1^R(\mathbf{x},t;\mathbf{x}',t') \;=\; \frac{1}{\hbar}\int\!d\mathbf{x}_1\!\int_{-\infty}^{\infty}\!dt_1\ G_0^R(\mathbf{x},t;\mathbf{x}_1,t_1)\,V(\mathbf{x}_1,t_1)\,G_0^R(\mathbf{x}_1,t_1;\mathbf{x}',t') \qquad (2.76)$$

as the retarded propagators restrict the time integration to the original time interval.

The first-order contribution to the propagator can be thought of as a product of three terms: the amplitude for free particle propagation from space-time point $(\mathbf{x}',t')$ to $(\mathbf{x}_1,t_1)$, where the particle experiences interaction with the potential, described by the factor $V(\mathbf{x}_1,t_1)$, and finally the amplitude for free particle propagation from $(\mathbf{x}_1,t_1)$ to $(\mathbf{x},t)$. Since the event of interaction with the potential, which we shall refer to as a scattering of the particle, can take place anywhere and at any time, we are summing over all these alternatives.

---

[8]Or immediately use the spectral representation of the potential operator in the interaction picture $\hat{V}_I(t) = V(\hat{\mathbf{x}}(t),t) = \int\!d\mathbf{x}\,|\mathbf{x},t> V(\mathbf{x},t)\,<\mathbf{x},t|$.

Graphically we represent the first-order term for the propagator by the diagram

$$G_1^R(\mathbf{x}, t; \mathbf{x}', t') \quad = \quad \underset{\mathbf{x}t}{\bullet} \xrightarrow{\quad R \quad} \underset{\mathbf{x}_1 t_1}{\times} \xleftarrow{\quad R \quad} \underset{\mathbf{x}'t'}{\bullet} \tag{2.77}$$

where a cross has been introduced to symbolize the interaction of the particle with the scalar potential

$$\underset{\mathbf{x}t}{\xrightarrow{\quad\times\quad}} \quad \equiv \quad \tfrac{1}{\hbar} V(\mathbf{x}, t) \tag{2.78}$$

and a thin line is used to represent the zeroth-order propagator

$$\underset{\mathbf{x}t}{\bullet} \xrightarrow{\quad R \quad} \underset{\mathbf{x}'t'}{\bullet} \quad \equiv \quad G_0^R(\mathbf{x}, t; \mathbf{x}', t') \tag{2.79}$$

in order to distinguish it from the propagator in the presence of the potential V

$$\underset{\mathbf{x}t}{\bullet} \xrightarrow{\quad R \quad} \underset{\mathbf{x}'t'}{\bullet} \quad \equiv \quad G^R(\mathbf{x}, t; \mathbf{x}', t') \tag{2.80}$$

depicted as a thick line. With this dictionary the analytical form, eq.(2.76), is obtained from the diagram, eq.(2.77), since integration is implied over the internal space-time point where interaction with the potential takes place.

Similarly we get for the second-order term by inserting complete sets of states

$$G_2^R(\mathbf{x}, t; \mathbf{x}', t') \quad = \quad -i\theta(t - t') \left(\frac{-i}{\hbar}\right)^2 \int_{t'}^{t} dt_2 \int_{t'}^{t_2} dt_1 \; <\mathbf{x}, t|\mathbf{x}_2, t_2> V(\mathbf{x}_2, t_2)$$

$$<\mathbf{x}_2, t_2|\mathbf{x}_1, t_1> V(\mathbf{x}_1, t_1) \; <\mathbf{x}_1, t_1|\mathbf{x}', t'>$$

$$= \quad \hbar^{-2} \int d\mathbf{x}_2 \int_{-\infty}^{\infty} dt_2 \int d\mathbf{x}_1 \int_{-\infty}^{\infty} dt_1 G_0^R(\mathbf{x}, t; \mathbf{x}_2, t_2) V(\mathbf{x}_2, t_2) G_0^R(\mathbf{x}_2, t_2; \mathbf{x}_1, t_1)$$

$$V(\mathbf{x}_1, t_1) G_0^R(\mathbf{x}_1, t_1; \mathbf{x}', t') \tag{2.81}$$

where we in the last equality have utilized that for the time relationship, $t' < t_1 < t_2 < t$, we have for the step function

$$\theta(t - t') \quad = \quad \theta(t - t_2)\, \theta(t_2 - t_1)\, \theta(t_1 - t') \tag{2.82}$$

and we can lift the time integration limitations as the step functions automatically limit the integration region to the original one.

The second-order term for the propagator is therefore represented diagrammatically by

$$G_2^R(\mathbf{x}, t; \mathbf{x}', t') \quad = \quad \underset{\mathbf{x}t}{\bullet} \xrightarrow{\quad R \quad} \underset{\mathbf{x}_2 t_2}{\times} \xleftarrow{\quad R \quad} \underset{\mathbf{x}_1 t_1}{\times} \xleftarrow{\quad R \quad} \underset{\mathbf{x}'t'}{\bullet} \tag{2.83}$$

corresponding to propagation governed by $\hat{H}_p$ in between the scattering twice by the potential.

Repeating this scheme of inserting complete sets of states, letting the system propagate through all the possible position property values at all possible times, we obtain that the n'th order term consists of n scatterings by the potential and $n+1$ propagators

$$G_n^R(\mathbf{x}, t; \mathbf{x}', t') = \frac{1}{\hbar^n} \int \prod_{m=1}^{n} d\mathbf{x}_m \int_{t'}^{t} \prod_{m=1}^{n} dt_m \, V(\mathbf{x}_n) V(\mathbf{x}_{n-1}) .. V(\mathbf{x}_2) V(\mathbf{x}_1)$$

$$G_0^R(\mathbf{x}, t; \mathbf{x}_n, t_n) \, G_0^R(\mathbf{x}_n, t_n; \mathbf{x}_{n-1}, t_{n-1}) .. G_0^R(\mathbf{x}_1, t_1; \tilde{\mathbf{x}}, t') \quad (2.84)$$

represented diagrammatically by

$$G_n^R(\mathbf{x}, t; \mathbf{x}', t') = \underset{xt}{\bullet} \xrightarrow{R} \underset{x_n t_n}{\times} \xleftarrow{R} \cdots \xrightarrow{R} \underset{x_1 t_1}{\times} \xleftarrow{R} \underset{x't'}{\bullet} \quad (2.85)$$

and the perturbative expression for the exact propagator is represented by the infinite sum of terms

$$G^R(\mathbf{x}, t; \mathbf{x}', t') = \underset{xt}{\bullet} \xrightarrow{R} \underset{x't'}{\bullet}$$

$$= \underset{xt}{\bullet} \xrightarrow{R} \underset{x't'}{\bullet} + \underset{xt}{\bullet} \xrightarrow{R} \underset{x_1 t_1}{\times} \xrightarrow{R} \underset{x't'}{\bullet} + \underset{xt}{\bullet} \xrightarrow{R} \underset{x_2 t_2}{\times} \xrightarrow{R} \underset{x_1 t_1}{\times} \xrightarrow{R} \underset{x't'}{\bullet}$$

$$+ \underset{xt}{\bullet} \xrightarrow{R} \underset{x_3 t_3}{\times} \xrightarrow{R} \underset{x_2 t_2}{\times} \xrightarrow{R} \underset{x_1 t_1}{\times} \xrightarrow{R} \underset{x't'}{\bullet} + \cdots . \quad (2.86)$$

Following Feynman we can elevate the diagrams to represent alternative physical scattering processes. The propagator in a potential being the sum of all possible scattering alternatives for the particle: not being scattered, being scattered once, being scattered twice, etc.

Such a series of scattering processes where each subsequent process has an extra scattering event and propagator (each subsequent diagram has an extra cross and propagator line) is iterative, and the propagator for the particle in a potential satisfies the diagrammatic equation

$$\underset{xt}{\bullet} \xrightarrow{R} \underset{x't'}{\bullet} = \underset{xt}{\bullet} \xrightarrow{R} \underset{x't'}{\bullet} + \underset{xt}{\bullet} \xrightarrow{R} \underset{\bar{x}\bar{t}}{\times} \xrightarrow{R} \underset{x't'}{\bullet} \quad (2.87)$$

as seen by iteration. Analytically we have the equation

$$G^R(\mathbf{x}, t; \mathbf{x}', t') = G_0^R(\mathbf{x}, t; \mathbf{x}', t') + \frac{1}{\hbar} \int d\bar{\mathbf{x}} \int_{-\infty}^{\infty} d\bar{t} \, G_0^R(\mathbf{x}, t; \bar{\mathbf{x}}, \bar{t}) V(\bar{\mathbf{x}}, \bar{t}) G^R(\bar{\mathbf{x}}, \bar{t}; \mathbf{x}', t') .$$

$$(2.88)$$

The equation for the advanced propagator we obtain by using its relationship to the retarded propagator, eq.(2.22),

$$G^A(\mathbf{x}, t; \mathbf{x}', t') = G_0^A(\mathbf{x}, t; \mathbf{x}', t') + \frac{1}{\hbar} \int d\bar{\mathbf{x}} \int_{-\infty}^{\infty} d\bar{t} \, G^A(\mathbf{x}, t; \bar{\mathbf{x}}, \bar{t}) V(\bar{\mathbf{x}}, \bar{t}) G_0^A(\bar{\mathbf{x}}, \bar{t}; \mathbf{x}', t')$$

(2.89)

or diagrammatically

(2.90)

with a convention for drawing the diagrams for the advanced propagator that makes explicit the backwards-in-time propagation.

Exercise 2.6 *Derive the perturbation expansion from the path integral formalism.*

Solution

From eq.(1.40) we obtain

$$K(\mathbf{x}, t; \mathbf{x}', t') = \int_{\mathbf{x}_{t'}=\mathbf{x}'}^{\mathbf{x}_t=\mathbf{x}} \mathcal{D}\mathbf{x}_{\bar{t}} \; e^{\frac{i}{\hbar} \int_{t'}^t d\bar{t} \, \frac{1}{2} m \dot{\mathbf{x}}_{\bar{t}}^2} e^{-\frac{i}{\hbar} \int_{t'}^t d\bar{t} \, V(\mathbf{x}_{\bar{t}}, \bar{t})}$$

$$= \int_{\mathbf{x}_{t'}=\mathbf{x}'}^{\mathbf{x}_t=\mathbf{x}} \mathcal{D}\mathbf{x}_{\bar{t}} \; e^{\frac{i}{\hbar} \int_{t'}^t d\bar{t} \, \frac{1}{2} m \dot{\mathbf{x}}_{\bar{t}}^2} \left(1 + \frac{-i}{\hbar} \int_{t'}^t dt_1 \, V(\mathbf{x}_{t_1}, t_1) \right.$$

$$\left. + \frac{1}{2!} \left(\frac{-i}{\hbar} \right)^2 \int_{t'}^t dt_2 \int_{t'}^t dt_1 \, V(\mathbf{x}_{t_2}, t_2) V(\mathbf{x}_{t_1}, t_1) + \dots \right) . \; (2.91)$$

Consider the first-order term. In the discretized form of the path integral we choose one of the intermediate times as the one dictated by the integration over t_1. There are N_1 and N_2 other internal moments of time, before and after the one singled out, respectively. The corresponding internal spatial integrations, and the number of "measure"-factors produces the product of the free propagators $K_0(\mathbf{x}_1, t_1; \mathbf{x}', t')$ and $K_0(\mathbf{x}, t; \mathbf{x}_1, t_1)$, and we obtain for the first-order correction to the propagator

$$K_1(\mathbf{x}, t; \mathbf{x}', t') = \frac{-i}{\hbar} \int_{t'}^t dt_1 \int d\mathbf{x}_1 \, K_0(\mathbf{x}, t; \mathbf{x}_1, t_1) V(\mathbf{x}_{t_1}, t_1) K_0(\mathbf{x}_1, t_1; \mathbf{x}', t') . \; (2.92)$$

In the second-order term we choose t_1 and t_2 as intermediate times. Since this can be done in two ways, $t_1 < t_2$ or $t_1 > t_2$, giving identical contributions and thereby canceling the factor $1/2!$, we obtain the expression in eq.(2.81). Similarly for the higher order terms, and we reproduce the perturbation series depicted diagrammatically in eq.(2.86).

2.6.1   Momentum Representation

For calculational purposes the momentum representation is often useful. In the momentum representation we encounter all the same manipulations as we did in the position representation except that we have $\mathbf{p}$'s instead of $\mathbf{x}$'s, and we have for the retarded propagator in the momentum representation

$$G^R(\mathbf{p}, t; \mathbf{p}', t') = -i\theta(t - t') < \mathbf{p}, t | \mathbf{p}', t' > = -i\theta(t - t') < \mathbf{p} | T e^{-\frac{i}{\hbar} \int_{t'}^{t} d\bar{t}\, \hat{H}_t} | \mathbf{p}' > \ . \tag{2.93}$$

For the n'th order term in the perturbative expansion of the propagator

$$G^R(\mathbf{p}, t; \mathbf{p}', t') = \sum_{n=0}^{\infty} G_n^R(\mathbf{p}, t; \mathbf{p}', t') \tag{2.94}$$

we have

$$G_n^R(\mathbf{p}, t; \mathbf{p}', t') = -i\theta(t - t') \frac{\left(\frac{-i}{\hbar}\right)^n}{n!} \int_{t'}^{t} \prod_{m=1}^{n} dt_m < \mathbf{p}, t | T(\hat{V}_I(t_n)\hat{V}_I(t_{n-1})..\hat{V}_I(t_1)) | \mathbf{p}', t' > \tag{2.95}$$

where the interaction picture momentum eigenstates

$$\hat{\mathbf{p}}_I(t) | \mathbf{p}, t > = \mathbf{p} | \mathbf{p}, t > \tag{2.96}$$

has been introduced.[9]

For the propagator in the absence of the potential V we have

$$G_0^R(\mathbf{p}, t; \mathbf{p}', t') = -i\theta(t - t') < \mathbf{p} | e^{-\frac{i}{\hbar} \hat{H}_p(t - t')} | \mathbf{p}' > \ . \tag{2.97}$$

In order to calculate the propagator to first order, we insert complete sets of momentum eigenstates and obtain

$$G_1^R(\mathbf{p}, t; \mathbf{p}', t') = \frac{1}{\hbar} \int_{-\infty}^{\infty} dt_1 \int_{\mathbf{p}_1', \mathbf{p}_1} G_0^R(\mathbf{p}, t; \mathbf{p}_1, t_1) < \mathbf{p}_1, t_1 | \hat{V}_I(t_1) | \mathbf{p}_1', t_1 > G_0^R(\mathbf{p}_1', t_1; \mathbf{p}', t') \tag{2.98}$$

---

[9]We suppress the index distinguishing these states from the states in eq.(2.93) defined by the momentum operator in the Heisenberg picture as it is clear from the context which states are involved.

where the interaction with the potential in the momentum representation is specified by (in three spatial dimensions)

$$<\mathbf{p}, t|\hat{V}_I(t)|\mathbf{p}', t> \ = \ <\mathbf{p}|\hat{V}_t|\mathbf{p}'> \ = \ \begin{cases} (2\pi\hbar)^{-3} \int d\mathbf{x} \, e^{-\frac{i}{\hbar}\mathbf{x}\cdot(\mathbf{p}-\mathbf{p}')} \, V(\mathbf{x}, t) \\[2mm] V^{-1} \int_V d\mathbf{x} \, e^{-\frac{i}{\hbar}\mathbf{x}\cdot(\mathbf{p}-\mathbf{p}')} \, V(\mathbf{x}, t) \end{cases} \tag{2.99}$$

depending on whether we have a particle in a box of volume V (exceeding the range of the potential), or in infinite space[10]

$$<\mathbf{x}, t|\mathbf{p}, t> \ = \ <\mathbf{x}|\mathbf{p}> \ = \ \begin{cases} (2\pi\hbar)^{-3/2} e^{\frac{i}{\hbar}\mathbf{x}\cdot\mathbf{p}} \\[2mm] V^{-1/2} e^{\frac{i}{\hbar}\mathbf{x}\cdot\mathbf{p}} \end{cases} \tag{2.100}$$

and we have in eq.(2.98) introduced the context-dependent notation

$$\sum_{\mathbf{P}} \leftrightarrow \int_{\mathbf{p}} \leftrightarrow \int_{-\infty}^{\infty} d\mathbf{p} \ . \tag{2.101}$$

In the momentum representation we therefore have the same first-order diagram as in the spatial representation

$$G_1^R(\mathbf{p}, t; \mathbf{p}', t') \ = \ \underset{\mathbf{p}t}{\bullet} \xleftarrow{\quad R \quad} \underset{\mathbf{p}_1 t_1 \mathbf{p}_1'}{\times} \xleftarrow{\quad R \quad} \underset{\mathbf{p}'t'}{\bullet} \tag{2.102}$$

however, with the momentum representation interpretation of the diagram: The propagator between momentum values $\mathbf{p}'$ and $\mathbf{p}$ in the absence of the potential V we represent diagrammatically by a thin line[11]

$$G_0^R(\mathbf{p}, t; \mathbf{p}', t') \ = \ -i\theta(t - t') < \mathbf{p}|e^{-\frac{i}{\hbar}\hat{H}_p(t-t')}|\mathbf{p}' >$$

$$\equiv \ \underset{\mathbf{p}t}{\bullet} \xleftarrow{\quad R \quad} \underset{\mathbf{p}'t'}{\bullet} \tag{2.103}$$

and in the momentum representation the cross designates the matrix element

$$\underset{\mathbf{p}\,t\ \ \mathbf{p}'}{\xleftarrow{\times}\!\!\leftarrow} \ = \ \frac{1}{\hbar} <\mathbf{p}|\hat{V}_t|\mathbf{p}'> \tag{2.104}$$

and signifies the momentum change due to the scattering by the potential at the time in question. Summation (integration) over all alternative intermediate momenta, and integration over time is implied according to eq.(2.98).

---

[10]We shall often interchange between the notation for a particle in a finite volume V (with appropriately imposed boundary conditions), and the continuum notation.

[11]The propagator is in the absence of the potential V invariant with respect to displacements in time as $\hat{H}_p$ was assumed time independent, a constraint we could easily relax.

For the second-order term we similarly obtain from eq.(2.95)

$$G_2^R(\mathbf{p},t;\mathbf{p}',t') = \frac{1}{\hbar^2} \int_{-\infty}^{\infty} dt_2 \int_{-\infty}^{\infty} dt_1 \int_{\mathbf{p}_1,\mathbf{p}_1',\mathbf{p}_2,\mathbf{p}_2'} G_0^R(\mathbf{p},t;\mathbf{p}_2,t_2) <\mathbf{p}_2|\hat{V}_{t_2}|\mathbf{p}_2'>$$

$$G_0^R(\mathbf{p}_2',t_2;\mathbf{p}_1,t_1) <\mathbf{p}_1|\hat{V}_{t_1}|\mathbf{p}_1'> G_0^R(\mathbf{p}_1',t_1;\mathbf{p}',t') \qquad (2.105)$$

and diagrammatically

$$G_2^R(\mathbf{p},t;\mathbf{p}',t') \quad = \quad \text{(diagram)} \qquad (2.106)$$

corresponding to propagation according to the Hamiltonian $\hat{H}_p$ in between the scattering by the potential where the momentum of the particle is changed.

Repeating the scheme of inserting complete sets of momentum eigenstates we obtain that the n'th order term consists of n scatterings by the potential and $n+1$ propagators

$$G_n^R(\mathbf{p},t;\mathbf{p}',t') \quad = \quad \text{(diagram)} \qquad (2.107)$$

and the exact propagator is represented by the infinite set of diagrams

$$G^R(\mathbf{p},t;\mathbf{p}',t') \quad = \quad \text{(diagrams)} \quad + \quad \cdots \quad + \quad \cdots$$

$$+ \quad \text{(diagram)} \quad + \quad \cdots \quad . \qquad (2.108)$$

The propagator in the momentum representation for a particle in a potential is therefore by iteration seen to satisfy the diagrammatic equation

$$\text{(diagram)} \quad = \quad \text{(diagram)} \quad + \quad \text{(diagram)} \qquad (2.109)$$

and analytically the equation

$$G^R(\mathbf{p},t;\mathbf{p}',t') = G_0^R(\mathbf{p},t;\mathbf{p}',t') + \frac{1}{\hbar} \int_{\bar{\mathbf{p}},\bar{\mathbf{p}}'} \int_{-\infty}^{\infty} d\bar{t}\, G_0^R(\mathbf{p},t;\bar{\mathbf{p}},\bar{t}) <\bar{\mathbf{p}}|\hat{V}_{\bar{t}}|\bar{\mathbf{p}}'> G^R(\bar{\mathbf{p}}',\bar{t};\mathbf{p}',t') .$$

$$(2.110)$$

In the case where the particle Hamiltonian, $\hat{H}_p$, represents a free particle, the zeroth-order propagator is the free propagator, eq.(2.33), and we introduce the diagrammatic representation for the amplitude for free propagation in momentum state **p**, eq.(2.35),

$$G_0^R(\mathbf{p}, t, t') \quad = \quad \overset{\text{R}}{\underset{t\ \ \mathbf{p}\ \ t'}{\bullet \longleftarrow \bullet}} \quad = \quad -i\theta(t - t')\, e^{-\frac{i}{\hbar}\epsilon_\mathbf{p}(t-t')}\,. \tag{2.111}$$

The first-order correction to the propagator due to the potential, eq.(2.98), reduces in this case to

$$G_1^R(\mathbf{p}, t; \mathbf{p}', t') = \frac{1}{\hbar} \int_{-\infty}^{\infty} d\bar{t}\, G_0^R(\mathbf{p}, t, \bar{t}) <\mathbf{p}|\hat{V}_{\bar{t}}|\mathbf{p}'> G_0^R(\mathbf{p}', \bar{t}, t') \tag{2.112}$$

corresponding diagrammatically to

$$G_1^R(\mathbf{p}, t; \mathbf{p}', t') \quad = \quad \overset{\text{R}\qquad\qquad\text{R}}{\underset{t\quad\ \mathbf{p}\quad\ \bar{t}\quad\ \mathbf{p}'\quad\ t'}{\bullet \longleftarrow \times \longleftarrow \bullet}} \,. \tag{2.113}$$

Similarly we get to second order in the potential

$$G_2^R(\mathbf{p}, t; \mathbf{p}', t') \quad = \quad \frac{1}{\hbar^2} \int_{\mathbf{p}''} \int_{-\infty}^{\infty} dt_2 \int_{-\infty}^{\infty} dt_1\, G_0^R(\mathbf{p}, t, t_2) <\mathbf{p}|\hat{V}_{t_2}|\mathbf{p}''> G_0^R(\mathbf{p}'', t_2, t_1)$$

$$<\mathbf{p}''|\hat{V}_{t_1}|\mathbf{p}'> G_0^R(\mathbf{p}', t_1, t')$$

$$= \quad \overset{\text{R}\qquad\quad\text{R}\qquad\quad\text{R}}{\underset{t\quad\ \mathbf{p}\quad\ t_2\quad\ \mathbf{p}''\quad\ t_1\quad\ \mathbf{p}'\quad\ t'}{\bullet \longleftarrow \times \longleftarrow \times \longleftarrow \bullet}} \tag{2.114}$$

and the momentum representation of the propagator in the potential is obtained by iterating the following equation

$$\overset{\text{R}}{\underset{\mathbf{p}t\qquad\ \mathbf{p}'t'}{\bullet \longleftarrow \bullet}} \quad = \quad \overset{\text{R}}{\underset{\mathbf{p}t\qquad\ \mathbf{p}'t'}{\bullet \longleftarrow \bullet}} \quad + \quad \overset{\text{R}\qquad\quad\text{R}}{\underset{t\quad\ \mathbf{p}\quad\ \bar{t}\,\mathbf{p}''\quad\ \mathbf{p}'t'}{\bullet \longleftarrow \times \longleftarrow \bullet}} \,. \tag{2.115}$$

Analytically we have the equation

$$G^R(\mathbf{p}, t; \mathbf{p}', t') = G_0^R(\mathbf{p}, t; \mathbf{p}', t') + \frac{1}{\hbar} \int_{\mathbf{p}''} \int d\bar{t}\, G_0^R(\mathbf{p}, t, \bar{t}) <\mathbf{p}|\hat{V}_{\bar{t}}|\mathbf{p}''> G^R(\mathbf{p}'', \bar{t}; \mathbf{p}', t')\,.$$
$$\tag{2.116}$$

2.6.2 Propagation in a Static Potential

In the case the system is isolated, the potential is time independent

$$\hat{H} = \hat{H}_p + V(\hat{\mathbf{x}}) \tag{2.117}$$

and the propagator only depends on the time difference. In the momentum representation, for example, we then have for the propagator

$$
\begin{aligned}
G^R(\mathbf{p}, t; \mathbf{p}', t') \;&=\; G^R(\mathbf{p}, \mathbf{p}'; t - t') \;=\; <\mathbf{p}, t | \mathbf{p}', t'> \\
&=\; <\mathbf{p}| e^{-\frac{i}{\hbar}\hat{H}(t-t')} |\mathbf{p}'> \; .
\end{aligned}
\tag{2.118}
$$

We therefore Fourier transform with respect to time[12]

$$G^R(\mathbf{p}, \mathbf{p}', E) \;\equiv\; \frac{1}{\hbar} \int\limits_{-\infty}^{\infty} d(t - t') \, e^{\frac{i}{\hbar}(E+i0)(t-t')} \, G^R(\mathbf{p}, t; \mathbf{p}', t') \tag{2.119}$$

and for the inverse transform we have

$$G^R(\mathbf{p}, \mathbf{p}', t - t') \;=\; \frac{1}{2\pi} \int\limits_{-\infty+i0}^{\infty+i0} dE \, e^{-\frac{i}{\hbar}E(t-t')} \, G^R(\mathbf{p}, \mathbf{p}', E) \; . \tag{2.120}$$

We shall call E the energy variable, emphasizing that the above Fourier transformation is not between property representations.

The invariance with respect to displacements in time is transparently reflected on a term-by-term basis in the perturbation expansion. The time convolutions of the propagators in the perturbative expansion will by introducing Fourier transformed propagators with respect to time produce delta functions in the energy variables leading to products of Fourier-transformed propagators all having the same energy variable, reflecting the energy conservation in elastic scattering.

In order to be specific let us assume that the particle is free in the absence of the potential V.[13] From the first-order term, eq.(2.112), we obtain by Fourier transformation

$$G_1^R(\mathbf{p}, \mathbf{p}', E) \;=\; G_0^R(\mathbf{p}, E) <\mathbf{p}|\hat{V}|\mathbf{p}'> G_0^R(\mathbf{p}', E) \tag{2.121}$$

as for a static potential we have

$$<\mathbf{p}, t|\hat{V}_I(t)|\mathbf{p}', t> \;=\; <\mathbf{p}|\hat{V}|\mathbf{p}'> \;=\; \underset{\mathbf{p} \quad \mathbf{p}'}{\overset{}{\longleftrightarrow\!\times\!\longleftrightarrow}} \tag{2.122}$$

---

[12]We observe that the retarded propagator is analytic in the upper half plane. The discussion of the analytic properties of propagators being deferred to the following section.

[13]It is of course not essential for exploiting the invariance with respect to displacements in time, that the particle in the absence of the potential V is assumed otherwise free, in which case the momentum and energy representations are identical. The particle could, for example, be exposed to a time-independent magnetic field. In that case, we would then just have to use the energy-representation specified by the eigenstates of the particle Hamiltonian $\hat{H}_p$ for which the propagator oscillates in time according to the energy value in question.

and we now represent this matrix element by a cross. The Fourier transform with respect to time of the free propagator, eq.(2.35), is given by

$$G_0^R(\mathbf{p}, E) \; = \; \frac{1}{E - \epsilon_{\mathbf{p}} + i0} \tag{2.123}$$

for which we introduce the diagrammatic notation

$$G_0^R(\mathbf{p}, E) \; \equiv \; \underset{\mathbf{p}E}{\overset{R}{\longleftarrow}} \; . \tag{2.124}$$

The first-order correction to the propagator is then specified diagrammatically by

$$G_1^R(\mathbf{p}, \mathbf{p}', E) \; = \; \underset{\mathbf{p}E}{\overset{R}{\longleftarrow}} \times \underset{\mathbf{p}'E}{\overset{R}{\longleftarrow}} \; . \tag{2.125}$$

Similarly, by Fourier transforming with respect to time, we get for the second-order term

$$G_2^R(\mathbf{p}, \mathbf{p}', E) \; = \; G_0^R(\mathbf{p}, E) \left(\sum_{\mathbf{p}''} <\mathbf{p}|\hat{V}|\mathbf{p}''> G_0^R(\mathbf{p}'', E) <\mathbf{p}''|\hat{V}|\mathbf{p}'> \right) G_0^R(\mathbf{p}', E) \tag{2.126}$$

and diagrammatically

$$G_2^R(\mathbf{p}, \mathbf{p}', E) \; = \; \underset{\mathbf{p}E}{\overset{R}{\longleftarrow}} \times \underset{\mathbf{p}''E}{\overset{R}{\longleftarrow}} \times \underset{\mathbf{p}'E}{\overset{R}{\longleftarrow}} \tag{2.127}$$

where a summation over all the possible alternative intermediate momentum values $\mathbf{p}''$ is implied.

For the propagator in a static potential

$$G^R(\mathbf{p}, \mathbf{p}', E) \; \equiv \; \underset{\mathbf{p} \qquad E \qquad \mathbf{p}'}{\bullet\!\!-\!\!\overset{R}{\longleftarrow}\!\!-\!\!\bullet}, \tag{2.128}$$

we then obtain the diagrammatic representation

$$\underset{\mathbf{p} \quad E \quad \mathbf{p}'}{\bullet\!\!-\!\!\overset{R}{\longrightarrow}\!\!-\!\!\bullet} = \underset{\mathbf{p} \quad E \quad \mathbf{p}'}{\bullet\!\!-\!\!\overset{R}{\longleftarrow}\!\!-\!\!\bullet} + \underset{\mathbf{p}E \quad \mathbf{p}'E}{\overset{R}{\longleftarrow}\times\overset{R}{\longleftarrow}} + \underset{\mathbf{p}E \quad \mathbf{p}''E \quad \mathbf{p}'E}{\overset{R}{\longleftarrow}\times\overset{R}{\longleftarrow}\times\overset{R}{\longleftarrow}}$$

$$+ \; \underset{\mathbf{p}E \quad \mathbf{p}'''E \quad \mathbf{p}''E \quad \mathbf{p}'E}{\overset{R}{\longleftarrow}\times\overset{R}{\longleftarrow}\times\overset{R}{\longleftarrow}\times\overset{R}{\longleftarrow}} + \; \dots \tag{2.129}$$

where we have introduced the diagrammatic notation

$$G_0^R(\mathbf{p}, \mathbf{p}', E) \; = \; G_0^R(\mathbf{p}, E)\, \delta_{\mathbf{p}\mathbf{p}'} \; = \; \underset{\mathbf{p} \quad E \quad \mathbf{p}'}{\bullet\!\!-\!\!\overset{R}{\longleftarrow}\!\!-\!\!\bullet} \; = \; \underset{\mathbf{p}E}{\overset{R}{\longleftarrow}}\, \delta_{\mathbf{p}\mathbf{p}'} \tag{2.130}$$

in order to absorb the Kronecker function in the free propagation term.

The full propagator is obtained by iterating the equation

$$\tag{2.131}$$

which analytically takes the form

$$
\begin{aligned}
G^R(\mathbf{p}, \mathbf{p}', E) &= G_0^R(\mathbf{p}, \mathbf{p}', E) + G_0^R(\mathbf{p}, E) \sum_{\mathbf{p}''} <\mathbf{p}|\hat{V}|\mathbf{p}''> G^R(\mathbf{p}'', \mathbf{p}', E) \\
&= G_0^R(\mathbf{p}, \mathbf{p}', E) + G_0^R(\mathbf{p}, E) \frac{1}{V} \sum_{\mathbf{p}''} V(\mathbf{p} - \mathbf{p}'') G^R(\mathbf{p}'', \mathbf{p}', E)
\end{aligned}
\tag{2.132}
$$

where we have introduced the Fourier transform of the potential

$$
V(\mathbf{p}) \equiv V <\mathbf{p}|\hat{V}|\mathbf{0}> = \int_V d\mathbf{x} \, e^{-\frac{i}{\hbar}\mathbf{x}\cdot\mathbf{p}} \, V(\mathbf{x})
\tag{2.133}
$$

for the case of a particle confined to a volume of size V.

2.7 Analytic Properties of Green's Functions

For an isolated system, where the Hamiltonian is time independent, we can for any complex number E with a positive imaginary part, transform the retarded Green's operator, eq.(2.24), according to

$$
\hat{G}_E^R = \frac{1}{\hbar} \int_{-\infty}^{\infty} d(t - t') \, e^{\frac{i}{\hbar}E(t-t')} \, \hat{G}^R(t - t') \, .
\tag{2.134}
$$

The Fourier transform is obtained as the analytic continuation from the upper half plane, $\Im m E > 0$. According to eq.(2.27) we have for $\Im m E > 0$ the equation

$$
\left(E - \hat{H}\right) \hat{G}_E^R = \hat{I} \, .
\tag{2.135}
$$

Analogously we obtain that the advanced Green's operator is the solution of the same equation

$$
\left(E - \hat{H}\right) \hat{G}_E = \hat{I}
\tag{2.136}
$$

for values of the energy variable E in the lower half-plane, $\Im m E < 0$, and by analytical continuation to the real axis

$$
\hat{G}_E^A \equiv \frac{1}{\hbar} \int_{-\infty}^{\infty} dt \, e^{\frac{i}{\hbar}Et} \, \hat{G}^A(t) \, .
\tag{2.137}
$$

We note the Fourier inversion formulas

$$
\hat{G}^{R(A)}(t) = \frac{1}{2\pi} \int_{-\infty \, (\pm) \, i0}^{\infty \, (\pm) \, i0} dE \, e^{-\frac{i}{\hbar}Et} \, \hat{G}_E^{R(A)}
\tag{2.138}
$$

and the hermitian property, eq.(2.26), leads to the relationship

$$\hat{G}^A_E = [\hat{G}^R_{E^*}]^\dagger . \qquad (2.139)$$

We introduce the Green's operator

$$\hat{G}_E \equiv \begin{cases} \hat{G}^R_E & \text{for } \Im m E > 0 \\[2mm] \hat{G}^A_E & \text{for } \Im m E < 0 \end{cases} \qquad (2.140)$$

for which we have the spectral representation

$$\hat{G}_E = \frac{1}{E - \hat{H}} = \sum_\lambda \frac{|\epsilon_\lambda><\epsilon_\lambda|}{E - \epsilon_\lambda} \qquad (2.141)$$

where $|\epsilon_\lambda>$ is the eigenstates of the Hamiltonian

$$\hat{H} |\epsilon_\lambda> = \epsilon_\lambda |\epsilon_\lambda> . \qquad (2.142)$$

The analytic properties of the retarded and advanced Green's operators leads, by an application of Cauchy's theorem, to the spectral representations

$$\hat{G}^{R(A)}_E = \int_{-\infty}^\infty \frac{dE'}{2\pi} \frac{\hat{A}_{E'}}{E - E' \, (\pm) \, i0} \qquad (2.143)$$

where we have introduced the spectral operator, the discontinuity of the Green's operator across the real axis

$$\hat{A}_E \equiv i(\hat{G}^R_E - \hat{G}^A_E) = i(\hat{G}_{E+i0} - \hat{G}_{E-i0})$$

$$= 2\pi \, \delta(E - \hat{H}) = 2\pi \sum_\lambda |\epsilon_\lambda><\epsilon_\lambda| \, \delta(E - \epsilon_\lambda) . \qquad (2.144)$$

Equivalently, we have the relationship between real and imaginary parts of, say, position representation matrix elements

$$\Re e \, G^R(\mathbf{x}, \mathbf{x}', E) = \mathcal{P} \int_{-\infty}^\infty \frac{dE'}{\pi} \frac{\Im m \, G^R(\mathbf{x}, \mathbf{x}', E')}{E' - E} \qquad (2.145)$$

and

$$\Im m \, G^R(\mathbf{x}, \mathbf{x}', E) = -\mathcal{P} \int_{-\infty}^\infty \frac{dE'}{\pi} \frac{\Re e \, G^R(\mathbf{x}, \mathbf{x}', E')}{E' - E} . \qquad (2.146)$$

The Kramers-Kronig relations due to the retarded propagator is analytic in the upper half-plane.

The perturbation expansion of the propagator in a static potential is seen to be equivalent to the operator expansion for the Green's operator

$$\hat{G}_E = \frac{1}{E - \hat{H}} = \frac{1}{E - \hat{H}_0 + \hat{V}} = \frac{1}{(E - \hat{H}_0)(1 - (E - \hat{H}_0)^{-1}\hat{V})}$$

$$= \frac{1}{1 - (E - \hat{H}_0)^{-1}\hat{V}} \frac{1}{E - \hat{H}_0}$$

$$= \left(1 + (E - \hat{H}_0)^{-1}\hat{V} + (E - \hat{H}_0)^{-1}\hat{V}(E - \hat{H}_0)^{-1}\hat{V} + ...\right) \frac{1}{E - \hat{H}_0}$$

$$= \hat{G}_0(E) + \hat{G}_0(E)\hat{V}\hat{G}_0(E) + \hat{G}_0(E)\hat{V}\hat{G}_0(E)\hat{V}\hat{G}_0(E) + ... \quad (2.147)$$

where

$$\hat{G}_0(E) = \frac{1}{E - \hat{H}_0} \quad (2.148)$$

is the free Green's operator.

The momentum representation of the retarded (advanced) propagator or Green's function in the energy variable can be expressed as the matrix element

$$G^{R(A)}(\mathbf{p}, \mathbf{p}', E) = <\mathbf{p}|\,\hat{G}_E^{R(A)}\,|\mathbf{p}'> \quad (2.149)$$

of the retarded (advanced) Green's operator

$$\hat{G}_E^{R(A)} = \frac{1}{E - \hat{H}\,{}_{(\mp)}^{\pm}\,i0} \equiv (E - \hat{H}\,{}_{(\mp)}^{\pm}\,i0)^{-1} \quad (2.150)$$

the analytical continuation from the various half-planes of the Green's operator. Other representations are obtained similarly, for example,

$$G^{R(A)}(\mathbf{x}, \mathbf{x}', E) = <\mathbf{x}|\,\hat{G}_E^{R(A)}\,|\mathbf{x}'> \,. \quad (2.151)$$

The hermitian property eq.(2.139) gives the relationship

$$[G^R(\mathbf{x}, \mathbf{x}', E)]^* = G^A(\mathbf{x}', \mathbf{x}, E^*) \quad (2.152)$$

and similarly in other representations.

Employing the resolution of the identity in terms of the eigenstates of $\hat{H}$

$$\hat{I} = \sum_\lambda |\epsilon_\lambda><\epsilon_\lambda| \quad (2.153)$$

we get the spectral representation in, for example, the position representation

$$G^{R(A)}(\mathbf{x}, \mathbf{x}', E) = \sum_\lambda \frac{\psi_\lambda(\mathbf{x})\psi_\lambda^*(\mathbf{x}')}{E - \epsilon_\lambda\,{}_{(\mp)}^{\pm}\,i0} \,. \quad (2.154)$$

The Green's functions thus have singularities at the energy eigenvalues (the energy spectrum), constituting a branch cut for the continuum part of the spectrum, and simple poles for the discrete part, the latter corresponding to states which are normalizable (possible bound states of the system).

Along a branch cut the spectral function measures the discontinuity in the Green's operator

$$
\begin{aligned}
A(\mathbf{x}, \mathbf{x}', E) &\equiv\ <\mathbf{x}|i(\hat{G}_{E+i0} - \hat{G}_{E-i0})|\mathbf{x}'> \\
&=\ i\left(G^R(\mathbf{x}, \mathbf{x}', E) - G^A(\mathbf{x}, \mathbf{x}', E)\right) \\
&=\ -2\,\Im m G^R(\mathbf{x}, \mathbf{x}', E) \\
&=\ 2\pi \sum_{\lambda} \psi_{\lambda}(\mathbf{x})\psi_{\lambda}^*(\mathbf{x}')\,\delta(E - \epsilon_{\lambda})\ .
\end{aligned}
\tag{2.155}
$$

From the expression

$$
A(\mathbf{x}, \mathbf{x}, E) = 2\pi\, Tr(\hat{P}(\mathbf{x})\delta(E - \hat{H})) = 2\pi \sum_{\lambda} |<\mathbf{x}|\epsilon_{\lambda}>|^2 \delta(E - \epsilon_{\lambda})
\tag{2.156}
$$

we note that the diagonal elements of the spectral function, $A(\mathbf{x}, \mathbf{x}, E)$, is the local density of states per unit volume: the unnormalized probability per unit energy to find the particle at position $\mathbf{x}$ with energy E (or vice versa, the probability density for the particle in energy state E to be found at position $\mathbf{x}$). Employing the resolution of the identity we have

$$
\int d\mathbf{x}\, A(\mathbf{x}, \mathbf{x}, E) =\ 2\pi \sum_{\lambda} \delta(E - \epsilon_{\lambda}) \equiv 2\pi \mathcal{N}(E)
\tag{2.157}
$$

where $\mathcal{N}(E)$ is seen to be the number of energy levels per unit energy, and eq.(2.157) is thus the statement that the relative probability of finding the particle somewhere in space with energy E is proportional to the number of states available at that energy.

We also note the completeness relation

$$
\int_{\sigma} \frac{dE}{2\pi}\, A(\mathbf{x}, \mathbf{x}', E) =\ \delta(\mathbf{x} - \mathbf{x}')
\tag{2.158}
$$

where the integration (and summation over discrete part) is over the energy spectrum.

The position and momentum representation matrix elements of any operator are related by Fourier transformation. For the spectral operator we have (assuming the system enclosed in a box of volume V)

$$
\begin{aligned}
A(\mathbf{x}, \mathbf{x}', E) &=\ \sum_{\mathbf{pp}'} <\mathbf{x}|\mathbf{p}> A(\mathbf{p}, \mathbf{p}', E) <\mathbf{p}'|\mathbf{x}'> \\
&=\ \frac{1}{V} \sum_{\mathbf{pp}'} e^{\frac{i}{\hbar}\mathbf{p}\cdot\mathbf{x} - \frac{i}{\hbar}\mathbf{p}'\cdot\mathbf{x}'} A(\mathbf{p}, \mathbf{p}', E)
\end{aligned}
\tag{2.159}
$$

and inversely we have

$$
A(\mathbf{p}, \mathbf{p}', E) =\ <\mathbf{p}|\hat{A}_E|\mathbf{p}'> =\ N^{-1} \int d\mathbf{x} \int d\mathbf{x}'\, e^{-\frac{i}{\hbar}\mathbf{p}\cdot\mathbf{x} + \frac{i}{\hbar}\mathbf{p}'\cdot\mathbf{x}'} A(\mathbf{x}, \mathbf{x}', E)
\tag{2.160}
$$

where the normalization depends on whether the particle is confined or not, $N = V, (2\pi\hbar)^d$.

For the diagonal momentum components of the spectral function we have

$$A(\mathbf{p}, \mathbf{p}, E) = 2\pi\, Tr(\hat{P}(\mathbf{p})\, \delta(E - \hat{H})) = 2\pi \sum_\lambda |<\mathbf{p}|\epsilon_\lambda>|^2 \delta(E - \epsilon_\lambda) \qquad (2.161)$$

describing the unnormalized probability for a particle with momentum $\mathbf{p}$ to have energy E (or vice versa). Analogously to the position representation we obtain

$$\sum_\mathbf{p} A(\mathbf{p}, \mathbf{p}, E) = 2\pi\,\mathcal{N}(E)\,. \qquad (2.162)$$

We have the momentum normalization condition

$$\int_\sigma \frac{dE}{2\pi}\, A(\mathbf{p}, \mathbf{p}', E) = \begin{cases} \delta(\mathbf{p} - \mathbf{p}') \\ \\ \delta_{\mathbf{p},\mathbf{p}'} \end{cases} \qquad (2.163)$$

depending on whether the particle is confined or not.

Let us finally discuss the analytical properties of the free propagator. Fourier transforming the free retarded propagator, eq.(2.123), we get (in three spatial dimensions for the pre-exponential factor to be correct), $\Im mE > 0$,

$$G_0^R(\mathbf{x}, \mathbf{x}', E) = \frac{-m}{2\pi\hbar^2} \frac{e^{\frac{i}{\hbar}p_E|\mathbf{x}-\mathbf{x}'|}}{|\mathbf{x}-\mathbf{x}'|} \quad , \quad p_E = \sqrt{2mE} \qquad (2.164)$$

the solution of the spatial representation of the operator equation, eq.(2.136),

$$\left(E - \frac{\hbar^2}{2m}\triangle_\mathbf{x}\right) G_0(\mathbf{x}, \mathbf{x}', E) = \delta(\mathbf{x} - \mathbf{x}') \qquad (2.165)$$

which is analytic in the upper half-plane.

The square root function, $\sqrt{E}$, has a half line branch cut, which according to the spectral representation, eq.(2.154), must be chosen along the positive real axis, the energy spectrum of a free particle, as we choose the lowest energy eigenvalue to have the value zero. In order for the Green's function to remain bounded for infinite separation of its spatial arguments, $|\mathbf{x} - \mathbf{x}'| \to \infty$, we must make the following choice of argument function

$$\sqrt{E} \equiv \begin{cases} \sqrt{E} & \text{for } \Re eE > 0 \\ \\ i\sqrt{|E|} & \text{for } \Re eE < 0 \end{cases}. \qquad (2.166)$$

rendering the free spectral function of the form

$$A_0(\mathbf{x}, \mathbf{x}', E) = \frac{m}{\pi\hbar^2} \frac{\sin(\frac{1}{\hbar}p_E|\mathbf{x} - \mathbf{x}'|)}{|\mathbf{x} - \mathbf{x}'|} \theta(E) \qquad (2.167)$$

and we can read off the free particle density of states, the number of energy levels per unit energy per unit volume,[14]

$$N_0(E) \equiv \frac{1}{2\pi} A_0(\mathbf{x}, \mathbf{x}; E) = \theta(E) \begin{cases} \sqrt{\frac{m}{2\pi^2 \hbar^2 E}} & d = 1 \\[2mm] \frac{m}{2\pi \hbar^2} & d = 2 \\[2mm] \frac{m \sqrt{2mE}}{2\pi^2 \hbar^3} & d = 3 \end{cases} \qquad (2.168)$$

where for completeness we have also listed the one- and two-dimensional cases.

The spectral function for a free particle in the momentum representation follows, for example, from eq.(2.123)

$$A_0(\mathbf{p}, E) \equiv A_0(\mathbf{p}, \mathbf{p}, E) = 2\pi \, \delta(E - \epsilon_{\mathbf{p}}) \qquad (2.169)$$

and describes that a free particle with momentum $\mathbf{p}$ with certainty has energy $E = \epsilon_{\mathbf{p}}$, or vice versa.

2.8   Scattering Cross Section

We shall consider the scattering of a particle by a static potential V. At time $t = 0$ we assume the particle to have momentum $\mathbf{p}'$ and are interested in the probability for finding the particle with momentum $\mathbf{p}$ at time t.

Let us first consider the Born approximation where we are only interested in the first-order correction to the propagator. According to eq.(2.98) we have

$$G_1^R(\mathbf{p}, t; \mathbf{p}', t' = 0) = -\frac{<\mathbf{p}|\hat{V}|\mathbf{p}'>}{\hbar} \int_0^t d\bar{t} \, e^{-\frac{i}{\hbar} \epsilon_{\mathbf{p}}(t - \bar{t})} \, e^{-\frac{i}{\hbar} \epsilon_{\mathbf{p}'} \bar{t}}$$

$$= -\frac{<\mathbf{p}|\hat{V}|\mathbf{p}'>}{\hbar} e^{-\frac{i}{\hbar} \epsilon_{\mathbf{p}} t} \frac{e^{\frac{i}{\hbar} t(\epsilon_{\mathbf{p}} - \epsilon_{\mathbf{p}'})} - 1}{\frac{i}{\hbar}(\epsilon_{\mathbf{p}} - \epsilon_{\mathbf{p}'})}. \qquad (2.170)$$

To lowest order in the potential we thus have that the probability to find the particle at a later time with momentum $\mathbf{p}$, given initially that the particle had momentum $\mathbf{p}'$, is given by

$$|G_1^R(\mathbf{p}, t; \mathbf{p}', 0)|^2 = 2 \, |<\mathbf{p}|\hat{V}|\mathbf{p}'>|^2 \, \frac{1 - \cos \frac{t}{\hbar}(\epsilon_{\mathbf{p}} - \epsilon_{\mathbf{p}'})}{(\epsilon_{\mathbf{p}} - \epsilon_{\mathbf{p}'})^2}. \qquad (2.171)$$

We are interested in the scattering of the particle into a momentum volume containing many states, and for large times, $t \gg \hbar/\epsilon_{\mathbf{p}}$, the wildly oscillating function

---

[14]This result is of course directly obtained by trivial counting of the momentum states in a given energy range, because for a free particle constrained to the volume L^d, there is one momentum state per momentum volume $(2\pi\hbar/L)^d$. However, the above argument makes no reference to a finite volume.

in eq.(2.171) is effectively a delta function, and we have

$$|G_1^R(\mathbf{p}, t; \mathbf{p}', 0)|^2 = t\, \frac{2\pi}{\hbar} |<\mathbf{p}|\hat{V}|\mathbf{p}'>|^2\, \delta(\epsilon_{\mathbf{p}} - \epsilon_{\mathbf{p}'})$$

$$\equiv t\, \Gamma_{\mathbf{pp}'} \tag{2.172}$$

and thereby for the transition probability per unit time

$$\Gamma_{\mathbf{pp}'} = \frac{2\pi}{\hbar} |<\mathbf{p}|\hat{V}|\mathbf{p}'>|^2\, \delta(\epsilon_{\mathbf{p}} - \epsilon_{\mathbf{p}'}) \tag{2.173}$$

i.e., Fermi's golden rule.

For the probability per unit time for the particle to be scattered into momentum states with values in the volume $\Delta\mathbf{p}$ around $\mathbf{p}$ we therefore have

$$\Gamma_{\mathbf{pp}'}\, \frac{\Delta\mathbf{p}}{\left(\frac{2\pi\hbar}{L}\right)^3} = L^3 \int_{\Delta\mathbf{p}} \frac{d\mathbf{p}}{(2\pi\hbar)^3}\, \Gamma_{\mathbf{pp}'} = \int_{\Delta\hat{\mathbf{p}}} \frac{d\hat{\mathbf{p}}}{4\pi} \int_{\Delta\epsilon_{\mathbf{p}}} d\epsilon_{\mathbf{p}}\, L^3 N_0(\epsilon_{\mathbf{p}})\Gamma_{\mathbf{pp}'}$$

$$= \frac{2\pi}{\hbar} |<\mathbf{p}|\hat{V}|\mathbf{p}'>|^2\, N_0(\epsilon_{\mathbf{p}})\, L^3\, \frac{\Delta\hat{\mathbf{p}}}{4\pi} \tag{2.174}$$

where it is understood that $|\mathbf{p}'| = |\mathbf{p}|$ as demanded by energy conservation. For the probability per unit time for the particle to be scattered into a unit solid angle in the $\hat{\mathbf{p}}$-direction, $\Gamma(\hat{\mathbf{p}})$, we thus have

$$\Gamma(\hat{\mathbf{p}}) = \frac{1}{2\hbar} |<\mathbf{p}|\hat{V}|\mathbf{p}'>|^2 N_0(\epsilon_{\mathbf{p}}) L^3 . \tag{2.175}$$

The probability current density at point $\mathbf{x}$ at time t for the given initial state is

$$\mathbf{j}(\mathbf{x}, t) = \frac{\hbar}{2im} \left(G^R(\mathbf{x}, t; \mathbf{p}', 0)\, \nabla_{\mathbf{x}}\, [G^R(\mathbf{x}, t; \mathbf{p}', 0)]^* \right.$$

$$\left. - [G^R(\mathbf{x}, t; \mathbf{p}', 0)]^*\, \nabla_{\mathbf{x}}\, G^R(\mathbf{x}, t; \mathbf{p}', 0) \right) . \tag{2.176}$$

In the absence of the potential, and thereby for the probability current density outside the range of the potential, we have the probability current density for a particle in state $|\mathbf{p}'>$ (recall that in eq.(2.174) box normalization is used, and the result then follows from eq.(1.336))

$$\mathbf{j}_0(\mathbf{x}, t) = \frac{\mathbf{p}'}{m} |G_0^R(\mathbf{x}, t; \mathbf{p}', 0)|^2 = \frac{\mathbf{p}'}{m} \frac{1}{L^3} . \tag{2.177}$$

The differential cross section, $d\sigma/d\hat{\mathbf{p}}$ is defined as the probability per unit time for scattering into a unit solid angle in the $\hat{\mathbf{p}}$-direction per unit incoming flux

$$\frac{d\sigma}{d\hat{\mathbf{p}}} \equiv \frac{\Gamma(\hat{\mathbf{p}})}{j_0} = \frac{\frac{1}{2\hbar}|<\mathbf{p}|\hat{V}|\mathbf{p}'>|^2 N_0(\epsilon_{\mathbf{p}}) L^3}{\frac{p}{mL^3}}$$

$$= L^6 \frac{mN_0(\epsilon_{\mathbf{p}})}{2\hbar p} \left| \frac{1}{L^3} \int_V d\mathbf{x}\, e^{-\frac{i}{\hbar}\mathbf{x}\cdot(\mathbf{p}-\mathbf{p}')} V(\mathbf{x}) \right|^2$$

$$= \left(\frac{m}{2\pi\hbar^2}\right)^2 |V(\mathbf{p}-\mathbf{p}')|^2 . \tag{2.178}$$

In a scattering experiment a beam of incoming particles is scattered by a target, and the number of particles flying off into different directions is counted. The differential cross section is therefore the quantity of interest, because it describes the relative flux of particles scattered into a given solid angle, i.e., the probability for an incoming particle per unit time to be scattered into a unit solid angle.

As an example we consider the Coulomb potential[15]

$$V_C(\mathbf{x}) = \frac{e^2}{4\pi\epsilon_0 |\mathbf{x}|} , \qquad V_C(\mathbf{p}) = \frac{e^2\hbar^2}{\epsilon_0\, \mathbf{p}^2} \tag{2.179}$$

for which in lowest order, the Born approximation, we obtain the cross section

$$\frac{d\sigma_C}{d\hat{\mathbf{p}}} = \left(\frac{m}{2\pi\hbar^2}\right)^2 \left(\frac{e^2}{\epsilon_0\, p^2}\right)^2 \frac{\hbar^4}{|\hat{\mathbf{p}} - \hat{\mathbf{p}}'|^4} . \tag{2.180}$$

Introducing the angle between the incoming and outgoing momentum directions, $\hat{\mathbf{p}} \cdot \hat{\mathbf{p}}' \equiv \cos\theta$, and noting that

$$|\hat{\mathbf{p}} - \hat{\mathbf{p}}'|^2 = 2\,(1-\cos\theta) = 4\sin^2\frac{\theta}{2} \tag{2.181}$$

we obtain for the differential cross section

$$\frac{d\sigma_C}{d\hat{\mathbf{p}}} = \left(\frac{e^2}{16\pi\epsilon_0}\right)^2 \frac{1}{\epsilon_{\mathbf{p}}^2\, \sin^4\frac{\theta}{2}} . \tag{2.182}$$

We note that the lowest order Born approximation for the Coulomb scattering cross section equals the classical Rutherford formula, which in fact is identical to the exact quantum mechanical result since the higher order terms only influence the phase of the propagator (see for example reference [6]).

For completeness we derive the expression for the differential cross section in general, i.e., beyond the Born approximation.[16] Consider the Hamiltonian $\hat{H} = \hat{H}_0 + \hat{V}$, where $\hat{H}_0$ describes a free particle $\hat{H}_0 |\mathbf{p}\!> = \epsilon_{\mathbf{p}} |\mathbf{p}\!>$, $\epsilon_{\mathbf{p}} = \mathbf{p}^2/2m$, and $\hat{V}$ is a time-independent potential. Assuming that $\hat{H}$ has the same spectrum as $\hat{H}_0$, we can label its eigenstates similarly $\hat{H} |\psi_{\mathbf{p}}\!> = \epsilon_{\mathbf{p}} |\psi_{\mathbf{p}}\!>$. For an exact eigenstate we have according to eq.(2.53)

$$e^{-\frac{i}{\hbar}\epsilon_{\mathbf{p}}t}|\psi_{\mathbf{p}}\!> = e^{-\frac{i}{\hbar}\epsilon_{\mathbf{p}}t}|\mathbf{p}\!> + \frac{1}{i\hbar}\int_{-\infty}^{t} dt'\, e^{-\frac{i}{\hbar}\hat{H}_0(t-t')}\, \hat{V}\, e^{-\frac{i}{\hbar}\epsilon_{\mathbf{p}}t'}|\psi_{\mathbf{p}}\!> . \tag{2.183}$$

---

[15]We adopt the standard SI units.

[16]Scattering theory beyond the Born approximation is discussed in the propagator langauge in section 3.7.

Expanding $\hat{V}|\psi_{\mathbf{p}}>$ on the complete set of momentum eigenstates, we get

$$|\psi_{\mathbf{p}}> \;=\; |\mathbf{p}> \,+\, \frac{1}{i\hbar}\int_{-\infty}^{t}dt'\int d\mathbf{p}'\, e^{\frac{i}{\hbar}(t-t')(\epsilon_{\mathbf{p}}-\epsilon_{\mathbf{p}'}+i\epsilon)}<\mathbf{p}'|\hat{V}|\psi_{\mathbf{p}}>\,|\mathbf{p}'> \quad (2.184)$$

where we have introduced a convergence factor. Performing the integration over time gives the Lippmann-Schwinger equation,

$$|\psi_{\mathbf{p}}> \;=\; |\mathbf{p}> \,+\, \int d\mathbf{p}'\,\frac{<\mathbf{p}'|\hat{V}|\psi_{\mathbf{p}}>}{\epsilon_{\mathbf{p}}-\epsilon_{\mathbf{p}'}+i\epsilon}\,|\mathbf{p}'> \;. \quad (2.185)$$

In the position representation we obtain

$$\psi_{\mathbf{p}}(\mathbf{x}) \;=\; \frac{e^{\frac{i}{\hbar}\mathbf{p}\cdot\mathbf{x}}}{(2\pi\hbar)^{d/2}} \,+\, \int d\mathbf{x}'\, G_0^R(\mathbf{x},\mathbf{x}',\epsilon_{\mathbf{p}})\,V(\mathbf{x}')\,\psi_{\mathbf{p}}(\mathbf{x}') \quad (2.186)$$

where the free retarded propagator was obtained in eq.(2.164). We are free to drop the normalization factor since $|\psi_{\mathbf{p}}>$ and $|\mathbf{p}>$ have the same normalization, because the time evolution, eq.(2.53), of course is unitary. Integrating eq.(2.183) with an envelope function we obtain

$$|\psi_g(t)> \;=\; |\phi_g(t)> \,+\, \frac{1}{i\hbar}\int_{-\infty}^{t}dt'\, e^{-\frac{i}{\hbar}\hat{H}_0(t-t')}\,\hat{V}\,e^{-\frac{i}{\hbar}\epsilon_{\mathbf{p}}t'}|\psi_g(t')> \quad (2.187)$$

where

$$|\psi_g(t)> \;\equiv\; \int_0^\infty d\epsilon_{\mathbf{p}}\,g(\epsilon_{\mathbf{p}})\,e^{-\frac{i}{\hbar}\epsilon_{\mathbf{p}}t}|\psi_{\mathbf{p}}> \;, \qquad |\phi_g(t)> \;\equiv\; \int_0^\infty d\epsilon_{\mathbf{p}}\,g(\epsilon_{\mathbf{p}})\,e^{-\frac{i}{\hbar}\epsilon_{\mathbf{p}}t}|\mathbf{p}> \quad (2.188)$$

and the envelope function g is assumed a smooth function peaked at some energy value. Far in the past the wave packet in eq.(2.187) therefore has free evolution toward the target potential as described by the first term on the right side of eq.(2.187), and at later times a scattered wave develops, the second term on the right side. Instead of performing the wave packet analysis of scattering, we note that we can calculate the scattering properties from the asymptotic form of the exact solution to the stationary Schrödinger equation, $|\mathbf{x}| \to \infty$, as easily obtained from eq.(2.186):

$$\psi_{\mathbf{p}}(\mathbf{x}) \;=\; e^{\frac{i}{\hbar}pz} \,+\, f(\theta)\,\frac{e^{\frac{i}{\hbar}|\mathbf{p}|\,|\mathbf{x}|}}{|\mathbf{x}|} \quad (2.189)$$

where

$$f(\theta) \;=\; -\frac{m}{2\pi\hbar^2}\int d\mathbf{x}'\, e^{-\frac{i}{\hbar}\frac{|\mathbf{P}|}{|\mathbf{x}|}\mathbf{x}\cdot\mathbf{x}'}\,V(\mathbf{x}')\,\psi_{\mathbf{p}}(\mathbf{x}') \quad (2.190)$$

and we have chosen the $\hat{z}$-direction along the direction of the momentum of the incoming particle, and the scattering angle, θ, is the angle between the incoming momentum and the direction to the point $\mathbf{x}$. Calculating the probability current density in the scattered wave relative to the incident wave, we get for the differential cross section[17]

$$\frac{d\sigma}{d\hat{\mathbf{p}}} = |f(\theta)|^2 .$$ (2.191)

2.9   Inversion and Time-Reversal Symmetry

If we change the sense of positive direction of the coordinate axes of a reference frame, we get an equivalent description of space in which points in space change label according to $\mathbf{x} \to -\mathbf{x}$.[18] The unitary operator relating the two descriptions obtained by spatial inversion through a point, here chosen as the origin, is specified by

$$\hat{U}_P |\mathbf{x}> = e^{i\varphi(\mathbf{x})}| - \mathbf{x}>$$ (2.192)

The inversion or reflection operator is equivalently, up to a phase transformation, specified by the transformation property of the position operator

$$\hat{U}_P^{-1} \hat{\mathbf{x}} \hat{U}_P = -\hat{\mathbf{x}}$$ (2.193)

or

$$\hat{\mathbf{x}} \hat{U}_P |\mathbf{x}> = -\mathbf{x} \hat{U}_P |\mathbf{x}> .$$ (2.194)

We could also define the inversion operator on an arbitrary state, with the phase choice $\hat{U}_P^2 = \hat{I}$, by

$$<\mathbf{x}|\hat{U}_P^\dagger |\psi> = <-\mathbf{x}|\psi> \quad \text{or equivalently} \quad \hat{U}_P \psi(\mathbf{x}) = \psi(-\mathbf{x}) .$$ (2.195)

By construction of the complementary operator (using the inversion changed basis $<x_i| \to <\underline{x_i}| = <x_i|\hat{U}_P^\dagger$, we encounter, compared to sections 1.4 and 1.5, the change $\hat{\mathcal{V}} \to \underline{\hat{\mathcal{V}}} = \exp\{-id\hat{p}\}$, i.e., $\hat{p} \to -\hat{p}$) we have for the momentum operator

$$\hat{\mathbf{p}} \hat{U}_P |\mathbf{p}> = -\mathbf{p} \hat{U}_P |\mathbf{p}> \quad \text{or equivalently} \quad \hat{U}_P^{-1} \hat{\mathbf{p}} \hat{U}_P = -\hat{\mathbf{p}} .$$ (2.196)

This is also immediately verified by exploiting the property of the transformation function $<\mathbf{p}| - \mathbf{x}> = <-\mathbf{p}|\mathbf{x}>$

$$<\mathbf{p}|\hat{U}_P^\dagger|\psi> = \int d\mathbf{x} <\mathbf{p}|\mathbf{x}><\mathbf{x}|\hat{U}_P^\dagger|\psi> = \int d\mathbf{x} <\mathbf{p}|\mathbf{x}><-\mathbf{x}|\psi>$$

$$= \int d\mathbf{x} <\mathbf{p}| - \mathbf{x}><\mathbf{x}|\psi> = \int d\mathbf{x} <-\mathbf{p}|\mathbf{x}><\mathbf{x}|\psi>$$

$$= <-\mathbf{p}|\psi>$$ (2.197)

---

[17]In the absence of spherical symmetry $f(\theta) \to f(\hat{\mathbf{p}}, \hat{\mathbf{p}}')$.

[18]The spatial inversion or reflection through a point interchanges right- and left-handed coordinate systems.

i.e.,

$$\hat{U}_P \, \psi(\mathbf{p}) \;=\; \psi(-\mathbf{p}) \, . \tag{2.198}$$

Position and momentum vectors change sign under reflection, and are called polar vectors, whereas angular momentum $\hat{\mathbf{L}} = \hat{\mathbf{x}} \times \hat{\mathbf{p}}$ is invariant, $\hat{U}_P^{-1} \, \hat{\mathbf{L}} \, \hat{U}_P = \hat{\mathbf{L}}$, and are called an axial vector or pseudo-vector. The spin up and down states of an electron are defined relative to a quantization axis, say the direction of the magnetic field in the Stern-Gerlach apparatus. Since the magnetic field is described by an axial vector it is invariant under space inversion, and consequently we have that the spin is invariant under space inversion

$$\hat{U}_P^{-1} \, \hat{\mathbf{s}} \, \hat{U}_P \;=\; \hat{\mathbf{s}}. \tag{2.199}$$

Since $\hat{U}_P^2$ commutes with both the position and momentum operators, $\hat{\mathbf{x}}$ and $\hat{\mathbf{p}}$, it is proportional to the identity operator, and since $\hat{U}_P$ is unitary the proportionality factor is just a phase factor. With the phase choice $\hat{U}_P^2 = \hat{I}$,[19] we have that the reflection or inversion operator is also hermitian, $\hat{U}_P^\dagger = \hat{U}_P = \hat{U}_P^{-1}$. For $\hat{U}_P^2 = \hat{I}$, the eigenvalues of the reflection operator is ± 1, and are called the parity.

Using eq.(1.333) we note that

$$\int \frac{d\mathbf{x}' d\mathbf{p}'}{(4\pi\hbar)^d} \; e^{\frac{i}{\hbar}\mathbf{x}'\cdot\hat{\mathbf{p}} - \frac{i}{\hbar}\mathbf{p}'\cdot\hat{\mathbf{x}}} \; |\mathbf{x}> \;=\; |-\mathbf{x}> \tag{2.200}$$

and similarly for momentum states

$$\int \frac{d\mathbf{x}' d\mathbf{p}'}{(4\pi\hbar)^d} \; e^{\frac{i}{\hbar}\mathbf{x}'\cdot\hat{\mathbf{p}} - \frac{i}{\hbar}\mathbf{p}'\cdot\hat{\mathbf{x}}} \; |\mathbf{p}> \;=\; |-\mathbf{p}> \tag{2.201}$$

and for the above phase choice we have

$$\hat{U}_P \;=\; \int \frac{d\mathbf{x}' d\mathbf{p}'}{(4\pi\hbar)^d} \; e^{\frac{i}{\hbar}\mathbf{x}'\cdot\hat{\mathbf{p}} - \frac{i}{\hbar}\mathbf{p}'\cdot\hat{\mathbf{x}}} \, . \tag{2.202}$$

For a Hamiltonian invariant under reflection[20]

$$\hat{U}_P \, \hat{H} \, \hat{U}_P^{-1} \;=\; \hat{H} \tag{2.203}$$

we have the properties of the transformation functions

$$<\mathbf{x}, t | \mathbf{x}', t'> \;=\; <-\mathbf{x}, t | -\mathbf{x}', t'> \quad , \quad <\mathbf{p}, t | \mathbf{p}', t'> \;=\; <-\mathbf{p}, t | -\mathbf{p}', t'> \, . \tag{2.204}$$

Finally we wish to derive the consequences of time-reversal invariance for the transformation functions. If the potential in the Hamiltonian eq.(1.26) is time independent, we can immediately infer that if $\psi(\mathbf{x}, t)$ is a solution of the Schrödinger

---

[19]The group of reflections in a point has only two elements, $\hat{U}_P$ and $\hat{I}$.

[20]In this case $\hat{U}_P$ is a constant of the motion.

equation, eq.(1.25), so is $\overline{\psi}(\mathbf{x}, t) \equiv \psi^*(\mathbf{x}, -t)$. Comparing the time evolution on integral form (eq.(1.7) or eq.(1.367)) of $\psi(\mathbf{x}, t)$ and $\overline{\psi}(\mathbf{x}, t)$, we discover that for the considered Hamiltonian the transformation function has the following property

$$<\mathbf{x}, t|\mathbf{x}', t'>^* = <\mathbf{x}, -t|\mathbf{x}', -t'> \qquad (2.205)$$

which is equivalent to

$$<\mathbf{x}, t|\mathbf{x}', t'> = <\mathbf{x}', t|\mathbf{x}, t'> . \qquad (2.206)$$

For the transformation function in the momentum representation we then get

$$<\mathbf{p}, t|\mathbf{p}', t'> = <-\mathbf{p}', t| - \mathbf{p}, t'> . \qquad (2.207)$$

Time reversal interchanges the initial and final states, and reverses the direction of motion.

The Schrödinger equation, eq.(1.25), is clearly not invariant with respect to time inversion $t \rightarrow -t$. However, by in addition subduing the wave function to complex conjugation a solution is generated, the motion-reversed solution,[21]

$$\hat{T} \psi(\mathbf{x}, t) = \psi^*(\mathbf{x}, -t) . \qquad (2.208)$$

In the position representation we can contemplate a motion picture of the time evolution of a system, say for simplicity of the probability density distribution for a particle $P(\mathbf{x}, t)$. The time-reversal invariance of a systems dynamics can then vividly be expressed in the active point of view[22] as the statement: the time evolution of the probability density obtained by watching the motion picture played backwards, the motion-reversed state, is a possible solution of the Schrödinger equation for the system. For the considered Hamiltonian this solution is given by $\overline{P}(\mathbf{x}, t) \equiv |\overline{\psi}(\mathbf{x}, t)|^2 = |\psi^*(\mathbf{x}, -t)|^2 = |\psi(\mathbf{x}, -t)|^2$, and represents the time-reversed motion of the probability density. We speak of $\psi^*(\mathbf{x}, -t)$ as the time-reversed solution of the original solution, $\psi(\mathbf{x}, t)$, of the Schrödinger equation, or the time-reversed state.[23]

In quantum mechanics we thus encounter a symmetry which falls outside the scheme of being represented by a unitary operator, and more importantly by a linear operator. This is the possible symmetry connected with the dynamics of the system, and we now give a general discussion of time-reversal invariance. A

---

[21] This is elucidated in exercise 4.6 on page 181.

[22] In reality, reversing the direction of time is not a viable option. The passive point of view corresponds to using backward-running clocks (reversed direction for measuring the progression of time), in which case the Schrödinger equation reads

$$-i\hbar \frac{d|\overline{\psi}(t)>}{dt} = \hat{H} |\overline{\psi}(t)> .$$

[23] Complex conjugation is not an option in classical mechanics. There wave equations are second order in time and the time reversed motion to the real wave $\psi(\mathbf{x}, t)$ is $\psi(\mathbf{x}, -t)$.

system is said to respect time-reversal symmetry if there exists an operator $\hat{T}$ for which ($|T\psi> \equiv \hat{T}|\psi>$)

$$<\psi_f|e^{-\frac{i}{\hbar}\hat{H}(t_f-t_i)}|\psi_i> \quad = \quad <T\psi_i|e^{-\frac{i}{\hbar}\hat{H}(t_f-t_i)}|T\psi_f> \ . \tag{2.209}$$

Equivalently it is said that the dynamics of a system is time-reversal invariant if the transition amplitude from state $|\psi_i>$ at time t_i to state $|\psi_f>$ at time t_f equals the transition amplitude from state $|T\psi_f>$ at time t_i to state $|T\psi_i>$ at time t_f.[24]

In view of the relation eq.(C.6), established in appendix C, applied to the linear operator $\exp\{-i\hat{H}(t_f - t_i)/\hbar\}$, the dynamics of a system is thus time-reversal invariant if there exists an antiunitary operator $\hat{T}$ which commutes with the Hamiltonian

$$\hat{T}\hat{H}\hat{T}^{-1} \quad = \quad \hat{H} \ . \tag{2.210}$$

Let us construct the time-reversal operator for the case of a particle in a time-independent potential V for which we have the Hamiltonian

$$\hat{H} \quad = \quad \frac{\hat{\mathbf{p}}^2}{2m} + V(\hat{\mathbf{x}}) \ . \tag{2.211}$$

We immediately find that in the position representation the antiunitary operation of complex conjugation[25]

$$\hat{K}_{(x)}\,\psi(\mathbf{x}) \quad \equiv \quad \psi^*(\mathbf{x}) \tag{2.212}$$

commutes with the Hamiltonian. In view of the wave function being the expansion coefficients on the position basis (recall eq.(1.148))

$$|\psi> \quad = \quad \int d\mathbf{x}\ \psi(\mathbf{x})\,|\mathbf{x}> \tag{2.213}$$

we have for the complex conjugate operator with respect to the position basis (see appendix C)

$$\hat{K}_{(x)}\,|\psi> \quad \equiv \quad \int d\mathbf{x}\ \psi^*(\mathbf{x})\,|\mathbf{x}> \tag{2.214}$$

and because

$$<\mathbf{x}|K_{(x)}\psi> \quad \equiv \quad <\mathbf{x}|\hat{K}_{(x)}|\psi> \quad = \quad \psi^*(\mathbf{x}) \ . \tag{2.215}$$

we have that for a spinless particle the time-reversal operator is simply the complex conjugation operator defined with respect to the position basis, $\hat{T} = \hat{K}_{(x)}$. We note that for a spinless particle we have $\hat{T}^2 = \hat{I}$.

Since the time-reversal operator in the position representation is the complex conjugation operator, we immediately obtain the transformation properties of the

---

[24]By proper phase choice of the states in eq.(2.209) the appearance of absolute value signs are superfluous.

[25]We are discussing the properties of the wave function at the moment in time of inversion. If we wish to discuss the action of the time-reversal operator at a different time we must also reverse the direction of time, see eq.(2.208).

position and momentum operators $(\hat{T}_{(x)}(-i\hbar\nabla_{\mathbf{x}})\hat{T}_{(x)}^{-1} = i\hbar\nabla_{\mathbf{x}})$ under time-reversal invariance

$$\hat{T}\hat{\mathbf{x}}\hat{T}^{-1} = \hat{\mathbf{x}} \quad , \quad \hat{T}\hat{\mathbf{p}}\hat{T}^{-1} = -\hat{\mathbf{p}} \, . \tag{2.216}$$

Transforming to the momentum representation (with the convention eq.(1.85) or eq.(1.336) for the phase factor) we find that the time-reversed state in the momentum representation is given by[26]

$$\hat{T}\psi(\mathbf{p}) = \psi^*(-\mathbf{p}) \, . \tag{2.217}$$

The time-reversed state is thus the motion-reversed state.

For the current density we have (see eq.(1.28) on page 11)

$$<\hat{\mathbf{j}}>_{\bar{\psi}} = -<\hat{\mathbf{j}}>_\psi \tag{2.218}$$

or equivalently, $\hat{T}\hat{\mathbf{j}}(\mathbf{x})\hat{T}^\dagger = -\hat{\mathbf{j}}(\mathbf{x})$. For a time-reversal invariant Hamiltonian we obtain for the current density operator in the Heisenberg picture the relation (we are inverting time with respect to $t = 0$)

$$\hat{T}\hat{\mathbf{j}}(\mathbf{x},t)\hat{T}^\dagger = -\hat{\mathbf{j}}(\mathbf{x},-t) \, . \tag{2.219}$$

Exercise 2.7 *Verify the transformation properties eq.(2.216) using the momentum representation.*

In particular we note that with the phase choice we have made (see appendix C) we have

$$\hat{T}|\mathbf{x}> = |\mathbf{x}> \tag{2.220}$$

and eq.(2.206) is a special case of eq.(2.209). Similarly, since

$$\hat{T}|\mathbf{p}> = \hat{T}\int d\mathbf{x} \, <\mathbf{x}|\mathbf{p}> |\mathbf{x}> = \int d\mathbf{x} \, <\mathbf{x}|\mathbf{p}>^* |\mathbf{x}> = |-\mathbf{p}> \tag{2.221}$$

we recover that eq.(2.207) is a special case of eq.(2.209).

Exercise 2.8 *Discuss time-reversal symmetry for a system exposed to an external magnetic field.*

---

[26]The time-reversal operator in the momentum representation is thus not simply the complex conjugation operator with respect to the momentum basis, but involves the substitution $\mathbf{p} \to -\mathbf{p}$.

The orbital angular momentum, $\hat{\mathbf{L}} = \hat{\mathbf{x}} \times \hat{\mathbf{p}}$, transforms under time reversal according to eq.(2.216) as

$$\hat{T}\,\hat{\mathbf{L}}\,\hat{T}^{\dagger} = -\hat{\mathbf{L}}\,. \tag{2.222}$$

Since the magnetic field changes sign under time reversal (the sources generating the field are supposed to have their motion reversed) we have for the transformation properties of the spin under time reversal

$$\hat{T}\,\hat{\mathbf{s}}\,\hat{T}^{\dagger} = -\hat{\mathbf{s}}\,. \tag{2.223}$$

Using the standard basis in the operator spin space (recall exercise 1.14 on page 53 for a spin-1/2 particle) whose matrix representation are the Pauli matrices, we have

$$\hat{K}_{(x)}\,\hat{s}_x\,\hat{K}_{(x)} = \hat{s}_x\,, \quad \hat{K}_{(x)}\,\hat{s}_y\,\hat{K}_{(x)} = -\hat{s}_y\,, \quad \hat{K}_{(x)}\,\hat{s}_z\,\hat{K}_{(x)} = \hat{s}_z \tag{2.224}$$

and for the spin part $\hat{T}_{\sigma}$ of the time-reversal operator, $\hat{T} = \hat{T}_{\sigma}\,\hat{K}_{(x)}$, we have

$$\hat{T}_{\sigma}\,\hat{\mathbf{x}}\,\hat{T}_{\sigma}^{\dagger} = \hat{\mathbf{x}}\,, \quad \hat{T}_{\sigma}\,\hat{\mathbf{p}}\,\hat{T}_{\sigma}^{\dagger} = \hat{\mathbf{p}} \tag{2.225}$$

and

$$\hat{T}_{\sigma}\,\hat{s}_x\,\hat{T}_{\sigma}^{\dagger} = -\hat{s}_x\,, \quad \hat{T}_{\sigma}\,\hat{s}_y\,\hat{T}_{\sigma}^{\dagger} = \hat{s}_y\,, \quad \hat{T}_{\sigma}\,\hat{s}_z\,\hat{T}_{\sigma}^{\dagger} = -\hat{s}_z\,. \tag{2.226}$$

The last set of equalities describes a rotation in spin space through the angle π around the y-axis, and according to exercise 1.16 we have (up to a phase factor)

$$\hat{T}_{\sigma} = e^{\frac{i}{\hbar}\pi\hat{s}_y}\,. \tag{2.227}$$

For the spin-1/2 case we have the matrix representation

$$\underline{\underline{T_{\sigma}}} = i\sigma_y\,. \tag{2.228}$$

The time-reversal operator for a spin-1/2 particle is seen to satisfy $\hat{T}^2 = -\hat{I}$ (independent of phase convention and choice of representation).

We note that the spin-orbit coupling, $\hat{\mathbf{s}} \cdot \hat{\mathbf{L}}$, does not break time-reversal symmetry.

2.10   The Density Matrix

When a particle interacts with an environment which has its own dynamics, i.e., its effect upon the particle can not be described by potentials, we need to develop the diagrammatic technique for the density matrix since statistical averages with respect to the environment are taken over the distribution function. In this section we develop the density-matrix formalism for a particle interacting with a potential, which we shall need for the treatment of a particle moving in a random potential. The treatment of a quantum environment is given in chapter 6.

We assume, that at time t' the particle is in the state ρ' described by the statistical operator $\hat{\rho}' \equiv \hat{\rho}(t')$. For the statistical operator at time t we have according to eq.(1.383)

$$\hat{\rho}(t) = \hat{U}(t,t')\,\hat{\rho}'\,\hat{U}^\dagger(t,t') \; . \tag{2.229}$$

The density matrix in the position representation is

$$\rho(\mathbf{x},\mathbf{x}',t) \equiv \,<\mathbf{x}|\hat{\rho}(t)|\mathbf{x}'> \,=\, Tr(\hat{\rho}(t)|\mathbf{x}'><\mathbf{x}|) \; . \tag{2.230}$$

The diagonal element $\rho(\mathbf{x},\mathbf{x},t)$ is the probability density to find the particle at position $\mathbf{x}$ at time t

$$\rho(\mathbf{x},\mathbf{x},t) = Tr(\hat{\rho}(t)\hat{P}(\mathbf{x})) = Tr(\hat{\rho}(t_r)\hat{P}(\mathbf{x},t)) \equiv P(\mathbf{x},t) \; . \tag{2.231}$$

Here $\hat{\rho}(t_r)$ is the statistical operator at the reference time t_r, where the Heisenberg and Schrödinger pictures are chosen to coincide. We can also express the diagonal elements of the density matrix in terms of the density operator

$$P(\mathbf{x},t) = \rho(\mathbf{x},\mathbf{x},t) = Tr(\hat{\rho}(t)\hat{n}(\mathbf{x})) = Tr(\hat{\rho}(t_r)\hat{n}(\mathbf{x},t)) \equiv n(\mathbf{x},t) \; . \tag{2.232}$$

The diagonal elements of the density matrix exemplify the simplest kind of a consistent family of histories, referring only to one moment in time. The propositions in the family are of the form (suppressing the reference to the initial state) *the particle is at position* $\mathbf{x}$ *at time* t, and have their associated probability density

$$p_\rho((\mathbf{x}t)) = \rho(\mathbf{x},\mathbf{x},t) = P(\mathbf{x,t}) \; . \tag{2.233}$$

We can also consider the two-time history, that the particle is at position $\mathbf{x}'$ at time t', and at position $\mathbf{x}$ at the later time t, given the state of the particle is known to be ρ at some moment in time in the past. The associated probability of this history is (choosing the moment in time where the state is ρ as our reference time $\hat{\rho} \equiv \hat{\rho}(t_r)$)

$$\begin{aligned}
p_\rho((\mathbf{x}t),(\mathbf{x}'t')) &= Tr(\hat{n}(\mathbf{x}',t')\,\hat{\rho}\,\hat{n}(\mathbf{x}',t')\,\hat{n}(\mathbf{x},t)) \\
&= \,<\mathbf{x}',t'|\hat{n}(\mathbf{x},t)|\mathbf{x}',t'> \rho(\mathbf{x}',\mathbf{x}',t') \\
&= |G^R(\mathbf{x},t;\mathbf{x}',t')|^2\,\rho(\mathbf{x}',\mathbf{x}',t') \\
&= P(\mathbf{x},t;\mathbf{x}',t')\,\rho(\mathbf{x}',\mathbf{x}',t') \\
&= P(\mathbf{x},t;\mathbf{x}',t')\,p_\rho(\mathbf{x}',t') \tag{2.234}
\end{aligned}$$

which of course is expressible as the probability that the particle in state ρ will be at position $\mathbf{x}'$ at time t' multiplied by the conditional probability for the particle to be at position $\mathbf{x}$ at time t given it was at position $\mathbf{x}'$ at time t'.

Exercise 2.9 *Show that for the conditional probability density we have the formula*

$$P(\mathbf{x},t;\mathbf{x}',t') = Tr(\delta(\mathbf{x}-\hat{\mathbf{x}}(t))\,\delta(\mathbf{x}'-\hat{\mathbf{x}}(t'))) \tag{2.235}$$

where $\hat{\mathbf{x}}(t)$ *is the position operator in the Heisenberg picture.*

Inserting complete sets of position eigenstates in eq.(2.229), we get the integral equation determining the time evolution of the density matrix

$$\rho(\mathbf{x}, \mathbf{x}', t) = \int d\tilde{\mathbf{x}} \int d\tilde{\mathbf{x}}' \ J(\mathbf{x}, \mathbf{x}', t; \tilde{\mathbf{x}}, \tilde{\mathbf{x}}', t') \, \rho(\tilde{\mathbf{x}}, \tilde{\mathbf{x}}', t') \tag{2.236}$$

which is specified in terms of the propagator of the density matrix

$$
\begin{aligned}
J(\mathbf{x}, \mathbf{x}', t; \tilde{\mathbf{x}}, \tilde{\mathbf{x}}', t') &= \ <\tilde{\mathbf{x}}'|\hat{U}^{\dagger}(t, t')|\mathbf{x}'> <\mathbf{x}|\hat{U}(t, t')|\tilde{\mathbf{x}}> \\[2mm]
&= \ <\tilde{\mathbf{x}}', t'|\mathbf{x}', t> <\mathbf{x}, t|\tilde{\mathbf{x}}, t'> \\[2mm]
&= \ G^*(\mathbf{x}', t; \tilde{\mathbf{x}}', t') \, G(\mathbf{x}, t; \tilde{\mathbf{x}}, t') \\[2mm]
&= \ G(\tilde{\mathbf{x}}', t'; \mathbf{x}', t) \, G(\mathbf{x}, t; \tilde{\mathbf{x}}, t') \tag{2.237}
\end{aligned}
$$

and the density matrix at time t'

$$\rho(\mathbf{x}, \mathbf{x}', t') = \ <\mathbf{x}|\mathring{\rho}(t')|\mathbf{x}'> \ \equiv \ \rho'(\mathbf{x}, \mathbf{x}') \,. \tag{2.238}$$

Expressing the density matrix propagator, J, in terms of the retarded and advanced propagators we have

$$
J(\mathbf{x}, \mathbf{x}', t; \tilde{\mathbf{x}}, \tilde{\mathbf{x}}', t') =
\begin{cases}
G^R(\mathbf{x}, t; \tilde{\mathbf{x}}, t') \, G^A(\tilde{\mathbf{x}}', t'; \mathbf{x}', t) & \text{for } t > t' \\[3mm]
G^A(\mathbf{x}, t; \tilde{\mathbf{x}}, t') \, G^R(\tilde{\mathbf{x}}', t'; \mathbf{x}', t) & \text{for } t < t' \,.
\end{cases}
\tag{2.239}
$$

Since

$$
\begin{aligned}
J(\mathbf{x}, \mathbf{x}, t; \mathbf{x}', \mathbf{x}', t') &= \ G^R(\mathbf{x}, t; \mathbf{x}', t') \, G^A(\mathbf{x}', t'; \mathbf{x}, t) \ = \ |G^R(\mathbf{x}, t; \mathbf{x}', t')|^2 \\[2mm]
&\equiv \ P(\mathbf{x}, t; \mathbf{x}', t') \tag{2.240}
\end{aligned}
$$

the spatial diagonal elements of the density matrix propagator, $J(\mathbf{x}, \mathbf{x}, t; \mathbf{x}', \mathbf{x}', t')$, have the simple physical interpretation: It is the conditional probability density for the particle to be found at position $\mathbf{x}$ at time t, given that it was at position $\mathbf{x}'$ at time t'.[27]

The probability distribution at time t, $P(\mathbf{x}, t) = \rho(\mathbf{x}, \mathbf{x}, t)$, can not, according to eq.(2.236), be expressed as a functional of the probability distribution at an earlier time, as off-diagonal elements of the density matrix are of importance. In particular we note the failure of the Markovian property in general for primitive histories

$$p_\rho((\mathbf{x}, t)) = \rho(\mathbf{x}, \mathbf{x}, t) \neq \int d\mathbf{x}' \ p_\rho((\mathbf{x}t), (\mathbf{x}'t')) = \int d\mathbf{x}' \ P(\mathbf{x}, t; \mathbf{x}', t') \, p_\rho((\mathbf{x}', t'))$$

$$\tag{2.241}$$

---

[27]This is also immediately obtained from eq.(2.236) by noting that the state for a particle at position $\mathbf{x}'$ is described by the statistical operator $\hat{\rho} = |\mathbf{x}'><\mathbf{x}'|$, and therefore by the density matrix $\rho(\tilde{\mathbf{x}}, \tilde{\mathbf{x}}') = \ <\tilde{\mathbf{x}}|\mathbf{x}'><\mathbf{x}'|\tilde{\mathbf{x}}'> = \delta(\tilde{\mathbf{x}} - \mathbf{x}') \, \delta(\tilde{\mathbf{x}}' - \mathbf{x}')$.

except for the case where the state ρ' corresponds to a state of definite position.

For the simplest of environments, that of an external potential, we have diagrammatically for the density matrix (we assume $t > t'$)

$$(2.242)$$

and the dictionary for transcribing the diagrams is according to eq.(2.236) as follows:

A stipulated vertical line represents the density matrix:

$$\rho(\mathbf{x}, \mathbf{x}', t) \quad \equiv \quad \begin{array}{c} \mathbf{x} \\ t \\ \mathbf{x}' \end{array} \qquad (2.243)$$

and we have introduced thick solid lines to represent the particle propagators in the presence of the potential, for example the advanced propagator is depicted as

$$\underset{\mathbf{x}'t' \quad \mathbf{x}t}{\xrightarrow{\text{A}}} \quad = \quad G^A(\mathbf{x}, t; \mathbf{x}', t') \quad . \qquad (2.244)$$

In accordance with eq.(2.236), spatial integrations over the initial density-matrix coordinates are implied.

The perturbative expansion and diagrammatic representation of the density-matrix propagator, and thereby also of the density matrix, is immediately obtained because we know the perturbative expansion for the retarded propagator (and thereby also the one for the advanced propagator)

$$J(\mathbf{x}, \mathbf{x}', t; \tilde{\mathbf{x}}, \tilde{\mathbf{x}}', t') \;=\; \sum_{n,m=0}^{\infty} \left(\frac{1}{\hbar}\right)^{n+m} \int \prod_{n'=1}^{n} dx_{n'} \int_{-\infty}^{\infty} \prod_{n'=1}^{n} dt_{n'} \int \prod_{m'=1}^{m} d\tilde{x}_{m'} \int_{-\infty}^{\infty} \prod_{m'=1}^{m} d\tilde{t}_{m'}$$

$$G_0^R(\mathbf{x}, t; \mathbf{x}_n, t_n) G_0^R(\mathbf{x}_n, t_n; \mathbf{x}_{n-1}, t_{n-1})..G_0^R(\mathbf{x}_1, t_1; \tilde{\mathbf{x}}, t')$$

$$G_0^A(\tilde{\mathbf{x}}', t'; \tilde{\mathbf{x}}_m, \tilde{t}_m) G_0^A(\tilde{\mathbf{x}}_m, \tilde{t}_m; \tilde{\mathbf{x}}_{m-1}, \tilde{t}_{m-1})..G_0^A(\tilde{\mathbf{x}}_1, \tilde{t}_1; \mathbf{x}', t)$$

$$V(\tilde{\mathbf{x}}_m, \tilde{t}_{m-1}) V(\mathbf{x}_{m-1}, \tilde{t}_{m-1})..V(\tilde{\mathbf{x}}_1, \tilde{t}_1)$$

$$V(\mathbf{x}_n, t_n) V(\mathbf{x}_{n-1}, t_{n-1})..V(\mathbf{x}_1, t_1) \qquad (2.245)$$

where we have assumed that the time t is later than t' (for the opposite sequence the advanced and retarded labels should be interchanged).

Diagrammatically we have for the density matrix for a particle in a potential the perturbative expansion

$$(2.246)$$

where we as usual use thin solid lines to represent the particle propagators in the absence of the potential; for example, the advanced free propagator is depicted as

$$= G_0^A(\mathbf{x}, t; \mathbf{x}', t') \,. \qquad (2.247)$$

In accordance with the derivation, integration over interaction space-time points should be performed, and spatial integrations over the initial density matrix coordinates. With the chosen conventions there are no additional factors, so with each diagram is associated the same trivial factor of $+1$.

The double line diagrams for the density matrix, with the retarded propagator exclusively appearing on the upper line, and the advanced propagator exclusively on the lower line, are generic to quantum dynamics, reflecting the presence of both $\hat{U}$ and $\hat{U}^\dagger$ in the time evolution of the density matrix. The diagonal elements of the density matrix, which are real numbers, are expressed as sums of complex numbers, but they come in pairs that are each other's complex conjugates as is characteristic of quantum mechanical interference.

If the density matrix at some point in time factorizes; i.e., the system is prepared in some pure state ψ, $\rho(\mathbf{x}, \mathbf{x}', t') = \psi(\mathbf{x}) \, \psi^*(\mathbf{x}')$, the motion of the particle in a potential is uniquely determined by the propagator.

For a statistical operator diagonal in the energy representation

$$\hat{\rho} = \sum_\lambda \rho(\epsilon_\lambda) \, |\epsilon_\lambda><\epsilon_\lambda| \qquad (2.248)$$

we obtain from eq.(2.156) the relation between the density matrix and the combination of the energy distribution function and the spectral weight

$$\int_{-\infty}^{\infty} \frac{dE}{2\pi} \, \rho(E) A(\mathbf{x}, \mathbf{x}'; E) = <\mathbf{x}|\hat{\rho}|\mathbf{x}'> = \rho(\mathbf{x}, \mathbf{x}') \,. \qquad (2.249)$$

For the diagonal elements, $\mathbf{x}' = \mathbf{x}$, the equation has the interpretation: the probability of finding the particle at position $\mathbf{x}$ is the probability to find the particle at position $\mathbf{x}$ *given* it has energy E, $A(\mathbf{x},\mathbf{x};E)$, times the probability it has energy E, $\rho(E)$, summed over all possible energy values.

Exercise 2.10 *Show, by taking the momentum matrix element of the von Neumann equation, eq.(1.384), that the density matrix for a free particle in the momentum representation satisfies the equation*

$$\frac{\partial \rho(\mathbf{p},\mathbf{p}',t)}{\partial t} \; - \; \frac{i}{\hbar}\left(\epsilon_{\mathbf{p}} - \epsilon_{\mathbf{p}'}\right)\rho(\mathbf{p},\mathbf{p}',t) \; = \; 0 \,. \tag{2.250}$$

Chapter 3

Particle in a Random Potential

In this chapter we shall introduce the diagrammatic impurity-averaging technique, which will be our basic tool for studying the physical properties of systems with quenched disorder. After introducing the concept of a random potential, we first study the average propagator order by order in perturbation theory, and finally the average density matrix. In the course of this we shall show how to partially sum perturbation expressions based on the topological structure of diagrams, and encounter the important concepts of self-energy and skeleton diagrams.

A metallic conductor exhibits at low temperatures a temperature-independent resistance, the value of which is called the residual resistance. The residual resistance is due to the deviation of the sample from that of an ideal crystal. A conductor always has imperfections: foreign atoms substituting for atoms of the crystal, vacancies due to missing atoms, dislocations in the crystal, grain boundaries, etc. These defects will scatter an electron in the conductor between the different current-carrying eigenstates of the ideal crystal Hamiltonian, and thus cause current degradation. The effect on the electron motion of such imperfections we can model as giving rise to a potential deviating from that of the ideal crystal. We shall call this potential the impurity potential, and view it as a result of *impurities*. We shall in the following assume static impurities characterized by an effective potential. This is in contrast to the case where a defect has several energy states available, between which it can make transitions either by thermal excitation or by quantum tunneling. We shall in chapter 11 discuss a case where the dynamics of the defects is of importance, viz. electron-phonon interaction in dirty metals.

3.1   Random Potential Model

Consider that in a given sample we have impurities located at some definite positions $\mathbf{r}_i$, $i = 1, .., N'$. The impurity potential felt by a particle is the sum of the individual impurity potentials, and we assume for simplicity that the impurities are identical:

$$V(\mathbf{x}) = \sum_{i=1}^{N'} V_{imp}(\mathbf{x} - \mathbf{r}_i) = \int d\mathbf{x}' \, V_{imp}(\mathbf{x}') \, \rho_{imp}(\mathbf{x} - \mathbf{x}') \qquad (3.1)$$

where

$$\rho_{imp}(\mathbf{x}) \equiv \rho_{imp}(\mathbf{x}; \mathbf{r}_1, .., \mathbf{r}_{N'}) = \sum_{i=1}^{N'} \delta(\mathbf{x} - \mathbf{r}_i) \qquad (3.2)$$

is the impurity density distribution, which has the Fourier transform

$$\rho_{imp}(\mathbf{k}) \equiv \rho_{imp}(\mathbf{k}; \mathbf{r}_1, .., \mathbf{r}_{N'}) = \int d\mathbf{x} \, e^{-i\mathbf{k}\cdot\mathbf{x}} \, \rho_{imp}(\mathbf{x}) = \sum_{i=1}^{N'} e^{-i\mathbf{k}\cdot\mathbf{r}_i} \, . \qquad (3.3)$$

The Fourier transform of the potential is

$$V(\mathbf{q}) = V_{imp}(\mathbf{q}) \sum_{i=1}^{N'} e^{-i\mathbf{q}\cdot\mathbf{r}_i} \qquad (3.4)$$

where we have introduced the Fourier transform of the potential of the individual impurity

$$V_{imp}(\mathbf{q}) = \int_V d\mathbf{x} \, e^{-i\mathbf{x}\cdot\mathbf{q}} \, V_{imp}(\mathbf{x}) \qquad (3.5)$$

V denoting the volume of the system.

In a large system, where the impurities on the average are evenly distributed, one might expect that the properties of the system would be characterized solely in terms of the macroscopic parameter, the mean impurity concentration $n_i = N'/V$. This is in accordance with the usual statistical description of macroscopic systems in equilibrium, where one assumes that the behavior of the system can be characterized as an average over its ensemble of macroscopically identical subsystems. The sample is said to be self-averaging. The precise conditions under which this macroscopic ensemble point of view provides a sufficient description of the transport properties of a sample is a question which has only been resolved recently. It has been realized that average values are not exhaustive for characterizing a conductor at sufficiently low temperatures. There are important quantum interference effects contributing to transport properties which do not behave in the fashion of thermodynamic fluctuations. For example, in two dimensions the quantum fluctuations in transport properties at zero temperature are independent of the size of the system, and hence do not vanish in the thermodynamic limit. In this case a transport property is not completely characterized by its average value, we need to know higher moments of the distribution as well. We shall discuss these so-called mesoscopic fluctuations in detail in chapter 11, where we will obtain quantitative criteria for the applicability of the macroscopic description.

In each subsystem of a disordered conductor, having N impurities and volume V, we contemplate a mesh of M cells of volumes $\Delta\mathbf{r}$, assumed to be so fine that the mean distance between the impurities, $n_i^{-1/d} = (V/N)^{1/d}$, is much larger than the cell size. The probability for having more than one impurity in each cell is therefore insignificant (of order $1/V^2$).

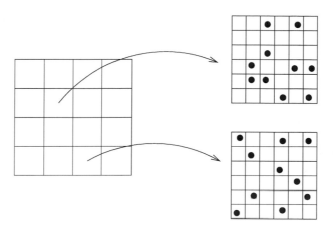

Figure 3.1 Individual impurity configurations of subsamples of a macroscopic sample.

A particular impurity configuration of any subsample is thus equivalent to specifying in which cells the impurities are located: one in $\Delta\mathbf{r}_{i_1}$,.., one in $\Delta\mathbf{r}_{i_N}$. In the limit of small cells the probability, $P(\Delta\mathbf{r}_{i_1}, .., \Delta\mathbf{r}_{i_N})$, for this configuration of impurities is specified in terms of a probability density

$$P(\Delta\mathbf{r}_{i_1}, .., \Delta\mathbf{r}_{i_N}) = P(\mathbf{r}_{i_1}, .., \mathbf{r}_{i_N})\,\Delta\mathbf{r}_{i_1} \cdots \Delta\mathbf{r}_{i_N} \tag{3.6}$$

where $\mathbf{r}_{i_k}$ is a point in cell k. Assuming that the positions of the impurities are distributed independently, we have for the probability density for having the impurities located at points $\mathbf{r}_1, .., \mathbf{r}_N$

$$P(\mathbf{r}_1, .., \mathbf{r}_N) = P(\mathbf{r}_1) \cdots P(\mathbf{r}_N) = \prod_{i=1}^{N} P(\mathbf{r}_i) \tag{3.7}$$

where $P(\mathbf{r}_i)$ is the probability density for having an impurity in the cell around point $\mathbf{r}_i$.

In the random case where we assume equal probability for an impurity to be located in any cell, we then have by normalization[1]

$$P(\mathbf{r}) = \frac{1}{V} \tag{3.8}$$

the random potential case.

---

[1]No confusion between the notation for the potential and the volume should arise.

In the self-averaging case, the value of a physical quantity F is represented by the average value over the ensemble of macroscopically identical subsystems. The average value is obtained by taking the average over systems where the location of impurities differ, but they have identical mean impurity concentration. In any given system of this ensemble, each with N impurities, the quantity F will depend on the actual positions of the impurities, $F = F(\mathbf{r}_1, \mathbf{r}_2, ..\mathbf{r}_N)$, and for the impurity average of F we have

$$< F > \; \equiv \; \int \prod_{i=1}^{N} d\mathbf{r}_i \; F(\mathbf{r}_1, \mathbf{r}_2, .., \mathbf{r}_N) \, P(\mathbf{r}_1, \mathbf{r}_2, .., \mathbf{r}_N)$$

$$= \; \int \prod_{i=1}^{N} \frac{d\mathbf{r}_i}{V} \; F(\mathbf{r}_1, \mathbf{r}_2, .., \mathbf{r}_N) \tag{3.9}$$

where the last equality is valid for the random case.

3.2   Propagation in a Random Potential

The all important ingredient in the further analysis is the impurity-averaged propagator. It is the basic building block we shall need when we discuss the motion of a particle in a random potential. Using the results of the previous chapter we study the impurity-averaged propagator order by order in perturbation theory.[2] In a given sample or subsystem the propagator (here displayed in the momentum representation)

$$G^R(\mathbf{p}, t; \mathbf{p}', t') \; \equiv \; G^R(\mathbf{p}, t; \mathbf{p}', t'; \mathbf{r}_1, \mathbf{r}_2, ..\mathbf{r}_N) \tag{3.10}$$

depends parametrically on the impurity positions.

The impurity-averaged propagator can be obtained by averaging each term in the perturbative expansion of the propagator in terms of the impurity potential. With each impurity potential we have in the momentum representation, according to eq.(3.4), associated the factor $V(\mathbf{p}-\mathbf{p}')$ depending on the incoming and outgoing momenta to the vertex (in accordance with eq.(2.132)).[3] Expressing the impurity potential in terms of the individual impurity potentials

$$V(\mathbf{p} - \mathbf{p}') \; \equiv \; V(\mathbf{p}, \mathbf{p}'; \mathbf{r}_1, \mathbf{r}_2, ..\mathbf{r}_N) = V_{imp}(\mathbf{p} - \mathbf{p}') \sum_{i=1}^{N} e^{-\frac{i}{\hbar}(\mathbf{p}-\mathbf{p}')\cdot\mathbf{r}_i} \tag{3.11}$$

we have associated with each potential term the impurity phase factor

$$\rho_{imp}(\mathbf{p} - \mathbf{p}'; \mathbf{r}_1, .., \mathbf{r}_N) \; = \; \sum_{i=1}^{N} \exp\left\{ -\frac{i}{\hbar}(\mathbf{p} - \mathbf{p}') \cdot \mathbf{r}_i \right\} \; . \tag{3.12}$$

---

[2]We follow the original presentations of references [19] and [20].

[3]We shall, when convenient, instead of the wave vector representation use the momentum representation of the previous chapter, $V(\mathbf{k} - \mathbf{k}') \to V(\mathbf{p} - \mathbf{p}')$, $\mathbf{p} \equiv \hbar\mathbf{k}$.

Averaging the first-order term, eq.(2.121), we get in the momentum representation[4]

$$<G_1^R(\mathbf{p}, \mathbf{p}', E)> \; = \; G_0^R(\mathbf{p}, E) \, V_{imp}(\mathbf{p}-\mathbf{p}') \, G_0^R(\mathbf{p}', E) \frac{1}{V} < \sum_{i=1}^{N} e^{-\frac{i}{\hbar}(\mathbf{p}-\mathbf{p}')\cdot\mathbf{r}_i} > \quad (3.13)$$

and we have to impurity average the Fourier transform of the impurity density. When $\mathbf{p} \neq \mathbf{p}'$, the oscillating exponents average to zero, and for $\mathbf{p} = \mathbf{p}'$ we get the number of impurities; i.e.,

$$< \sum_{i=1}^{N} e^{-\frac{i}{\hbar}(\mathbf{p}-\mathbf{p}')\cdot\mathbf{r}_i} > \; = \; N \, \delta_{\mathbf{p},\mathbf{p}'} \quad (3.14)$$

because there are N terms in the sum giving identical contributions. The overall factor in eq.(3.13) is therefore the average impurity density. As expected we have recovered translation invariance for the first-order impurity-averaged propagator

$$< G_1^R(\mathbf{p}, \mathbf{p}', E) > \; = \; n_i \, V_{imp}(\mathbf{p} = \mathbf{0}) \, [G_0^R(\mathbf{p}, E)]^2 \, \delta_{\mathbf{p},\mathbf{p}'} \; . \quad (3.15)$$

The factor in front of the Kronecker function

$$G_1^R(\mathbf{p}, E) \; = \; n_i \, V_{imp}(\mathbf{p} = \mathbf{0}) [G_0^R(\mathbf{p}, E)]^2 \quad (3.16)$$

we can depict diagrammatically

$$G_1^R(\mathbf{p}, E) = \quad \underset{\mathbf{p}E}{\overset{R}{\underline{\qquad}}} \underset{\mathbf{p}E}{\overset{R}{\underline{\qquad}}} \quad (3.17)$$

as we introduce

$$\mathbf{p} \; \longleftarrow \; \mathbf{p}' \quad = \quad n_i \, V_{imp}(\mathbf{p} - \mathbf{p}') \quad (3.18)$$

where the cross designates the impurity concentration, n_i, and the dashed line the Fourier transform of the impurity potential, $V_{imp}(\mathbf{p} - \mathbf{p}')$, where the argument is the outgoing minus the incoming momentum.

The first-order term is proportional to $V_{imp}(\mathbf{p} = \mathbf{0})$, the spatial average of the impurity potential, whose value is arbitrary, and just acts as a constant, $n_i V_{imp}(\mathbf{p} = \mathbf{0})$, added to the Hamiltonian. This term therefore has no observable consequences, and we can assume it to be zero, or redefine the reference for measuring energy.

---

[4]We shall for simplicity assume the free electron model for the conduction electrons. Formally, however, all formulas are identical for Bloch electrons; we just have a suppressed band index. Eventually, one must usually in order to obtain analytical results resort to a simple Fermi surface, and neglect interband scattering.

The latter case is a simple example of renormalizing the term away; i.e., we add the constant $-n_i V_{imp}(\mathbf{0})$ to the Hamiltonian, and this term will generate the first-order term

$$\begin{array}{ccc} \underset{\substack{R \\ \mathbf{p}E}}{\xleftarrow{}} \;\;\; \underset{\substack{R \\ \mathbf{p}E}}{\xleftarrow{}} & = & G_0^R(\mathbf{p}, E)(-n_i V_{imp}(\mathbf{0}))G_0^R(\mathbf{p}, E) \end{array} \qquad (3.19)$$

canceling the previous term. The extra diagrams generated by the added term to the Hamiltonian will exactly cancel all the dangling impurity-scattering diagrams as

$$\underset{\substack{\mathbf{p} \;\;\;\;\;\; \mathbf{p}}}{\xleftarrow{}} \;\; = \;\; - \;\; \underset{\substack{\mathbf{p} \;\;\;\;\;\; \mathbf{p}}}{\xleftarrow{}} \qquad . \qquad (3.20)$$

For the second-order term, eq.(2.127),

$$G_2^R(\mathbf{p}, \mathbf{p}', E) \;\; = \;\; \underset{\substack{R \\ \mathbf{p}E}}{\xleftarrow{}} \times \underset{\substack{R \\ \mathbf{p}''E}}{\xleftarrow{}} \times \underset{\substack{R \\ \mathbf{p}'E}}{\xleftarrow{}} \qquad (3.21)$$

we get, upon impurity averaging,

$$< G_2^R(\mathbf{p}, \mathbf{p}', E) > \;\; = \;\; G_0^R(\mathbf{p}, E) \sum_{\mathbf{p}''} V_{imp}(\mathbf{p} - \mathbf{p}'') \, G_0^R(\mathbf{p}'', E) \, V_{imp}(\mathbf{p}'' - \mathbf{p}')$$

$$G_0^R(\mathbf{p}', E) \frac{1}{V^2} \langle \sum_{i,j=1}^{N} e^{-\frac{i}{\hbar}(\mathbf{p} - \mathbf{p}'')\cdot \mathbf{r}_i - \frac{i}{\hbar}(\mathbf{p}'' - \mathbf{p}')\cdot \mathbf{r}_j} \rangle \qquad (3.22)$$

and we have to average the sum over the impurity positions. The terms in the sum corresponding to scattering off different impurities, $i \neq j$, gives the factor

$$\frac{1}{V^2} \langle \sum_{i \neq j}^{N} e^{-\frac{i}{\hbar}(\mathbf{p} - \mathbf{p}'')\cdot \mathbf{r}_i - \frac{i}{\hbar}(\mathbf{p}'' - \mathbf{p}')\cdot \mathbf{r}_j} \rangle \;\; = \;\; \frac{N(N-1)}{V^2} \delta_{\mathbf{p}, \mathbf{p}''} \, \delta_{\mathbf{p}'', \mathbf{p}'} \qquad (3.23)$$

as there are $N(N-1)$ terms in the sum giving identical contributions. We shall be interested in the thermodynamic limit where we let V and N approach infinity in such a way that the ratio is kept fixed, $n_i \equiv N/V$, and we get

$$\frac{1}{V^2} \langle \sum_{i \neq j}^{N} e^{-\frac{i}{\hbar}(\mathbf{p} - \mathbf{p}'')\cdot \mathbf{r}_i - \frac{i}{\hbar}(\mathbf{p}'' - \mathbf{p}')\cdot \mathbf{r}_j} \rangle \;\; = \;\; n_i^2 \, \delta_{\mathbf{p}, \mathbf{p}''} \, \delta_{\mathbf{p}'', \mathbf{p}'} \;\; = \;\; n_i^2 \, \delta_{\mathbf{p}, \mathbf{p}''} \, \delta_{\mathbf{p}, \mathbf{p}'} \qquad (3.24)$$

and thereby the contribution from scattering off different impurities

$$< G_2^R(\mathbf{p}, \mathbf{p}', E) >^{i \neq j} \;\; = \;\; \delta_{\mathbf{p}, \mathbf{p}'} \, n_i^2 \, [V_{imp}(\mathbf{p} = 0)]^2 \, [G_0^R(\mathbf{p}, E)]^3$$

$$\equiv \;\; G_2^R(\mathbf{p}, E)^{i \neq j} \, \delta_{\mathbf{p}, \mathbf{p}'} \qquad . \qquad (3.25)$$

The prefactor in this term has the diagrammatic representation

$$G_2^R(\mathbf{p}, E)^{i \neq j} \;\; = \;\; \underset{\mathbf{p}E}{\mathrm{R}} \;\; \underset{\mathbf{p}E}{\mathrm{R}} \;\; \underset{\mathbf{p}E}{\mathrm{R}}$$

$$= \;\; [G_0^R(\mathbf{p}, E)]^3 \, n_i^2 \, [V_{imp}(\mathbf{p} = \mathbf{0})]^2 \tag{3.26}$$

with two dangling impurity lines. The term can be kept or renormalized away by the above mentioned prescription.

The term corresponding to scattering off the same impurity $i = j$ gives the factor in eq.(3.14), and we get the contribution to the impurity-averaged propagator

$$< G_2^R(\mathbf{p}, \mathbf{p}', E) >^{i=j} \;\; = \;\; n_i \, \delta_{\mathbf{p}, \mathbf{p}'} \, [G_0^R(\mathbf{p}, E)]^2 \, \frac{1}{V} \sum_{\mathbf{p}''} |V_{imp}(\mathbf{p} - \mathbf{p}'')|^2 \, G_0^R(\mathbf{p}'', E)$$

$$\equiv \;\; G_2^R(\mathbf{p}, E)^{i=j} \, \delta_{\mathbf{p}, \mathbf{p}'} \;\; . \tag{3.27}$$

The prefactor in this term has the diagrammatic representation

$$G_2^R(\mathbf{p}, E)^{i=j} \;\; = \;\; \underset{\mathbf{p}E}{\mathrm{R}} \;\; \underset{\mathbf{p}''E}{\mathrm{R}} \;\; \underset{\mathbf{p}E}{\mathrm{R}} \tag{3.28}$$

where we have introduced the impurity correlator in the momentum representation

$$\equiv \;\; n_i \, \delta_{\mathbf{p}_1 + \mathbf{p}_2, \mathbf{p}'_1 + \mathbf{p}'_2} \, V_{imp}(\mathbf{p}_1 - \mathbf{p}'_1) \, V_{imp}(\mathbf{p}_2 - \mathbf{p}'_2)$$

$$= \;\; n_i \, \delta_{\mathbf{p}_1 + \mathbf{p}_2, \mathbf{p}'_1 + \mathbf{p}'_2} \, |V_{imp}(\mathbf{p}_1 - \mathbf{p}'_1)|^2 \;\; . \tag{3.29}$$

The Kronecker function reflects that translation invariance is recovered upon impurity averaging, and expresses that the sum of the incoming momenta in the impurity correlator equals the sum of the outgoing momenta. In the above second-order case the momentum conservation is trivially expressed as $1 = \delta_{\mathbf{p},\mathbf{p}} = \delta_{\mathbf{p}+\mathbf{p}'',\mathbf{p}+\mathbf{p}''}$. Summation (and an inverse volume factor) over the internal momentum variable is implied as a Feynman rule.

To second order in the impurity potential the impurity-averaged propagator has recovered translation invariance and we have in the continuum limit (for the $i = j$ term)

$$G_2^R(\mathbf{p}, E) = [G_0^R(\mathbf{p}, E)]^2 \, n_i \int \frac{d\mathbf{p}'}{(2\pi\hbar)^3} \, |V_{imp}(\mathbf{p} - \mathbf{p}')|^2 G_0^R(\mathbf{p}', E) \, . \qquad (3.30)$$

For the third-order diagram

$$G_3^R(\mathbf{p}, \mathbf{p}', E) \quad = \quad \bullet \underset{\mathbf{p}E}{\overset{R}{\leftarrow}} \times \underset{\mathbf{p}_2 E}{\overset{R}{\leftarrow}} \times \underset{\mathbf{p}_1 E}{\overset{R}{\leftarrow}} \times \underset{\mathbf{p}'E}{\overset{R}{\leftarrow}} \bullet \qquad (3.31)$$

we have the sum of position-dependent impurity phase factors

$$< \sum_{i,j,k=1}^{N} e^{-\frac{i}{\hbar}(\mathbf{p}-\mathbf{p}_2)\cdot\mathbf{r}_i - \frac{i}{\hbar}(\mathbf{p}_2-\mathbf{p}_1)\cdot\mathbf{r}_j - \frac{i}{\hbar}(\mathbf{p}_1-\mathbf{p}')\cdot\mathbf{r}_k} > \qquad (3.32)$$

which for triple scattering off the same impurity, $i = j = k$, gives the factor $N\delta_{\mathbf{p},\mathbf{p}'}$ and thereby the diagram

$$\underset{\mathbf{p}E}{\overset{R}{\leftarrow}} \ \underset{\mathbf{p}_2 E}{\overset{R}{\leftarrow}} \ \underset{\mathbf{p}_1 E}{\overset{R}{\leftarrow}} \ \underset{\mathbf{p}E}{\overset{R}{\leftarrow}} \quad = \quad n_i \, [G_0^R(\mathbf{p}, E)]^2 \frac{1}{V^2} \sum_{\mathbf{p}_1,\mathbf{p}_2} V_{imp}(\mathbf{p} - \mathbf{p}_2) G_0^R(\mathbf{p}_2, E)$$

$$V_{imp}(\mathbf{p}_2 - \mathbf{p}_1) G_0^R(\mathbf{p}_1, E) V_{imp}(\mathbf{p}_1 - \mathbf{p}) \qquad (3.33)$$

where the three-leg represents the three impurity potential factors, and the cross, as before, the impurity concentration. Terms with three or more scatterings off the same impurity, we shall refer to as multiple scattering.

The terms with double scattering off the same impurity, the $i \neq j = k$, $i = j \neq k$ terms, contains a dangling impurity line, as the average gives

$$\frac{N(N-1)}{V^2} \delta_{\mathbf{p},\mathbf{p}_2} \delta_{\mathbf{p}_2,\mathbf{p}'} \simeq \frac{N^2}{V^2} \delta_{\mathbf{p},\mathbf{p}_2} \delta_{\mathbf{p},\mathbf{p}'} \qquad (3.34)$$

where the last expression is an identity in the large volume limit. For example we have the contribution

$$\text{[diagram]} \quad = \quad n_i V_{imp}(\mathbf{p}=0)[G_0^R(\mathbf{p}, E)]^3$$

$$n_i \frac{1}{V} \sum_{\mathbf{p}'} |V_{imp}(\mathbf{p}-\mathbf{p}')|^2 G_0^R(\mathbf{p}', E) \qquad (3.35)$$

which can be kept or renormalized to zero.

The fourth-order diagram

$$\text{[diagram]} \qquad (3.36)$$

has the impurity phase factor

$$< \sum_{i,j,k,l=1}^{N} e^{-\frac{i}{\hbar}(\mathbf{p}-\mathbf{p_3})\cdot\mathbf{r}_i - \frac{i}{\hbar}(\mathbf{p_3}-\mathbf{p_2})\cdot\mathbf{r}_j - \frac{i}{\hbar}(\mathbf{p_2}-\mathbf{p_1})\cdot\mathbf{r}_k - \frac{i}{\hbar}(\mathbf{p_1}-\mathbf{p}')\cdot\mathbf{r}_l} > . \qquad (3.37)$$

The term corresponding to quadruple scattering off the same impurity, $i = j = k = l$, contains the factor

$$< \sum_{i=j=k=l}^{N} e^{-\frac{i}{\hbar}(\mathbf{p}-\mathbf{p_3})\cdot\mathbf{r}_i - \frac{i}{\hbar}(\mathbf{p_3}-\mathbf{p_2})\cdot\mathbf{r}_j - \frac{i}{\hbar}(\mathbf{p_2}-\mathbf{p_1})\cdot\mathbf{r}_k - \frac{i}{\hbar}(\mathbf{p_1}-\mathbf{p}')\cdot\mathbf{r}_l} > = N \delta_{\mathbf{p},\mathbf{p}'} \qquad (3.38)$$

giving the multiple scattering term corresponding to the diagram

$$G_4^R(\mathbf{p}, E)^{i=j=k=l} \quad = \quad \text{[diagram]}$$

$$= \quad n_i \int \frac{d\mathbf{p_3}}{(2\pi\hbar)^3} \int \frac{d\mathbf{p_2}}{(2\pi\hbar)^3} \int \frac{d\mathbf{p_1}}{(2\pi\hbar)^3} V_{imp}(\mathbf{p}-\mathbf{p_3})$$

$$V_{imp}(\mathbf{p_3}-\mathbf{p_2})V_{imp}(\mathbf{p_2}-\mathbf{p_1})V_{imp}(\mathbf{p_1}-\mathbf{p})$$

$$G_0^R(\mathbf{p_3}, E)G_0^R(\mathbf{p_2}, E)G_0^R(\mathbf{p_1}, E)[G_0^R(\mathbf{p}, E)]^2 . \qquad (3.39)$$

We also get fourth-order terms corresponding to diagrams with dangling impurity lines. For example the $N(N-1)(N-2)(N-3)$ terms where the scattering is off different impurities $i \neq j \neq k \neq l$,

$$(3.40)$$

or terms with two dangling lines, for example the term where $i \neq j = l \neq k$,

$$(3.41)$$

All such dangling impurity line diagrams we can keep or renormalize to zero.

Interesting terms arise when we have double scattering off two different impurities. For the case where we consider the term $i = j \neq k = l$ in the sum, the impurity phase factor is

$$< e^{-\frac{i}{\hbar}(\mathbf{p}-\mathbf{p}_2)\cdot\mathbf{r}_i} e^{-\frac{i}{\hbar}(\mathbf{p}_2-\mathbf{p}')\cdot\mathbf{r}_k} > \;=\; \delta_{\mathbf{p},\mathbf{p}_2}\delta_{\mathbf{p}_2,\mathbf{p}'} \;=\; \delta_{\mathbf{p},\mathbf{p}_2}\delta_{\mathbf{p},\mathbf{p}'}$$

$$=\; \delta_{\mathbf{p}_3+\mathbf{p}_2,\mathbf{p}_3+\mathbf{p}}\;\delta_{\mathbf{p}_2+\mathbf{p}_1,\mathbf{p}'+\mathbf{p}_1} \qquad (3.42)$$

and we get

$$G_4^R(\mathbf{p},E)^{i=j\neq k=l} \;=\; n_i^2[G_0^R(\mathbf{p},E)]^3 \frac{1}{V^2}\sum_{\mathbf{p}_1,\mathbf{p}_3} V_{imp}(\mathbf{p}-\mathbf{p}_3)G_0^R(\mathbf{p}_3,E)$$

$$V_{imp}(\mathbf{p}_3-\mathbf{p})G_0^R(\mathbf{p},E)V_{imp}(\mathbf{p}-\mathbf{p}_1)G_0^R(\mathbf{p}_1,E)V_{imp}(\mathbf{p}_1-\mathbf{p})$$

$$(3.43)$$

For the case where $i = k \neq j = l$, the impurity phase factor is

$$< e^{-\frac{i}{\hbar}((\mathbf{p}-\mathbf{p}_3)+(\mathbf{p}_2-\mathbf{p}_1))\cdot\mathbf{r}_i} e^{-\frac{i}{\hbar}((\mathbf{p}_3-\mathbf{p}_2)+(\mathbf{p}_1-\mathbf{p}'))\cdot\mathbf{r}_j} > \;=\; \delta_{\mathbf{p},\mathbf{p}'}\,\delta_{\mathbf{p}_3-\mathbf{p}_2,\mathbf{p}-\mathbf{p}_1} \quad (3.44)$$

and we obtain

$$G_4^R(\mathbf{p},E)^{i=k\neq j=l} \;=\; n_i^2[G_0^R(\mathbf{p},E)]^2 \frac{1}{V^2}\sum_{\mathbf{p}_1,\mathbf{p}_3} V_{imp}(\mathbf{p}-\mathbf{p}_3)G_0^R(\mathbf{p}_3,E)V_{imp}(\mathbf{p}-\mathbf{p}_1)$$

$$G_0^R(\mathbf{p}_3+\mathbf{p}_1-\mathbf{p},E)V_{imp}(\mathbf{p}_3-\mathbf{p})G_0^R(\mathbf{p}_1,E)V_{imp}(\mathbf{p}_1-\mathbf{p})$$

$$= \quad \underset{\mathbf{p}E \quad \mathbf{p}_3 E \quad \mathbf{p}_2 E \quad \mathbf{p}_1 E \quad \mathbf{p}E}{\text{R} \quad \text{R} \quad \text{R} \quad \text{R} \quad \text{R}} \tag{3.45}$$

where the momentum conservation in the impurity correlator constrains $\mathbf{p}_2$ to the value $\mathbf{p}_2 = \mathbf{p}_3 + \mathbf{p}_1 - \mathbf{p}$.

For the last possibility where $i = l \neq j = k$, the impurity phase factor is

$$< e^{-\frac{i}{\hbar}((\mathbf{p}-\mathbf{p}_3)+(\mathbf{p}_1-\mathbf{p}'))\cdot\mathbf{r}_i} e^{-\frac{i}{\hbar}(\mathbf{p}_3-\mathbf{p}_1)\cdot\mathbf{r}_j} > \quad = \quad \delta_{\mathbf{p},\mathbf{p}'}\,\delta_{\mathbf{p}_3,\mathbf{p}_1} \tag{3.46}$$

and we obtain

$$G_4^R(\mathbf{p}, E)^{i=l\neq j=k} \quad = \quad n_i^2 [G_0^R(\mathbf{p}, E)]^2 \frac{1}{V^2} \sum_{\mathbf{p}_2,\mathbf{p}_1} V_{imp}(\mathbf{p} - \mathbf{p}_1) G_0^R(\mathbf{p}_1, E) V_{imp}(\mathbf{p}_1 - \mathbf{p}_2)$$

$$G_0^R(\mathbf{p}_2, E) V_{imp}(\mathbf{p}_2 - \mathbf{p}_1) G_0^R(\mathbf{p}_1, E) V_{imp}(\mathbf{p}_1 - \mathbf{p})$$

$$= \quad \underset{\mathbf{p}E \quad \mathbf{p}_1 E \quad \mathbf{p}_2 E \quad \mathbf{p}_1 E \quad \mathbf{p}E}{\text{R} \quad \text{R} \quad \text{R} \quad \text{R} \quad \text{R}} \tag{3.47}$$

Double scattering off two impurities can occur in three different ways, and the impurity correlators lead in each term to different momentum-conservation constraints.

Continuing impurity averaging the higher-order terms, the impurity-averaged propagator is expressed as a perturbative expansion in the impurity concentration or the impurity potential, and the bare propagator, G_0^R, the propagator in the absence of the impurity potential.

The approximation where multiple scatterings can be neglected we shall call the Born approximation, and we establish the quantitative criterion for its validity in section 3.5. As we discuss in section 3.7, the multiple scattering terms can be taken into account without qualitative changes for the cases we shall have in mind.[5] In the Born approximation we therefore have an easy prescription for obtaining all the diagrams for the n'th order impurity averaged propagator: Tie all the impurity crosses pairwise together in all possible ways! We note, that in the no-dangling Born approximation the perturbation series is an expansion in the parameter $n_i |V_{imp}|^2$.

---

[5]When the potential can give rise to resonances, multiple scattering can give rise to effects not included in the Born approximation.

Any internal momentum appears in two impurity-phase-factors, once with a plus sign and once with a minus sign. The Kronecker functions appearing upon impurity averaging will thus allways be proportional to $\delta \mathbf{p}, \mathbf{p}'$.[6] To each order in perturbation theory we thus have translation invariance of the impurity-averaged propagator, and we have analytically for the impurity-averaged propagator

$$< G^R(\mathbf{p}, \mathbf{p}', E) > \; = \; G^R(\mathbf{p}, E)\, \delta_{\mathbf{p},\mathbf{p}'} \; . \tag{3.48}$$

The translation invariance of the impurity-averaged propagator is obvious from a physical point of view because the averaging procedure distinguishes no point. We have verified this property to each order in perturbation theory, as the momentum conservation at the impurity correlator, eq.(3.29), assures that the total momentum flow through a diagram is conserved.[7]

The diagrams of third order in the impurity concentration are in the Born approximation (leaving out all dangling impurity line diagrams) the following

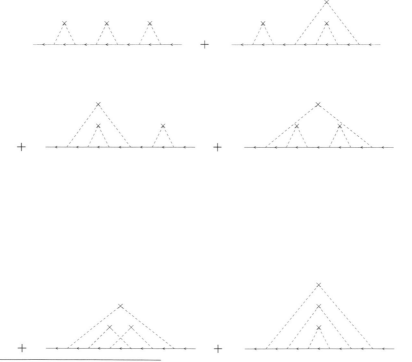

---

[6]Similarly, an internal line which when cut separates a diagram in two (like the one in eq.(3.43) carries the external momentum $\mathbf{p}$.

[7]Such an identification of a property valid in each order of perturbation theory can forcibly be turned around to allow statements valid beyond perturbation theory.

$+$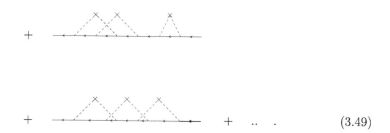

$+$ $+$ .. . (3.49)

Exercise 3.1 *Draw the rest of the 15 third-order diagrams.*

3.3 The Self-energy

We have previously derived diagrammatic formulas from formal expressions. Now we shall argue directly in the diagrammatic language in order to generate new diagrammatic expressions from previous ones, and thereby diagrammatically derive new equations.

In order to get a grasp of the totality of diagrams for the impurity-averaged propagator we shall use their topology for classification. We introduce the one-particle irreducible (1PI) propagator, corresponding to all the diagrams which can not be cut in two by cutting an internal particle line. In the example

1PI   1PR

the first diagram is one-particle irreducible,, 1PI, whereas the second is one-particle reducible, 1PR.

Amputating the external legs of the one-particle irreducible diagrams for the impurity-averaged propagator, we get an object we call the self-energy:

$$\Sigma^R(E, \mathbf{p}) \quad \equiv \quad {}_{\mathbf{p}E} \blacktriangleleft \left(\Sigma^R(E,\mathbf{p}) \right) \blacktriangleleft {}_{\mathbf{p}E}$$

$$(3.51)$$

consisting, by construction, of all amputated diagrams which can not be cut in two by cutting one bare propagator line.

We can now go on and uniquely classify all diagrams of the impurity-averaged propagator according to whether they can be cut in two by cutting an internal particle line at only one place, or at two, three, etc. places. By construction we uniquely exhaust all the possible diagrams for the propagator

$$(3.52)$$

By iteration, this equation is seen to be equivalent to the equation[8]

----

[8]In the last term we can interchange the free and full propagator, because iterating from the left generates the same series as iterating from the right.

$$\underset{\mathbf{p}E}{\overset{R}{\longrightarrow}} \;=\; \underset{\mathbf{p}E}{\overset{R}{\longleftarrow}} \;+\; \underset{\mathbf{p}E}{\overset{R}{\longleftarrow}} \left(\Sigma^R(E,\mathbf{p})\right) \underset{\mathbf{p}E}{\overset{R}{\longleftarrow}} \;. \tag{3.53}$$

In terms of the self-energy we therefore have for the impurity-averaged propagator the equation

$$G^R(\mathbf{p}, E) \;=\; G_0^R(\mathbf{p}, E) \;+\; G_0^R(\mathbf{p}, E)\, \Sigma^R(E, \mathbf{p})\, G^R(\mathbf{p}, E) \tag{3.54}$$

which we can solve to get

$$G^R(\mathbf{p}, E) \;=\; \frac{1}{G_0^{-1}(\mathbf{p}, E) - \Sigma^R(E, \mathbf{p})} \;=\; \frac{1}{E - \epsilon_{\mathbf{p}} - \Sigma^R(E, \mathbf{p})} \;. \tag{3.55}$$

The self-energy determines the analytic structure of the propagator, the location of the poles of the analytically continued propagator, and thereby the lifetime of (in the present case) momentum states. The effect of the random potential is clearly to give momentum states a finite lifetime (see also exercise 3.3 on page 152).

Fourier transforming eq.(3.55) we get in our isotropic model (in three spatial dimensions for the prefactor to be correct)

$$G_E^R(\mathbf{x} - \mathbf{x}') \;\equiv\; <G_E^R(\mathbf{x}, \mathbf{x}')> \;=\; \frac{-m}{2\pi\hbar^2}\, \frac{e^{\frac{i}{\hbar}|\mathbf{x}-\mathbf{x}'|\sqrt{2m(E-\Sigma^R(E,\, p_E \hat{\mathbf{p}}))}}}{|\mathbf{x}-\mathbf{x}'|} \tag{3.56}$$

where p_E is the solution of the equation

$$p_E = \sqrt{2m(E - \Sigma^R(E, p_E \hat{\mathbf{p}}))} \;. \tag{3.57}$$

3.4 Skeleton Diagrams

So far we only have a perturbative description of the self-energy; i.e., we have a representation of the self-energy as a functional of the free propagator $\Sigma_{\mathbf{p},E}[G_0^R]$. In a realistic description of a physical system, we always need to invoke the specifics of the problem in order to implement a controlled approximation. To this end we must study the actual correlations in the system, and it is necessary to have the self-energy expressed in terms of the impurity-averaged propagator. Coherent quantum processes correspond to an infinite repetition of bare processes, and the diagrammatic approach is precisely useful for capturing this feature, as irreducible summations are easily described diagrammatically. In order to achieve a description of the self-energy in terms of the full propagator, let us consider the perturbative expansion of the self-energy.

For any given self-energy diagram in the perturbative expansion, eq.(3.51), we also encounter self-energy diagrams with all possible self-energy decorations on internal lines; for example

$$\to \quad + \quad + \quad + \ldots$$

$$= \quad + \quad \Sigma^R(E,\mathbf{p}) \quad + \ldots$$

$$\to \quad + \quad \Sigma^R(E,\mathbf{p})$$

$$+ \quad \Sigma^R(E,\mathbf{p}) \quad \Sigma^R(E,\mathbf{p}) \quad + \quad \cdots$$

$$= \quad \mathbf{p}E \quad \longleftrightarrow \quad \mathbf{p}E . \tag{3.58}$$

We can uniquely classify all these self-energy decorations in the perturbative expansion according to whether the particle line can be cut into two, three, or more pieces by cutting particle lines (the step indicated by the second arrow in eq.(3.58)). We can therefore partially sum the self-energy diagrams according to the unique prescription: for a given self-energy diagram, remove all internal self-energy insertions, and substitute for the remaining bare particle propagator lines the full impurity-averaged propagator lines.[9] Through this partial summation of the origi-

---

[9]Synonymous names for the full Green's function or propagator are renormalized or dressed propagator.

nal perturbative expansion of the self-energy only so-called skeleton diagrams containing the full propagator will then appear. Since in the skeleton expansion we have removed self-energy insertions (decorations) which allowed a 1PI self-energy diagram to be cut in two by cutting two lines, we can characterize the skeleton expansion of the self-energy as the set of skeleton diagrams which can not be cut in two by cutting two lines (2PI-diagrams).[10]

By construction, only self-energy skeleton diagrams which can not be cut in two by cutting only two full propagator lines appear, and we have the partially summed diagrammatic expansion for the self-energy

$$\Sigma^R(\mathbf{p}, E) = \Sigma^R_{\mathbf{p}, E}[G^R] . \tag{3.60}$$

The summations are unique, since the initial and final impurity correlator lines are attached internally in different ways in each class of summed diagrams. No double counting of diagrams thus take place due to the different topology of the skeleton self-energy diagrams, and all diagrams in the perturbative expansion of the self-energy, eq.(3.51), are by construction contained in the skeleton diagrams of eq.(3.59).

What has been achieved by the partial summation, where each diagram corresponds to an infinite sum of terms in perturbation theory, is that the self-energy is expressed as a functional of the exact impurity-averaged propagator

$$\Sigma^R(\mathbf{p}, E) = \Sigma^R_{\mathbf{p}, E}[G^R] . \tag{3.60}$$

We can continue this topological classification, and introduce the higher order vertex functions; however, we defer this until chapter 8.

---

[10]Since propagator and impurity lines appear topologically equivalently, we can restate: the skeleton self-energy expansion consists of all the two-line irreducible skeleton diagrams.

Exercise 3.2 *Draw the rest of the 4 skeleton self-energy diagrams with three impurity correlators.*

3.5　Impurity-Averaged Propagator

In the next section we show that, for sufficiently high energies and momenta, we can for the self-energy neglect skeleton diagrams where impurity lines cross. We are thereby left with a single self-energy diagram[11]

$$\Sigma^R_{NCA}(E, \mathbf{p}) \quad \equiv \quad \text{[diagram]} \qquad (3.61)$$

in the Born approximation.[12] Analytically we have for the self-energy in the Born approximation

$$\Sigma^R_{NCA}(\mathbf{p}, E) = n_i \int \frac{d\mathbf{p}'}{(2\pi\hbar)^3} \, |V_{imp}(\mathbf{p} - \mathbf{p}')|^2 \, G^R(\mathbf{p}', E) \qquad (3.62)$$

which is an implicit expression since the propagator is in turn specified in terms of the self-energy.

In order to orient ourselves as to the effect of impurity scattering, we insert the free propagator into eq.(3.62) and obtain

$$\tilde{\Sigma}^R_{NCA}(\mathbf{p}, E) \quad \equiv \quad \text{[diagram]} \quad = \quad -i \, \frac{\hbar}{2\,\tau(E, \mathbf{p})} \quad + \quad \Re e \tilde{\Sigma}^R_{NCA}(\mathbf{p}, E) \qquad (3.63)$$

where we have introduced the momentum relaxation time (in the Born approximation).

$$\frac{\hbar}{\tau(E, \mathbf{p})} \quad = \quad 2\pi n_i N_0(E) \int \frac{d\hat{\mathbf{p}}'}{4\pi} \, |V_{imp}(\mathbf{p} - p_E \hat{\mathbf{p}}')|^2 \qquad (3.64)$$

where $p_E = \sqrt{2mE}$, and the real part of the self-energy is the principal value integral

$$\Re e \tilde{\Sigma}^R_{NCA}(\mathbf{p}, E) \quad = \quad n_i \, P \int_0^\infty d\epsilon_{p'} \, N_0(\epsilon_{p'}) \, \frac{1}{E - \epsilon_{p'}} \int \frac{d\hat{\mathbf{p}}'}{4\pi} \, |V_{imp}(\mathbf{p} - \sqrt{2m\epsilon_{p'}} \, \hat{\mathbf{p}}')|^2 \qquad (3.65)$$

---

[11]Here the index NCA simply stands for noncrossing approximation, but as mentioned, we establish its validity in the next section.

[12]In section 3.6 we show that inclusion of multiple scattering is handled with equal care.

where the P signifies that the principal value of the integral is to be taken. In this first iteration we thus have for the impurity-averaged propagator

$$\tilde{G}_0^R(E, \mathbf{p}) = \frac{1}{E - \epsilon_p - \Re\tilde{\Sigma}_{NCA}^R(\mathbf{p}, E) + i\hbar/2\tau(E, \mathbf{p})} . \qquad (3.66)$$

In the following we shall only be interested in the region of large energies $E \simeq E_F \gg \hbar/\tau$ and large momenta $p \simeq p_F \equiv \sqrt{2mE_F} \gg \hbar/l$, where $\tau \equiv \tau(E_F, \mathbf{p}_F)$, and we have introduced the impurity mean free path $l \equiv v_F\tau$, and the velocity $v_F \equiv p_F/m$.[13] We shall further assume that the impurity potential has a range, a, much shorter than the mean free path, $a \ll l$, say[14]

$$V_{imp}(\mathbf{r}) \propto \frac{1}{r} e^{-r/a} \qquad (3.67)$$

or for the Fourier transform

$$V_{imp}(\mathbf{k}) \propto \frac{4\pi}{\mathbf{k}^2 + a^{-2}} . \qquad (3.68)$$

In the region of energies $|E - E_F| \ll E_F$, and momenta $|p - p_F| \ll p_F$, the other relevant quantities for the calculation of the self-energy, density of states and impurity potential, are essentially constant, since they vary only on the large scales E_F and p_F, say $a \sim \hbar/p_F$. In the region of interest, the real part of the self-energy is thus essentially constant, only giving rise to an irrelevant shift in the reference for measuring the energy.[15] The first iterated propagator expression, eq.(3.66), is therefore as a function of momentum sharply peaked at the value p_F with the small width $\hbar/l$. The result for the imaginary part of the self-energy, eq.(3.64), is therefore unchanged, to order $\hbar/E_F\tau$, by the substitution of the improved propagator, $\tilde{G}_0^R$, instead of the free propagator. We therefore have, to order $\hbar/E_F\tau$, for the self-energy in the Born approximation, $|E - E_F| \ll E_F$, $|p - p_F| \ll p_F$,

$$\Sigma^R(E, \mathbf{p}) \equiv$$

$$= -\frac{i\hbar}{2\tau} . \qquad (3.69)$$

where

$$\frac{\hbar}{\tau} = 2\pi n_i N_0(E_F) \int \frac{d\hat{\mathbf{p}}'}{4\pi} \left| V_{imp}\left(\sqrt{2mE_F} (\hat{\mathbf{p}} - \hat{\mathbf{p}}') \right) \right|^2 . \qquad (3.70)$$

[13]The present single-particle problem has of course no built-in energy scale. However, when we eventually shall discuss the transport properties of degenerate fermions, such as electrons in a metal, say, a large energy scale will be provided, viz. the Fermi energy.

[14]We show in section that this is indeed the relevant case.

[15]For a delta-correlated random potential $<V(\mathbf{x})V(\mathbf{x}')> = u^2\,\delta(\mathbf{x} - \mathbf{x}')$ (see also eq.(3.103)), the real part of the self-energy $\Re\Sigma^R(E, \mathbf{p})$, though divergent, can be absorbed into an irrelevant renormalization of the energy.

For the impurity-averaged propagator we therefore obtain, $|E - E_F| \ll E_F$, $|p - p_F| \ll p_F$,

$$< G^R(E, \mathbf{p}, \mathbf{p}') > \; = \; G^R(E, \mathbf{p}) \, \delta_{\mathbf{p}, \mathbf{p}'} \qquad (3.71)$$

where

$$G^R(E, \mathbf{p}) = \frac{1}{E - \epsilon_\mathbf{p} + i\hbar/2\tau} \; . \qquad (3.72)$$

We note that this form of the propagator can be used for all values of E and $\mathbf{p}$ without violating the sum rules eq.(2.157) and eq.(2.158).

In the region of interest, $E \simeq E_F$, we have according to eq.(3.56) that the impurity-averaged propagator decays exponentially as a function of its spatial variable with the scale of the mean free path, and is in three spatial dimensions given by the expression

$$G_E^{R(A)}(\mathbf{x} - \mathbf{x}') \; \equiv \; < G_E^{R(A)}(\mathbf{x}, \mathbf{x}') > \; = \; -\frac{m}{2\pi\hbar^2} \frac{\exp\{|\mathbf{x} - \mathbf{x}'|\,(\overset{+}{\scriptstyle -})\, ik_E - 1/2l_E)\}}{|\mathbf{x} - \mathbf{x}'|}$$

$$= \; G_0^{R(A)}(\mathbf{x} - \mathbf{x}', E) \, e^{-|\mathbf{x} - \mathbf{x}'|/2l} \qquad (3.73)$$

where $k_E = \sqrt{2mE}/\hbar$ is the electronic wave vector, and $l_E = \hbar k_E \tau_E/m$ the impurity mean free path. The imaginary part of the self-energy leads to exponential damping of the propagator as a function of spatial separation. In the present case the damping of the propagator is due to the directional scattering of a plane wave (the spatial representation of the momentum eigenstate) due to the impurities, (say by passing a slab of material with impurities the amplitude of a plane wave is damped).

According to eq.(3.73) and eq.(2.157) the density of states is unchanged by the presence of weak disorder[16]:

$$N(E_F) \; = \; N_0(E_F) \left(1 + \mathcal{O}\left(\hbar/E_F\tau\right)\right) \; . \qquad (3.74)$$

Exercise 3.3 *Show that for $p \simeq p_F$ we have (to order $\hbar/E_F\tau$)*

$$G^R(\mathbf{p}, t) \; = \; \int_{-\infty}^{\infty} \frac{dE}{2\pi} \, e^{-\frac{i}{\hbar}Et} \, \frac{1}{E - \epsilon_\mathbf{p} + \frac{i\hbar}{2\tau(E,p)}} \; = \; G_0^R(\mathbf{p}, t) \, e^{-t/2\tau} \qquad (3.75)$$

i.e., τ is the momentum relaxation time (see also section 4.8.1 on page 193).

Exercise 3.4 *Obtain the expressions for the impurity averaged propagator in two spatial dimensions.*

---

[16]The specific heat of a degenerate Fermi gas is proportional to the density of states at the Fermi surface (see exercise 5.3 on page 208). The presence of weak disorder does therefore not change this result, which is not surprising in view of the scattering being elastic.

3.6 Diagram Estimation

In the skeleton expansion all internal propagators are the exact impurity-averaged propagator instead of the bare propagator as in the naive perturbation expansion in the potential. This is advantageous since it is the full propagator that reflects the physical properties of the system. The properties of the full propagator are determined by its analytic structure, its poles. When we wish to estimate the order of magnitude of the contribution of various diagrams, the quantity of interest to estimate is therefore the self-energy. This is the quantity determining the pole in the propagator, the singular point where a small change in variables makes a huge difference.

We shall only be interested in estimating the various contributions to the propagator for large energies and momenta, $p \simeq p_F \gg \hbar/l$, $E \simeq E_F \gg \hbar/\tau$. For $E \simeq E_F$, the impurity-averaged propagator, eq.(3.72), is therefore only large near the momentum value p_F where we have for the order of magnitude of the propagator $G \sim \tau/\hbar$, as we have in the Born approximation for the order of magnitude of the self-energy $\Sigma \sim \hbar/\tau$, a small value compared to E_F. For conduction electrons in a metal the large momentum value p_F is the Fermi momentum, and we shall in the following refer to the surface in the space of momentum values at the large momentum value p_F as the Fermi surface. The large contribution to a diagram therefore comes from the internal momentum integration regions where the momenta of the propagators are all on the Fermi surface.

Let us start the diagram estimation by establishing the criterion for the validity of neglecting multiple scattering. Consider for example the fourth-order (in the impurity potential) diagram for the skeleton self-energy

$$
\begin{aligned}
pE \quad \underset{\mathbf{p}_3 E \quad \mathbf{p}_2 E \quad \mathbf{p}_1 E}{\longleftarrow\!\!\!\!\bullet\!\!\!\!-\!\!\!\!\bullet\!\!\!\!-\!\!\!\!\bullet\!\!\!\!\longrightarrow} \quad pE \;=\; & n_i \int \frac{d\mathbf{p}_3}{(2\pi\hbar)^3} \int \frac{d\mathbf{p}_2}{(2\pi\hbar)^3} \int \frac{d\mathbf{p}_1}{(2\pi\hbar)^3} V_{imp}(\mathbf{p} - \mathbf{p}_3) \\
& V_{imp}(\mathbf{p}_3 - \mathbf{p}_2) V_{imp}(\mathbf{p}_2 - \mathbf{p}_1) V_{imp}(\mathbf{p}_1 - \mathbf{p}) \\
& G^R(\mathbf{p}_3, E) G^R(\mathbf{p}_2, E) G^R(\mathbf{p}_1, E) \qquad (3.76)
\end{aligned}
$$

The impurity concentration appears to first order, giving for one of the integrations a self-energy type contribution, and since the Fourier transform of the impurity potential is assumed slowly varying we have the estimate

$$
pE \quad \underset{\mathbf{p}_3 E \quad \mathbf{p}_2 E \quad \mathbf{p}_1 E}{\longleftarrow\!\!\!\!\bullet\!\!\!\!-\!\!\!\!\bullet\!\!\!\!-\!\!\!\!\bullet\!\!\!\!\longrightarrow} \quad pE \;\sim\; \left[V_{imp}(\mathbf{p} = 0) \int \frac{d\mathbf{p}'}{(2\pi\hbar)^3} G^R(\mathbf{p}', E) \right]^2
$$

$$
n_i \int \frac{d\mathbf{p}_1}{(2\pi\hbar)^3} |V_{imp}(\mathbf{p}_1 - \mathbf{p})|^2 G^R(\mathbf{p}_1, E) \; . \quad (3.77)
$$

For the momentum integral of the impurity-averaged propagator we have

$$\int \frac{d\mathbf{p}}{(2\pi\hbar)^3} \, G^R(\mathbf{p}, E) \ \sim \ iN_0(E_F) \ \sim \ \frac{mp_F}{\hbar^3} \tag{3.78}$$

and we obtain the estimate

$$\sim \ \left(\frac{V_{imp}(\mathbf{p}=0)p_F^2}{\hbar^3 v_F} \right)^2 \frac{\hbar}{\tau} . \tag{3.79}$$

In the case where the range of the impurity potential is specified in terms of p_F, $a \sim \hbar/p_F$, the Born criterion is (see for example reference [13])

$$p_F^3 \, V_{imp}(\mathbf{p}=0) \ll \hbar^3 \, E_F . \tag{3.80}$$

We therefore have that a multiple scattering term is small relative to the Born term when the Born criterion is satisfied

$$\sim \left(\frac{V_{imp}(\mathbf{p}=0)p_F^2}{\hbar^3 v_F} \right)^2 \ll \left(\frac{E_F}{v_F p_F} \right)^2 \sim 1 . \tag{3.81}$$

To show that the crossed self-energy diagrams can be neglected when calculating the propagator, we note that the intermediate momentum integrations in a self-energy diagram with crossed impurity lines ($E \simeq E_F, p \simeq p_F$)

$$= n_i^2 \int \frac{d\mathbf{p}_3}{(2\pi\hbar)^3} \int \frac{d\mathbf{p}_1}{(2\pi\hbar)^3} \, |V_{imp}(\mathbf{p} - \mathbf{p}_3)|^2 |V_{imp}(\mathbf{p}_1 - \mathbf{p})|^2$$

$$G^R(\mathbf{p}_3, E)G^R(\mathbf{p}_1 + \mathbf{p}_3 - \mathbf{p}, E)G^R(\mathbf{p}_1, E) \tag{3.82}$$

for the region of large contribution, are not free. In order for all the momenta of the propagators to be in the thin shell of extension $\hbar/l$ around the sphere in momentum space with radius p_F, where the large contribution arises, one of the angular integrations is restricted to a cone of angle $\hbar/p_F l$. The crossed diagrams will therefore be relatively smaller by the same factor compared to the diagram where impurity lines do not cross and no angular restriction occurs[17]

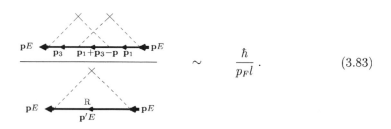

$$\sim \quad \frac{\hbar}{p_F l} \, . \qquad (3.83)$$

Using the diagrammatic technique we have thus achieved the goal of identifying a parameter on which we can base a perturbation theory. A nontrivial perturbation expansion as the expansion parameter $\hbar/p_F l$ is not a parameter in the Hamiltonian.

3.7 Multiple Scattering

If we relax the Born restriction, eq.(3.80), we would to order $\hbar/p_F l \ll 1$ have for the impurity-averaged propagator

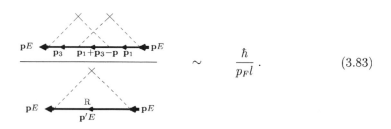

The result beyond the Born approximation is simply obtained by summing all the multiple scatterings, and we will everywhere instead of the Born amplitude

---

[17]The argument being based on an angular integration restriction is thus not valid for the case of one spatial dimension.

$V_{imp}(\mathbf{p} - \mathbf{p}')$ encounter the t-matrix for scattering off a single impurity, where the t-matrix satisfies the equation

$$t^R_{\mathbf{pp}'}(E) = n_i V_{imp}(\mathbf{p} - \mathbf{p}') + \frac{1}{V} \sum_{\mathbf{p}''} V_{imp}(\mathbf{p} - \mathbf{p}'')\, G^R(\mathbf{p}'', E)\, t^R_{\mathbf{p}''\mathbf{p}'}(E) \qquad (3.85)$$

corresponding to the diagrams for the t-matrix

$$(3.86)$$

The above t-matrix equation differs from the one for scattering off a single impurity in free space in that the impurity-averaged propagator appears instead of the free propagator, reflecting the presence of the other impurities.

Except for the obvious connection to the previous notation there is no logic to tying the potential lines together since we are concerned with multiple scattering off a single impurity, so we could equally well use the depiction[18]

$$= V_{imp}(\mathbf{p} - \mathbf{p}') + \frac{1}{V} \sum_{\mathbf{p}_1} V_{imp}(\mathbf{p} - \mathbf{p}_1)\, G^R(\mathbf{p}_1, E)\, V_{imp}(\mathbf{p}_1 - \mathbf{p}')$$

---

[18]If the propagator is interpreted as the free propagator, the t-matrix describes the scattering of a particle by a potential.

$$+ \ \frac{1}{V^2} \sum_{\mathbf{p}_1, \mathbf{p}_2} V_{imp}(\mathbf{p} - \mathbf{p}_2) \, G^R(\mathbf{p}_2, E) \, V_{imp}(\mathbf{p}_2 - \mathbf{p}_1)$$

$$G^R(\mathbf{p}_1, E) \, V_{imp}(\mathbf{p}_1 - \mathbf{p}') \ + \ \dots \tag{3.87}$$

and extract the factor of n_i from the previous definition of the t-matrix. For the self-energy we then have in the noncrossing approximation

$$\Sigma^R(\mathbf{p}, E) \ = \ \frac{n_i}{V} \sum_{\mathbf{p}'} \ V_{imp}(\mathbf{p} - \mathbf{p}') \, G^R(\mathbf{p}', E) \, t^R_{\mathbf{p}'\mathbf{p}}(E)$$

$$= \quad pE \xleftarrow{\quad} \underset{\mathbf{p}'E \quad E}{\overset{R \qquad R}{\longleftarrow}} pE \tag{3.88}$$

and for the imaginary part (recall eq.(3.85))

$$\Im m \Sigma^R(\mathbf{p}, E) \ = \ n_i \, \Im m \, t^R_{\mathbf{p}\mathbf{p}}(E) \ . \tag{3.89}$$

Similarly we encounter when we consider the advanced propagator

$$t^A_{\mathbf{p}\mathbf{p}'}(E) = n_i \, V_{imp}(\mathbf{p} - \mathbf{p}') + \frac{1}{V} \sum_{\mathbf{p}''} \ t^A_{\mathbf{p}\mathbf{p}''}(E) \, G^A(\mathbf{p}'', E) \, V_{imp}(\mathbf{p}'' - \mathbf{p}') \tag{3.90}$$

corresponding to the same set of diagrams, but with the impurity-averaged advanced propagator appearing instead of the retarded

$$p \xleftarrow{\ \ \overset{A}{\ } \ } p' \ = \ p \xleftarrow{\ \ \underset{E}{\times} \ } p' \ + \ p \xleftarrow{\ \ \overset{A}{\underset{\mathbf{p}''E}{E}} \ } p' \quad . \tag{3.91}$$

We note the hermitian property

$$t^A_{\mathbf{p}\mathbf{p}'}(E) = [t^R_{\mathbf{p}'\mathbf{p}}(E)]^* \tag{3.92}$$

and as always

$$\Sigma^A(\mathbf{p}, E) = [\Sigma^R(\mathbf{p}, E)]^* \ . \tag{3.93}$$

Combining the two t-matrix equations we get (extracting the factor n_i from the definition of the t-matrix)

$$t^R_{\mathbf{p}\mathbf{p}'}(E) \ = \ V_{imp}(\mathbf{p} - \mathbf{p}') + \frac{1}{V} \sum_{\mathbf{p}''} \ t^A_{\mathbf{p}\mathbf{p}''}(E) \, G^R(\mathbf{p}'', E) \, t^R_{\mathbf{p}''\mathbf{p}'}(E)$$

$$- \ \frac{1}{V^2} \sum_{\mathbf{p}_1 \mathbf{p}_2} t^A_{\mathbf{p}\mathbf{p}_1}(E) \, G^A(\mathbf{p}_1, E) \, V_{imp}(\mathbf{p}_1 - \mathbf{p}_2) \, G^R(\mathbf{p}_2, E) \, t^R_{\mathbf{p}_2\mathbf{p}'}(E) \tag{3.94}$$

and similarly

$$t_{\mathbf{pp'}}^A(E) = V_{imp}(\mathbf{p} - \mathbf{p'}) + \frac{1}{V}\sum_{\mathbf{p''}} t_{\mathbf{pp''}}^A(E)\, G^A(\mathbf{p''}, E)\, t_{\mathbf{p''p'}}^R(E)$$

$$- \frac{1}{V^2}\sum_{\mathbf{p_1 p_2}} t_{\mathbf{pp_1}}^A(E)\, G^A(\mathbf{p_1}, E)\, V_{imp}(\mathbf{p_1} - \mathbf{p_2})\, G^R(\mathbf{p_2}, E)\, t_{\mathbf{p_2 p'}}^R(E)\ . \quad (3.95)$$

Subtracting the diagonal terms we get

$$\Im m\, t_{\mathbf{pp}}^R(E) = -\frac{i}{2}[t_{\mathbf{pp}}^R(E) - t_{\mathbf{pp}}^A(E)] = \frac{1}{V}\sum_{\mathbf{p'}} |t_{\mathbf{pp'}}^R(E)|^2\, \Im m G^R(\mathbf{p'}, E) \quad (3.96)$$

and thereby for the imaginary part of the self-energy

$$\Im m\Sigma^R(\mathbf{p}, E) = \frac{n_i}{V}\sum_{\mathbf{p'}} |t_{\mathbf{pp'}}^R(E)|^2\, \Im m G^R(\mathbf{p'}, E)\ . \quad (3.97)$$

Assuming that the t-matrix is a slowly varying function of momentum on the scale $\hbar/l$ (i.e., no resonances), we get to lowest order in $\hbar/pl$[19]

$$\Im m\Sigma^R(\mathbf{p}, E) = -\frac{\hbar}{2\tau(\mathbf{p}, E)} \quad (3.98)$$

where ($p' = \sqrt{2mE}$)

$$\frac{\hbar}{\tau(\mathbf{p}, E)} = 2\pi n_i N_0(E) \int\frac{d\hat{\mathbf{p}}'}{4\pi}\, |t_{\mathbf{pp'}}^R(E)|^2\ . \quad (3.99)$$

The only change from the Born approximation, eq.(3.64), being that the exact differential cross section for scattering off an impurity appears instead of the Born expression

$$\equiv\ n_i\, |V_{imp}(\mathbf{p_1} - \mathbf{p_1'})|^2\, \delta_{\mathbf{p_1} + \mathbf{p_2}, \mathbf{p_1'} + \mathbf{p_2'}}$$

---

[19]For the case of scattering of a particle off a potential, i.e., the propagator is free, we obtain from eq.(3.99) and exercise 3.3 on page 152 that $|t_{\mathbf{pp'}}^R(\epsilon_{\mathbf{p}})m/2\pi\hbar^2|^2$ is the exact differential cross section. The imaginary part of the diagonal part of the t-matrix, the forward scattering amplitude, is thus proportional to the total elastic scattering cross section σ, $\Im m t_{\mathbf{pp}}^R(\epsilon_{\mathbf{p}}) = \sigma p/4\pi\hbar$, the optical theorem.

$$\rightarrow \qquad \equiv \quad n_i \, |t^R_{\mathbf{p}_1-\mathbf{p}'_1}(E)|^2 \, \delta_{\mathbf{p}_1+\mathbf{p}_2,\mathbf{p}'_1+\mathbf{p}'_2} \, . \qquad (3.100)$$

Going beyond the Born approximation thus adds nothing qualitatively new if there are no resonances. Furthermore, the topological structure of the diagrams of the disorder problem is also unchanged by including multiple scattering off the same impurity, as the t-matrix correlator simply appears instead of the Gaussian impurity correlator.

3.8   Gaussian Approximation

According to the analysis of impurity averaging the propagator in momentum space, we realize that if we can neglect multiple scattering, and are considering the thermodynamic limit, $(N-1)/V \simeq N/V$, the impurity average corresponds to tying potential vertices pairwise together, through the impurity correlator, in all possible ways. In this approximation the impurity average is a Gaussian average in the random variable, the potential. Fourier-transforming, we find that in the position representation the Gaussian average is done according to the specification (for an even number of potential terms)

$$< V(\mathbf{x}_1)V(\mathbf{x}_2) .. V(\mathbf{x}_{2N-1})V(\mathbf{x}_{2N}) >$$

$$= \sum_P < V(\mathbf{x}_{P(1)})V(\mathbf{x}_{P(2)}) > \ .. .. < V(\mathbf{x}_{P(2N-1)})V(\mathbf{x}_{P(2N)}) > \qquad (3.101)$$

where we sum over all permutations P. We have assumed that the mean value of the potential is zero, $< V(\mathbf{x}) > \, = 0$, so that only fully contracted terms are nonvanishing (diagrams with dangling impurity lines vanish). For an odd number of potential vertices, we always encounter a dangling impurity line, and the contribution is proportional to the mean value of the potential, which we have chosen to be zero.

When impurity averaging a diagram in the position representation, we therefore get the sum of diagrams with all possible pairings of potential vertices by the impurity correlator in real space

$$= \frac{1}{\hbar^2} < V(\mathbf{x})V(\mathbf{x}') > \qquad (3.102)$$

which is specified by

$$
\begin{aligned}
< V(\mathbf{x})V(\mathbf{x}') > \; &= \; \frac{1}{V^2} \sum_{\mathbf{p},\mathbf{p}'} e^{\frac{i}{\hbar}(\mathbf{p}\cdot\mathbf{x}+\mathbf{p}'\cdot\mathbf{x}')} \, V_{imp}(\mathbf{p}) \, V_{imp}(\mathbf{p}') \, \langle \, \sum_{i,j=1}^{N} e^{-\frac{i}{\hbar}(\mathbf{p}\cdot\mathbf{r}_i+\mathbf{p}'\cdot\mathbf{r}_j)} \, \rangle \\[2mm]
&= \; \frac{N}{V^2} \sum_{\mathbf{p}} e^{\frac{i}{\hbar}\mathbf{p}\cdot(\mathbf{x}-\mathbf{x}')}|V_{imp}(\mathbf{p})|^2 \; + \; \frac{N(N-1)}{V^2}|V_{imp}(\mathbf{p}=0)|^2 \\[2mm]
&= \; \frac{N}{V^2} \sum_{\mathbf{p}} e^{\frac{i}{\hbar}\mathbf{p}\cdot(\mathbf{x}-\mathbf{x}')}|V_{imp}(\mathbf{p})|^2 \\[2mm]
&= \; n_i \int d\mathbf{r} \, V_{imp}(\mathbf{x}-\mathbf{r}) V_{imp}(\mathbf{x}'-\mathbf{r}) \; .
\end{aligned}
\tag{3.103}
$$

where we in the third equality have used that our reference for measuring energies is such that the spatial average of the potential vanishes. Inversely we have

$$
\begin{aligned}
< V(\mathbf{p}_1-\mathbf{p}_1')V(\mathbf{p}_2-\mathbf{p}_2') > \; &= \; \int d\mathbf{x} \int d\mathbf{x}' \, e^{-\frac{i}{\hbar}\mathbf{x}\cdot(\mathbf{p}_1-\mathbf{p}_1')-\frac{i}{\hbar}\mathbf{x}'\cdot(\mathbf{p}_2-\mathbf{p}_2')} < V(\mathbf{x})V(\mathbf{x}') > \\[2mm]
&= \; N \, |V_{imp}(\mathbf{p}_1-\mathbf{p}_1')|^2 \, \delta_{\mathbf{p}_1+\mathbf{p}_2,\mathbf{p}_1'+\mathbf{p}_2'}
\end{aligned}
\tag{3.104}
$$

the impurity correlator in the momentum representation.

A delta-correlated random potential

$$
< V(\mathbf{x})\, V(\mathbf{x}') > \; = \; u^2 \, \delta(\mathbf{x}-\mathbf{x}')
\tag{3.105}
$$

corresponds, according to eq.(3.103), to the limit of dense point scatterers.

The self-energy in the Born approximation is in the position representation specified in terms of the impurity correlator

$$
\Sigma^R(\mathbf{x},\mathbf{x}'; E) \; = \; \Sigma^R(\mathbf{x}-\mathbf{x}'; E) \; \equiv \; \mathbf{x} \underset{E}{\overset{R}{\longleftrightarrow}} \mathbf{x}'
$$

$$
= \; < V(\mathbf{x})V(\mathbf{x}') > G_E^R(\mathbf{x},\mathbf{x}') \; .
\tag{3.106}
$$

For the case where the impurity potential is short ranged compared to the mean free path, $a \sim \hbar/p_F$, the impurity correlator is, according to eq.(3.103), essentially a delta function, and

$$
\Sigma^R(\mathbf{x},\mathbf{x}'; E) = \Sigma^R(\mathbf{x}-\mathbf{x}'; E) \; = \; u^2 \, \delta(\mathbf{x}-\mathbf{x}') \, G_E^R(\mathbf{x},\mathbf{x}')
\tag{3.107}
$$

where

$$
u^2 \; = \; n_i \int \frac{d\hat{\mathbf{p}}}{4\pi} \, \left| V_{imp}\left(\sqrt{2mE_F}\,\hat{\mathbf{p}}\right) \right|^2 \; .
\tag{3.108}
$$

At large energies, $E \simeq E_F$, we then have according to eq.(3.73), that the imaginary part of the retarded self-energy (the real part is an irrelevant infinite constant) is given by

$$\Im m \Sigma^R(\mathbf{x}, \mathbf{x}'; E) \quad = \quad -\frac{\hbar}{2\tau} \, \delta(\mathbf{x} - \mathbf{x}') \; . \tag{3.109}$$

In the momentum representation we encountered products of propagators and self-energies. Upon Fourier-transforming, these turn into convolutions in the position variable.

3.9 Motion in a Random Potential

We now have all the ingredients enabling us to perform perturbative calculations for the motion of a particle in a random potential. We introduce the impurity-averaged density matrix in the position representation

$$f(\mathbf{x}, \mathbf{x}', t) \quad \equiv \quad <\rho(\mathbf{x}, \mathbf{x}', t)> \; = \; <<\mathbf{x}|\hat{\rho}(t)|\mathbf{x}'>> \; . \tag{3.110}$$

A diagonal element has the interpretation: Start a particle off at the same point in each impurity sample; then $f(\mathbf{x}, \mathbf{x}, t)$ gives the probability that we by an arbitrary pick of sample will find a particle at position $\mathbf{x}$ at time t.

In the following we shall obtain a perturbative description of the time evolution of the impurity-averaged density matrix in terms of an integral equation. Let us assume that at time t' the particle is described by the statistical operator $\hat{\rho}(t')$. The impurity-averaged density matrix at time t

$$f(\mathbf{x}, \mathbf{x}', t) \quad = \quad \int d\tilde{\mathbf{x}} \int d\tilde{\mathbf{x}}' < J(\mathbf{x}, \mathbf{x}', t; \tilde{\mathbf{x}}, \tilde{\mathbf{x}}', t') > \rho'(\tilde{\mathbf{x}}, \tilde{\mathbf{x}}') \tag{3.111}$$

is then expressed in terms of the density matrix at time t'

$$\rho'(\tilde{\mathbf{x}}, \tilde{\mathbf{x}}') \quad \equiv \quad \rho(\tilde{\mathbf{x}}, \tilde{\mathbf{x}}', t') \; = \; <\tilde{\mathbf{x}}|\hat{\rho}(t')|\tilde{\mathbf{x}}'> \tag{3.112}$$

and the impurity-averaged density-matrix propagator equals

$$< J(\mathbf{x}, \mathbf{x}', t; \tilde{\mathbf{x}}, \tilde{\mathbf{x}}', t') > \quad = \quad < G_0^R(\mathbf{x}, t; \tilde{\mathbf{x}}, t') \, G_0^A(\tilde{\mathbf{x}}', t'; \mathbf{x}', t) >$$

$$= \quad \sum_{n=0}^{\infty} (\frac{1}{\hbar})^n \sum_{m=0}^{\infty} (\frac{1}{\hbar})^m \int \prod_{n'=1}^{n} dx_{n'} \int_{-\infty}^{\infty} \prod_{n'=1}^{n} dt_{n'} \int \prod_{m'=1}^{m} d\tilde{x}_{m'} \int_{-\infty}^{\infty} \prod_{m'=1}^{m} d\tilde{t}_{m'}$$

$$G_0^R(\mathbf{x}, t; \mathbf{x}_n, t_n) G_0^R(\mathbf{x}_n, t_n; \mathbf{x}_{n-1}, t_{n-1})..G_0^R(\mathbf{x}_1, t_1; \tilde{\mathbf{x}}, t')$$

$$G_0^A(\tilde{\mathbf{x}}', t'; \tilde{\mathbf{x}}_m, \tilde{t}_m) G_0^A(\tilde{\mathbf{x}}_m, \tilde{t}_m; \tilde{\mathbf{x}}_{m-1}, \tilde{t}_{m-1})..G_0^A(\tilde{\mathbf{x}}_1, \tilde{t}_1; \mathbf{x}', t)$$

$$< V(\tilde{\mathbf{x}}_m) V(\mathbf{x}_{m-1})..V(\tilde{\mathbf{x}}_1) V(\mathbf{x}_n) V(\mathbf{x}_{n-1})..V(\mathbf{x}_1) > \tag{3.113}$$

where we have assumed $t > t'$.[20] A virtue of the diagrammatic approach is that we shall not have to deal with unwieldy perturbative expressions as the above!

Before the impurity average has been performed, the perturbative expansion of the density matrix corresponds to the sum of diagrams with any number of potential vertices on the upper and lower lines as depicted in eq.(2.246). In the approximation where we neglect multiple scattering we get the impurity-averaged density matrix by tying all the potential vertices pairwise together, the Gaussian average, and obtain for the impurity-averaged density matrix the diagrammatic expansion

$$(3.114)$$

In the diagrammatic language the perturbative structure of the impurity-averaged density matrix is thus easy to grasp. We recall that on the upper line only retarded propagators appear, and on the lower line only advanced propagators.

We have drawn impurity correlators connecting upper and lower particle lines vertically, although their space-time points are different, as the presence of the retarded and advanced propagators makes the distinct topology of a diagram correspond to a unique virtual physical process. The fifth and eight diagram on the right-hand side of eq.(3.114) thus represents two physically different processes. The first contributes to the classical probability, while the other, as we shall see, represents a quantum interference process with which we shall be particularly concerned later.

Every particle line can be dressed by self-energy insertions so that the exact impurity-averaged propagator appears everywhere, and we get the skeleton dia-

---

[20]For $t < t'$ the retarded and advanced labels are interchanged, and we are concerned with retrodicting the state of the particle at previous times consistent with our knowledge of the present.

grammatic expansion for the impurity-averaged density matrix

We shall return to study this integral equation for the impurity-averaged density matrix in chapter 8.

In order to include multiple scattering, we only need to substitute the t-matrix for the impurity correlator

$$< V(\mathbf{x})V(\mathbf{x}') > = \quad \rightarrow \quad = t_E^R(\mathbf{x}_1, \mathbf{x}_1')t_{E'}^A(\mathbf{x}_2, \mathbf{x}_2')$$

(3.116)

here specified in terms of the Fourier transform with respect to time, and the t-matrix is carrying either an R or an A label depending on the attachment.

Chapter 4

Kinetics in a Random Potential

In this chapter the motion of a particle in a random potential is studied. The general quantum kinetic equation is discussed in terms of diagrams. The Wigner function is introduced, and a criterion for the validity of the classical kinetic equation is established. This Boltzmann equation for a particle in a random potential is then used to calculate the mobility.

At the end of the preceding chapter, we examined the time evolution of the impurity-averaged density matrix on integral form, and described the diagrammatic expansion in the spatial representation. In the following we shall pursue the kinetic approach, which amounts to studying the differential time evolution of the impurity-averaged density matrix. This will permit us to eliminate the explicit appearance of the initial density matrix, and thereby pave the way for an easy description of steady-state situations, so important to transport theory. In order to do this, we now display the spatial representation of the von Neumann equation and interpret it diagrammatically.

4.1 Density Matrix Equation of Motion

From the von Neumann equation for the statistical operator, eq.(1.384), for a particle in a potential V

$$i\hbar \frac{d\hat{\rho}(t)}{dt} = [\hat{H}_p + \hat{V}, \hat{\rho}(t)] \tag{4.1}$$

we obtain in the spatial representation the equation of motion for the density matrix

$$\frac{\partial \rho(\mathbf{x}, \mathbf{x}', t)}{\partial t} - \frac{i}{\hbar} <\mathbf{x}|[\hat{\rho}(t), \hat{H}_p]|\mathbf{x}'> = -\frac{i}{\hbar}\left(\hat{V}(\mathbf{x}) - \hat{V}(\mathbf{x}')\right)\rho(\mathbf{x}, \mathbf{x}', t) \tag{4.2}$$

where we assume the potential V (later to be taken random) to be static. In the case where the particle in addition is exposed to a time-dependent field U we have the particle Hamiltonian

$$\hat{H}_p = \hat{H}_t^p = \frac{\hat{\mathbf{p}}^2}{2m} - U(\hat{\mathbf{x}}, t) \tag{4.3}$$

and we get for the terms due to the kinetic and potential energy of the particle, the drift and driving terms,

$$<\mathbf{x}|[\hat{\rho}(t), \hat{H}_t^p]|\mathbf{x}'> \;=\; \int d\bar{\mathbf{x}} \left(\rho(\mathbf{x}, \bar{\mathbf{x}}, t) <\bar{\mathbf{x}}|\hat{H}_t^p|\mathbf{x}'> - <\mathbf{x}|\hat{H}_t^p|\bar{\mathbf{x}}> \rho(\bar{\mathbf{x}}, \mathbf{x}', t) \right)$$

$$=\; \left(\frac{\hbar^2}{2m} (\triangle_\mathbf{x} - \triangle_{\mathbf{x}'}) - U(\mathbf{x}, t) + U(\mathbf{x}', t) \right) \rho(\mathbf{x}, \mathbf{x}', t) \;. \quad (4.4)$$

Expressing the density matrix on the right-hand side of the spatial representation of the von Neumann equation, eq.(4.2), in terms of the propagator of the density matrix, eq.(2.236), we get the integro-differential equation for the density matrix

$$\left(\frac{\partial}{\partial t} - \frac{i}{\hbar} \left(\frac{\hbar^2}{2m} (\triangle_\mathbf{x} - \triangle_{\mathbf{x}'} - U(\mathbf{x}, t) + U(\mathbf{x}', t)) \right) \right) \rho(\mathbf{x}, \mathbf{x}', t) = -iF_{\mathbf{x}, \mathbf{x}', t}[\rho'] \quad (4.5)$$

where

$$F_{\mathbf{x}, \mathbf{x}', t}[\rho'] \equiv \int d\tilde{\mathbf{x}} \int d\tilde{\mathbf{x}}' \, \bar{J}(\mathbf{x}, \mathbf{x}', t; \tilde{\mathbf{x}}, \tilde{\mathbf{x}}', t') \, \rho(\tilde{\mathbf{x}}, \tilde{\mathbf{x}}', t') \quad (4.6)$$

is a functional of the density matrix at time t', and we have introduced

$$\bar{J}(\mathbf{x}, \mathbf{x}', t; \tilde{\mathbf{x}}, \tilde{\mathbf{x}}', t') \;\equiv\; \frac{1}{\hbar} \left(V(\mathbf{x}) - V(\mathbf{x}') \right) J(\mathbf{x}, \mathbf{x}', t; \tilde{\mathbf{x}}, \tilde{\mathbf{x}}', t') \;. \quad (4.7)$$

At this point there is of course no reason to treat the static potential on a special footing, and it could equally well appear on the left-hand side through the substitution $U \to U + V$. However, as anticipated we shall shortly treat the static potential as a random variable.

Using the perturbative expansion of the density-matrix propagator, eq.(2.245), we get the same perturbative structure for $\bar{J}$ as for J, except for an additional potential factor

$$\bar{J}(\mathbf{x}, \mathbf{x}', t; \tilde{\mathbf{x}}, \tilde{\mathbf{x}}', t') \;=\; \sum_{n,m=0}^{\infty} \frac{1}{\hbar^{n+m+1}} \int \prod_{n'=1}^{n} d\mathbf{x}_{n'} \int_{-\infty}^{\infty} \prod_{n'=1}^{n} dt_{n'} \, G_0^R(\mathbf{x}, t; \mathbf{x}_n, t_n)$$

$$G_0^R(\mathbf{x}_n, t_n; \mathbf{x}_{n-1}, t_{n-1}) \;..\; G_0^R(\mathbf{x}_1, t_1; \tilde{\mathbf{x}}, t')$$

$$\int \prod_{m'=1}^{m} d\tilde{\mathbf{x}}_{m'} \int_{-\infty}^{\infty} \prod_{m'=1}^{m} d\tilde{t}_{m'} \, G_0^A(\tilde{\mathbf{x}}', t'; \tilde{\mathbf{x}}_m, \tilde{t}_m)$$

$$G_0^A(\tilde{\mathbf{x}}_m, \tilde{t}_m; \tilde{\mathbf{x}}_{m-1}, \tilde{t}_{m-1}) \;..\; G_0^A(\tilde{\mathbf{x}}_1, \tilde{t}_1; \mathbf{x}', t)$$

$$V(\tilde{\mathbf{x}}_m) \;..\; V(\tilde{\mathbf{x}}_1) V(\mathbf{x}) V(\mathbf{x}_n) \;..\; V(\mathbf{x}_1)$$

$$-\; \sum_{n,m=0}^{\infty} \frac{1}{\hbar^{n+m+1}} \int \prod_{m'=1}^{m} d\tilde{\mathbf{x}}_{m'} \int_{-\infty}^{\infty} \prod_{m'=1}^{m} d\tilde{t}_{m'} \int \prod_{n'=1}^{n} d\mathbf{x}_{n'} \int_{-\infty}^{\infty} \prod_{n'=1}^{n} dt_{n'}$$

$$G_0^R(\mathbf{x}, t; \mathbf{x}_n, t_n) G_0^R(\mathbf{x}_n, t_n; \mathbf{x}_{n-1}, t_{n-1}) \ .. \ G_0^R(\mathbf{x}_1, t_1; \tilde{\mathbf{x}}, t')$$

$$G_0^A(\tilde{\mathbf{x}}', t'; \tilde{\mathbf{x}}_m, \tilde{t}_m) G_0^A(\tilde{\mathbf{x}}_m, \tilde{t}_m; \tilde{\mathbf{x}}_{m-1}, \tilde{t}_{m-1}) \ .. \ G_0^A(\tilde{\mathbf{x}}_1, \tilde{t}_1; \mathbf{x}', t)$$

$$V(\tilde{\mathbf{x}}_m) \ .. \ V(\tilde{\mathbf{x}}_1) V(\mathbf{x}') V(\mathbf{x}_n) \ .. \ V(\mathbf{x}_1) \ . \tag{4.8}$$

The additional potential factor can diagrammatically be depicted as appearing first on the upper line for the first term, representing $V(\mathbf{x})/\hbar$, and for the second term as last on the lower line, representing $V(\mathbf{x}')/\hbar$, (we shall in the following avoid repeating identical space-time labeling, and be dropping dots for indication of space-time points)

$$\bar{J}(\mathbf{x}, \mathbf{x}', t; \tilde{\mathbf{x}}, \tilde{\mathbf{x}}, t') =$$

The presence of both plus and minus signs simply reflects the commutator in the von Neumann equation, or explicitly the sign in eq.(4.7). In the last equality we have introduced the perturbative expansions of the propagators.

The diagrammatic expansion of F is, according to eq.(4.6), obtained by attaching the density matrix at time t' according to the defining equations, eq.(4.6) and eq.(4.7),

$$F_{\mathbf{x},\mathbf{x}',t}[\rho'] =$$

$$\tag{4.10}$$

where we in the last equality have displayed the perturbative expansion in the potential V.

4.2   Impurity-Averaged Density Matrix

We now treat the potential as a random variable. Taking the impurity average of the von Neumann equation we get for the time evolution of the impurity-averaged density matrix

$$\left(\frac{\partial}{\partial t} - \frac{i}{\hbar}\left(\frac{\hbar^2}{2m}\left(\triangle_{\mathbf{x}} - \triangle_{\mathbf{x}'}\right) - U(\mathbf{x},t) + U(\mathbf{x}',t)\right)\right) < \rho(\mathbf{x},\mathbf{x}',t) >$$

$$= -i \int d\tilde{\mathbf{x}} \int d\tilde{\mathbf{x}}' < \bar{J}(\mathbf{x},\mathbf{x}',t;\tilde{\mathbf{x}},\tilde{\mathbf{x}}',t') > f_i(\tilde{\mathbf{x}},\tilde{\mathbf{x}}') \tag{4.11}$$

where the bracket, recall eq.(3.9), denotes the impurity average and

$$< \bar{J}(\mathbf{x},\mathbf{x}',t;\tilde{\mathbf{x}},\tilde{\mathbf{x}}',t') > \equiv \frac{1}{\hbar} < (V(\mathbf{x}) - V(\mathbf{x}')) J(\mathbf{x},\mathbf{x}',t;\tilde{\mathbf{x}},\tilde{\mathbf{x}}',t') > \tag{4.12}$$

and we have introduced the notation for the density matrix at some initial time t'

$$f_i(\mathbf{x},\mathbf{x}') \equiv \rho(\mathbf{x},\mathbf{x}',t') \equiv \rho'(\mathbf{x},\mathbf{x}') . \tag{4.13}$$

Denoting the impurity-averaged density matrix by f

$$f(\mathbf{x},\mathbf{x}',t) \equiv < \rho(\mathbf{x},\mathbf{x}',t) > \tag{4.14}$$

the equation of motion for the impurity-averaged density matrix becomes

$$\left(\frac{\partial}{\partial t} - \frac{i}{\hbar}\left(\frac{\hbar^2}{2m}\left(\triangle_{\mathbf{x}} - \triangle_{\mathbf{x}'}\right) - U(\mathbf{x},t) + U(\mathbf{x}',t)\right)\right) f(\mathbf{x},\mathbf{x}',t) = -i < F_{\mathbf{x},\mathbf{x}',t} > [f_i]$$

$$\tag{4.15}$$

where on the right-hand side

$$< F_{\mathbf{x},\mathbf{x}',t} > [f_i] \equiv \frac{1}{\hbar}\left(< V(\mathbf{x})\rho(\mathbf{x},\mathbf{x}',t) > - < V(\mathbf{x}')\rho(\mathbf{x},\mathbf{x}',t) >\right)$$

$$= \int d\tilde{\mathbf{x}} \int d\tilde{\mathbf{x}}' < \bar{J}(\mathbf{x},\mathbf{x}',t;\tilde{\mathbf{x}},\tilde{\mathbf{x}}',t') > f_i(\tilde{\mathbf{x}},\tilde{\mathbf{x}}') \tag{4.16}$$

is a functional of the density matrix at the initial time t', and a function of $\mathbf{x}, \mathbf{x}', t$. For the impurity average of $\bar{J}$ we have the perturbative expansion

$$
\begin{aligned}
<\bar{J}(\mathbf{x}, \mathbf{x}', t; \tilde{\mathbf{x}}, \tilde{\mathbf{x}}', t')> \; &= \; \sum_{n,m=0}^{\infty} \frac{1}{\hbar^{n+m+1}} \int \prod_{m'=1}^{m} d\tilde{\mathbf{x}}_{m'} \int \prod_{m'=1}^{m} d\tilde{t}_{m'} \int \prod_{n'=1}^{n} d\mathbf{x}_{n'} \int \prod_{n'=1}^{n} dt_{n'} \\
& \quad G_0^R(\mathbf{x}, t; \mathbf{x}_n, t_n) G_0^R(\mathbf{x}_n, t_n; \mathbf{x}_{n-1}, t_{n-1}) .. G_0^R(\mathbf{x}_1, t_1; \tilde{\mathbf{x}}, t') \\
& \quad G_0^A(\tilde{\mathbf{x}}', t'; \tilde{\mathbf{x}}_m, \tilde{t}_m) G_0^A(\tilde{\mathbf{x}}_m, \tilde{t}_m; \tilde{\mathbf{x}}_{m-1}, \tilde{t}_{m-1}) .. G_0^A(\tilde{\mathbf{x}}_1, \tilde{t}_1; \mathbf{x}', t) \\
& \quad < V(\tilde{\mathbf{x}}_m) .. V(\tilde{\mathbf{x}}_1) V(\mathbf{x}) V(\mathbf{x}_n) .. V(\mathbf{x}_1) > \\[4pt]
& - \; \sum_{n,m=0}^{\infty} \frac{1}{\hbar^{n+m+1}} \int \prod_{m'=1}^{m} d\tilde{\mathbf{x}}_{m'} \int \prod_{m'=1}^{m} d\tilde{t}_{m'} \int \prod_{n'=1}^{n} d\mathbf{x}_{n'} \int \prod_{n'=1}^{n} dt_{n'} \\
& \quad G_0^R(\mathbf{x}, t; \mathbf{x}_n, t_n) G_0^R(\mathbf{x}_n, t_n; \mathbf{x}_{n-1}, t_{n-1}) .. G_0^R(\mathbf{x}_1, t_1; \tilde{\mathbf{x}}, t') \\
& \quad G_0^A(\tilde{\mathbf{x}}', t'; \tilde{\mathbf{x}}_m, \tilde{t}_m) G_0^A(\tilde{\mathbf{x}}_m, \tilde{t}_m; \tilde{\mathbf{x}}_{m-1}, \tilde{t}_{m-1}) .. G_0^A(\tilde{\mathbf{x}}_1, \tilde{t}_1; \mathbf{x}', t) \\
& \quad < V(\tilde{\mathbf{x}}_m) .. V(\tilde{\mathbf{x}}_1) V(\mathbf{x}') V(\mathbf{x}_n) .. V(\mathbf{x}_1) > \qquad (4.17)
\end{aligned}
$$

where all time integrations are from minus to plus infinity. In order to perform the impurity average, we thus have to average the following product of impurity potentials $< V(\tilde{\mathbf{x}}_m) .. V(\tilde{\mathbf{x}}_1) V(\mathbf{x}) V(\mathbf{x}_n) .. V(\mathbf{x}_1) >$, and in performing the Gaussian average[1] we tie all the potential vertices pairwise together, and obtain for the impurity-averaged $\bar{J}$

---

[1]Inclusion of multiple scattering is straightforward, in a diagram we simply have to allow for the additional feature of the impurity line to branch into multiple scattering, and we will encounter the t-matrix instead of the Gaussian impurity correlator.

where the box signifies that the leftmost impurity correlator line can enter into any entanglement, and the doublets of diagrams that the first impurity correlator can end up on either the upper or lower line, as indicated by the two different depictions of the impurity correlator.

The functional $<F>$ is obtained by attaching the initial reduced density matrix onto $<\bar{J}>$ at the initial time. We therefore see that the impurity-averaged functional $<F>$ has a diagrammatic representation which is obtained from the diagrams for the impurity-averaged density matrix, eq.(3.114), by the following prescription: Remove the final external retarded or advanced particle line on the diagrams with impurity correlators, and change the prefactor from plus to minus according to whether the last impurity correlator attaches to the upper or lower particle line.

Every particle line can be dressed by self-energy insertions so that the exact impurity-averaged propagator appears everywhere in the skeleton diagrammatic expansion.

4.3   Quantum Kinetic Equation

For problems that are not concerned with transient behavior, such as nonequilibrium steady-state problems, it would be preferable to have a description in which the initial condition does not appear explicitly. In order to establish such a formulation we shall exploit a topological feature of the diagrams. In this case for the diagrams in the expansion of $<F>$, and we shall obtain new diagrammatic expressions from the naive perturbative ones, and thereby diagrammatically derive new equations. The desired formulation is provided by introducing the concept of equal-time two-line irreducibility [21]. An $<F>[f_i]$ - diagram can be cut in two by cutting vertically only an upper and lower particle line, as any $\bar{J}$-diagram can eventually be cut prior to the initial time t'. However, some diagrams can be cut in a less trivial way, viz. when they can be cut at intermediate times, i.e., not at the extremity next to the initial point in time. A diagram which in this fashion can not be cut in two by cutting vertically two internal particle lines is said to be equal-time two-line irreducible.

Using the concept of equal-time two-line irreducibility, we can now eliminate the explicit appearance of the initial-time density matrix by observing the following summation of diagrams. Take an arbitrary diagram from the perturbative expansion of $< F >$. It can be either equal-time two-line reducible (ETLR) or irreducible (ETLI). Let us first consider the former case, and say we encounter the diagram

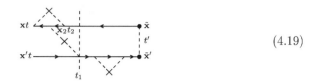

$$(4.19)$$

a typical diagram containing three impurity correlators. It is reducible at an intermediate time labeled t_1. To check whether an arbitrary diagram is reducible or not, we can use the following mechanical scanning procedure: Move back in time along the particle lines until the first time t_1, where the diagram might be cut in two by cutting vertically (at equal time) only the particle lines, as illustrated above. In the above example we insert a complete set of states on the retarded line at the cut, and rewrite

$$G_0^R(\mathbf{x}_2, t_2; \tilde{\mathbf{x}}, t') = i \int d\mathbf{x}_1\, G_0^R(\mathbf{x}_2, t_2; \mathbf{x}_1, t_1) G_0^R(\mathbf{x}_1, t_1; \tilde{\mathbf{x}}, t') \ . \qquad (4.20)$$

At times prior to t_1, any process can take place as witnessed by the diagrammatic expansion of $<F>[f_i]$. Along with the above diagram, with its unique equal-time two-line irreducible configuration of the first two impurity correlator lines, there will be additional reducible diagrams (depicted in the parenthesis below) corresponding to all possible scattering processes that can take place prior to t_1, i.e., all the diagrams for the density matrix propagated to time t_1

$$< F^d_{\mathbf{x},\mathbf{x}',t} > [f_i] \quad = \quad \qquad\qquad\qquad\qquad\qquad\qquad\qquad\qquad$$

$$\rightarrow \quad i \qquad\qquad \left(\qquad + \qquad + \ldots \right)$$

$$= \quad i \qquad\qquad \equiv i \tilde{F}^d_{\mathbf{x},\mathbf{x}',t}[f] \qquad\qquad (4.21)$$

where the factor $+i$ stems from cutting the retarded propagator into two in accordance with eq.(4.20) (a factor $-i$ will appear when the advanced line is cut). Had the initially picked diagram been equal-time two-line irreducible, we also would have additional reducible diagrams in the perturbation expansion corresponding to all possible additional prior scattering processes. We can thus group all diagrams according to their equal-time two-line irreducible part, their unique irreducible starting configuration of impurity correlators. The total sum of processes prior to t_1 is the same as those for the density matrix at time t_1, so that the initial time density matrix f_i is propagated to time t_1. We can therefore resum all diagrams according to their equal-time two-line reducibility and obtain the identity

$$-i < F_{\mathbf{x},\mathbf{x}',t} > [f_i] \; = \; \tilde{F}_{\mathbf{x},\mathbf{x}',t}[f] \tag{4.22}$$

where the diagrammatic expansion of $\tilde{F}$, consists of all equal-time two-line irreducible diagrams of which only the lowest order ones are shown explicitly in the figure

The sign attributed to a diagram is determined by whether a propagator on the upper or lower line is split into two due to the cutting.[2]

We can express the functional $\tilde{F}_{\mathbf{x},\mathbf{x}',t}$

$$\tilde{F}_{\mathbf{x},\mathbf{x}',t}[f] = \int d\tilde{\mathbf{x}} \int d\tilde{\mathbf{x}}' \int_{t'}^{t} d\tilde{t} \; \tilde{J}(\mathbf{x},\mathbf{x}',t;\tilde{\mathbf{x}},\tilde{\mathbf{x}}',\tilde{t}) f(\tilde{\mathbf{x}},\tilde{\mathbf{x}}',\tilde{t}) \tag{4.24}$$

in terms of the irreducible propagator of the density matrix, $\tilde{J}$, consisting of all the equal-time two-line irreducible $\bar{J}$-diagrams, i.e., the ones which can not be cut in two by cutting vertically only an internal upper and lower particle line. Since the cutting is at an internal vertex, an interaction time which is integrated over, we have integration over all times in between the initial time t' and t. The

---

[2]We shall not at this point bother to go into further details with the sign ascription, as we shall presently only study lowest-order diagrams. Once we want to discuss diagrams of arbitrary order we shall introduce a generalized density matrix and a stringer concept of irreducibility for which this issue disappears.

functional $\tilde{F}_{\mathbf{x},\mathbf{x}',t}[f]$ is therefore a functional in all the parameters of the impurity-averaged density matrix, as well as a function of the coordinates of the reduced density matrix appearing on the left-hand side of the von Neumann equation for the density matrix.

For the diagrammatic expansion of the irreducible density matrix propagator we have

$$\tilde{J}(\mathbf{x},\mathbf{x}',t;\tilde{\mathbf{x}},\tilde{\mathbf{x}}',\tilde{t}) \quad = \quad \pm \quad \pm \quad \pm \quad \pm$$

$$= \quad - \quad - \quad + \quad + \quad \pm ... \quad (4.25)$$

where we only display explicitly the lowest-order diagrams.

Every particle line can be dressed by self-energy insertions so that the exact impurity-averaged propagator appears everywhere in the skeleton diagrammatic expansion. Again, inclusion of multiple scattering is irrelevant for the topological arguments used to analyze the diagrammatic structure, and we can in the skeleton diagrammatic expansion everywhere substitute the t-matrix for the Gaussian impurity correlator.

We note that equal-time two-line irreducibility of a diagram is determined not by topology alone, but also by the relative time-ordering between interaction times on the upper and lower branches, as illustrated below,

ETLR   ETLI

$$(4.26)$$

so that beyond lowest order, equal-time two-line irreducibility of the perturbative expansion of $<F>[f]$ is nontrivial from a topological point of view.[3]

---

[3]In section 6.7 we develop the concept of irreducibility further in order to make it trivial

The motion of a particle in a random potential is therefore determined by the nonlocal integro-differential equation, the quantum kinetic equation,

$$\left(\frac{\partial}{\partial t} - \frac{i}{\hbar} \left(\frac{\hbar^2}{2m} (\triangle_{\mathbf{x}} - \triangle_{\mathbf{x}'}) - U(\mathbf{x}, t) + U(\mathbf{x}', t) \right) \right) f(\mathbf{x}, \mathbf{x}', t)$$

$$= \int d\tilde{\mathbf{x}} \int d\tilde{\mathbf{x}}' \int_{t'}^{t} d\tilde{t} \, \tilde{J}(\mathbf{x}, \mathbf{x}', t; \tilde{\mathbf{x}}, \tilde{\mathbf{x}}', \tilde{t}) f(\tilde{\mathbf{x}}, \tilde{\mathbf{x}}', \tilde{t}) . \tag{4.27}$$

Instead of pursuing the general discussion at this point, and in order to gain familiarity with the Feynman rules and the general features of the diagrammatic method, let us look at the lowest-order contribution to the irreducible propagator of the density matrix $\tilde{J}$. We observe that all lowest-order diagrams are two-line irreducible. For illustration, we note that the first first-order diagram in eq.(4.25) corresponds to the analytical expression $\tilde{J}^1$ given by

$$\tilde{J}^{(1)}(\mathbf{x}, \mathbf{x}', t; \tilde{\mathbf{x}}, \tilde{\mathbf{x}}', \tilde{t}) = -\frac{1}{\hbar^2} < V(\mathbf{x}) V(\tilde{\mathbf{x}}) > G^R(\mathbf{x}, t; \tilde{\mathbf{x}}, \tilde{t}) \, G^A(\tilde{\mathbf{x}}', \tilde{t}; \mathbf{x}', t) \quad (4.28)$$

by application of the Feynman rules.

The present spatial representation allows inclusion of boundary conditions, and the general formalism presented is therefore of importance for a full quantum statistical description of finite size effects of physical systems.

4.4   The Wigner Function

In the following, we shall pursue the treatment of quantum transport that resembles the kinetic description of the dynamics of classical gases due to Boltzmann [22]. The virtues of a transport description in terms of kinetic equations are the simple physical interpretation it can produce, and its ability to describe nonlinear behavior. Furthermore, the kinetic equation allows a straightforward description of thermal properties, such as heat conduction, whereas a treatment of nonmechanical induced nonequilibrium states in linear-response theory is less immediate.

We introduce new coordinates

$$\mathbf{R} = \frac{\mathbf{x} + \mathbf{x}'}{2} \qquad\qquad \mathbf{r} = \mathbf{x} - \mathbf{x}' \tag{4.29}$$

and define the function

$$f(\mathbf{R}, \mathbf{r}, t) \equiv f(\mathbf{x}, \mathbf{x}', t) . \tag{4.30}$$

The central object to the Boltzmann approach to nonequilibrium classical statistical mechanics is the one-particle probability distribution function on phase

---

from a topological point of view, and recognize that in this stringer sense the reducibility versus irreducibility feature corresponds to different physical phenomena, viz. renormalization and dissipation.

space. In quantum mechanics Heisenberg's uncertainty principle excludes the existence of a probability distribution with such a physical interpretation, but not, however, the introduction of a function with formal resemblance to it. This, so-called Wigner function [23], is definable in terms of the density matrix by Fourier transformation with respect to the relative spatial coordinate [4]:

$$f(\mathbf{R}, \mathbf{p}, t) \equiv \int d\mathbf{r}\, e^{-\frac{i}{\hbar}\mathbf{r}\cdot\mathbf{p}} f(\mathbf{R}, \mathbf{r}, t) = \int d\mathbf{r}\, e^{-\frac{i}{\hbar}\mathbf{r}\cdot\mathbf{p}} <\mathbf{R}+\mathbf{r}/2|\hat{\rho}(t)|\mathbf{R}-\mathbf{r}/2> \quad (4.31)$$

and inversely we have

$$f(\mathbf{R}, \mathbf{r}, t) = \frac{1}{V}\sum_{\mathbf{p}} e^{\frac{i}{\hbar}\mathbf{p}\cdot\mathbf{r}} f(\mathbf{R}, \mathbf{p}, t) = \int \frac{d\mathbf{p}}{(2\pi\hbar)^3} e^{\frac{i}{\hbar}\mathbf{p}\cdot\mathbf{r}} f(\mathbf{R}, \mathbf{p}, t) \,. \quad (4.32)$$

We shall call $(\mathbf{R}, \mathbf{p})$ the Wigner coordinates.[5]

For an arbitrary physical quantity A we can define a function, called the corresponding phase space function,

$$A(\mathbf{R}, \mathbf{p}) \equiv \int d\mathbf{x} \int d\mathbf{x}'\, e^{-\frac{i}{\hbar}(\mathbf{x}-\mathbf{x}')\cdot\mathbf{p}} <\mathbf{x}|\hat{A}|\mathbf{x}'> \delta(\mathbf{R} - (\mathbf{x}+\mathbf{x}')/2) \quad (4.33)$$

and we see that the Wigner function corresponds to the phase space function of the statistical operator $\hat{\rho}(t)$.

The density distribution is obtained by integrating the Wigner function over the *momentum*-coordinate

$$n(\mathbf{R}, t) = \int \frac{d\mathbf{p}}{(2\pi\hbar)^3} f(\mathbf{R}, \mathbf{p}, t) = f(\mathbf{R}, \mathbf{r}=0, t) = <\mathbf{R}|\hat{\rho}(t)|\mathbf{R}> = Tr(\hat{\rho}(t_r)\hat{P}(\mathbf{R}, t))$$
$$(4.34)$$

and vice versa for the momentum distribution, as integrating the Wigner function over the *position*-coordinate we obtain the momentum probability distribution

$$f_{\mathbf{p}}(t) = \int \frac{d\mathbf{R}}{(2\pi\hbar)^3} f(\mathbf{R}, \mathbf{p}, t) \quad (4.35)$$

since the probability density for the particle to have momentum $\mathbf{p}$ at time t is

$$Tr(\hat{\rho}(t_r)\hat{P}(\mathbf{p}, t)) = <\mathbf{p}|\hat{\rho}(t)|\mathbf{p}>$$
$$= \int d\mathbf{x} \int d\mathbf{x}' <\mathbf{p}|\mathbf{x}> f(\mathbf{x}, \mathbf{x}', t) <\mathbf{x}'|\mathbf{p}>$$
$$= \frac{1}{(2\pi\hbar)^3} \int d\mathbf{R} \int d\mathbf{r}\, e^{-\frac{i}{\hbar}\mathbf{r}\cdot\mathbf{p}} f(\mathbf{R}, \mathbf{r}, t) = \int \frac{d\mathbf{R}}{(2\pi\hbar)^3} f(\mathbf{R}, \mathbf{p}, t)$$
$$\equiv f_{\mathbf{p}}(t) \quad (4.36)$$

---

[4]We first discuss the case where there is no magnetic field.

[5]We shall avoid the phrase Wigner representation, because the Wigner coordinates do not correspond to some mixed representation since it is forbidden by Heisenberg's uncertainty principle.

where in the shift of integration variables we have used that the coordinate transformation, eq.(4.29), has Jacobian 1.

For a spatially homogeneous state f is independent of $\mathbf{R}$, and the Wigner function is thus proportional to the momentum distribution function, and for a spatially localized state (f independent of $\mathbf{p}$), the Wigner function is proportional to the density distribution function.

Exercise 4.1 *Show that for a free particle the Wigner function at time t is expressed in terms of the Wigner function at a previous time t' through*

$$f(\mathbf{R}, \mathbf{p}, t) = f\left(\mathbf{R} - \frac{\mathbf{p}}{m}(t - t'), \mathbf{p}, t'\right) . \tag{4.37}$$

Solution

Introducing Wigner coordinates we obtain from eq.(2.236) for the case of a free particle

$$
\begin{aligned}
f(\mathbf{R}, \mathbf{p}, t) &= \int d\mathbf{r}\, e^{-\frac{i}{\hbar}\mathbf{r}\cdot\mathbf{p}} \int d\tilde{\mathbf{R}} \int d\tilde{\mathbf{r}}\, G_0^R(\mathbf{R} + \mathbf{r}/2, t; \tilde{\mathbf{R}} + \tilde{\mathbf{r}}/2, t') \\
&\quad G_0^A(\tilde{\mathbf{R}} - \tilde{\mathbf{r}}/2, t'; \mathbf{R} - \mathbf{r}/2, t)\, \rho(\tilde{\mathbf{R}} + \tilde{\mathbf{r}}/2, \tilde{\mathbf{R}} - \tilde{\mathbf{r}}/2, t') \\
&= \left(\frac{m}{2\pi\hbar(t - t')}\right)^2 \int d\mathbf{r}\, e^{-\frac{i}{\hbar}\mathbf{r}\cdot\mathbf{p}} \int d\tilde{\mathbf{R}} \int d\tilde{\mathbf{r}}\, e^{\frac{im}{2\hbar(t-t')}(\mathbf{R}-\tilde{\mathbf{R}})\cdot(\mathbf{r}-\tilde{\mathbf{r}})} \\
&\quad \int \frac{d\tilde{\mathbf{p}}}{(2\pi\hbar)^3}\, e^{\frac{i}{\hbar}\tilde{\mathbf{p}}\cdot\tilde{\mathbf{r}}} f(\tilde{\mathbf{R}}, \tilde{\mathbf{p}}, t') .
\end{aligned}
\tag{4.38}
$$

The integrations over $\mathbf{r}$ and $\tilde{\mathbf{r}}$ can now be performed and leads to the appearance of delta functions, which thereupon makes the $\tilde{\mathbf{R}}$ and $\tilde{\mathbf{p}}$ integrations trivial, and we obtain the stated result.

The time evolution of the Wigner function for a free particle (or a particle moving in homogeneous fields) is thus identical to that of the classical distribution function. If we have a probabilistic interpretation of the Wigner function at time t', say a simultaneous measurement of position and momentum in conformity with Heisenberg's uncertainty principle, the Wigner function for a particle moving in homogeneous fields will at later times evolve identically to the distribution function of classical mechanics. The above exact equation for the Wigner function for a free particle has the simple interpretation in the classical limit: in order for a particle with momentum $\mathbf{p}$ to arrive at position $\mathbf{R}$ at time t it must at time t' have been at position $\mathbf{R} - \mathbf{p}(t - t')/m$. In the following exercises we demonstrate the analogous conclusion for motion in a homogeneous field, and the case of a magnetic field is discussed in appendix A.

Exercise 4.2 *Show that the propagator for a particle in an external homogeneous field, $\mathbf{F}$, in terms of the Wigner coordinates is given by the expression*

$$G_{\mathbf{F}}^{R}(\mathbf{R}, \mathbf{p}, t, t') = -i\theta(t - t') \exp\left\{ -\frac{i}{\hbar}(t - t')(\epsilon_{\mathbf{p}} + \frac{\mathbf{F}^2(t - t')^2}{24m} - \mathbf{F} \cdot \mathbf{R}) \right\}$$

$$= G_0^{R}(\mathbf{p}, t, t') \exp\left\{ -\frac{i}{\hbar}(t - t')\left(\frac{\mathbf{F}^2(t - t')^2}{24m} - \mathbf{F} \cdot \mathbf{R} \right) \right\} . \quad (4.39)$$

Solution

In terms of the coordinates $\mathbf{R}$ and $\mathbf{r}$ we obtain from eq. (A.19) for the propagator

$$G_{\mathbf{F}}^{R}(\mathbf{R}, t; \mathbf{r}, t') = G_0^{R}(\mathbf{r}, t, t') \, e^{\frac{i}{\hbar}(t - t')(\mathbf{F} \cdot \mathbf{R} - \frac{1}{24m}\mathbf{F}^2(t - t')^2)} . \quad (4.40)$$

and by Fourier transformation to Wigner coordinates we obtain the above result.

Exercise 4.3 *Show that for a particle in an external homogeneous field, $\mathbf{F}$, the Wigner function at time t is expressed in terms of the Wigner function at a previous time t' through*

$$f(\mathbf{R}, \mathbf{p}, t) = f(\mathbf{R} - \frac{\mathbf{p}}{m}(t - t') + \frac{\mathbf{F}}{2m}(t - t')^2, \ \mathbf{p} - \mathbf{F}(t - t'), \ t') \quad (4.41)$$

and interpret the result.

Solution

According to eq. (A.19) we have

$$f(\mathbf{R}, \mathbf{p}, t) = \int d\mathbf{r} \, e^{-\frac{i}{\hbar}\mathbf{r} \cdot \mathbf{p}} \int d\tilde{\mathbf{R}} \int d\tilde{\mathbf{r}} \, G_{\mathbf{F}}^{R}(\mathbf{R} + \mathbf{r}/2, t; \tilde{\mathbf{R}} + \tilde{\mathbf{r}}/2, t')$$

$$G_{\mathbf{F}}^{A}(\tilde{\mathbf{R}} - \tilde{\mathbf{r}}/2, t'; \mathbf{R} - \mathbf{r}/2, t) \, \rho(\tilde{\mathbf{R}} + \tilde{\mathbf{r}}/2, \tilde{\mathbf{R}} - \tilde{\mathbf{r}}/2, t')$$

$$= \left(\frac{m}{2\pi\hbar(t-t')}\right)^2 \int d\mathbf{r}\; e^{-\frac{i}{\hbar}\mathbf{r}\cdot\mathbf{p}} \int d\tilde{\mathbf{R}} \int d\tilde{\mathbf{r}}\; e^{\frac{im}{2\hbar(t-t')}(\mathbf{R}-\tilde{\mathbf{R}})\cdot(\mathbf{r}-\tilde{\mathbf{r}})}$$

$$\int \frac{d\tilde{\mathbf{p}}}{(2\pi\hbar)^3}\; e^{\frac{i}{\hbar}\tilde{\mathbf{p}}\cdot\tilde{\mathbf{r}}} f(\tilde{\mathbf{R}},\tilde{\mathbf{p}},t')$$

$$= \left(\frac{m}{2\pi\hbar(t-t')}\right)^2 \int d\mathbf{r}\; e^{-\frac{i}{\hbar}\mathbf{r}\cdot\mathbf{p}} \int d\tilde{\mathbf{R}} \int d\tilde{\mathbf{r}}\; e^{\frac{im}{2\hbar(t-t')}(\mathbf{R}-\tilde{\mathbf{R}})\cdot(\mathbf{r}-\tilde{\mathbf{r}})}$$

$$e^{\frac{i(t-t')}{2\hbar}\mathbf{F}\cdot(\mathbf{r}+\tilde{\mathbf{r}})} \int \frac{d\tilde{\mathbf{p}}}{(2\pi\hbar)^3}\; e^{\frac{i}{\hbar}\tilde{\mathbf{p}}\cdot\tilde{\mathbf{r}}} f(\tilde{\mathbf{R}},\tilde{\mathbf{p}},t') . \qquad (4.42)$$

Upon integration over $\mathbf{r}$ *and* $\tilde{\mathbf{r}}$ *we get delta functions rendering the rest of the integrations trivial, and the result eq.(4.41) follows.*

The value of the Wigner function at time t *of the Wigner coordinates* $\mathbf{R}$ *and* $\mathbf{p}$ *is the same as the value the Wigner function at time* t' *attains for the Wigner coordinates related to* $\mathbf{R}$ *and* $\mathbf{p}$ *through the classical equation of motion. The time evolution of the Wigner function for a particle moving in a homogeneous field is thus identical to that of the classical distribution function.*

Exercise 4.4 *Show that the expectation value of a quantity* A *is given by*

$$<\hat{A}> = \int d\mathbf{R} \int d\mathbf{p}\; f(\mathbf{R},\mathbf{p},t)\, A(\mathbf{R},\mathbf{p}) \qquad (4.43)$$

i.e., formally identical to the expression in classical physics.

4.4.1   Classical Distribution Function

The Wigner function reduces in the classical limit to the classical distribution function. This is conveniently seen by introducing Gaussian wave packet states. In the position representation such a state is specified by (in three spatial dimensions)

$$\psi_{\mathbf{x}_0\mathbf{p}_0}(\mathbf{x}) \equiv <\mathbf{x}|\mathbf{x}_0,\mathbf{p}_0> \equiv \left(\frac{1}{2\pi\Delta x^2}\right)^{3/4} \exp\left\{-\frac{(\mathbf{x}-\mathbf{x}_0)^2}{4\Delta x^2} + \frac{i}{\hbar}\mathbf{p}_0\cdot(\mathbf{x}-\mathbf{x}_0)\right\} . \qquad (4.44)$$

Being a Gaussian wave packet centered around the value $\mathbf{x}_0$, the spatial variance and average characterizes the state[6]

$$\Delta x^2 \equiv \frac{1}{3} < (\hat{\mathbf{x}} - <\hat{\mathbf{x}}>_{\mathbf{x}_0\mathbf{p}_0})^2 >_{\mathbf{x}_0\mathbf{p}_0} \quad , \quad <\hat{\mathbf{x}}>_{\mathbf{x}_0\mathbf{p}_0} = \mathbf{x}_0 . \qquad (4.45)$$

---

[6]For reason of notational simplicity we assume equal variances in all three spatial directions, $\Delta x^2 = \Delta y^2 = \Delta z^2$.

In the momentum representation the Gaussian wave packet is also a Gaussian function

$$\psi_{\mathbf{x}_0\mathbf{p}_0}(\mathbf{p}) \equiv <\mathbf{p}|\mathbf{x}_0,\mathbf{p}_0> = \left(\frac{2\Delta x^2}{\pi\hbar^2}\right)^{3/4} \exp\left\{-\frac{(\mathbf{p}-\mathbf{p}_0)^2\Delta x^2}{\hbar^2} - \frac{i}{\hbar}\mathbf{p}\cdot\mathbf{x}_0\right\}$$

(4.46)

centered around the average momentum $\mathbf{p}_0$

$$<\hat{\mathbf{p}}>_{\mathbf{x}_0\mathbf{p}_0} = \mathbf{p}_0$$

(4.47)

with a momentum variance

$$\Delta p^2 \equiv \frac{1}{3}<(\hat{\mathbf{p}}-<\hat{\mathbf{p}}>_{\mathbf{x}_0\mathbf{p}_0})^2>_{x_0 p_0} = \frac{\hbar^2}{4\Delta x^2}.$$

(4.48)

The variances minimize Heisenberg's uncertainty relation

$$\Delta x^2 \Delta p^2 = \frac{\hbar^2}{4}$$

(4.49)

and since the Gaussian wave packet states are the only states that minimize Heisenberg's uncertainty relation, they are also referred to as minimal-uncertainty wave packet states.[7] The Gaussian wave packet states provide a resolution of the identity

$$\int_{-\infty}^{\infty} \frac{d\mathbf{x}_0 d\mathbf{p}_0}{(2\pi\hbar)^3} |\mathbf{x}_0,\mathbf{p}_0><\mathbf{x}_0,\mathbf{p}_0| = \hat{I}$$

(4.50)

which for instance can be seen by recalling that $\hat{I}$ is the only operator for which $<\mathbf{x}|\hat{I}|\mathbf{x}'> = \delta(\mathbf{x}-\mathbf{x}')$, but they constitute an overcomplete set because the states are not orthogonal

$$<\mathbf{x}_0,\mathbf{p}_0|\mathbf{x}_0',\mathbf{p}_0'> = \exp\left\{-\frac{(\mathbf{x}_0-\mathbf{x}_0')^2}{8\Delta x^2} - \frac{(\mathbf{p}_0-\mathbf{p}_0')^2}{8\Delta p^2} + \frac{i}{2\hbar}(\mathbf{p}_0+\mathbf{p}_0')\cdot(\mathbf{x}_0-\mathbf{x}_0')\right\}$$

$$\neq \delta(\mathbf{x}_0-\mathbf{x}_0')\,\delta(\mathbf{p}_0-\mathbf{p}_0').$$

(4.51)

Exercise 4.5 *Consider the free evolution of a Gaussian wave packet which at time $t=0$ is centered around position $x=0$ (we consider the one-dimensional case for simplicity), and momentum p:*

$$\psi_{0p}(x,t=0) = \left(\frac{1}{2\pi\Delta x^2}\right)^{1/4} \exp\left\{-\frac{x^2}{4\Delta x^2} + \frac{i}{\hbar}px\right\}.$$

(4.52)

Find the wave functions and probability densities at times $\pm t$. Consider the variances at times $\pm t$.

---

[7] In the context of the harmonic oscillator they are also referred to as coherent states.

Solution

Since the integration is Gaussian we immediately obtain (t > 0)

$$\psi_{0p}(x,t) = i \int_{-\infty}^{\infty} dx' \, G_0^R(x,t;x',0) \, \psi_{0p}(x',0)$$

$$= \left(\frac{\Delta x^2}{2\pi \delta x_t^4}\right)^{1/4} \exp\left\{\frac{im}{2\hbar t}x^2\right\} \exp\left\{-\frac{im\Delta x^2}{2\hbar t}\frac{(x-x_0(t))^2}{\delta x_t^2}\right\} \quad (4.53)$$

where

$$\delta x_t^2 = \Delta x^2 \left(1 + \frac{i\hbar t}{2m\Delta x^2}\right) = \Delta x^2 \left(1 + \frac{2i\Delta p^2}{m\hbar}t\right) \quad (4.54)$$

and the center of the Gaussian wave packet, $x_0(t) = pt/m$, is seen to move with the velocity $p/m = \partial\epsilon/\partial p$, the group velocity.

For the probability density at time t we then get

$$P_{0p}(x,t) = |\psi_{0p}(x,t)|^2 = \sqrt{\frac{\Delta x^2}{2\pi|\delta x_t^2|^2}} \exp\left\{-\frac{\Delta x^2(x-x_0(t))^2}{2|\delta x_t^2|^2}\right\} . \quad (4.55)$$

At time −t the wave function was

$$\psi_{0p}(x,-t) = -i \int_{-\infty}^{\infty} dx' \, G_0^A(x,-t;x',0) \, \psi_{0p}(x',0)$$

$$= \left(\frac{\Delta x^2}{2\pi \delta x_{-t}^4}\right)^{1/4} \exp\left\{-\frac{im}{2\hbar t}x^2\right\} \exp\left\{\frac{im\Delta x^2}{2\hbar t}\frac{(x-x_0(-t))^2}{\delta x_{-t}^2}\right\} \quad (4.56)$$

and the probability density

$$P_{0p}(x,-t) = |\psi_{0p}(x,-t)|^2 = \sqrt{\frac{\Delta x^2}{2\pi|\delta x_t^2|^2}} \exp\left\{-\frac{\Delta x^2(x-x_0(-t))^2}{2|\delta x_t^2|^2}\right\} . \quad (4.57)$$

The position uncertainty thus shrinks until it reaches its minimal value at time $t = 0$, whereupon it increases. For the position variance at times $\pm t$ we get

$$\Delta x_{\pm t}^2 \equiv \, < (\hat{x}- <\hat{x}>_{\pm t})^2 >_{\pm t} = \Delta x^2 \left(1 + \left(\frac{\pm t\hbar}{2m\Delta x^2}\right)^2\right)^{1/2} = |\delta x_t^2| \quad (4.58)$$

and, since the momentum of a free particle is conserved, we get that the momentum variance stays constant $\Delta p_{\pm t}^2 \equiv \, < (\hat{p}- <\hat{p}>_{\pm t})^2 >_{\pm t} = \hbar^2/4\Delta x^2$, and we have

$$\Delta x_{\pm t}^2 \Delta p_{\pm t}^2 = \frac{\hbar^2}{4\Delta x^2}|\delta x_t^2| = \frac{\hbar^2}{4}\left(1 + \left(\frac{\hbar t}{2m\Delta x^2}\right)^2\right)^{1/2} . \quad (4.59)$$

If in the expression for the probability density, eq.(4.55), $\hbar$ and then Δx^2 is set to zero we obtain

$$P_{0p}(x,t) \;=\; \delta(x - x_0(t)) \tag{4.60}$$

i.e., the classical trajectory expressed in the language of probability densities: the probability for the particle to be anywhere but on the classical trajectory is zero.

Exercise 4.6 *Consider the motion of the time-reversed state of the previous exercise.*

Solution

The time-reversed state at time $-t$ is specified by the wave function

$$\overline{\psi}_{0p}(x,-t) \equiv \psi_{0p}^*(x,t) = \left(\frac{\Delta x^2}{2\pi(\delta x_t^4)^*}\right)^{1/4} e^{-\frac{im}{2\hbar t}x^2} \exp\left\{\frac{im\Delta x^2}{2\hbar t}\frac{(x - x_0(t))^2}{(\delta x_t^2)^*}\right\} \tag{4.61}$$

and its free evolution gives for time $t = 0$ the wave function (t > 0)

$$\overline{\psi}_{0p}(x,0) \;=\; i\int_{-\infty}^{\infty} dx'\, G_0^R(x,0;x',-t)\, \overline{\psi}_{0p}(x',-t)$$

$$= \left(\frac{1}{2\pi\Delta x^2}\right)^{1/4} \exp\left\{-\frac{x^2}{4\Delta x^2} - \frac{i}{\hbar}px\right\}$$

$$= \psi_{0-p}(x,0) \;=\; \psi_{0p}^*(x,0)\,. \tag{4.62}$$

The time-reversed state is seen to be the motion-reversed state $p \to -p$. At time t the time-reversed wave function is

$$\overline{\psi}_{0p}(x,t) \;=\; i\int_{-\infty}^{\infty} dx'\, G_0^R(x,t;x',-t)\, \overline{\psi}_{0p}(x',-t)$$

$$= \left(\frac{\Delta x^2}{2\pi(\delta x_{-t}^4)^*}\right)^{1/4} \exp\left\{\frac{im}{2\hbar t}x^2\right\} \exp\left\{\frac{im\Delta x^2}{2\hbar t}\frac{(x - x_0(-t))^2}{(\delta x_{-t}^2)^*}\right\}$$

$$= \psi_{0p}^*(x,-t)\,. \tag{4.63}$$

The probability density profile for the time-reversed state thus evolves as motion reversed.

The Wigner function for the Gaussian wave packet state $|x_0, p_0 >$ is immediately obtained by performing a Gaussian integral

$$f_{\mathbf{x_0 p_0}}(\mathbf{R}, \mathbf{p}) = \left(\frac{\hbar}{\sqrt{\Delta x^2 \Delta p^2}}\right)^3 \exp\left\{-\frac{(\mathbf{R} - \mathbf{x_0})^2}{2\Delta x^2} - \frac{(\mathbf{p} - \mathbf{p_0})^2}{2\Delta p^2}\right\} \qquad (4.64)$$

and is a Gaussian function in both the position and momentum variables. In the classical limit, $\hbar \to 0$, where we can let Δx and Δp approach zero simultaneously, we get the classical distribution function corresponding to the particle being at a definite position with a definite momentum

$$f^{cl}_{\mathbf{x_0 p_0}}(\mathbf{R}, \mathbf{p}) \equiv \frac{f_{\mathbf{x_0 p_0}}(\mathbf{R}, \mathbf{p})}{(2\pi\hbar)^3} = \delta(\mathbf{R} - \mathbf{x_0})\,\delta(\mathbf{p} - \mathbf{p_0})\ . \qquad (4.65)$$

In the classical limit the Wigner function thus reduces to the distribution function for the particle to be in a phase space volume of size $(2\pi\hbar)^3$ around $(\mathbf{R}, \mathbf{p})$. We can thus anticipate that in the classical limit the Wigner function satisfies the classical kinetic equation.

However, caution must be exercised in interpreting the Wigner function in a probabilistic sense. Although f is always a real function, since the statistical operator is hermitian, there is no general physical principle guaranteeing that it be nonnegative. In any event, the Wigner function is a valid construction, and all physical quantities are expressible in terms of it. For instance, for the probability current density we have in terms of the Wigner function

$$\mathbf{j}(\mathbf{R}, t) = \int \frac{d\mathbf{p}}{(2\pi\hbar)^3}\ \mathbf{v_p}\ f(\mathbf{R}, \mathbf{p}, t)\ . \qquad (4.66)$$

In the absence of the random potential V we get from eq.(4.27), upon transforming to the Wigner coordinates, and assuming a spatially constant force, $U(\mathbf{x}, t) = \mathbf{F}(t) \cdot \mathbf{x}$, Liouville's equation[8]

$$\frac{\partial f(\mathbf{R}, \mathbf{p}, t)}{\partial t} + \mathbf{F}(t) \cdot \frac{\partial f(\mathbf{R}, \mathbf{p}, t)}{\partial \mathbf{p}} + \mathbf{v_p} \cdot \frac{\partial f(\mathbf{R}, \mathbf{p}, t)}{\partial \mathbf{R}} = 0 \qquad (4.67)$$

as differentiation is specified by the chain rule

$$\frac{\partial}{\partial \mathbf{x}} = \frac{1}{2}\frac{\partial}{\partial \mathbf{R}} + \frac{\partial}{\partial \mathbf{r}} \qquad\qquad \frac{\partial}{\partial \mathbf{x'}} = \frac{1}{2}\frac{\partial}{\partial \mathbf{R}} - \frac{\partial}{\partial \mathbf{r}}\ . \qquad (4.68)$$

The group velocity

$$\mathbf{v_p} = \frac{\partial \epsilon_{\mathbf{p}}}{\partial \mathbf{p}} \qquad (4.69)$$

equals in the considered model $\mathbf{p}/m$.

---

[8]The conclusion of exercise 4.1 4.3, that the time evolution of the Wigner function in homogeneous fields is identical to that of the classical distribution function, follows immediately from Liouville's equation.

In the classical limit Liouville's equation is simply the continuity equation in phase space, expressing that a change in time of the probability density equals the net flow of probability current out of the unit phase space volume in question. The flow is either drift due to the inertia of the particle, or the particle is driven by the presence of an external force. The equation for the Wigner function is the same provided the forces are spatially homogeneous, and the dispersion quadratic. This is to be expected in view of exercise 4.3 on page 177 as the classical and quantum dynamics in this case (see also exercise 4.9 on page 186 for the case of motion in a homogeneous magnetic field) are identical.

We note, that for a particle launched at time $t = 0$ in the Gaussian wave packet state $|\mathbf{x}_0, \mathbf{p}_0>$, we obtain from eq.(4.37) for the Wigner function at later times

$$f_{\mathbf{x}_0\mathbf{p}_0}(\mathbf{R}, \mathbf{p}, t) = 2^d \, e^{-\frac{(\mathbf{R} - \frac{\mathbf{p}t}{m} - \mathbf{x}_0)^2}{2\Delta x^2} - \frac{(\mathbf{P} - \mathbf{P}_0)^2}{2\Delta p^2}} . \qquad (4.70)$$

Exercise 4.7 *Suppose we launch a particle in free space in the Gaussian wave packet state $|0, \mathbf{p}_0>$. Find the probability for the particle again to be at the origin after the time span t.*

Solution

The return probability is according to eq.(4.34) and eq.(4.37) given by

$$P_R(0, t) = \int \frac{d\mathbf{p}}{(2\pi\hbar)^3} f_{\mathbf{x}_0\mathbf{p}_0}(0, \mathbf{p}, t) = \int \frac{d\mathbf{p}}{(2\pi\hbar)^3} f_{\mathbf{x}_0\mathbf{p}_0}\left(-\frac{\mathbf{p}}{m}t, \mathbf{p}, 0\right)$$

$$= \prod_{\alpha=x,y,z} 2 \int \frac{dp_\alpha}{2\pi\hbar} \exp\left\{ -\frac{\left(\frac{p_\alpha t}{m}\right)^2}{2\Delta x_\alpha^2} - \frac{(p_\alpha - p_\alpha^{(0)})^2}{2\Delta p_\alpha^2} \right\}$$

$$= \prod_{\alpha=x,y,z} \sqrt{\frac{2\Delta p_\alpha^2}{\hbar^2\pi}} \left(1 + \frac{4\Delta p_\alpha^4 t^2}{\hbar^2 m^2}\right)^{-1/2}$$

$$\exp\left\{ -\frac{2\Delta p_\alpha^2 (p_\alpha^{(0)} t)^2}{\hbar^2 m^2} \frac{1}{1 + \frac{4\Delta p_\alpha^4 t^2}{\hbar^2 m^2}} \right\} . \qquad (4.71)$$

Assuming for simplicity an isotropic initial state, $\Delta p_x = \Delta p_y = \Delta p_z \equiv \Delta p$, we obtain in d spatial dimensions the limiting behaviors

$$P_R(0, t) = \left(\frac{2\Delta p^2}{\hbar^2\pi}\right)^{d/2} \begin{cases} e^{-dt^2/t_p^2} & t \ll \frac{\sqrt{2}p_0}{\Delta p} t_b \\ \left(\frac{m\hbar}{2\Delta p^2 t}\right)^d \exp\left\{ -\frac{dp_0^2}{2\Delta p^2} \right\} & t \gg \frac{\sqrt{2}p_0}{\Delta p} t_b \end{cases} \qquad (4.72)$$

where the ballistic decay time $t_b \equiv m\hbar/p_0\sqrt{2\Delta p^2}$ separates two regimes: initially we have ballistic motion, but eventually the wave packet spreading is dominating.

Exercise 4.8 *Show that the projection operator on the Gaussian wave packet state, $\hat{P}_{x_0,p_0} \equiv |x_0,p_0 > < x_0,p_0|$, has the integral representation (we consider the one-dimensional case, the generalization to any dimension is trivial)*

$$\hat{P}_{x_0,p_0} \equiv \int_{-\infty}^{\infty} \frac{dxdp}{2\pi\hbar} \, e^{-\frac{\Delta x^2}{2\hbar^2}p^2 - \frac{\Delta p^2}{2\hbar^2}x^2 + \frac{i}{\hbar}p(\hat{x}-x_o) + \frac{i}{\hbar}x(\hat{p}-p_o)} \ . \tag{4.73}$$

Solution

Using the identity eq.(1.333), we get for the mixed representation matrix elements

$$< p' | \int_{-\infty}^{\infty} \frac{dxdp}{2\pi\hbar} \, e^{-\frac{\Delta x^2}{2\hbar^2}p^2 - \frac{\Delta p^2}{2\hbar^2}x^2 + \frac{i}{\hbar}p(\hat{x}-x_o) + \frac{i}{\hbar}x(\hat{p}-p_o)} | x' >$$

$$= \sqrt{\frac{1}{\pi\hbar}} \exp\left\{ -\frac{(x_0 - x')^2}{4\Delta x^2} - \frac{(p_0 - p')^2\Delta x^2}{\hbar^2} + \frac{i}{\hbar}(p_0 - p')x_0 - \frac{i}{\hbar}p_0x' \right\}$$

$$= < p'|x_0,p_0 > < x_0,p_0|x' > \ = \ < p'|\hat{P}_{x_0,p_0}|x' > \ . \tag{4.74}$$

4.4.2   Magnetic Field Driving Term

The Hamiltonian for a particle with charge e in electromagnetic fields generated by a vector potential $\mathbf{A}(\mathbf{x},t)$ ($\mathbf{B}(\mathbf{x},t) = \nabla \times \mathbf{A}(\mathbf{x},t)$, $\mathbf{E}(\mathbf{x},t) = -\dot{\mathbf{A}}(\mathbf{x},t)$), is (recall exercise 1.4 on page 15)

$$\hat{H} = \frac{(\hat{\mathbf{p}}_t^{kin})^2}{2m} = \frac{1}{2m}\left(\hat{\mathbf{p}}_{can} - e\mathbf{A}(\hat{\mathbf{x}},t)\right)^2 = \hat{H}_0 + \hat{H}_{A(t)} \tag{4.75}$$

where

$$\hat{H}_0 = \frac{\hat{\mathbf{p}}_{can}^2}{2m} \tag{4.76}$$

is the Hamiltonian in the absence of the electromagnetic fields. In the presence of a vector potential the kinematic momentum operator is related to the canonical momentum operator through the minimal coupling relation

$$\hat{\mathbf{p}}_t^{kin} = \hat{\mathbf{p}}_{can} - e\mathbf{A}(\hat{\mathbf{x}},t) \ . \tag{4.77}$$

The additional terms in the Hamiltonian due to the coupling to the vector potential

$$\hat{H}_{A(t)} = \frac{-e}{2m} \left(\hat{\mathbf{p}}_{can} \cdot \mathbf{A}(\hat{\mathbf{x}}, t) + \mathbf{A}(\hat{\mathbf{x}}, t) \cdot \hat{\mathbf{p}}_{can} \right) + \frac{e^2}{2m} \mathbf{A}^2(\hat{\mathbf{x}}, t) \qquad (4.78)$$

generate in the equation of motion for the density matrix the terms

$$<\mathbf{x}|[\hat{H}_{A(t)}, \hat{\rho}(t)]|\mathbf{x}'> \;=\; \left(\frac{-e\hbar}{im} (\mathbf{A}(\mathbf{x}, t) \cdot \nabla_{\mathbf{x}} + \mathbf{A}(\mathbf{x}', t) \cdot \nabla_{\mathbf{x}'}) \right.$$

$$- \frac{e\hbar}{2im} (\nabla_{\mathbf{x}} \cdot \mathbf{A}(\mathbf{x}, t) + \nabla_{\mathbf{x}'} \cdot \mathbf{A}(\mathbf{x}', t))$$

$$+ \left. \frac{e^2}{2m} \left(\mathbf{A}^2(\mathbf{x}, t) - \mathbf{A}^2(\mathbf{x}', t) \right) \right) \rho(\mathbf{x}, \mathbf{x}', t) . \qquad (4.79)$$

Assuming a spatially homogeneous magnetic field, and choosing the symmetric gauge $\mathbf{A}(\mathbf{x}, t) = \frac{1}{2} \mathbf{B}(t) \times \mathbf{x}$, we obtain

$$\left\langle \mathbf{R} + \frac{\mathbf{r}}{2} \left| [\hat{H}, \hat{\rho}(t)] \right| \mathbf{R} - \frac{\mathbf{r}}{2} \right\rangle$$

$$= -\frac{\hbar^2}{m} \left(\nabla_{\mathbf{R}} + \frac{ie}{\hbar} \mathbf{A}(\mathbf{r}, t) \right) \cdot \left(\nabla_{\mathbf{r}} - \frac{ie}{\hbar} \mathbf{A}(\mathbf{R}, t) \right) \rho(\mathbf{R}, \mathbf{r}, t) . \qquad (4.80)$$

In a kinetic description we must use the kinematic momentum, not the gauge-dependent canonical one, so in the presence of a vector potential we shall define the Wigner function according to

$$f(\mathbf{p}, \mathbf{R}, t) \;=\; \int d\mathbf{r} \, e^{-\frac{i}{\hbar}\mathbf{r} \cdot (\mathbf{p} + e\mathbf{A}(\mathbf{R}, t))} \left\langle \mathbf{R} + \frac{\mathbf{r}}{2} \right| \hat{\rho}(t) \left| \mathbf{R} - \frac{\mathbf{r}}{2} \right\rangle \qquad (4.81)$$

and we get the driving terms

$$\frac{i}{\hbar} \int d\mathbf{r} \, e^{-\frac{i}{\hbar}\mathbf{r} \cdot (\mathbf{p} + e\mathbf{A}(\mathbf{R}, t))} \left\langle \mathbf{R} + \frac{\mathbf{r}}{2} \left| [\hat{H}, \hat{\rho}(t)] \right| \mathbf{R} - \frac{\mathbf{r}}{2} \right\rangle \;=\; e(\mathbf{v}_{\mathbf{p}} \times \mathbf{B}(t)) \cdot \frac{\partial f(\mathbf{R}, \mathbf{p}, t)}{\partial \mathbf{p}}$$

$$+ \; \mathbf{v}_{\mathbf{p}} \cdot \frac{\partial f(\mathbf{R}, \mathbf{p}, t)}{\partial \mathbf{R}} \qquad (4.82)$$

where $\mathbf{v}_{\mathbf{p}}$ is the group velocity, eq.(4.69). The driving term due to a homogeneous magnetic field is thus the Lorentz force term.[9]

---

[9] The driving term due to the electric field is generated by the time derivative term in eq.(4.2).

Exercise 4.9 *Consider a charged particle of mass m moving in a spatially homogeneous and time-independent magnetic field (whose direction is chosen as the z-direction). Show that the Wigner function at time t is related to the Wigner function at time t = 0 through the same relationship as the classical distribution function*

$$f(\mathbf{R}, \mathbf{p}, t) \;=\; f\left(\mathbf{R}_\perp(-t),\; R_z - \frac{p_z}{m}t,\; \mathbf{p}_\perp(-t),\; p_z, 0\right) \tag{4.83}$$

where $\mathbf{R}_\perp(-t)$ and $\mathbf{p}_\perp(-t)$ are determined by the equations

$$\mathbf{R}_\perp - \mathbf{R}_\perp(-t) = \frac{1}{m\omega_c}\,\hat{\mathbf{B}} \times (\mathbf{p}_\perp - \mathbf{p}_\perp(-t)) \tag{4.84}$$

and

$$\mathbf{p}_\perp - \mathbf{p}_\perp(-t) = -m\omega_c\,\hat{\mathbf{B}} \times (\mathbf{R}_\perp - \mathbf{R}_\perp(-t)) \tag{4.85}$$

the solution of the classical equation of motion, describing the circular motion (with the frequency $\omega_c = eB/m$ in the plane perpendicular to the magnetic field) for which the particle at time t is in the phase space point $(\mathbf{R}_\perp, \mathbf{p}_\perp)$.

Solution

Introducing Wigner coordinates, we get for the Wigner function

$$f(\mathbf{R}, \mathbf{p}, t) \;=\; \int d\mathbf{r}\, e^{-\frac{i}{\hbar}\mathbf{r}\cdot(\mathbf{p}+e\mathbf{A}(\mathbf{R}))} \int d\tilde{\mathbf{R}} \int d\tilde{\mathbf{r}} \int \frac{d\tilde{\mathbf{p}}}{(2\pi\hbar)^3}\, e^{\frac{i}{\hbar}(\tilde{\mathbf{p}}+e\mathbf{A}(\mathbf{R}))\cdot\tilde{\mathbf{r}}} f(\tilde{\mathbf{R}}, \tilde{\mathbf{p}}, 0)$$

$$G_{\mathbf{B}}^R(\mathbf{R}+\mathbf{r}/2, t; \tilde{\mathbf{R}}+\tilde{\mathbf{r}}/2, 0)\, G_{\mathbf{B}}^A(\tilde{\mathbf{R}}-\tilde{\mathbf{r}}/2, 0; \mathbf{R}-\mathbf{r}/2, t) \tag{4.86}$$

where in the symmetric gauge $\mathbf{A}(\mathbf{R}) = \mathbf{B} \times \mathbf{R}/2$.

Using the result of appendix A, eq.(A.25), for the propagators in the presence of a magnetic field (the motion in the z-direction is the same as in the absence of the magnetic field) we obtain

$$f(\mathbf{R}_\perp, R_z, \mathbf{p}_\perp, p_z, t) \;=\; \left(\frac{\frac{\omega_c t}{2}}{\sin\frac{\omega_c t}{2}}\right)^2 \int d\mathbf{r}_\perp\, e^{-\frac{i}{\hbar}\mathbf{r}_\perp\cdot(\mathbf{p}+e\mathbf{A}(\mathbf{R}))} \int d\tilde{\mathbf{R}}_\perp \int d\tilde{\mathbf{r}}_\perp$$

$$e^{\frac{i}{\hbar}\frac{m\omega_c}{2}(R_x\tilde{r}_y + \tilde{R}_y r_x - R_y\tilde{r}_x - \tilde{R}_x r_y)}$$

$$e^{\frac{i}{\hbar}\frac{m\omega_c}{2}\cot\frac{\omega_c t}{2}[(R_x-\tilde{R}_x)(r_x-\tilde{r}_x)+(R_y-\tilde{R}_y)(r_y-\tilde{r}_y)]}$$

$$e^{\frac{i}{\hbar}\frac{m\omega_c}{2}(-r_x R_y + R_x r_y + \tilde{r}_x\tilde{R}_y - \tilde{r}_y\tilde{R}_x)}$$

$$\int \frac{d\tilde{\mathbf{p}}_\perp}{(2\pi\hbar)^3}\, e^{\frac{i}{\hbar}(\tilde{\mathbf{p}}+e\mathbf{A}(\mathbf{R}))\cdot\tilde{\mathbf{r}}_\perp} f(\tilde{\mathbf{R}}_\perp, R_z - p_z t/m, \tilde{\mathbf{p}}_\perp, p_z, 0). \tag{4.87}$$

Upon performing the $\mathbf{r}_\perp$ and $\tilde{\mathbf{r}}_\perp$ integrations, we obtain delta functions which enforce time evolution according to the classical equations of motion.

4.5 Wigner Function Dynamics

We now turn to the description of the term on the right side of the equation of motion for the density matrix, eq.(4.27), the term due to the scattering by the random potential. This integro-differential equation for the density matrix can be turned into an equation for the Wigner function by Fourier transformation, and we obtain, assuming a spatially constant force, $\mathbf{F}(t)$, the equation $(U(\mathbf{x}, t) = \mathbf{F}(t) \cdot \mathbf{x})$

$$\frac{\partial f(\mathbf{R}, \mathbf{p}, t)}{\partial t} + \mathbf{v_p} \cdot \frac{\partial f(\mathbf{R}, \mathbf{p}, t)}{\partial \mathbf{R}} + \mathbf{F}(t) \cdot \frac{\partial f(\mathbf{R}, \mathbf{p}, t)}{\partial \mathbf{p}} = \tilde{F}_{\mathbf{R}, \mathbf{p}, t}[f] \quad (4.88)$$

where

$$\tilde{F}_{\mathbf{R}, \mathbf{p}, t}[f] = \int \frac{d\tilde{\mathbf{p}}}{(2\pi\hbar)^3} \int d\tilde{\mathbf{R}} \int_{t'}^{t} d\tilde{t} \, \tilde{J}\left(\mathbf{p}, \mathbf{R}, t; \tilde{\mathbf{p}}, \tilde{\mathbf{R}}, \tilde{t}\right) f(\tilde{\mathbf{R}}, \tilde{\mathbf{p}}, \tilde{t}) \quad (4.89)$$

is expressed in terms of the irreducible propagator of the Wigner function

$$\tilde{J}(\mathbf{R}, \mathbf{p}, t; \tilde{\mathbf{R}}, \tilde{\mathbf{p}}, \tilde{t}) = \int d\mathbf{r} \int d\tilde{\mathbf{r}} \, e^{-\frac{i}{\hbar}\mathbf{p}\mathbf{r} + \frac{i}{\hbar}\tilde{\mathbf{p}}\cdot\tilde{\mathbf{r}}} \, \tilde{J}\left(\mathbf{R} + \frac{1}{2}\mathbf{r}, \mathbf{R} - \frac{1}{2}\mathbf{r}, t; \tilde{\mathbf{R}} + \frac{1}{2}\tilde{\mathbf{r}}; \tilde{\mathbf{R}} - \frac{1}{2}\tilde{\mathbf{r}}, \tilde{t}\right) . \quad (4.90)$$

Each diagram for the irreducible propagator in the spatial representation can, by simply Fourier transforming, be labeled by the Wigner coordinates. We note, that the quantum kinetic equation, eq.(4.88), is nonlocal in the space and time variables due to the term on the right-hand side.

A diagrammatic technique is only useful for calculational purposes if the diagrammatic structure has easily recognizable features to ensure that fundamental properties of physical quantities are respected. For example, what in the diagrammatic structure reflects that the Wigner function is real? As we shall see shortly, it is an easily recognizable symmetry between the diagrams of the irreducible propagator $\tilde{J}$. For any diagram, d, for the irreducible propagator of the density matrix, with its analytical expression denoted by $\tilde{J}^d(\mathbf{x}, \mathbf{x}', t; \tilde{\mathbf{x}}, \tilde{\mathbf{x}}', \tilde{t})$, labeled by its external points, there is a symmetric diagram, $\bar{d}$, obtained by mirror reflection of the impurity lines in a line that is parallel to and in between the upper and lower particle lines. The diagrams in the last line of eq.(4.25) are an example containing two such pairs of mirror diagrams, and the general structure of the diagrams, as displayed, clearly allows for such a symmetry operation. Using the relationship between the retarded and advanced propagators, eq.(2.22), and the trivial property

$$< V(\mathbf{x})V(\mathbf{x}') >^* = < V(\mathbf{x}')V(\mathbf{x}) > \quad (4.91)$$

since V is real, it follows that

$$\tilde{J}^d(\mathbf{x}, \mathbf{x}', t; \tilde{\mathbf{x}}, \tilde{\mathbf{x}}', \tilde{t}) = \tilde{J}^{\bar{d}}(\mathbf{x}', \mathbf{x}, t; \tilde{\mathbf{x}}', \tilde{\mathbf{x}}, \tilde{t}) . \quad (4.92)$$

Transforming to Wigner coordinates, we therefore have the property

$$[\tilde{J}^d(\mathbf{p}, \mathbf{R}, t; \tilde{\mathbf{p}}, \tilde{\mathbf{R}}, \tilde{t})]^* = \tilde{J}^{\bar{d}}(\mathbf{p}, \mathbf{R}, t; \tilde{\mathbf{p}}, \tilde{\mathbf{R}}, \tilde{t}) . \quad (4.93)$$

The fact that the Wigner function is real is thus reflected in the diagrammatic structure by the two symmetric classes the diagrams fall into, as allowed in general by the two classes of diagrams in eq.(4.25).

4.6   Weak-Coupling Kinetic Equation

In this section we shall study the time evolution of the density matrix or equivalently the Wigner function, in the weak-coupling limit; i.e., we shall treat the random potential to lowest order. The irreducible propagator of the impurity-averaged density matrix is in the weak-coupling limit given by the four lowest-order diagrams depicted in the last line of eq.(4.25).

Let us assume that we initially start from a spatially homogeneous state, i.e., the Wigner function is independent of the spatial variable, $f(\mathbf{p}, \mathbf{R}, t') = f(\mathbf{p}, t')$, and the Wigner function thus has the interpretation of being proportional to the momentum probability distribution. The Wigner function will then be independent of its spatial coordinate at all times since we assume the external force, $\mathbf{F}(t)$, to be spatially homogeneous. Along with the integrations over $\mathbf{r}$ and $\tilde{\mathbf{r}}$, we can therefore perform the integration over $\tilde{\mathbf{R}}$ on the right-hand side of the weak-coupling kinetic equation, as a consequence of which the right-hand side becomes independent of $\mathbf{R}$.

Since we study the effect of the random potential to lowest order, we substitute the free particle propagators for the full propagators

$$G_0^{R(A)}(\mathbf{x}, t; \mathbf{x}', t') = -i\theta(t - t') \left(\frac{m}{(\overset{+}{-})2\pi\hbar i (t - t')} \right)^{3/2} e^{\frac{i}{\hbar} \frac{m(\mathbf{x}-\mathbf{x}')^2}{2(t-t')}} \qquad (4.94)$$

because for a start, we neglect the effect of the external force on the right-hand side of eq.(4.88).[10]

As dictated by the introduction of the Wigner function, we need the propagator in the Wigner coordinates (obtainable by simple Fourier transformation)

$$\begin{aligned} G_0^R(\mathbf{p}, t - t') &= \int d\mathbf{r} \, e^{-\frac{i}{\hbar}\mathbf{p}\cdot\mathbf{r}} \, G_0^R(\mathbf{r}, t, t') \\ &= -i\theta(t - t') \, e^{-\frac{i}{\hbar}\epsilon_\mathbf{p}(t-t')} \end{aligned} \qquad (4.95)$$

which for the considered free case leaves the propagator independent of the spatial coordinate. In general we note the relationship

$$G^A(\mathbf{R}, \mathbf{p}; t, t') = [G^R(\mathbf{R}, \mathbf{p}; t', t)]^* \qquad (4.96)$$

between the advanced and retarded propagator in the Wigner coordinates.

We start by considering the first first-order diagram in eq.(4.23)

$$-\qquad \begin{array}{c} \times \\ \mathbf{x}=\mathbf{R}+\mathbf{r}/2 \longleftarrow \bullet \quad \tilde{\mathbf{x}}=\tilde{\mathbf{R}}+\tilde{\mathbf{r}}/2 \\ \mid \quad \tilde{t} \\ \mathbf{x}'=\mathbf{R}-\mathbf{r}/2 \longrightarrow \bullet \quad \tilde{\mathbf{x}}'=\tilde{\mathbf{R}}-\tilde{\mathbf{r}}/2 \ . \end{array} \qquad (4.97)$$

---

[10]In the subsequent exercise we include the effect of the homogeneous external force to all orders.

All the first-order diagrams contain the product of the two propagators

$$G_0^R(\mathbf{x}, t; \tilde{\mathbf{x}}, \tilde{t})\, G_0^A(\mathbf{x}', t; \tilde{\mathbf{x}}', \tilde{t}) \;=\; \theta(t - \tilde{t}) \int \frac{d\mathbf{p}_1}{(2\pi\hbar)^3} \int \frac{d\mathbf{p}_2}{(2\pi\hbar)^3}\; e^{\frac{i}{\hbar}(\mathbf{R} - \tilde{\mathbf{R}}) \cdot (\mathbf{p}_1 - \mathbf{p}_2)}$$

$$e^{-\frac{i}{\hbar}(t - \tilde{t})(\epsilon_{\mathbf{p}_1} - \epsilon_{\mathbf{p}_2}) + \frac{i}{\hbar}(\mathbf{p}_1 + \mathbf{p}_2) \cdot \frac{\mathbf{r} - \tilde{\mathbf{r}}}{2}} \tag{4.98}$$

and introducing the Fourier transform of the impurity correlator

$$<V(\mathbf{x})V(\tilde{\mathbf{x}})> \;=\; \frac{n_i}{V} \sum_{\mathbf{p}'} e^{\frac{i}{\hbar}\mathbf{p} \cdot (\mathbf{R} - \tilde{\mathbf{R}} + \frac{\mathbf{r} - \tilde{\mathbf{r}}}{2})}\; |V_{imp}(\mathbf{p}')|^2 \tag{4.99}$$

all spatial dependences are through simple phase factors providing three delta functions in the momentum variables. The subsequent momentum integrations are therefore trivial, and we obtain for the diagram in question, in the Wigner coordinates, the contribution

$$\tilde{F}_{out}^{(1)}[f] = -\frac{n_i}{V\hbar^2} \sum_{\mathbf{p}'} \int_{t'}^{t} d\tilde{t}\; \theta(t - \tilde{t})\, e^{\frac{i}{\hbar}(\tilde{t} - t)(\epsilon_{\mathbf{p}'} - \epsilon_{\mathbf{p}})}\, |V_{imp}(\mathbf{p} - \mathbf{p}')|^2\, f(\mathbf{p}, \tilde{t}) . \tag{4.100}$$

In the calculation of the contribution from the diagram

$$\tag{4.101}$$

the only change, as compared to the above calculation, involves the sign attributed to the diagram, and the change in the impurity correlator coordinates

$$<V(\mathbf{x})V(\tilde{\mathbf{x}})> \;\rightarrow\; -<V(\tilde{\mathbf{x}}')V(\mathbf{x})> \;=\; -<V(\tilde{\mathbf{R}} - \tilde{\mathbf{r}}/2)V(\mathbf{R} + \mathbf{r}/2)> \tag{4.102}$$

and we obtain in the Wigner coordinates the contribution

$$\tilde{F}_{in}^{(1)}[f] \;=\; \frac{n_i}{V\hbar^2} \sum_{\mathbf{p}'} \int_{t'}^{t} d\tilde{t}\; \theta(t - \tilde{t})\, e^{\frac{i}{\hbar}(\tilde{t} - t)(\epsilon_{\mathbf{p}'} - \epsilon_{\mathbf{p}})}\, |V_{imp}(\mathbf{p} - \mathbf{p}')|^2\, f(\mathbf{p}', \tilde{t}) . \tag{4.103}$$

The two terms have been indexed *in* and *out* since in the classical limit they correspond, as we shall see, to the scattering in and out terms in the Boltzmann description.

Using that symmetric diagrams give complex conjugate contributions, eq.(4.93), we obtain the collision-type integral $\tilde{F}^{(1)}$ for the lowest-order diagrams

$$\tilde{F}_t^{(1)}[f] \;=\; 2\mathrm{Re}\,(\tilde{F}_{in}^{(1)}[f] + \tilde{F}_{out}^{(1)}[f])$$

$$=\; -\frac{2n_i}{V\hbar^2} \sum_{\mathbf{p}'} \int_{t'}^{t} d\tilde{t}\; \theta(t - \tilde{t}) \cos\left(\frac{1}{\hbar}(\tilde{t} - t)(\epsilon_{\mathbf{p}'} - \epsilon_{\mathbf{p}})\right)$$

$$|V_{imp}(\mathbf{p} - \mathbf{p}')|^2\, [f(\mathbf{p}, \tilde{t}) - f(\mathbf{p}', \tilde{t})] \tag{4.104}$$

leading to the weak-coupling equation

$$\frac{\partial f(\mathbf{p}, t)}{\partial t} + \mathbf{F}(t) \cdot \frac{\partial f(\mathbf{p}, t)}{\partial \mathbf{p}} = \tilde{F}_t^{(1)}[f] \ . \tag{4.105}$$

We note that this equation is local in the spatial coordinate, but still nonlocal in time.

Exercise 4.10 *Show that if we include the effect of a spatially homogeneous time-independent external force, $\mathbf{F}$, to arbitrary order, we get in the weak disorder limit the collision integral*

$$\tilde{F}_t^{(1)}[f] = -\sum_{\mathbf{p}'} \int_{t'}^{t} d\tilde{t}\, \theta(t - \tilde{t}) \cos\left(\frac{1}{\hbar}(\tilde{t} - t)\left\{\epsilon\left(\mathbf{p}' - \frac{\mathbf{F}(t - \tilde{t})}{2}\right) - \epsilon\left(\mathbf{p} - \frac{\mathbf{F}(t - \tilde{t})}{2}\right)\right\}\right)$$

$$\frac{2 n_i}{V \hbar^2} |V_{imp}(\mathbf{p} - \mathbf{p}')|^2 \left[f(\mathbf{p} - \mathbf{F}(t - \tilde{t}), \tilde{t}) - f(\mathbf{p}' - \mathbf{F}(t - \tilde{t}), \tilde{t}) \right] . \tag{4.106}$$

Solution

We insert the propagator in the presence of the field, eq.(A.19) or eq.(4.39), and since the integration in eq.(4.89) is still over simple exponential factors they are immediately performed, leading to the above result. The collision integral takes into account the effect of the field during the collision, the intracollisional field effect. Furthermore, the result is straightforwardly generalized to a time-dependent field by the substitution

$$\mathbf{p} \to \mathbf{p} - \int_{t'}^{t} dt'' \, \mathbf{F}(t'') \ . \tag{4.107}$$

4.7   Classical Kinetics

We now study the collision integral, eq.(4.105), in the classical limit, which we formally obtain by letting the quantum of action approach zero, $\hbar \to 0$. We can then make use of the identity proved in appendix B, eq.(B.27),

$$\theta(t - \tilde{t}) \lim_{\hbar \to 0} \frac{1}{\hbar} \cos\left\{\frac{1}{\hbar}(t - \tilde{t})(\epsilon_{\mathbf{p}'} - \epsilon_{\mathbf{p}})\right\} = \pi\, \delta(t - \tilde{t})\, \delta(\epsilon_{\mathbf{p}'} - \epsilon_{\mathbf{p}}) \ . \tag{4.108}$$

In the classical limit we can therefore perform the time integration to obtain the local in time expression

$$I_t^{(1)}[f] = -\sum_{\mathbf{p'}} \{W(\mathbf{p'}, \mathbf{p})f(\mathbf{p}, t) - W(\mathbf{p}, \mathbf{p'})f(\mathbf{p'}, t)\} \tag{4.109}$$

where

$$\begin{aligned} W(\mathbf{p'}, \mathbf{p}) &= \frac{2\pi n_i}{\hbar V} |V_{imp}(\mathbf{p} - \mathbf{p'})|^2 \delta(\epsilon_{\mathbf{p'}} - \epsilon_{\mathbf{p}}) \\ &= \frac{2\pi}{\hbar} n_i V |<\mathbf{p}|V_{imp}(\hat{\mathbf{x}})|\mathbf{p'}>|^2 \delta(\epsilon_{\mathbf{p'}} - \epsilon_{\mathbf{p}}) \ . \end{aligned} \tag{4.110}$$

We note that in the Born approximation we always have $W(\mathbf{p'}, \mathbf{p}) = W(\mathbf{p}, \mathbf{p'})$, but this is generally true due to time-reversal symmetry and space inversion symmetry.[11]

We then have the following equation for the homogeneous Wigner function for a particle in a weak random potential and a spatially homogeneous field, $\mathbf{F}(t)$,

$$\frac{\partial f(\mathbf{p}, t)}{\partial t} + \mathbf{F}(t) \cdot \frac{\partial f(\mathbf{p}, t)}{\partial \mathbf{p}} = I_{\mathbf{p}, t}^{(1)}[f] \tag{4.111}$$

with the collision integral given by

$$I_t^{(1)}[f] = -\sum_{\mathbf{p'}} W(\mathbf{p'}, \mathbf{p}) (f(\mathbf{p}, t) - f(\mathbf{p'}, t)) \ . \tag{4.112}$$

We have arrived at the classical kinetic equation describing the motion of a particle in a weakly disordered system, the Boltzmann equation for a particle in a random potential.[12] The derived equation is called a kinetic equation because the collision integral is not a functional in time (or space), i.e., local in both the space and time variable, and only a functional with respect to the momentum.

We note that the expression $W(\mathbf{p'}, \mathbf{p})$ is Fermi's golden rule expression for the transition probability per unit time from momentum state $\mathbf{p}$ to momentum state $\mathbf{p'}$ (or vice versa) caused by the scattering off an impurity, times the number of impurities. The two terms in the collision integral thus have a simple interpretation because they describe the scattering in and out of a momentum state. For example, the first term in the collision integral of the Boltzmann equation, eq.(4.112), is a loss term, and gives the rate of change of occupation of a phase space volume due to the scattering of the particle from momentum $\mathbf{p}$ to momentum $\mathbf{p'}$ by the random potential. The probability per unit time to be scattered out of the phase space

---

[11]Potential scattering is time-reversal invariant, so according to eq.(2.207) we always have $W(\mathbf{p'}, \mathbf{p}) = W(-\mathbf{p}, -\mathbf{p'})$. If in addition the potential is invariant with respect to space inversion, we have according to eq.(2.204) $W(\mathbf{p'}, \mathbf{p}) = W(-\mathbf{p'}, -\mathbf{p})$, and thereby $W(\mathbf{p'}, \mathbf{p}) = W(\mathbf{p}, \mathbf{p'})$.

[12]In fact, the Boltzmann equation, eq.(4.111), has a much wider range of applicability than the above consideration suggests. Furthermore, Planck's constant is a *constant* and the above argument should be improved, as done in section 6.7 where we show that the corrections to the Boltzmann kinetics are of order $\hbar/\epsilon_{\mathbf{p}}\tau$.

volume around $\mathbf{p}$, and into a volume around $\mathbf{p}'$ is the product of three probabilities: (the probability that the particle is in that phase space volume to be available for scattering) $\times$ (the transition probability per unit time for the transition from state $\mathbf{p}$ to $\mathbf{p}'$) $\times$ (the probability that there is an impurity in the space volume to scatter). Similarly we have the interpretation of the other term as a scattering-in term.

The obtained equation is a quasi-classical equation, because in between collisions the particle moves along the straight line just as in classical mechanics, but the scattering cross section is the quantum mechanical one.[13]

The weak-disorder kinetic equation for a particle in a random potential is of course immediately obtained from classical mechanics, granted the stochastic treatment of the impurity scattering. In classical mechanics the distribution function concept is unproblematic because we can simultaneously specify position and momentum. and the left-hand side is simply the streaming terms in phase space for the situation in question. However, the classical cross section for scattering will appear in the equation.

The presented diagrammatic method to derive transport equations is capable of going beyond the Markov process described by the classical kinetic equation, to include quantum effects (the topic of section 8.9).

For the sole purpose of obtaining the weak-disorder kinetic equation, the use of Feynman diagrams is not necessary. However, it allows us in a simple way, by using the result of section 3.6, to assess a validity criterion for the classical kinetic description. In view of the neglected diagrams, the validity of the Boltzmann equation requires for the momentum of the particle to satisfy $p \gg \hbar/l$; i.e., the de Broglie wavelength of the particle must be much smaller than the mean free path.

Let us study the simplest nonequilibrium situation where the distribution is out of momentum equilibrium only for a single momentum value

$$f_{\mathbf{p}'}(t) = f(\epsilon_{\mathbf{p}'}) + \delta f_{\mathbf{p}}(t)\, \delta_{\mathbf{p},\mathbf{p}'} \tag{4.113}$$

and we assume no external fields. The Boltzmann equation then reduces to

$$\frac{\partial \delta f_{\mathbf{p}}(t)}{\partial t} = -\frac{\delta f_{\mathbf{p}}}{\tau_{\mathbf{p}}} \tag{4.114}$$

whose solution describes the exponential relaxation to equilibrium

$$f_{\mathbf{p}}(t) = f(\epsilon_{\mathbf{p}}) + \delta f_{\mathbf{p}}(t=0)\, e^{-t/\tau_{\mathbf{p}}} \tag{4.115}$$

and the momentum relaxation time

$$\frac{1}{\tau_{\mathbf{p}}} = \sum_{\mathbf{p}'(\neq \mathbf{p})} W_{\mathbf{p}',\mathbf{p}} \tag{4.116}$$

is seen to be identical to the imaginary part of the self-energy, eq.(3.64), for $E = \epsilon_{\mathbf{p}}$

$$\frac{1}{\tau_{\mathbf{p}}} = \frac{1}{\tau(E = \epsilon_{\mathbf{p}}, \mathbf{p})} . \tag{4.117}$$

---

[13]If we go beyond the considered Born approximation, and include multiple scattering, we obtain the exact cross-section for scattering off an impurity as expressed by the t-matrix.

4.8 Validity of the Boltzmann Equation

In the derivation of the weak-coupling kinetic equation, eq.(4.111), we assumed a spatially homogeneous field and a spatially homogeneous state, and did not need to invoke any spatial properties of the propagator of the Wigner function. To discuss spatially nonhomogeneous fields, we must take into account the intrinsic length scale in the problem, the mean free path. This is easily done as we observe that in the skeleton diagram expansion all propagator lines are the exact impurity propagators which, according to eq.(3.73), have the range of the mean free path. Obtaining an equation local in the spatial coordinate can now be justified for a spatially nonhomogeneous state if the spatial scale of $\tilde{J}_1$, i.e., l, is much larger than the de Broglie wavelength of the particle, $\lambda \equiv \hbar/p$. We simply note, that when expanding the Wigner function in the collision integral around the point $\mathbf{R}$, specified by the argument of the Wigner function on the left-hand side of the kinetic equation, the spatial scale of the integration region is set by the range of the propagators, and on comparison with the drift term the nonlocal terms can be neglected in the limit where $p \gg \hbar/l$.[14] We then obtain the kinetic equation for the case of a spatially varying field

$$\frac{\partial f(\mathbf{R}, \mathbf{p}, t)}{\partial t} + \mathbf{F}(\mathbf{R}, t) \cdot \frac{\partial f(\mathbf{R}, \mathbf{p}, t)}{\partial \mathbf{p}} + \mathbf{v_p} \cdot \frac{\partial f(\mathbf{R}, \mathbf{p}, t)}{\partial \mathbf{R}} = I_{\mathbf{R}t}[f] \qquad (4.118)$$

an equation local in both space and time.

A general discussion of the validity of the Boltzmann equation is given in section 6.7.

Exercise 4.11 *Show that the continuity equation is obtained by integrating the kinetic equation, eq.(4.118), with respect to the momentum variable.*

4.8.1 The Master Equation

We noted above that the collision integral rendered the kinetic equation a stochastic equation for the momentum. We can easily establish the emergence of such an equation. Let us assume that the density matrix at time t is diagonal

$$<\mathbf{p}|\hat{\rho}(t)|\mathbf{p}'> = P_{\mathbf{p}}(t)\, \delta_{\mathbf{p}, \mathbf{p}'} \qquad (4.119)$$

---

[14] This so-called Landau criterion is not sufficient for the applicability of the Boltzmann equation in low-dimensional systems, $d \leq 2$. A subject we shall discuss in detail in chapters 9 and 11.

where the probability $P_{\mathbf{p}}(t) = Tr(\hat{\rho}(t)\hat{P}(\mathbf{p}))$ for finding the particle in momentum state $\mathbf{p}$ at time t, according to eq.(4.36), is related to the Wigner function by

$$P_{\mathbf{p}}(t) = \int \frac{d\mathbf{R}}{(2\pi\hbar)^3} \, f(\mathbf{R}, \mathbf{p}, t) = f_{\mathbf{p}}(t) \, . \qquad (4.120)$$

At a later time the probability for the particle to have momentum $\mathbf{p}$ has changed to

$$P_{\mathbf{p}}(t + \Delta t) = <\mathbf{p}| \, \hat{\rho}(t + \Delta t) \, |\mathbf{p}> = <\mathbf{p}|\hat{U}(t + \Delta t, t) \, \hat{\rho}(t) \, \hat{U}^{\dagger}(t + \Delta t, t)|\mathbf{p}>$$

$$= \sum_{\mathbf{p}'} T_{\mathbf{p},\mathbf{p}'}(\Delta t, t) \, P_{\mathbf{p}'}(t) \qquad (4.121)$$

where the last line is only obtained because the momentum representation of the statistical operator is assumed diagonal at time t, and

$$T_{\mathbf{p},\mathbf{p}'}(\Delta t, t) = |<\mathbf{p}|\hat{U}(t + \Delta t, t)|\mathbf{p}'>|^2 = |G^R(\mathbf{p}, t + \Delta t; \mathbf{p}', t)|^2 \qquad (4.122)$$

is the probability that the momentum state of the particle in time span Δt is changed from state $\mathbf{p}'$ to $\mathbf{p}$. Assuming the elementary process which gives rise to scattering is without history (as is the case for time-independent Hamiltonians), $T_{\mathbf{p},\mathbf{p}'}(\Delta t, t)$ is independent of the time t in question. Conservation of probability, unitarity of the evolution operator, requires the normalization $\sum_{\mathbf{p}} T_{\mathbf{p},\mathbf{p}'}(\Delta t) = 1$, and for a short time interval we have

$$T_{\mathbf{p},\mathbf{p}'}(\Delta t) = \begin{cases} W_{\mathbf{p},\mathbf{p}'} \, \Delta t & \mathbf{p} \neq \mathbf{p}' \\ \\ 1 - W_{\mathbf{p}',\mathbf{p}'} \, \Delta t & \mathbf{p} = \mathbf{p}' \end{cases} \qquad (4.123)$$

where we have introduced the transition probability per unit time

$$W_{\mathbf{p},\mathbf{p}'} \equiv \lim_{\Delta t \to 0} \frac{T_{\mathbf{p},\mathbf{p}'}(\Delta t)}{\Delta t} \qquad \mathbf{p} \neq \mathbf{p}' \qquad (4.124)$$

and

$$W_{\mathbf{p}',\mathbf{p}'} = \sum_{\mathbf{p}(\neq \mathbf{p}')} W_{\mathbf{p},\mathbf{p}'} \, . \qquad (4.125)$$

Furthermore, we note from eq.(4.122) that $T_{\mathbf{p}',\mathbf{p}}(\Delta t) = T_{\mathbf{p},\mathbf{p}'}(\Delta t)$. Then, using $\sum_{\mathbf{p}'} T_{\mathbf{p}',\mathbf{p}}(\Delta t) = 1$, we obtain for the rate of change of the momentum probability distribution

$$\dot{P}_{\mathbf{p}}(t) = \sum_{\mathbf{p}'} (W_{\mathbf{p},\mathbf{p}'} P_{\mathbf{p}'}(t) - W_{\mathbf{p}',\mathbf{p}} P_{\mathbf{p}}(t)) \qquad (4.126)$$

Pauli's master equation.

We note, that in our derivation of the master equation we only had to assume that the density matrix is diagonal at the initial time. The impurity average done in the previous sections thus effectively corresponds to that after each collision all nondiagonal elements of the density matrix are phase-averaged to zero. The question as to whether there are important quantum coherence effects neglected by the phase averaging we shall address in chapters 9 and 11.

The transition rate for making a transition from $\mathbf{p}$ to any other $\mathbf{p}'$ is

$$\frac{1}{\tau(\mathbf{p})} = \sum_{\mathbf{p}'(\neq\mathbf{p})} W_{\mathbf{p}',\mathbf{p}} . \qquad (4.127)$$

In the case where the momentum relaxation time, $\tau(\mathbf{p})$, can be considered independent of $\mathbf{p}$, τ is the phenomenological parameter of the Drude theory of conduction, $\Delta t/\tau(\mathbf{p})$ giving the probability that a particle with momentum $\mathbf{p}$ in the time span Δt will suffer a collision with total loss of momentum direction memory [24]. Such an assumption is not valid in the quantum mechanical description as the scattering of a wave sets up correlations that can not lead to a total memory loss in general, as we shall discuss in detail in chapters 9 and 11.

4.9 Classical Mobility

The classical motion of a particle in a random potential under the influence of a constant force $\mathbf{F}$ is, in the weak disorder limit, described by the kinetic equation

$$\frac{\partial f(\mathbf{R}, \mathbf{p}, t)}{\partial t} + \mathbf{F} \cdot \frac{\partial f(\mathbf{R}, \mathbf{p}, t)}{\partial \mathbf{p}} + \mathbf{v_p} \cdot \frac{\partial f(\mathbf{R}, \mathbf{p}, t)}{\partial \mathbf{R}} = I_{\mathbf{R},t}[f] \qquad (4.128)$$

where, as derived above, we have the collision integral

$$I_{\mathbf{R},\mathbf{p},t}[f] = -\int d\mathbf{p}' \, \tilde{W}(\mathbf{p}, \mathbf{p}')[f(\mathbf{R}, \mathbf{p}, t) - f(\mathbf{R}, \mathbf{p}', t)] . \qquad (4.129)$$

We let f denote the classical distribution function, and we have introduced

$$\tilde{W}(\mathbf{p}, \mathbf{p}') \equiv \frac{W(\mathbf{p}, \mathbf{p}')V}{(2\pi\hbar)^3} \qquad (4.130)$$

the conversion factor reflecting, that for each momentum volume $\Delta\mathbf{p} = (2\pi\hbar)^3/V$ we shall count one quantum state in accordance with Heisenberg's uncertainty principle. We shall in the following assume that we consider a particle of charge e in an electromagnetic field so that the force is $\mathbf{F} = e(\mathbf{E} + \mathbf{v_p} \times \mathbf{B})$, see eq.(4.82).

From normalization it follows that $f(\mathbf{p}, |\mathbf{R}| = \infty, t) = 0$, so that by integrating the kinetic equation with respect to $\mathbf{R}$ we get the kinetic equation for the momentum distribution

$$\frac{\partial f_{\mathbf{p}}(t)}{\partial t} + \mathbf{F} \cdot \frac{\partial f_{\mathbf{p}}(t)}{\partial \mathbf{p}} = -\int d\mathbf{p}' \, \tilde{W}(\mathbf{p}, \mathbf{p}')[f_{\mathbf{p}}(t) - f_{\mathbf{p}'}(t)] . \qquad (4.131)$$

We shall here just consider the isotropic scattering model,[15] where the transition probability

$$\tilde{W}(\mathbf{p}, \mathbf{p}') = \tilde{W}(\hat{\mathbf{p}} \cdot \hat{\mathbf{p}}', \epsilon_{\mathbf{p}}) \, \delta(\epsilon_{\mathbf{p}} - \epsilon_{\mathbf{p}'}) \qquad (4.132)$$

---

[15]The potential of an impurity is thus assumed to be a delta function $V_{imp}(\mathbf{x}) \propto \delta(\mathbf{x})$. We consider the case of a spherical symmetric impurity potential, $V_{imp}(\mathbf{x}) = V_{imp}(|\mathbf{x}|)$, in section 5.4.

is independent of the angle between $\mathbf{p}$ and $\mathbf{p}'$, so that $\tilde{W}(\hat{\mathbf{p}} \cdot \hat{\mathbf{p}}', \epsilon_\mathbf{p}) = \tilde{W}_0(\epsilon_\mathbf{p})$. We then obtain the kinetic equation

$$\frac{\partial f_\mathbf{p}(t)}{\partial t} + \mathbf{F} \cdot \frac{\partial f_\mathbf{p}(t)}{\partial \mathbf{p}} = -\frac{f_\mathbf{p}(t)}{\tau(\epsilon_\mathbf{p})} + \frac{1}{\tau(\epsilon_\mathbf{p})} \int \frac{d\hat{\mathbf{p}}'}{4\pi} f_{\mathbf{p}'}(t) \tag{4.133}$$

where in the last integral it is understood that the length of the momentum $\mathbf{p}'$ is the same as that of $\mathbf{p}$, $|\mathbf{p}'| = |\mathbf{p}|$, and

$$\frac{1}{\tau(\epsilon_\mathbf{p})} = N_0(\epsilon_\mathbf{p})\tilde{W}_0(\epsilon_\mathbf{p}) . \tag{4.134}$$

The average momentum at time t is given by

$$\mathbf{P}(t) \equiv \int d\mathbf{p} \, \mathbf{p} \, f_\mathbf{p}(t) \tag{4.135}$$

and for the time derivative we have

$$\dot{\mathbf{P}}(t) \equiv \frac{d\mathbf{P}(t)}{dt} = \int d\mathbf{p} \, \mathbf{p} \, \frac{\partial f_\mathbf{p}(t)}{\partial t} \tag{4.136}$$

which upon using the kinetic equation can be rewritten

$$\dot{\mathbf{P}}(t) = \int d\mathbf{p} \, \mathbf{p} \left(-e(\mathbf{E}(t) + \mathbf{v}_\mathbf{p} \times \mathbf{B}(t)) \cdot \frac{\partial f_\mathbf{p}(t)}{\partial \mathbf{p}} - \frac{f_\mathbf{p}(t)}{\tau(\epsilon_\mathbf{p})} + \frac{1}{\tau(\epsilon_\mathbf{p})} \int \frac{d\hat{\mathbf{p}}'}{4\pi} f_{\mathbf{p}'}(t) \right) . \tag{4.137}$$

The last term is proportional to the angular integral over $\mathbf{p}$ and accordingly vanishes.

Noting that from normalization

$$\int d\mathbf{p} \, f_\mathbf{p}(t) = 1 \tag{4.138}$$

it follows that $f_{|\mathbf{p}|=\infty}(t) = 0$, and in fact as the average momentum is assumed well-defined, the stronger statement

$$[\mathbf{p} f_\mathbf{p}(t)] \Big|_{|\mathbf{p}|=\infty} = 0 \tag{4.139}$$

is valid, and we can immediately integrate the force terms by parts to obtain

$$\dot{\mathbf{P}}(t) = \mathbf{F}(\mathbf{P}(t)) - \int d\mathbf{p} \, \mathbf{p} \frac{f_\mathbf{p}(t)}{\tau(\epsilon_\mathbf{p})} . \tag{4.140}$$

Assuming that the deviation $\delta f_\mathbf{p}(t)$ from an equilibrium distribution, $f_\mathbf{p}(t) = f(\epsilon_\mathbf{p}) + \delta f_\mathbf{p}(t)$, is sufficiently peaked as a function of the length of the momentum, peaked at the momentum value $p_F = \sqrt{2mE_F}$, so that we can use the approximation

$$\int d\mathbf{p} \, \mathbf{p} \frac{1}{\tau(\epsilon_\mathbf{p})} f_\mathbf{p}(t) \simeq \frac{1}{\tau(E_F)} \int d\mathbf{p} \, \mathbf{p} \, f_\mathbf{p}(t) \tag{4.141}$$

we obtain for the equation of motion for the average momentum

$$\dot{\mathbf{P}}(t) + \frac{\mathbf{P}(t)}{\tau} = e\mathbf{E}(t) + \frac{e}{m}\mathbf{P}(t) \times \mathbf{B}(t) \qquad (4.142)$$

where we have introduced the notation $\tau \equiv \tau(E_F)$. The effect of the random potential on the average momentum is thus described by a friction term, the Drude theory.

In the case where the magnetic field is absent, and the force spatially homogeneous and time independent, we get from the preceding equation the expression for the average momentum

$$\mathbf{P}(t) = \mathbf{F}\tau(1 - e^{-t/\tau}) + \mathbf{P}_0 e^{-t/\tau} \qquad (4.143)$$

where $\mathbf{P}_0$ is the average momentum at time $t = 0$. The memory of the initial velocity direction will, due to scattering, be lost after the time span τ, and in the presence of the constant force the particle will get the average drift velocity

$$\mathbf{V}_f = \frac{\mathbf{P}(t \gg \tau)}{m} = \frac{\tau}{m}\mathbf{F} \qquad (4.144)$$

corresponding to a diagonal mobility tensor

$$\underline{\underline{\mu}} = \mu\,\delta_{\alpha\beta} \qquad (4.145)$$

where the mobility is given by

$$\mu = \frac{\tau}{m}\,. \qquad (4.146)$$

Thus the stationary solution of eq.(4.142), $\dot{\mathbf{P}} = \mathbf{0}$, is given by

$$\mathbf{P} = \tau\mathbf{F}\,. \qquad (4.147)$$

We can also consider the average kinetic energy $E(t)$ at time t

$$E(t) \equiv Tr(\hat{H}_p\,\hat{\rho}(t)) = \int d\mathbf{p}\;\epsilon_{\mathbf{p}}f_{\mathbf{p}}(t) \qquad (4.148)$$

The rate of change of the kinetic energy is

$$\dot{E}(t) \equiv \frac{dE(t)}{dt} = \int d\mathbf{p}\;\epsilon_{\mathbf{p}}\frac{\partial f_{\mathbf{p}}(t)}{\partial t} \qquad (4.149)$$

and using the kinetic equation we get, the magnetic field makes no contribution,

$$\dot{E}(t) = \int d\mathbf{p}\;\epsilon_{\mathbf{p}}\left(-e\mathbf{E}\cdot\frac{\partial f_{\mathbf{p}}(t)}{\partial \mathbf{p}} - \frac{f_{\mathbf{p}}(t)}{\tau(\epsilon_{\mathbf{p}})} + \frac{1}{\tau(\epsilon_{\mathbf{p}})}\int\frac{d\hat{\mathbf{p}}'}{4\pi}\,f_{\mathbf{p}'}(t)\right) \qquad (4.150)$$

where, as usual for elastic scattering, it is understood that in the last term we have $|\mathbf{p}'| = |\mathbf{p}|$. The two terms from the collision integral cancel, reflecting that no energy change of the electron can take place in a random elastic potential. By noting that as the average kinetic energy is assumed well-defined, we have

$$[\mathbf{p}^2 f_\mathbf{p}(t)]\Big|_{|\mathbf{p}|=\infty} = 0 \tag{4.151}$$

so that, by partial integration, the electric field term gives the following rate of change in the kinetic energy:

$$\dot{E}(t) = \frac{e}{m}\int d\mathbf{p}\ \mathbf{E}\cdot\mathbf{p}\, f_\mathbf{p}(t) = \frac{e}{m}\mathbf{E}\cdot\mathbf{P}(t) = \frac{e}{m}\mathbf{E}\cdot(e\mathbf{E}\tau(1-e^{-t/\tau})+\mathbf{P}_0\, e^{-t/\tau})\,. \tag{4.152}$$

If at a given time, say $t = 0$, the average kinetic energy is E_i, the average kinetic energy will, after a transient of time span τ, grow linearly in time

$$E(t \gg \tau) = E_i + \frac{e^2\tau}{m}\mathbf{E}^2 t \tag{4.153}$$

reflecting that there is no sink for the energy pumped into the system, since the scattering is elastic. The particle gains energy from the external field at the Joule heating rate, $e^2\mu\mathbf{E}^2$. Only on time scales where heating can be neglected is the assumption eq.(4.141) valid in the present model. In practice, a steady state is maintained because the generated heat is carried off by other degrees of freedom with which the particle interacts.

Chapter 5

Fermi Gas in a Random Potential

We have hitherto considered the kinetics of a single particle in a random potential. However, our main interest shall be conduction electrons in a metal or semiconductor which constitute an assembly of interacting fermions. The Coulomb interaction between conduction electrons is effectively weak due to screening (as we discuss in chapter 10), and as far as the effect of disorder on transport properties the mutual Coulomb interaction can to a first approximation usually be neglected. We shall therefore study the Sommerfeld model of noninteracting electrons in a random potential. We first demonstrate that the kinetics of an assembly of noninteracting identical particles in a random potential is formally identical to the kinetics of a single particle in a random potential. This is due to the fact that the indistinguishability of the particles is not being probed by elastic scattering, and the conclusion of the following analysis is simply stated: The weak-disorder kinetic equation for noninteracting fermions is identical to the one-particle case, except that the distribution function respects the Pauli principle! A reader not interested in the details of the derivation of this result can proceed right to section 5.4 where the Boltzmann theory for an electron gas in a random potential is presented.

5.1 N-Particle Kinematics

If the particles in an assembly are distinguishable, an orthonormal basis in the N-particle state space $H^{(N)} = H_1 \otimes H_2 \otimes \, .. \, \otimes H_N$ is the (tensor) product states, for example specified in terms of the momentum quantum numbers of the particles[1]

$$|\mathbf{p}_1, \mathbf{p}_2, \, .. \, , \mathbf{p}_N> \; \equiv \; |\mathbf{p}_1> \otimes |\mathbf{p}_2> \otimes \, .. \, \otimes |\mathbf{p}_N> \; \equiv \; |\mathbf{p}_1> \, |\mathbf{p}_2> \, .. \, |\mathbf{p}_N> \; . \quad (5.1)$$

We follow the custom of suppressing the tensorial notation.

For an assembly of identical particles, however, we must respect their quantum statistics; assemblies of fermions and bosons are described by states which

---

[1] We shall in the next section use the momentum basis, and refer in the following to the quantum numbers labeling the one-particle states as momentum. The N-tuple $(\mathbf{p}_1, \mathbf{p}_2, \, .. \, , \mathbf{p}_N)$ is a complete description of the N-particle system, since we neglect internal degrees of freedom. However, any complete set of quantum numbers could equally well be used.

with respect to interchange of pairs of identical particles are antisymmetric and symmetric, respectively.[2]

Any N-particle state can be split into a superposition of its symmetric and antisymmetric part with respect to interchange of any two particles. For example

$$|\mathbf{p}_1, \mathbf{p}_2, .., \mathbf{p}_N> \ = \ (\hat{\mathcal{S}} + \hat{\mathcal{A}}) \ |\mathbf{p}_1, \mathbf{p}_2, .., \mathbf{p}_N> \tag{5.2}$$

where the symmetrization operator $\hat{\mathcal{S}}$ symmetrizes an N-particle state

$$\hat{\mathcal{S}} \ |\mathbf{p}_1, \mathbf{p}_2, .., \mathbf{p}_N> \ = \ \frac{1}{N!} \sum_P \ |\mathbf{p}_{P_1}> \otimes |\mathbf{p}_{P_2}> \otimes .. \ \otimes |\mathbf{p}_{P_N}> \tag{5.3}$$

and the antisymmetrization operator $\hat{\mathcal{A}}$ antisymmetrizes

$$\hat{\mathcal{A}} \ |\mathbf{p}_1, \mathbf{p}_2, .., \mathbf{p}_N> \ = \ \frac{1}{N!} \sum_P \ (-1)^{\zeta_P} |\mathbf{p}_{P_1}> \otimes |\mathbf{p}_{P_2}> \otimes .. \ \otimes |\mathbf{p}_{P_N}> \ . \tag{5.4}$$

The summations are over all permutations P of the particles, and ζ_P counts the number of transpositions in the permutation P, or equivalently we have for the sign of the permutation[3]

$$(-1)^{\zeta_P} \ = \ \prod_{1 \leq i < j \leq N} \frac{j - i}{P_j - P_i} \ . \tag{5.5}$$

We note that if any two single-particle states are identical, the antisymmetrized state vector equals the zero vector, i.e., Pauli's exclusion principle for fermions: No two fermions can occupy the same state!

The symmetrization operators are normalized so that they are projectors, $\hat{\mathcal{A}}^2 = \hat{\mathcal{A}}$, $\hat{\mathcal{S}}^2 = \hat{\mathcal{S}}$, as redoing the above symmetrization or antisymmetrization leaves a state unchanged on its already symmetrized form. Furthermore,

$$\hat{\mathcal{A}} \hat{\mathcal{S}} = \hat{0} = \hat{\mathcal{S}} \hat{\mathcal{A}} \tag{5.6}$$

since, for example, symmetrizing an antisymmetric state gives the zero vector. The operators $\hat{\mathcal{S}}$ and $\hat{\mathcal{A}}$ project a state onto either of the two orthogonal subspaces of symmetric or antisymmetric states. We note, that eq.(5.2) expresses the identity operator on the N-particle state space

$$\hat{I}^{(N)} = \hat{I}_1 \otimes \hat{I}_2 \otimes \hat{I}_3 \otimes .. \otimes \hat{I}_N \ = \ \hat{\mathcal{A}} + \hat{\mathcal{S}} \ . \tag{5.7}$$

---

[2]Quantum statistics and the spin degree of a particle is intimately connected as relativistic quantum field theory demands that bosons have integer spin, whereas particles with half-integer spin are fermions. In the following we suppress the spin labeling, or simply assume it absorbed in the momentum labeling.

[3]It is customary to introduce the notation $(\zeta)^P \equiv (-1)^{\zeta_P}$, allowing for a unfied notation for bosons and fermions corresponding to $\zeta = \pm 1$, respectively. However, we shall concentrate on the fermionic case.

The symmetrization operators are hermitian, $\hat{\mathcal{A}}^\dagger = \hat{\mathcal{A}}$, $\hat{\mathcal{S}}^\dagger = \hat{\mathcal{S}}$, as verified for example for $\hat{\mathcal{A}}$ by first noting that

$$<\mathbf{p}_1, \ldots, \mathbf{p}_N|\hat{\mathcal{A}}^\dagger|\mathbf{p}_1', \mathbf{p}_2', \ldots, \mathbf{p}_N'> \;=\; <\mathbf{p}_1', \ldots, \mathbf{p}_N'|\hat{\mathcal{A}}|\mathbf{p}_1, \mathbf{p}_2, \ldots, \mathbf{p}_N>^*$$

$$=\; \frac{1}{N!}\sum_P (-1)^{\zeta_P} <\mathbf{p}_1'|\mathbf{p}_{P_1}>^* \;\ldots\; <\mathbf{p}_N'|\mathbf{p}_{P_N}>^*$$

$$=\; \frac{(-1)^{\zeta_S}}{N!} <\mathbf{p}_{S_1}|\mathbf{p}_1'> \;\ldots\; <\mathbf{p}_{S_N}|\mathbf{p}_N'> \qquad (5.8)$$

the matrix element only being nonzero if the set $\{\mathbf{p}_i'\}_{i=1,..,N}$ is a permutation of the set $\{\mathbf{p}_i\}_{i=1,..,N}$, S being the permutation that brings the set $\{\mathbf{p}_i\}_{i=1,..,N}$ into the set $\{\mathbf{p}_i'\}_{i=1,..,N}$, $\mathbf{p}_{S_i} = \mathbf{p}_i'$. Permuting both sets of indices by the inverse permutation S^{-1} of S, and using that a permutation and its inverse have the same sign, $\zeta_{S^{-1}} = \zeta_S$, we get

$$<\mathbf{p}_1, \ldots, \mathbf{p}_N|\hat{\mathcal{A}}^\dagger|\mathbf{p}_1', \mathbf{p}_2', \ldots, \mathbf{p}_N'> \;=\; \frac{1}{N!}(-1)^{\zeta_{S^{-1}}} <\mathbf{p}_1|\mathbf{p}_{S_1^{-1}}'> \;\ldots\; <\mathbf{p}_N|\mathbf{p}_{S_N^{-1}}'>$$

$$=\; \frac{1}{N!}\sum_P (-1)^{\zeta_P} <\mathbf{p}_1, \ldots, \mathbf{p}_N|\mathbf{p}_{P_1}', \ldots, \mathbf{p}_{P_N}'>$$

$$=\; <\mathbf{p}_1, \ldots, \mathbf{p}_N|\hat{\mathcal{A}}|\mathbf{p}_1', \ldots, \mathbf{p}_N'> \;. \qquad (5.9)$$

We introduce the complete orthogonal set of momentum states for the anti-symmetric state space

$$|\mathbf{p}_1 \wedge \mathbf{p}_2 \wedge \ldots \wedge \mathbf{p}_N> \;\equiv\; \sqrt{N!}\, \hat{\mathcal{A}}\left(|\mathbf{p}_1> \otimes |\mathbf{p}_2> \otimes \ldots \otimes |\mathbf{p}_N>\right)$$

$$=\; \frac{1}{\sqrt{N!}}\sum_P (-1)^{\zeta_P} |\mathbf{p}_{P_1}> \otimes |\mathbf{p}_{P_2}> \otimes \ldots \otimes |\mathbf{p}_{P_N}> \qquad (5.10)$$

In order to demonstrate their orthogonality we consider

$$<\mathbf{p}_1 \wedge \ldots \wedge \mathbf{p}_N|\mathbf{p}_1' \wedge \ldots \wedge \mathbf{p}_N'> \;=\; N! <\mathbf{p}_1, \ldots, \mathbf{p}_N|\hat{\mathcal{A}}^\dagger\hat{\mathcal{A}}|\mathbf{p}_1', \ldots, \mathbf{p}_N'>$$

$$=\; N! <\mathbf{p}_1, \ldots, \mathbf{p}_N|\hat{\mathcal{A}}|\mathbf{p}_1', \ldots, \mathbf{p}_N'>$$

$$=\; <\mathbf{p}_1, \ldots, \mathbf{p}_N|\sum_P (-1)^{\zeta_P} |\mathbf{p}_{P_1}', \ldots, \mathbf{p}_{P_N}'>$$

$$=\; \begin{cases} (-1)^{\zeta_S} & \{\mathbf{p}'\}_i \equiv \{\mathbf{p}\}_i \\[2mm] 0 & \text{otherwise} \end{cases} \qquad (5.11)$$

where $\{\mathbf{p}_i'\}_{i=1,..,N} \equiv \{\mathbf{p}_i\}_{i=1,..,N}$ is short for *the labels $\{\mathbf{p}_i'\}_{i=1,..,N}$ is a permutation of the labels $\{\mathbf{p}_i\}_{i=1,..,N}$*, and S is the permutation that takes the set $\{\mathbf{p}_i\}_{i=1,..,N}$

into $\{\mathbf{p}'_i\}_{i=1,..,N}$, $\mathbf{p}_{S_i} = \mathbf{p}'_i$. Equivalently we have

$$
<\mathbf{p}_1 \wedge \mathbf{p}_2, \wedge \, .. \, \wedge \mathbf{p}_N | \, \mathbf{p}'_1, \mathbf{p}'_2, .., \mathbf{p}'_N > \; = \; \begin{cases} \frac{1}{\sqrt{N!}} (-1)^{\zeta_s} & \{\mathbf{p}'\}_i \equiv \{\mathbf{p}\}_i \\ \\ 0 & \text{otherwise} \end{cases} . \quad (5.12)
$$

We can also write the matrix element as a determinant of the matrix with entries $<\mathbf{p}_i|\mathbf{p}'_j>$

$$
<\mathbf{p}_1 \wedge \, .. \, \wedge \mathbf{p}_N | \mathbf{p}'_1 \wedge \, .. \, \wedge \mathbf{p}'_N > \; = \; \det(<\mathbf{p}_i|\mathbf{p}'_j>) \quad (5.13)
$$

the Slater determinant.

The phase factor can always be chosen to equal 1 by considering proper orderings in the definition of the basis states. For example, if we choose the basis vectors according to the ordering $|\mathbf{p}_1| < |\mathbf{p}_2| < \, .. \, < |\mathbf{p}_N|$, we have an orthonormal basis. We shall use the notation

$$
|\{\mathbf{p}_i\}> \; \equiv \; |\mathbf{p}_1 \wedge \mathbf{p}_2 \wedge .. \wedge \mathbf{p}_N > \quad (5.14)
$$

for an antisymmetrized ordered basis vector.

The resolution of the identity on the N-particle state space splits into resolutions on the two orthogonal symmetrized subspaces

$$
\hat{I}^{(N)} \; = \; (\hat{A} + \hat{S}) \, \hat{I}^{(N)} \, (\hat{A} + \hat{S}) = \hat{I}^{(N)}_A + \hat{I}^{(N)}_S . \quad (5.15)
$$

The resolution on the antisymmetric state space is seen to be given by

$$
\begin{aligned}
\hat{I}^{(N)}_A \; &= \; \hat{A} \, \hat{I}^{(N)} \, \hat{A} = \hat{A} \left(\hat{I}_1 \otimes \hat{I}_2 \otimes \hat{I}_3 \otimes .. \otimes \hat{I}_N \right) \hat{A}^\dagger \\[4pt]
&= \; \hat{A} \sum_{\mathbf{p}_1,..,\mathbf{p}_N} |\mathbf{p}_1><\mathbf{p}_1| \otimes |\mathbf{p}_2><\mathbf{p}_2| \otimes .. \otimes |\mathbf{p}_N><\mathbf{p}_N| \; \hat{A}^\dagger \\[4pt]
&= \; \hat{A} \sum_{\mathbf{p}_1,..,\mathbf{p}_N} |\mathbf{p}_1, \mathbf{p}_2, .., \mathbf{p}_N><\mathbf{p}_1, \mathbf{p}_2, .., \mathbf{p}_N| \; \hat{A}^\dagger \\[4pt]
&= \; \frac{1}{N!} \sum_{\mathbf{p}_1,..,\mathbf{p}_N} |\mathbf{p}_1 \wedge \mathbf{p}_2 \wedge .. \wedge \mathbf{p}_N><\mathbf{p}_1 \wedge \mathbf{p}_2 \wedge .. \wedge \mathbf{p}_N| \\[4pt]
&= \; \sum_{\mathbf{p}_1<\mathbf{p}_2<..<\mathbf{p}_N} |\mathbf{p}_1 \wedge \mathbf{p}_2 \wedge .. \wedge \mathbf{p}_N><\mathbf{p}_1 \wedge \mathbf{p}_2 \wedge .. \wedge \mathbf{p}_N| . \quad (5.16)
\end{aligned}
$$

Exercise 5.1 *Derive the analogous results for the symmetric state space, in particular that the set of ordered vectors, choosing, for example, the ordering according to $\mathbf{p}_1 \leq \mathbf{p}_2 \leq \, .. \, \leq \mathbf{p}_N$,*

$$
|\mathbf{p}_1 \vee \mathbf{p}_2 \vee .. \vee \mathbf{p}_N > \; \equiv \; \sqrt{\frac{N!}{n_1! \, n_2! \, ..}} \; \hat{S} \, |\mathbf{p}_1, \mathbf{p}_2, .., \mathbf{p}_N > \quad (5.17)
$$

constitute an orthonormal basis. Here n_i is the number of times $\mathbf{p}_i$ occurs among the $\mathbf{p}_1, \mathbf{p}_2, .., \mathbf{p}_N$.

The state symbol for the one-particle momentum state $\mathbf{p}$

$$\hat{P}(\mathbf{p}) = |\mathbf{p}><\mathbf{p}| = \delta_{\hat{\mathbf{p}},\mathbf{p}} \equiv \hat{n}_{\mathbf{p}} \tag{5.18}$$

could as discussed in chapter 1 (see eq.(1.166)) be considered the operator describing whether the single-particle state $\mathbf{p}$ is occupied or not

$$\hat{n}_{\mathbf{p}} |\mathbf{p}'> = \delta_{\mathbf{p},\mathbf{p}'} |\mathbf{p}'> \tag{5.19}$$

or as we shall say, the operator counting whether the state in question is occupied or not. Similarly, the operator on the N-particle state space that counts the extend to which a single particle momentum state $\mathbf{p}$ is occupied, the number operator, is

$$\hat{n}_{\mathbf{p}}^{(N)} = \hat{n}_{\mathbf{p}} \otimes \hat{I}_2 \otimes .. \otimes \hat{I}_N + \hat{I}_1 \otimes \hat{n}_{\mathbf{p}} \otimes \hat{I}_3 \otimes .. \otimes \hat{I}_N + .. + \hat{I}_1 \otimes .. \otimes \hat{I}_{N-1} \otimes \hat{n}_{\mathbf{p}}$$

$$\equiv \sum_{i=1}^{N} \hat{n}_{\mathbf{p}}^{(i)} \tag{5.20}$$

because

$$\hat{n}_{\mathbf{p}}^{(N)} |\mathbf{p}_1, \mathbf{p}_2, .., \mathbf{p}_N> = (\delta_{\mathbf{p}_1,\mathbf{p}} + \delta_{\mathbf{p}_2,\mathbf{p}} + .. + \delta_{\mathbf{p}_N,\mathbf{p}}) |\mathbf{p}_1, \mathbf{p}_2, .., \mathbf{p}_N> . \tag{5.21}$$

5.2 Fermi Gas Kinematics

We shall in the following assume the N identical particles to be fermions. The state space is thus spanned by the antisymmetric basis states. A basis state is also an eigenstate of $\hat{n}_{\mathbf{p}}^{(N)}$ with eigenvalues one or zero depending on whether $\mathbf{p}$ equals exactly one of the $\mathbf{p}_i$'s or not, i.e., whether state $\mathbf{p}$ is occupied or not,

$$\hat{n}_{\mathbf{p}}^{(N)} |\mathbf{p}_1 \wedge \mathbf{p}_2 \wedge .. \wedge \mathbf{p}_N> = (\delta_{\mathbf{p}_1,\mathbf{p}} + \delta_{\mathbf{p}_2,\mathbf{p}} + .. + \delta_{\mathbf{p}_N,\mathbf{p}}) |\mathbf{p}_1 \wedge \mathbf{p}_2 \wedge .. \wedge \mathbf{p}_N> . \tag{5.22}$$

A mixture of states for N identical fermions is described by a statistical operator $\hat{\rho}(t)$

$$\hat{\rho}(t') = \sum_{\mathbf{p}_1,..,\mathbf{p}_N} \sum_{\mathbf{p}'_1,..,\mathbf{p}'_N} \rho^{(N)}(\mathbf{p}_1, .., \mathbf{p}_N; \mathbf{p}'_1, .., \mathbf{p}'_N, t') |\mathbf{p}_1 \wedge .. \wedge \mathbf{p}_N><\mathbf{p}'_1.. \wedge \mathbf{p}'_N|$$

$$= N!^2 \sum_{\mathbf{p}_1<..<\mathbf{p}_N, \mathbf{p}'_1<..<\mathbf{p}'_N} \rho^{(N)}(\mathbf{p}_1, .., \mathbf{p}_N; \mathbf{p}'_1, .., \mathbf{p}'_N, t') |\mathbf{p}_1 \wedge .. \wedge \mathbf{p}_N><\mathbf{p}'_1.. \wedge \mathbf{p}'_N| \tag{5.23}$$

where the N-fermion density matrix

$$\rho^{(N)}(\mathbf{p}_1, .., \mathbf{p}_N; \mathbf{p}'_1, .., \mathbf{p}'_N, t) \equiv <\mathbf{p}_1 \wedge .. \wedge \mathbf{p}_N|\hat{\rho}(t)|\mathbf{p}'_1 \wedge .. \wedge \mathbf{p}'_N> \tag{5.24}$$

is antisymmetric in each set of variables. For example,

$$\rho^{(N)}(\mathbf{p}_1, \mathbf{p}_2, \mathbf{p}_3, .., \mathbf{p}_N; \mathbf{p}'_1, .., \mathbf{p}'_N, t) = \rho^{(N)}(\mathbf{p}_2, \mathbf{p}_1, \mathbf{p}_3, .., \mathbf{p}_N; \mathbf{p}'_2, \mathbf{p}'_1, \mathbf{p}'_3, .., \mathbf{p}'_N, t)$$

$$= -\rho^{(N)}(\mathbf{p}_2, \mathbf{p}_1, \mathbf{p}_3, .., \mathbf{p}_N; \mathbf{p}'_1, .., \mathbf{p}'_N, t) \tag{5.25}$$

and since the statistical operator is hermitian, we have

$$\rho^{(N)}(\mathbf{p}_1, .., \mathbf{p}_N; \mathbf{p}'_1, .., \mathbf{p}'_N, t) = [\rho^{(N)}(\mathbf{p}'_1, .., \mathbf{p}'_N; \mathbf{p}_1, .., \mathbf{p}_N, t)]^* . \qquad (5.26)$$

The state symbol representing the event *one fermion has momentum* $\mathbf{p}_1$, *another fermion has momentum* $\mathbf{p}_2$, .., *the N'th fermion has momentum* $\mathbf{p}_N$ is

$$\hat{P}(\mathbf{p}_1, \mathbf{p}_2, \mathbf{p}_3, , .., \mathbf{p}_N) \equiv |\mathbf{p}_1 \wedge \mathbf{p}_2 \wedge .. \wedge \mathbf{p}_N> <\mathbf{p}_1 \wedge \mathbf{p}_2 \wedge .. \wedge \mathbf{p}_N| \qquad (5.27)$$

as

$$\hat{P}(\mathbf{p}_1, .., \mathbf{p}_N) |\mathbf{p}'_1 \wedge \mathbf{p}'_2 \wedge .. \wedge \mathbf{p}'_N> = \begin{cases} (-1)^{\zeta_s} |\mathbf{p}'_1 \wedge \mathbf{p}'_2 \wedge .. \wedge \mathbf{p}'_N> & \{\mathbf{p}'_i\}_i \equiv \{\mathbf{p}_i\}_i \\ \\ |0> & \text{otherwise} \end{cases}$$

$$(5.28)$$

The phase factor can always be chosen to be 1 by using identical ordering conventions in the definition of the basis states and state symbols.

The probability $P(\mathbf{p}_1, \mathbf{p}_2, \mathbf{p}_3, .., \mathbf{p}_N)$ to find one fermion with momentum $\mathbf{p}_1$, another fermion with momentum $\mathbf{p}_2$, .., and the N'th fermion with momentum $\mathbf{p}_N$ for an N-fermion system in state $\rho(t)$ can be expressed in terms of the trace over the N-fermion state space

$$\begin{aligned} P(\mathbf{p}_1, \mathbf{p}_2, \mathbf{p}_3, .., \mathbf{p}_N) &= Tr(\hat{\rho}(t)\hat{P}(\mathbf{p}_1, \mathbf{p}_2, \mathbf{p}_3, .., \mathbf{p}_N)) \\ \\ &= <\mathbf{p}_1 \wedge .. \wedge \mathbf{p}_N| \hat{\rho}(t) |\mathbf{p}_1 \wedge .. \wedge \mathbf{p}_N> \\ \\ &= \rho^{(N)}(\mathbf{p}_1, .., \mathbf{p}_N; \mathbf{p}_1, .., \mathbf{p}_N, t) . \qquad (5.29) \end{aligned}$$

The diagonal elements of the N-fermion density matrix have their expected interpretation as the probability for the event *finding one fermion with momentum* $\mathbf{p}_1$, *another fermion with momentum* $\mathbf{p}_2$, .., *and the N'th fermion with momentum* $\mathbf{p}_N$.

Fixing any pair of variables to the values $\mathbf{p}$ and $\mathbf{p}'$, and tracing in the $(N-1)$-fermion state space we get the so-called one-fermion density matrix

$$\begin{aligned} \rho(\mathbf{p}, \mathbf{p}', t) &\equiv N \sum_{\mathbf{p}_2, \mathbf{p}_3, .., \mathbf{p}_N} \rho^{(N)}(\mathbf{p}, \mathbf{p}_2, .., \mathbf{p}_N; \mathbf{p}', \mathbf{p}_2, .., \mathbf{p}_N, t) \\ \\ &= N \sum_{\mathbf{p}_1, \mathbf{p}_3, .., \mathbf{p}_N} \rho^{(N)}(\mathbf{p}_1, \mathbf{p}, \mathbf{p}_3, .., \mathbf{p}_N; \mathbf{p}_1, \mathbf{p}', \mathbf{p}_3, .., \mathbf{p}_N, t) \\ \\ &= N \sum_{\mathbf{p}_1, \mathbf{p}_2, .., \mathbf{p}_{N-1}} \rho^{(N)}(\mathbf{p}_1, .., \mathbf{p}_{N-1}, \mathbf{p}; \mathbf{p}_1, .., \mathbf{p}_{N-1}, \mathbf{p}', t) \\ \\ &= \sum_{i=1}^{N} \sum_{\{\mathbf{p}\}_j \{\mathbf{p}'\}_j} \delta_{\mathbf{p}_i, \mathbf{p}} \delta_{\mathbf{p}'_i, \mathbf{p}'} \prod_{j \neq i} \delta_{\mathbf{p}_j, \mathbf{p}'_j} \rho^{(N)}(\mathbf{p}_1, .., \mathbf{p}_N; \mathbf{p}'_1, .., \mathbf{p}'_N, t) \\ \\ &\equiv Tr_{(N-1)}(\hat{\rho}(t))\Big|_{\mathbf{p}, \mathbf{p}'} \equiv <\mathbf{p}|Tr_{(N-1)}(\hat{\rho}(t))|\mathbf{p}'> \qquad (5.30) \end{aligned}$$

where the symmetry of the state makes it irrelevant which pair of variables we choose not to trace over. When using the ordered basis we have, for example,

$$\rho(\mathbf{p}, \mathbf{p}', t) \equiv N! \sum_{\mathbf{p}_2 < \mathbf{p}_3 < .. < \mathbf{p}_N} \rho^{(N)}(\mathbf{p}, \mathbf{p}_2, .., \mathbf{p}_N; \mathbf{p}', \mathbf{p}_2, .., \mathbf{p}_N, t) . \qquad (5.31)$$

The one-fermion density matrix is hermitian

$$\rho(\mathbf{p}, \mathbf{p}', t) = [\rho(\mathbf{p}', \mathbf{p}, t)]^* \qquad (5.32)$$

simply because the statistical operator is hermitian.

The probability $f_{\mathbf{p}}(t)$ of finding one of the fermions with momentum $\mathbf{p}$ given the N-fermion state $\rho(t)$ is

$$\begin{aligned}
f_{\mathbf{p}}(t) &= \sum_{\mathbf{p}_1, .., \mathbf{p}_N} (\delta_{\mathbf{p}_1, \mathbf{p}} + \delta_{\mathbf{p}_2, \mathbf{p}} + .. + \delta_{\mathbf{p}_N, \mathbf{p}}) \rho^{(N)}(\mathbf{p}_1, \mathbf{p}_2, .., \mathbf{p}_N; \mathbf{p}_1, \mathbf{p}_2, .., \mathbf{p}_N, t) \\[2mm]
&= \rho(\mathbf{p}, \mathbf{p}, t) \\[2mm]
&= \sum_{\mathbf{p}_1, .., \mathbf{p}_N} (\delta_{\mathbf{p}_1, \mathbf{p}} + \delta_{\mathbf{p}_2, \mathbf{p}} + .. + \delta_{\mathbf{p}_N, \mathbf{p}}) < \mathbf{p}_1 \wedge .. \wedge \mathbf{p}_N | \hat{\rho}(t) | \mathbf{p}_1 \wedge .. \wedge \mathbf{p}_N > \\[2mm]
&= \sum_{\mathbf{p}_1, .., \mathbf{p}_N} < \mathbf{p}_1 \wedge .. \wedge \mathbf{p}_N | \hat{\rho}(t) \, \hat{n}_{\mathbf{p}}^{(N)} | \mathbf{p}_1 \wedge .. \wedge \mathbf{p}_N > \\[2mm]
&= Tr(\hat{\rho}(t) \, \hat{n}_{\mathbf{p}}^{(N)}) \qquad (5.33)
\end{aligned}$$

the expectation value of the number operator.

The Hamiltonian for N identical fermions in an external potential is

$$\hat{H} = \hat{H}_0 + \hat{V}_N \qquad (5.34)$$

where $\hat{H}_0$ is the kinetic energy operator for the N fermions

$$\hat{H}_0 = \frac{\hat{\mathbf{p}}_1^2}{2m} \otimes \hat{I}_2 \otimes .. \otimes \hat{I}_N + .. + \hat{I}_1 \otimes .. \otimes \hat{I}_{N-1} \otimes \frac{\hat{\mathbf{p}}_N^2}{2m} \equiv \sum_{i=1}^{N} \frac{\hat{\mathbf{p}}_i^2}{2m} \qquad (5.35)$$

and $\hat{V}_N$ denotes the interaction with the external potential

$$\hat{V}_N = \sum_{i=1}^{N} V(\hat{\mathbf{x}}_i) = \sum_{i=1}^{N} \hat{V} . \qquad (5.36)$$

Both operators are sums of identical one-body operators since the fermions are identical. For the kinetic energy we have the eigenvalue equation

$$\hat{H}_0 |\mathbf{p}_1 \wedge .. \wedge \mathbf{p}_N> = \left(\sum_{i=1}^{N} \frac{\mathbf{p}_i}{2m} \right) |\mathbf{p}_1 \wedge .. \wedge \mathbf{p}_N> . \qquad (5.37)$$

We can introduce the orthogonal position basis vectors for the N-fermion state space analogously to the momentum representation

$$|\mathbf{x}_1 \wedge \mathbf{x}_2 \wedge \,..\, \wedge \mathbf{x}_N> \;\equiv\; \sqrt{N!}\; \hat{A}\; |\mathbf{x}_1> \otimes |\mathbf{x}_2> \otimes .. \otimes |\mathbf{x}_N>$$

$$= \frac{1}{\sqrt{N!}} \sum_P (-1)^{\zeta_P} |\mathbf{x}_{P_1}> \otimes |\mathbf{x}_{P_2}> \otimes \,..\, \otimes |\mathbf{x}_{P_N}> \quad (5.38)$$

and similarly as for the momentum states we have the orthogonality relation

$$<\mathbf{x}_1 \wedge \,..\, \wedge \mathbf{x}_N | \mathbf{x}'_1 \wedge .. \wedge \mathbf{x}'_N> \;=\; (-1)^{\zeta_S} <\mathbf{x}_1, \,..\,, \mathbf{x}_N | \mathbf{x}'_{S_1}, \,..\,, \mathbf{x}'_{S_N}>$$

$$= (-1)^{\zeta_S}\, \delta(\mathbf{x}_1 - \mathbf{x}'_{S_1}) \,..\, \delta(\mathbf{x}_N - \mathbf{x}'_{S_N}) \quad (5.39)$$

which by proper ordering can be turned into the orthonormality relation.

For the transformation function between the antisymmetric position and momentum basis vectors we have

$$<\mathbf{x}_1 \wedge \,..\, \wedge \mathbf{x}_N | \mathbf{p}_1 \wedge .. \wedge \mathbf{p}_N>$$

$$= \frac{1}{N!} \sum_{P_x, P_p} (-1)^{\zeta_{P_x}} (-1)^{\zeta_{P_p}} <\mathbf{x}_{P_x(1)} | \mathbf{p}_{P_p(1)}> \,..\, <\mathbf{x}_{P_x(N)} | \mathbf{p}_{P_p(N)}>$$

$$= \frac{1}{N!} \sum_{P_x, P_p} (-1)^{\zeta_{P_x} + \zeta_{P_p}} <\mathbf{x}_1 | \mathbf{p}_{P_p P_x^{-1}(1)}> \,..\, <\mathbf{x}_N | \mathbf{p}_{P_p P_x^{-1}(N)}> \quad (5.40)$$

where in the last equality we have permuted the factors by the permutation P_x^{-1}. Noting that $\zeta_{P_p P_x^{-1}} = \zeta_{P_x^{-1}} + \zeta_{P_p} = \zeta_{P_x} + \zeta_{P_p}$ (the product of two even or odd permutations is an even permutation, and the product of an even and an odd permutation is an odd permutation), we obtain $N!$ identical terms giving

$$<\mathbf{x}_1 \wedge \,..\, \wedge \mathbf{x}_N | \mathbf{p}_1 \wedge .. \wedge \mathbf{p}_N> \;=\; \sum_P (-1)^{\zeta_{P_p}} <\mathbf{x}_1 | \mathbf{p}_{P_p(1)}> \,..\, <\mathbf{x}_N | \mathbf{p}_{P_p(N)}>$$

$$= \begin{vmatrix} <\mathbf{x}_1|\mathbf{p}_1> & <\mathbf{x}_1|\mathbf{p}_2> & .. & <\mathbf{x}_1|\mathbf{p}_N> \\ <\mathbf{x}_2|\mathbf{p}_1> & <\mathbf{x}_2|\mathbf{p}_2> & .. & <\mathbf{x}_2|\mathbf{p}_N> \\ \vdots & \vdots & \vdots & \vdots \\ <\mathbf{x}_N|\mathbf{p}_1> & <\mathbf{x}_N|\mathbf{p}_2> & .. & <\mathbf{x}_N|\mathbf{p}_N> \end{vmatrix}$$

$$= \det(<\mathbf{x}_i|\mathbf{p}_j>) \quad (5.41)$$

the Slater determinant.

The potential operator is in the position representation given by

$$\hat{V}_N = \frac{1}{N!} \int d\mathbf{x}_1 \,..\, \int d\mathbf{x}_N\; |\mathbf{x}_1 \wedge \,..\, \wedge \mathbf{x}_N> \left(\sum_{i=1}^N V(\mathbf{x}_i) \right) <\mathbf{x}_1 \wedge \,..\, \wedge \mathbf{x}_N| \quad (5.42)$$

whose operation on a position basis state is simply given by the eigenvalue equation

$$\hat{V}_N \,|\mathbf{x}_1 \wedge \,..\, \wedge \mathbf{x}_N> \;\; = \;\; \left(\sum_{i=1}^{N} V(\mathbf{x}_i) \right) |\mathbf{x}_1 \wedge \,..\, \wedge \mathbf{x}_N> \,. \qquad (5.43)$$

The density operator for the N-particle system is the sum of the density operators (we leave out the tensor products with the $(N-1)$ identity operators)

$$\hat{n}^{(N)}(\mathbf{x}) \;\; = \;\; \sum_{i=1}^{N} \hat{n}_i(\mathbf{x}) \;\; = \;\; \sum_{i=1}^{N} \delta(\hat{\mathbf{x}}_i - \mathbf{x}) \qquad (5.44)$$

and we have that

$$\hat{n}^{(N)}(\mathbf{x}) \,|\mathbf{x}_1 \wedge \,..\, \wedge \mathbf{x}_N> \;\; = \;\; \Big(\delta(\mathbf{x} - \mathbf{x}_1) + .. + \delta(\mathbf{x} - \mathbf{x}_N) \Big) |\mathbf{x}_1 \wedge \,..\, \wedge \mathbf{x}_N> \qquad (5.45)$$

equals $\delta(\mathbf{x} - \mathbf{x}_i) \,|\mathbf{x}_1 \wedge \,..\, \wedge \mathbf{x}_N>$ if $\mathbf{x}$ equals exactly one of the $\mathbf{x}_i$'s, and the zero vector otherwise, i.e., counting the extent to which there is a particle at position $\mathbf{x}$. We can then write for the potential operator in the position representation

$$\hat{V}_N(t) \;\; = \;\; \int d\mathbf{x} \, \hat{n}^{(N)}(\mathbf{x}) \, V(\mathbf{x}, t) \,. \qquad (5.46)$$

Exercise 5.2 *Consider the thermal equilibrium state of a large number of non-interacting fermions specified by the temperature T*

$$\hat{\rho}_T \;\; = \;\; \frac{e^{-\hat{H}_0/kT}}{Tr e^{-\hat{H}_0/kT}} \,. \qquad (5.47)$$

Show that the one-fermion equilibrium density matrix (suppressing the irrelevant spin degree of freedom as the Hamiltonian is assumed spin independent)

$$\rho_T(\mathbf{p}, \mathbf{p}') \;\; = \;\; \delta_{\mathbf{p}, \mathbf{p}'} \, f_0(\epsilon_{\mathbf{p}}) \qquad (5.48)$$

is specified in terms of the mean occupation number of the levels, the Fermi function

$$f_0(\epsilon_{\mathbf{p}}) \;\; = \;\; \frac{1}{\exp\{(\epsilon_{\mathbf{p}} - \mu)/kT\} + 1} \qquad (5.49)$$

where the chemical potential μ of a Fermi gas is essentially the Fermi energy ϵ_F

$$\mu \;\; = \;\; \epsilon_F \left(1 - \frac{\pi^2}{12} \left(\frac{T}{T_F} \right)^2 \right) \,. \qquad (5.50)$$

The Fermi energy is specified by the density n of the fermions, as $\epsilon_F = \hbar^2 k_F^2/2m$, where the magnitude of the Fermi wave vector is given by $k_F = (3\pi^2 n)^{1/3}$ (in three dimensions). We have introduced the Fermi temperature, $T_F \equiv \epsilon_F/k$, which for a metal is huge, typically $T_F \sim 10^4 K$.

Exercise 5.3 *Show that the specific heat of a spin-1/2 Fermi gas is given by*

$$c = \frac{\pi^2 k^2 T}{3} N_0(\epsilon_F) = \frac{\pi^2 n k^2 T}{2\epsilon_F} . \qquad (5.51)$$

5.3 Fermi Gas Kinetics

Instead of introducing at this stage a field theoretic description of assemblies of arbitrary numbers of identical fermions, and thereupon considering appropriate circumstances where for their kinetics will lead to the Boltzmann equation, we shall instead assume appropriate circumstances and obtain the Boltzmann equation for a Fermi gas in a random potential by an explicit consideration of N identical fermions.

In order to obtain the kinetic equation for a Fermi gas in a random potential we start from the von Neumann equation

$$\frac{d\hat{\rho}(t)}{dt} + \frac{i}{\hbar}[\hat{H}_0, \hat{\rho}(t)] = -\frac{i}{\hbar}[\hat{V}_N, \hat{\rho}(t)] . \qquad (5.52)$$

The external potential is assumed caused by a set of scattering centers which we later shall assume randomly positioned.

Taking the matrix element in the state space, we obtain the equation of motion for the N-fermion density matrix

$$\frac{d\rho^{(N)}(\mathbf{p}_1, .., \mathbf{p}_N; \mathbf{p}'_1, .., \mathbf{p}'_N, t)}{dt} + \frac{i}{\hbar} <\mathbf{p}_1, \wedge .. \wedge \mathbf{p}_N [\hat{H}_0, \hat{\rho}(t)] | \mathbf{p}'_1 \wedge .. \wedge \mathbf{p}'_N >$$

$$= -\frac{i}{\hbar} <\{\mathbf{p}_i\} | [\hat{V}_N, \hat{\rho}(t)] | \{\mathbf{p}'_i\} > . \qquad (5.53)$$

For the kinetic energy term we get

$$<\{\mathbf{p}_i\} | \hat{H}_0 \, \hat{\rho}(t) | \{\mathbf{p}'_i\} > = \left(\sum_{i=1}^{N} \frac{\mathbf{p}_i^2}{2m} \right) \rho^{(N)}(\mathbf{p}_1, .., \mathbf{p}_N; \mathbf{p}'_1, .., \mathbf{p}'_N, t) \qquad (5.54)$$

and by tracing the expression over the $(N-1)$-fermion space, and relabeling the untraced pair of variables, $\mathbf{p}_i \equiv \mathbf{p}$ and $\mathbf{p}'_i \equiv \mathbf{p}'$, gives

$$Tr_{(N-1)}(\hat{H}_0 \, \hat{\rho}(t)) \Big|_{\mathbf{p}, \mathbf{p}'} = \epsilon_{\mathbf{p}} \, \rho(\mathbf{p}, \mathbf{p}', t) + \sum_{\tilde{\mathbf{p}}} \epsilon_{\tilde{\mathbf{p}}} \, \rho^{(2)}(\mathbf{p}, \tilde{\mathbf{p}}; \mathbf{p}', \tilde{\mathbf{p}}, t) \qquad (5.55)$$

where we have introduced the two-particle density matrix

$$\rho^{(2)}(\mathbf{p}, \tilde{\mathbf{p}}; \mathbf{p}', \tilde{\mathbf{p}}', t) = \frac{N!}{(N-2)!} \sum_{\mathbf{p}_3, .., \mathbf{p}_N} \rho^{(N)}(\mathbf{p}, \tilde{\mathbf{p}}, \mathbf{p}_3, .., \mathbf{p}_N; \mathbf{p}', \tilde{\mathbf{p}}', \mathbf{p}_3, .., \mathbf{p}_N, t)$$

$$= \frac{N!}{(N-2)!} \sum_{\mathbf{p}_2,..,\mathbf{p}_{N-1}} \rho^{(N)}(\mathbf{p},\mathbf{p}_2,..,\mathbf{p}_{N-1},\tilde{\mathbf{p}};\mathbf{p}',\mathbf{p}_2,..,\mathbf{p}_{N-1},\tilde{\mathbf{p}}',t)$$

$$= N! \sum_{\mathbf{p}_3<..<\mathbf{p}_N} \rho^{(N)}(\mathbf{p},\tilde{\mathbf{p}},\mathbf{p}_3,..,\mathbf{p}_N;\mathbf{p}',\tilde{\mathbf{p}}',\mathbf{p}_3,..,\mathbf{p}_N,t) \quad (5.56)$$

and noted that the symmetry of the state makes it irrelevant which two pairs of variables we choose not to trace over.

Noting that

$$<\mathbf{p}_1\wedge ..\wedge\mathbf{p}_N|\hat{H}_0\,\hat{\rho}(t)|\mathbf{p}'_1\wedge ..\wedge\mathbf{p}'_N> = <\mathbf{p}'_1\wedge ..\wedge\mathbf{p}'_N|\hat{\rho}(t)\,\hat{H}_0|\mathbf{p}_1\wedge ..\wedge\mathbf{p}_N>^* \quad (5.57)$$

we immediately obtain for the other term in the commutator in eq.(5.53)

$$Tr_{(N-1)}(\hat{\rho}(t)\,\hat{H}_0)\bigg|_{\mathbf{p},\mathbf{p}'} = \epsilon_{\mathbf{p}'}\,\rho(\mathbf{p},\mathbf{p}',t) + \sum_{\tilde{\mathbf{p}}}\epsilon_{\tilde{\mathbf{p}}}\,\rho^{(2)}(\mathbf{p},\tilde{\mathbf{p}};\mathbf{p}',\tilde{\mathbf{p}},t) \quad (5.58)$$

and thereby the equation

$$\frac{\partial\rho(\mathbf{p};\mathbf{p}',t)}{\partial t} - \frac{i}{\hbar}(\epsilon_{\mathbf{p}}-\epsilon_{\mathbf{p}'})\,\rho(\mathbf{p};\mathbf{p}',t) = -\frac{i}{\hbar}Tr_{N-1}([\hat{V}_N,\hat{\rho}(t)])\bigg|_{\mathbf{p},\mathbf{p}'} . \quad (5.59)$$

We note that the drift term is analogous to the one for the one-particle case (see exercise 2.10 on page 132).

Since there will be no linear order effect of the impurity potential on the impurity-averaged density matrix, we consider the second-order term. In order to calculate the effect of the potential to second order in eq.(5.53) we only need the evolution operator to lowest order in the potential

$$\hat{\rho}(t) = \hat{\rho}_0(t) + \frac{i}{\hbar}\hat{U}_0(t,t')\,\hat{\rho}(t')\,\hat{U}_0(t',t_r)\int_{t'}^{t}d\bar{t}\,\hat{V}_N(\bar{t})\,\hat{U}_0^\dagger(t,t_r)$$

$$- \frac{i}{\hbar}\hat{U}_0(t,t_r)\int_{t'}^{t}d\bar{t}\,\hat{V}_N(\bar{t})\,\hat{U}_0^\dagger(t',t_r)\,\hat{\rho}(t')\,\hat{U}_0^\dagger(t,t') \quad (5.60)$$

where

$$\hat{V}_N(t) = \hat{U}_0^\dagger(t,t_r)\,\hat{V}_N\,\hat{U}_0(t,t_r) \quad (5.61)$$

is the potential operator in the interaction picture, and

$$\hat{U}_0(t,t_r) = e^{-\frac{i}{\hbar}\hat{H}_0(t-t_r)} \quad (5.62)$$

is the evolution operator in the absence of the potential. We choose the arbitrary reference time conveniently, $t_r = t'$, and have to lowest order in the potential

$$\hat{\rho}(t) = \hat{\rho}_0(t) + \hat{\rho}^{(1)}(t) \quad (5.63)$$

where

$$\hat{\rho}^{(1)}(t) = \frac{i}{\hbar}\hat{U}_0(t,t')\,\hat{\rho}(t')\int_{t'}^{t}d\bar{t}\,\hat{V}_N(\bar{t})\,\hat{U}_0^{\dagger}(t,t') + h.c. \quad . \tag{5.64}$$

Considering one of the four second-order terms, we obtain by inserting complete sets of states:

$$<\mathbf{p}_1 \wedge .. \wedge \mathbf{p}_N|\hat{V}_N\,\hat{U}_0(t,t')\,\hat{\rho}(t')\,\hat{V}_N(\bar{t})\,\hat{U}_0^{\dagger}(t,t')|\mathbf{p}_1' \wedge .. \wedge \mathbf{p}_N'>$$

$$= \frac{1}{N!^2}\sum_{\mathbf{p}_1'',...,\mathbf{p}_N''} e^{-\frac{i}{\hbar}(t-t')\sum_{i=1}^{N}\epsilon_{\mathbf{p}_i''}} <\mathbf{p}_1 \wedge .. \wedge \mathbf{p}_N|\hat{V}_N|\mathbf{p}_1'' \wedge .. \wedge \mathbf{p}_N''>$$

$$\sum_{\mathbf{p}_1''',...,\mathbf{p}_N'''} e^{\frac{i}{\hbar}(\bar{t}-t')\sum_{i=1}^{N}\epsilon_{\mathbf{p}_i'''}} e^{-\frac{i}{\hbar}(\bar{t}-t)\sum_{i=1}^{N}\epsilon_{\mathbf{p}_i'}} <\mathbf{p}_1'' \wedge .. \wedge \mathbf{p}_N''|\hat{\rho}(t')|\mathbf{p}_1''' \wedge .. \wedge \mathbf{p}_N'''>$$

$$<\mathbf{p}_1''' \wedge .. \wedge \mathbf{p}_N'''|\hat{V}_N|\mathbf{p}_1' \wedge .. \wedge \mathbf{p}_N'> \quad . \tag{5.65}$$

In kinetics we are interested in how the scattering changes the momentum distribution. Assuming that at time t' the density matrix is diagonal,

$$\hat{\rho}(t') = \sum_{\mathbf{p}_1,...,\mathbf{p}_N} \rho^{(N)}(\mathbf{p}_1,..,\mathbf{p}_N;\mathbf{p}_1,..,\mathbf{p}_N,t')\,|\mathbf{p}_1 \wedge .. \wedge \mathbf{p}_N><\mathbf{p}_1 \wedge .. \wedge \mathbf{p}_N|$$

$$= N!\sum_{\mathbf{p}_1<\mathbf{p}_2<..<\mathbf{p}_N} \rho^{(N)}(\mathbf{p}_1,..,\mathbf{p}_N;\mathbf{p}_1,..,\mathbf{p}_N,t')\,|\{\mathbf{p}_i\}><\{\mathbf{p}_i\}| \tag{5.66}$$

we obtain for the matrix element of interest

$$<\{\mathbf{p}_i\}|\hat{V}_N\,\hat{U}_0(t,t')\,\hat{\rho}(t')\,\hat{V}_N(\bar{t})\,\hat{U}_0^{\dagger}(t,t')|\{\mathbf{p}_i'\}>$$

$$= N!\sum_{\mathbf{p}_1''<..<\mathbf{p}_N''} e^{\frac{i}{\hbar}(\bar{t}-t')\sum_{i=1}^{N}(\epsilon_{\mathbf{p}_i''}-\epsilon_{\mathbf{p}_i'})} <\{\mathbf{p}_i\}|\hat{V}_N|\{\mathbf{p}_i''\}>$$

$$\rho^{(N)}(\mathbf{p}_1'',..,\mathbf{p}_N'';\mathbf{p}_1'',..,\mathbf{p}_N'',t')\,<\{\mathbf{p}_i''\}|\hat{V}_N|\{\mathbf{p}_i'\}> \tag{5.67}$$

where we now use ordered basis states. Collecting all four second-order terms we have

$$<\{\mathbf{p}_i\}|[\hat{V}_N,\hat{\rho}^{(1)}(t)]|\{\mathbf{p}_i'\}>$$

$$= \frac{i}{\hbar}\int_{t'}^{t}d\bar{t}\,N!\sum_{\mathbf{p}_1''<..<\mathbf{p}_N''} <\{\mathbf{p}_i\}|\hat{V}_N|\{\mathbf{p}_i''\}><\{\mathbf{p}_i''\}|\hat{V}_N|\{\mathbf{p}_i'\}>$$

$$\left\{ \left(e^{\frac{i}{\hbar}(t-\bar{t}) \sum_i (\epsilon_{\mathbf{p}_i'} - \epsilon_{\mathbf{p}_i''})} + e^{\frac{i}{\hbar}(t-\bar{t}) \sum_i (\epsilon_{\mathbf{p}_i''} - \epsilon_{\mathbf{p}_i})} \right) \rho^{(N)}(\mathbf{p}_1'', .., \mathbf{p}_N''; \mathbf{p}_1'', .., \mathbf{p}_N'', t') \right.$$

$$- \left(e^{\frac{i}{\hbar}(t-\bar{t}) \sum_i (\epsilon_{\mathbf{p}_i'} - \epsilon_{\mathbf{p}_i''})} \rho^{(N)}(\mathbf{p}_1', .., \mathbf{p}_N'; \mathbf{p}_1', .., \mathbf{p}_N', t') \right.$$

$$\left. \left. + \ e^{\frac{i}{\hbar}(t-\bar{t}) \sum_i (\epsilon_{\mathbf{p}_i''} - \epsilon_{\mathbf{p}_i})} \rho^{(N)}(\mathbf{p}_1, .., \mathbf{p}_N; \mathbf{p}_1, .., \mathbf{p}_N, t') \right) \right\} . \tag{5.68}$$

We choose to measure the energy from a reference value such that the spatial average of the potential vanishes, $<\mathbf{p}|\hat{V}|\mathbf{p}> = 0$, and the matrix element

$$<\mathbf{p}_1 \wedge .. \wedge \mathbf{p}_N|\hat{V}_N|\mathbf{p}_1'' \wedge .. \wedge \mathbf{p}_N''> \ = \ \sum_{i=1}^{N} <\mathbf{p}_1 \wedge .. \wedge \mathbf{p}_N|V(\hat{\mathbf{x}}_i)|\mathbf{p}_1'' \wedge .. \wedge \mathbf{p}_N''>$$

$$= \sum_{i=1}^{N} \sum_{P} (-1)^{\zeta_P} <\mathbf{p}_{P_i}|V(\hat{\mathbf{x}}_i)|\mathbf{p}_i''> \prod_{k \neq i} \delta_{\mathbf{p}_{P_k}, \mathbf{p}_k''} \tag{5.69}$$

is then only nonzero if the $\mathbf{p}_i''$'s is a set of vectors for which exactly one of the vectors differ from the vectors in the set of $\mathbf{p}_i$'s. This is simply the statement that the momentum of a particle that is not scattered is conserved. Let us denote the vector in the set of $\{\mathbf{p}\}_i$ which is different from all the vectors in the set of $\{\mathbf{p}_i''\}$ for $\mathbf{p}_s$.

In the sum of terms

$$<\{\mathbf{p}_i\}|\hat{V}_N|\{\mathbf{p}_i''\}><\{\mathbf{p}_i''\}|\hat{V}_N|\{\mathbf{p}_i'\}>$$

$$= \sum_{i,j=1}^{N} \sum_{P,Q} (-1)^{\zeta_P + \zeta_Q} <\mathbf{p}_{P_i}|V(\hat{\mathbf{x}}_i)|\mathbf{p}_i''><\mathbf{p}_j''|V(\hat{\mathbf{x}}_j)|\mathbf{p}_{Q_j}'> \prod_{k \neq i} \delta_{\mathbf{p}_{P_k}, \mathbf{p}_k''} \prod_{l \neq j} \delta_{\mathbf{p}_l'', \mathbf{p}_{Q_l}'}$$

$$\tag{5.70}$$

we have two possibilities, $i = j$ or $i \neq j$, i.e., $\mathbf{p}_i'' = \mathbf{p}_j''$ or $\mathbf{p}_i'' \neq \mathbf{p}_j''$. In the latter case we encounter two products of $(N-1)$ Kronecker deltas, and we will unavoidably in eq.(5.70) have a potential matrix element between two equal momenta rendering this type of term equal to zero. Due to the ordering of the basis vectors, we therefore only get a nonzero contribution when P and Q are the identity permutations so that

$$<\{\mathbf{p}_i\}|\hat{V}_N|\{\mathbf{p}_i''\}><\{\mathbf{p}_i''\}|\hat{V}_N|\{\mathbf{p}_i'\}>$$

$$= \ <\mathbf{p}_s|V(\hat{\mathbf{x}}_s)|\mathbf{p}_s''><\mathbf{p}_s''|V(\hat{\mathbf{x}}_s)|\mathbf{p}_s'> \prod_{k,l \neq s} \delta_{\mathbf{p}_k, \mathbf{p}_k''} \, \delta_{\mathbf{p}_l'', \mathbf{p}_l'} . \tag{5.71}$$

The presence of the Kronecker functions makes the summation in eq.(5.68) over the momenta $\mathbf{p}_i''$ for which $\mathbf{p}_i'' \neq \mathbf{p}_s''$ trivial, and when performing the $(N-1)$-trace, we set the $(N-1)$ momenta in the set $\{\mathbf{p}_i'\}$ for which $\mathbf{p}_i' \neq \mathbf{p}_s'$ equal to the $(N-1)$ momenta, $\mathbf{p}_i \neq \mathbf{p}_s$, in the set $\{\mathbf{p}_i\}$, and obtain

$$
Tr_{N-1}([\hat{V}_N, \hat{\rho}^{(1)}(t)])\bigg|_{\mathbf{p}_s, \mathbf{p}_s'} = \frac{i}{\hbar} \int_{t'}^{t} d\bar{t} \sum_{\mathbf{p}_s''} <\mathbf{p}_s|V(\hat{\mathbf{x}}_s)|\mathbf{p}_s''><\mathbf{p}_s''|V(\hat{\mathbf{x}}_s)|\mathbf{p}_s'>
$$

$$
\left\{ \left(e^{\frac{i}{\hbar}(t-\bar{t})(\epsilon_{\mathbf{p}_s'} - \epsilon_{\mathbf{p}_s''})} + e^{\frac{i}{\hbar}(t-\bar{t})(\epsilon_{\mathbf{p}_s''} - \epsilon_{\mathbf{p}_s})} \right) \rho(\mathbf{p}_s'', \mathbf{p}_s'', t') \right.
$$

$$
- \left(e^{\frac{i}{\hbar}(t-\bar{t})(\epsilon_{\mathbf{p}_s'} - \epsilon_{\mathbf{p}_s''})} \rho(\mathbf{p}_s', \mathbf{p}_s', t') \right)
$$

$$
+ \left. e^{\frac{i}{\hbar}(t-\bar{t})(\epsilon_{\mathbf{p}_s''} - \epsilon_{\mathbf{p}_s})} \rho(\mathbf{p}_s, \mathbf{p}_s, t') \right) \right\} . \tag{5.72}
$$

In the absence of the potential the diagonal elements of the one-particle density matrix stay, according to the von Neumann equation, eq.(5.52), constant in time. We can therefore to lowest order in the potential substitute $t' \to t$ on the right side of eq.(5.72). In the following we are only interested in the collision term, so we assume a spatially homogeneous state, $\rho(\mathbf{p}, \mathbf{p}', t) = \rho(\mathbf{p}, \mathbf{p}, t)\, \delta_{\mathbf{p}, \mathbf{p}'}$. The drift term then vanishes. Setting $\mathbf{p}_s' = \mathbf{p} = \mathbf{p}_s$ in the collision term, we obtain for the one-fermion distribution function the kinetic equation

$$
\frac{\partial f_{\mathbf{p}}(t)}{\partial t} = I_{\mathbf{p}}[f] \tag{5.73}
$$

where

$$
I_{\mathbf{p}}[f] \equiv -\frac{i}{\hbar} Tr_{N-1}([\hat{V}_N, \hat{\rho}^{(1)}(t)])\bigg|_{\mathbf{p}, \mathbf{p}}
$$

$$
= \frac{2}{\hbar^2} \int_{t'}^{t} d\bar{t} \sum_{\mathbf{p}''} |<\mathbf{p}|\hat{V}|\mathbf{p}''>|^2 \cos(\frac{1}{\hbar}(t-\bar{t})(\epsilon_{\mathbf{p}} - \epsilon_{\mathbf{p}''}))\left\{ f_{\mathbf{p}''}(t') - f_{\mathbf{p}}(t') \right\} . \tag{5.74}
$$

Taking the classical limit as in section 4.7 (see eq.(4.108)) we obtain the equation for the one-fermion distribution function[4]

$$
\frac{\partial f_{\mathbf{p}}(t)}{\partial t} = -\frac{2\pi}{\hbar} \sum_{\mathbf{p}'} |<\mathbf{p}|\hat{V}|\mathbf{p}'>|^2\, \delta(\epsilon_{\mathbf{p}} - \epsilon_{\mathbf{p}'})\, (f_{\mathbf{p}}(t) - f_{\mathbf{p}'}(t)) . \tag{5.75}
$$

Upon impurity averaging we have

$$
\overline{< |<\mathbf{p}|\hat{V}|\mathbf{p}'>|^2 >} = \frac{n_i}{V} |V_{imp}(\mathbf{p} - \mathbf{p}')|^2 \tag{5.76}
$$

_____

[4]Or employing the quasi-classical criterion, $\hbar/\epsilon_{\mathbf{p}}\tau$, of 6.7.

and the same collision integral as in the one-particle case, eq.(4.112), except that $f_\mathbf{p}(t)$ is the one-fermion distribution obtained by $(N-1)$-tracing the antisymmetric N-fermion density matrix.[5]

For the density of fermions at point $\mathbf{x}$ at time t for a system in state $\rho(t)$ we have

$$
\begin{aligned}
n(\mathbf{x}, t) &= Tr(\hat{\rho}(t)\,\hat{n}^{(N)}(\mathbf{x})) = \sum_{i=1}^{N} \int_{\mathbf{x}_1 < .. < \mathbf{x}_N} \prod_{j=1}^{N} d\mathbf{x}_j \; < \{\mathbf{x}_j\}|\hat{\rho}(t)\,\hat{n}^{(N)}(\mathbf{x})|\{\mathbf{x}_j\} > \\
&= \sum_{i=1}^{N} \frac{1}{N} \int d\mathbf{x}_i \, \delta(\mathbf{x}_i - \mathbf{x}) < \{\mathbf{x}_i\}|Tr_{(N-1)}\hat{\rho}(t)|\{\mathbf{x}_i\} > \\
&= \rho(\mathbf{x}, \mathbf{x}, t)
\end{aligned}
\tag{5.77}
$$

the diagonal element of the one-fermion density matrix.

Similarly we obtain from the current density operator

$$
\hat{\mathbf{j}}^{(N)}(\mathbf{x}) = \sum_{i=1}^{N} \hat{\mathbf{j}}_i(\mathbf{x}) = \frac{1}{2}\sum_{i=1}^{N} \{\hat{n}_i(\mathbf{x}), \hat{\mathbf{v}}_i\}
\tag{5.78}
$$

the current density[6]

$$
\mathbf{j}(\mathbf{x}, t) = Tr(\hat{\rho}(t)\hat{\mathbf{j}}^{(N)}(\mathbf{x})) = \frac{e\hbar}{2mi}(\nabla_\mathbf{x} - \nabla_{\mathbf{x}'})\,\rho(\mathbf{x}, \mathbf{x}', t)\Big|_{\mathbf{x}'=\mathbf{x}}
\tag{5.79}
$$

again in complete analogy with the one-particle case. We can introduce the Wigner function as in section 4.4, and obtain the same expression for the current density as in eq.(4.66).

Recalling the calculation in exercise 1.21 on page 71, we obtain that the N-fermion density operator in the Heisenberg picture satisfies the continuity equation

$$
\frac{\partial\,\hat{n}^{(N)}(\mathbf{x}, t)}{\partial t} + \nabla_\mathbf{x} \cdot \hat{\mathbf{j}}^{(N)}(\mathbf{x}, t) = 0 \, .
\tag{5.80}
$$

5.4 Boltzmann Theory

In this section we shall consider the kinetics of the conduction electrons in a metal, assuming only interaction with impurities, and neglecting band structure effects;[7] i.e., we shall study the Boltzmann theory for a Fermi gas in a random potential. Assuming an isotropic model, the Boltzmann equation can be solved to linear order in the external fields. We start by considering the electric conductivity before turning to thermal conductivity and thermoelectric effects.

---

[5]In case we include multiple scattering, the exact cross section appears instead of the Born expression.

[6]In the presence of a vector potential the formula must be amended with the diamagnetic term, recall exercise 1.4 on page 15 and exercise 1.21 on page 71, or section 7.3.1.

[7]We only expect alkali metals to be properly described by such an isotropic model. As for band structure effects, we refer to reference [25].

5.4.1   Classical Conductivity

At sufficiently low temperatures the temperature-independent residual resistance of a metal (or a heavily doped semiconductor) is determined by scattering of conduction electrons due to impurities. In the following we calculate the classical residual resistance by solving the Boltzmann equation[8]

$$\frac{\partial f(\mathbf{x}, \mathbf{p}, t)}{\partial t} + e\mathbf{E}(\mathbf{x}, t) \cdot \frac{\partial f(\mathbf{x}, \mathbf{p}, t)}{\partial \mathbf{p}} + \mathbf{v_p} \cdot \frac{\partial f(\mathbf{x}, \mathbf{p}, t)}{\partial \mathbf{x}} = I_{\mathbf{x}, \mathbf{p}, t}[f] \qquad (5.81)$$

where the impurity collision integral is given by

$$I_{\mathbf{x}, \mathbf{p}, t}[f] = -\int \frac{d\mathbf{p}'}{(2\pi\hbar)^3} W(\mathbf{p}, \mathbf{p}') [f(\mathbf{x}, \mathbf{p}, t) - f(\mathbf{x}, \mathbf{p}', t)] . \qquad (5.82)$$

The electron-impurity collision integral has the property

$$\int \frac{d\mathbf{p}}{(2\pi\hbar)^3} \epsilon_{\mathbf{p}} I_{\mathbf{x}, \mathbf{p}, t}[f] = 0 \qquad (5.83)$$

since the integrand is antisymmetric with respect to interchange of $\mathbf{p}$ and $\mathbf{p}'$. This property of the collision integral reflects that the energy of an electron is conserved in a collision with an impurity. Multiplying the Boltzmann equation by $\epsilon_{\mathbf{p}}$, and integrating over the momentum therefore gives

$$\frac{\partial E(\mathbf{x}, t)}{\partial t} + \nabla_{\mathbf{x}} \cdot \mathbf{j}_\epsilon(\mathbf{x}, t) = \mathbf{E}(\mathbf{x}, t) \cdot \mathbf{j}(\mathbf{x}, t) \qquad (5.84)$$

where (the factors of two accounting for the spin degree of freedom of the electron)

$$E(\mathbf{x}, t) = 2 \int \frac{d\mathbf{p}}{(2\pi\hbar)^3} \epsilon_{\mathbf{p}} f(\mathbf{x}, \mathbf{p}, t) \qquad (5.85)$$

is the energy density of the electrons, and

$$\mathbf{j}_\epsilon(\mathbf{x}, t) = 2 \int \frac{d\mathbf{p}}{(2\pi\hbar)^3} \mathbf{v_p} \epsilon_{\mathbf{p}} f(\mathbf{x}, \mathbf{p}, t) \qquad (5.86)$$

the energy current density and

$$\mathbf{j}(\mathbf{x}, t) = 2e \int \frac{d\mathbf{p}}{(2\pi\hbar)^3} \mathbf{v_p} f(\mathbf{x}, \mathbf{p}, t) \qquad (5.87)$$

the charge current density. The energy equation, eq.(5.84), has a source term describing the nonconservation of the energy of the electrons due to the work performed by the electric field.

---

[8]The criterion for the validity of the Boltzmann equation for a spatially inhomogeneous field is identical to the one-particle case of section 4.8.

In the isotropic model, the impurity potential is assumed spherically symmetric, and the transition probability has the form

$$W(\mathbf{p}, \mathbf{p}') = W(\hat{\mathbf{p}} \cdot \hat{\mathbf{p}}', \epsilon_{\mathbf{p}}) \, \delta(\epsilon_{\mathbf{p}} - \epsilon_{\mathbf{p}'}) \tag{5.88}$$

and furthermore a spherical Fermi surface, $\epsilon_{\mathbf{p}} = \epsilon(|\mathbf{p}|)$, is assumed. To be specific we consider the free electron model, $\epsilon_{\mathbf{p}} = \mathbf{p}^2/2m$, where the group velocity is $v_{\mathbf{p}} = \mathbf{p}/m$.

To calculate the classical conductivity of a disordered conductor, we assume that the distribution function has a Taylor expansion in terms of the applied field

$$f_{\mathbf{p}} = f_{\mathbf{p}}^{(0)} + f_{\mathbf{p}}^{(1)} + f_{\mathbf{p}}^{(2)} + \cdots . \tag{5.89}$$

The random potential due to the quenched disorder can not relax the energy, and any function independent of the direction of the momentum, $f_{\mathbf{p}}^{(0)} = f_0(\epsilon_{\mathbf{p}})$, makes the collision integral vanish, an equilibrium function with respect to impurity scattering. However, an electron interacts with its environment; say, in a metal the electrons have mutual Coulomb interaction and interaction with the lattice vibrations, and are as such in thermal contact with a heat reservoir. We can therefore assume that the distribution function in the absence of the electric field is the thermal equilibrium distribution, the Fermi function,

$$f_{\mathbf{p}}^{(0)} = f_0(\epsilon_{\mathbf{p}}) = \frac{1}{e^{\frac{\epsilon_{\mathbf{p}} - \mu}{kT}} + 1} . \tag{5.90}$$

In a metal the chemical potential μ is essentially the Fermi energy, recall eq.(5.50), and in a degenerate Fermi system, states below the Fermi energy are occupied and states above unoccupied, except in the thin thermal layer of size kT around the Fermi surface where states are only partially occupied.

We first consider a spatially homogeneous and time-independent electric field. In the steady state the first order in the electric field part of the distribution function is determined by the linearized kinetic equation

$$e\mathbf{E} \cdot \mathbf{v}_{\mathbf{p}} \frac{\partial f_0(\epsilon_{\mathbf{p}})}{\partial \epsilon_{\mathbf{p}}} = -\frac{1}{\tau(\epsilon_{\mathbf{p}})} f_{\mathbf{p}}^{(1)} + N_0(\epsilon_{\mathbf{p}}) \int \frac{d\hat{\mathbf{p}}'}{4\pi} W(\hat{\mathbf{p}} \cdot \hat{\mathbf{p}}', \epsilon_{\mathbf{p}}) f_{\mathbf{p}\hat{\mathbf{p}}'}^{(1)} \tag{5.91}$$

where the momentum relaxation time in the isotropic model is a function only of the energy

$$\frac{1}{\tau(\epsilon_{\mathbf{p}})} \equiv \int \frac{d\mathbf{p}'}{(2\pi\hbar)^3} W(\mathbf{p}, \mathbf{p}') = N_0(\epsilon_{\mathbf{p}}) \int \frac{d\hat{\mathbf{p}}'}{4\pi} W(\hat{\mathbf{p}} \cdot \hat{\mathbf{p}}', \epsilon_{\mathbf{p}}) = N_0(\epsilon_{\mathbf{p}}) W_0(\epsilon_{\mathbf{p}}) . \tag{5.92}$$

Here W_0 is the $l = 0$ polar angle component (with respect to the direction $\hat{\mathbf{p}}$) of the transition probability in the expansion on the Legendre polynomials P_l

$$W(\hat{\mathbf{p}} \cdot \hat{\mathbf{p}}', \epsilon_{\mathbf{p}}) = \sum_{l=0}^{\infty} W_l(\epsilon_{\mathbf{p}}) P_l(\hat{\mathbf{p}} \cdot \hat{\mathbf{p}}') \tag{5.93}$$

or inversely

$$W_l(\epsilon_{\mathbf{p}}) = (2l+1) \int \frac{d\hat{\mathbf{p}}'}{4\pi} P_l(\hat{\mathbf{p}} \cdot \hat{\mathbf{p}}') W(\hat{\mathbf{p}} \cdot \hat{\mathbf{p}}', \epsilon_{\mathbf{p}}) \tag{5.94}$$

and the density of states is in the three-dimensional case given by (see eq.(2.168))

$$N_0(\epsilon) = \frac{\sqrt{2m^3\epsilon}}{2\pi^2\hbar^3} . \tag{5.95}$$

We expand the linear-order distribution function on the complete set of Legendre polynomials in the polar angle with respect to the direction of the electric field

$$f_{\mathbf{p}}^{(1)} = \sum_{l=1}^{\infty} f_l^{(1)}(\epsilon_{\mathbf{p}}) P_l(\hat{\mathbf{p}} \cdot \hat{\mathbf{E}}) \tag{5.96}$$

where the $l = 0$ component is absent as $f_{\mathbf{p}}^{(1)}$ vanishes in the absence of the electric field. The linearized kinetic equation, eq.(5.91), then appears as

$$0 = \frac{e\mathbf{E} \cdot \mathbf{p}}{m} \frac{\partial f_0(\epsilon_{\mathbf{p}})}{\partial \epsilon_{\mathbf{p}}} + \frac{1}{\tau(\epsilon_{\mathbf{p}})} \sum_{l=1}^{\infty} f_l^{(1)}(\epsilon_{\mathbf{p}}) P_l(\hat{\mathbf{p}} \cdot \hat{\mathbf{E}})$$

$$- N_0(\epsilon_{\mathbf{p}}) \int \frac{d\hat{\mathbf{p}}'}{4\pi} \sum_{l=0}^{\infty} W_l(\epsilon_{\mathbf{p}}) P_l(\hat{\mathbf{p}} \cdot \hat{\mathbf{p}}') \sum_{l'=1}^{\infty} f_{l'}^{(1)}(\epsilon_{\mathbf{p}}) P_{l'}(\hat{\mathbf{p}}' \cdot \hat{\mathbf{E}}) . \tag{5.97}$$

Using the completeness of the Legendre polynomials

$$\int \frac{d\hat{\mathbf{p}}'}{4\pi} P_l(\hat{\mathbf{p}} \cdot \hat{\mathbf{p}}') P_{l'}(\hat{\mathbf{p}} \cdot \hat{\mathbf{p}}') = \frac{1}{2l+1} \delta_{l,l'} \tag{5.98}$$

and the addition theorem

$$\int \frac{d\hat{\mathbf{p}}'}{4\pi} P_l(\hat{\mathbf{p}} \cdot \hat{\mathbf{p}}') P_{l'}(\hat{\mathbf{p}}' \cdot \hat{\mathbf{E}}) = \frac{1}{2l+1} \delta_{l,l'} P_l(\hat{\mathbf{p}} \cdot \hat{\mathbf{E}}) \tag{5.99}$$

we obtain, by taking the angular average of eq.(5.97), the equation

$$0 = \sum_{l=0}^{\infty} P_l(\hat{\mathbf{p}} \cdot \hat{\mathbf{E}}) \left(f_l^{(1)}(\epsilon_{\mathbf{p}}) \left(\frac{W_l(\epsilon_{\mathbf{p}})}{2l+1} - W_0(\epsilon_{\mathbf{p}}) \right) N_0(\epsilon_{\mathbf{p}}) - \delta_{l,1} eEv_{\mathbf{p}} \frac{\partial f_0(\epsilon_{\mathbf{p}})}{\partial \epsilon_{\mathbf{p}}} \right) . \tag{5.100}$$

By the uniqueness of the expansion coefficients we thereby obtain an infinite set of equations which we immediately can solve

$$f_{l=1}^{(1)}(\epsilon_{\mathbf{p}}) = \left(\frac{W_l(\epsilon_{\mathbf{p}})}{3} - W_0(\epsilon_{\mathbf{p}}) \right)^{-1} \frac{eEv_{\mathbf{p}}}{N_0(\epsilon_{\mathbf{p}})} \frac{\partial f_0(\epsilon_{\mathbf{p}})}{\partial \epsilon_{\mathbf{p}}} \quad , \quad f_l^{(1)}(\epsilon_{\mathbf{p}}) = 0 \quad l \neq 1 \tag{5.101}$$

or equivalently

$$f_{\mathbf{p}}^{(1)} = -e\mathbf{E} \cdot \mathbf{v}_{\mathbf{p}} \, \tau_{tr}(\epsilon_{\mathbf{p}}) \frac{\partial f_0(\epsilon_{\mathbf{p}})}{\partial \epsilon_{\mathbf{p}}} \tag{5.102}$$

where the transport relaxation time is given by

$$\frac{1}{\tau_{tr}(\epsilon_{\mathbf{p}})} \equiv N_0(\epsilon_{\mathbf{p}}) \left(W_0(\epsilon_{\mathbf{p}}) - \frac{1}{3} W_1(\epsilon_{\mathbf{p}}) \right)$$

$$= \int \frac{d\mathbf{p}'}{(2\pi\hbar)^3} W(\mathbf{p} \cdot \mathbf{p}')[1 - \hat{\mathbf{p}} \cdot \hat{\mathbf{p}}']$$

$$= N_0(\epsilon_{\mathbf{p}}) \int \frac{d\hat{\mathbf{p}}'}{4\pi} W(\hat{\mathbf{p}} \cdot \hat{\mathbf{p}}', \epsilon_{\mathbf{p}})[1 - \hat{\mathbf{p}} \cdot \hat{\mathbf{p}}'] . \tag{5.103}$$

As to be expected in the isotropic model, we have demonstrated that to linear order in the electric field the angular symmetry of the solution to the kinetic equation is solely determined by the form of the driving term $\mathbf{E} \cdot \mathbf{v_p}$. We have shown that for the spherical model the solution to the kinetic equation has to linear order the solution

$$f(\mathbf{p}) = f_0(\epsilon_{\mathbf{p}}) + \mathbf{p} \cdot \mathbf{f}(\epsilon_{\mathbf{p}}) \tag{5.104}$$

where

$$\mathbf{f}(\epsilon_{\mathbf{p}}) = -e\mathbf{E} \frac{\tau_{tr}(\epsilon_{\mathbf{p}})}{m} \frac{\partial f_0(\epsilon_{\mathbf{p}})}{\partial \epsilon_{\mathbf{p}}} . \tag{5.105}$$

Using the assumed spherical symmetry of the scattering potential, we note, by expanding $\mathbf{p}'$ along $\mathbf{p}$, $\mathbf{p}' = (\mathbf{p}' \cdot \hat{\mathbf{p}})\hat{\mathbf{p}} - \mathbf{p}'_{\perp}$, that the perpendicular component $\mathbf{p}'_{\perp}$, does not contribute to the angular average

$$\int \frac{d\hat{\mathbf{p}}'}{4\pi} \mathbf{p}' \, W(\mathbf{p}, \mathbf{p}') = \int \frac{d\hat{\mathbf{p}}'}{4\pi} ((\mathbf{p}' \cdot \hat{\mathbf{p}})\hat{\mathbf{p}} - \mathbf{p}'_{\perp}) \, W(\hat{\mathbf{p}} \cdot \hat{\mathbf{p}}') \, \delta(\epsilon_{\mathbf{p}} - \epsilon_{\mathbf{p}'})$$

$$= \mathbf{p} \int \frac{d\hat{\mathbf{p}}'}{4\pi} W(\mathbf{p}, \mathbf{p}') \, \hat{\mathbf{p}}' \cdot \hat{\mathbf{p}} \tag{5.106}$$

and thereby that the linearized collision integral reduces to

$$I_{\mathbf{p}}[f] = I_{\mathbf{p}}[f_0 + \mathbf{p} \cdot \mathbf{f}] = -\frac{\mathbf{p} \cdot \mathbf{f}(\epsilon_{\mathbf{p}})}{\tau_{tr}(\epsilon_{\mathbf{p}})} = -\frac{f(\epsilon_{\mathbf{p}}) - f_0(\epsilon_{\mathbf{p}})}{\tau_{tr}(\epsilon_{\mathbf{p}})} . \tag{5.107}$$

In the linear approximation the impurity collision integral in the isotropic model thus reduces to a simple relaxation time form. Thus in the isotropic model the function $\mathbf{p} \cdot \mathbf{f}$ is an eigenfunction of the collision operator with the eigenvalue $-1/\tau_{tr}(\epsilon_{\mathbf{p}})$

$$I_{\mathbf{p}}[\mathbf{p} \cdot \mathbf{f}] = -\frac{\mathbf{p} \cdot \mathbf{f}(\epsilon_{\mathbf{p}})}{\tau_{tr}(\epsilon_{\mathbf{p}})} . \tag{5.108}$$

For a distribution function of the form eq.(5.104), we get the current density

$$\mathbf{j} = \frac{2e}{m} \int \frac{d\mathbf{p}}{(2\pi\hbar)^3} \mathbf{p} \, f(\mathbf{p}) = \frac{2e}{3m} \int \frac{d\mathbf{p}}{(2\pi\hbar)^3} \mathbf{p}^2 \, \mathbf{f}(\epsilon_{\mathbf{p}})$$

$$= \frac{4e^2}{3m} \int_0^{\infty} d\epsilon \, N_0(\epsilon) \, \epsilon \, \tau_{tr}(\epsilon) \left(-\frac{\partial f_0(\epsilon)}{\partial \epsilon} \right) \mathbf{E} . \tag{5.109}$$

The derivative of the Fermi function is peaked at the Fermi surface with a width kT.[9] Since the rest of the integrand only varies on the scale ϵ_F, those quantities can be taken at the Fermi surface and we have for the current density

$$j_\alpha = \sum_\beta \sigma_{\alpha\beta} E_\beta \tag{5.110}$$

where the conductivity tensor

$$\sigma_{\alpha\beta} = \sigma_0 \, \delta_{\alpha\beta} \tag{5.111}$$

in the isotropic model is specified in terms of the conductivity (with small corrections at most of order $(T/T_F)^2$)

$$\sigma_0 = \frac{ne^2 \tau_{tr}(\epsilon_F)}{m} . \tag{5.112}$$

For the current I through a cube of linear dimension L we have, Ohm's law,

$$I = \int_S d\mathbf{s} \cdot \mathbf{j} = jL^{d-1} = L^{d-1}\sigma_0 E = R^{-1}V \tag{5.113}$$

where $V = EL$ is the voltage drop over the cube, and for the resistance of a cube of linear dimension L we have

$$R = \frac{m}{ne^2 \tau_{tr} L} . \tag{5.114}$$

A resistance measurement thus allows one to extract the transport cross section for an electron on the Fermi surface, $\tau_{tr} \equiv \tau_{tr}(\epsilon_F)$. The transport mean free path $l_{tr} = v_F \tau_{tr}$ is the characteristic distance an electron with the Fermi velocity, $\epsilon_F = m v_F^2/2$, can travel before the direction of its velocity is randomized.

In a time-dependent electric field, $\mathbf{E}(t) = \Re e(\mathbf{E}(\omega)e^{-i\omega t})$, the deviation from the equilibrium distribution, $\delta f(\mathbf{p}, t) \equiv f(\mathbf{p}, t) - f_0(\epsilon_{\mathbf{p}})$, will in the steady state have the form $\delta f(\mathbf{p}, t) = \Re e(\delta f(\mathbf{p}, \omega)e^{-i\omega t})$, and is in the isotropic model to linear order determined by the kinetic equation Fourier-transformed with respect to time

$$-i\omega \, \delta f(\mathbf{p}, \omega) + e\mathbf{E}(\omega) \cdot \mathbf{v_p} \frac{\partial f_0}{\partial \epsilon_{\mathbf{p}}} = -\frac{\delta f(\mathbf{p}, \omega)}{\tau_{tr}(\epsilon_{\mathbf{p}})} . \tag{5.115}$$

For the frequency-dependent conductivity

$$\mathbf{j}(\omega) = \sigma(\omega) \, \mathbf{E}(\omega) \tag{5.116}$$

we then get the expression

$$\sigma(\omega) = \frac{\sigma_0}{1 - i\omega\tau_{tr}} . \tag{5.117}$$

---

[9]We note, that the distribution function to linear order precisely satisfies the criterion of having the peaked derivatives required in section 4.9. The results of that section can therefore be taken over with the quantity p_F now being played by the Fermi momentum.

The current

$$\mathbf{j}(t) = 2e \int \frac{d\mathbf{p}}{(2\pi\hbar)^3} \, \mathbf{v_p} \, \delta f(\mathbf{p},t) = \Re e(\sigma(\omega) \, \mathbf{E}(\omega) \, e^{-i\omega t}) \tag{5.118}$$

has a dissipative component, and an inductive component due to inertia, and the average work done by the electric field on the electrons in a unit volume per unit time is ($T = 2\pi/\omega$ is the period of the electric field)

$$\frac{1}{T} \int_0^T dt \, \mathbf{j}(t) \cdot \mathbf{E}(t) = \frac{1}{2} \frac{\sigma_0}{1 + (\omega\tau_{tr})^2} \, |\mathbf{E}(\omega)|^2 . \tag{5.119}$$

5.4.2   Classical Kinetics in a Magnetic Field

In this section we shall assume that our isotropic metal is placed in crossed electric and magnetic fields, both spatially homogeneous and time independent.[10] In view of the previous section we shall seek the solution of the Boltzmann equation linearized with respect to the electric field

$$e\mathbf{E} \cdot \mathbf{v_p} \frac{\partial f_0(\epsilon_\mathbf{p})}{\partial \epsilon_\mathbf{p}} + e\mathbf{v_p} \times \mathbf{B} \cdot \frac{\partial f(\mathbf{p})}{\partial \mathbf{p}} = - \int \frac{d\mathbf{p}'}{(2\pi\hbar)^3} \, W(\mathbf{p}, \mathbf{p}')[f(\mathbf{p}) - f(\mathbf{p}')] \tag{5.120}$$

on the form

$$f(\mathbf{p}) = f_0(\epsilon_\mathbf{p}) + \mathbf{p} \cdot \mathbf{f}(\epsilon_\mathbf{p}) + ... \tag{5.121}$$

and for the linear-order term with respect to the electric field we have the equation

$$e\mathbf{E} \cdot \mathbf{v_p} \frac{\partial f_0(\epsilon_\mathbf{p})}{\partial \epsilon_\mathbf{p}} + e((\mathbf{v_p} \times \mathbf{B}) \cdot \nabla_\mathbf{p})(\mathbf{p} \cdot \mathbf{f}(\epsilon_\mathbf{p})) = - \frac{\mathbf{p} \cdot \mathbf{f}(\epsilon_\mathbf{p})}{\tau_{tr}(\epsilon_\mathbf{p})} . \tag{5.122}$$

Noting the identity in the isotropic model

$$\nabla_\mathbf{p}(\mathbf{p} \cdot \mathbf{f}(\epsilon_\mathbf{p})) = \mathbf{f}(\epsilon_\mathbf{p}) + \mathbf{p} \left(\mathbf{v_p} \cdot \frac{\partial \mathbf{f}(\epsilon_\mathbf{p})}{\partial \epsilon_\mathbf{p}} \right) \tag{5.123}$$

we have

$$(\mathbf{v_p} \times \mathbf{B}) \cdot \frac{\partial f(\mathbf{p})}{\partial \mathbf{p}} = (\mathbf{v_p} \times \mathbf{B}) \cdot \mathbf{f}(\epsilon_\mathbf{p}) . \tag{5.124}$$

The solution of eq.(5.122) is then seen to be given by[11]

$$\mathbf{f}(\epsilon_\mathbf{p}) = - \frac{e\tau_{tr}(\epsilon_\mathbf{p})}{m} \frac{\partial f_0(\epsilon_\mathbf{p})}{\partial \epsilon_\mathbf{p}} \frac{1}{1 + (\omega_c \tau_{tr}(\epsilon_\mathbf{p}))^2} \left[\mathbf{E} + (\omega_c \tau_{tr}(\epsilon_\mathbf{p}))^2 \mathbf{E}_{\parallel} + \frac{e\tau_{tr}(\epsilon_\mathbf{p})}{m} \mathbf{E} \times \mathbf{B} \right] \tag{5.125}$$

---

[10]Band structure and magnetic field can give rise to many interesting effects; for these we refer to reference [25].

[11]Since the magnetic field is an axial (or pseudo) vector, the vector $\mathbf{f}$, which is linear in the electric field, has the form $\mathbf{f}(\epsilon_\mathbf{p}) = c_1(\epsilon_\mathbf{p}) \mathbf{E} + c_2(\epsilon_\mathbf{p}) \mathbf{E} \times \mathbf{B} + c_3(\epsilon_\mathbf{p}) (\mathbf{E} \cdot \mathbf{B})\mathbf{B}$.

where $\mathbf{E}_{\parallel}$ is the component of the electric field along the direction of the magnetic field, and $\omega_c = |e|B/m$ is the Larmor or cyclotron frequency.

From the current density formula, eq.(5.109), we then obtain for the conductivity tensor (choosing the $\hat{\mathbf{z}}$-direction along the magnetic field, $\mathbf{B} = B\,\hat{\mathbf{z}}$)

$$\underline{\sigma} = \frac{\sigma_0}{1 + (\omega_c \tau_{tr})^2} \begin{pmatrix} 1 & -\omega_c \tau_{tr} & 0 \\ \omega_c \tau_{tr} & 1 & 0 \\ 0 & 0 & 1 + (\omega_c \tau_{tr})^2 \end{pmatrix} . \tag{5.126}$$

We observe explicitly the Onsager symmetry relation[12]

$$\sigma_{\alpha\beta}(\mathbf{B}) = \sigma_{\beta\alpha}(-\mathbf{B}) . \tag{5.127}$$

Inverting, we get the resistivity tensor

$$\underline{\rho} = [\underline{\sigma}]^{-1} = \rho_0 \begin{pmatrix} 1 & \omega_c \tau_{tr} & 0 \\ -\omega_c \tau_{tr} & 1 & 0 \\ 0 & 0 & 1 \end{pmatrix} \tag{5.128}$$

where $\rho_0 = \sigma_0^{-1}$. We note that the isotropic model does not display magnetoresistance; i.e., ρ_{xx} does not depend on the magnetic field due to the isotropic dispersion, $\epsilon_{\mathbf{p}} = \epsilon(|\mathbf{p}|)$. We further note that $\omega_c \tau_{tr}$ sets the scale for the classical magnetoresistance effects.

In a Hall-bar setup, the conduction electrons are deflected by the magnetic field, and an electric field in the direction transverse to the current (say the y-direction in which the sample is finite) develops in addition to the driving field E_x, or equivalently, a Hall voltage appears. The transverse electric field, E_y, is determined by the equilibrium condition $j_y = 0$, and as a consequence we obtain the relation (we neglect fringe fields due to the finite size of the sample; i.e., $E_z = 0$)

$$\frac{E_y}{E_x} = -\frac{\sigma_{yx}}{\sigma_{yy}} . \tag{5.129}$$

For the Hall coefficient we then get

$$R_H \equiv \frac{E_y}{B j_x} = \frac{1}{ne} \tag{5.130}$$

and for the Hall voltage

$$V_H \equiv E_y L_y = L_y B j_x \frac{1}{ne} . \tag{5.131}$$

A Hall experiment thus determines the sign and density of the charge carriers.

---

[12] For a discussion of the properties of transport coefficients under time-reversal symmetry, see section 7.7.

Exercise 5.4 *The energy dependence of the transport relaxation time $\tau_{tr}(\epsilon_{\mathbf{p}})$ has so far not played any role. Let us therefore consider a model where it is a constant $\tau_{tr}(\epsilon_{\mathbf{p}}) \equiv \tau$. Show by multiplying the Boltzmann equation by $e v_{\mathbf{p}}$ and integrating with respect to the momentum that*

$$\mathbf{E} = \rho_0 \mathbf{j} + R_H \, \mathbf{B} \times \mathbf{j} . \tag{5.132}$$

We note the absence on the right-hand side of the terms $B^2 \mathbf{j}$ and $(\mathbf{B} \cdot \mathbf{j})\mathbf{B}$ in the isotropic model.

5.4.3   Thermal and Thermoelectric Effects

In this section we consider the kinetics of the conduction electrons in the isotropic model of a metal experiencing a time-independent electric field $\mathbf{E}$, and a temperature gradient ∇T, and a chemical potential gradient $\nabla \mu$. We are thus considering a macroscopic description where the distribution function is assumed a local equilibrium function; i.e., a function of the local temperature and density of the electrons, which are assumed to vary slowly through space. We shall again assume low temperatures so that the kinetics of the conduction electrons is described by the Boltzmann equation with the electron-impurity collision integral, eq.(5.82), and a steady state. As already commented upon at the end of section 4.9 and in section 5.4.1, a steady state can not be attained when only elastic scattering is considered. However, in reality the steady state of the conduction electrons is, for example, maintained due to their coupling to the lattice vibrations, which in turn interact with the surroundings of the sample. We shall not here consider the inelastic processes explicitly, but instead assume they are efficient enough to allow the introduction of a local temperature of the conduction electrons (exceeding, but in a metal typically close to the temperature of the lattice).[13]

Referring to the analysis of section 5.4.1, the solution to the Boltzmann equation for a small deviation from equilibrium has the form

$$f(\mathbf{x}, \mathbf{p}) = f^{\text{l.e.}}(\mathbf{x}, \mathbf{p}) + \mathbf{p} \cdot \mathbf{f}(\epsilon_{\mathbf{p}}) \tag{5.133}$$

where

$$f^{\text{l.e.}}(\mathbf{x}, \mathbf{p}) = \frac{1}{e^{\frac{\epsilon_{\mathbf{p}} - \mu(\mathbf{x})}{kT(\mathbf{x})}} + 1} \tag{5.134}$$

is the local equilibrium Fermi distribution function, specified by the local temperature, $T(\mathbf{x}) = T + \delta T(\mathbf{x})$, and the local chemical potential $\mu(\mathbf{x}) = \mu + \delta\mu(\mathbf{x})$. Here $T = T(\mathbf{x_a})$ and $\mu = \mu(\mathbf{x_a})$ is the temperature and chemical potential at some arbitrary point in the system.

---

[13]Electron-electron interaction and electron-phonon interaction are considered in chapter 10.

Linearizing the driving terms with respect to $\mathbf{E}$ and ∇T and $\nabla \mu$, we obtain

$$e\mathbf{E} \cdot \frac{\partial f(\mathbf{x}, \mathbf{p})}{\partial \mathbf{p}} \; + \; \mathbf{v_p} \cdot \frac{\partial f(\mathbf{x}, \mathbf{p})}{\partial \mathbf{x}} \quad \rightarrow$$

$$e\mathbf{E} \cdot \frac{\partial f_0(\epsilon_\mathbf{p})}{\partial \mathbf{p}} \; + \; \nabla T \cdot \mathbf{v_p} \frac{\partial f_0(\epsilon_\mathbf{p})}{\partial T} \; + \; \mathbf{v_p} \cdot \nabla \mu \frac{\partial f_0(\epsilon_\mathbf{p})}{\partial \mu}$$

$$= \; \left((e\mathbf{E} - \nabla \mu) - \frac{\epsilon_\mathbf{p} - \mu}{T} \nabla T \right) \cdot \mathbf{v_p} \, \frac{\partial f_0(\epsilon_\mathbf{p})}{\partial \epsilon_\mathbf{p}} \; . \tag{5.135}$$

Solving the linearized kinetic equation

$$\left((e\mathbf{E} - \nabla \mu) \; - \; \frac{\epsilon_\mathbf{p} - \mu}{T} \nabla T \right) \cdot \mathbf{v_p} \, \frac{\partial f_0(\epsilon_\mathbf{p})}{\partial \epsilon_\mathbf{p}} \; = \; -\frac{\mathbf{p} \cdot \mathbf{f}(\epsilon_\mathbf{p})}{\tau_{tr}(\epsilon_\mathbf{p})} \tag{5.136}$$

we get

$$\mathbf{f}(\epsilon_\mathbf{p}) \; = \; \mathbf{f}_\mathcal{E}(\epsilon_\mathbf{p}) \; + \; \mathbf{f}_T(\epsilon_\mathbf{p}) \tag{5.137}$$

where

$$\mathbf{f}_\mathcal{E}(\epsilon_\mathbf{p}) \; = \; -\tau_{tr}(\epsilon_\mathbf{p}) \frac{\partial f_0}{\partial \epsilon_\mathbf{p}} \frac{e\mathcal{E}}{m} \tag{5.138}$$

and

$$\mathbf{f}_T(\epsilon_\mathbf{p}) \; = \; \tau_{tr}(\epsilon_\mathbf{p}) \frac{\partial f_0}{\partial \epsilon_\mathbf{p}} \frac{\epsilon_\mathbf{p} - \mu}{T} \frac{\nabla T}{m} \tag{5.139}$$

and we have introduced the gradient of the electrochemical potential

$$\mathcal{E} \; \equiv \; \mathbf{E} - \frac{\nabla \mu}{e} \; . \tag{5.140}$$

In the particle and energy current densities, the $\mathbf{f}_\mathcal{E}$-part of the nonequilibrium distribution function will give rise to susceptibilities which only depend on the value of $N_0(\epsilon)$ and $\tau_{tr}(\epsilon)$ at the Fermi energy, whereas the terms arising from the $\mathbf{f}_T$-part of the nonequilibrium distribution function will depend on how these quantities vary across the Fermi surface. In the latter case numerical factors are thus model dependent. Let us in the following consider the case of isotropic scattering

$$W(\mathbf{p}, \mathbf{p}') \; = \; W \, \delta(\epsilon_\mathbf{p} - \epsilon_{\mathbf{p}'}) \tag{5.141}$$

where W is a constant. The impurity potential is thus assumed of zero range, i.e., it is a delta function.

Inserting the solution into the current expressions, and using the Sommerfeld expansion

$$-\int_{-\infty}^{\infty} d\epsilon \, G(\epsilon) \frac{\partial f_0(\epsilon)}{\partial \epsilon} \; = \; G(\mu) + \frac{\pi^2}{6} (kT)^2 \frac{\partial^2 G(\mu)}{\partial \mu^2} + \mathcal{O}((T/T_F)^4) \tag{5.142}$$

we get the constitutive equations

$$\mathbf{j} = L_{11}\,\mathcal{E} + L_{12}\,\nabla\mathbf{T} \tag{5.143}$$

and

$$\mathbf{j}_\epsilon = L_{21}\,\mathcal{E} + L_{22}\,\nabla\mathbf{T} \tag{5.144}$$

where (noting that $N_0(\epsilon)\,\tau_{tr}(\epsilon)$ is a constant in the case of isotropic scattering)

$$L_{11} = \frac{4e^2}{3m}\,\mu N_0(\mu)\tau_{tr}(\mu) = \sigma(\mu) = \sigma(\epsilon_F) \equiv \sigma_0 \tag{5.145}$$

up to corrections of order $(T/T_F)^2$, and similarly for

$$L_{12} = -\frac{\pi^2}{3}\,\frac{k^2 T}{\epsilon_F}\,\frac{\sigma_0}{e} \tag{5.146}$$

and

$$L_{21} = \frac{\epsilon_F}{e}\sigma_0 \tag{5.147}$$

and

$$L_{22} = \frac{-2\pi^2}{3e^2}\,k^2 T\sigma_0 \;. \tag{5.148}$$

The dissipative energy current density, the heat current density, is the energy current in the absence of particle current

$$\mathbf{j}_h \underset{\mathbf{j}=0}{\equiv} \mathbf{j}_\epsilon \tag{5.149}$$

and the proportionality constant, κ, between the heat current and the temperature gradient is called the thermal conductivity

$$\mathbf{j}_h \underset{\kappa}{\equiv} -\kappa\,\nabla\mathbf{T} \;. \tag{5.150}$$

Expressing κ in terms of the L_{ij}'s we have

$$\kappa = \frac{L_{21}L_{12}}{L_{11}} - L_{22} = -\frac{1}{2}L_{22} \tag{5.151}$$

and thereby the Wiedemann-Franz relation

$$\kappa = \frac{\pi^2}{3}\,k^2 T\,\frac{\sigma_0}{e^2} \tag{5.152}$$

between the thermal and electric conductivities.

We can also express the thermal conductivity in terms of the specific heat of the degenerate Fermi gas, $c = \pi^2 n k^2 T/2\epsilon_F$ (recall exercise 5.3 on page 208), and obtain the simple kinetic result[14]

$$\kappa = \frac{1}{3}\,c\,v_F^2\,\tau_{tr} = \frac{1}{3}\,c\,v_F\,l_{tr} \tag{5.153}$$

---

[14] The specific heat for fixed volume c_V and for fixed pressure c_P differ for a degenerate Fermi system only to order $(T/T_F)^2$.

where $l_{tr} \equiv v_F \tau_{tr}$ is the transport mean free path.

Using the constitutive equations, we note

$$\mathbf{j}_\epsilon = \frac{L_{21}}{L_{11}}\mathbf{j} - \left(\frac{L_{21}L_{12}}{L_{11}} - L_{22}\right)\nabla T = \frac{\epsilon_F}{e}\mathbf{j} - \kappa\nabla T + \mathcal{O}\left((T/T_F)^2\right) \quad (5.154)$$

so that, to order $(T/T_F)^2$, we can rewrite the heat current in terms of the energy and particle currents as

$$\mathbf{j}_h = \mathbf{j}_\epsilon - \frac{\mu}{e}\mathbf{j} = 2\int\frac{d\mathbf{p}}{(2\pi\hbar)^3}\,\mathbf{v_p}\,(\epsilon_\mathbf{p} - \mu)f(\mathbf{x},\mathbf{p})\,. \quad (5.155)$$

The interpretation of this result is simply that the shift in the distribution induced by the electrochemical field

$$\delta f_\mathcal{E}(\mathbf{p}) \equiv f_0(\mathbf{p}) + \mathbf{p}\cdot\mathbf{f}_\mathcal{E}(\epsilon_\mathbf{p}) \simeq f_0(\mathbf{p} + e\tau_{tr}\mathcal{E}) \quad (5.156)$$

corresponds to a rigid shift $\delta\mathbf{p}_\mathcal{E} = e\tau_{tr}\mathcal{E}$ of the Fermi sphere, and the corresponding energy transport is not dissipated and thus not part of the entropy or heat current.[15]

The off-diagonal element L_{12} describes according to eq.(5.143) a thermoelectric effect: the particle current response to a temperature gradient. The thermoelectric coefficient is seen to be smaller by the factor T/T_F compared to the electric conductivity.

Calculating the expression

$$\mathbf{j}_\epsilon - \frac{\mu}{e}\mathbf{j} = 2\int\frac{d\mathbf{p}}{(2\pi\hbar)^3}\,\mathbf{v_p}\,(\epsilon_\mathbf{p} - \mu)f(\mathbf{x},\mathbf{p}) \quad (5.157)$$

to linear order in the disturbances gives for the heat current density

$$\mathbf{j}_h = 2\int\frac{d\mathbf{p}}{(2\pi\hbar)^3}\,\mathbf{v_p}\,(\epsilon_\mathbf{p} - \mu)\,\tau_{tr}(\epsilon_\mathbf{p})\,\mathbf{v_p}\cdot\left(-e\mathcal{E} + \nabla T\,\frac{\epsilon_\mathbf{p} - \mu}{T}\right)\frac{\partial f_0}{\partial\epsilon_\mathbf{p}}$$

$$= -L_{12}\,T\,\mathcal{E} - \kappa\nabla T = \frac{T\kappa e}{\epsilon_F}\mathcal{E} - \kappa\nabla T\,, \quad (5.158)$$

from which we get

$$\mathbf{j}_h = \tilde{L}_{21}\mathcal{E} + \tilde{L}_{22}\nabla\mathbf{T} \quad (5.159)$$

where

$$\tilde{L}_{21} = -TL_{12} = e\kappa\,\frac{T}{\mu} \quad (5.160)$$

and

$$\tilde{L}_{22} = -\frac{\pi^2}{3}\frac{k^2T}{e^2}L_{11} = -\kappa\,. \quad (5.161)$$

---

[15]Neglecting the energy dependence in $\tau_{tr}(\epsilon_\mathbf{p})$ we note that the deviation from the equilibrium distribution $\delta f_\mathcal{E}(\mathbf{p}) \equiv \mathbf{p}\cdot\mathbf{f}_\mathcal{E}(\epsilon_\mathbf{p})$ is odd in $\mathbf{p}$ and even in $(\epsilon_\mathbf{p} - \mu)$, whereas $\delta f_T(\mathbf{p}) \equiv \mathbf{p}\cdot\mathbf{f}_T(\epsilon_\mathbf{p})$ is odd in $\mathbf{p}$ and odd in $(\epsilon_\mathbf{p} - \mu)$.

The other thermoelectric effect, the heat current response to an electric field, is described by the coefficient $\tilde{L}_{21}$, and is seen to be smaller by the factor T/T_F compared to the thermal conductivity.

The thermoelectric effects are only nonzero because quantities vary as a function of energy across the Fermi surface. If we neglect this dependence, the thermoelectric effects would vanish.[16] In a degenerate Fermi system, thermoelectric effects are small due to the large Fermi energy, the scale on which quantities vary across the Fermi surface.

We note that we can express all the transport coefficients in terms of the conductivity and its derivative

$$
\begin{pmatrix} \mathbf{j} \\ \mathbf{j}_h \end{pmatrix} = \begin{pmatrix} \sigma_0 & -\frac{\pi^2}{3e}k^2 T \sigma_0' \\ \frac{\pi^2}{3e}(kT)^2 \sigma_0' & -\frac{\pi^2}{3e}k^2 T \sigma_0 \end{pmatrix} \begin{pmatrix} \mathcal{E} \\ \nabla T \end{pmatrix}
\tag{5.162}
$$

where

$$
\sigma_0' \equiv \sigma'(\epsilon_F) \equiv \frac{d\sigma(\mu)}{d\mu}\bigg|_{\mu=\epsilon_F} .
\tag{5.163}
$$

If a piece of material with mobile charges is subjected to a temperature gradient ∇T, an electrochemical field $\mathcal{E}$ will in equilibrium ($\mathbf{j} = 0$) be induced in order to counter the thermoelectric current. The linear relationship between the two is the thermopower of the material

$$
\mathcal{E} = Q \nabla T
\tag{5.164}
$$

and we obtain from eq.(5.143) that the thermopower or the Seebeck coefficient is given by $Q = -L_{12}/L_{11}$. For the model under consideration we have[17]

$$
Q = \frac{\pi^2 k^2 T}{3e} \frac{d \ln \sigma(\mu)}{d\mu}\bigg|_{\mu=\epsilon_F} = \frac{\pi^2}{3} \frac{k^2 T}{e \epsilon_F} = \frac{2}{3} \frac{s}{ne}
\tag{5.165}
$$

where s is the entropy density of the electron gas, which in a degenerate Fermi gas, $T \ll T_F$, is identical to the specific heat, $s = c$. In the considered model the mean free path does not enter the expression for the thermopower, and, although a nonequilibrium quantity, it is expressed in terms of a thermodynamic quantity.

Exercise 5.5 *Show that in the model where τ_{tr} is assumed energy independent, the thermopower is given by*

$$
Q = \frac{s}{ne} .
\tag{5.166}
$$

---

[16]Neglecting this energy dependence is customarily referred to as particle-hole symmetry.

[17]For sodium and potassium the isotropic model result is in fair agreement with the measured low temperature thermopower, but for other alkali metals, for example lithium, not even the sign is correct, and more realistic descriptions of band structure and collisions are needed.

Using the Boltzmann equation and eq.(5.155), we obtain the equation

$$\frac{\partial q(\mathbf{x}, t)}{\partial t} + \nabla_\mathbf{x} \cdot \mathbf{j}_h(\mathbf{x}, t) = \mathcal{E}(\mathbf{x}, t) \cdot \mathbf{j}(\mathbf{x}, t) \tag{5.167}$$

governing the rate of change of the heat density

$$q(\mathbf{x}, t) = 2 \int \frac{d\mathbf{p}}{(2\pi\hbar)^3} (\epsilon_\mathbf{p} - \mu) f(\mathbf{x}, \mathbf{p}, t) . \tag{5.168}$$

We also note the identity

$$\frac{\partial E(\mathbf{x}, t)}{\partial t} - \mu \frac{\partial n(\mathbf{x}, t)}{\partial t} = \frac{\partial q(\mathbf{x}, t)}{\partial t} \tag{5.169}$$

or alternatively the thermodynamic relation, relating the change in heat to the change in energy and the change in density $\Delta q = \Delta E - \mu \, \Delta n$.

Let us consider the case where the temperature varies in space. Using eq.(5.167) and eq.(5.159), and assuming conditions of uniform current flow, $\nabla \cdot \mathbf{j} = 0$, we obtain for the rate of change in the heat density[18]

$$\frac{\partial q}{\partial t} = \rho \mathbf{j}^2 - T \frac{dQ}{dT} (\mathbf{j} \cdot \nabla T) + \kappa \nabla^2 T + \frac{d\kappa}{dT} (\nabla T)^2 \tag{5.170}$$

where ρ, κ and Q are the resistivity, the thermal conductivity and the thermopower, respectively. A nonequilibrium steady state can thus be sustained when the Joule heating is carried away by the thermal conductivity.

In a system with a spatially slowly varying temperature, $q(\mathbf{x}, t) = q(T(\mathbf{x}, t))$, we obtain from eq.(5.167) and eq.(5.159), assuming a constant electrochemical potential,

$$\frac{\partial T(\mathbf{x}, t)}{\partial t} \frac{dq}{dT} - \kappa \triangle_\mathbf{x} T(\mathbf{x}, t) = 0 . \tag{5.171}$$

Since $dq/dT = c$ is the specific heat, and $\kappa/c = v_F^2 \tau_{tr}/3 = D_0$, we obtain the heat equation

$$\frac{\partial T(\mathbf{x}, t)}{\partial t} - D_0 \triangle_\mathbf{x} T(\mathbf{x}, t) = 0 . \tag{5.172}$$

As discussed in detail in section 5.6 this simply reflects that on length scales larger than the mean free path the motion of the electrons is diffusive. The heat at a hot spot diffuses away.

5.5 Boltzmannian Motion in a Random Potential

In later chapters we shall discuss quantum corrections to classical transport. However, in many cases we often still only need to know the classical kinetics of the

---

[18]We note that measuring the change in heating of a sample for reversed current directions, determines the derivative of the thermopower with respect to the temperature.

particle motion. We therefore discuss the Boltzmannian motion of a particle scattered by impurities at this opportunity, although we shall not need these results before we discuss destruction of phase coherence due to electron-phonon interaction in chapter 11. The Boltzmann theory is a stochastic description of the classical motion of a particle in a weakly disordered potential. At each instant the particle has attributed a probability for a certain position and velocity (or momentum). In the absence of external fields the Boltzmann equation for a particle in a random potential has the form

$$\frac{\partial f(\mathbf{x}, \mathbf{p}, t)}{\partial t} + \mathbf{v} \cdot \frac{\partial f(\mathbf{x}, \mathbf{p}, t)}{\partial \mathbf{x}} = -\int \frac{d\mathbf{p}'}{(2\pi\hbar)^3} W(\mathbf{p}, \mathbf{p}') \left[f(\mathbf{x}, \mathbf{p}, t) - f(\mathbf{x}, \mathbf{p}', t) \right]$$

(5.173)

where we have introduced the notation $\mathbf{v} = \mathbf{v_p} = \mathbf{p}/m$.

The Boltzmann equation is first order in time (the state of a particle is completely determined in classical mechanics by specifying its position and momentum), and the solution for such a Markovian process can be expressed in terms of the conditional probability F for the particle to have position $\mathbf{x}$ and momentum $\mathbf{p}$ at time t *given* it had position $\mathbf{x}'$ and momentum $\mathbf{p}'$ at time t'

$$f(\mathbf{x}, \mathbf{p}, t) = \int \frac{d\hat{\mathbf{p}}'}{4\pi} \int d\mathbf{x}' \, F(\mathbf{x}, \mathbf{p}, t; \mathbf{x}', \mathbf{p}', t') \, f(\mathbf{x}', \mathbf{p}', t') \,.$$

(5.174)

For elastic scattering only the direction of momentum can change, and consequently we need only integrate over the direction of the momentum. In the absence of external fields the motion in between scattering events is along straight lines, and the conditional probability describes how the particle by impurity scattering, is thrown between different straight-line segments, i.e., a Boltzmannian path.

We define the Boltzmann propagator as the conditional probability for the initial condition that it vanishes for times $t < t'$, the retarded Green's function for the Boltzmann equation. The equation obeyed by the Boltzmann propagator is thus, assuming isotropic scattering,

$$\left(\frac{\partial}{\partial t} + \mathbf{v_p} \cdot \frac{\partial}{\partial \mathbf{x}} + \frac{1}{\tau} \right) F(\mathbf{p}, \mathbf{x}, t; \mathbf{p}', \mathbf{x}', t') - \frac{1}{\tau} \int \frac{d\hat{\mathbf{p}}}{4\pi} F(\mathbf{p}, \mathbf{x}, t; \mathbf{p}', \mathbf{x}', t')$$

$$= \hat{\delta}(\hat{\mathbf{p}} - \hat{\mathbf{p}}') \, \delta(\mathbf{x} - \mathbf{x}') \, \delta(t - t')$$

(5.175)

where $\hat{\delta}$ is the spherical delta function

$$\int \frac{d\hat{\mathbf{p}}'}{4\pi} \hat{\delta}(\hat{\mathbf{p}} - \hat{\mathbf{p}}') \, f(\mathbf{p}') = f(\mathbf{p}) \,.$$

(5.176)

The equation for the Boltzmann propagator is solved by Fourier transformation, and we obtain

$$F(\mathbf{p}, \mathbf{x}, t; \mathbf{p}', \mathbf{x}', t') = \int \frac{d\mathbf{q} \, d\omega}{(2\pi)^4} \, e^{i\mathbf{q} \cdot (\mathbf{x} - \mathbf{x}') - i\omega(t - t')} \, F(\mathbf{p}, \mathbf{p}'; \mathbf{q}, \omega)$$

(5.177)

where

$$F(\mathbf{p}, \mathbf{p}'; \mathbf{q}, \omega) = \frac{1}{-i\omega + \mathbf{p} \cdot \mathbf{q}/m + 1/\tau} \left(\frac{1/\tau}{-i\omega + \mathbf{p}' \cdot \mathbf{q}/m + 1/\tau} I(q, \omega) + \hat{\delta}(\hat{\mathbf{p}} - \hat{\mathbf{p}}') \right)$$
(5.178)

and

$$I(q, \omega) = \frac{ql}{ql - \arctan ql/(1 - i\omega\tau)}$$
(5.179)

where $l = v\tau$ is the mean free path.

We note, by direct integration, the property

$$F(\mathbf{x}, \mathbf{p}, t; \mathbf{x}', \mathbf{p}', t') = \int \frac{d\hat{\mathbf{p}}''}{4\pi} \int d\mathbf{x}'' \, F(\mathbf{x}, \mathbf{p}, t; \mathbf{x}'', \mathbf{p}'', t'') \, F(\mathbf{x}'', \mathbf{p}'', t''; \mathbf{x}', \mathbf{p}', t')$$
(5.180)

the signature of a Markovian process.[19] This property will be utilized in section 11.3.1 in the calculation of the dephasing rate in weak localization due to electron-phonon interaction.

5.6   Brownian Motion

If we are only interested in the long-time and large-distance behavior of the particle motion, $|\mathbf{x} - \mathbf{x}'| \gg l$, $t - t' \gg \tau$, the wave vectors and frequencies of importance in the Boltzmann propagator, eq.(5.178), satisfy $ql, \omega\tau \ll 1$, and we obtain the diffusion approximation

$$I(q, \omega) \simeq \frac{1/\tau}{-i\omega + D_0 q^2}$$
(5.181)

where $D_0 = vl/3$ is the diffusion constant in the considered case of three dimensions (and isotropic scattering). Fourier-transforming we find that in the diffusion approximation the dependence on the magnitude of the momentum (velocity) in the momentum directional averaged Boltzmann propagator only appears through the diffusion constant, $t > t'$,

$$D(\mathbf{x}, t; \mathbf{x}, t') \equiv \int \frac{d\hat{\mathbf{p}}d\hat{\mathbf{p}}'}{(4\pi)^2} F(\mathbf{p}, \mathbf{x}, t; \mathbf{p}', \mathbf{x}', t') = \int \frac{d\mathbf{q}d\omega}{(2\pi)^4} \frac{e^{i\mathbf{q}\cdot(\mathbf{x}-\mathbf{x}')-i\omega(t-t')}}{-i\omega + D_0 q^2}$$

$$= \frac{e^{-(\mathbf{x}-\mathbf{x}')^2/4D_0(t-t')}}{(4\pi D_0(t - t'))^{d/2}} .$$
(5.182)

This diffusion propagator describes the diffusive or Brownian motion of the particle, the conditional probability for the particle to diffuse from point $\mathbf{x}'$ to $\mathbf{x}$ in time span $t - t'$, described by the one parameter, the diffusion constant. The absence of the explicit appearance of the magnitude of the velocity reflects the fact that the

---

[19]For a Markovian process, the *future* is independent of the *past* when the *present* is known, i.e., the causality principle of classical physics in the context of a stochastic dynamic system, here the process in question is Boltzmannian motion.

local velocity is a meaningless quantity in Brownian motion, a point we discuss further in section 8.7.

Exercise 5.6 *Show that*

$$< \mathbf{x}^2 >_{t,\mathbf{x}',t'} \equiv \int d\mathbf{x} \, \mathbf{x}^2 \, D(\mathbf{x}, t; \mathbf{x}', t') = \mathbf{x}'^2 + 2dD_0(t - t') \qquad (5.183)$$

where d is the spatial dimension.

If we are only interested in the long-time and large-distance behavior of the Boltzmannian motion we can, as noted above, get a simplified description of the classical motion of a particle in a random potential. We are thus not interested in the zigzag Boltzmannian trajectories, but only in the smooth large-scale behavior. It is instructive to relate the large-scale behavior to the velocity (or momentum) moments of the distribution function, and the corresponding physical quantities, density and current density. Expanding the distribution function on spherical harmonics

$$f(\mathbf{x}, \mathbf{p}, t) = f_0(\epsilon_\mathbf{p}, \mathbf{x}, t) + \mathbf{p} \cdot \mathbf{f}(\epsilon_\mathbf{p}, \mathbf{x}, t) + \dots \qquad (5.184)$$

we have that the particle current density is given in terms of the first moment

$$\mathbf{j}(\mathbf{x}, t) = \frac{1}{m} \int \frac{d\mathbf{p}}{(2\pi\hbar)^3} \, \mathbf{p} \, \mathbf{p} \cdot \mathbf{f}(\epsilon_\mathbf{p}, \mathbf{x}, t) = \frac{1}{3m} \int \frac{d\mathbf{p}}{(2\pi\hbar)^3} \, p^2 \, \mathbf{f}(\epsilon_\mathbf{p}, \mathbf{x}, t) \qquad (5.185)$$

and the density in terms of the zeroth moment

$$n(\mathbf{x}, t) = \int \frac{d\mathbf{p}}{(2\pi\hbar)^3} \, f_0(\epsilon_\mathbf{p}, \mathbf{x}, t) \, . \qquad (5.186)$$

Taking the spherical average

$$< \dots > \equiv \int \frac{d\hat{\mathbf{p}}}{4\pi} \, \dots \qquad (5.187)$$

of the force-free Boltzmann equation, eq.(5.173), we obtain the zeroth moment equation

$$\frac{\partial f_0(\epsilon_\mathbf{p}, \mathbf{x}, t)}{\partial t} + \frac{p^2}{3m} \nabla_\mathbf{x} \cdot \mathbf{f}(\epsilon_\mathbf{p}, \mathbf{x}, t) = 0 \, . \qquad (5.188)$$

Integrating this equation with respect to momentum gives the continuity equation

$$\frac{\partial n(\mathbf{x}, t)}{\partial t} + \nabla_\mathbf{x} \cdot \mathbf{j}(\mathbf{x}, t) = 0 \, . \qquad (5.189)$$

This result is of course independent of whether external fields are present or not. This is seen directly from the Boltzmann equation, eq.(4.118), by integrating with respect to momentum as we have the identity

$$\int \frac{d\hat{\mathbf{p}}}{4\pi} \, I_{\mathbf{x},\mathbf{p},t}[f] = 0 \qquad (5.190)$$

simply reflecting that the collision integral respects particle conservation.

Taking the first moment of the Boltzmann equation, $< \mathbf{p} \ldots >$,

$$\int \frac{d\hat{\mathbf{p}}}{4\pi} \, \mathbf{p} \left(\frac{\partial f(\mathbf{x}, \mathbf{p}, t)}{\partial t} + \mathbf{v_p} \cdot \frac{\partial f(\mathbf{x}, \mathbf{p}, t)}{\partial \mathbf{x}} - I_{\mathbf{x}, \mathbf{p}, t}[f] \right) = 0 \qquad (5.191)$$

we obtain the first moment equation

$$\frac{p^2}{3} \left(\frac{\partial}{\partial t} + \frac{1}{\tau(\epsilon_{\mathbf{p}})} \right) \mathbf{f}(\mathbf{x}, \mathbf{p}, t) + \frac{p^2}{3m} \frac{\partial f_0(\mathbf{x}, \mathbf{p}, t)}{\partial \mathbf{x}} = 0 \qquad (5.192)$$

where we have used the angular average formulas

$$\int \frac{d\hat{\mathbf{p}}}{4\pi} \, p_\alpha \, p_\beta = \frac{p^2}{3} \delta_{\alpha, \beta} \quad , \quad \int \frac{d\hat{\mathbf{p}}}{4\pi} \, p_\alpha \, p_\beta \, p_\gamma = 0 \qquad (5.193)$$

repeatedly.

We have thus reduced the kinetic equation to a closed set of equations relating the two lowest moments of the distribution function,[20] f_0 and $\mathbf{f}$, and we get the equation satisfied by the zeroth moment f_0:

$$\left(\frac{\partial}{\partial t} + \frac{1}{\tau(\epsilon_{\mathbf{p}})} \right) \frac{\partial f_0(\mathbf{x}, \mathbf{p}, t)}{\partial t} - \frac{p^2}{3m^2} \triangle_{\mathbf{x}} f_0(\mathbf{x}, \mathbf{p}, t) = 0 \, . \qquad (5.194)$$

In a metal the derivatives of the zeroth harmonic of the distribution function for the conduction electrons, $\partial_t f_0(\epsilon_{\mathbf{p}}, \mathbf{x}, t)$ and $\triangle_{\mathbf{x}} f_0(\epsilon_{\mathbf{p}}, \mathbf{x}, t)$, are peaked at the Fermi energy, and we can use the approximations

$$\int \frac{d\mathbf{p}}{(2\pi\hbar)^3} \, p^2 \triangle_{\mathbf{x}} f_0(\epsilon_{\mathbf{p}}, \mathbf{x}, t) \simeq p_F^2 \int \frac{d\mathbf{p}}{(2\pi\hbar)^3} \, \triangle_{\mathbf{x}} f_0(\epsilon_{\mathbf{p}}, \mathbf{x}, t) \qquad (5.195)$$

and

$$\int \frac{d\mathbf{p}}{(2\pi\hbar)^3} \left(\frac{\partial}{\partial t} + \frac{1}{\tau(\epsilon_{\mathbf{p}})} \right) \frac{\partial f_0(\epsilon_{\mathbf{p}}, \mathbf{x}, t)}{\partial t} \simeq \left(\frac{\partial}{\partial t} + \frac{1}{\tau} \right) \frac{\partial n(\mathbf{x}, t)}{\partial t} \qquad (5.196)$$

where as usual $\tau \equiv \tau(\epsilon_{p_F})$. Assuming that we only have low-frequency oscillations in the density, $\omega\tau \ll 1$,

$$\left| \frac{\partial^2 n}{\partial t^2} \right| \ll \frac{1}{\tau} \left| \frac{\partial n}{\partial t} \right| \qquad (5.197)$$

we obtain from eq.(5.194) the continuity equation on diffusive form

$$\left(\frac{\partial}{\partial t} - D_0 \triangle_{\mathbf{x}} \right) n(\mathbf{x}, t) = 0 \, . \qquad (5.198)$$

Since $\nabla_{\mathbf{x}} f_0(\epsilon_{\mathbf{p}}, \mathbf{x}, t)$ is peaked at the Fermi energy, we can use the approximation

$$\int \frac{d\mathbf{p}}{(2\pi\hbar)^3} \, p^2 \nabla_{\mathbf{x}} f_0(\epsilon_{\mathbf{p}}, \mathbf{x}, t) \simeq p_F^2 \int \frac{d\mathbf{p}}{(2\pi\hbar)^3} \, \nabla_{\mathbf{x}} f_0(\epsilon_{\mathbf{p}}, \mathbf{x}, t) \qquad (5.199)$$

---

[20]This is only possible due to the relaxation time form, eq.(5.107), of the collision integral.

and assuming only low-frequency current oscillations

$$\left|\frac{\partial \mathbf{j}(\mathbf{x}, t)}{\partial t}\right| \ll \frac{1}{\tau} |\mathbf{j}(\mathbf{x}, t)| \tag{5.200}$$

we obtain from the first moment equation, eq.(5.192), the diffusion expression for the current density

$$\mathbf{j}(\mathbf{x}, t) = -D_0 \frac{\partial n(\mathbf{x}, t)}{\partial \mathbf{x}} . \tag{5.201}$$

If we assume that the particle is absent prior to time t', at which time the particle is created at point $\mathbf{x}'$, the diffusion equation, eq.(5.198), gets a source term, and we obtain for the conditional probability or diffusion propagator $D(\mathbf{x}, t; \mathbf{x}', t')$

$$n(\mathbf{x}, t) = \int d\mathbf{x}' \, D(\mathbf{x}, t; \mathbf{x}', t') \, n(\mathbf{x}', t') \tag{5.202}$$

the equation

$$\left(\frac{\partial}{\partial t} - D_0 \triangle_\mathbf{x}\right) D(\mathbf{x}, t; \mathbf{x}', t') = \delta(\mathbf{x} - \mathbf{x}') \, \delta(t - t') \tag{5.203}$$

with the initial condition

$$D(\mathbf{x}, t; \mathbf{x}', t') = 0 , \quad \text{for} \quad t < t' . \tag{5.204}$$

We can solve the equation for the diffusion propagator, the retarded Green's function for the diffusion equation, by referring to the solution of the free particle Schrödinger Green's function equation, eq.(2.37), and let $it \to t$, and $\hbar/2m \to D_0$, and we obtain

$$D(\mathbf{x}, t; \mathbf{x}', t') = \theta(t - t') \frac{e^{-\frac{(\mathbf{x} - \mathbf{x}')^2}{4D_0(t - t')}}}{(4\pi D_0(t - t'))^{d/2}} . \tag{5.205}$$

Chapter 6

Particle in Oscillator Environment

In this chapter we shall consider the interaction between a single degree of freedom and an environment. We shall assume that the environment can be represented by a set of harmonic oscillators, and that the interaction between the system of interest and the environment is linear in the coupling to the oscillator degrees of freedom.[1] There are systems where the linear coupling to the environment degrees of freedom is exact, such as the fundamental interaction of a charged particle interacting with the electromagnetic field, electron-photon interaction. Furthermore, many situations are adequately described by such an interaction, because it is sufficient to consider each environment degree of freedom only weakly perturbed. This is often the case for an electron interacting with the lattice vibrations of a crystal.[2] We shall mainly be interested in the important case where the environment is supposed to represent a heat bath in which case the oscillator model with linear coupling represents the proper phenomenology.

6.1 Particle-Oscillator Coupling

Let us consider a single degree of freedom, referred to as the particle, interacting with some other degrees of freedom which we refer to as the environment. The Hamiltonian for such a system has the form

$$\hat{H} = \hat{H}_0 + \hat{H}_i \tag{6.1}$$

where the Hamiltonian for the noninteracting subsystems

$$\hat{H}_0 = \hat{H}_p + \hat{H}_E \tag{6.2}$$

---

[1] This representation of the environment is appropriate when any single degree of freedom of the environment is only weakly perturbed by the system. For a detailed discussion of this qualitative assertion see references [26] and [27].

[2] We consider this case in detail in chapter 10.

consists of a term $\hat{H}_p$ describing the particle and a term $\hat{H}_E$ the environment. The environment is modeled by a set of harmonic oscillator degrees of freedom represented by the position and momentum operators $\hat{x}_q$ and $\hat{p}_q$,[3] labeled by q, so that the environment Hamiltonian is

$$\hat{H}_E = \sum_q \left(\frac{\hat{p}_q^2}{2M_q} + \frac{1}{2} M_q \omega_q^2 \hat{x}_q^2 \right). \tag{6.3}$$

The oscillators are characterized by their mass M_q and frequency ω_q.

In the case where the particle degree of freedom is continuous, such as is the case for position, and the particle is subject to an external spatially homogeneous force $F(t)$, we have the particle Hamiltonian

$$\hat{H}_p = \frac{\hat{p}^2}{2m} - F(t)\,\hat{x} \tag{6.4}$$

where m denotes the mass of the particle, and $\hat{x}$ and $\hat{p}$ the position and momentum operators for the particle, respectively.

The interaction we assume linear in the coupling to the environment, i.e., to the position and momentum of the oscillators[4]

$$\hat{H}_i = i \sum_q \{\alpha_q(\hat{x})\hat{a}_q - \alpha_q^*(\hat{x})\hat{a}_q^\dagger\} \equiv H_i(\hat{x}, \{\hat{a}_q, \hat{a}_q^\dagger\}_q) \equiv H_i(\hat{x}, \{\hat{x}_q, \hat{p}_q\}_q) \tag{6.5}$$

where $\alpha_q(\hat{x})$ describes the coupling between the particle and the q'th oscillator, and $\hat{a}_q^\dagger$ and $\hat{a}_q$ are the creation and annihilation operators for the q'th harmonic oscillator

$$\hat{a}_q = \sqrt{\frac{M_q \omega_q}{2\hbar}} \left(\hat{x}_q + \frac{i}{M_q \omega_q} \hat{p}_q \right) \quad , \quad \hat{a}_q^\dagger = \sqrt{\frac{M_q \omega_q}{2\hbar}} \left(\hat{x}_q - \frac{i}{M_q \omega_q} \hat{p}_q \right). \tag{6.6}$$

The commutation relations for position and momentum lead to the following commutation relations for the creation and annihilation operators

$$[\hat{a}_q, \hat{a}_{q'}] = 0 = [\hat{a}_q^\dagger, \hat{a}_{q'}^\dagger] \quad , \quad [\hat{a}_q, \hat{a}_{q'}^\dagger] = \delta_{qq'} . \tag{6.7}$$

Inverting the transformation

$$\begin{pmatrix} \hat{a}_q \\ \hat{a}_q^\dagger \end{pmatrix} = \sqrt{\frac{M_q \omega_q}{2\hbar}} \begin{pmatrix} 1 & \frac{i}{M_q \omega_q} \\ 1 & \frac{-i}{M_q \omega_q} \end{pmatrix} \begin{pmatrix} \hat{x}_q \\ \hat{p}_q \end{pmatrix} \tag{6.8}$$

---

[3]It is immaterial for the formal developments of this chapter that we refer to position and momentum, we could equally well consider any complementary properties characterizing the oscillator, which would be appropriate for discussing the degrees of freedom of the electromagnetic field.

[4]For systems with nonlinear coupling, the applicability of the model thus assumes that each oscillator degree of freedom is only weakly perturbed. In that case the model neglects anharmonicity, and is only the starting point for a perturbative study of such effects.

in order to express the position and momentum operators in terms of the creation and annihilation operators

$$\begin{pmatrix} \hat{x}_q \\ \hat{p}_q \end{pmatrix} = \sqrt{\frac{\hbar}{2M_q \omega_q}} \begin{pmatrix} 1 & 1 \\ -iM_q \omega_q & iM_q \omega_q \end{pmatrix} \begin{pmatrix} \hat{a}_q \\ \hat{a}_q^\dagger \end{pmatrix} \tag{6.9}$$

we get the normal mode representation of the environment Hamiltonian

$$\hat{H}_E = \sum_q \hat{h}_q \ , \ \ \hat{h}_q \equiv \hbar \omega_q \left(\hat{n}_q + \frac{1}{2} \right) = \hbar \omega_q \left(\hat{a}_q^\dagger \hat{a}_q + \frac{1}{2} \right) . \tag{6.10}$$

An energy eigenstate of an oscillator is simultaneously an eigenstate of the number operator $\hat{n}_q = \hat{a}_q^\dagger \hat{a}_q$

$$\hat{n}_q \, |n_q> \ = \ n_q \, |n_q> \qquad\qquad n_q = 0, 1, 2, 3, ... \tag{6.11}$$

counting the number of quanta in the q'th oscillator, and the corresponding representation is referred to as the number representation.

Exercise 6.1 *Show that (the eigenstates are normalized)*

$$\hat{a}_q^\dagger \, |n_q> \ = \ \sqrt{n_q + 1} \, |n_q + 1> \ , \ \ \hat{a}_q \, |n_q> \ = \ \sqrt{n_q} \, |n_q - 1> \ , \ \ n_q = 0, 1, 2, 3, ... \tag{6.12}$$

and in particular that the annihilation operator turns the ground state vector into the zero vector, i.e., $\hat{a}_q \, |0> \ = \ 0$. Show that

$$|n_q> \ = \ \frac{1}{\sqrt{n_q!}} \, (\hat{a}_q^\dagger)^{n_q} \, |n_q = 0> \qquad\qquad n_q = 0, 1, 2, 3, ... \tag{6.13}$$

6.2 Reduced Density Matrix

We shall in the following consider situations where only observation of the particle degree of freedom is made, and are therefore interested in operators of the form

$$\hat{A} = \hat{I}_{osc} \otimes \hat{A}_p \equiv \hat{A}_p . \tag{6.14}$$

The last identification suppresses the trivial identity action of the operator on the oscillator degrees of freedom

$$\hat{A}_p \, |osc> \, |\psi> \ = \ (\hat{I}_{osc} \otimes \hat{A}_p)(|osc> \otimes |\psi>) \ = \ \hat{I}_{osc} |osc> \otimes \hat{A}_p |\psi> \ = \ |osc> \, \hat{A}_p |\psi> \tag{6.15}$$

for an arbitrary state ψ of the particle, and an arbitrary state of the oscillators as represented by the state vector $|osc>$.

In the first chapter we showed quite generally that any experimental outcome can be expressed as a weighted trace of the operator representing the corresponding physical quantity. At present there is no preferred basis with which to take the environmental trace, but later on it is convenient to take the trace with respect to the oscillator degrees of freedom in the eigenbasis of the environment Hamiltonian. We shall therefore use this basis explicitly in the following.

For a bunch of oscillators we have the energy eigenstates, the product states,

$$|n_{q_1}, n_{q_2}, n_{q_3}, ..> \equiv |n_{q_1} > \otimes |n_{q_2} > \otimes |n_{q_3} > .. \equiv |n_{q_1} > |n_{q_2} > |n_{q_3} > ..$$

$$\equiv |\{n_q\}_q > \qquad (6.16)$$

where each number, n_{q_n}, is an integer number, $n_{q_n} = 0, 1, 2, 3, ...$, specifying the number of quanta in oscillator mode q_n. The product eigenstates is the complete set of eigenvectors for the environment Hamiltonian, or equivalently for the total number operator $\hat{N} = \sum_q \hat{n}_q$, and instead of the equivalent energy representation we speak of the number representation. We have in eq.(6.16) chosen a definite ordering of the oscillators, and can therefore dismiss the tensor symbol $\otimes$ as we have an unambiguous rule as to which operators operate on which state vectors. For example, for the bunch of only two oscillators we have

$$\hat{B}_1 \otimes \hat{B}_2 \ |n_{q_1}, n_{q_2} > = \hat{B}_1 |n_{q_1} > \hat{B}_2 |n_{q_2} > \ . \qquad (6.17)$$

If for the particle we choose the eigenvectors for the quantity of interest, A_p, we have as a complete set of states for the combined system of particle and environment $\{|a> |\{n_q\}_q >\}_{a,\{n_q\}_q}$. For the expectation value of the observable $\hat{A}_p$ we have (recall eq.(1.389) or confer section 6.2.2)

$$<\hat{A}_p(t)> = Tr(\hat{\rho}(t)\hat{A}_p) = \int_a \sum_{\{n_q\}_q} <\{n_q\}_q|<a| \ \hat{\rho}(t)\hat{A}_p \ |a>|\{n_q\}_q > \ . \qquad (6.18)$$

The trace of an operator equals the sum of the diagonal elements in a basis, and in each term we perform the matrix element for each degree of freedom separately, since the involved states are product states. Making only a distinction between the particle and oscillator degrees of freedom we can therefore split the total trace into partial traces

$$Tr = tr_p \, tr = tr \, tr_p \qquad (6.19)$$

where we have introduced tr for the trace over the oscillator states, and tr_p denotes the trace with respect to the particle. The multiplicative nature of the trace over multiple degrees of freedom, allows for an arbitrary order of performance.

For the expectation value of interest, we perform the trace over the oscillators first, and since $\hat{A}_p$ only operates on the particle degree of freedom, the operator can be moved outside the trace over the oscillator degrees of freedom to give

$$<\hat{A}_p(t)> \equiv Tr(\hat{\rho}(t)\hat{A}_p) = \int_a <a|\hat{A}_p \sum_{\{n_q\}_q} <\{n_q\}_q| \ \hat{\rho}(t) \ |\{n_q\}_q >|a>$$

$$= tr_p(\hat{A}_p \sum_{\{n_q\}_q} <\{n_q\}_q| \hat{\rho}(t) |\{n_q\}_q>) . \qquad (6.20)$$

Introducing the reduced statistical operator for the particle

$$\hat{f}(t) \equiv \sum_{\{n_q\}_q} <\{n_q\}_q| \hat{\rho}(t) |\{n_q\}_q> \equiv tr\hat{\rho}(t) \qquad (6.21)$$

as the statistical operator traced over the oscillator degrees of freedom, we can write for expectation values relating only to the particle properties

$$<\hat{A}_p(t)> = tr_p(\hat{f}(t) \hat{A}_p) = \int_a <a|\hat{A}_p \hat{f}(t)|a> . \qquad (6.22)$$

We note, that since $\hat{\rho}(t)$ is hermitian so is $\hat{f}(t)$.

The reduced statistical operator is only an operator with respect to the particle degree of freedom, but contains all the information of the influence of the unobserved oscillators on the particle. The matrix elements of the reduced statistical operator, the reduced density matrix, is in the a-representation

$$f(a, a', t) \equiv <a|\hat{f}(t)|a'> = Tr(\hat{\rho}(t)|a'><a|) = tr <a|\hat{\rho}(t)|a'> . \qquad (6.23)$$

For the expectation value of the property A_p we have in terms of the reduced density matrix

$$<\hat{A}_p(t)> = \int_a a <a|\hat{f}(t)|a> = \int_a a f(a, a, t) . \qquad (6.24)$$

Knowledge of the reduced density matrix allows us all information regarding measurements only on the subsystem, the particle. For example, the probability density to find the particle with property value a at time t is

$$P_a(t) = <\delta(\hat{A}_p(t) - a)> = Tr(\hat{\rho}(t)\delta(\hat{A}_p - a)) = f(a, a, t) \qquad (6.25)$$

the diagonal elements of the reduced density matrix in the property representation in question. Equivalently stated, the diagonal elements of the reduced density matrix, $f(a, a, t)$, gives the probability for the particle to be in state a *times* the probability to find the oscillators in *some* state, the latter probability being equal to one.

6.2.1 Particle Density and Current

In the case where the degree of freedom of interest is the position of a particle, we have for the matrix elements of the reduced statistical operator

$$f(\mathbf{x}, \mathbf{x}', t) = <\mathbf{x}|\hat{f}(t)|\mathbf{x}'> = Tr(\hat{\rho}(t)|\mathbf{x}'><\mathbf{x}|) = tr <\mathbf{x}|\hat{\rho}(t)|\mathbf{x}'> \qquad (6.26)$$

the reduced density matrix in the position representation. The diagonal elements specify the probability density, which we now demonstrate explicitly by asking for the probability density to find the particle at position $\mathbf{x}$ at time t, irrespective of the state of the oscillator degrees of freedom. This probability, $P_{\mathbf{x}}^{\rho}(t)$, is given by (recall eq.(1.378) or see the discussion in section 6.2.2)

$$
\begin{aligned}
P_{\mathbf{x}}^{\rho}(t) &\equiv \ <\hat{P}(\mathbf{x},t)>_{\rho} \ = Tr(\hat{\rho}_i \hat{P}(\mathbf{x},t)) = Tr(\hat{\rho}(t)\hat{P}(\mathbf{x})) = tr_p(\hat{f}(t)\hat{P}(\mathbf{x})) \\
&= f(\mathbf{x},\mathbf{x},t)
\end{aligned}
\tag{6.27}
$$

where $\hat{P}(\mathbf{x},t)$ is short for $\hat{I}_{\text{osc}} \otimes \hat{P}(\mathbf{x},t)$, i.e., the identity operator for the oscillator degrees of freedom times the position projector at time t in the Heisenberg picture

$$
\hat{P}(\mathbf{x},t) \ = \ |\mathbf{x},t><\mathbf{x},t| \ = \ e^{\frac{i}{\hbar}\hat{H}t}\hat{P}(\mathbf{x})e^{-\frac{i}{\hbar}\hat{H}t} \ .
\tag{6.28}
$$

Here we have taken the reference time to be zero, and $\hat{\rho}_i \equiv \hat{\rho}(t=0)$.

The probability density to find the particle at position $\mathbf{x}$ at time t, which we also refer to as the particle density, we have thus shown to be given by the diagonal element of the reduced density matrix

$$
n(\mathbf{x},t) \ = \ f(\mathbf{x},\mathbf{x},t) \ = \ <\mathbf{x}|\hat{f}(t)|\mathbf{x}>
\tag{6.29}
$$

for the case where the environment state is left unobserved as represented by the identity operator for the oscillators.[5]

Another example, say, the probability current density $\mathbf{j}(\mathbf{x},t)$ for the particle at time t and at position $\mathbf{x}$ is given by the expectation value of the current density operator

$$
\hat{\mathbf{j}}(\mathbf{x}) \ = \ \frac{1}{2}\{\hat{\mathbf{v}}, \hat{n}(\mathbf{x})\}
\tag{6.30}
$$

which gives for the expectation value for the case where the oscillator degrees of freedom are left unobserved[6]

$$
\mathbf{j}(\mathbf{x},t) \ = \ Tr(\hat{\rho}(t)\hat{\mathbf{j}}(\mathbf{x})) \ = \ \frac{\hbar}{2mi}\left(\frac{\partial}{\partial \mathbf{x}} - \frac{\partial}{\partial \mathbf{x}'}\right)f(\mathbf{x},\mathbf{x}',t)\bigg|_{\mathbf{x}'=\mathbf{x}} \ .
\tag{6.31}
$$

Particle conservation is reflected in the continuity equation

$$
\frac{\partial n(\mathbf{x},t)}{\partial t} \ + \ \nabla_{\mathbf{x}} \cdot \mathbf{j}(\mathbf{x},t) \ = \ 0
\tag{6.32}
$$

which is immediately obtained using the equation of motion for the density operator, eq.(1.369), and tracing over the oscillator degrees of freedom.

---

[5]This is in accordance with the physical interpretation of the identity operator in section 1.3.1.

[6]In the presence of a vector potential the formula must be amended with the diamagnetic term, as discussed in section 7.3.1.

6.2.2 Partial Trace

When calculating the probability distribution for outcomes for the case where some
degrees of freedom are left unobserved, we must sum over all possible final states
of the unobserved degrees of freedom. It is useful to consider this for the specific
case of interest in order to gain familiarity with the partial trace, and at the same
time we shall take the opportunity to introduce thermal equilibrium notation.

Let us to this end assume, that at some time, t', we have prepared the particle
in state $|\psi>$ and the oscillators in the energy eigenstate $|\{n_q\}_q>$ whereby the
state of the total system is

$$|i> \ = \ |\psi> \otimes |\{n_q\}_q> \ = \ |\psi> \ |\{n_q\}_q> \qquad (6.33)$$

and, say, we are interested in the probability for finding the total system at time t
in the state corresponding to finding the particle at position $\mathbf{x}$ and the oscillators
in state $|\{m_q\}_q>$

$$|f> \ = \ |\mathbf{x}> \ |\{m_q\}_q> \ . \qquad (6.34)$$

The conditional amplitude to find the system in state $|f>$ at time t, given it was
in state $|i>$ at time t', is

$$A_{fi} \ = \ <f,t|i,t'> \ = \ <f|\hat{U}(t,t')|i> \qquad (6.35)$$

with the corresponding probability

$$P_{fi} \ = \ |A_{fi}|^2 \ = \ <i|\hat{U}^\dagger(t,t')|f><f|\hat{U}(t,t')|i>$$

$$= \ <\{n_q\}_q|<\psi|\,\hat{U}^\dagger(t,t')\,|\mathbf{x}>|\{m_q\}_q><\{m_q\}_q|<\mathbf{x}|\,\hat{U}(t,t')\,|\psi>|\{n_q\}_q> . (6.36)$$

Now, let us renounce precise knowledge of the initial state, and assume that
the initial state is a mixture where we only know that the probability for the envi-
ronment to be in state $|\{n_q\}_q>$ is $P(\{n_q\}_q)$. The statistical operator representing
this mixture is specified through its eigenvalues on the complete oscillator set of
states

$$\hat{\rho}_i^E \ |\{n_q\}_q> \ = \ P(\{n_q\}_q) \ |\{n_q\}_q> \ . \qquad (6.37)$$

In order to calculate the probability for finding the final state $|f>$ conforming with
this initial information, we must take the statistical average (i.e., relative sum) over
the initial distribution

$$\bar{P}_{fi} \ = \ \sum_{\{n_q\}_q} P_{fi} \, P(\{n_q\}_q) \ . \qquad (6.38)$$

Suppose no detection of the oscillators is done. They can then have propagated
to any final state. We must therefore, to obtain the final probability distribution
for the particle, sum over all these final environment alternatives in order to be in
concordance with the supposed complete ignorance of the final environment state.
Given that initially the particle is prepared in state ψ, and the oscillators initially
in the mixture described by the initial statistical operator $\hat{\rho}_i^E$, we thus have for the

probability to find the particle at position $\mathbf{x}$ at time t without any detection of the oscillators

$$
\begin{aligned}
P_{|\psi>\to|\mathbf{x}>} &= \sum_{\{m_q\}_q} \bar{P}_{fi} \\
&= \sum_{\{n_q\}_q, \{m_q\}_q} P(\{n_q\}_q) <\{n_q\}_q|<\psi|\,\hat{U}^\dagger(t,t')\,|\mathbf{x}>|\{m_q\}_q> \\
&\qquad <\{m_q\}_q|<\mathbf{x}|\,\hat{U}(t,t')\,|\psi>|\{n_q\}_q> \ .
\end{aligned}
\tag{6.39}
$$

Using the completeness of the oscillator states

$$
\hat{I}_{\text{osc}} = \sum_{\{m_q\}_q} |\{m_q\}_q><\{m_q\}_q|
\tag{6.40}
$$

we obtain for the transition probability

$$
\begin{aligned}
P_{|\psi>\to|\mathbf{x}>} &= \sum_{\{n_q\}_q} <\{n_q\}_q|\hat{\rho}_i^E <\psi|\hat{U}^\dagger(t,t')|\mathbf{x}> <\mathbf{x}|\hat{U}(t,t')|\psi> |\{n_q\}_q> \\
&= tr(\hat{\rho}_i^E <\psi|\hat{U}^\dagger(t,t')\,\hat{P}(\mathbf{x})\,\hat{U}(t,t')|\psi>) \ .
\end{aligned}
\tag{6.41}
$$

We of course recognize the above expression for the transition probability as identical to the one in eq.(6.27) for the initial state

$$
\hat{\rho}_i = \hat{P}_\psi \otimes \hat{\rho}_i^E = |\psi><\psi| \otimes \hat{\rho}_i^E \ .
\tag{6.42}
$$

Employing the spectral representation of the operator representing a physical quantity, eq.(1.157), we immediately arrive at the expression for the expectation value eq.(6.18).

If we assume that the environment initially is in thermal equilibrium at temperature T, and therefore acts as a heat bath for the particle, then the initial distribution is specified by

$$
P(\{n_q\}_q) = \frac{1}{Z}\, e^{-E(\{n_q\}_q)/kT}
\tag{6.43}
$$

where

$$
E(\{n_q\}_q) = \sum_q \hbar\omega_q \left(n_q + \frac{1}{2}\right)
\tag{6.44}
$$

is the energy of the oscillator state $|\{n_q\}_q>$

$$
\hat{H}_E\,|\{n_q\}_q> = E(\{n_q\}_q)\,|\{n_q\}_q>
\tag{6.45}
$$

and the normalization factor

$$
Z = \sum_{\{n_q\}_q} e^{-E(\{n_q\}_q)/kT} = tr(\exp\{-\hat{H}_E/kT\}) = \prod_q \frac{1}{2\sinh\frac{\hbar\omega_q}{2kT}}
\tag{6.46}
$$

is the partition function for the oscillator environment (for the last explicit expression recall exercise 1.23 on page 76). The thermal environment statistical operator is thus specified by[7]

$$\hat{\rho}_T = Z^{-1} \exp\{-\hat{H}_E/kT\} = Z^{-1} e^{-E(\{n_q\}_q)/kT} |\{n_q\}_q><\{n_q\}_q| . \tag{6.47}$$

If the degrees of freedom, besides the one of the particle, are considered representing a heat bath, they are rendered unobserved and we are only interested in the measurable properties of the particle.[8] This is the situation pertaining to the measurement of the electrical conductivity of a metal which we for instance shall have in mind. We must therefore perform the trace over the bath degrees of freedom, i.e., allow for all possible final states of the bath, and are led to study the reduced statistical operator $\hat{f}$ for the particle.

6.3 Particle-Heat Bath Diagrammatics

In the following we are not so much interested in transient properties as in steady states, where the dependence on the initial density matrix is lost, and any time dependence is due to external forces. We shall therefore assume that at some time t' the statistical operator describing the state of the total system is separable

$$\hat{\rho}_i = \hat{f}_i \otimes \hat{\rho}_E . \tag{6.48}$$

Such an initial condition corresponds to the physical condition, that prior to time t' the two subsystems have not interacted.

With a separable statistical operator at time t', the formal solution of the von Neumann equation, eq.(1.383), gives that the reduced density matrix

$$<\mathbf{x}|\hat{f}(t)|\mathbf{x}'> = <\mathbf{x}|tr(\hat{U}(t,t')\,\hat{\rho}_i\,\hat{U}^\dagger(t,t'))|\mathbf{x}'> \tag{6.49}$$

can be rewritten on the form[9]

$$f(\mathbf{x},\mathbf{x}',t) = \int d\tilde{\mathbf{x}}\int d\tilde{\mathbf{x}}'\, J(\mathbf{x},\mathbf{x}',t;\tilde{\mathbf{x}},\tilde{\mathbf{x}}',t')\, f_i(\tilde{\mathbf{x}},\tilde{\mathbf{x}}') \tag{6.50}$$

where J is the reduced density matrix propagator

$$J(\mathbf{x},\mathbf{x}',t;\tilde{\mathbf{x}},\tilde{\mathbf{x}}',t') = tr(\hat{\rho}_T < \tilde{\mathbf{x}}'|\hat{U}^\dagger(t,t')|\mathbf{x}' > < \mathbf{x}|\hat{U}(t,t')|\tilde{\mathbf{x}} >)$$

$$\equiv\; << \tilde{\mathbf{x}}'|\hat{U}^\dagger(t,t')|\mathbf{x}' > < \mathbf{x}|\hat{U}(t,t')|\tilde{\mathbf{x}} >> \tag{6.51}$$

---

[7]We have considered the oscillators distinguishable. However, for symmetric states such as the thermal equilibrium state we obtain identical results in the following if we treat the oscillators as identical bosons.

[8]In cases where nonequilibrium states of the oscillator degrees of freedom need to be considered (as, for instance, in the case of the phonon drag effect), we must in addition introduce the reduced-density matrices for these degrees of freedom, which in turn leads to a set of equations coupling the density matrices of all the degrees of freedom.

[9]We of course include all transient effects for the chosen initial condition eq.(6.48). Whether this choice is appropriate for the study of transient effects depends on the given physical situation.

and the initial reduced density matrix is the matrix elements of the initial statistical operator for the particle

$$f_i(\mathbf{x}, \mathbf{x}') \equiv <\mathbf{x}|\hat{f}_i|\mathbf{x}'> . \tag{6.52}$$

We assume that the oscillator degrees of freedom are in thermal equilibrium at the temperature T, so that initially the state of the environment is the thermal mixture described by the statistical operator $\hat{\rho}_E = \hat{\rho}_T$, and we have introduced the bracket

$$< \ \dots \ > \equiv tr(\hat{\rho}_T \ \dots) \tag{6.53}$$

to denote the thermally weighted trace over the oscillators.

The spatial diagonal elements, $J(\mathbf{x}, \mathbf{x}, t; \mathbf{x}', \mathbf{x}', t')$, of the reduced density matrix propagator have the simple physical interpretation (recall section 2.10): It is the conditional probability for the particle to be found at position $\mathbf{x}$ at time t, given that it was at position $\mathbf{x}'$ at time t', and the oscillators initially in the thermal state, and no signature distinguishing the final state of the oscillators is available in principle.

We shall be interested in the perturbative structure, in the coupling to the environment, of the reduced density matrix, or equivalently the reduced density matrix propagator J. We therefore express the evolution operator in the interaction picture with respect to $\hat{H}_0 = \hat{H}_p + \hat{H}_E$

$$\hat{U}(t, t') = \exp\left\{-\frac{i}{\hbar}\hat{H}_0 t\right\} T \exp\left\{-\frac{i}{\hbar}\int_{t'}^{t}d\bar{t}\hat{H}_i(\bar{t})\right\} \exp\left\{\frac{i}{\hbar}\hat{H}_0 t'\right\}$$

$$= \exp\left\{-\frac{i}{\hbar}\hat{H}_0 t\right\} \hat{U}_I(t, t') \exp\left\{\frac{i}{\hbar}\hat{H}_0 t'\right\} \tag{6.54}$$

where

$$\hat{H}_i(t) = \exp\left\{\frac{i}{\hbar}\hat{H}_0 t\right\} \hat{H}_i \exp\left\{-\frac{i}{\hbar}\hat{H}_0 t\right\} \tag{6.55}$$

is the interaction Hamiltonian in the interaction picture. The reference time is here chosen to be zero, $t_r = 0$, but could of course be chosen arbitrarily without any consequences for the following discussion.

We obtain for the reduced density matrix propagator

$$J(\mathbf{x}, \mathbf{x}', t; \tilde{\mathbf{x}}, \tilde{\mathbf{x}}', t') = tr(\hat{\rho}_T <\tilde{\mathbf{x}}'|\hat{U}^\dagger(t, t')|\mathbf{x}'><\mathbf{x}|\hat{U}(t, t')|\tilde{\mathbf{x}}>)$$

$$= tr\left(\hat{\rho}_T \left\langle \tilde{\mathbf{x}}', t' \left| \bar{T} \exp\left\{\frac{i}{\hbar}\int_{t'}^{t}d\bar{t}\hat{H}_i(\bar{t})\right\} \right| \mathbf{x}', t \right\rangle\right.$$

$$\left.\left\langle \mathbf{x}, t \left| T \exp\left\{-\frac{i}{\hbar}\int_{t'}^{t}d\bar{t}\hat{H}_i(\bar{t})\right\} \right| \tilde{\mathbf{x}}, t' \right\rangle\right) \tag{6.56}$$

where we have used the cyclic property of the trace, and that $\hat{H}_E$ and $\hat{\rho}_T$ commute, whereby the explicit appearance of temporal dependence due to $\hat{H}_E$ cancel out, and

$$|\mathbf{x}, t> = e^{\frac{i}{\hbar}\hat{H}_p t} |\mathbf{x}> \tag{6.57}$$

denotes a particle position state in the interaction picture.

In order to grasp the structure of the reduced density matrix propagator J we first study it order by order. To zeroth order in the interaction the two exponentials referring to the interaction with the environment are just identity operators, and the weighted oscillator trace is one by normalization, so that we of course get propagation as in the absence of the environment

$$J_0(\mathbf{x}, \mathbf{x}', t; \tilde{\mathbf{x}}, \tilde{\mathbf{x}}', t') \;=\; <\tilde{\mathbf{x}}', t'|\mathbf{x}', t> <\mathbf{x}, t|\tilde{\mathbf{x}}, t'> \;. \tag{6.58}$$

The position states are determined by the particle evolution operator in the interaction picture, so that

$$<\mathbf{x}, t|\mathbf{x}', t'> \;=\; <\mathbf{x}|e^{-\frac{i}{\hbar}\hat{H}_p(t-t')}|\mathbf{x}'> \tag{6.59}$$

is the amplitude for propagation of the particle between the space-time points in question in the absence of the environment.

The propagator of the reduced density matrix in the absence of coupling to the environment, J_0, is thus as expected just the product of the retarded and advanced particle propagators which we depicted by the following diagram

$$J_0(\mathbf{x}, \mathbf{x}', t; \tilde{\mathbf{x}}, \tilde{\mathbf{x}}', t') \;= \tag{6.60}$$

where it is understood that the two amplitudes represented by the two propagator lines are being multiplied together. To zeroth order we therefore have for the reduced density matrix

$$f^{(0)}(\mathbf{x}, \mathbf{x}', t) \;=\; \int d\tilde{\mathbf{x}} \int d\tilde{\mathbf{x}}' \, G_0^R(\mathbf{x}, t; \tilde{\mathbf{x}}, t') \, G_0^A(\tilde{\mathbf{x}}', t'; \mathbf{x}', t) \, f_i(\tilde{\mathbf{x}}, \tilde{\mathbf{x}}') \;. \tag{6.61}$$

The subscript on the propagator indicates that it is the propagator for the particle in the absence of the environment. We have assumed that $t > t'$. If $t < t'$, we should interchange R and A (recall eq.(2.239) on page 129).

For the zeroth-order contribution to the reduced density matrix we introduce the diagrammatic representation

$$f^{(0)}(\mathbf{x}, \mathbf{x}', t) \;= \tag{6.62}$$

describing the free propagation of the initial particle-density matrix, depicted as the dashed vertical line.

The first-order terms in the interaction vanish, $<\hat{H}_i(t)> = 0$, as the oscillators are initially chosen to have zero average displacement, $<\hat{a}_q> = 0 = <\hat{a}_q^\dagger>$, a point we elaborate in section 6.5. The lowest-order terms in the interaction are thus of second order. A second-order term can come about in three ways: Expanding $\hat{U}$ to second order in the interaction and $\hat{U}^\dagger$ to zeroth order, or vice versa, or expanding each to first order.

Consider expanding $\hat{U}(t, t')$ to zeroth order

$$\hat{U}(t, t') \to e^{-\frac{i}{\hbar}\hat{H}_0(t-t')} \qquad \text{as} \qquad \hat{U}_I(t, t') \to \hat{I} \qquad (6.63)$$

and taking in $\hat{U}^\dagger(t, t')$ the second order term

$$\hat{U}^\dagger(t, t') \to e^{-\frac{i}{\hbar}\hat{H}_0 t'} \left(\frac{i}{\hbar}\right)^2 \int_{t'}^t dt_2 \int_{t_2}^t dt_1 \; \hat{H}_i(t_2)\hat{H}_i(t_1) \, e^{\frac{i}{\hbar}\hat{H}_0 t} \qquad (6.64)$$

or

$$\hat{U}_I^\dagger(t, t') \to \left(\frac{i}{\hbar}\right)^2 \int_{t'}^t dt_2 \int_{t_2}^t dt_1 \; \hat{H}_i(t_2)\hat{H}_i(t_1) \, . \qquad (6.65)$$

Inserting the resolution of the identity in terms of the complete set of particle position eigenstates

$$\hat{I} = \int d\mathbf{x}_n \, |\mathbf{x}_n, t_n><\mathbf{x}_n, t_n| \qquad (6.66)$$

where $\{|\mathbf{x}, t>\}_{\mathbf{x}\in R}$ is the complete set of eigenstates for the position operator, $\hat{\mathbf{x}}(t)$, in the interaction picture

$$\hat{\mathbf{x}}(t) = e^{\frac{i}{\hbar}\hat{H}_0 t} \hat{\mathbf{x}} e^{-\frac{i}{\hbar}\hat{H}_0 t} = e^{\frac{i}{\hbar}\hat{H}_p t} \hat{\mathbf{x}} e^{-\frac{i}{\hbar}\hat{H}_p t} \qquad (6.67)$$

we turn the position operator in $\hat{H}_i(t_n)$ into a number

$$\hat{H}_i(t_n)|\mathbf{x}_n, t_n> = H_i(\hat{\mathbf{x}}(t_n), \{\hat{x}_q(t_n), \hat{p}_q(t_n)\}_q)|\mathbf{x}_n, t_n>$$

$$= H_i(\mathbf{x}_n, \{\hat{x}_q(t_n), \hat{p}_q(t_n)\}_q)|\mathbf{x}_n, t_n> \, . \qquad (6.68)$$

The operator in the interaction picture $H_i(\mathbf{x}, \{\hat{a}_q(t), \hat{a}_q^\dagger(t)\}_q)$ is now only an operator with respect to the oscillator degrees of freedom and has the explicit form

$$H_i(\mathbf{x}, \{\hat{x}_q(t), \hat{p}_q(t)\}_q) = i \sum_q [\alpha_q(\mathbf{x}) \, \hat{a}_q(t) - \alpha_q^*(\mathbf{x}) \, \hat{a}_q^\dagger(t)] \qquad (6.69)$$

where $\hat{a}_q(t)$ is the annihilation operator in the interaction picture

$$\hat{a}_q(t) = \exp\left\{\frac{i}{\hbar}\hat{H}_0 t\right\} \hat{a}_q \exp\left\{-\frac{i}{\hbar}\hat{H}_0 t\right\} = \exp\left\{\frac{i}{\hbar}\hat{H}_E t\right\} \hat{a}_q \exp\left\{-\frac{i}{\hbar}\hat{H}_E t\right\} \, . \qquad (6.70)$$

We therefore obtain products of terms of the form

$$< \mathbf{x}_3, t_3 | \hat{H}_i(\hat{\mathbf{x}}(t_2), \{\hat{x}_q(t_2), \hat{p}_q(t_2)\}_q) | \mathbf{x}_2 t_2 >$$

$$= H_i(\mathbf{x}_2, \{\hat{x}_q(t_2), \hat{p}_q(t_2)\}_q) < \mathbf{x}_3, t_3 | \mathbf{x}_2, t_2 >$$

$$= i \sum_q \{\alpha_q(\mathbf{x}_2) \hat{a}_q(t_2) - \alpha_q^*(\mathbf{x}_2) \hat{a}_q^\dagger(t_2)\} < \mathbf{x}_3, t_3 | \mathbf{x}_2, t_2 > . \qquad (6.71)$$

and we can therefore write the considered contribution to $J(\mathbf{x}, \mathbf{x}', t; \tilde{\mathbf{x}}, \tilde{\mathbf{x}}', t')$ as

$$\left(\frac{i}{\hbar}\right)^2 \int_{t'}^t d\tilde{t}_2 \int_{t_2}^t d\tilde{t}_1 \, tr(\hat{\rho}_T < \tilde{\mathbf{x}}', t' | \hat{H}_i(\tilde{t}_2) \, \hat{H}_i(\tilde{t}_1) | \mathbf{x}', t >) < \mathbf{x}, t | \tilde{\mathbf{x}}, t' >$$

$$= \frac{-1}{\hbar^2} \int d\tilde{\mathbf{x}}_2 \int_{t'}^t d\tilde{t}_2 \int d\tilde{\mathbf{x}}_1 \int_{t_2}^t d\tilde{t}_1 < \tilde{\mathbf{x}}_2, \tilde{t}_2 | \tilde{\mathbf{x}}', t' > < \tilde{\mathbf{x}}_1, \tilde{t}_1 | \tilde{\mathbf{x}}_2, \tilde{t}_2 > < \mathbf{x}', t | \tilde{\mathbf{x}}_1, \tilde{t}_1 >$$

$$< \mathbf{x}, t | \tilde{\mathbf{x}}, t' > \ll H_i(\tilde{\mathbf{x}}_2, \{\hat{a}_q(\tilde{t}_2), \hat{a}_q^\dagger(\tilde{t}_2)\}_q) \, H_i(\tilde{\mathbf{x}}_1, \{\hat{a}_q(\tilde{t}_1), \hat{a}_q^\dagger(\tilde{t}_1)\}_q) >$$

$$= \int d\tilde{\mathbf{x}}_2 \int_{-\infty}^\infty d\tilde{t}_2 \int d\tilde{\mathbf{x}}_1 \int_{-\infty}^\infty d\tilde{t}_1 \, G_0^R(\mathbf{x}, t; \tilde{\mathbf{x}}, t') \, G_0^A(\tilde{\mathbf{x}}', t'; \tilde{\mathbf{x}}_2, \tilde{t}_2) \, D(\tilde{\mathbf{x}}_2, \tilde{t}_2; \tilde{\mathbf{x}}_1, \tilde{t}_1)$$

$$G_0^A(\tilde{\mathbf{x}}_2, \tilde{t}_2; \tilde{\mathbf{x}}_1, \tilde{t}_1) G_0^A(\tilde{\mathbf{x}}_1, \tilde{t}_1; \mathbf{x}', t) \qquad (6.72)$$

where the step functions in the propagators have lifted the time integration restrictions, and we have introduced the heat bath correlation function[10]

$$D(\mathbf{x}, t; \mathbf{x}', t') = \frac{1}{\hbar^2} < H_i(\mathbf{x}, \{\hat{a}_q(t), \hat{a}_q^\dagger(t)\}_q) \, H_i(\mathbf{x}', \{\hat{a}_{q'}(t'), \hat{a}_{q'}^\dagger(t')\}_{q'}) >$$

$$= -\frac{1}{\hbar^2} < \sum_{q,q'} \{\alpha_q(\mathbf{x}) \hat{a}_q(t) - \alpha_q^*(\mathbf{x}) \hat{a}_q^\dagger(t)\}$$

$$\{\alpha_{q'}(\mathbf{x}') \hat{a}_{q'}(t') - \alpha_{q'}^*(\mathbf{x}') \hat{a}_{q'}^\dagger(t')\} > \qquad (6.73)$$

describing the interaction energy correlations of the heat bath when the particle propagates between space-time points $(\mathbf{x}', t')$ and $(\mathbf{x}, t)$.

The considered second-order contribution to the propagator of the density matrix we depict as

$$(6.74)$$

---

[10]No confusion with the notation for the diffusion propagator in later chapters should arise.

For the analogous term where we take the second order term in $\hat{U}(t,t')$ and $\hat{U}^\dagger(t,t')$ to zeroth order we get the contribution

$$\frac{-1}{\hbar^2}\int d\mathbf{x}_2\int_t^{t'}dt_2\int d\mathbf{x}_1\int_{t'}^{t_2}dt_1 <\mathbf{x},t|\mathbf{x}_2,t_2><\mathbf{x}_2,t_2|\mathbf{x}_1,t_1><\mathbf{x}_1,t_1|\tilde{\mathbf{x}},t'>$$

$$<H_i(\mathbf{x}_2,\{\hat{a}_q(t_2),\hat{a}_q^\dagger(t_2)\}_q)H_i(\mathbf{x}_1,\{\hat{a}_q(t_1),\hat{p}_q^\dagger(t_1)\}_q)><\tilde{\mathbf{x}}',t'|\mathbf{x}',t>$$

$$=\int d\mathbf{x}_2\int_{-\infty}^{\infty}dt_2\int d\mathbf{x}_1\int_{-\infty}^{\infty}dt_1\,G_0^R(\mathbf{x},t;\mathbf{x}_2,t_2)\,G_0^R(\mathbf{x}_2,t_2;\mathbf{x}_1,t_1)\,G_0^R(\mathbf{x}_1,t_1;\tilde{\mathbf{x}},t')$$

$$D(\mathbf{x}_2,t_2;\mathbf{x}_1,t_1)\,G_0^A(\tilde{\mathbf{x}}',t';\mathbf{x}',t) \tag{6.75}$$

which we depict as

$$\tag{6.76}$$

When we take the first order terms in both $\hat{U}(t,t')$ and $\hat{U}^\dagger(t,t')$, we get

$$\frac{1}{\hbar^2}\int d\tilde{\mathbf{x}}_1\int_{t_2}^t d\tilde{t}_1\int d\mathbf{x}_1\int_{t_2}^t dt_1 <\mathbf{x},t|\mathbf{x}_1,t_1><\mathbf{x}_1,t_1|\tilde{\mathbf{x}},t'><\tilde{\mathbf{x}}_1,\tilde{t}_1|\tilde{\mathbf{x}}',t'>$$

$$<\mathbf{x}',t|\tilde{\mathbf{x}}_1,\tilde{t}_1><H_i(\tilde{\mathbf{x}}_1,\{\hat{a}_q(\tilde{t}_1),\hat{a}_q^\dagger(\tilde{t}_1)\}_q)H_i(\mathbf{x}_1,\{\hat{a}_q(t_1),\hat{p}_q^\dagger(t_1)\}_q)>$$

$$=\int d\tilde{\mathbf{x}}_1\int_{-\infty}^{\infty}d\tilde{t}_1\int d\mathbf{x}_1\int_{-\infty}^{\infty}dt_1\,G_0^R(\mathbf{x},t;\mathbf{x}_1,t_1)\,G_0^R(\mathbf{x}_1,t_1;\tilde{\mathbf{x}},t')$$

$$D(\tilde{\mathbf{x}}_1,\tilde{t}_1;\mathbf{x}_1,t_1)\,G_0^A(\tilde{\mathbf{x}}',t';\tilde{\mathbf{x}}_1,\tilde{t}_1)\,G_0^A(\tilde{\mathbf{x}}_1,\tilde{t}_1;\mathbf{x}',t) \tag{6.77}$$

which we depict as

$$\tag{6.78}$$

We have now derived the lowest-order diagrammatic expressions for the reduced density matrix for a particle interacting with a heat bath

$$+ \quad \ldots \tag{6.79}$$

where the dashed line represents the reduced density matrix

$$\vdots \quad = \quad f(\mathbf{x}, \mathbf{x}', t) \ . \tag{6.80}$$

We now proceed to get an understanding of the complete perturbative structure of the particle-heat bath problem. For this purpose we shall again find a diagrammatic representation useful.

6.4   Particle-Heat Bath Perturbation Theory

To obtain the full diagrammatic expansion for the reduced density matrix propagator, we introduce the iterative solution for the evolution operators; i.e., we expand the time-ordered and anti-time-ordered exponentials, for example,

$$< \mathbf{x}, t | \hat{U}_I(t, t') | \mathbf{x}', t' > \quad \equiv \quad \left\langle \mathbf{x}, t \left| T \exp \left\{ -\frac{i}{\hbar} \int_{t'}^{t} d\bar{t} \hat{H}_i(\bar{t}) \right\} \right| \mathbf{x}', t' \right\rangle$$

$$= \quad \sum_{n=0}^{\infty} (i\hbar)^{-n} \int_{t'}^{t} dt_n \int_{t'}^{t_n} dt_{n-1} .. \int_{t'}^{t_3} dt_2 \int_{t'}^{t_2} dt_1 \ <\mathbf{x}, t | \hat{H}_i(t_n) .. \hat{H}_i(t_1) | \mathbf{x}', t' > \ . \tag{6.81}$$

Inserting the resolution of the identity repeatedly so as to turn the position operator in the $\hat{H}_i(t_j)$'s into numbers, and noting that the n'th order term will have $n + 1$ transition amplitudes we get

$$\left\langle \mathbf{x}, t \left| T \exp \left\{ -\frac{i}{\hbar} \int_{t'}^{t} d\bar{t} \hat{H}_i(\bar{t}) \right\} \right| \tilde{\mathbf{x}}, t' \right\rangle \quad = \quad i \sum_{n=0}^{\infty} \frac{1}{\hbar^n} \int d\mathbf{x}_n \int_{t'}^{t} dt_n \int d\mathbf{x}_{n-1} \int_{t'}^{t} dt_{n-1} .. \int d\mathbf{x}_1 \int_{t'}^{t} dt_1$$

$$G_0^R(\mathbf{x}, t; \mathbf{x}_n, t_n) \, G_0^R(\mathbf{x}_n, t_n; \mathbf{x}_{n-1}, t_{n-1}) .. \, G_0^R(\mathbf{x}_1, t_1; \tilde{\mathbf{x}}, t')$$

$$H_i(\mathbf{x}_n, \{\hat{x}_q(t_n), \hat{p}_q(t_n)\}_q) .. \, H_i(x_1, \{\hat{x}_q(t_1), \hat{p}_q(t_1)\}_q) \tag{6.82}$$

where the step functions in the propagators have lifted the time integration restrictions.

For the anti-time-ordered exponential ($\tilde{t}_m < \tilde{t}_{m-1} < .. \ \tilde{t}_2 < \tilde{t}_1$)

$$\left\langle \tilde{\mathbf{x}}', t' \left| \tilde{T} \exp\left\{ \frac{i}{\hbar} \int_{t'}^t d\tilde{t} \hat{H}_i(\tilde{t}) \right\} \right| \mathbf{x}', t \right\rangle = \sum_{m=0}^{\infty} \left(\frac{i}{\hbar} \right)^n \int_{t'}^t d\tilde{t}_m \int_{\tilde{t}_m}^t d\tilde{t}_{m-1} .. \int_{\tilde{t}_3}^t d\tilde{t}_2 \int_{\tilde{t}_2}^t d\tilde{t}_1$$

$$< \tilde{\mathbf{x}}', t' | H_i(\tilde{t}_m) \ .. \ H_i(\tilde{t}_1) | \mathbf{x}', t > \qquad (6.83)$$

we obtain a similar expression, the difference being that the advanced particle propagator G_0^A appears instead of the retarded.

Using the cyclic property of the trace, and that $\hat{H}_E$ and $\hat{\rho}_T$ commute, thereby canceling out the explicit appearance of $\hat{H}_E$, we obtain for the reduced density matrix propagator

$$J(\mathbf{x}, \mathbf{x}', t; \tilde{\mathbf{x}}, \tilde{\mathbf{x}}', t') = \sum_{n=0}^{\infty} \left(\frac{1}{\hbar} \right)^n \sum_{m=0}^{\infty} \left(\frac{1}{\hbar} \right)^m \int d\tilde{\mathbf{x}}_m \int_{-\infty}^{\infty} d\tilde{t}_m .. \int d\tilde{\mathbf{x}}_1 \int_{-\infty}^{\infty} d\tilde{t}_1 \int d\mathbf{x}_n \int_{-\infty}^{\infty} d t_n .. \int d\mathbf{x}_1 \int_{-\infty}^{\infty} d t_1$$

$$G_0^R(\mathbf{x}, t; \mathbf{x}_n, t_n) \, G_0^R(\mathbf{x}_n, t_n; \mathbf{x}_{n-1}, t_{n-1}) \ .. \ G_0^R(\mathbf{x}_1, t_1; \tilde{\mathbf{x}}, t')$$

$$G_0^A(\tilde{\mathbf{x}}', t'; \tilde{\mathbf{x}}_m, \tilde{t}_m) G_0^A(\tilde{\mathbf{x}}_m, \tilde{t}_m; \tilde{\mathbf{x}}_{m-1}, \tilde{t}_{m-1}) \ .. \ G_0^A(\tilde{\mathbf{x}}_1, \tilde{t}_1; \mathbf{x}', t)$$

$$tr(\hat{\rho}_T H_i(\tilde{\mathbf{x}}_m, \{\hat{x}_q(\tilde{t}_m), \hat{p}_q(\tilde{t}_m)\}_q) \ .. \ H_i(\tilde{\mathbf{x}}_1, \{\hat{x}_q(\tilde{t}_1), \hat{p}_q(\tilde{t}_1)\}_q)$$

$$H_i(\mathbf{x}_n, \{\hat{x}_q(t_n), \hat{p}_q(t_n)\}_q) \ .. \ H_i(\mathbf{x}_1, \{\hat{x}_q(t_1), \hat{p}_q(t_1)\}_q)) \ . \qquad (6.84)$$

The only operation left to perform in eq.(6.84) is then the trace over the bath states. Keeping track of the time-ordering and anti-time-ordering is done by the presence of the retarded and advanced propagators; i.e., we only get contribution from the integration region, $t > t_n > .. > t_1 > t'$ and $t > \tilde{t}_1 > .. > \tilde{t}_m > t'$. We can therefore for free throw in the time-ordering and anti-time-ordering operations for the sequence of bath operators occurring in eq.(6.84)

$$tr\left(\hat{\rho}_T \hat{H}_i(\tilde{\mathbf{x}}_m, \{\hat{x}_q(\tilde{t}_m), \hat{p}_q(\tilde{t}_m)\}_q) \ .. \ \hat{H}_i(\tilde{\mathbf{x}}_1, \{\hat{x}_q(\tilde{t}_1), \hat{p}_q(\tilde{t}_1)\}_q) \right.$$

$$\left. \hat{H}_i(\mathbf{x}_n, \{\hat{x}_q(t_n), \hat{p}_q(t_n)\}_q) \ .. \ \hat{H}_i(\mathbf{x}_1, \{\hat{x}_q(t_1), \hat{p}_q(t_1)\}_q) \right)$$

$$= \ tr\left(\hat{\rho}_T \tilde{T}[\hat{H}_i(\tilde{\mathbf{x}}_m, \{\hat{x}_q(\tilde{t}_m), \hat{p}_q(\tilde{t}_m)\}_q) \ .. \ \hat{H}_i(\tilde{\mathbf{x}}_1, \{\hat{x}_q(\tilde{t}_1), \hat{p}_q(\tilde{t}_1)\}_q)] \right.$$

$$\left. T[\hat{H}_i(\mathbf{x}_n, \{\hat{x}_q(t_n), \hat{p}_q(t_n)\}_q) \ .. \ \hat{H}_i(\mathbf{x}_1, \{\hat{x}_q(t_1), \hat{p}_q(t_1)\}_q)] \right) \ . \qquad (6.85)$$

For any order in perturbation theory, we have in eq.(6.84) a corresponding string S of bath operators which are explicitly time-ordered or anti-time-ordered

$$S = tr(\hat{\rho}_T \ \tilde{T}[\hat{c}_{\tilde{q}_m}(\tilde{t}_m) \ .. \ \hat{c}_{\tilde{q}_1}(\tilde{t}_1)] \, T[\hat{c}_{q_n}(t_n) \ .. \ \hat{c}_{q_1}(t_1)]) \qquad (6.86)$$

where $\hat{c}_q$ denotes either a creation or an annihilation operator. This expression is most easily evaluated by introducing the closed time-path description. Consider the time-ordered and anti-time-ordered times as lying on a pair of different real axes – the time-ordered ones on the forward and the anti-time-ordered ones, distinguished by a tilde, on the return part of the closed time-path contour c_t starting and ending at t_i, as depicted in figure 6.1.[11]

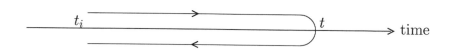

Figure 6.1 Closed time-path contour c_t.

This is the essence in the real-time formulation of nonequilibrium quantum statistical mechanics due to Schwinger [28]: letting the quantum dynamics do the doubling of the degrees of freedom necessary for describing nonequilibrium states.

The expression for the time-ordered and anti-time-ordered string of operators can then be subsumed under one contour ordering along the closed time path contour c_t

$$S = tr(\hat{\rho}_T \, T_{c_t}[\hat{c}_{\tilde{q}_m}(\tilde{t}_m) \, .. \, \hat{c}_{\tilde{q}_1}(\tilde{t}_1) \, \hat{c}_{q_n}(t_n) \, .. \, \hat{c}_{q_1}(t_1)])$$

$$\equiv \; <T_{c_t}[\hat{c}_{\tilde{q}_m}(\tilde{t}_m) \, .. \, \hat{c}_{\tilde{q}_1}(\tilde{t}_1) \, \hat{c}_{q_n}(t_n) \, .. \, \hat{c}_{q_1}(t_1)]> \qquad (6.87)$$

where T_{c_t} orders the operators according to their position on the contour c_t (earliest contour position to the right) so that, for example, for two operators indexed by contour times τ and τ'

$$T_{c_t}(\hat{c}(\tau)\,\hat{c}(\tau')) \;=\; \begin{cases} \hat{c}(\tau)\,\hat{c}(\tau') & \text{for} \quad \tau >_c \tau' \\[2mm] \hat{c}(\tau')\,\hat{c}(\tau) & \text{for} \quad \tau' >_c \tau \end{cases} \qquad (6.88)$$

where the upper identity is for contour time τ being further along the contour than τ' and the lower identity being the ordering for the opposite case.

Such an ordered expression can now be decomposed according to Wick's theorem, which relies only on the simple property

$$[\hat{c}_q, \hat{\rho}_T] \;=\; \hat{\rho}_T \, \hat{c}_q \, [\exp\{\lambda_c \hbar \omega_q / k_B T\} - 1] \qquad (6.89)$$

---

[11]The closed time-path contour can be shifted out into the complex time plane provided analyticity of expressions is respected.

valid for a quadratic bath Hamiltonian, eq.(6.10) ($\lambda_c = \pm 1$, depending upon whether $\hat{c}_q$ is a creation or an annihilation operator). We now turn to prove Wick's theorem, which is the statement that the harmonically weighted trace of a contour-ordered string of interaction picture creation and annihilation operators can be decomposed into a sum over all possible pairwise products

$$<T_{c_t}[\hat{c}_{\bar{q}_m}(\bar{t}_m)..\hat{c}_{\bar{q}_1}(\bar{t}_1)\hat{c}_{q_n}(t_n)..\hat{c}_{q_1}(t_1)]> \;=\; \sum_{a.p.p.} \prod_{q,q'} <T_{c_t}(\hat{c}_q(\tau)\hat{c}_{q'}(\tau'))> \quad (6.90)$$

where the sum is over all possible ways of picking pairs (*a.p.p.*) among the $n + m$ operators. Equivalently, Wick's theorem states that the trace of a contour-ordered string of creation and annihilation operators weighted with a quadratic Hamiltonian has the Gaussian property.

6.5   Wick's Theorem

Before proving Wick's theorem and the relation eq.(6.89), we first observe some preliminary results. Different q-label's describe different oscillator degrees of freedom, so operators for different q's commute, and algebraic manipulations with commuting operators is just as for usual numbers giving for example

$$\hat{\rho}_T \;=\; \prod_q \hat{\rho}_q^T \quad (6.91)$$

where we have introduced the thermal statistical operator for each oscillator

$$\hat{\rho}_q^T \;=\; z_q^{-1}\, e^{-\hat{h}_q/kT} \quad (6.92)$$

and the partition function for the single oscillator

$$z_q \;=\; tr_q\, e^{-\hat{h}_q/kT} \;=\; \sum_{n_q=0}^{\infty} <n_q|e^{-\hat{h}_q/kT}|n_q> \;=\; \sum_{n_q=0}^{\infty} e^{-\hbar\omega_q(n_q+1/2)/kT}$$

$$=\; \frac{e^{-\hbar\omega_q/2kT}}{1 - e^{-\hbar\omega_q/kT}} \;. \quad (6.93)$$

The independence of each oscillator degree of freedom, as expressed by the commutation of operators corresponding to different degrees of freedom, gives

$$\hat{c}_q\, \hat{\rho}_T \;=\; \left(\prod_{q'(\neq q)} \hat{\rho}_{q'}^T \right) \hat{c}_q\, \hat{\rho}_q^T \;. \quad (6.94)$$

Now, using the commutation relations for the creation and annihilation operators we have

$$\hat{c}_q\hat{h}_q \;=\; (\hat{h}_q - \lambda_c\hbar\omega_q)\, \hat{c}_q \quad (6.95)$$

where

$$\lambda_c = \begin{cases} +1 & \text{for} \quad \hat{c}_q = \hat{a}_q^\dagger \\ -1 & \text{for} \quad \hat{c}_q = \hat{a}_q \end{cases} . \tag{6.96}$$

Using eq.(6.95) repeatedly gives

$$\hat{c}_q \hat{h}_q^n = (\hat{h}_q - \lambda_c \hbar \omega_q)^n \hat{c}_q \tag{6.97}$$

and upon Taylor-expanding and re-exponentiating we can commute through to get

$$\hat{c}_q \hat{\rho}_T = e^{\lambda_c \hbar \omega_q / kT} \hat{\rho}_T \hat{c}_q \tag{6.98}$$

so that for the commutator of interest we have the property stated in eq.(6.89)

$$[\hat{c}_q, \hat{\rho}_T] = (e^{\lambda_c \hbar \omega_q / kT} - 1) \hat{\rho}_T \hat{c}_q . \tag{6.99}$$

We then prove for an arbitrary operator $\hat{A}$ that

$$< [\hat{c}_q, \hat{A}] > = (1 - e^{\lambda_c \hbar \omega_q / kT}) < \hat{c}_q \hat{A} > \tag{6.100}$$

as we first note, by using the cyclic property of the trace, that

$$< [\hat{c}_q, \hat{A}] > = -tr([\hat{c}_q, \hat{\rho}_T] \hat{A}) \tag{6.101}$$

and then by using eq.(6.99) we get eq.(6.100).

Employing eq.(6.100) with $\hat{A} = 1, 1, \hat{a}, \hat{a}^\dagger$, respectively, we observe that all the following averages vanish

$$0 = <\hat{a}(t)> = <\hat{a}^\dagger(t)> = <\hat{a}(t)\hat{a}(t')> = <\hat{a}^\dagger(t)\hat{a}^\dagger(t')> \tag{6.102}$$

and as a consequence $<\hat{H}_i(t)> = 0$ as announced on page 244.

Repeating the algebraic manipulations leading to eq.(6.98), or by analytical continuation of the result, we have

$$\hat{c}_q(t) = \hat{c}_q e^{-\frac{i}{\hbar}\hat{H}_E t} = e^{i\lambda_c \omega_q t} e^{-\frac{i}{\hbar}\hat{H}_E t} \hat{c}_q \tag{6.103}$$

from which we get that the creation and annihilation operators in the interaction picture has a simple time dependence in terms of a phase factor

$$\hat{c}_q(t) = e^{\frac{i}{\hbar}\hat{H}_E t}\hat{c}_q e^{-\frac{i}{\hbar}\hat{H}_E t} = \hat{c}_q e^{i\lambda_c \omega_q t} . \tag{6.104}$$

The commutators formed by creation and annihilation operators in the interaction picture are thus c-numbers, the only nonvanishing one being specified by

$$[\hat{a}_q(t), \hat{a}_{q'}^\dagger(t')] = \delta_{q,q'} e^{-i\omega_q(t-t')} . \tag{6.105}$$

According to eq.(6.100) we thereby have

$$<\hat{a}_q(t) \hat{a}_{q'}^\dagger(t')> = (1 - e^{-\hbar \omega_q / kT})^{-1} < [\hat{a}_q(t), \hat{a}_{q'}^\dagger(t')] >$$

$$= \delta_{q,q'} (1 - e^{-\hbar \omega_q / kT})^{-1} e^{-i\omega_q(t-t')}$$

$$= \delta_{q,q'} (n(\omega_q) + 1) e^{-i\omega_q(t-t')}$$

$$\equiv D_{qq'}^>(t, t') \tag{6.106}$$

where n denotes the Bose-Einstein distribution function

$$n(\omega_q) = \frac{1}{e^{\hbar\omega_q/kT} - 1} . \tag{6.107}$$

Exercise 6.2 *Show that for the opposite ordering of the operators, the correlation function is*

$$D^<_{q'q}(t',t) \equiv <\hat{a}^\dagger_q(t)\hat{a}_{q'}(t')> = tr(\hat{\rho}_T\,\hat{a}^\dagger_q(t)\,\hat{a}_{q'}(t'))$$

$$= n(\omega_q)\,\delta_{q,q'}\,e^{i\omega_q(t-t')} . \tag{6.108}$$

If the string contains an odd number of operators, i.e., $n+m$ is odd, the expression for S equals zero since the expectation value is with respect to the equilibrium bath state (and we assumed no linear term in the oscillator Hamiltonian). For an odd number of operators we namely encounter a matrix element between states with different number of quanta; for example,

$$< \hat{a}_q \hat{a}^\dagger_q \hat{a}_q > = Z^{-1} \sum_{\{n_q\}_q} e^{-E(\{n_q\}_q)/kT}(\sqrt{n_q})^3 <n_q|n_q - 1> = 0 \tag{6.109}$$

which is zero by orthogonality of the different energy eigenstates.

As an example of using Wick's theorem we write down the term we encounter at fourth order in the coupling to the oscillators (we suppress, for the present consideration, the immaterial q labels)

$$tr(\rho_T T_{c_t}[a(\tau_1)a^\dagger(\tau_2)a(\tau_3)a^\dagger(\tau_4)]) = <T_{c_t}(a(\tau_1)a^\dagger(\tau_2))> <T_{c_t}(a(\tau_3)a^\dagger(\tau_4))> $$

$$+ <T_{c_t}(a(\tau_1)a^\dagger(\tau_4))> <T_{c_t}(a(\tau_3)a^\dagger(\tau_4))> .$$

$$\tag{6.110}$$

Here we have deleted terms which do not pair creation and annihilation operators as such terms, just as above, leads to matrix elements between orthogonal states:

$$< T_{c_t}(a(\tau)a(\tau')) > = 0 = < T_{c_t}(a^\dagger(\tau)a^\dagger(\tau')) > . \tag{6.111}$$

At the fourth-order level the ordered Gaussian decomposition can of course be obtained by noting that only by pairing equal numbers of creation and annihilation operators can the number of quanta stay conserved and the matrix element be non-zero as we have the expression

$$tr(\rho_T T_{c_t}(a(\tau_1)a^\dagger(\tau_2)a(\tau_3)a^\dagger(\tau_4)))$$

$$= \sum_{\{n_q\}_q} e^{-E(\{n_q\}_q)/kT} <\{n_q\}_q|T_{c_t}(a(\tau_1)a^\dagger(\tau_2)a(\tau_3)a^\dagger(\tau_4))|\{n_q\}_q> . \tag{6.112}$$

Wick's theorem is the generalization of this simple observation.

Wick's theorem is trivially true for $N = 1$ (and for $N = 2$ according to the above consideration), and we now turn to prove Wick's theorem by induction.[12] Let us therefore consider an N-string with $2N$ operators

$$S_N = < T_{c_t}(\hat{c}(\tau_{2N})\hat{c}(\tau_{2N-1})..\hat{c}(\tau_2)\hat{c}(\tau_1)) > . \qquad (6.113)$$

We can assume that the contour-time labeling already corresponds to the contour-ordered one, otherwise we just relabel the indices, and we have

$$S_N = < \prod_{n=1}^{2N} \hat{c}(\tau_n) > = < \hat{c}(\tau_{2N}) \prod_{n=1}^{2N-1} \hat{c}(\tau_n) > . \qquad (6.114)$$

We then use the above proved relation, eq.(6.100), to rewrite

$$S_N = (1 - e^{\lambda_c \hbar \omega_q / kT})^{-1} < [\hat{c}(\tau_{2N}), \prod_{n=1}^{2N-1} \hat{c}(\tau_n)] > . \qquad (6.115)$$

In the first term in the commutator we commute $\hat{c}(\tau_{2N})$ to the right

$$[\hat{c}(\tau_{2N}), \prod_{n=1}^{2N-1} \hat{c}(\tau_n)] = \hat{c}(\tau_{2N-1})\hat{c}(\tau_{2N}) \prod_{n=1}^{2N-2} \hat{c}(\tau_n) + [\hat{c}(\tau_{2N}), \hat{c}(\tau_{2N-1})] \prod_{n=1}^{2N-2} \hat{c}(\tau_n)$$

$$- \left(\prod_{n=1}^{2N-1} \hat{c}(\tau_n) \right) \hat{c}(\tau_{2N}) . \qquad (6.116)$$

We now keep commuting $\hat{c}(\tau_{2N})$ through in the first term repeatedly, each time generating a commutator, and eventually ending up with canceling the last term in eq.(6.116), so that

$$[\hat{c}(\tau_{2N}), \prod_{n=1}^{2N-1} \hat{c}(\tau_n)] = \sum_{n=1}^{2N-1} [\hat{c}(\tau_{2N}), \hat{c}(\tau_n)] \prod_{\substack{m=1 \\ m \neq n}}^{2N-1} \hat{c}(\tau_m) . \qquad (6.117)$$

Then we use that the commutator, eq.(6.106), is a c-number which we according to eq.(6.106) can rewrite

$$[\hat{c}_q(\tau_{2N}), \hat{c}_{q'}(\tau_n)] = \delta_{q,q'} (1 - e^{-\hbar\omega_q/kT}) <\hat{c}_q(\tau_{2N}) \hat{c}_{q'}(\tau_n)> \qquad (6.118)$$

and being a c-number it can be taken outside the thermal average in eq.(6.115), and we obtain

$$S_N = \sum_{n=1}^{2N-1} < \hat{c}(\tau_{2N})\hat{c}(\tau_n) > \left\langle \prod_{\substack{m=1 \\ m \neq n}}^{2N-1} \hat{c}(\tau_m) \right\rangle$$

$$= \sum_{n=1}^{2N-1} < T_{c_t}(\hat{c}(\tau_{2N})\hat{c}(\tau_n)) > \left\langle T_{c_t} \left(\prod_{\substack{m=1 \\ m \neq n}}^{2N-1} \hat{c}(\tau_m) \right) \right\rangle \qquad (6.119)$$

---

[12]We follow the elegant and simple proof given by Mills [29].

where we reintroduce the contour ordering. By assumption the second factor can be written as a sum over all possible pairs (on a.p.p.-form), and by induction the N case is then precisely seen to be of that form too. We note, that to prove Wick's theorem we have only exploited that the weight was a quadratic form.[13]

Regrouping terms we have shown

$$tr(\hat{\rho}_T H_i(\tilde{\mathbf{x}}_m, \{\hat{a}_q(\tilde{t}_m), \hat{a}_q^\dagger(\tilde{t}_m)\}_q) \ .. \ H_i(\tilde{\mathbf{x}}_1, \{\hat{a}_q(\tilde{t}_1), \hat{a}_q^\dagger(\tilde{t}_1)\}_q)$$

$$H_i(\mathbf{x}_n, \{\hat{a}_q(t_n), \hat{a}_q^\dagger(t_n)\}_q) \ .. \ H_i(\mathbf{x}_1, \{\hat{a}_q(t_1), \hat{a}_q^\dagger(t_1)\}_q))$$

$$= \ tr(\hat{\rho}_T \tilde{T}[\hat{H}_i(\tilde{\mathbf{x}}_m, \{\hat{x}_q(\tilde{t}_m), \hat{p}_q(\tilde{t}_m)\}_q) \ .. \ \hat{H}_i(\tilde{\mathbf{x}}_1, \{\hat{x}_q(\tilde{t}_1), \hat{p}_q(\tilde{t}_1)\}_q)]$$

$$T[\hat{H}_i(\mathbf{x}_n, \{\hat{x}_q(t_n), \hat{p}_q(t_n)\}_q) \ .. \ \hat{H}_i(\mathbf{x}_1, \{\hat{x}_q(t_1), \hat{p}_q(t_1)\}_q)])$$

$$= \ < \tilde{T}[\hat{H}_i(\tilde{\mathbf{x}}_m, \{\hat{a}_q(\tilde{t}_m), \hat{a}_q^\dagger(\tilde{t}_m)\}_q) \ .. \ \hat{H}_i(\tilde{\mathbf{x}}_1, \{\hat{a}_q(\tilde{t}_1), \hat{a}_q^\dagger(\tilde{t}_1)\}_q)]$$

$$T[\hat{H}_i(\mathbf{x}_n, \{\hat{a}_q(t_n), \hat{a}_q^\dagger(t_n)\}_q) \ .. \ \hat{H}_i(\mathbf{x}_1, \{\hat{a}_q(t_1), \hat{a}_q^\dagger(t_1)\}_q))] >$$

$$= \ \sum_{a.p.p.} \prod_{i \neq j} < T_{c_t}(H_i(\mathbf{x}_i, \{\hat{a}_q(t_i), \hat{a}_q^\dagger(t_i)\}_q) H_i(\mathbf{x}_j, \{\hat{a}_q(t_j), \hat{a}_q^\dagger(t_j)\}_q)) >$$

$$\equiv \ \sum_{a.p.p.} \prod_{i \neq j} \hbar^2 \, D(\mathbf{x}_i, t_i; \mathbf{x}_j, t_j) \ . \tag{6.120}$$

Performing the trace over the bath states in eq.(6.84) therefore corresponds to pairing the interactions in all possible ways.

It follows from the construction, that in the bath correlator D the time ordering is inherited from the original ordering on the contour c_t. The notation of distinguishing times on the backward part of the contour by a tilde is therefore superfluous and was henceforth omitted in the last equality in eq.(6.120).

We therefore have a simple prescription for performing the trace of the object of interest in eq.(6.84). The interaction operators we depict as dangling bonds, and for an N'th order diagram we have $2N$ such interaction operators[14]

$$\bar{J} \quad = \Big\langle \qquad \Big\rangle \tag{6.121}$$

where

---

[13]If the weight were not quadratic, we would have encountered correlations which must be handled additionally.

[14]We now label the order of a diagram according to its number of bath correlators.

$$\equiv \quad H_i(\mathbf{x}, \{\hat{x}_q(t), \hat{p}_q(t)\}_q) \tag{6.122}$$

designates the interaction operator, and the prescription for performing the average is: Tie the dangling bonds together in pairs in all possible ways!

For example, for the fourth-order contribution where all interactions stems from $\hat{U}$ we have[15]

$$\bar{J} \quad = \quad \left\langle \rule{0pt}{1em} \right\rangle \tag{6.123}$$

and this term can be paired together in three different ways to give the diagrams:

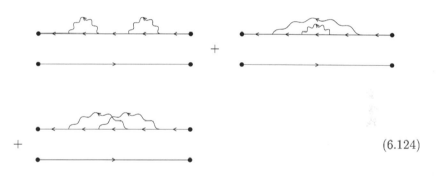

$$\tag{6.124}$$

Exercise 6.3 *Draw the rest of the 15 second-order diagrams for the propagator of the reduced-density matrix. Show, that the number d_n of perturbative diagrams of n'th order, i.e., with n bath correlators, is $d_n = (2n+1)!! \equiv (2n+1) \cdot (2n-1) \cdots 3 \cdot 1$.*

Exercise 6.4 *Show that we have for the generating functional ($\hat{x}_q(t)$ is the q'th oscillator operator in the interaction picture with respect to $\hat{h}_q$)*

$$Z[f_t] \quad \equiv \quad \left\langle Te^{\,i\int\limits_{-\infty}^{\infty} dt\, f_t\, \hat{x}_q(t)} \right\rangle \quad = \quad tr_q \left(\hat{\rho}_q^T\, Te^{\,i\sqrt{\frac{\hbar}{2M_q\omega_q}}\int\limits_{-\infty}^{\infty} dt\, f_t(\hat{a}_q(t)+\hat{a}_q^\dagger(t))} \right)$$

---

[15]Note, that the number of diagrams of any order is the same as in the previous Gaussian impurity average case of chapter 3. Also, in both cases we have trivial combinatorics, the prefactor of all diagrams is 1.

$$= e^{-\frac{1}{2} \int\limits_{-\infty}^{\infty} dt \int\limits_{-\infty}^{\infty} dt'\, f_t <T(\hat{x}_q(t)\,\hat{x}_q(t'))> f'_t} . \tag{6.125}$$

Solution

Using Wick's theorem we obtain

$$
\begin{aligned}
Z[f_t] &= \sum_{n=0}^{\infty} \frac{(-i)^n}{n!} \int\limits_{-\infty}^{\infty} dt_1 \,..\, \int\limits_{-\infty}^{\infty} dt_n\, f_{t_1} \,..\, f_{t_n} \; <T(\hat{x}_q(t_1) \,..\, \hat{x}_q(t_n))> \\
&= \sum_{n=0}^{\infty} \sum_{a.p.p.} \frac{(-i)^{2n}}{(2n)!} \int\limits_{-\infty}^{\infty} dt_1 \,..\, \int\limits_{-\infty}^{\infty} dt_{2n}\, f_{t_1} \,..\, f_{t_{2n}} \prod_{i,i'} <T(\hat{x}_q(t_i)\,\hat{x}_q(t_{i'}))> \\
&= \sum_{n=0}^{\infty} \frac{(-\frac{1}{2})^n}{(n)!} \int\limits_{-\infty}^{\infty} dt_1 \,..\, \int\limits_{-\infty}^{\infty} dt_{2n}\, f_{t_1} \,..\, f_{t_{2n}} \prod_{i,i'} <T(\hat{x}_q(t_i)\,\hat{x}_q(t_{i'}))> \quad (6.126)
\end{aligned}
$$

where in the last equality we have used that all the terms in the " all possible pair-sum" are equal, and that their number equals the number of ways of picking n pairs out of a set with $2n$ members, which is $(2n-1)!! = (2n)!/2^n n!$.

A unique feature of a diagram is its topology, which is a property easy to spot. A very useful aspect of the diagrammatic technique is that we can get a complete description of the perturbative structure of the theory in terms of a simple prescription.

We can now state the Feynman diagrammatics in the spatial representation for the reduced density matrix for a particle interacting with a heat bath: The Feynman diagrams for the reduced density matrix $f(\mathbf{x}, \mathbf{x}', t)$ comprises all the topologically different diagrams of the type depicted in the figure (showing only the lowest-order terms)

$$+ \qquad\qquad\qquad\qquad + \quad \cdots \qquad\qquad (6.127)$$

The dictionary for transcribing the diagrams is according to the derivation as follows: The reduced density matrix, being a single-time object, is represented as a vertical dashed line

$$f(\mathbf{x}, \mathbf{x}', t) \equiv \begin{array}{c} \mathbf{x} \; \bullet \\ t \; \vdots \\ \mathbf{x}' \; \bullet \end{array} \qquad\qquad (6.128)$$

An upper solid line represents the retarded particle propagator in the absence of the oscillator environment

$$\underset{\mathbf{x}t}{\bullet}\xrightarrow{\quad R \quad}\underset{\mathbf{x}'t'}{\bullet} \quad \equiv \quad G_0^R(\mathbf{x}, t; \mathbf{x}', t') \qquad\qquad (6.129)$$

and a lower solid line represents the advanced particle propagator in the absence of the oscillator environment

$$\underset{\mathbf{x}'t'}{\bullet}\xrightarrow{\quad A \quad}\underset{\mathbf{x}t}{\bullet} \quad \equiv \quad G_0^A(\mathbf{x}, t; \mathbf{x}', t') \qquad\qquad (6.130)$$

A wavy line represents the heat bath correlation function

$$\underset{\mathbf{x}t}{\bullet}\!\!\!\sim\!\!\!\sim\!\!\!\sim\!\!\!\sim\!\!\!\underset{\mathbf{x}'t'}{\bullet} \quad \equiv \quad D(\mathbf{x}, t; \mathbf{x}', t') \;. \qquad\qquad (6.131)$$

In accordance with the derivation (and a consequence of the superposition principle), integration over interaction space-time points should be performed, and spatial integration over the initial reduced density matrix coordinates. With the chosen convention, as illustrated in the diagrams of eq.(6.127), for assigning direction to the heat bath correlator line, we only need to introduce one type of heat bath correlation function.[16]

We note, that the interaction of the particle with a heat bath can be simulated by a Gaussian-distributed time-dependent fluctuating potential specified by the correlator

$$< V(\mathbf{x}, t) V(\mathbf{x}', t') > \; \equiv \; D(\mathbf{x}, t; \mathbf{x}', t') \;. \qquad\qquad (6.132)$$

---

[16]The reason for the absence of the explicitly time-ordered oscillator correlation function in the theory, as the Wick decomposition suggests, is the fact that we are dealing with (besides the oscillator bath) only one particle, so that no particle-hole (pair) creation can take place as one encounters in the many-body case for electrons in solids, or electron-positron pair creation in quantum electrodynamics, or quark-antiquark creation in quantum chromodynamics.

The final ingredient needed to turn a diagram for the reduced density matrix into an analytical expression is the knowledge of the particular analytical form of the heat bath correlation function, eq.(6.73), which can be calculated readily using eq.(6.100) as only the equilibrium average with respect to the bath state is involved (see eq.(6.106) and eq.(6.108)). For convenience, we have included the particle-oscillator coupling constant $\alpha_{\mathbf{q}}$ in the definition of the heat bath correlation function so that the analytical expression following from eq.(6.73) is

$$D(\mathbf{x}, t; \mathbf{x}', t') = \sum_{\mathbf{q}} \left(\alpha_{\mathbf{q}}(\mathbf{x}) \, D_{\mathbf{q}}^{>}(t, t') \, \alpha_{\mathbf{q}}^{*}(\mathbf{x}') + \alpha_{\mathbf{q}}^{*}(\mathbf{x}) \, D_{\mathbf{q}}^{<}(t', t) \, \alpha_{\mathbf{q}}(\mathbf{x}') \right) \quad (6.133)$$

where

$$D_{\mathbf{q}}^{>}(t, t') = (n(\omega_{\mathbf{q}}) + 1) \, e^{-i\omega_{\mathbf{q}}(t-t')} \quad (6.134)$$

and

$$D_{\mathbf{q}}^{<}(t', t) = n(\omega_{\mathbf{q}}) \, e^{i\omega_{\mathbf{q}}(t-t')} \ . \quad (6.135)$$

The heat bath correlation function is thus completely specified by the temperature and the oscillator frequencies.

We now have all the ingredients to perform perturbative calculations for the particle-heat bath problem, a task we shall shortly embark on.

6.6   Kinetic Approach

In the preceding section, we examined the time evolution of the reduced density matrix on integral form, and described the diagrammatic expansion in the spatial representation. In the following we shall pursue the kinetic approach for studying the motion of a particle interacting with an oscillator environment, We shall therefore first display the spatial representation of the von Neumann equation and establish its diagrammatic interpretation. In order to do so, we must look at the change in time of the reduced density matrix. From the von Neumann equation, eq.(1.384), we obtain by taking trace over the environment degrees of freedom

$$i\hbar \frac{d\hat{f}(t)}{dt} = [\hat{H}_p, \hat{f}(t)] + tr([\hat{H}_i, \hat{\rho}(t)]) \ . \quad (6.136)$$

We have noticed that since the trace is with respect to the environment, the trace of a commutator where one of the hermitian operators acts only on the environment degrees of freedom has zero trace

$$tr([\hat{H}_E, \hat{\rho}(t)]) = \sum_{\{n_q\}_q} < \{n_q\}_q | [\hat{H}_E, \hat{\rho}(t)] | \{n_q\}_q >$$

$$= \sum_{\{n_q\}_q} (E(\{n_q\}_q) - E(\{n_q\}_q)) < \{n_q\}_q | \hat{\rho}(t) | \{n_q\}_q >$$

$$= 0 \ . \quad (6.137)$$

The spatial representation of the operator equation for the reduced statistical operator, eq.(6.136), leads to the following equation for the reduced density matrix

$$\frac{\partial f(\mathbf{x}, \mathbf{x}', t)}{\partial t} - \frac{i}{\hbar} <\mathbf{x}|[\hat{f}(t), \hat{H}_p]|\mathbf{x}'> = -iF_{\mathbf{x},\mathbf{x}',t}[f_i] \qquad (6.138)$$

where

$$F_{\mathbf{x},\mathbf{x}',t}[f_i] = \frac{1}{\hbar} tr(<\mathbf{x}|[\hat{H}_i, \hat{\rho}(t)]|\mathbf{x}'>) . \qquad (6.139)$$

Inserting complete sets of position states, we see that $F_{\mathbf{x},\mathbf{x}',t}$ is a functional of the initial time particle density matrix $f_i(\tilde{\mathbf{x}}, \tilde{\mathbf{x}}')$, and a function of $\mathbf{x}, \mathbf{x}'$ and t

$$F_{\mathbf{x},\mathbf{x}',t}[f_i] = \frac{1}{\hbar} <\mathbf{x}|tr([\hat{H}_i, \hat{\rho}(t)])|\mathbf{x}'> = \int d\tilde{\mathbf{x}} \int d\tilde{\mathbf{x}}' \, \bar{J}(\mathbf{x}, \mathbf{x}', t; \tilde{\mathbf{x}}, \tilde{\mathbf{x}}', t') f_i(\tilde{\mathbf{x}}, \tilde{\mathbf{x}}')$$
$$(6.140)$$

and the reduced density matrix propagator, $\bar{J}$, is given by

$$\bar{J}(\mathbf{x}, \mathbf{x}', t; \tilde{\mathbf{x}}, \tilde{\mathbf{x}}', t') = \frac{1}{\hbar} tr(\hat{\rho}_T <\tilde{\mathbf{x}}'|\hat{U}^\dagger(t, t')|\mathbf{x}'><\mathbf{x}|\hat{H}_i \hat{U}(t, t')|\tilde{\mathbf{x}}>)$$

$$- \frac{1}{\hbar} tr(\hat{\rho}_T <\tilde{\mathbf{x}}'|\hat{U}^\dagger(t, t') \hat{H}_i|\mathbf{x}'><\mathbf{x}|\hat{U}(t, t')|\tilde{\mathbf{x}}>)$$

$$= \frac{1}{\hbar} tr(\hat{\rho}_T <\tilde{\mathbf{x}}', t'|\hat{U}_I^\dagger(t, t')|\mathbf{x}', t><\mathbf{x}, t|\hat{H}_i(t)\hat{U}_I(t, t')|\tilde{\mathbf{x}}, t'>)$$

$$- \frac{1}{\hbar} tr(\hat{\rho}_T <\tilde{\mathbf{x}}', t'|\hat{U}_I^\dagger(t, t')\hat{H}_i(t)|\mathbf{x}', t><\mathbf{x}, t|\hat{U}_I(t, t')|\tilde{\mathbf{x}}, t'>)$$

$$= \frac{1}{\hbar} tr\left(\hat{\rho}_T \left\langle \tilde{\mathbf{x}}', t' \left| \tilde{T} \exp\left\{\frac{i}{\hbar} \int_{t'}^t d\bar{t} \hat{H}_i(\bar{t})\right\} \right| \mathbf{x}', t \right\rangle\right.$$

$$H_i(\mathbf{x}, \{\hat{x}_q(t), \hat{p}_q(t)\}_q) \left\langle \mathbf{x}, t \left| T \exp\left\{-\frac{i}{\hbar} \int_{t'}^t d\bar{t} \hat{H}_i(\bar{t})\right\} \right| \tilde{\mathbf{x}}, t' \right\rangle\right)$$

$$- \frac{1}{\hbar} tr\left(\hat{\rho}_T \left\langle \tilde{\mathbf{x}}', t' \left| \tilde{T} \exp\left\{\frac{i}{\hbar} \int_{t'}^t d\bar{t} \hat{H}_i(\bar{t})\right\} \right| \mathbf{x}', t \right\rangle\right.$$

$$H_i(\mathbf{x}', \{\hat{x}_q(t), \hat{p}_q(t)\}_q)$$

$$\left\langle \mathbf{x}, t \left| T \exp\left\{-\frac{i}{\hbar} \int_{t'}^t d\bar{t} \hat{H}_i(\bar{t})\right\} \right| \tilde{\mathbf{x}}, t' \right\rangle\right) . \qquad (6.141)$$

This expression is thus analogous to the one we encountered when we derived the diagrammatic structure for the reduced density matrix. We only have an additional interaction operator compared to the previous case. We consequently insert complete sets of states into the expression for $\bar{J}$ just as we did in the case of J, and we get the perturbative expansion[17]

$$\bar{J}(\mathbf{x}, \mathbf{x}', t; \tilde{\mathbf{x}}, \tilde{\mathbf{x}}', t') =$$

---

[17]The presence of the time-ordering and anti-time-ordering operations is optional as it is automatically taken care of by the presence of the retarded and advanced propagators.

$$\sum_{n,m=0}^{\infty} \frac{1}{\hbar^{n+m+1}} \int d\tilde{\mathbf{x}}_m \int_{-\infty}^{\infty} d\tilde{t}_m .. \int d\tilde{\mathbf{x}}_1 \int_{-\infty}^{\infty} d\tilde{t}_1 \int d\mathbf{x}_n \int_{-\infty}^{\infty} d\tilde{t}_n .. \int d\mathbf{x}_1 \int_{-\infty}^{\infty} d\tilde{t}_1$$

$$G_0^R(\mathbf{x}, t; \mathbf{x}_n, t_n)\, G_0^R(\mathbf{x}_n, t_n; \mathbf{x}_{n-1}, t_{n-1}) .. G_0^R(\mathbf{x}_2, t_2; \mathbf{x}_1, t_1)\, G_0^R(\mathbf{x}_1, t_1; \tilde{\mathbf{x}}, t')$$

$$G_0^A(\tilde{\mathbf{x}}', t'; \tilde{\mathbf{x}}_m, \tilde{t}_m) G_0^A(\tilde{\mathbf{x}}_m, \tilde{t}_m; \tilde{\mathbf{x}}_{m-1}, \tilde{t}_{m-1}) .. G_0^A(\tilde{\mathbf{x}}_2, \tilde{t}_2; \tilde{\mathbf{x}}_1, \tilde{t}_1)\, G_0^A(\tilde{\mathbf{x}}_1, \tilde{t}_1; \mathbf{x}', t)$$

$$tr(\hat{\rho}_T\, \tilde{T}\{H_i(\tilde{\mathbf{x}}_m, \{\hat{x}_q(\tilde{t}_m), \hat{p}_q(\tilde{t}_m)\}_q) .. H_i(\tilde{\mathbf{x}}_1, \{\hat{x}_q(\tilde{t}_1), \hat{p}_q(\tilde{t}_1)\}_q)\}$$

$$T\{H_i(\mathbf{x}, \{\hat{x}_q(t), \hat{p}_q(t)\}_q) H_i(\mathbf{x}_n, \{\hat{x}_q(t_n), \hat{p}_q(t_n)\}_q) .. H_i(\mathbf{x}_1, \{\hat{x}_q(t_1), \hat{p}_q(t_1)\}_q)\})$$

$$- \sum_{n,m=0}^{\infty} \frac{1}{\hbar^{n+m+1}} \int d\tilde{\mathbf{x}}_m \int_{-\infty}^{\infty} d\tilde{t}_m .. \int d\tilde{\mathbf{x}}_1 \int_{-\infty}^{\infty} d\tilde{t}_1 \int d\mathbf{x}_n \int_{-\infty}^{\infty} d\tilde{t}_n .. \int d\mathbf{x}_1 \int_{-\infty}^{\infty} d\tilde{t}_1$$

$$G_0^R(\mathbf{x}, t; \mathbf{x}_n, t_n) .. G_0^R(\mathbf{x}_1, t_1; \tilde{\mathbf{x}}, t')\, G_0^A(\tilde{\mathbf{x}}', t'; \tilde{\mathbf{x}}_m, \tilde{t}_m) .. G_0^A(\tilde{\mathbf{x}}_1, \tilde{t}_1; \mathbf{x}', t)$$

$$tr(\hat{\rho}_T\, \tilde{T}\{H_i(\tilde{\mathbf{x}}_m, \{\hat{x}_q(\tilde{t}_m), \hat{p}_q(\tilde{t}_m)\}_q) .. H_i(\tilde{\mathbf{x}}_1, \{\hat{x}_q(\tilde{t}_1), \hat{p}_q(\tilde{t}_1)\}_q)$$

$$H_i(\mathbf{x}' \{\hat{x}_q(t), \hat{p}_q(t)\}_q)\} T\{H_i(\mathbf{x}_n, \{\hat{x}_q(t_n), \hat{p}_q(t_n)\}_q) .. H_i(\mathbf{x}_1, \{\hat{x}_q(t_1), \hat{p}_q(t_1)\}_q)\}). \quad (6.142)$$

We therefore have to average, or trace, the following objects (whose structure is easy to grasp diagrammatically in comparison to the unwieldy long expression above!) where we only display two typical members

$$(6.143)$$

Compared to the case of J we only have an additional interaction operator, which diagrammatically for the first type of terms can be depicted as appearing first on the upper line, representing $H_i(\mathbf{x}, \{\hat{x}_q(t), \hat{p}_q(t)\}_q)$, and for the second type of terms as last on the lower line, representing $H_i(\mathbf{x}', \{\hat{x}_q(t), \hat{p}_q(t)\}_q)$.

Now we use Wick's theorem to tie all the dangling interaction operators pairwise together, and we obtain for $\bar{J}$ the diagrammatic expansion

$$\text{(6.144)}$$

where the box signifies that the final oscillator correlator line can enter into any entanglement. The diagrams appear in doublets because the final oscillator correlator can end up on either the upper or lower particle line (uniquely specified by the direction of the environment correlator arrow). The signs of the diagrams just reflect their origin in a commutator expression.

The functional F has the diagrammatic form obtained by attaching the initial reduced density matrix at the initial time. We therefore see that the functional F has a diagrammatic representation which is obtained from the diagrams for the reduced density matrix f by the same prescription as in the random potential case: Remove the final external retarded or advanced particle line on the diagrams with bath correlators, and change the prefactor from plus to minus according to whether the last interaction line attaches to the upper or lower particle line. The diagrammatic expansion is therefore as follows:

$$\text{(6.145)}$$

where the box signifies that we can have arbitrary entanglement of environment lines.

6.6.1   Quantum Kinetic Equation

The partial summation of diagrams we performed in section 4.3 when considering the kinetics of a particle in a random potential relied only upon the topology of the diagrams. Since the topological structure of the diagrams in the particle-heat bath problem is the same as for the random potential problem (just wiggly lines instead of impurity lines) we can therefore sum all the diagrams according to their equal-time two-line reducibility to get the identity

$$-iF_{\mathbf{x},\mathbf{x}',t}[f_i] \quad = \quad \tilde{F}_{\mathbf{x},\mathbf{x}',t}[f] \tag{6.146}$$

where $\tilde{F}_{\mathbf{x},\mathbf{x}',t}$ is the functional

$$\tilde{F}_{\mathbf{x},\mathbf{x}',t}[f] \quad = \quad \int d\tilde{\mathbf{x}} \int d\tilde{\mathbf{x}}' \int_{t'}^{t} d\tilde{t}\, \tilde{J}(\mathbf{x},\mathbf{x}',t;\tilde{\mathbf{x}},\tilde{\mathbf{x}}',\tilde{t})\, f(\tilde{\mathbf{x}},\tilde{\mathbf{x}}',\tilde{t}) \tag{6.147}$$

described by the irreducible reduced density matrix propagator, $\tilde{J}_{\mathbf{x},\mathbf{x}',t}[f]$, consisting of all the equal-time two-line irreducible diagrams, i.e., the ones which can not be cut in two by cutting only an upper and lower internal particle line at equal time. At time t' the density matrix is the initial one f_i.

We thus have the integro-differential equation determining the time evolution of the reduced density matrix

$$\left(\frac{\partial}{\partial t} - \frac{i}{\hbar}\left(\frac{\hbar^2}{2m}(\triangle_{\mathbf{x}} - \triangle_{\mathbf{x}'}) - U(\mathbf{x},t) + U(\mathbf{x}',t)\right)\right)\rho(\mathbf{x},\mathbf{x}',t) \quad = \quad \tilde{F}_{x,x',t}[f]\,. \tag{6.148}$$

The diagrammatic expansion of the irreducible reduced density matrix propagator, $\tilde{J}$, is in one-to-one correspondence with the equal-time two-line irreducible diagrams of J

where the sign is determined by whether a propagator on the upper or lower line is split into two. Here only the lowest-order diagrams are shown explicitly.[18]

The diagrammatic expansion of $\tilde{F}$ consists of all equal-time two-line irreducible diagrams

---

[18]We shall not at this point bother to go into further details with the sign ascription, as we shall presently only study lowest-order diagrams. At the end of this chapter we shall introduce a generalized density matrix and a stringer concept of irreducibility for which this issue disappears.

$$\tilde{F}[f] = \pm \quad \pm \quad \pm \quad \pm$$

$$= - \quad - \quad + \quad + \quad \pm \dots \text{(6.150)}$$

and only the lowest-order ones are shown explicitly in the figure (the diagrams are identical to the $\tilde{J}$-diagrams, except that the intermediate reduced density matrix is attached at the earliest interaction line).[19]

Instead of pursuing the general discussion at this point, and in order to gain familiarity with the Feynman rules and the general features of the method, let us look at the lowest-order contribution to the irreducible reduced density matrix propagator $\tilde{J}$. This contribution is represented by the diagrams

$$\tilde{J}^{(1)} = - \quad - \quad + \quad + \quad . \quad \text{(6.151)}$$

As illustration, we note that the first diagram corresponds to the analytical expression $\tilde{J}^1$ given by

$$\tilde{J}^1(\mathbf{x}, \mathbf{x}', t; \tilde{\mathbf{x}}, \tilde{\mathbf{x}}', \tilde{t}) = - G_0^R(\mathbf{x}, t; \tilde{\mathbf{x}}, \tilde{t}) \, G_0^A(\tilde{\mathbf{x}}', \tilde{t}; \mathbf{x}') \, D(\mathbf{x}, t; \tilde{\mathbf{x}}, \tilde{t}) \qquad \text{(6.152)}$$

by application of the Feynman rules. The full calculation of the lowest-order irreducible propagator shall be deferred to the next section where the kinetic approach to transport is considered. We observe that for all lowest-order diagrams, equal-time two-line irreducibility is trivial.

6.6.2 Weak-Coupling Kinetic Equation

Just as we did for the particle in a random potential in section 4.6, we shall study the time evolution of the density matrix or, equivalently, the Wigner function in the weak-coupling limit, where the coupling to the heat bath is treated to lowest order.

---

[19]We recall that the introduced equal-time two-line irreducibility of a diagram, beyond lowest order, is determined not by topology alone, but also by the relative time-ordering between interaction times on the upper and lower branches, just as in the random potential case of section 4.3. This undesirable feature is circumvented at the end of this section by considering a generalization of the density matrix.

Let us assume that we initially start from a spatially homogeneous state; i.e., the Wigner function is independent of the spatial variable, $f(\mathbf{p}, \mathbf{R}, t') = f(\mathbf{p}, t')$. The Wigner function thus has the interpretation of being proportional to the momentum probability distribution. Assuming that the external force, $\mathbf{F}(t)$, is spatially homogeneous ($U(\mathbf{x}, t) = \mathbf{F}(t) \cdot \mathbf{x}$), the Wigner function will then be independent of its spatial coordinate at all times. Along with the integration over $\mathbf{r}$ and $\tilde{\mathbf{r}}$, we can therefore perform the integration over $\tilde{\mathbf{R}}$ on the right-hand side of the weak-coupling kinetic equation for the heat bath case, analogously to the impurity case treated in section 4.6. As a consequence, the right-hand side becomes independent of $\mathbf{R}$, just as in the random potential case.

Performing the integrations over the propagator variables, just as we did in the case of the random potential, we obtain for the diagram

$$-\qquad \qquad \qquad \text{(6.153)}$$

the following contribution:

$$\tilde{F}_{\text{out}}^{(1)}[f] \;=\; -\frac{1}{V} \sum_{\mathbf{p}'} \int_{t'}^{t} d\tilde{t}\; e^{\frac{i}{\hbar}(\tilde{t}-t)(\epsilon_{\mathbf{p}'}-\epsilon_{\mathbf{p}})} D(\mathbf{p}-\mathbf{p}', t, \tilde{t}) f(\mathbf{p}, \tilde{t}) \qquad \text{(6.154)}$$

where we have introduced momentum labeling for the oscillator states, and anticipated a three-dimensional labeling, $\mathbf{p} \equiv \hbar\mathbf{q}$.

In calculating the contribution from the diagram

$$\qquad \qquad \qquad \text{(6.155)}$$

the only change as compared to the above calculation involves the sign attributed to the diagram, and the change in the heat bath correlator

$$[D(\mathbf{x}, t; \tilde{\mathbf{x}}, \tilde{t})]^* \quad \rightarrow \quad - D(\tilde{\mathbf{x}}, \tilde{t}; \mathbf{x}, t) \qquad \text{(6.156)}$$

and we obtain from this diagram the contribution

$$\tilde{F}_{\text{in}}^{(1)}[f] \;=\; \frac{1}{V} \sum_{\mathbf{p}'} \int_{t'}^{t} d\tilde{t}\; e^{\frac{i}{\hbar}(\tilde{t}-t)(\epsilon_{\mathbf{p}'}-\epsilon_{\mathbf{p}})} D(\mathbf{p}'-\mathbf{p}, t, \tilde{t}) f(\mathbf{p}', \tilde{t}) \;. \qquad \text{(6.157)}$$

The two terms have been indexed *in* and *out* since they correspond, as we realize shortly, to the scattering in and out terms in the Boltzmann description.

Using the relationship between the retarded and advanced propagators, eq.(2.22), and the property

$$[D(\mathbf{x}, t; \mathbf{x}', t')]^* \;=\; D(\mathbf{x}', t'; \mathbf{x}, t) \qquad \text{(6.158)}$$

it follows as in the impurity case that

$$\tilde{J}^d(\mathbf{x}, \mathbf{x}', t; \tilde{\mathbf{x}}, \tilde{\mathbf{x}}', \tilde{t}) \;\; = \;\; \tilde{J}^d(\mathbf{x}', \mathbf{x}, t; \tilde{\mathbf{x}}', \tilde{\mathbf{x}}, \tilde{t}) \; . \tag{6.159}$$

Transforming to the Wigner representation, we then have the property

$$[\tilde{J}^d(\mathbf{p}, \mathbf{R}, t; \tilde{\mathbf{p}}, \tilde{\mathbf{R}}, \tilde{t})]^* \;\; = \;\; \tilde{J}^d(\mathbf{p}, \mathbf{R}, t; \tilde{\mathbf{p}}, \tilde{\mathbf{R}}, \tilde{t}) \; . \tag{6.160}$$

Just as in the case of the kinetics of a particle in a random potential, the fact that the Wigner function is real is reflected in the mirror symmetry of the diagrams. The diagrams come in two symmetric classes obtained by mirror reflection of the bath correlator lines in a line that is parallel to and in between the upper and lower particle lines.

Utilizing that symmetric diagrams thus give complex conjugate contributions, as just demonstrated, we obtain the collision-type integral $\tilde{F}^{(1)}$ for the lowest-order diagrams

$$\tilde{F}_{\mathbf{p}t}^{(1)}[f] \;\; = \;\; 2\,\Re e(\tilde{F}_{in}^{(1)}[f] \, + \, \tilde{F}_{out}^{(1)}[f])$$

$$= \;\; -2\,\frac{1}{V}\,\Re e \sum_{\mathbf{p}'} \int_{t'}^{t} d\tilde{t}\; e^{\frac{i}{\hbar}(\tilde{t}-t)(\epsilon_{\mathbf{p}'} - \epsilon_{\mathbf{p}})} D(\mathbf{p} - \mathbf{p}', t, \tilde{t})\,[f(\mathbf{p}, \tilde{t}) - f(\mathbf{p}', \tilde{t})] . \tag{6.161}$$

Exercise 6.5 *Show that if we include the effect of a spatially homogeneous time-independent external force, $\mathbf{F}$, to arbitrary order we get the weak-coupling collision integral [30],*

$$\tilde{F}_{\mathbf{p}t}^{(1)}[f] \;\; = \;\; -\sum_{\mathbf{p}'} \int_{t'}^{t} d\tilde{t}\; \theta(t - \tilde{t}) \cos\left(\frac{1}{\hbar}(\tilde{t} - t) \left\{ \epsilon\left(\mathbf{p}' - \frac{\mathbf{F}(t - \tilde{t})}{2} \right) - \epsilon\left(\mathbf{p} - \frac{\mathbf{F}(t - \tilde{t})}{2} \right) \right\} \right)$$

$$\frac{2}{V}\, D(\mathbf{p} - \mathbf{p}')\,[f(\mathbf{p} - \mathbf{F}(t - \tilde{t}), \tilde{t}) - f(\mathbf{p}' - \mathbf{F}(t - \tilde{t}), \tilde{t})] \; . \tag{6.162}$$

Solution

We insert the propagators in the presence of the field, eq.(A.19) or eq.(4.39), into the expressions corresponding to the diagrams of eq.(6.151), and since the resulting integrals are still simple exponential factors they are immediately performed, leading to the above result. The collision integral takes into account the effect of the field during the collision, the intracollisional field effect. Furthermore, the result is straightforwardly generalized to a time-dependent field by the substitution

$$\mathbf{p} \;\; \rightarrow \;\; \mathbf{p} - \int_{t'}^{t} dt''\, \mathbf{F}(t'') \; . \tag{6.163}$$

6.6.3   Classical Kinetics

We take the classical limit formally as we did for the impurity case in section 4.7,[20] and obtain the following equation for the homogeneous Wigner function, the Boltzmann equation:

$$\frac{\partial f}{\partial t} + \mathbf{F} \cdot \frac{\partial f}{\partial \mathbf{p}} = I_{\mathbf{p}t}[f] \tag{6.164}$$

with the collision integral

$$I_{\mathbf{p},t}[f] = -\int \frac{d\mathbf{p}'}{(2\pi\hbar)^3} \left\{ W^+(\mathbf{p}',\mathbf{p}) \left[f(\mathbf{p},t)(1+n_{\mathbf{p}-\mathbf{p}'}) - f(\mathbf{p}',t)n_{\mathbf{p}-\mathbf{p}'} \right] \right.$$

$$\left. + W^-(\mathbf{p}',\mathbf{p}) \left[f(\mathbf{p},t)n_{\mathbf{p}'-\mathbf{p}} - f(\mathbf{p}',t)(1+n_{\mathbf{p}'-\mathbf{p}}) \right] \right\} \tag{6.165}$$

where[21]

$$W^\pm(\mathbf{p}',\mathbf{p}) = \frac{2\pi}{\hbar} |\alpha_{\mathbf{p}'-\mathbf{p}}|^2 \, \delta(\epsilon_{\mathbf{p}'} - \epsilon_{\mathbf{p}} \pm \hbar\omega_{\pm(\mathbf{p}-\mathbf{p}')}) \,. \tag{6.166}$$

We note that $W^\pm(\mathbf{p}',\mathbf{p})$ is Fermi's golden rule expression for the transition probability per unit time from momentum state $\mathbf{p}$ to momentum state $\mathbf{p}'$ caused by the interaction with the oscillators. Each of the four terms has thus a simple interpretation. As an example let us calculate the transition probability for scattering due to absorption of an oscillator quantum. Let us look at scattering out of momentum state $\mathbf{p}$ by absorption: We thus have the initial state and initial energy

$$|i> = |\mathbf{p}> |n_{\mathbf{q}} = 1> , \qquad \epsilon_i = \epsilon_{\mathbf{p}} + \hbar\omega_{\mathbf{q}} \tag{6.167}$$

where $|n_{\mathbf{q}} = 1>$ is the state where all oscillators are in the ground state except the one labeled $\mathbf{q}$, which is assumed to contain one quanta. In the final state all the oscillators are in their ground state

$$|f> = |\mathbf{p}'> |0> , \qquad \epsilon_f = \epsilon_{\mathbf{p}'} \,. \tag{6.168}$$

The transition probability per unit time for the transition $|i> \rightarrow |f>$ is given in lowest-order perturbation theory by (Fermi's golden rule)

$$P_{fi} = \frac{2\pi}{\hbar} |<f|\hat{H}_i|i>|^2 \, \delta(\epsilon_f - \epsilon_i) \,. \tag{6.169}$$

Since no track of the oscillators are kept, the absorption of a quantum can be from any of the oscillators, and we must therefore sum the probabilities for all the possible transitions to get the relevant transition probability per unit time

$$W_{fi} = \sum_{\mathbf{q}} P_{fi} = \frac{2\pi}{\hbar} \sum_{\mathbf{q}} |<f|\alpha_{\mathbf{q}}(\hat{x})\hat{a}_{\mathbf{q}} - \alpha_{\mathbf{q}}^*(\hat{x})\hat{a}_{\mathbf{q}}^\dagger|i>|^2 \, \delta(\epsilon_f - \epsilon_i)$$

---

[20]As discussed in section 6.7, the Boltzmann equation is valid in the quasi-classical regime which for the case of a degenerate electron gas only amounts to the Fermi energy being much larger than the Debye energy, $\epsilon_F \gg \hbar\omega_D$.

[21]We note that the phonon dispersion satisfies $\omega_{-\mathbf{p}} = \omega_{\mathbf{p}}$ due to time-reversal symmetry.

$$= \sum_{\mathbf{q}} \left| \int d\mathbf{x} \, \frac{1}{V} \, e^{\frac{i}{\hbar}\mathbf{x}(\mathbf{p}+\mathbf{q}-\mathbf{p}')} \, \alpha_{\mathbf{q}} < n_{\mathbf{q}} = 0|\hat{a}_{\mathbf{q}}|n_{\mathbf{q}} = 1> \right|^2 \delta(\epsilon_f - \epsilon_i)$$

$$= \frac{2\pi}{\hbar} |\alpha_{\mathbf{p}-\mathbf{p}'}|^2 \, \delta(\epsilon_{\mathbf{p}'} - \epsilon_{\mathbf{p}} - \hbar\omega_{-(\mathbf{p}-\mathbf{p}')})$$

$$= W^-(\mathbf{p}',\mathbf{p}) \, . \tag{6.170}$$

We therefore have the following interpretation of the third term in the collision integral of the Boltzmann equation, eq.(6.165), for a particle interacting with an oscillator environment. It is a loss term and gives the rate of change of particle density in a phase space volume due to the scattering of a particle with momentum $\mathbf{p}$ into momentum state $\mathbf{p}'$ by absorption of a quanta of momentum $\mathbf{p}' - \mathbf{p}$. This probability to be scattered out of the phase space volume around $\mathbf{p}$, and into the volume around $\mathbf{p}'$ by emission is the product of three probabilities $f(\mathbf{p},t)$, $W^-(\mathbf{p}',\mathbf{p})$, and $n_{\mathbf{p}'-\mathbf{p}}$: (the probability that a particle in the phase space volume in question is available for scattering) $\times$ (the transition probability for the transition from $\mathbf{p}$ to $\mathbf{p}'$) $\times$ (the probability that there is a quantum to be absorbed). We have analogous interpretations of the other three terms reflecting that the interaction Hamiltonian consists of a creation and an annihilation part which give rise to scattering processes due to emission and absorption of oscillator quanta. The different terms in the collision integral correspond to scattering in and out due to emission or absorption of quanta. We can therefore memorize the four scattering terms easily by drawing the following Boltzmann diagrams

$$\tag{6.171}$$

with the Boltzmann dictionary for occupation probabilities of quanta (wiggly lines)

$$\equiv \quad n_{\mathbf{p}-\mathbf{p}'} \quad , \qquad\qquad \equiv \quad 1 + n_{\mathbf{p}-\mathbf{p}'} \quad (6.172)$$

and occupation probabilities for the particle

$$\equiv \quad f_{\mathbf{p}} \quad , \qquad\qquad \equiv \quad 1 \qquad\qquad (6.173)$$

and we use for the transition probability the notation

$$\textcircled{\pm} \equiv W^{\pm}(\mathbf{p}, \mathbf{p}') \, . \qquad\qquad (6.174)$$

The classical limit we took above only involved the particle kinematics. As regards the oscillator statistics an important quantum feature is apparent, namely that of stimulated emission, the appearance of the factor $1 + n$ instead of simply the factor one for spontaneous emission. The favoring of emission of the type of quanta already present, is fundamental to the operation of lasers and masers.

We note, that the electron-impurity collision integral is the following limit of the particle-oscillator model: Take the temperature to be zero, so that the Bose function equals zero, $n_{\mathbf{q}} = 0$ for all $\mathbf{q}$, corresponding to no quanta being present, and assume that $\omega_{\mathbf{q}} = 0$ for all $\mathbf{q}$, no energy transfer between the particle and environment is possible, making the interaction elastic, and identify $|\alpha_{\mathbf{p}-\mathbf{p}'}|^2 \rightarrow n_i |V_{imp}(\mathbf{p} - \mathbf{p}')|^2$.

Exercise 6.6 *Show that the equilibrium distribution f_0, i.e., $I_{\mathbf{p}}[f_0] = 0$, is the Maxwell-Boltzmann distribution*

$$f_0(\epsilon_{\mathbf{p}}) = f_{MB}(\epsilon_{\mathbf{p}}) = \frac{1}{V} \left(\frac{2\pi\hbar^2}{mkT} \right)^{3/2} e^{-\epsilon_{\mathbf{p}}/kT} \quad . \qquad\qquad (6.175)$$

6.7 Kinetic Propagator

We have noticed in section 4.3 that the previously introduced topological concept of equal-time two-line reducibility is impractical for topological classification beyond lowest-order perturbation theory, because it does not refer exclusively to the topology of a density matrix diagram. As we now demonstrate, defining reducibility by the ability to cut either an R-line or an A-line is the concept referring only to topology, and separates the two different aspects of interactions, the virtual processes leading to renormalization, and the real dissipative processes.

When we sum density-matrix diagrams, we encounter reduced amplitudes which do not in general have a physical interpretation, but nevertheless enter the skeleton perturbation theory as basic building blocks. An example is the full retarded propagator

$$G^R(\mathbf{x}, t; \mathbf{x}', t') \equiv -i\theta(t - t')\, tr(\hat{\rho}_T <\mathbf{x}|\hat{U}(t, t')|\mathbf{x}' >)\,. \tag{6.176}$$

Since the environment is assumed to be in the thermodynamic equilibrium state, we also have, $t > t'$,

$$G^R(\mathbf{x}, t; \mathbf{x}', t') = \frac{-i}{Z} \sum_{\{n_q\}_q} e^{-E(\{n_q\}_q)/kT} <\{n_q\}_q|\otimes<\mathbf{x}|\hat{U}(t, t')|\mathbf{x}'>\otimes|\{n_q\}_q>\,. \tag{6.177}$$

The full retarded propagator has the diagrammatic perturbative representation given by the upper line segments in the diagrams of eq.(6.127)

$$\tag{6.178}$$

The insertion which iteratively generates the full propagator is the self-energy, the sum of amputated diagrams which cannot be cut in two by cutting only an internal particle line[22]

$$\tag{6.179}$$

---

[22]The topological argument is identical to the impurity case, section 3.3.

In terms of the retarded self-energy Σ^R we thus have the integral equation for the full retarded propagator

$$G^R(\mathbf{x},t;\mathbf{x}',t') = \int d\mathbf{x}_2 \int_{-\infty}^{\infty} dt_2 \int d\mathbf{x}_1 \int_{-\infty}^{\infty} dt_1\, G_0^R(\mathbf{x},t;\mathbf{x}_2,t_2)\, \Sigma^R(\mathbf{x}_2,t_2;\mathbf{x}_1,t_1)\, G^R(\mathbf{x}_1,t_1;\mathbf{x}',t')$$

$$+ \ G_0^R(\mathbf{x},t;\mathbf{x}',t') \ . \tag{6.180}$$

We introduce the notation $\otimes$ for matrix multiplication in the space and time variables. For example,

$$(\Sigma^R \otimes G^R)(\mathbf{x},t;\mathbf{x}',t') \equiv \int d\bar{\mathbf{x}} \int d\bar{t}\, \Sigma^R(\mathbf{x},t;\bar{\mathbf{x}},\bar{t})\, G^R(\bar{\mathbf{x}},\bar{t};\mathbf{x}',t') \ . \tag{6.181}$$

In the matrix notation we have the equation for the full propagator

$$G^R = G_0^R + G_0^R \otimes \Sigma^R \otimes G^R \ . \tag{6.182}$$

Iterating from the other side we get the equivalent equation

$$G^R = G_0^R + G^R \otimes \Sigma^R \otimes G_0^R \ . \tag{6.183}$$

The inverse matrix to the free particle propagator

$$(G_0^{-1} \otimes G_0^R)(\mathbf{x},t;\mathbf{x}',t') = \hbar\,\delta(\mathbf{x}-\mathbf{x}')\,\delta(t-t') = (G_0^R \otimes G_0^{-1})(\mathbf{x},t;\mathbf{x}',t') \tag{6.184}$$

is, according to eq.(2.15), given by

$$G_0^{-1}(\mathbf{x},t;\mathbf{x}',t') = \left\langle \mathbf{x} \left| i\hbar\frac{\partial}{\partial t} - \hat{H}_p \right| \mathbf{x}' \right\rangle \delta(t-t')$$

$$= \left(i\hbar\frac{\partial}{\partial t}\,\delta(\mathbf{x}-\mathbf{x}') - <\mathbf{x}|\hat{H}_p|\mathbf{x}'> \right) \delta(t-t') \ . \tag{6.185}$$

Multiplying eq.(6.182) by the inverse propagator from the left, we get the equation

$$(G_0^{-1} \otimes G^R)(\mathbf{x},t;\mathbf{x}',t') = \hbar\,\delta(\mathbf{x}-\mathbf{x}')\delta(t-t') + (\Sigma^R \otimes G^R)(\mathbf{x},t;\mathbf{x}',t') \tag{6.186}$$

and multiplying eq.(6.183) from the right, we get the equation

$$(G^R \otimes G_0^{-1})(\mathbf{x},t;\mathbf{x}',t') = \hbar\,\delta(\mathbf{x}-\mathbf{x}')\delta(t-t') + (G^R \otimes \Sigma^R)(\mathbf{x},t;\mathbf{x}',t') \ . \tag{6.187}$$

Similarly we have for the full advanced propagator

$$G^A(\mathbf{x},t;\mathbf{x}',t') \equiv i\theta(t'-t)\,tr(\hat{\rho}_T <\mathbf{x}|\hat{U}^\dagger(t,t')|\mathbf{x}'> = [G^R(\mathbf{x}',t';\mathbf{x},t)]^* \tag{6.188}$$

the equation

$$(G_0^{-1} - \Sigma^A) \otimes G^A = \hbar\,\delta \tag{6.189}$$

and

$$G^A \otimes (G_0^{-1} - \Sigma^A) = \hbar \delta \tag{6.190}$$

where we have introduced the unit matrix

$$\delta \equiv \delta(\mathbf{x} - \mathbf{x}') \delta(t - t') . \tag{6.191}$$

Combining the above equations, we obtain for the nonequilibrium spectral function

$$A(\mathbf{x}, t; \mathbf{x}', t') = i \left(G^R(\mathbf{x}, t; \mathbf{x}', t') - G^A(\mathbf{x}, t; \mathbf{x}', t') \right) \tag{6.192}$$

the equation

$$[G_0^{-1} \, \substack{\otimes \\ ,} \, A]_- = i\Sigma^R \otimes G^R - i\Sigma^A \otimes G^A - iG^R \otimes \Sigma^R + iG^A \otimes \Sigma^A \tag{6.193}$$

where we have introduced the notation

$$[G_0^{-1} \, \substack{\otimes \\ ,} \, A]_- = G_0^{-1} \otimes A - A \otimes G_0^{-1} . \tag{6.194}$$

Introducing further the notation for real and imaginary parts

$$\Gamma(\mathbf{x}, t; \mathbf{x}', t') = i \left(\Sigma^R(\mathbf{x}, t; \mathbf{x}', t') - \Sigma^A(\mathbf{x}, t; \mathbf{x}', t') \right) = -2 \, \Im m \Sigma^R(\mathbf{x}, t; \mathbf{x}', t') \tag{6.195}$$

and

$$\Re e \Sigma^R(\mathbf{x}, t; \mathbf{x}', t') = \frac{1}{2} \left(\Sigma^R(\mathbf{x}, t; \mathbf{x}', t') + \Sigma^A(\mathbf{x}, t; \mathbf{x}', t') \right) = \Re e \Sigma^A(\mathbf{x}, t; \mathbf{x}', t') \tag{6.196}$$

and

$$\Re e G^R(\mathbf{x}, t; \mathbf{x}', t') = \frac{1}{2} \left(G^R(\mathbf{x}, t; \mathbf{x}', t') + G^A(\mathbf{x}, t; \mathbf{x}', t') \right) = \Re e G^A(\mathbf{x}, t; \mathbf{x}', t') \tag{6.197}$$

the equation for the nonequilibrium spectral function can be rewritten

$$[G_0^{-1} - \Re e \Sigma \, \substack{\otimes \\ ,} \, A]_- - [\Gamma \, \substack{\otimes \\ ,} \, \Re e G]_- = 0 . \tag{6.198}$$

In order to obtain a symmetric formulation in the time variables, we introduce the kinetic propagator, the double-time reduced density matrix,

$$f(\mathbf{x}, t; \mathbf{x}', t') \equiv tr(<\mathbf{x}|\hat{\rho}(t, t')|\mathbf{x}' >) = <\mathbf{x}|\hat{f}(t, t')|\mathbf{x}' > \tag{6.199}$$

where

$$\hat{\rho}(t, t') \equiv \hat{U}(t, t_i) \, \hat{\rho}_i \, \hat{U}^\dagger(t', t_i) \tag{6.200}$$

and

$$\hat{f}(t, t') \equiv tr(\hat{\rho}(t, t')) . \tag{6.201}$$

The equal-time kinetic propagator

$$f(\mathbf{x}, \mathbf{x}', t) = f(\mathbf{x}, t; \mathbf{x}', t) \tag{6.202}$$

equals the reduced density matrix as

$$\hat{\rho}(t,t) = \hat{\rho}(t) .$$ (6.203)

and all physical quantities can thus be expressed in terms of the kinetic propagator.

Let us assume that the initial state of the particle and environment is separable

$$\hat{\rho}_i = \hat{f}_i \, \hat{\rho}_T .$$ (6.204)

In the absence of interaction, we have for the free kinetic propagator the diagrammatic expansion (we choose the reference time equal to the initial time, $t_r = t_i$)

$$f_0(\mathbf{x}, t; \mathbf{x}', t') = \;<\mathbf{x}, t|\hat{f}_i|\mathbf{x}', t'>$$

$$\equiv \quad \underset{\mathbf{x}t}{\bullet} - - - \underset{\mathbf{x}'t'}{\bullet}$$ (6.205)

where we have introduced a thin horizontal dashed line to represent the free kinetic propagator.

We note the identities

$$G_0^{-1} \otimes f_0 = 0 = f_0 \otimes G_0^{-1} .$$ (6.206)

For an initial state of the particle of the form

$$\hat{f}_i = \sum_{\mathbf{p}} \rho(\epsilon_{\mathbf{p}}) \, |\mathbf{p}><\mathbf{p}|$$ (6.207)

we have for the Fourier transform of the free kinetic propagator

$$f_0(\mathbf{p}, E) = i \, (G_0^R(\mathbf{p}, E) - G_0^A(\mathbf{p}, E)) \, \rho(E) .$$ (6.208)

The diagrammatic expansion of the kinetic propagator is identical to the diagrammatic expansion of the reduced density matrix, eq.(6.127), except that the advanced propagator starts at space-time point $(\mathbf{x}', t')$ and not at $(\mathbf{x}', t)$. Whenever the free kinetic propagator is attached to a retarded or advanced particle line, a D-line is also attached at the vertex. To the left of f_0 is attached a G_0^R-line and at the right a G_0^A-line. From a topological point of view the free kinetic propagator is thus on an equal footing with the free retarded and advanced propagators. The diagrammatic expansion for the kinetic propagator is according to its definition obtained by the procedure: first draw any sequence of propagator lines where only one free kinetic propagator line appears, and G_0^R-lines only appears to the left of the f_0-line, and G_0^A-lines only at the right. At each internal point is attached a dangling interaction line (recall the diagram in eq.(6.121)). The diagrams for the kinetic propagator (the kinetic propagator is represented diagrammatically by a horizontal fat dashed line) are then obtained by connecting the dangling bonds together in pairs in all possible ways

$$(6.209)$$

Exercise 6.7 *Draw the rest of the 15 second-order diagrams for the kinetic propagator.*

In order to obtain the one-particle irreducible equation for the kinetic propagator we consider instead of the above diagrams the self-energy type diagrams where we amputate the two outermost lines from the diagrams. If such a diagram can not be cut in two by cutting only one f_0-, G_0^R-, or G_0^A-line, it is said to be irreducible. Such a self-energy diagram can be of three types: it contains only G_0^R-lines or only G_0^A-lines, or exactly one f_0-line (and arbitrary many G_0^R or G_0^A-lines). In accordance with our previous notation, the three classes of self-energy diagrams are denoted Σ^R, Σ^A and Σ, respectively.

Exercise 6.8 *Draw all Σ-diagrams containing one and two D-lines.*

The integral equation determining the kinetic propagator

$$f = f_0 + G_0^R \otimes \Sigma^R \otimes f + f_0 \otimes \Sigma^A \otimes G^A + G_0^R \otimes \Sigma \otimes G^A \qquad (6.210)$$

can be obtained by studying the topology of its diagrams, see eq.(6.209). In a diagram with a Σ^R-piece appearing, it is always attached at the left to G_0^R, and at the right to the f_0-line or a G_0^R-line, and the string to the right of a Σ^R-piece is thus one of the terms in the diagrammatic expansion of f. We can therefore generate a unique class of kinetic propagator diagrams by the second term on the right-hand side of the preceding equation. A Σ^A-piece in a diagram must be attached to an f-type diagram on the left (or a G_0^A-line, but such terms are generated by the Σ^A self-energy iteration). However, in order not to generate diagrams already included in the $G_0^R \otimes \Sigma^R \otimes f$ term, it must be attached to f_0, and we have the third term in the preceding equation. We have now exhausted all kinetic propagator diagrams containing Σ^R and Σ^A pieces. In $G_0^R \otimes \Sigma^R \otimes f$ we have diagrams containing Σ-pieces, viz. those which always have attached to the left a $\Sigma^R \otimes G_0^R$-piece. The remaining Σ-terms are thus given by the last term in the above equation.

We could equally well have established that the kinetic propagator satisfies the equation

$$f = f_0 + f \otimes \Sigma^A \otimes G_0^A + G^R \otimes \Sigma^R \otimes f_0 + G^R \otimes \Sigma \otimes G_0^A . \qquad (6.211)$$

To get the equation for the kinetic propagator on a form without explicit appearance of the initial condition, i.e., the free kinetic propagator, we study its equation of motion on differential form (subtracting eq.(6.211) multiplied by the inverse propagator, G_0^{-1}, from the right from eq.(6.210) multiplied by the inverse propagator from the left)

$$[G_0^{-1} \,\overset{\otimes}{,}\, f]_- = \Sigma^R \otimes f - f \otimes \Sigma^A - G^R \otimes \Sigma + \Sigma \otimes G^A \qquad (6.212)$$

which constitute the general quantum kinetic equation. Introducing for arbitrary functions A and B the notation

$$[A \,\overset{\otimes}{,}\, B]_+ = A \otimes B + B \otimes A \qquad (6.213)$$

the quantum kinetic equation can be rewritten

$$[G_0^{-1} - \Re e \Sigma^R \,\overset{\otimes}{,}\, f]_- - [\Sigma \,\overset{\otimes}{,}\, \Re e G^R]_- = \frac{i}{2}[\Sigma \,\overset{\otimes}{,}\, A]_+ - \frac{i}{2}[\Gamma \,\overset{\otimes}{,}\, f]_+ . \qquad (6.214)$$

We introduce the coordinates

$$\mathbf{R} = \frac{\mathbf{x}_1 + \mathbf{x}_{1'}}{2} \qquad \mathbf{r} = \mathbf{x}_1 - \mathbf{x}_{1'} \qquad T = \frac{t_1 + t_{1'}}{2} \qquad t = t_1 - t_{1'} \qquad (6.215)$$

and the mixed variables with respect to both space and time; i.e., we Fourier transform with respect to relative coordinates. For example, we have for the kinetic propagator in the mixed variables

$$f(E, \mathbf{p}, \mathbf{R}, T) = \int d\mathbf{r} \int dt \, e^{-\frac{i}{\hbar}\mathbf{r}\cdot\mathbf{p} + \frac{i}{\hbar}tE} f(\mathbf{r}, \mathbf{R}, t, T) . \qquad (6.216)$$

The matrix product becomes in the mixed variables

$$(A \otimes B)(E, \mathbf{p}, \mathbf{R}, T) = e^{\frac{i}{2}(\partial_X^A \partial_P^B - \partial_P^A \partial_X^B)} A(E, \mathbf{p}, \mathbf{R}, T) \, B(E, \mathbf{p}, \mathbf{R}, T) \qquad (6.217)$$

where we have introduced the notation

$$\partial_X^A \partial_P^B \equiv -\frac{\partial^A}{\partial T} \frac{\partial^B}{\partial E} + \frac{\partial^A}{\partial \mathbf{R}} \cdot \frac{\partial^B}{\partial \mathbf{p}} \qquad (6.218)$$

and the superscript indicates on which function the differentiation takes place.

In equilibrium the kinetic equation reduces to

$$\Sigma(E, \mathbf{p}) \, A(E, \mathbf{p}) = \Gamma(E, \mathbf{p}) \, f(E, \mathbf{p}) \qquad (6.219)$$

determining the equilibrium distribution function, the one for which the collision integral vanishes.

The dependence of any function on $\mathbf{R}$ and T is due to a nonequilibrium situation, for example caused by the coupling of the particle to external fields. If this dependence is slowly varying only the lowest order variation needs be retained, leading to the gradient expansion

$$[A \, ? \, B]_+ \rightarrow 2 \, A(E, \mathbf{p}, \mathbf{R}, T) \, B(E, \mathbf{p}, \mathbf{R}, T) \;, \quad -i[A \, ? \, B]_- \rightarrow [A, B]_p \qquad (6.220)$$

where we have introduced a generalization of the Poisson bracket notation

$$[A, B]_p = \partial_X^A A \, \partial_P^B B - \partial_P^A A \, \partial_X^B B = (\partial_E^A \partial_T^B - \partial_T^A \partial_E^B$$
$$- \nabla_\mathbf{p}^A \cdot \nabla_\mathbf{R}^B + \nabla_\mathbf{R}^A \cdot \nabla_\mathbf{p}^B) \, A(E, \mathbf{p}, \mathbf{R}, T) \, B(E, \mathbf{p}, \mathbf{R}, T) \;. \qquad (6.221)$$

Let us assume that the nonequilibrium situation is created by coupling the particle to a scalar potential U.[23] In the gradient approximation, applicaple to slowly varying disturbances, we then obtain the quantum kinetic equation[24]

$$\left(\frac{\partial}{\partial T} + \frac{\mathbf{p}}{m} \cdot \nabla_\mathbf{R} - \nabla_\mathbf{R} U(\mathbf{R}, T) \cdot \nabla_\mathbf{p} + \frac{\partial U(\mathbf{R}, T)}{\partial T} \frac{\partial}{\partial E} \right) f(E, \mathbf{p}, \mathbf{R}, T)$$

$$- [\Re e \Sigma^R, f]_p - [\Re e G^R, \Sigma]_p = \Sigma \, A - \Gamma \, f \qquad (6.222)$$

where all propagators and self-energies have $(E, \mathbf{p}, \mathbf{R}, T)$ as argument. The last two terms on the left side describe renormalization effects due to virtual processes,[25] and the terms on the right side describe the dissipative effect due to the interaction of the particle with the oscillators.

---

[23]The coupling to a vector potential can be handled with equal ease, recall section 4.4.2.

[24]The first term is not dependent on the gradient approximation, but as usual exact, simply due to the von Neumann equation being first order in time, and similarly for the second term for the case of quadratic dispersion.

[25]Self-energy renormalization is for example important for the Polaron.

The preceding equation can be rewritten

$$[E - \epsilon_{\mathbf{p}} - U(\mathbf{R}, T) - \Re e\Sigma^R, f]_p + [\Re eG^R, \Sigma]_p$$

$$= \Sigma(E, \mathbf{p}, \mathbf{R}, T) \, A(E, \mathbf{p}, \mathbf{R}, T) - \Gamma(E, \mathbf{p}, \mathbf{R}, T) \, f(E, \mathbf{p}, \mathbf{R}, T) \,. \quad (6.223)$$

In the gradient approximation, the nonequilibrium spectral function satiesfies the equation (recall eq.(6.198))

$$[E - \epsilon_{\mathbf{p}} - U(\mathbf{R}, T) - \Re e\Sigma^R, A]_p + [\Re eG^R, \Gamma]_p = 0 \,. \quad (6.224)$$

We note that

$$A(E, \mathbf{p}, \mathbf{R}, T) = \frac{\Gamma(E, \mathbf{p}, \mathbf{R}, T)}{(E - \epsilon_{\mathbf{p}} - U(\mathbf{R}, T) - \Re e\Sigma^R(E, \mathbf{p}, \mathbf{R}, T))^2 + \left(\frac{\Gamma(E, \mathbf{p}, \mathbf{R}, T)}{2}\right)^2} \quad (6.225)$$

solves eq.(6.224) since, using that $[A, B]_p = -[B, A]_p$, and noting that

$$\Re e \left(G^R(E, \mathbf{p}, \mathbf{R}, T)\right)^{-1} = E - \epsilon_{\mathbf{p}} - U(\mathbf{R}, T) - \Re e\Sigma^R(E, \mathbf{p}, \mathbf{R}, T) \quad (6.226)$$

the left side of the equation can then be rewritten on the form

$$-i\left[\Re e\left(G^R\right)^{-1} - \frac{i}{2}\Gamma, \left(\Re e\left(G^R\right)^{-1} - \frac{i}{2}\Gamma\right)^{-1}\right]_p$$

$$+ \; i\left[\Re e\left(G^R\right)^{-1} + \frac{i}{2}\Gamma, \left(\Re e\left(G^R\right)^{-1} + \frac{i}{2}\Gamma\right)^{-1}\right]_p \quad (6.227)$$

which vanishes since for any function F, we have $[A, F(A)]_p = 0$. In the far past, where the system is assumed undisturbed, i.e., U vanishes, the presented solution, eq.(6.225), reduces to the equilibrium spectral function, and is therefore the sought solution since it satisfies the correct initial condition.

Let us finally see how the impurity Boltzmann equation for a degenerate electron gas emerges from the corresponding equation for the kinetic propagator.[26] To lowest order in $\hbar/\epsilon_F\tau$ we have according to section 3.6 for the nonequilibrium self-energies (in the Born approximation)

$$\Sigma^{R(A)}(E, \mathbf{p}, \mathbf{R}, T) = n_i \int \frac{d\mathbf{p}'}{(2\pi\hbar)^3} \, |V_{imp}(\mathbf{p} - \mathbf{p}')|^2 \, G^{R(A)}(E, \mathbf{p}', \mathbf{R}, T) \quad (6.228)$$

and

$$\Sigma(E, \mathbf{p}, \mathbf{R}, T) = n_i \int \frac{d\mathbf{p}'}{(2\pi\hbar)^3} \, |V_{imp}(\mathbf{p} - \mathbf{p}')|^2 \, f(E, \mathbf{p}', \mathbf{R}, T) \,. \quad (6.229)$$

---

[26]The irreducible diagrammatic classification for the random potential case is identical to the one for the case of an oscillator environment, and all the above results are immediately taken over. The universality of diagrammatics is a powerful feature.

The nonequilibrium spectral function is according to eq.(6.225) a strongly peaked function in the energy variable, and has to zeroth order in the disorder the form

$$A(E, \mathbf{p}, \mathbf{R}, T) = 2\pi\, \delta(E - \epsilon_\mathbf{p} - U(\mathbf{R}, T)) \,. \tag{6.230}$$

We can therefore integrate the quantum kinetic equation, eq.(6.223), with respect to the energy variable, obtaining an equation for the Wigner distribution function

$$f(\mathbf{p}, \mathbf{R}, T) = \int_{-\infty}^{\infty} \frac{dE}{2\pi}\, f(E, \mathbf{p}, \mathbf{R}, T) \tag{6.231}$$

by inserting in the self-energy term for the kinetic propagator the approximation

$$f(E, \mathbf{p}, \mathbf{R}, T) = 2\pi\, \delta(E - \epsilon_\mathbf{p} - U(\mathbf{R}, T))\, f(\mathbf{p}, \mathbf{R}, T) \tag{6.232}$$

i.e., eq.(6.208) for the case where a potential U is present. To the desired order of accuracy, and since the disturbance is slowly varying, i.e., its frequency and wavelength are small compared to the characteristic energy of the system, the Fermi energy ϵ_F, the renormalization terms on the left side of the quantum kinetic equation vanishes. Integrating the quantum kinetic equation, eq.(6.223), with respect to the energy variable therefore gives an equation in terms of the equal-time kinetic progator, i.e., the Wigner function, which is exactly the Boltzmann equation for impurity scattering eq.(5.120). We have thus established that the criterion for the validity of the Boltzmann equation is $\hbar/\epsilon_F \tau \ll 1$, the Landau criterion.[27]

---

[27]Similarly, the Boltzmann equation for a degenerate electron gas interacting with phonons is valid if the Fermi energy is much larger than the Debye energy, $\epsilon_F \gg \hbar\omega_D$, since then diagrams with crossing phonon lines can be neglected and the self-energy is given by a single diagram.

Chapter 7

Linear Response Theory

In the three preceding chapters we have shown how to obtain kinetic equations. The kinetic-equation approach to transport is a general method, and allows in principle nonlinear effects to be considered. However, in many practical situations one is interested only in the linear response of the system to the external force. The linear response limit is a tremendous simplification in comparison with general nonequilibrium conditions, since the linear response is uniquely determined by the equilibrium properties of the system.

In this chapter linear response is discussed, and in particular the density and current response of an electron gas. The symmetry properties of response functions, and the fluctuation-dissipation theorem are established. Last, we demonstrate how correlation functions can be measured in scattering experiments, as illustrated by considering neutron scattering from matter.

7.1 Linear Response

In this section we consider the response of a system to a perturbation. The Hamiltonian consists of two parts:

$$\hat{H} = \hat{H}_0 + \hat{H}'_t \tag{7.1}$$

where $\hat{H}_0$ governs the dynamics in the absence of the perturbation $\hat{H}'_t$.

For the expectation value of a quantity A for a system in state ρ we have

$$A(t) = Tr(\hat{\rho}(t)\,\hat{A}) = Tr(\hat{U}(t,t')\,\hat{\rho}(t')\,\hat{U}^\dagger(t,t')\,\hat{A})\,. \tag{7.2}$$

Expanding the time-evolution operator to linear order in the applied perturbation we get

$$\hat{U}(t,t') = \hat{U}_0(t,t') - \hat{U}_0(t,t_r)\frac{i}{\hbar}\int_{t'}^{t}d\bar{t}\,\hat{H}'_I(\bar{t})\,\hat{U}_0^\dagger(t',t_r) + \mathcal{O}(\hat{H}'^2_t) \tag{7.3}$$

where the perturbation is in the interaction picture with respect to $\hat{H}_0$

$$\hat{H}'_I(t) = \hat{U}_0^\dagger(t,t_r)\hat{H}'_t\,\hat{U}_0(t,t_r)\,. \tag{7.4}$$

For the statistical operator we thus have the perturbative expansion in terms of the perturbation

$$\hat{\rho}(t) = \hat{\rho}_0(t) + \hat{\rho}_1(t) + \mathcal{O}(H_t'^2) \tag{7.5}$$

where

$$\hat{\rho}_0(t) = \hat{U}_0(t, t_i)\,\hat{\rho}_i\,\hat{U}_0^\dagger(t, t_i) = \hat{U}_0(t, t_r)\,\hat{\rho}_0(t_r)\,\hat{U}_0^\dagger(t, t_r) \tag{7.6}$$

and the linear correction in the applied potential is given by

$$\hat{\rho}_1(t) = \frac{i}{\hbar}\hat{U}_0(t, t_i)\,\hat{\rho}_i\,\hat{U}_0(t_i, t_r)\int_{t_i}^t d\bar{t}\,\hat{H}_I'(\bar{t})\,\hat{U}_0^\dagger(t, t_r)$$

$$- \frac{i}{\hbar}\hat{U}_0(t, t_r)\int_{t_i}^t d\bar{t}\,\hat{H}_I'(\bar{t})\,\hat{U}_0^\dagger(t_i, t_r)\,\hat{\rho}_i\,\hat{U}_0^\dagger(t, t_i)\,. \tag{7.7}$$

We have assumed that prior to time t_i, the applied field is absent, and the system is in state $\hat{\rho}_i$. For the expectation value we then get to linear order

$$A(t) = Tr(\hat{\rho}_0(t)\,\hat{A}) + \frac{i}{\hbar}\int_{t_i}^t d\bar{t}\,Tr(\hat{\rho}_0(t_r)\,[\hat{H}_I'(\bar{t}), \hat{A}_I(t)])\,. \tag{7.8}$$

We first discuss the density response to an external scalar potential, and afterwards the current response to a vector potential.

7.2 Density Response

In this section we consider the density response to an applied external field. The external field is represented by the potential $V(\mathbf{x}, t)$, and the Hamiltonian consists of two parts:

$$\hat{H} = \hat{H}_0 + \hat{H}_{V(t)} \tag{7.9}$$

where $\hat{H}_0$ governs the dynamics in the absence of the applied potential, and the applied potential couples to the density operator of the system

$$\hat{H}_{V(t)} = \int d\mathbf{x}\,\hat{n}(\mathbf{x})\,V(\mathbf{x}, t) \tag{7.10}$$

where for a system of N-particles we have

$$\hat{n}(\mathbf{x}) = \sum_{i=1}^N \hat{n}_i(\mathbf{x}) = \sum_{i=1}^N \delta(\hat{\mathbf{x}}_i - \mathbf{x})\,. \tag{7.11}$$

The density will adjust to the applied potential, and according to eq.(7.8) the deviation from equilibrium is to linear order

$$\delta n(\mathbf{x}, t) = n(\mathbf{x}, t) - n_0(\mathbf{x}, t) = \int d\mathbf{x}' \int_{t_i}^\infty d t'\,\chi(\mathbf{x}, t; \mathbf{x}', t')V(\mathbf{x}', t') \tag{7.12}$$

where

$$n_0(\mathbf{x}, t) = Tr(\hat{\rho}_0(t)\hat{n}(\mathbf{x})) \tag{7.13}$$

is the density in the absence of the potential, and the linear density response function can be specified in the various ways:

$$\chi(\mathbf{x}, t; \mathbf{x}', t') = -\frac{i}{\hbar}\theta(t - t')Tr(\hat{\rho}_0(t_r)[\hat{n}(\mathbf{x}, t), \hat{n}(\mathbf{x}', t')])$$

$$\equiv -\frac{i}{\hbar}\theta(t - t') < [\hat{n}(\mathbf{x}, t), \hat{n}(\mathbf{x}', t')] >_0$$

$$= -\frac{i}{\hbar}\theta(t - t')Tr(\hat{\rho}_0(t_r)[\delta\hat{n}(\mathbf{x}, t), \delta\hat{n}(\mathbf{x}', t')])$$

$$\equiv \chi^R(\mathbf{x}, t; \mathbf{x}', t') . \qquad (7.14)$$

The density operator is in the interaction picture with respect to $\hat{H}_0$

$$\hat{n}(\mathbf{x}, t) = \hat{U}_0^\dagger(t, t_r)\,\hat{n}(\mathbf{x})\,\hat{U}_0(t, t_r) = \sum_{i=1}^{N} \delta(\hat{\mathbf{x}}_i(t) - \mathbf{x}) \qquad (7.15)$$

and we have introduced the density deviation operator $\delta\hat{n}(\mathbf{x}, t) \equiv \hat{n}(\mathbf{x}, t) - n_0(\mathbf{x}, t)$. The retarded density response function appears in eq.(7.12) in order to respect causality; i.e., a change in the density at time t can only occur as a cause of the applied potential prior to that time.

Before the external potential is applied we assume a stationary state with respect to the unperturbed Hamiltonian $\hat{H}_0$, and the initial state is described by a statistical operator of the form

$$\hat{\rho}_i = \rho_i(\hat{H}_0) = \sum_\lambda \rho_\lambda |\lambda><\lambda| \qquad (7.16)$$

where the $|\lambda>$'s are the eigenstates of $\hat{H}_0$,

$$\hat{H}_0|\lambda> = \epsilon_\lambda|\lambda> \qquad (7.17)$$

and $\rho_\lambda = \rho_i(\epsilon_\lambda)$ is the probability for finding the unperturbed system with energy ϵ_λ. The unperturbed statistical operator is then time independent, $\hat{\rho}_0(t) = \hat{\rho}_i$, and the equilibrium density profile is time independent, $n_0(\mathbf{x}, t) = n_0(\mathbf{x}) = Tr(\hat{\rho}_i\,\hat{n}(\mathbf{x}))$. The response function will then only depend on the time difference:

$$\chi(\mathbf{x}, t; \mathbf{x}', t') = \chi(\mathbf{x}, \mathbf{x}'; t - t') =$$

$$-\frac{i}{\hbar}\theta(t - t')\sum_{\lambda\lambda'}(\rho_\lambda - \rho_{\lambda'}) <\lambda|\hat{n}(\mathbf{x})|\lambda'><\lambda'|\hat{n}(\mathbf{x}')|\lambda> e^{\frac{i}{\hbar}(\epsilon_\lambda - \epsilon_{\lambda'})(t - t')} . \qquad (7.18)$$

In linear response, each Fourier component contributes additively, so without loss of generality we just need to seek the response at one driving frequency, say ω,

$$V(\mathbf{x}, t) = V_\omega(\mathbf{x})\,e^{-i\omega t} . \qquad (7.19)$$

For any ω in the the upper half plane, $\Im m\,\omega > 0$, the applied potential vanishes in the far past, $V(t \to -\infty) = 0$, and the state of the system in the far past becomes smoothly independent of the applied potential. For ω real we are thus interested in the analytic continuation from the upper half plane of the frequency-dependent response function.

Since we shall be interested in steady-state properties, the time integration in eq.(7.12) can be performed by letting the arbitrary initial time t_i be taken in the remote past. By letting t_i approach minus infinity, transients are absent, and there is then only a linear density response at the driving frequency

$$\delta n(\mathbf{x}, t) = n(\mathbf{x}, t) - n_0(\mathbf{x}) = \delta n(\mathbf{x}, \omega)\, e^{-i\omega t} . \tag{7.20}$$

We obtain for the Fourier transform of the linear density response

$$\delta n(\mathbf{x}, \omega') = \delta n(\mathbf{x}, \omega)\, \delta(\omega - \omega') \tag{7.21}$$

where

$$\delta n(\mathbf{x}, \omega) = \int d\mathbf{x}'\, \chi(\mathbf{x}, \mathbf{x}'; \omega)\, V_\omega(\mathbf{x}') \tag{7.22}$$

and

$$\chi(\mathbf{x}, \mathbf{x}'; \omega) = \sum_{\lambda\lambda'} \frac{\rho_\lambda - \rho_{\lambda'}}{\epsilon_\lambda - \epsilon_{\lambda'} + \hbar\omega + i0} <\lambda|\hat{n}(\mathbf{x})|\lambda'><\lambda'|\hat{n}(\mathbf{x}')|\lambda> \tag{7.23}$$

is the Fourier transform of the time-dependent linear response function for a steady state. The positive infinitesimal stems from the theta-function; i.e., causality causes the response function $\chi_\omega \equiv \chi_\omega^R$ to be an analytic function in the upper half plane.

We now assume, that the Hamiltonian, $\hat{H}_0$, describes the dynamics of a single particle, and in terms of its eigenstates we have for the linear response function

$$\chi(\mathbf{x}, \mathbf{x}'; t - t') = -\frac{i}{\hbar}\theta(t - t') \sum_{\lambda\lambda'}(\rho_\lambda - \rho_{\lambda'})e^{\frac{i}{\hbar}(\epsilon_\lambda - \epsilon_{\lambda'})(t - t')}\psi_\lambda^*(\mathbf{x})\psi_{\lambda'}(\mathbf{x})\psi_{\lambda'}^*(\mathbf{x}')\psi_\lambda(\mathbf{x}')$$
$$\tag{7.24}$$

where $\psi_\lambda(\mathbf{x}) = <\mathbf{x}|\lambda>$ is the energy eigenfunction corresponding to the energy eigenvalue ϵ_λ. For the Fourier transform we have

$$\chi(\mathbf{x}, \mathbf{x}'; \omega) = \sum_{\lambda\lambda'} \frac{\rho_\lambda - \rho_{\lambda'}}{\epsilon_\lambda - \epsilon_{\lambda'} + \hbar\omega + i0}\, \psi_\lambda^*(\mathbf{x})\psi_{\lambda'}(\mathbf{x})\psi_{\lambda'}^*(\mathbf{x}')\psi_\lambda(\mathbf{x}') . \tag{7.25}$$

Recalling eq.(2.155), we can express the Fourier transform of the density response function in terms of the spectral function

$$\chi(\mathbf{x}, \mathbf{x}'; \omega) = \int_{-\infty}^\infty \frac{dE}{2\pi} \int_{-\infty}^\infty \frac{dE'}{2\pi} \frac{\rho_i(E') - \rho_i(E)}{E' - E + \hbar\omega + i0}\, A(\mathbf{x}, \mathbf{x}'; E)\, A(\mathbf{x}', \mathbf{x}, E') . \tag{7.26}$$

Introducing the propagators instead of the spectral functions

$$A(\mathbf{x}, \mathbf{x}'; E) = i[G^R(\mathbf{x}, \mathbf{x}', E) - G^A(\mathbf{x}, \mathbf{x}', E)] \tag{7.27}$$

we have expressed the response function in terms of the particle propagators, quantities we know how to handle well.[1]

---

[1]If the particle has coupling to other degrees of freedom the propagators are still operators with respect to these.

7.2.1 Scalar Potential Coupling

In order to familiarize ourselves with the diagram technique, we derive the above result for the density response function using the diagrammatic technique for the density matrix for a single particle. Inserting complete sets of position states, we obtain for the density matrix

$$
\begin{aligned}
\rho(\mathbf{x}, \mathbf{x}', t) &= \ <\mathbf{x}|\hat{U}(t, t_i)\,\hat{\rho}_i\,\hat{U}^\dagger(t, t_i)|\mathbf{x}'> \\
&= \int d\tilde{\mathbf{x}} \int d\tilde{\mathbf{x}}' \ <\mathbf{x}|\hat{U}(t, t_i)|\tilde{\mathbf{x}}><\tilde{\mathbf{x}}|\hat{\rho}_i|\tilde{\mathbf{x}}'><\tilde{\mathbf{x}}'|\hat{U}^\dagger(t, t_i)|\mathbf{x}'> \\
&= \int d\tilde{\mathbf{x}} \int d\tilde{\mathbf{x}}' \ <\mathbf{x}, t|T e^{-\frac{i}{\hbar}\int_{t'}^{t} d\bar{t}\,\hat{H}_V(\bar{t})}|\tilde{\mathbf{x}}, t_i><\tilde{\mathbf{x}}', t_i|\tilde{T} e^{\frac{i}{\hbar}\int_{t'}^{t} d\bar{t}\,\hat{H}_V(\bar{t})}|\mathbf{x}', t> \\
&\quad \rho_i(\tilde{\mathbf{x}}, \tilde{\mathbf{x}}') \ .
\end{aligned}
\tag{7.28}
$$

For the linear correction to the density matrix we have the expression given by the two diagrams (the second and third diagrams of eq.(2.246))

$$
\equiv \ \rho^u(\mathbf{x}, \mathbf{x}', t) + \rho^l(\mathbf{x}, \mathbf{x}', t) \ .
\tag{7.29}
$$

We note the hermitian relationship between the two terms:

$$
\rho^l(\mathbf{x}, \mathbf{x}', t) = [\rho^u(\mathbf{x}', \mathbf{x}, t)]^* \ .
\tag{7.30}
$$

The potential vertex describes the coupling of the particle to the applied potential at the space-time point in question according to the standard Feynman rule:

$$
\overset{\times}{\underset{\mathbf{x}t}{}} \ \equiv \ \frac{1}{\hbar}\,V(\mathbf{x}, t)
\tag{7.31}
$$

and for the first diagram we have the expression (the superscript indicates that the perturbation acts on the upper, retarded propagator, line)

$$
\begin{aligned}
\rho^u(\mathbf{x}, \mathbf{x}', t) &= \ \frac{1}{\hbar}\int d\tilde{\mathbf{x}}\int d\tilde{\mathbf{x}}'\int d\mathbf{x}_1 \int_{t_i}^{t} dt_1 \ G^R(\mathbf{x}, t; \mathbf{x}_1, t_1)\,V(\mathbf{x}_1, t_1) \\
&\quad G^R(\mathbf{x}_1, t_1; \tilde{\mathbf{x}}, t_i)\,\rho_i(\tilde{\mathbf{x}}, \tilde{\mathbf{x}}')\,G^A(\tilde{\mathbf{x}}', t_i; \mathbf{x}', t) \ .
\end{aligned}
\tag{7.32}
$$

In linear response the initial density matrix is evolved in time according to the Hamiltonian in the absence of the external field, $\hat{H}_0$, up until the time of interaction with the external field, and we obtain the set of diagrams (where the $\pm i$ factors stem from cutting the retarded or advanced propagator into two in accordance with eq.(4.20) on page 171)

$$\rho^{(1)}(\mathbf{x}, \mathbf{x}', t) \quad = \quad -i \quad \begin{array}{c} \text{R} \\ \text{A} \end{array} t_1 \quad + \quad i \quad \begin{array}{c} \text{R} \\ \text{A} \end{array} t_1 \ . \tag{7.33}$$

or by application of the Feynman rules to the diagrams the expression

$$\rho^{(1)}(\mathbf{x}, \mathbf{x}', t) \quad = \quad \frac{-i}{\hbar} \int d\mathbf{x}_1 \int d\mathbf{x}_1' \int_{t_i}^{t} dt_1 \ G^R(\mathbf{x}, t; \mathbf{x}_1, t_1) \, G^A(\mathbf{x}'_1, t_1; \mathbf{x}', t)$$

$$[V(\mathbf{x}_1, t_1) - V(\mathbf{x}_1', t_1)] \, \rho_0(\mathbf{x}_1, \mathbf{x}'_1, t_1) \ . \tag{7.34}$$

Since we initially assume a stationary state with respect to the unperturbed Hamiltonian $\hat{H}_0$, eq.(7.16), the unperturbed density matrix stays time independent

$$\rho_0(\mathbf{x}, \mathbf{x}', t) \ = \ \rho_i(\mathbf{x}, \mathbf{x}') \ = \ <\mathbf{x}|\hat{\rho}_i|\mathbf{x}'> \ = \ \sum_\lambda \rho_\lambda \, \psi_\lambda(\mathbf{x}) \, \psi_\lambda^*(\mathbf{x}') \ . \tag{7.35}$$

The final ingredient we need in order to extract the temporal Fourier component at frequency ω of the density matrix is to make use of the spectral representations, eq.(2.154), of the retarded and advanced propagators in terms of the complete set of eigenfunctions of the Hamiltonian $\hat{H}_0$. In the ρ^u-term the integration over the spatial variable $\mathbf{x}_1'$ can then be done using the completeness relation for the eigenfunctions

$$\int d\mathbf{x} \, \psi_\lambda^*(\mathbf{x}) \, \psi_{\lambda'}(\mathbf{x}) \ = \ \delta_{\lambda\lambda'} \tag{7.36}$$

and we obtain

$$\rho^u(\mathbf{x}, \mathbf{x}', t) \quad = \quad -\frac{i}{\hbar} \int d\mathbf{x}_1 \int_{-\infty}^{\infty} \frac{dE_1}{2\pi} \int_{-\infty}^{\infty} \frac{dE_2}{2\pi} \sum_{\lambda_1 \lambda_2} \int_{t_i}^{t} dt_1 \ e^{-\frac{i}{\hbar} E_1(t-t_1) - \frac{i}{\hbar} E_2(t_1 - t)}$$

$$\rho_i(\epsilon_{\lambda_2}) \frac{\psi_{\lambda_1}^*(\mathbf{x}_1)\psi_{\lambda_1}(\mathbf{x})\psi_{\lambda_2}(\mathbf{x}_1)\psi_{\lambda_2}^*(\mathbf{x}')}{(E_1 - \epsilon_{\lambda_1} + i0)(E_2 - \epsilon_{\lambda_2} - i0)} \, V(\mathbf{x}_1, t_1) \ . \tag{7.37}$$

In linear response, each Fourier component contributes additively, so without loss of generality we can concentrate on a specific frequency ω of the potential

$$V(\mathbf{x}, t) \ = \ V_\omega(\mathbf{x}) \, e^{-i\omega t} \ + \ V_\omega^*(\mathbf{x}) \, e^{i\omega t} \ . \tag{7.38}$$

Since we are interested in the steady state, we can let the initial time at which the external field is switched on be in the remote past; i.e., we let t_i approach minus infinity, whereupon we can perform the time and energy variable integrations, and we obtain that we only get a response at the driving frequency, ω,

$$\rho^u(\mathbf{x}, \mathbf{x}', t) = \rho^u_{\mathbf{x}\mathbf{x}'\omega}[V_\omega] \, e^{-i\omega t} + \rho^u_{\mathbf{x}\mathbf{x}'-\omega}[V^*_\omega] \, e^{i\omega t} \qquad (7.39)$$

where

$$\rho^u_{\mathbf{x}\mathbf{x}'\omega}[V_\omega] = -\sum_{\lambda_1 \lambda_2} \frac{\rho_i(\epsilon_{\lambda_2})}{\epsilon_{\lambda_1} - \epsilon_{\lambda_2} - \hbar\omega - i0} \, \psi_{\lambda_1}(\mathbf{x}) \psi^*_{\lambda_2}(\mathbf{x}')$$

$$\int d\mathbf{x}_1 \, \psi^*_{\lambda_1}(\mathbf{x}_1) \, V_\omega(\mathbf{x}_1) \, \psi_{\lambda_2}(\mathbf{x}_1)$$

$$= \int d\mathbf{x}_1 \int_{-\infty}^{\infty} \frac{dE_1}{2\pi} \int_{-\infty}^{\infty} \frac{dE_2}{2\pi} \frac{\rho_i(E_2)}{E_1 - E_2 - \hbar\omega - i0}$$

$$[G^R(\mathbf{x}, \mathbf{x}_1; E_1) - G^A(\mathbf{x}, \mathbf{x}_1; E_1)] \, V_\omega(\mathbf{x}_1)$$

$$[G^R(\mathbf{x}_1, \mathbf{x}', E_2) - G^A(\mathbf{x}_1, \mathbf{x}', E_2)] \, . \qquad (7.40)$$

Using the hermitian relationship, eq.(7.30), we find for the linear correction to the density matrix

$$\rho^{(1)}(\mathbf{x}, \mathbf{x}', t) = \rho^{(1)}_{\mathbf{x}\mathbf{x}'\omega}[V_\omega] \, e^{-i\omega t} + \rho^{(1)}_{\mathbf{x}\mathbf{x}'-\omega}[V^*_\omega] \, e^{i\omega t} \qquad (7.41)$$

where the coefficients are the functionals of the applied potential

$$\rho^{(1)}_{\mathbf{x}\mathbf{x}'\omega}[V_\omega] = \int d\mathbf{x}_1 \int_{-\infty}^{\infty} \frac{dE_1}{2\pi} \int_{-\infty}^{\infty} \frac{dE_2}{2\pi} \frac{\rho_i(E_2) - \rho_i(E_1)}{E_1 - E_2 - \hbar\omega - i0}$$

$$[G^R(\mathbf{x}, \mathbf{x}_1; E_1) - G^A(\mathbf{x}, \mathbf{x}_1; E_1)] \, V_\omega(\mathbf{x}_1)$$

$$[G^R(\mathbf{x}_1, \mathbf{x}', E_2) - G^A(\mathbf{x}_1, \mathbf{x}', E_2)] \, . \qquad (7.42)$$

We then consider the diagonal elements of the density matrix and obtain the formula for the density response

$$\delta n(\mathbf{x}, t) = \delta n(\mathbf{x}, \omega) \, e^{-i\omega t} + \delta n^*(\mathbf{x}, \omega) \, e^{i\omega t} \qquad (7.43)$$

where

$$\delta n(\mathbf{x}, \omega) = \int d\mathbf{x}' \, \chi(\mathbf{x}, \mathbf{x}', \omega) \, V_\omega(\mathbf{x}') \qquad (7.44)$$

and we have for the Fourier transform of the density response function the expression

$$\chi(\mathbf{x}, \mathbf{x}', \omega) = \int_{-\infty}^{\infty} \frac{dE_1}{2\pi} \int_{-\infty}^{\infty} \frac{dE_2}{2\pi} \frac{\rho_i(E_2) - \rho_i(E_1)}{E_1 - E_2 - \hbar\omega - i0}$$

$$[G^R(\mathbf{x}, \mathbf{x}'; E_1) - G^A(\mathbf{x}, \mathbf{x}'; E_1)]$$

$$[G^R(\mathbf{x}', \mathbf{x}; E_2) - G^A(\mathbf{x}', \mathbf{x}; E_2)] \qquad (7.45)$$

as also arrived at earlier, eq.(7.26).

7.2.2 Fermi Gas Response

We could continue to discuss the linear response of a single particle as in the previous section. This is appropriate, for example, when describing nondegenerate electron dynamics in semiconductors. In that case we have for the thermal state

$$\rho_\lambda = Z_T^{-1} e^{-\epsilon_\lambda/kT} \tag{7.46}$$

where the normalization factor is the partition function for the particle

$$Z_T = \sum_\lambda e^{-\epsilon_\lambda/kT} . \tag{7.47}$$

However, since we wish to consider the dynamics of, say, the conduction electrons in a metal, we must bear in mind the quantum statistics of the particles. According to the analysis in section 5.3, the linear response function for a Fermi gas is given by the expression eq.(7.45), except for the occupation probabilities now being in accordance with Pauli's exclusion principle. For the case of the thermal equilibrium state at temperature T we have that the mean occupation number for a level is specified by the Fermi function

$$\rho_\lambda = f_0(\epsilon_\lambda) = \frac{1}{\exp\{(\epsilon_\lambda - \mu)/kT\} + 1} . \tag{7.48}$$

In that case we have for the relationship between the one-particle density matrix and the combination of the Fermi function and the spectral weight[2]

$$\int_{-\infty}^{\infty} \frac{dE}{2\pi} f_0(E) A(\mathbf{x}, \mathbf{x}'; E) = \rho_0(\mathbf{x}, \mathbf{x}') . \tag{7.49}$$

We can also straightforwardly handle the spin degree of freedom of the electron. However, we shall presently not allow the spin to play any dynamic role, and the spin of the electron therefore simply results in multiplying sums over quantum numbers, say momentum, by the spin degeneracy factor of 2.[3] Inserting into eq.(7.45) the energy distribution function for the electrons, we therefore have for the density response function for an electron gas (the factor of 2 accounts for the spin of the electron)

$$\chi(\mathbf{x}, \mathbf{x}', \omega) = \frac{2}{(2\pi)^2} \int_{-\infty}^{\infty} dE_1 \int_{-\infty}^{\infty} dE_2 \frac{f_0(E_1) - f_0(E_2)}{E_1 - E_2 - \hbar\omega - i0} A(\mathbf{x}, \mathbf{x}'; E_1) A(\mathbf{x}', \mathbf{x}; E_2). \tag{7.50}$$

In the expression for the response function we can in each term perform one of the energy integrations, and exploiting the analytical properties of the propagators half of the terms are seen not to contribute, and we obtain for the density response function for an electron gas

$$\chi(\mathbf{x}, \mathbf{x}', \omega) = \chi^{RA}(\mathbf{x}, \mathbf{x}', \omega) + \chi^{RR}(\mathbf{x}, \mathbf{x}', \omega) + \chi^{AA}(\mathbf{x}, \mathbf{x}', \omega) \tag{7.51}$$

---

[2]We recall the interpretation of this equation, see page 131.

[3]In section 11.6, electron scattering off magnetic impurities and the effect of spin-orbit scattering in disordered systems are considered.

where

$$\chi^{RA}(\mathbf{x}, \mathbf{x}', \omega) = -\frac{i}{\pi} \int_{-\infty}^{\infty} dE \; f_0(E) \, G^R(\mathbf{x}, \mathbf{x}'; E + \hbar\omega) \, G^A(\mathbf{x}', \mathbf{x}; E)$$

$$+ \frac{i}{\pi} \int_{-\infty}^{\infty} dE \; f_0(E) \, G^R(\mathbf{x}, \mathbf{x}'; E) \, G^A(\mathbf{x}', \mathbf{x}; E - \hbar\omega) \quad (7.52)$$

and

$$\chi^{RR}(\mathbf{x}, \mathbf{x}', \omega) = \frac{i}{\pi} \int_{-\infty}^{\infty} dE \; f_0(E) \, G^R(\mathbf{x}, \mathbf{x}'; E + \hbar\omega) \, G^R(\mathbf{x}', \mathbf{x}; E) \quad (7.53)$$

and

$$\chi^{AA}(\mathbf{x}, \mathbf{x}', \omega) = -\frac{i}{\pi} \int_{-\infty}^{\infty} dE \; f_0(E) \, G^A(\mathbf{x}, \mathbf{x}'; E) \, G^A(\mathbf{x}', \mathbf{x}; E - \hbar\omega) . \quad (7.54)$$

All three functions are seen individually to be analytic in the upper half plane, and we note the relationship

$$\chi^{AA}(\mathbf{x}, \mathbf{x}', -\omega^*) = [\chi^{RR}(\mathbf{x}, \mathbf{x}', \omega)]^* . \quad (7.55)$$

7.2.3 Momentum Representation Response

The position and momentum representations are related by Fourier transformation. For example, for the spectral function we have

$$A(\mathbf{x}, \mathbf{x}', E) = \sum_{\mathbf{p}\mathbf{p}'} <\mathbf{x}|\mathbf{p}> A(\mathbf{p}, \mathbf{p}', E) <\mathbf{p}'|\mathbf{x}'>$$

$$= \frac{1}{V} \sum_{\mathbf{p}\mathbf{p}'} e^{\frac{i}{\hbar}\mathbf{p}\cdot\mathbf{x} - \frac{i}{\hbar}\mathbf{p}'\cdot\mathbf{x}'} A(\mathbf{p}, \mathbf{p}', E) . \quad (7.56)$$

In linear response the wave vector components contribute additively, and we need only to consider the response to, say, wave vector $\mathbf{q}$

$$V_\omega(\mathbf{x}) = V(\mathbf{q}, \omega) \, e^{i\mathbf{q}\cdot\mathbf{x}} . \quad (7.57)$$

The spatial integration in the expression for the density response, eq.(7.22), can then be performed, and we obtain for the density response at wave vector $\mathbf{q}'$ in response to an external field with wave vector $\mathbf{q}$ the expression

$$\delta n(\mathbf{q}', \omega) \equiv \frac{1}{V} \int d\mathbf{x} \; e^{-i\mathbf{x}\cdot\mathbf{q}'} \, \delta n(\mathbf{x}, \omega) = \chi(\mathbf{q}', \mathbf{q}, \omega) \, V(\mathbf{q}, \omega) \quad (7.58)$$

where the response function

$$\chi(\mathbf{q}', \mathbf{q}, \omega) = \int d\mathbf{x} \int d\mathbf{x}' \; <\hbar\mathbf{q}'|\mathbf{x}> \chi(\mathbf{x}, \mathbf{x}', \omega) <\mathbf{x}'|\hbar\mathbf{q}> \quad (7.59)$$

is given by

$$\chi(\mathbf{q}', \mathbf{q}, \omega) = \chi^{RA}(\mathbf{q}', \mathbf{q}, \omega) + \chi^{RR}(\mathbf{q}', \mathbf{q}, \omega) + \chi^{AA}(\mathbf{q}', \mathbf{q}, \omega) . \tag{7.60}$$

Here

$$\chi^{RA}(\mathbf{q}', \mathbf{q}, \omega) = -\frac{i}{\pi} \frac{1}{V} \sum_{\mathbf{p}\mathbf{p}'} \int_{-\infty}^{\infty} dE \, f_0(E) \, G^R(\mathbf{p}_+, \mathbf{p}'_+; E + \hbar\omega) G^A(\mathbf{p}'_-, \mathbf{p}_-; E)$$

$$+ \frac{i}{\pi} \frac{1}{V} \sum_{\mathbf{p}\mathbf{p}'} \int_{-\infty}^{\infty} dE \, f_0(E) \, G^R(\mathbf{p}_+, \mathbf{p}'_+; E) G^A(\mathbf{p}'_-, \mathbf{p}_-; E - \hbar\omega) \tag{7.61}$$

and

$$\chi^{RR}(\mathbf{q}', \mathbf{q}, \omega) = \frac{i}{\pi} \frac{1}{V} \sum_{\mathbf{p}\mathbf{p}'} \int_{-\infty}^{\infty} dE \, f_0(E) \, G^R(\mathbf{p}_+, \mathbf{p}'_+; E + \hbar\omega) G^R(\mathbf{p}'_-, \mathbf{p}_-; E) \tag{7.62}$$

and

$$\chi^{AA}(\mathbf{q}', \mathbf{q}, \omega) = -\frac{i}{\pi} \frac{1}{V} \sum_{\mathbf{p}\mathbf{p}'} \int_{-\infty}^{\infty} dE \, f_0(E) G^A(\mathbf{p}_+, \mathbf{p}'_+; E) G^A(\mathbf{p}'_-, \mathbf{p}_-; E - \hbar\omega) . \tag{7.63}$$

We have introduced the notation

$$\mathbf{p}_\pm = \mathbf{p} \pm \frac{\hbar\mathbf{q}'}{2} , \qquad \mathbf{p}'_\pm = \mathbf{p}' \pm \frac{\hbar\mathbf{q}}{2} . \tag{7.64}$$

To linear order we have for the Fourier transform of the density

$$n(\mathbf{q}', \omega) = n_0(\mathbf{q}') \, \delta(\omega) + \delta n(\mathbf{q}', \omega) \tag{7.65}$$

where

$$n_0(\mathbf{q}') = \frac{1}{V} \int d\mathbf{x} \, e^{-i\mathbf{x} \cdot \mathbf{q}'} \rho_0(\mathbf{x}, \mathbf{x}) \tag{7.66}$$

is the Fourier transform of the density in the absence of the applied potential. For a spatially homogeneous equilibrium density we have

$$n_0(\mathbf{q}) = n_0 \, \delta_{\mathbf{q}, 0} . \tag{7.67}$$

7.3 Current Response

In this section, we shall discuss the linear current response. We shall specifically discuss the electric current response to an applied electric field $\mathbf{E}$. In the following we shall represent the external electric field $\mathbf{E}$ by a time-dependent vector potential $\mathbf{A}$

$$\mathbf{E} = -\frac{\partial \mathbf{A}}{\partial t} \tag{7.68}$$

and not by a scalar potential as in the previous section. The two cases can be handled with an equal amount of labor and are equivalent by gauge invariance.

7.3.1 Particle Interacting with a Vector Potential

We wish to calculate the current response of a system to lowest order in an applied electric field. For a start we shall have in mind a single particle with charge e, say an electron, moving in an impurity potential in addition to the applied field. The Hamiltonian in question is therefore (recall exercise 1.4 on page 15)

$$\hat{H} = \frac{(\hat{\mathbf{p}}_t^{kin})^2}{2m} + V(\hat{\mathbf{x}}) = \frac{1}{2m}(\hat{\mathbf{p}}_{can} - e\mathbf{A}(\hat{\mathbf{x}}, t))^2 + V(\hat{\mathbf{x}}) = \hat{H}_0 + \hat{H}_{A(t)}, \quad (7.69)$$

where

$$\hat{H}_0 = \frac{\hat{\mathbf{p}}_{can}^2}{2m} + V(\hat{\mathbf{x}}) \quad (7.70)$$

is the Hamiltonian for the particle in the impurity potential, V, and we have the perturbation due to the applied field

$$\hat{H}_{A(t)} = -\frac{e}{2m}(\hat{\mathbf{p}}_{can} \cdot \mathbf{A}(\hat{\mathbf{x}}, t) + \mathbf{A}(\hat{\mathbf{x}}, t) \cdot \hat{\mathbf{p}}_{can}) + \frac{e^2}{2m}\mathbf{A}^2(\hat{\mathbf{x}}, t). \quad (7.71)$$

The charge current density operator is specified in terms of the kinematic momentum operator

$$\hat{\mathbf{p}}_t^{kin} = \hat{\mathbf{p}}_{can} - e\mathbf{A}(\hat{\mathbf{x}}, t). \quad (7.72)$$

and the density operator (recall exercise 1.4 on page 15 and exercise 1.21 on page 71)

$$\hat{\mathbf{j}}_t(\mathbf{x}) = \frac{e}{2m}\{\hat{\mathbf{p}}_t^{kin}, \hat{n}(\mathbf{x})\} \quad (7.73)$$

and satisfies in the Heisenberg picture the (charge) continuity equation, eq.(1.369).

The current density operator has two distinct parts

$$\hat{\mathbf{j}}_t(\mathbf{x}) = \hat{\mathbf{j}}_p(\mathbf{x}) + \hat{\mathbf{j}}_t^d(\mathbf{x}) \quad (7.74)$$

consisting of the so-called paramagnetic current density operator

$$\hat{\mathbf{j}}_p(\mathbf{x}) = \frac{e}{2m}\{\hat{\mathbf{p}}_{can}, \hat{n}(\mathbf{x})\} \quad (7.75)$$

and in the present case a time-dependent so-called diamagnetic current density operator

$$\hat{\mathbf{j}}_t^d(\mathbf{x}) = -\frac{e^2}{2m}\{\mathbf{A}(\hat{\mathbf{x}}, t), \hat{n}(\mathbf{x})\} = -\frac{e^2}{m}\hat{n}(\mathbf{x})\mathbf{A}(\hat{\mathbf{x}}, t). \quad (7.76)$$

The last equality sign follows from the fact that the two operators commute.

The interaction between the particle and the vector potential can be written in terms of the current density operator and the density operator

$$\hat{H}_{A(t)} = -\int d\mathbf{x}\,\hat{\mathbf{j}}_t(\mathbf{x}) \cdot \mathbf{A}(\mathbf{x}, t) - \frac{e^2}{2m}\int d\mathbf{x}\,\hat{n}(\mathbf{x})\mathbf{A}^2(\mathbf{x}, t)$$

$$= -\int d\mathbf{x}\,\hat{\mathbf{j}}_p(\mathbf{x}) \cdot \mathbf{A}(\mathbf{x}, t) + \frac{e^2}{2m}\int d\mathbf{x}\,\hat{n}(\mathbf{x})\mathbf{A}^2(\mathbf{x}, t). \quad (7.77)$$

We note that the discussion is quite general once the current and density operators represent all the particles of interest. For a system of N particles we simply have to add the contribution from each particle, and the density operator $\hat{n}(\mathbf{x})$ appearing in the above formulas is the total density of particles, eq.(7.11).

In the presence of a vector potential, the average current density at time t, in the state specified by the statistical operator $\hat{\rho}(t)$, is given by

$$\mathbf{j}(\mathbf{x}, t) = Tr(\hat{\rho}(t)\,\hat{\mathbf{j}}_t(\mathbf{x}))$$

$$= \frac{e\hbar}{2im}\left(\frac{\partial}{\partial \mathbf{x}} - \frac{\partial}{\partial \mathbf{x}'}\right)\rho(\mathbf{x}, \mathbf{x}', t)\Bigg|_{\mathbf{x}'=\mathbf{x}} - \frac{e^2}{m}\mathbf{A}(\mathbf{x}, t)\,\rho(\mathbf{x}, \mathbf{x}, t) \quad (7.78)$$

where in the last line we have introduced the one-particle density matrix.

Since the second, diamagnetic, term in the expression for the current density, eq.(7.78), is explicitly linear in the external field, we can to linear order replace the density matrix $\rho(\mathbf{x}, \mathbf{x}', t)$ by $\rho_0(\mathbf{x}, \mathbf{x}', t)$, the density matrix evolved by the unperturbed Hamiltonian $\hat{H}_0$.

7.3.2 Current Correlation Function

To calculate the current density to linear order in the external electric field, we write the evolution operator in the interaction picture with respect to the unperturbed Hamiltonian $\hat{H}_0$

$$\hat{U}(t, t') = \hat{U}_0(t, t_r)\, T \exp\left\{-\frac{i}{\hbar}\int_{t'}^{t}d\bar{t}\,\hat{H}_A(\bar{t})\right\}\hat{U}_0^{\dagger}(t', t_r) \quad (7.79)$$

where $\hat{U}_0(t, t')$ is the time-evolution operator in the absence of the applied field. For the perturbation in the interaction picture we have

$$\hat{H}_A(t) = \hat{U}_0^{\dagger}(t, t_r)\hat{H}_{A(t)}\,\hat{U}_0(t, t_r) = \frac{e^2}{2m}\mathbf{A}^2(\hat{\mathbf{x}}(t), t)$$

$$- \frac{e}{2m}\left(\hat{\mathbf{p}}_{can}(t)\cdot\mathbf{A}(\hat{\mathbf{x}}(t), t) + \mathbf{A}(\hat{\mathbf{x}}(t), t)\cdot\hat{\mathbf{p}}_{can}(t)\right)$$

$$\equiv \hat{H}_A^{(1)}(t) + \frac{e^2}{2m}\mathbf{A}^2(\hat{\mathbf{x}}(t), t)\,. \quad (7.80)$$

Here $\hat{H}_A^{(1)}(t)$ denotes the part of $\hat{H}_A(t)$ which is linear in the field.

Expanding the time-evolution operator to linear order in the applied electric field we get

$$\hat{U}(t, t') = \hat{U}_0(t, t') - \hat{U}_0(t, t_r)\frac{i}{\hbar}\int_{t'}^{t}d\bar{t}\,\hat{H}_A^{(1)}(\bar{t})\,\hat{U}_0^{\dagger}(t', t_r) + \mathcal{O}(E^2) \quad (7.81)$$

and for the statistical operator we then obtain

$$\hat{\rho}(t) = \hat{\rho}_0(t) + \hat{\rho}^{(1)}(t) + \mathcal{O}(\mathbf{E}^2) \quad (7.82)$$

where the linear correction is given by

$$\hat{\rho}^{(1)}(t) = \frac{i}{\hbar}\hat{U}_0(t, t_i)\,\hat{\rho}_i\,\hat{U}_0(t_i, t_r)\int_{t_i}^{t}\!\!d\bar{t}\,\hat{H}_A^{(1)}(\bar{t})\,\hat{U}_0^\dagger(t, t_r)$$

$$- \frac{i}{\hbar}\hat{U}_0(t, t_r)\int_{t_i}^{t}\!\!d\bar{t}\,\hat{H}_A^{(1)}(\bar{t})\,\hat{U}_0^\dagger(t_i, t_r)\,\hat{\rho}_i\,\hat{U}_0^\dagger(t, t_i)\,. \tag{7.83}$$

For the current density we therefore obtain to linear order

$$\mathbf{j}(\mathbf{x}, t) = Tr(\hat{\rho}_0(t)\hat{\mathbf{j}}_t(\mathbf{x}))$$

$$- \frac{i}{\hbar}\int_{t_i}^{t}\!\!d\bar{t}\,Tr(\hat{\rho}_0(t_r)[\hat{\mathbf{j}}_p(\mathbf{x}, t), \hat{H}_A^{(1)}(\bar{t})]) + \mathcal{O}(E^2) \tag{7.84}$$

where $\hat{\mathbf{j}}_p(\mathbf{x}, t)$ is just the paramagnetic part of the current density operator in the interaction representation

$$\hat{\mathbf{j}}_p(\mathbf{x}, t) = \hat{U}_0^\dagger(t, t_r)\,\hat{\mathbf{j}}_p(\mathbf{x})\,\hat{U}_0(t, t_r)\,. \tag{7.85}$$

To linear order in the external electric field we therefore have that the current density

$$j_\alpha(\mathbf{x}, t) = Tr(\hat{\rho}_0(t)\hat{j}_\alpha^p(\mathbf{x})) + \sum_\beta \int\!d\mathbf{x}'\!\!\int_{t_i}^{\infty}\!\!dt'\,Q_{\alpha\beta}(\mathbf{x}, t; \mathbf{x}', t')A_\beta(\mathbf{x}', t') \tag{7.86}$$

is determined by the current response function

$$Q_{\alpha\beta}(\mathbf{x}, t; \mathbf{x}', t') = K_{\alpha\beta}(\mathbf{x}, t; \mathbf{x}', t')$$

$$- \frac{e^2\rho_0(\mathbf{x}, \mathbf{x}, t)}{m}\,\delta_{\alpha\beta}\,\delta(\mathbf{x} - \mathbf{x}')\,\delta(t - t') \tag{7.87}$$

where we have introduced the current-current response function

$$K_{\alpha\beta}(\mathbf{x}, t; \mathbf{x}', t') = \frac{i}{\hbar}\theta(t - t')\,Tr(\hat{\rho}_0(t_r)[\hat{j}_\alpha^p(\mathbf{x}, t), \hat{j}_\beta^p(\mathbf{x}', t')])$$

$$\equiv \frac{i}{\hbar}\theta(t - t') < [\hat{j}_\alpha^p(\mathbf{x}, t), \hat{j}_\beta^p(\mathbf{x}', t')] >_0 \tag{7.88}$$

and $Tr(\hat{\rho}_0(t)\hat{j}_\alpha^p(\mathbf{x}))$ is a possible current density in the absence of the field. We shall not consider superconductivity or magnetism, and can therefore in the following assume that this term vanishes.

Assuming that we have a stationary state with respect to the unperturbed Hamiltonian before the external field is applied, the response function only depends on the relative time

$$K_{\alpha\beta}(\mathbf{x}, t; \mathbf{x}', t') = \frac{i}{\hbar}\theta(t - t')\sum_{\lambda\lambda'}(\rho_\lambda - \rho_{\lambda'}) < \lambda|\hat{j}_\alpha^p(\mathbf{x})|\lambda' > < \lambda'|\hat{j}_\beta^p(\mathbf{x}')|\lambda > e^{\frac{i}{\hbar}(\epsilon_\lambda - \epsilon_{\lambda'})(t - t')}. \tag{7.89}$$

In linear response each frequency contributes additively so we just need to seek the response at one driving frequency, say ω,

$$\mathbf{A}(\mathbf{x}, t) = \mathbf{A}(\mathbf{x}, \omega) e^{-i\omega t} . \tag{7.90}$$

The time integration in eq.(7.86) can then be performed by letting the arbitrary initial time, t_i, be taken in the remote past (letting t_i approach minus infinity), and we only get a current response at the driving frequency

$$j_\alpha(\mathbf{x}, t) = j_\alpha(\mathbf{x}, \omega) e^{-i\omega t} . \tag{7.91}$$

For the Fourier transform of the current density we then have

$$j_\alpha(\mathbf{x}, \omega) = \sum_\beta \int d\mathbf{x}' \, Q_{\alpha\beta}(\mathbf{x}, \mathbf{x}'; \omega) A_\beta(\mathbf{x}', \omega) + \mathcal{O}(\mathbf{E}^2) \tag{7.92}$$

where

$$Q_{\alpha\beta}(\mathbf{x}, \mathbf{x}'; \omega) = K_{\alpha\beta}(\mathbf{x}, \mathbf{x}'; \omega) - \frac{\rho_0(\mathbf{x}, \mathbf{x}) e^2}{m} \delta_{\alpha\beta} \, \delta(\mathbf{x} - \mathbf{x}') \tag{7.93}$$

and

$$K_{\alpha\beta}(\mathbf{x}, \mathbf{x}'; \omega) = \sum_{\lambda\lambda'} \frac{\rho_{\lambda'} - \rho_\lambda}{\epsilon_\lambda - \epsilon_{\lambda'} + \hbar\omega + i0} <\lambda|\hat{j}_\alpha^p(\mathbf{x})|\lambda'><\lambda'|\hat{j}_\beta^p(\mathbf{x}')|\lambda> . \tag{7.94}$$

The paramagnetic current density matrix element is given by

$$<\lambda|\hat{\mathbf{j}}^p(\mathbf{x})|\lambda'> = \frac{e\hbar}{2im}\psi_\lambda^*(\mathbf{x}) \left(\frac{\overrightarrow{\partial}}{\partial\mathbf{x}} - \frac{\overleftarrow{\partial}}{\partial\mathbf{x}} \right) \psi_{\lambda'}(\mathbf{x}) \tag{7.95}$$

where the arrows indicate whether we differentiate to the left or the right.
Introducing the spectral weight, we can rewrite the response function

$$K_{\alpha\beta}(\mathbf{x}, \mathbf{x}'; \omega) = \left(\frac{e\hbar}{m} \right)^2 \int_{-\infty}^\infty \frac{dE}{2\pi} \int_{-\infty}^\infty \frac{dE'}{2\pi} \frac{\rho_i(E') - \rho_i(E)}{E' - E + \hbar\omega + i0}$$

$$[G^R(\mathbf{x}, \mathbf{x}'; E) - G^A(\mathbf{x}, \mathbf{x}'; E)]$$

$$\overset{\leftrightarrow}{\nabla}_{x_\alpha} \overset{\leftrightarrow}{\nabla}_{x'_\beta} [G^R(\mathbf{x}', \mathbf{x}, E') - G^A(\mathbf{x}', \mathbf{x}, E')] . \tag{7.96}$$

We have introduced the abbreviated notation

$$\overset{\leftrightarrow}{\nabla}_{\mathbf{x}} = \frac{1}{2} \left(\frac{\overrightarrow{\partial}}{\partial\mathbf{x}} - \frac{\overleftarrow{\partial}}{\partial\mathbf{x}} \right) \tag{7.97}$$

for the differential operator associated with the current vertex in the position representation. We have now achieved the goal of expressing the current response function in terms of propagators, the quantities that we know how to handle well.

In the expression for the current response kernel we can, as previously for the density response kernel, perform one of the energy integrations, and exploiting the analytical properties of the propagators half of the terms are seen not to contribute, and we obtain for the current response function

$$
K_{\alpha\beta}(\mathbf{x}, \mathbf{x}', \omega) = -2 \left(\frac{e\hbar}{m}\right)^2 \int_{-\infty}^{\infty} \frac{dE}{2\pi} \, \rho_i(E) \left(A(\mathbf{x}, \mathbf{x}'; E) \, \overleftrightarrow{\nabla}_{x_\alpha} \overleftrightarrow{\nabla}_{x'_\beta} \, G^A(\mathbf{x}', \mathbf{x}; E - \hbar\omega) \right.
$$

$$
\left. + \; G^R(\mathbf{x}, \mathbf{x}'; E + \hbar\omega) \, \overleftrightarrow{\nabla}_{x_\alpha} \overleftrightarrow{\nabla}_{x'_\beta} \, A(\mathbf{x}', \mathbf{x}; E) \right) . \tag{7.98}
$$

For the current response function for an electron gas we obtain (the factor of 2 accounts for spin)

$$
K_{\alpha\beta}(\mathbf{x}, \mathbf{x}', \omega) = -2 \left(\frac{e\hbar}{m}\right)^2 \int_{-\infty}^{\infty} \frac{dE}{2\pi} \, f_0(E) \left(A(\mathbf{x}, \mathbf{x}'; E) \, \overleftrightarrow{\nabla}_{x_\alpha} \overleftrightarrow{\nabla}_{x'_\beta} \, G^A(\mathbf{x}', \mathbf{x}; E - \hbar\omega) \right.
$$

$$
\left. + \; G^R(\mathbf{x}, \mathbf{x}'; E + \hbar\omega) \, \overleftrightarrow{\nabla}_{x_\alpha} \overleftrightarrow{\nabla}_{x'_\beta} \, A(\mathbf{x}', \mathbf{x}; E) \right) . \tag{7.99}
$$

7.3.3 Gauge Invariance

Gauge invariance implies a useful expression for the longitudinal part of the current response function, i.e., the current response to a longitudinal electric field, $\nabla \times \mathbf{E} = 0$.

The static paramagnetic current density response

$$
\mathbf{j}^p_\Lambda(\mathbf{x}, \omega = 0) = \frac{e\hbar}{2im} \left(\frac{\partial}{\partial \mathbf{x}} - \frac{\partial}{\partial \mathbf{x}'}\right) \rho_\Lambda(\mathbf{x}, \mathbf{x}', \omega = 0) \Big|_{\mathbf{x}'=\mathbf{x}} \tag{7.100}
$$

due to a pure gauge field

$$
\mathbf{A}(\mathbf{x}) = \nabla \Lambda(\mathbf{x}) \qquad \mathbf{E}(\mathbf{x}) = 0 \qquad \mathbf{B}(\mathbf{x}) = 0 \tag{7.101}
$$

is simply determined by the gauge transformation property of the density matrix

$$
\rho_\Lambda(\mathbf{x}, \mathbf{x}', \omega = 0) = e^{\frac{ie}{\hbar}\Lambda(\mathbf{x})} \rho_{\Lambda=0}(\mathbf{x}, \mathbf{x}', \omega = 0) e^{-\frac{ie}{\hbar}\Lambda(\mathbf{x}')} \tag{7.102}
$$

as gauge invariance implies the transformation properties

$$
\mathbf{A}(\mathbf{x}) \to \mathbf{A}(\mathbf{x}) + \nabla \Lambda(\mathbf{x}) \quad , \quad |\mathbf{x}\rangle \to e^{-\frac{ie}{\hbar}\Lambda(\mathbf{x})} |\mathbf{x}\rangle \; . \tag{7.103}
$$

The paramagnetic contribution to the static current density in the presence of the pure gauge field is therefore

$$
\mathbf{j}^p_\Lambda(\mathbf{x}, \omega = 0) = \frac{e\hbar}{2im} \left(\frac{\partial}{\partial \mathbf{x}} - \frac{\partial}{\partial \mathbf{x}'}\right) e^{\frac{ie}{\hbar}\Lambda(\mathbf{x})} \rho_{\Lambda=0}(\mathbf{x}, \mathbf{x}', \omega = 0) e^{-\frac{ie}{\hbar}\Lambda(\mathbf{x}')} \Big|_{\mathbf{x}'=\mathbf{x}} \tag{7.104}
$$

or equivalently

$$\mathbf{j}^p_\Lambda(\mathbf{x}, \omega = 0) \;=\; \mathbf{j}^p_{\Lambda=0}(\mathbf{x}, \omega = 0) \;+\; \frac{e^2 \rho_{\Lambda=0}(\mathbf{x}, \mathbf{x}, \omega = 0)}{m} \, \nabla \Lambda(\mathbf{x}) \;. \quad (7.105)$$

where $\mathbf{j}^p_{\Lambda=0}(\mathbf{x}, \omega = 0)$ is the paramagnetic current density in the system in the absence of the pure gauge field.[4]

From perturbation theory, on the other hand, we have to lowest order in Λ

$$j^{p(\Lambda)}_\alpha(\mathbf{x}, \omega=0) \;=\; j^{p(\Lambda=0)}_\alpha(\mathbf{x}, \omega=0) \;+\; \sum_\beta \int d\mathbf{x}' \, K_{\alpha\beta}(\mathbf{x}, \mathbf{x}'; \omega=0) \nabla_{x'_\beta} \Lambda(\mathbf{x}') . \; (7.106)$$

from which we obtain the relation

$$K_{\alpha\beta}(\mathbf{x}, \mathbf{x}'; \omega = 0) \;=\; \frac{e^2 \rho_{\Lambda=0}(\mathbf{x}, \mathbf{x}, \omega = 0)}{m} \, \delta_{\alpha\beta} \, \delta(\mathbf{x} - \mathbf{x}') \quad (7.107)$$

and the longitudinal part of the current response function can be written on the form

$$Q_{\alpha\beta}(\mathbf{x}, \mathbf{x}'; \omega) = K_{\alpha\beta}(\mathbf{x}, \mathbf{x}'; \omega) - K_{\alpha\beta}(\mathbf{x}, \mathbf{x}'; \omega=0) \;. \quad (7.108)$$

We can therefore express the longitudinal current density response solely in terms of the paramagnetic response function[5]

$$j_\alpha(\mathbf{x}, \omega) = \sum_\beta \int d\mathbf{x}' \, [K_{\alpha\beta}(\mathbf{x}, \mathbf{x}'; \omega) - K_{\alpha\beta}(\mathbf{x}, \mathbf{x}'; \omega=0)] A_\beta(\mathbf{x}', \omega)] \;. \quad (7.109)$$

7.3.4   Vector Potential Coupling

In order to familiarize ourselves with the Feynman rules for the coupling of a charged particle to a vector potential, we derive the above result for the current response function using the diagrammatic technique for the density matrix.

For the density matrix in the case of the perturbation eq.(7.71), we obtain by inserting complete sets

$$\rho(\mathbf{x}, \mathbf{x}', t) \;=\; <\mathbf{x}|\hat{U}(t, t_i) \, \hat{\rho}_i \, \hat{U}^\dagger(t, t_i)|\mathbf{x}'>$$

$$=\; \int d\tilde{\mathbf{x}} \int d\tilde{\mathbf{x}}' <\mathbf{x}|\hat{U}(t, t_i)|\tilde{\mathbf{x}}><\tilde{\mathbf{x}}|\hat{\rho}_i|\tilde{\mathbf{x}}'><\tilde{\mathbf{x}}'|\hat{U}^\dagger(t, t_i)|\mathbf{x}'>$$

$$=\; \int d\tilde{\mathbf{x}} \int d\tilde{\mathbf{x}}' <\mathbf{x}, t|Te^{-\frac{i}{\hbar}\int_{t'}^t d\bar{t}\, \hat{H}_A(\bar{t})}|\tilde{\mathbf{x}}, t_i><\tilde{\mathbf{x}}', t_i|\tilde{T}e^{\frac{i}{\hbar}\int_{t'}^t d\bar{t}\, \hat{H}_A(\bar{t})}|\mathbf{x}', t>$$

$$\rho_i(\tilde{\mathbf{x}}, \tilde{\mathbf{x}}') \quad (7.110)$$

---

[4]The total current response to a pure gauge field is of course zero. A pure gauge transformation corresponds to a phase transformation, and leaves, according to section 1.3.5, quantum mechanics invariant.

[5]In appendix E we relate this result to the equation of motion, and to the causal and dissipative character of linear response.

where time-labeled position eigenstates are in the interaction picture.

We obtain the density matrix to linear order in the external electric field by expanding the time-ordered exponential, and omit the quadratic term in $\hat{H}_A(t)$

$$
\begin{aligned}
\rho^{(1)}(\mathbf{x}, \mathbf{x}', t) &= -\frac{i}{\hbar} \int d\tilde{\mathbf{x}} \int d\tilde{\mathbf{x}}' \int_{t_i}^{t} d\bar{t} \, (<\mathbf{x}, t | \hat{H}_A^{(1)}(\bar{t}) | \tilde{\mathbf{x}}, t_i > < \tilde{\mathbf{x}}', t_i | \mathbf{x}', t> \\
&\quad - <\mathbf{x}, t | \tilde{\mathbf{x}}, t_i > < \tilde{\mathbf{x}}', t_i | \hat{H}_A^{(1)}(\bar{t}) | \mathbf{x}', t>) \rho_i(\tilde{\mathbf{x}}, \tilde{\mathbf{x}}') \\
&= \rho^u(\mathbf{x}, \mathbf{x}', t) + \rho^l(\mathbf{x}, \mathbf{x}', t) \, .
\end{aligned} \tag{7.111}
$$

The two terms are related according to eq.(7.30), reflecting the hermitian property of the statistical operator.

Inserting a complete set of states we obtain for the matrix element

$$
\begin{aligned}
<\mathbf{x}, t | \hat{H}_A^{(1)}(t) | \mathbf{x}', t> &= -<\mathbf{x}, t | \int d\tilde{\mathbf{x}} \, \hat{\mathbf{j}}_p(\tilde{\mathbf{x}}, t) \cdot \mathbf{A}(\tilde{\mathbf{x}}, t) | \mathbf{x}', t> \\
&= -\frac{e}{2m} <\mathbf{x}, t | \hat{\mathbf{p}}_{can}^{(t)} \cdot \mathbf{A}(\hat{\mathbf{x}}(t), t) + \mathbf{A}(\hat{\mathbf{x}}(t), t) \cdot \hat{\mathbf{p}}_{can}^{(t)} | \mathbf{x}', t> \\
&= \frac{ie\hbar}{2m} \{\mathbf{A}(\mathbf{x}, t) + \mathbf{A}(\mathbf{x}', t)\} \cdot \frac{\partial}{\partial \mathbf{x}} \delta(\mathbf{x} - \mathbf{x}') \, .
\end{aligned} \tag{7.112}
$$

The first term of eq.(7.111) contains the linear correction in the vector potential to the retarded propagator

$$
\begin{aligned}
G_1^R(\mathbf{x}, t; \tilde{\mathbf{x}}, t_i) &= \frac{e}{2im} \int d\bar{\mathbf{x}} \int_{-\infty}^{\infty} d\bar{t} \, G^R(\mathbf{x}, t; \bar{\mathbf{x}}, \bar{t}) \left\{ \frac{\overleftarrow{\partial}}{\partial \bar{\mathbf{x}}} \cdot \mathbf{A}(\bar{\mathbf{x}}, \bar{t}) \right. \\
&\quad \left. - \mathbf{A}(\bar{\mathbf{x}}, \bar{t}) \cdot \frac{\overrightarrow{\partial}}{\partial \bar{\mathbf{x}}} \right\} G^R(\bar{\mathbf{x}}, \bar{t}; \tilde{\mathbf{x}}, t_i)
\end{aligned} \tag{7.113}
$$

where we by G^R denote the retarded propagator in the absence of the vector potential.

We introduce the current vertex describing the coupling of the charged particle to the vector potential at the space-time point in question according to the Feynman rule

$$
\begin{array}{c} \text{R} \genfrac{}{}{0pt}{}{\gtrless}{} \text{R} \\ \overrightarrow{} \underset{\mathbf{x}t}{} \overleftarrow{} \end{array} = \frac{e}{2im} \left(\frac{\overleftarrow{\partial}}{\partial \mathbf{x}} \cdot \mathbf{A}(\mathbf{x}, t) - \mathbf{A}(\mathbf{x}, t) \cdot \frac{\overrightarrow{\partial}}{\partial \mathbf{x}} \right) \tag{7.114}
$$

where the upper arrows is meant to designate whether the differential operator operates on the propagator to the left or right, i.e., on the outgoing or incoming particle line.

Similarly one has the perturbative expansion of the advanced propagator, which is expressed in terms of the vertex

$$\overset{\mathbf{xt}}{\underset{\mathbf{A}\;\S\;\mathbf{A}}{\longrightarrow}} \quad = \quad \frac{e}{2im}\left(\mathbf{A}(\mathbf{x},t)\cdot\frac{\overset{\rightarrow}{\partial}}{\partial\mathbf{x}} - \frac{\overset{\leftarrow}{\partial}}{\partial\mathbf{x}}\cdot\mathbf{A}(\mathbf{x},t)\right) \tag{7.115}$$

i.e., the same sign for the operations on incoming and outgoing lines, irrespective of whether they are retarded or advanced.

Diagrammatically we have the perturbation expansion of the retarded propagator in terms of the vector potential vertex

$$\tag{7.116}$$

and similarly for the advanced propagator

$$\tag{7.117}$$

In the expansion of the density matrix propagator to lowest order in the field, we encounter precisely the terms explicitly displayed in eq.(7.116) and eq.(7.117), and for the linear corrections to the density matrix we obtain the expression given by the two diagrams

$$\rho^{(1)}(\mathbf{x},\mathbf{x}',t) \quad = \quad$$

$$\tag{7.118}$$

The first diagram corresponds to the expression

$$\rho^{u}(\mathbf{x},\mathbf{x}',t) = \frac{e}{2im}\int d\tilde{\mathbf{x}}\int d\tilde{\mathbf{x}}'\int d\mathbf{x}_1\int_{t_i}^{t}dt_1 \,[G^R(\mathbf{x},t;\mathbf{x}_1,t_1)\left\{\frac{\overset{\leftarrow}{\partial}}{\partial\mathbf{x}_1}\cdot\mathbf{A}(\mathbf{x}_1,t_1)\right.$$

$$\left. - \;\mathbf{A}(\mathbf{x}_1,t_1)\cdot\frac{\overset{\rightarrow}{\partial}}{\partial\mathbf{x}_1}\right\}G^R(\mathbf{x}_1,t_1;\tilde{\mathbf{x}},t_i)\,\rho_i(\tilde{\mathbf{x}},\tilde{\mathbf{x}}')\,G^A(\tilde{\mathbf{x}}',t_i;\mathbf{x}',t) \tag{7.119}$$

and analogously for the second diagram.

In linear response the initial density matrix is evolved in time by $\hat{H}_0$ up until the time of interaction with the external field, and we obtain the second set of diagrams

$$\rho^{(1)}(\mathbf{x},\mathbf{x}',t) \quad = \quad -i$$

$$+ \; i$$

$$\tag{7.120}$$

and the linear correction to the density matrix is therefore, according to the application of the Feynman rules to the diagrams, determined by the expression

$$\rho^u(\mathbf{x}, \mathbf{x}', t) = \frac{e}{2m} \int d\mathbf{x}_1 \int d\mathbf{x}'_1 \int_{t_i}^{t} dt_1 \; [G^R(\mathbf{x}, t; \mathbf{x}_1, t_1) \left\{ \frac{\overleftarrow{\partial}}{\partial \mathbf{x}_1} \cdot \mathbf{A}(\mathbf{x}_1, t_1) \right.$$

$$\left. - \mathbf{A}(\mathbf{x}_1, t_1) \cdot \frac{\overrightarrow{\partial}}{\partial \mathbf{x}_1} \right\} \rho_0(\mathbf{x}_1, \mathbf{x}'_1, t_1) \, G^A(\mathbf{x}'_1, t_1; \mathbf{x}', t) \; . \tag{7.121}$$

We now assume that in the remote past the system was in a stationary state with respect to the unperturbed Hamiltonian, and we can, just as in the density response case in section 7.2.1, perform one of the spatial integrations

$$\rho^u(\mathbf{x}, \mathbf{x}', t) = -\frac{e\hbar^2}{2m} \int d\mathbf{x}_1 \int_{-\infty}^{\infty} \frac{dE_1}{2\pi\hbar} \int_{-\infty}^{\infty} \frac{dE_2}{2\pi\hbar} \sum_{\lambda_1 \lambda_2} \int_{t_i}^{t} dt_1 e^{-\frac{i}{\hbar} E_1(t - t_1) - \frac{i}{\hbar} E_2(t_1 - t)}$$

$$\frac{f_0(\epsilon_{\lambda_2}) \psi_{\lambda_1}(\mathbf{x}) \psi^*_{\lambda_2}(\mathbf{x}')}{(E_1 - \epsilon_{\lambda_1} + i0)(E_2 - \epsilon_{\lambda_2} - i0)}$$

$$\psi^*_{\lambda_1}(\mathbf{x}_1) \left\{ \frac{\overleftarrow{\partial}}{\partial \mathbf{x}_1} \cdot \mathbf{A}(\mathbf{x}_1, t_1) - \mathbf{A}(\mathbf{x}_1, t_1) \cdot \frac{\overrightarrow{\partial}}{\partial \mathbf{x}_1} \right\} \psi^*_{\lambda_2}(\mathbf{x}_1) \; . \tag{7.122}$$

In linear response, each frequency contributes additively, so without loss of generality we can concentrate on a specific frequency ω of the vector potential

$$\mathbf{A}(\mathbf{x}, t) = \mathbf{A}_\omega(\mathbf{x}) \, e^{-i\omega t} + \mathbf{A}^*_\omega(\mathbf{x}) \, e^{i\omega t} \tag{7.123}$$

related to the electric field by

$$\mathbf{E}(\mathbf{x}, t) = i\omega \left(\mathbf{A}_\omega(\mathbf{x}) \, e^{-i\omega t} - \mathbf{A}^*_\omega(\mathbf{x}) \, e^{i\omega t} \right) = \mathbf{E}_\omega(\mathbf{x}) \, e^{-i\omega t} + \mathbf{E}^*_\omega(\mathbf{x}) \, e^{i\omega t} \; . \tag{7.124}$$

We then obtain (letting t_i approach $-\infty$ and performing the time and energy variable integrations) that we only get a response at the driving frequency, ω,

$$\rho^u(\mathbf{x}, \mathbf{x}', t) = \rho^u_{\mathbf{x}\mathbf{x}'\omega}[\mathbf{A}_\omega] \, e^{-i\omega t} + \rho^u_{\mathbf{x}\mathbf{x}' - \omega}[\mathbf{A}^*_\omega] \, e^{i\omega t} \tag{7.125}$$

where

$$\rho^u_{\mathbf{x}\mathbf{x}'\omega}[\mathbf{A}_\omega] = \frac{e\hbar}{2im} \sum_{\lambda_1 \lambda_2} \frac{f_0(\epsilon_{\lambda_2})}{\epsilon_{\lambda_1} - \epsilon_{\lambda_2} - \hbar\omega - i0} \psi_{\lambda_1}(\mathbf{x}) \psi^*_{\lambda_2}(\mathbf{x}')$$

$$\int d\mathbf{x}_1 \psi^*_{\lambda_1}(\mathbf{x}_1) \left\{ \frac{\overleftarrow{\partial}}{\partial \mathbf{x}_1} \cdot \mathbf{A}_\omega(\mathbf{x}_1) - \mathbf{A}_\omega(\mathbf{x}_1) \cdot \frac{\overrightarrow{\partial}}{\partial \mathbf{x}_1} \right\} \psi_{\lambda_2}(\mathbf{x}_1)$$

$$= \frac{ie\hbar}{2m} \int d\mathbf{x}_1 \int_{-\infty}^{\infty} \frac{dE_1}{2\pi} \int_{-\infty}^{\infty} \frac{dE_2}{2\pi} \frac{f_0(E_2)}{E_1 - E_2 - \hbar\omega - i0}$$

$$[G^R(\mathbf{x}, \mathbf{x}_1; E_1) - G^A(\mathbf{x}, \mathbf{x}_1; E_1)] \left\{ \frac{\overleftarrow{\partial}}{\partial \mathbf{x}_1} \cdot \mathbf{A}_\omega(\mathbf{x}_1) - \mathbf{A}_\omega(\mathbf{x}_1) \cdot \frac{\overrightarrow{\partial}}{\partial \mathbf{x}_1} \right\}$$

$$[G^R(\mathbf{x}_1, \mathbf{x}', E_2) - G^A(\mathbf{x}_1, \mathbf{x}', E_2)] . \tag{7.126}$$

Using the hermitian relation, eq.(7.30), between the terms with the vector potential inserted on either the retarded or advanced line, we obtain that the linear change in the density matrix due to a vector potential is given by

$$\rho^{(1)}(\mathbf{x}, \mathbf{x}', t) = \rho^{(1)}_{\mathbf{x}\mathbf{x}'\omega}[\mathbf{A}_\omega] \, e^{-i\omega t} + \rho^{(1)}_{\mathbf{x}\mathbf{x}'-\omega}[\mathbf{A}^*_\omega] \, e^{i\omega t} \tag{7.127}$$

where the coefficients are functionals of the vector potential

$$\rho^{(1)}_{\mathbf{x}\mathbf{x}'\omega}[\mathbf{A}_\omega] = \frac{ie\hbar}{2m} \int d\mathbf{x}_1 \int_{-\infty}^{\infty} \frac{dE_1}{2\pi} \int_{-\infty}^{\infty} \frac{dE_2}{2\pi} \frac{f_0(E_2) - f_0(E_1)}{E_1 - E_2 - \hbar\omega - i0}$$

$$[G^R(\mathbf{x}, \mathbf{x}_1; E_1) - G^A(\mathbf{x}, \mathbf{x}_1; E_1)] \left\{ \frac{\overleftarrow{\partial}}{\partial \mathbf{x}_1} \cdot \mathbf{A}_\omega(\mathbf{x}_1) - \mathbf{A}_\omega(\mathbf{x}_1) \cdot \frac{\overrightarrow{\partial}}{\partial \mathbf{x}_1} \right\}$$

$$[G^R(\mathbf{x}_1, \mathbf{x}', E_2) - G^A(\mathbf{x}_1, \mathbf{x}', E_2)] . \tag{7.128}$$

Upon inserting the linearized expression for the density matrix into the expression for the current density, eq.(7.78), and Fourier transforming with respect to the frequency we obtain the current density response

$$j_\alpha(\mathbf{x}, t) = j_\alpha(\mathbf{x}, \omega) \, e^{-i\omega t} + j^*_\alpha(\mathbf{x}, \omega) \, e^{i\omega t} \tag{7.129}$$

where

$$j_\alpha(\mathbf{x}, \omega) = -\frac{e^2 \rho_0(\mathbf{x}, \mathbf{x})}{m} A_\alpha(\mathbf{x}, \omega) + \sum_\beta \int d\mathbf{x}' \, K_{\alpha\beta}(\mathbf{x}, \mathbf{x}', \omega) \, A_\beta(\mathbf{x}', \omega) \tag{7.130}$$

and (a factor of 2 accounts for spin)

$$K_{\alpha\beta}(\mathbf{x}, \mathbf{x}', \omega) = -2 \left(\frac{e\hbar}{2\pi m} \right)^2 \int_{-\infty}^{\infty} dE_1 \int_{-\infty}^{\infty} dE_2 \frac{f_0(E_2) - f_0(E_1)}{E_1 - E_2 - \hbar\omega - i0}$$

$$[G^R(\mathbf{x}, \mathbf{x}'; E_1) - G^A(\mathbf{x}, \mathbf{x}'; E_1)]$$

$$\overleftrightarrow{\nabla}_{x_\alpha} \overleftrightarrow{\nabla}_{x'_\beta} [G^R(\mathbf{x}', \mathbf{x}; E_2) - G^A(\mathbf{x}', \mathbf{x}; E_2)] \tag{7.131}$$

the current response function previously obtained for the electron gas, eq.(7.96).

7.3.5   Conductivity Tensor

Introducing the applied electric field, we express the current density

$$j_\alpha(\mathbf{x}, \omega) = \sum_\beta \int d\mathbf{x}' \, \sigma_{\alpha\beta}(\mathbf{x}, \mathbf{x}'; \omega) E_\beta(\mathbf{x}', \omega) + \mathcal{O}(\mathbf{E}^2) \tag{7.132}$$

in terms of the conductivity tensor,

$$\sigma_{\alpha\beta}(\mathbf{x}, \mathbf{x}'; \omega) = \frac{Q_{\alpha\beta}(\mathbf{x}, \mathbf{x}'; \omega)}{i\omega} \tag{7.133}$$

or equivalently for the longitudinal part,

$$\sigma_{\alpha\beta}(\mathbf{x}, \mathbf{x}', \omega) = \frac{K_{\alpha\beta}(\mathbf{x}, \mathbf{x}', \omega) - K_{\alpha\beta}(\mathbf{x}, \mathbf{x}', \omega = 0)}{i\omega}. \tag{7.134}$$

We note that the conductivity tensor is analytic in the upper half plane as causality demands, and as a consequence the real and imaginary parts are related through principal value integrals, Kramers-Kronig relations,

$$\Re e\, \sigma_{\alpha\beta}(\mathbf{x}, \mathbf{x}', \omega) = \frac{1}{\pi} P \int_{-\infty}^{\infty} d\omega' \, \frac{\Im m\, \sigma_{\alpha\beta}(\mathbf{x}, \mathbf{x}'; \omega')}{\omega' - \omega} \tag{7.135}$$

and

$$\Im m\, \sigma_{\alpha\beta}(\mathbf{x}, \mathbf{x}', \omega) = -\frac{1}{\pi} P \int_{-\infty}^{\infty} d\omega' \, \frac{\Re e\, \sigma_{\alpha\beta}(\mathbf{x}, \mathbf{x}'; \omega')}{\omega' - \omega}. \tag{7.136}$$

The time average of the response function, $K_{\alpha\beta}(\mathbf{x}, \mathbf{x}'; \omega = 0)$, is a real function according to eq.(7.131), and we have (for ω real)

$$\Re e\, \sigma_{\alpha\beta}(\mathbf{x}, \mathbf{x}'; \omega) = \Re e \left(\frac{-i}{\omega} K_{\alpha\beta}(\mathbf{x}, \mathbf{x}'; -\omega) \right) = \frac{1}{\omega} \Im m\, K_{\alpha\beta}(\mathbf{x}, \mathbf{x}'; \omega). \tag{7.137}$$

The real part of the conductivity tensor is according to eq.(7.131) given by

$$\Re e\, \sigma_{\alpha\beta}(\mathbf{x}, \mathbf{x}', \omega) = \frac{1}{\pi} \left(\frac{e}{m} \right)^2 \int_{-\infty}^{\infty} dE \, \frac{f_0(E) - f_0(E + \hbar\omega)}{\omega}$$

$$[G^R(\mathbf{x}, \mathbf{x}'; E + \hbar\omega) - G^A(\mathbf{x}, \mathbf{x}'; E + \hbar\omega)]$$

$$\overset{\leftrightarrow}{\nabla}_{x_\alpha} \overset{\leftrightarrow}{\nabla}_{x'_\beta} [G^R(\mathbf{x}', \mathbf{x}; E) - G^A(\mathbf{x}', \mathbf{x}; E)]. \tag{7.138}$$

In the case where the electron gas in the absence of the applied field is in the thermal state, only electrons occupying levels in the thermal layer around the Fermi surface contribute to the real part of the longitudinal conductivity, as expected.

7.3.6 Conductance

Often we are only interested in the total average current through the system (S denotes a cross-sectional surface through the system)

$$I(\omega) = \int_S d\mathbf{s} \cdot \mathbf{j}(\mathbf{x}, \omega) \tag{7.139}$$

and a proper description is in terms of the conductance, the inverse of the resistance. Let us consider a hypercube of volume L^d, and choose the surface S

perpendicular to the direction of the current flow, say, the α-direction. In terms of the conductivity we have (where ds_α denotes the infinitesimal area on the surface S):

$$I_\alpha(\omega) \;=\; \sum_\beta \int_S \!\!ds_\alpha \!\!\int\! d\mathbf{x}' \; \sigma_{\alpha\beta}(\mathbf{x}, \mathbf{x}', \omega) E_\beta(\mathbf{x}', \omega) \;. \tag{7.140}$$

Since the current, by particle conservation, is independent of the position of the cross section we get

$$I_\alpha(\omega) \;=\; L^{-1} \int\! d\mathbf{x} \; \mathbf{j}(\mathbf{x}, \omega) \;=\; L^{-1} \sum_\beta \int\! d\mathbf{x} \int\! d\mathbf{x}' \; \sigma_{\alpha\beta}(\mathbf{x}, \mathbf{x}', \omega) E_\beta(\mathbf{x}', \omega) \;.$$
$$\tag{7.141}$$

For the case of a spatially homogeneous external field in the β-direction, $E_\alpha(\mathbf{x}, \omega) = \delta_{\alpha\beta} E(\omega)$, we have in terms of the applied voltage across the system, $V_\beta(\omega) = E(\omega)\,L$,

$$I_\alpha(\omega) = G_{\alpha\beta}(\omega) V_\beta(\omega) \tag{7.142}$$

where we have introduced the conductance tensor

$$G_{\alpha\beta}(\omega) \;=\; L^{-2} \int\! d\mathbf{x} \int\! d\mathbf{x}' \; \sigma_{\alpha\beta}(\mathbf{x}, \mathbf{x}', \omega) \tag{7.143}$$

the inverse of the resistance tensor.

7.4   Properties of the Response Function

Response functions must satisfy certain relationships. In order to be specific, we illustrate these relationships by considering the current response function. We have already noted that causality causes the response function to be analytic in the upper half ω-plane. The current response function therefore has the representation in terms of the current spectral function, $\Im m K_{\alpha\beta}$ (the current response function vanishes in the limit of large ω),

$$K_{\alpha\beta}(\mathbf{x}, \mathbf{x}'; \omega) \;=\; \int_{-\infty}^{\infty} \frac{d\omega'}{\pi} \frac{\Im m K_{\alpha\beta}(\mathbf{x}, \mathbf{x}'; \omega')}{\omega' - \omega - i0} \;. \tag{7.144}$$

Since $K_{\alpha\beta}(\mathbf{x}, t; \mathbf{x}', t')$ contains a commutator of hermitian operators multiplied by the imaginary unit, it is real, and we have the property of the response function (ω real)

$$[K_{\alpha\beta}(\mathbf{x}, \mathbf{x}'; \omega)]^* \;=\; K_{\alpha\beta}(\mathbf{x}, \mathbf{x}'; -\omega) \tag{7.145}$$

and the real part of the response function is even[6]

$$\Re e K_{\alpha\beta}(\mathbf{x}, \mathbf{x}'; -\omega) \;=\; \Re e K_{\alpha\beta}(\mathbf{x}, \mathbf{x}'; \omega) \tag{7.146}$$

---

[6]In the presence of a magnetic field $\mathbf{B}$, we must also reverse the direction of the field, for example, $\Im m K_{\alpha\beta}(\mathbf{x}, \mathbf{x}'; \omega, \mathbf{B}) = -\Im m K_{\alpha\beta}(\mathbf{x}, \mathbf{x}'; -\omega, -\mathbf{B})$.

and the imaginary part odd

$$\Im m\, K_{\alpha\beta}(\mathbf{x}, \mathbf{x}'; -\omega) \;=\; -\Im m K_{\alpha\beta}(\mathbf{x}, \mathbf{x}'; \omega)\,. \tag{7.147}$$

From the spectral representation, eq.(7.94), we have

$$\Im m K_{\alpha\alpha}(\mathbf{x}, \mathbf{x}; \omega) = \pi \sum_{\lambda\lambda'} \rho(\epsilon_\lambda) |{<}\lambda|\, \hat{j}_\alpha^p(\mathbf{x}) |\lambda'{>}|^2 (\delta(\epsilon_\lambda - \epsilon_{\lambda'} - \hbar\omega) - \delta(\epsilon_\lambda - \epsilon_{\lambda'} + \hbar\omega))\,.$$
$$\tag{7.148}$$

For the thermal equilibrium state, where

$$\rho(\epsilon_{\lambda'}) \;=\; \rho(\epsilon_\lambda)\, e^{\frac{\epsilon_\lambda - \epsilon_{\lambda'}}{kT}} \tag{7.149}$$

we then obtain

$$\Im m\, K_{\alpha\alpha}(\mathbf{x}, \mathbf{x}; \omega) \;=\; \pi \left(1 - e^{-\hbar\omega/kT}\right) \sum_{\lambda\lambda'} \rho(\epsilon_\lambda) |{<}\lambda|\, \hat{j}_\alpha^p(\mathbf{x}) |\lambda'{>}|^2 \, \delta(\epsilon_\lambda - \epsilon_{\lambda'} + \hbar\omega)\,.$$
$$\tag{7.150}$$

For a state where the probability distribution, $\rho(\epsilon_\lambda)$, is a decreasing function of the energy, such as in the case of the thermal equilibrium state, the imaginary part of the diagonal response function is therefore positive for positive frequencies

$$\Im m\, K_{\alpha\alpha}(\mathbf{x}, \mathbf{x}; \omega \geq 0) \;\geq\; 0\,. \tag{7.151}$$

For the imaginary part of the diagonal part of the response function K we therefore have[7]

$$\omega\, \Im m K_{\alpha\alpha}(\mathbf{x}, \mathbf{x}, \omega) \;\geq\; 0\,. \tag{7.152}$$

From the spectral representation, eq.(7.144), we then find that the real part of the response function at zero frequency is larger than zero, $\Re e K_{\alpha\alpha}(\mathbf{x}, \mathbf{x}, \omega = 0) > 0$. The diagonal part of the real part of the response function is therefore positive for small frequencies. Since for large frequencies, the integral in eq.(7.144) is controlled by the singularity in the denominator, and as $\Im m K_{\alpha\alpha}(\mathbf{x}, \mathbf{x}, \omega)$ is a decaying function, the real part of the response function is negative for large frequencies, eventually approaching zero.

7.5 Stability of the Thermal Equilibrium State

In this section we shall show that the thermal equilibrium state is stable; i.e., manipulating the system by coupling its physical properties to a weak classical field which vanishes in the past and future can only increase the energy of the system. The average energy of a system is

$$E(t) \;=\; <\hat{H}(t)> \;=\; Tr(\hat{\rho}(t)\hat{H}(t))\,. \tag{7.153}$$

---

[7]We stress the important role played by the canonical ensemble.

The rate of change of the expectation value for the energy, for the Hamiltonian in question, eq.(7.69) (the term appearing when differentiating the statistical operator with respect to time vanishes, as seen by using the von-Neumann equation, eq.(1.384), and the cyclic property of the trace),

$$\frac{dE}{dt} \;=\; Tr\left(\hat{\rho}(t)\frac{d\hat{H}}{dt}\right) \;=\; -\int d\mathbf{x}\, Tr(\hat{\rho}(t)\hat{\mathbf{j}}_t(\mathbf{x}))\cdot\dot{\mathbf{A}}(\mathbf{x},t) \qquad (7.154)$$

has the perturbation expansion in the time-dependent external field

$$\frac{dE}{dt} \;=\; \frac{-i}{\hbar}\sum_{\alpha\beta}\int d\mathbf{x}\int d\mathbf{x}'\int_{t_i}^{t}dt'\,\dot{A}_\alpha(\mathbf{x},t)<[\hat{j}_\alpha^p(\mathbf{x},t),\hat{j}_\beta^p(\mathbf{x}',t')]>_0 A_\beta(\mathbf{x}',t')$$

$$-\int d\mathbf{x} <\hat{\mathbf{j}}_t(\mathbf{x})>_0 \cdot\dot{\mathbf{A}}(\mathbf{x},t) + \mathcal{O}(A^3)$$

$$=\; -\sum_{\alpha\beta}\int d\mathbf{x}\int d\mathbf{x}'\int_{t_i}^{\infty}dt'\,\dot{A}_\alpha(\mathbf{x},t)\,Q_{\alpha\beta}(\mathbf{x},t;\mathbf{x}',t')\,A_\beta(\mathbf{x}',t')$$

$$-\int d\mathbf{x} <\hat{\mathbf{j}}_t(\mathbf{x})>_0 \cdot\dot{\mathbf{A}}(\mathbf{x},t) + \mathcal{O}(A^3)\,. \qquad (7.155)$$

The dot signifies differentiation with respect to time. We recall that the equilibrium current, $<\hat{\mathbf{j}}_t(\mathbf{x})>_0$, is in fact time independent.

An external field therefore performs, in the time span between t_i and t_f, the work

$$W \;\equiv\; E(t_f) - E(t_i)$$

$$=\; -\sum_{\alpha\beta}\int_{t_i}^{t_f}dt\int d\mathbf{x}\int d\mathbf{x}'\int_{t_i}^{\infty}dt'\,\dot{A}_\alpha(\mathbf{x},t)\,K_{\alpha\beta}(\mathbf{x},t;\mathbf{x}',t')\,A_\beta(\mathbf{x}',t')$$

$$=\; \sum_{\alpha\beta}\int_{t_i}^{t_f}dt\int d\mathbf{x}\int d\mathbf{x}'\int_{t_i}^{\infty}dt'\,A_\alpha(\mathbf{x},t)\,\frac{dK_{\alpha\beta}(\mathbf{x},t;\mathbf{x}',t')}{dt}\,A_\beta(\mathbf{x}',t')$$

$$+\;\mathcal{O}(A^3)\,. \qquad (7.156)$$

In the first equality we have noticed that the diamagnetic term in the response function Q does not contribute. For the second equality we have assumed that the vector potential vanishes in the past and in the future (i.e., the time average of the electric field is zero), so that the boundary terms vanish, and we observe that in that case there is no linear contribution; to linear order the energy of the system is unchanged.

For an isotropic system we have

$$K_{\alpha\beta}(\mathbf{x}, \mathbf{x}', \omega) = K(\mathbf{x}, \omega)\, \delta(\mathbf{x} - \mathbf{x}')\, \delta_{\alpha\beta} \qquad (7.157)$$

and we obtain, in view of eq.(7.152), that the mean change in energy of the system to second order is positive

$$\Delta E \equiv W = \int d\mathbf{x} \int_{-\infty}^{\infty} \frac{d\omega}{2\pi}\, \omega\, \Im m\, K(\mathbf{x}, \omega)\, \mathbf{A}(\mathbf{x}, \omega) \cdot \mathbf{A}^*(\mathbf{x}, -\omega) \geq 0 . \qquad (7.158)$$

Interacting weakly with the physical quantities of a system in thermal equilibrium through a classical field, which vanishes in the past and in the future, can thus only lead to an increase in the energy of the system; the energy never decreases. The thermodynamic equilibrium state is thus a stable state.[8]

In the case of a monochromatic field

$$\mathbf{A}(\mathbf{x}, t) = \frac{1}{2}\left(\mathbf{A}(\mathbf{x}, \omega)e^{-i\omega t} + \mathbf{A}^*(\mathbf{x}, \omega)e^{i\omega t}\right) = \Re e\left(\mathbf{A}(\mathbf{x}, \omega)e^{-i\omega t}\right) \qquad (7.159)$$

we have for the mean rate of change of the energy to second order in the applied field, $T \equiv 2\pi/\omega$,

$$\overline{\frac{dE_{\omega}}{dt}}^T \equiv \frac{1}{T}\int_0^T dt\, \frac{dE}{dt} = \frac{-i}{4\hbar}\frac{1}{T}\sum_{\alpha\beta}\int_0^T dt \int_{t_i}^{t} dt' \int d\mathbf{x} \int d\mathbf{x}'\, \dot{A}_{\alpha}(\mathbf{x}, t)$$

$$< [\hat{j}_{\alpha}^p(\mathbf{x}, t), \hat{j}_{\beta}^p(\mathbf{x}', t')] >_0 A_{\beta}(\mathbf{x}', t') \qquad (7.160)$$

as the diamagnetic term averages in time to zero. Turning the field on in the far past, $t_i \to -\infty$, we have in terms of the response function

$$\overline{\frac{dE_{\omega}}{dt}}^T = \frac{-i\omega}{4}\sum_{\alpha\beta}\int d\mathbf{x} \int d\mathbf{x}'\, A_{\alpha}^*(\mathbf{x}, \omega)\, (K_{\alpha\beta}(\mathbf{x}, \mathbf{x}', \omega) - K_{\beta\alpha}(\mathbf{x}', \mathbf{x}, -\omega))\, A_{\beta}(\mathbf{x}', \omega)$$

$$= \frac{\omega}{2}\sum_{\alpha\beta}\int d\mathbf{x} \int d\mathbf{x}'\, A_{\alpha}^*(\mathbf{x}, \omega)\, \Im m K_{\alpha\beta}(\mathbf{x}, \mathbf{x}', \omega) A_{\beta}(\mathbf{x}', \omega) . \qquad (7.161)$$

We can, according to eq.(7.147), rewrite the average work performed by the external field on the form

$$\overline{\frac{dE_{\omega}}{dt}}^T = \sum_{\lambda} \hbar\omega\, \rho(\epsilon_{\lambda})\, (P_{\lambda}(\hbar\omega) - P_{\lambda}(-\hbar\omega)) , \qquad (7.162)$$

where

$$P_{\lambda}(\hbar\omega) = \frac{2\pi}{\hbar}\sum_{\lambda'} \left| \frac{1}{2}\int d\mathbf{x} <\lambda'|\hat{\mathbf{j}}_p(\mathbf{x}) \cdot \mathbf{A}(\mathbf{x}, \omega)|\lambda> \right|^2 \delta(\epsilon_{\lambda} - \epsilon_{\lambda'} + \hbar\omega) \qquad (7.163)$$

---

[8]It is important to stress the crucial role of the canonical (or grand canonical) ensemble for the validity of eq.(7.152).

is Fermi's golden rule expression for the probability for the transition from state λ to any state λ' in which the system absorbs the amount $\hbar\omega$ of energy from the external field, and $P_\lambda(-\hbar\omega)$ is the transition probability for emission of the amount $\hbar\omega$ of energy to the external field. The equation for the change in energy is thus a master equation for the energy, and we infer that the energy exchange between a system and a classical field oscillating at frequency ω takes place in lumps of magnitude $\hbar\omega$.

At each frequency we have for the average work done on the system by the external field:

$$\overline{\frac{dE_\omega}{dt}}^T = \frac{1}{2}\sum_{\alpha\beta}\int d\mathbf{x}\int d\mathbf{x}'\, E_\alpha^*(\mathbf{x},\omega)\,\Re e\sigma_{\alpha\beta}(\mathbf{x},\mathbf{x}',\omega)E_\beta(\mathbf{x}',\omega) \qquad (7.164)$$

where we have utilized eq.(7.137) to introduce the real part of the conductivity tensor.

For a translational invariant system we have

$$\sigma_{\alpha\beta}(\mathbf{x},\mathbf{x}',\omega) = \sigma_{\alpha\beta}(\mathbf{x}-\mathbf{x}',\omega) = \frac{1}{V}\sum_\mathbf{q} e^{i\mathbf{q}\cdot(\mathbf{x}-\mathbf{x}')}\,\sigma_{\alpha\beta}(\mathbf{q},\omega) \qquad (7.165)$$

and we get for each wave vector

$$E_\alpha(\mathbf{x},\omega) = E_\alpha(\mathbf{q},\omega)\,e^{i\mathbf{q}\cdot\mathbf{x}} \qquad (7.166)$$

the contribution

$$\overline{\frac{dE_{\mathbf{q}\omega}}{dt}}^T = \frac{V}{2}\sum_{\alpha\beta} E_\alpha^*(\mathbf{q},\omega)\,\Re e\sigma_{\alpha\beta}(\mathbf{q},\omega)E_\beta(\mathbf{q},\omega)\,. \qquad (7.167)$$

Each harmonic contributes additively, and we get for the average energy absorption for arbitrary spatial dependence of the electric field the expression

$$\overline{\frac{dE_\omega}{dt}}^T = \frac{V}{2}\sum_{\alpha\beta}\sum_\mathbf{q} E_\alpha^*(\mathbf{q},\omega)\,\Re e\sigma_{\alpha\beta}(\mathbf{q},\omega)E_\beta(\mathbf{q},\omega)\,. \qquad (7.168)$$

For an isotropic system the conductivity tensor is diagonal

$$\sigma_{\alpha\beta}(\mathbf{x},\mathbf{x}',\omega) = \delta_{\alpha\beta}\,\sigma(\mathbf{x}-\mathbf{x}',\omega) \qquad (7.169)$$

and we have

$$\overline{\frac{dE_\omega}{dt}}^T = \frac{V}{2}\sum_\alpha |E_\alpha(\mathbf{q},\omega)|^2\,\Re e\sigma_{\alpha\alpha}(\mathbf{q},\omega)\,. \qquad (7.170)$$

For the spatially homogeneous field case, $E_\alpha(\mathbf{q}\neq 0,\omega) = 0$, we then obtain

$$\overline{\frac{dE_\omega}{dt}}^T = \frac{V}{2}\sum_\alpha |E_\alpha^*(\mathbf{q}=0,\omega)|^2 \sum_\mathbf{q}\Re e\,\sigma_{\alpha\alpha}(\mathbf{q},\omega)\,. \qquad (7.171)$$

Since

$$\frac{1}{V} \sum_q \Re e \, \sigma_{\alpha\alpha}(\mathbf{q}, \omega) \; = \; \Re e \, \sigma_{\alpha\alpha}(\mathbf{x}, \mathbf{x}, \omega) \; = \; \frac{1}{\omega} \Im m K_{\alpha\alpha}(\mathbf{x}, \mathbf{x}, \omega) \geq 0 \qquad (7.172)$$

we obtain the result, that for a system in thermal equilibrium the average change in energy can only be increased by interaction with a weak periodic external field[9]

$$\frac{\overline{dE_\omega}^T}{dt} \geq 0 \,. \qquad (7.173)$$

The thermal state is stable against a weak periodic perturbation.[10]

Considering the isotropic dc case we get directly from eq.(7.164) the familiar Joule heating expression for the energy absorbed per unit time in a resistor

$$\frac{\overline{dE}^T}{dt} \; = \; \frac{1}{2} G V^2 \; = \; \frac{1}{2} R I^2 \qquad (7.174)$$

where R is the resistance, the inverse conductance, $R \equiv G^{-1}$, and we have used that in the dc case the imaginary part of the conductance tensor vanishes.

The absorbed energy of a system in thermal equilibrium interacting with an external field, is dissipated in the system, and we thus note that $\Re e \, \sigma$ or equivalently $\Im m \, K$ describes the dissipation in the system.

7.6 Fluctuation-Dissipation Theorem

The most important hallmark of linear response is the relation between equilibrium fluctuations and dissipation. We shall illustrate this feature by again considering the current response function; however, the argument is equivalent for any correlation function. We introduce the current correlation function in the thermal equilibrium state

$$\tilde{K}^{(j)}_{\alpha\beta}(\mathbf{x}, t; \mathbf{x}', t') \; \equiv \; \frac{1}{2} < \{ \delta \hat{\jmath}^p_\alpha(\mathbf{x}, t), \delta \hat{\jmath}^p_\beta(\mathbf{x}', t') \} >_0 \qquad (7.175)$$

where

$$\delta \hat{\jmath}^p_\alpha(\mathbf{x}, t) \; \equiv \; \hat{\jmath}^p_\alpha(\mathbf{x}, t) - < \hat{\jmath}^p_\alpha(\mathbf{x}, t) >_0 \qquad (7.176)$$

is the deviation from a possible equilibrium current, $<\hat{\mathbf{j}}^p(\mathbf{x}, t)>_0$, which in fact is independent of time. However, for notational simplicity we assume in the following that the equilibrium current density vanishes. By taking the anticommutator, we have symmetrized the correlation function, and since the current operator is hermitian, the correlation function is a real function.

---

[9]In fact, we have from the positivity of $\Re e \, \sigma_{\alpha\alpha}(\mathbf{q}, \omega)$ for arbitrary wave vector that the conclusion is valid for arbitrary spatially varying external field.

[10]Since this result is valid at any frequency, we again obtain the result that a system in thermal equilibrium is stable.

Since the statistical average is with respect to the equilibrium state (for an arbitrary Hamiltonian $\hat{H}_0$), we have on account of the cyclic property of the trace

$$
\begin{aligned}
K^>_{\alpha\beta}(\mathbf{x},t;\mathbf{x}',t') &\equiv Tr\left(e^{-\hat{H}_0/kT}\,\hat{j}^p_\alpha(\mathbf{x},t)\,\hat{j}^p_\beta(\mathbf{x}',t')\right) \equiv \; <\hat{j}^p_\alpha(\mathbf{x},t)\,\hat{j}^p_\beta(\mathbf{x}',t')>_0 \\
&= Tr\left(e^{-\hat{H}_0/kT}\,\hat{j}^p_\beta(\mathbf{x}',t')\,\hat{j}^p_\alpha(\mathbf{x},t+i\hbar/kT)\right) \\
&= K^<_{\alpha\beta}(\mathbf{x},t+i\hbar/kT;\mathbf{x}',t')
\end{aligned}
\tag{7.177}
$$

as we define

$$
K^<_{\alpha\beta}(\mathbf{x},t;\mathbf{x}',t') \equiv Tr\left(e^{-\hat{H}_0/kT}\hat{j}^p_\beta(\mathbf{x}',t')\hat{j}^p_\alpha(\mathbf{x},t)\right) \equiv \; <\hat{j}^p_\beta(\mathbf{x}',t')\hat{j}^p_\alpha(\mathbf{x},t)>_0 \; .
\tag{7.178}
$$

We note the crucial role played by the assumption of a canonical ensemble.

We assume the canonical ensemble average exists for all real times t and t', and consequently $K^<$ is an analytic function in the region $0 < \Im m(t-t') < \hbar/kT$, and $K^>$ is analytic in the region $-\hbar/kT < \Im m(t-t') < 0$. For the Fourier transforms we therefore obtain the relation

$$
K^>_{\alpha\beta}(\mathbf{x},\mathbf{x}';\omega) = e^{-\hbar\omega/kT}\,K^<_{\alpha\beta}(\mathbf{x},\mathbf{x}';\omega) \; .
\tag{7.179}
$$

We observe the following relation of the commutator to the retarded and advanced correlation functions

$$
\begin{aligned}
K^>_{\alpha\beta}(\mathbf{x},t;\mathbf{x}',t') - K^<_{\alpha\beta}(\mathbf{x},t;\mathbf{x}',t') &= \; <[\hat{j}^p_\alpha(\mathbf{x},t),\hat{j}^p_\beta(\mathbf{x}',t')]>_0 \\
&= -i\hbar\left(K^R_{\alpha\beta}(\mathbf{x},t;\mathbf{x}',t') - K^A_{\alpha\beta}(\mathbf{x},t;\mathbf{x}',t')\right)
\end{aligned}
\tag{7.180}
$$

where we have introduced the advanced correlation function

$$
K^A_{\alpha\beta}(\mathbf{x},t;\mathbf{x}',t') = -\frac{i}{\hbar}\,\theta(t'-t)\;<[\hat{j}^p_\alpha(\mathbf{x},t),\hat{j}^p_\beta(\mathbf{x}',t')]>_0
\tag{7.181}
$$

corresponding to the retarded one appearing in the current response, eq.(7.88),

$$
K^R_{\alpha\beta}(\mathbf{x},t;\mathbf{x}',t') \equiv K_{\alpha\beta}(\mathbf{x},t;\mathbf{x}',t') \; .
\tag{7.182}
$$

Since the response function involves the commutator of two hermitian operators we immediately verify that (for ω real)

$$
K^{R(A)}_{\alpha\beta}(\mathbf{x},\mathbf{x}',-\omega) = [K^{R(A)}_{\alpha\beta}(\mathbf{x},\mathbf{x}',\omega)]^* \; .
\tag{7.183}
$$

Analogous to eq.(7.178) we have for the correlation function, the anticommutator,

$$
<\{\hat{j}^p_\alpha(\mathbf{x},t),\hat{j}^p_\beta(\mathbf{x}',t')\}>_0 = K^>_{\alpha\beta}(\mathbf{x},t;\mathbf{x}',t') + K^<_{\alpha\beta}(\mathbf{x},t;\mathbf{x}',t') \; .
\tag{7.184}
$$

Using eq.(7.178) we can rewrite

$$\tilde{K}_{\alpha\beta}^{(j)}(\mathbf{x}, \mathbf{x}', \omega) = \frac{1}{2} K_{\alpha\beta}^{>}(\mathbf{x}, \mathbf{x}', \omega) \left(1 + e^{\hbar\omega/kT}\right)$$

$$= \left(\frac{1}{2}\left(K_{\alpha\beta}^{>}(\mathbf{x}, \mathbf{x}', \omega) + K_{\alpha\beta}^{<}(\mathbf{x}, \mathbf{x}', \omega)\right) - \frac{1}{2}\left(K_{\alpha\beta}^{>}(\mathbf{x}, \mathbf{x}', \omega) - K_{\alpha\beta}^{<}(\mathbf{x}, \mathbf{x}', \omega)\right)\right)$$

$$\frac{1}{2}\left(1 + e^{\hbar\omega/kT}\right) \qquad (7.185)$$

and thereby

$$\tilde{K}_{\alpha\beta}^{(j)}(\mathbf{x}, \mathbf{x}', \omega) = \left(K_{\alpha\beta}^{R}(\mathbf{x}, \mathbf{x}', \omega) - K_{\alpha\beta}^{A}(\mathbf{x}, \mathbf{x}', \omega)\right) \frac{\hbar}{2i} \coth\frac{\hbar\omega}{2kT} \qquad (7.186)$$

Using eq.(7.186), and noting that for omega real (we establish this as a consequence of time-reversal invariance in the next section)

$$K_{\alpha\beta}^{A}(\mathbf{x}, \mathbf{x}', \omega) = [K_{\alpha\beta}^{R}(\mathbf{x}, \mathbf{x}', \omega)]^* \qquad (7.187)$$

we then get the relation between the correlation function and the imaginary part of the response function

$$\tilde{K}_{\alpha\beta}^{(j)}(\mathbf{x}, \mathbf{x}', \omega) = \hbar \coth\frac{\hbar\omega}{2kT} \, \Im m K_{\alpha\beta}(\mathbf{x}, \mathbf{x}', \omega) \,. \qquad (7.188)$$

We have established the relationship between the imaginary part of the linear response function, governing according to eq.(7.161) the dissipation in the system, and the equilibrium fluctuations, the fluctuation-dissipation theorem.[11]

According to the fluctuation-dissipation theorem we can express the change in energy of a system in an external field of frequency ω, eq.(7.161), in terms of the current fluctuations

$$\overline{\frac{dE_\omega}{dt}}^T = \frac{1}{2\hbar\omega \coth\frac{\hbar\omega}{2kT}} \sum_{\alpha\beta} \int d\mathbf{x} \int d\mathbf{x}' \, E_\alpha^*(\mathbf{x}, \omega) \, \tilde{K}_{\alpha\beta}^{j}(\mathbf{x}, \mathbf{x}', \omega) E_\beta(\mathbf{x}', \omega) \,. \qquad (7.189)$$

For the current fluctuations we have (recall eq.(7.137))

$$\frac{1}{2} < \{\hat{j}_\alpha^p(\mathbf{x}, \omega), \hat{j}_\beta^p(\mathbf{x}', -\omega)\} > = \tilde{K}_{\alpha\beta}^{(j)}(\mathbf{x}, \mathbf{x}', \omega)$$

$$= \hbar\omega \coth\frac{\hbar\omega}{2kT} \, \Re e \, \sigma_{\alpha\beta}(\mathbf{x}, \mathbf{x}', \omega) \qquad (7.190)$$

---

[11]Formally the fluctuation-dissipation theorem expresses the relationship between a commutator and anticommutator canonical equilibrium average. The fluctuation-dissipation relation, eq.(7.188), is also readily established by comparing the spectral representation of the imaginary part of the retarded current response function, eq.(7.94), with that of $\tilde{K}^{(j)}$. The fluctuation-dissipation relationship expresses that the system is in equilibrium and described by the canonical ensemble.

and the equal-time current fluctuations are specified by

$$\tilde{K}_{\alpha\beta}^{(j)}(\mathbf{x},t;\mathbf{x}',t) = \int\limits_{-\infty}^{\infty}\frac{d\omega}{2\pi}\, \tilde{K}_{\alpha\beta}^{(j)}(\mathbf{x},\mathbf{x}',\omega) = \hbar\int\limits_{-\infty}^{\infty}\frac{d\omega}{2\pi}\,\coth\frac{\hbar\omega}{2kT}\,\Im m K_{\alpha\beta}(\mathbf{x},\mathbf{x}',\omega)$$

(7.191)

and eq.(7.152) guarantees the positivity of the equal-time and space current density fluctuations.

In a macroscopic description we have a local relationship between field and current density, Ohm's law,

$$j_\alpha(\mathbf{x},\omega) = \sigma_{\alpha\beta}(\mathbf{x},\omega)E_\beta(\mathbf{x},\omega)$$

(7.192)

or equivalently

$$\sigma_{\alpha\beta}(\mathbf{x},\mathbf{x}',\omega) = \sigma_{\alpha\beta}(\mathbf{x},\omega)\,\delta(\mathbf{x}-\mathbf{x}')\,.$$

(7.193)

The equilibrium current density fluctuations at point $\mathbf{x}$ are then specified by

$$<j_\alpha^2>_{\mathbf{x}\omega} \equiv \frac{1}{2V}\int d(\mathbf{x}-\mathbf{x}')\,<\{\hat{j}_\alpha^p(\mathbf{x},\omega),\hat{j}_\alpha^p(\mathbf{x}',-\omega)\}>_0$$

$$= \frac{1}{V}\,\tilde{K}_{\alpha\alpha}^{(j)}(\mathbf{x},\omega) = \frac{1}{V}\,\hbar\omega\coth\frac{\hbar\omega}{2kT}\,\Re e\,\sigma_{\alpha\alpha}(\mathbf{x},\omega)\,.$$ (7.194)

We note, that the factor

$$\frac{\hbar\omega}{2}\coth\frac{\hbar\omega}{2kT} = \hbar\omega\left(n(\omega)+\frac{1}{2}\right)$$

(7.195)

is the average energy of a harmonic oscillator, with frequency ω, in the thermal state. The average energy consists of a thermal contribution described by the Bose function, and a zero-point quantum fluctuation contribution.

In the high-temperature limit where relevant frequencies are small compared to the temperature, $\hbar\omega \ll kT$, we get for the current fluctuations in a homogeneous conductor with conductivity σ, Johnson noise,

$$<j_\alpha^2>_\omega = \frac{2kT\sigma}{V}$$

(7.196)

independent of the specific nature of the conductor.

In the linear response treatment we have assumed the field fixed, and studied the fluctuations in the current density. However, fluctuations in the current (or charge) density gives rise to fluctuations in the electromagnetic field as well. As an example of using the fluctuation-dissipation theorem we therefore turn the point of view around, and consider the external current as specified, and obtain that the (longitudinal) electric field fluctuations are given by

$$<E_\alpha^2>_{\mathbf{x}\omega} = \frac{1}{|\sigma(\mathbf{x},\omega)|^2}<j_\alpha^2>_{\mathbf{x}\omega}\,.$$

(7.197)

According to eq.(7.194) we then obtain for the (longitudinal) electric field fluctuations

$$< E_\alpha^2 >_{\mathbf{x}\omega} \;=\; \frac{1}{V} \, \hbar\omega \coth \frac{\hbar\omega}{2kT} \, \frac{\Re e \sigma_{\alpha\alpha}(\mathbf{x}, \omega)}{|\sigma_{\alpha\alpha}(\mathbf{x}, \omega)|^2} \; . \tag{7.198}$$

In the high temperature-limit, $\hbar\omega \ll kT$, we have for the (longitudinal) electric field fluctuations, Nyquist noise,

$$< E_\alpha^2 >_\omega \;=\; \frac{2kT}{\sigma V} \; . \tag{7.199}$$

7.7 Time-Reversal Symmetry

We showed in section 2.9 that hermitian operators by suitable phase choice have a definite sign under time reversal, position and electric field have positive sign, and velocity and magnetic field have negative sign. The following considerations can be performed for any pair of operators (see exercise 7.1), but we shall for definiteness consider the current operator, and show that eq.(7.187) is a consequence of time-reversal invariance.

Recalling eq.(2.219) on page 126 we have in case the Hamiltonian is time-reversal invariant

$$\left(\hat{T} \, [\hat{j}_\alpha^p(\mathbf{x}, t), \hat{j}_\beta^p(\mathbf{x}', t')] \, \hat{T}^\dagger \right)^\dagger \;=\; [\hat{T} \, \hat{j}_\beta^p(\mathbf{x}', t') \, \hat{T}^\dagger, \hat{T} \, \hat{j}_\alpha^p(\mathbf{x}, t) \, \hat{T}^\dagger]$$

$$=\; -\, [\hat{j}_\alpha^p(\mathbf{x}, -t), \hat{j}_\beta^p(\mathbf{x}', -t')] \tag{7.200}$$

Assuming that the ensemble of states are invariant with respect to time reversal, we have according to eq.(C.7) of appendix C

$$<\psi | [\hat{j}_\alpha^p(\mathbf{x}, t), \hat{j}_\beta^p(\mathbf{x}', t')] | \psi> \;=\; <T\psi | \hat{T} \, [\hat{j}_\alpha^p(\mathbf{x}, t), \hat{j}_\beta^p(\mathbf{x}', t')]^\dagger \hat{T}^\dagger | T\psi> \tag{7.201}$$

and thereby that

$$Tr(\rho(\hat{H}_0)[\hat{j}_\alpha^p(\mathbf{x}, -t), \hat{j}_\beta^p(\mathbf{x}', -t')]) \;=\; -Tr(\rho(\hat{H}_0)[\hat{j}_\alpha^p(\mathbf{x}, t), \hat{j}_\beta^p(\mathbf{x}', t')]) \tag{7.202}$$

and we therefore find that time-reversal invariance implies

$$K_{\alpha\beta}^R(\mathbf{x}, \mathbf{x}'; \omega) \;=\; [K_{\alpha\beta}^A(\mathbf{x}, \mathbf{x}'; \omega)]^* \tag{7.203}$$

i.e., we have established eq.(7.187).

Exercise 7.1 *Consider two physical quantities represented by the operators $\hat{A}_1(\mathbf{x}, t)$ and $\hat{A}_2(\mathbf{x}, t)$ which transform under time reversal according to*

$$\hat{T} \, \hat{A}_i(\mathbf{x}, t) \, \hat{T}^\dagger = s_i \, \hat{A}_i(\mathbf{x}, -t) \; , \quad s_i = \pm 1, \; i = 1, 2. \tag{7.204}$$

Show that when the Hamiltonian is invariant under time reversal, the response function

$$A_{ij}(\mathbf{x}, \mathbf{x}', t - t') \equiv Tr(\rho(\hat{H})[\hat{A}_i(\mathbf{x}, t), \hat{A}_j(\mathbf{x}, t)]) \qquad (7.205)$$

satisfies the relations

$$A_{ij}(\mathbf{x}, \mathbf{x}', t - t') = -s_i s_j A_{ij}(\mathbf{x}, \mathbf{x}', t' - t) = s_i s_j A_{ij}(\mathbf{x}', \mathbf{x}, t - t') \qquad (7.206)$$

and thereby[12]

$$A_{ij}(\mathbf{x}, \mathbf{x}', \omega) = -s_i s_j A_{ij}(\mathbf{x}, \mathbf{x}', -\omega) = s_i s_j A_{ij}(\mathbf{x}', \mathbf{x}, \omega) . \qquad (7.207)$$

7.8   Momentum Representation Response

Transforming to the momentum representation, the spatial differentiations representing the current vertices simply brings down momentum factors. Again, using that in linear response the wave vector components contribute additively, so that we only need to consider the response to, say, wave vector $\mathbf{q}$

$$\mathbf{A}_\omega(\mathbf{x}) \equiv \mathbf{A}(\mathbf{q}, \omega)e^{i\mathbf{q}\cdot\mathbf{x}} \qquad (7.208)$$

the spatial integration in eq.(7.92) can be performed, and we obtain for the current density at wave vector $\mathbf{q}'$ in response to an external field with wave vector $\mathbf{q}$ the expression

$$j_\alpha(\mathbf{q}', \omega) = \frac{1}{V} \int d\mathbf{x}\, e^{-i\mathbf{x}\cdot\mathbf{q}'} j_\alpha(\mathbf{x}, \omega) = \sum_\beta Q_{\alpha\beta}(\mathbf{q}', \mathbf{q}, \omega) A_\beta(\mathbf{q}, \omega) \qquad (7.209)$$

where

$$Q_{\alpha\beta}(\mathbf{q}', \mathbf{q}, \omega) = \int d\mathbf{x} \int d\mathbf{x}' <\hbar\mathbf{q}'|\mathbf{x}> [K_{\alpha\beta}(\mathbf{x}, \mathbf{x}', \omega) - K_{\alpha\beta}(\mathbf{x}, \mathbf{x}', 0)] <\mathbf{x}'|\hbar\mathbf{q}>$$

$$= K_{\alpha\beta}(\mathbf{q}', \mathbf{q}, \omega) - K_{\alpha\beta}(\mathbf{q}', \mathbf{q}, 0) = K_{\alpha\beta}(\mathbf{q}', \mathbf{q}, \omega) - \frac{e^2}{m} n_0(\mathbf{q}' - \mathbf{q}) \delta_{\alpha\beta} \quad (7.210)$$

and the response function for the electron gas (the factor of 2 accounts for spin) is given by

$$K_{\alpha\beta}(\mathbf{q}', \mathbf{q}, \omega) = 2 \left(\frac{e}{2\pi m}\right)^2 \int_{-\infty}^{\infty} dE_1 \int_{-\infty}^{\infty} dE_2 \frac{f_0(E_2) - f_0(E_1)}{E_1 - E_2 - \hbar\omega - i0}$$

$$\frac{1}{V} \sum_{\mathbf{p}\mathbf{p}'} p_\alpha p'_\beta A(\mathbf{p}_+, \mathbf{p}'_+; E_1) A(\mathbf{p}'_-, \mathbf{p}_-; E_2) . \qquad (7.211)$$

---

[12]If the Hamiltonian contains a term coupling to a magnetic field, the symmetry of the correlation function is $A_{ij}(\mathbf{x}, \mathbf{x}', \omega, \mathbf{B}) = -s_i s_j A_{ij}(\mathbf{x}, \mathbf{x}', -\omega, -\mathbf{B}) = s_i s_j A_{ji}(\mathbf{x}', \mathbf{x}, \omega, -\mathbf{B})$.

We can perform one of the energy integrations to obtain the expression

$$K_{\alpha\beta}(\mathbf{q}', \mathbf{q}, \omega) = \frac{-e^2}{\pi m^2 V} \sum_{\mathbf{p}\mathbf{p}'} p_\alpha p'_\beta \int_{-\infty}^{\infty} dE\, f_0(E) \left(G^R(\mathbf{p}_+, \mathbf{p}'_+; E + \hbar\omega)\, A(\mathbf{p}'_-, \mathbf{p}_-; E) \right.$$

$$\left. + A(\mathbf{p}_+, \mathbf{p}'_+; E)\, G^A(\mathbf{p}'_-, \mathbf{p}_-; E - \hbar\omega) \right) . \tag{7.212}$$

For the real part of the longitudinal conductivity tensor we have according to eq.(7.138) in the momentum representation

$$\Re e\sigma_{\alpha\beta}(\mathbf{q}, \mathbf{q}', \omega) = \left(\frac{e}{m} \right)^2 \frac{1}{\pi} \int_{-\infty}^{\infty} dE\, \frac{f_0(E_-) - f_0(E_+)}{\omega}$$

$$\frac{1}{V} \sum_{\mathbf{p}\mathbf{p}'} p_\alpha p'_\beta\, A(\mathbf{p}_+, \mathbf{p}'_+; E_+)\, A(\mathbf{p}'_-, \mathbf{p}_-; E_-) \tag{7.213}$$

where we have introduced the notation $E_\pm = E \pm \hbar\omega/2$ (ω assumed to be real). If the Fermi gas is degenerate, $\hbar\omega, kT \ll \epsilon_F$, the gate function is sharp

$$\frac{f_0(E_-) - f_0(E_+)}{\hbar\omega} \simeq \delta(E - \epsilon_F) \tag{7.214}$$

and sets the energy variable to the Fermi energy $E \simeq \epsilon_F$; i.e., only electrons on the Fermi surface contribute to the real part of the longitudinal conductivity, as expected.

Exercise 7.2 *We could equally well right from the start have performed the calculation of the current response in the momentum representation in which case we have for the paramagnetic part of the current density*

$$\mathbf{j}_p(\mathbf{x}, \omega) = 2 \frac{e}{2m} \frac{1}{V} \sum_{\mathbf{p}\mathbf{p}'} (\mathbf{p} + \mathbf{p}')\, e^{\frac{i}{\hbar}\mathbf{x}\cdot(\mathbf{p} - \mathbf{p}')}\, \rho(\mathbf{p}, \mathbf{p}', \omega) . \tag{7.215}$$

Show that to linear order in the field we obtain for the density matrix in the momentum representation

$$\rho^{(1)}(\mathbf{p}, \mathbf{p}', t) = \rho^{(1)}_{\mathbf{p}\mathbf{p}'\omega}[\mathbf{A}_\omega]\, e^{-i\omega t} + \rho^{(1)}_{\mathbf{p}\mathbf{p}'-\omega}[\mathbf{A}^*_\omega]\, e^{i\omega t} \tag{7.216}$$

where

$$\rho^{(1)}_{\mathbf{p}\mathbf{p}'\omega}[\mathbf{A}_\omega] = \frac{e}{2m} \int d\mathbf{x}_1 \int_{-\infty}^{\infty} \frac{dE_1}{2\pi} \int_{-\infty}^{\infty} \frac{dE_2}{2\pi} \frac{f_0(E_2) - f_0(E_1)}{E_1 - E_2 - \hbar\omega - i0} \frac{1}{V} \sum_{\mathbf{p}_1\mathbf{p}_2} e^{-\frac{i}{\hbar}\mathbf{x}_1\cdot(\mathbf{p}_1 - \mathbf{p}_2)}$$

$$(\mathbf{p}_1 + \mathbf{p}_2) \cdot \mathbf{A}_\omega(\mathbf{x}_1)\, A(\mathbf{p}, \mathbf{p}_1; E_1) A(\mathbf{p}_2, \mathbf{p}'; E_2) . \tag{7.217}$$

This is the momentum representation of the expression in eq.(7.128), and we obtain the same result for the response function as previously, eq.(7.211).

7.9   Scattering and Correlation Functions

In this section we shall consider transport of particles (neutrons, photons, etc.) through matter. To be specific we consider the scattering of slow neutrons by a piece of matter.[13] A neutron interacts with the nuclei of the substance (all assumed identical). The interaction potential is short ranged, and we take for the interaction with the nucleus at position $\mathbf{R}_N$[14]

$$V(\mathbf{r}_n - \mathbf{R}_N) = a\,\delta(\mathbf{r}_n - \mathbf{R}_N)\,. \tag{7.218}$$

We have thus neglected the spin of the nuclei (or consider the case of spinless bosons).[15] For the interaction of a neutron with the nuclei of the substance we then have

$$V(\mathbf{r}_n) = \sum_N V(\mathbf{r}_n - \mathbf{R}_N) = a\sum_N \delta(\mathbf{r}_n - \mathbf{R}_N) \tag{7.219}$$

The interaction is weak, and the scattering can be treated in the Born approximation. For the transition rate between initial and final states we then have

$$\Gamma_{fi} = \frac{2\pi}{\hbar}\,|<f|\sum_N V(\hat{\mathbf{r}}_n - \hat{\mathbf{R}}_N)|i>|^2\,\delta(E_f - E_i)\,. \tag{7.220}$$

For simplicity we assume that the states of the substance can be labeled solely by their energy

$$|i> = |\mathbf{p}', E_S^{(i)}> = |\mathbf{p}'>|E_S^{(i)}>\,, \qquad |f> = |\mathbf{p}, E_S^{(f)}> = |\mathbf{p}>|E_S^{(f)}> \tag{7.221}$$

where the initial and final energies are

$$E_f = E_S^{(f)} + \frac{\mathbf{p}^2}{2m_n}\,, \qquad E_i = E_S^{(i)} + \frac{\mathbf{p}'^2}{2m_n} \tag{7.222}$$

and m_n is the mass of the neutron. We introduce the energy transfer from the neutron to the material

$$\hbar\omega = \frac{\mathbf{p}'^2}{2m_n} - \frac{\mathbf{p}^2}{2m_n} = E_S^{(f)} - E_S^{(i)} \tag{7.223}$$

and we have for the transition probability per unit time

$$\Gamma_{fi} = \frac{2\pi}{\hbar}|<\mathbf{p}, E_S^{(f)}|a\sum_N \delta(\hat{\mathbf{r}}_n - \hat{\mathbf{R}}_N)|\mathbf{p}', E_S^{(i)}>|^2\,\delta(E_S^{(f)} - E_S^{(i)} - \hbar\omega)\,. \tag{7.224}$$

Since the interaction is inelastic, the differential cross section of interest is the fraction of incident neutrons with momentum $\mathbf{p}'$ being scattered into a unit solid angle $d\hat{\mathbf{p}}$ with energy in the range between ϵ and $\epsilon + d\epsilon$, $d^2\sigma/d\hat{\mathbf{p}}\,d\epsilon$. Noting that

$$\Delta\mathbf{p} = p^2\,dp\,d\hat{\mathbf{p}} = m_n\,p\,d\epsilon\,d\hat{\mathbf{p}} \tag{7.225}$$

---

[13]In chapter 10 we consider electrons scattering off phonons or electron density fluctuations.

[14]We thus exclude the possibility of any nuclear reaction taking place.

[15]However, it is precisely the magnetic moment of the neutron that makes it an ideal tool to investigate the magnetic properties of matter. The subject of neutron scattering is thus vast, and for a general reference we refer to [31].

we obtain from eq.(2.178) for the inelastic differential cross section for neutron scattering off the substance

$$\frac{d^2\sigma}{d\hat{\mathbf{p}}\,d\epsilon} = \frac{m_n^2\,L^6}{(2\pi\hbar)^3}\,\frac{p}{p'}\,\frac{2\pi}{\hbar}\,|<\mathbf{p},E_S^{(f)}|a\sum_N \delta(\hat{\mathbf{r}}_n - \hat{\mathbf{R}}_N)|\mathbf{p}',E_S^{(i)}>|^2$$

$$\delta(E_S^{(f)} - E_S^{(i)} - \hbar\omega) \tag{7.226}$$

which we can express as

$$\frac{d^2\sigma}{d\hat{\mathbf{p}}\,d\epsilon} = \frac{m_n^2\,a^2}{(2\pi\hbar)^3}\,\frac{p}{p'}\,\frac{2\pi}{\hbar}\int d\mathbf{x}\int d\mathbf{x}'\,e^{-\frac{i}{\hbar}(\mathbf{x}-\mathbf{x}')\cdot(\mathbf{p}-\mathbf{p}')}\int_{-\infty}^{\infty}\frac{d(t-t')}{2\pi\hbar}\,e^{-i(t-t')\omega}$$

$$<E_S^{(f)}|\hat{n}(\mathbf{x},t)|E_S^{(i)}> \; <E_S^{(i)}|\hat{n}(\mathbf{x}',t')|E_S^{(f)}> \tag{7.227}$$

where $\hat{n}(\mathbf{x},t)$ is the density operator for the nuclei of the material in the Heisenberg picture with respect to the substance Hamiltonian $\hat{H}_S$, the density operator in the Schrödinger picture being $\hat{n}(\mathbf{x}) = \sum_N \delta(\mathbf{x} - \hat{\mathbf{R}}_N)$.

Exercise 7.3 *Show that for scattering off a single heavy nucleus, $M \gg m_n$, we have for the total cross section*

$$\sigma = \int_{4\pi} d\hat{\mathbf{p}}\int_0^\infty d\epsilon\,\frac{d^2\sigma}{d\hat{\mathbf{p}}\,d\epsilon} = 4\pi\left(\frac{m_n\,a}{2\pi\hbar^2}\right)^2. \tag{7.228}$$

In the scattering experiment we only know the probability distribution for the initial state of the material, which we shall assume to be the thermal equilibrium state

$$\hat{\rho}_S = \sum_\lambda |E_S(\lambda)> P(E_S(\lambda)) <E_S(\lambda)| \tag{7.229}$$

where

$$P(E_S(\lambda)) = \frac{e^{-E_S(\lambda)/kT}}{Z_S}, \qquad \hat{H}_S\,|E_S(\lambda)> = E_S(\lambda)\,|E_S(\lambda)> . \tag{7.230}$$

For the transition rate weighted over the thermal mixture of initial states of the substance we have

$$\Gamma_{f\mathbf{p}'} \equiv \sum_\lambda P(E_S^{(i)}(\lambda))\,\Gamma_{fi}$$

$$= \frac{m_n^2\,a^2}{(2\pi\hbar)^3}\,\frac{p}{p'}\,\frac{2\pi}{\hbar}\sum_\lambda P(E_S(\lambda))\int d\mathbf{x}\int d\mathbf{x}'\int_{-\infty}^{\infty}\frac{d(t-t')}{2\pi\hbar}\,e^{-i(t-t')\omega}$$

$$e^{-\frac{i}{\hbar}(\mathbf{x}-\mathbf{x}')\cdot(\mathbf{p}-\mathbf{p}')}\,<E_S^{(f)}|\hat{n}(\mathbf{x},t)|E_S(\lambda)> <E_S(\lambda)|\hat{n}(\mathbf{x}',t')|E_S^{(f)}> \tag{7.231}$$

and we obtain for the weighted differential cross section (we use the same notation)

$$\frac{d^2\sigma}{d\hat{\mathbf{p}}\,d\epsilon} = \sum_\lambda P(E_S^{(i)}(\lambda))\,\frac{d^2\sigma}{d\hat{\mathbf{p}}\,d\epsilon}$$

$$= \frac{m_n^2\,a^2}{(2\pi\hbar)^3}\,\frac{p}{p'}\,\frac{2\pi}{\hbar}\sum_\lambda P(E_S(\lambda))\int dx \int dx' \int_{-\infty}^{\infty} \frac{d(t-t')}{2\pi\hbar}\,e^{-i(t-t')\omega}$$

$$e^{-\frac{i}{\hbar}(\mathbf{x}-\mathbf{x}')\cdot(\mathbf{p}-\mathbf{p}')} < E_S^{(f)}|\hat{n}(\mathbf{x},t)|E_S(\lambda)> < E_S(\lambda)|\hat{n}(\mathbf{x}',t')|E_S^{(f)}> . \qquad (7.232)$$

Furthermore, in the experiment the final state of the substance is not measured, and we must sum over all possible final states of the substance, and we obtain finally for the observed differential cross section (we use the same notation)

$$\frac{d^2\sigma}{d\hat{\mathbf{p}}\,d\epsilon} = \frac{m_n^2\,a^2}{(2\pi\hbar)^3}\,\frac{p}{p'}\,\frac{2\pi}{\hbar}\int dx \int dx'\, e^{-\frac{i}{\hbar}(\mathbf{x}-\mathbf{x}')\cdot(\mathbf{p}'-\mathbf{p})}\int_{-\infty}^{\infty} \frac{d(t-t')}{2\pi\hbar}\,e^{i(t-t')\omega}$$

$$< \hat{n}(\mathbf{x},t)\,\hat{n}(\mathbf{x}',t')> \qquad\qquad (7.233)$$

where the bracket denotes the weighted trace with respect to the state of the substance

$$< \hat{n}(\mathbf{x},t)\,\hat{n}(\mathbf{x}',t')> \equiv tr_S(\hat{\rho}_S\,\hat{n}(\mathbf{x},t)\,\hat{n}(\mathbf{x}',t')) . \qquad (7.234)$$

We thus obtain the formula

$$\frac{d^2\sigma}{d\hat{\mathbf{p}}\,d\epsilon} = \frac{p}{p'}\,\frac{m_n^2\,a^2}{(2\pi\hbar)^3\hbar^2}\,V\,S(\mathbf{q},\omega) \qquad (7.235)$$

where $S(\mathbf{q},\omega)$ is the Fourier transform of the space-time density correlation function $S(\mathbf{x},t;\mathbf{x}',t') \equiv < \hat{n}(\mathbf{x},t)\,\hat{n}(\mathbf{x}',t') >$, and $\hbar\mathbf{q} \equiv \mathbf{p}' - \mathbf{p}$ and $\hbar\omega$ is the momentum and energy transfer from the neutron to the substance. We note that $S(-\mathbf{q},-\omega) = S(\mathbf{q},\omega)$. This correlation function is often referred to as the dynamic structure factor.[16] The dynamic structure factor gives the number of density excitations of the system with a given energy and momentum. A scattering experiment is thus a measurement of the density correlation function.

Exercise 7.4 *Show that for a target consisting of a single nucleus of mass M in the thermal state, the dynamic structure factor is given by*

$$S(\mathbf{q},\omega) = \frac{1}{V}\sqrt{\frac{2\pi M}{kTq^2}}\,e^{-\frac{M}{2kTq^2}\left(\omega - \frac{\hbar q^2}{2M}\right)} . \qquad (7.236)$$

---

[16]We here follow the conventional notation, although in the standard notation of this chapter we have $S(\mathbf{x},\mathbf{x}',\omega) = \chi^>(\mathbf{x},\mathbf{x}',\omega)$. According to the fluctuation-dissipation theorem, the structure function is related to the density response function according to $S(\mathbf{x},\mathbf{x}',\omega) = 2\hbar n(\omega)\,\Im m\,\chi(\mathbf{x},\mathbf{x}',\omega)$.

Solution

The dynamic structure factor is then the Fourier transform with respect to time of

$$S(\mathbf{q}, t) = \frac{Tr\left(e^{-\frac{\hat{\mathbf{P}}_N^2}{2MkT}} e^{-i\mathbf{q}\cdot\hat{\mathbf{R}}_N(t)} e^{i\mathbf{q}\cdot\hat{\mathbf{R}}_N}\right)}{V\, Tr\left(e^{-\frac{\hat{\mathbf{P}}_N^2}{2MkT}}\right)} \equiv \frac{1}{V} < e^{-i\mathbf{q}\cdot\hat{\mathbf{R}}_N(t)} e^{i\mathbf{q}\cdot\hat{\mathbf{R}}_N} > . \quad (7.237)$$

Using the analog of eq. (1.310)

$$e^{-i\mathbf{q}\cdot\hat{\mathbf{R}}_N} \hat{\mathbf{p}}_N e^{i\mathbf{q}\cdot\hat{\mathbf{R}}_N} = \hat{\mathbf{p}}_N + \hbar\mathbf{q}\,\hat{I} \quad (7.238)$$

we obtain

$$S(\mathbf{q}, t) = \frac{1}{V} e^{-\frac{i\hbar t}{2M}q^2} < e^{-\frac{it}{M}\mathbf{q}\cdot\hat{\mathbf{p}}_N} > \quad (7.239)$$

and by completing the square

$$S(\mathbf{q}, t) = \frac{1}{V} e^{-\frac{q^2}{2M}\left(t^2 kT - i\hbar t\right)} \quad (7.240)$$

and by Fourier transformation the stated result.

For the differential cross section of a Boltzmann gas of N noninteracting nuclei we have according to eq. (7.236)

$$\frac{d^2\sigma}{d\hat{\mathbf{p}}\, d\epsilon} = N \frac{m_n^2 a^2}{(2\pi\hbar)^3} \frac{p}{p'} \frac{2\pi}{\hbar} S(\mathbf{q}, \omega)$$

$$= N \frac{m_n^2 a^2}{(2\pi\hbar)^3} \frac{p}{p'} \frac{2\pi}{\hbar} \int_{-\infty}^{\infty} dt\, e^{-i\omega t} e^{-\frac{q^2}{2M}\left(t^2 kT - i\hbar t\right)} . \quad (7.241)$$

Exercise 7.5 *Show that the limiting behavior of the total cross section for a Boltzmann gas is*

$$\sigma = \int_{4\pi} d\hat{\mathbf{p}} \int_0^\infty d\epsilon\, \frac{d^2\sigma}{d\hat{\mathbf{p}}\, d\epsilon}$$

$$= \begin{cases} 4\pi N \left(\frac{m_n a}{2\pi\hbar^2}\right)^2 \dfrac{2}{\sqrt{\pi}\left(1 + \frac{m_n}{M}\right)^2 \sqrt{\frac{M p'^2}{2m_n^2 kT}}} & \text{for} \quad \frac{M p'^2}{2m_n^2 kT} \ll 1 \\[20pt] 4\pi N \left(\frac{m_n a}{2\pi\hbar^2}\right)^2 \dfrac{1}{\left(1 + \frac{m_n}{M}\right)^2} & \text{for} \quad \frac{M p'^2}{2m_n^2 kT} \gg 1 . \end{cases} \quad (7.242)$$

The divergent result for low energies is due to the almost vanishing flux of incoming neutrons being scattered by the moving nuclei in the gas, and in the opposite limit we recover the result for scattering off N free and noninteracting nuclei.

If the energy of the incident neutron is far in excess of the energy transfer, $\epsilon_{\mathbf{p}'} \gg \hbar\omega$, we obtain for the quasi-elastic cross section (equivalent to replacing $\delta(E_S^{(f)} - E_S^{(i)} - \hbar\omega) \to \delta(\hbar\omega)$ in eq.(7.224))

$$\frac{d\sigma}{d\hat{\mathbf{p}}} = \int_{\Delta\epsilon} d\epsilon \, \frac{d\sigma}{d\hat{\mathbf{p}}d\epsilon} \simeq \frac{m_n^2 a^2}{(2\pi\hbar)^3\hbar} V \int_{\Delta\omega} d\omega \, S(\mathbf{q},\omega) \simeq \frac{m_n^2 a^2}{(2\pi\hbar)^3\hbar} V \, S(\mathbf{q}) \quad (7.243)$$

where

$$S(\mathbf{q}) = \int_{-\infty}^{\infty} d\omega \, S(\mathbf{q},\omega) = \int d(\mathbf{x} - \mathbf{x}') \, e^{-i(\mathbf{x}-\mathbf{x}')\cdot\mathbf{q}} <\hat{n}(\mathbf{x})\,\hat{n}(\mathbf{x}')> \quad (7.244)$$

is the static structure factor.

In the following we shall have a gas or a liquid in mind[17] and consider a classical treatment of the nuclei in the material where we can ascribe the heavy nuclei positions. The average density of nuclei (or particles as we shall refer to) we denote by n, $n \equiv <n(\mathbf{x})>$. Consider a volume $\Delta\mathbf{x}$ of the substance that is so small that it contains either one particle or zero particles (in other words, the probability for more than one particle in volume $\Delta\mathbf{x}$ is negligible). We thus have the probabilities $n\,\Delta\mathbf{x}$ and $1 - n\,\Delta\mathbf{x}$ for the presence and absence of a particle in a volume $\Delta\mathbf{x}$, respectively. We introduce the conditional probability $p(\mathbf{x};\mathbf{x}')\,\Delta\mathbf{x}$ for a particle to be in the volume $\Delta\mathbf{x}$ given that there is a particle in volume $\Delta\mathbf{x}'$. Since we assume a spatially homogeneous state, we have $p(\mathbf{x};\mathbf{x}') = p(\mathbf{x}-\mathbf{x}') = p(|\mathbf{x}-\mathbf{x}'|)$. Defining $\mathbf{r} = \mathbf{x} - \mathbf{x}'$, we observe

$$p(r) = \begin{cases} 0 & \text{for } r = 0 \\ n & \text{for } r = \infty \end{cases} . \quad (7.245)$$

as particles separated by a large distance are uncorrelated, and the probability for two particles arbitrarily close to each other is negligible.

Since a volume of size $\Delta\mathbf{x}$ can only contain one or zero particles, we have for different points, $\mathbf{x}' \neq \mathbf{x}$,

$$<n(\mathbf{x})\,n(\mathbf{x}')> = n\,p(\mathbf{x};\mathbf{x}') . \quad (7.246)$$

However, as the small volume can only contain one or no particle, we have

$$\int_{\Delta\mathbf{x}} d\mathbf{x} \int_{\Delta\mathbf{x}} d\mathbf{x}' \, <n(\mathbf{x})\,n(\mathbf{x}')> = <\left(\int_{\Delta\mathbf{x}} d\mathbf{x}\, n(\mathbf{x})\right)^2> = n\,\Delta\mathbf{x} \quad (7.247)$$

[17]The treatment of a solid would give us the Bragg peak structure.

and we have, valid for all choices of positions,

$$< n(\mathbf{x})\, n(\mathbf{x}') > \; = \; n\, \delta(\mathbf{x} - \mathbf{x}') + n\, p(\mathbf{x}; \mathbf{x}') \,. \qquad (7.248)$$

Introducing the pair correlation function

$$g(\mathbf{x} - \mathbf{x}') \; = \; p(\mathbf{x} - \mathbf{x}') - n \qquad (7.249)$$

we can write for the variance of the density fluctuations

$$< (n(\mathbf{x}) - n)\,(n(\mathbf{x}') - n) > \; = \; n\, \delta(\mathbf{x} - \mathbf{x}') + n\, g(\mathbf{x} - \mathbf{x}') \,. \qquad (7.250)$$

Integrating this expression over a large volume V we get

$$\int_V d\mathbf{x} \int_V d\mathbf{x}' \; < (n(\mathbf{x}) - n)\,(n(\mathbf{x}') - n) > \; = \; < (N_V - <N_V>)\,(N_V - <N_V>) >$$

$$\equiv \; < (\Delta N_V)^2 > \qquad (7.251)$$

and therefore, if the volume V is much larger than the range of g, we have for the fluctuations in the number of particles in volume V

$$< (\Delta N_V)^2 > \; = \; n V \left(1 + \int_V d\mathbf{r}\, g(\mathbf{r}) \right) \qquad (7.252)$$

or equivalently

$$\int_V d\mathbf{r}\, g(\mathbf{r}) \; = \; \frac{< (\Delta N_V)^2 >}{< N_V >} - 1 \,. \qquad (7.253)$$

The fluctuations in the number of particles in a fixed volume is related to the compressibility

$$\frac{< (\Delta N_V)^2 >}{< N_V >} \; = \; -\frac{kT < N_V >}{V^2} \left(\frac{\partial V}{\partial P} \right)_T \,. \qquad (7.254)$$

For an ideal gas we thus obtain from eq.(7.253)

$$\int_V d\mathbf{r}\, g(\mathbf{r}) \; = \; 0 \,. \qquad (7.255)$$

Exercise 7.6 *Sketch the form of the pair correlation function, $g(r)$, for a gas of inert atoms of size a.*

Exercise 7.7 *Verify the limiting behavior of the static structure factor*

$$
S(\mathbf{q}) = \begin{cases} -\frac{kTn^2}{V}\left(\frac{\partial V}{\partial P}\right)_T & \text{for } q = 0 \\[2ex] n & \text{for } q = \infty . \end{cases} \tag{7.256}
$$

Approaching the liquid-gas transition, the compressibility of a substance diverges, and near the transition the density fluctuations become anomalously large. The free energy

$$
F = \int_V d\mathbf{x}\, f(\mathbf{x}) \tag{7.257}
$$

fluctuates, and we have for a fluctuation in the free energy

$$
\Delta F = \int_V d\mathbf{x}\, (f(\mathbf{x}) - <f(\mathbf{x})>) . \tag{7.258}
$$

We can express the free energy fluctuations in terms of the density fluctuations. In the mean field approximation we have for the free energy for a density profile $n(\mathbf{x})$[18]:

$$
\Delta F[n(\mathbf{x})] = \int_V d\mathbf{x}\, \left(\frac{1}{2}\alpha\,(n(\mathbf{x}) - n)^2 + \frac{1}{2}\beta\,(\nabla n(\mathbf{x}))^2\right) \tag{7.259}
$$

where α and β are functions of the temperature.

We introduce the Fourier transform of the density deviation

$$
\delta n_{\mathbf{q}} = \frac{1}{V}\int d\mathbf{x}\, e^{i\mathbf{q}\cdot\mathbf{x}}\,(n(\mathbf{x}) - n) \tag{7.260}
$$

and obtain

$$
\Delta F_{tot}[n(\mathbf{x})] = \frac{1}{2}V\sum_{\mathbf{q}\neq 0}(\alpha + \beta\,q^2)\,|\delta n_{\mathbf{q}}|^2 \tag{7.261}
$$

i.e., the fluctuations at different wavelengths are independent.

The probability for a density fluctuation is given by

$$
P[n(\mathbf{x})] \propto e^{-\frac{\Delta F[n_{\mathbf{x}}]}{kT}} = e^{-\frac{V}{2kT}\sum_{\mathbf{q}\neq 0}(\alpha+\beta\,q^2)\,|\delta n_{\mathbf{q}}|^2} . \tag{7.262}
$$

As the fluctuations at different wave vectors are independent, we can deal with them independently, and since $\delta n_{\mathbf{q}}^* = \delta n_{-\mathbf{q}}$, a density fluctuation

$$
\delta n(\mathbf{x}) = \sum_{\mathbf{q}\neq 0}\delta n_{\mathbf{q}}e^{i\mathbf{q}\cdot\mathbf{x}} = \frac{1}{2}\sum_{\mathbf{q}\neq 0}\left(\delta n_{\mathbf{q}}e^{i\mathbf{q}\cdot\mathbf{x}} + \delta n_{\mathbf{q}}^* e^{-i\mathbf{q}\cdot\mathbf{x}}\right) \tag{7.263}
$$

---

[18]The fluctuations in the density and in the temperature are independent and the temperature can be taken constant.

is completely specified by either $\delta n_{\mathbf{q}}$ or $\delta n_{-\mathbf{q}}$. Since $|\delta n_{-\mathbf{q}}|^2 = |\delta n_{\mathbf{q}}|^2$, $|\delta n_{\mathbf{q}}|^2$ appears twice in eq.(7.262), and we have for the probability for the fluctuation at wave length $\lambda = 2\pi/q$

$$P(\delta n_{\mathbf{q}}) \;=\; \frac{1}{\mathcal{N}_q} \, e^{-\frac{V}{kT}(\alpha+\beta q^2)|\,\delta n_{\mathbf{q}}|^2}$$

$$=\; \frac{1}{\mathcal{N}_q} \, e^{-\frac{V}{kT}(\alpha+\beta q^2)\,(\Re e\,\delta n_{\mathbf{q}})^2} \, e^{-\frac{V}{kT}(\alpha+\beta q^2)\,(\Im m\,\delta n_{\mathbf{q}})^2} \qquad (7.264)$$

where we have for the normalization constant

$$\mathcal{N}_q \;=\; \int\limits_{-\infty}^{\infty} d(\Re e\,n_{\mathbf{q}}) \int\limits_{-\infty}^{\infty} d(\Im m\,n_{\mathbf{q}}) \, P(\delta n_{\mathbf{q}}) \;=\; \frac{\pi}{\frac{kT}{V}(\alpha+\beta q^2)} \,. \qquad (7.265)$$

For the mean square fluctuation we get

$$<|\delta n_{\mathbf{q}}|^2> \;\equiv\; \int\limits_{-\infty}^{\infty} d(\Re e\,n_{\mathbf{q}}) \int\limits_{-\infty}^{\infty} d(\Im m\,n_{\mathbf{q}}) \, |\delta n_{\mathbf{q}}|^2 \, P(\delta n_{\mathbf{q}}) \;=\; \frac{kT}{V(\alpha+\beta q^2)} \,. \qquad (7.266)$$

Since

$$g(\mathbf{q}) \;=\; \int d\mathbf{r}\, e^{-i\mathbf{r}\cdot\mathbf{q}}\, g(\mathbf{r}) \;=\; \frac{V<|\delta n_{\mathbf{q}}|^2>}{n} \;-\; 1 \qquad (7.267)$$

we obtain for the pair correlation function, as the last term above can be neglected near the liquid-gas transition,

$$g(\mathbf{r}) \;=\; \frac{kT}{4\pi n\beta} \, \frac{e^{-r/\xi}}{r} \qquad (7.268)$$

where

$$\xi \;=\; \sqrt{\frac{\beta}{\alpha}} \qquad (7.269)$$

is the correlation length for density fluctuations. For the static structure factor we thus have

$$S(\mathbf{q}) \;=\; n\left(\frac{kT}{n\beta} \frac{1}{q^2 + \xi^{-2}} + N\,\delta_{\mathbf{q},0} \right) \qquad (7.270)$$

Under ordinary conditions the correlation length is of interatomic distance, but close to the liquid-gas transition, near the critical point, the correlation between the positions of different particles decays very slowly, we have long-range density correlations. At the critical point we have $\alpha_{\mathrm{cr}} = 0$ and $\xi_{\mathrm{cr}} = \infty$. Near the liquid-gas transition, small angle scattering is thus strongly enhanced (in contrast to the gas phase where the static structure factor, in the present model, is independent of $\mathbf{q}$), the phenomenon of critical opalescence.

Chapter 8

Linear Response of Disordered Conductors

We shall now consider the impurity-averaged current and density responses. We shall base our presentation on the diagrammatic technique developed previously, and develop the diagrammatic method directly for the correlation functions which describe the linear response of a system. We will thereby benefit from the transparent physical picture provided by the diagrammatic representation. Following the prescription of chapter 3, we shall develop the diagrammatic impurity-averaging technique for the transport coefficients of a disordered system. The existence of a small parameter in the problem, $\lambda_F/l \ll 1$, will make a systematic treatment possible. We will first discuss in general the density response, and afterwards the current response. We then establish what the Boltzmann conductivity amounts to in terms of conductivity diagrams, and describe the diffusion propagator of section 5.6 in terms of diagrams.

A physical property of a given sample depends on its specific impurity potential, as given by the impurity positions $\{\mathbf{r}_1, \mathbf{r}_2, \ldots, \mathbf{r}_N\}$. When we average over all possible impurity configurations, the average of a physical quantity will have the symmetry pertaining to the system in the absence of impurities. In the following we shall consider the Sommerfeld model. For a given realization of the impurity potential, the system does not posses translation invariance, and the response at point $\mathbf{x}$ to an external field applied at point $\mathbf{x}'$ will depend on both spatial variables $\mathbf{x}$ and $\mathbf{x}'$ separately. In contrast, when we average over all possible impurity configurations, the average response only depends on the separation between the points, $\mathbf{x} - \mathbf{x}'$, as translational invariance is recovered.[1]

8.1 Impurity-Averaged Density Response

The impurity-averaged density response is, to linear order in the applied potential, described by the impurity average of the product of two propagators, see eq.(7.60). Performing the impurity average of the density response function is analogous to

---

[1]In the model with spherical Fermi surface we also regain rotational invariance.

the average we performed for the density matrix: expand the propagators in terms of the impurity potential, and make all possible impurity correlation connections. When impurity averaging a product of two propagators of any type, the momentum conservation in the diagrammatic expansion at each impurity correlator leads to a nonzero result for any such diagram only when the sum of the external incoming and outgoing momenta are equal, since only then can the Kronecker functions associated with the impurity correlators be satisfied. This is illustrated in the equation below:

$$
\left\langle \begin{array}{c} \mathbf{p}+\hbar\mathbf{q}'/2 \xrightarrow[E_+]{R} \mathbf{p}'+\hbar\mathbf{q}/2 \\ \\ \mathbf{p}-\hbar\mathbf{q}'/2 \xrightarrow[E]{A} \mathbf{p}'-\hbar\mathbf{q}/2 \end{array} \right\rangle = \delta_{\mathbf{q}\mathbf{q}'} \left\langle \begin{array}{c} \mathbf{p}_+ \xrightarrow[E_+]{R} \mathbf{p}'_+ \\ \\ \mathbf{p}_- \xrightarrow[E]{A} \mathbf{p}'_- \end{array} \right\rangle \tag{8.1}
$$

where we have introduced the notation

$$
\mathbf{p}_\pm = \mathbf{p} \pm \frac{\hbar\mathbf{q}}{2} \qquad \mathbf{p}'_\pm = \mathbf{p}' \pm \frac{\hbar\mathbf{q}}{2} \qquad E_+ = E + \hbar\omega . \tag{8.2}
$$

In the retarded and advanced case, for example, eq.(8.1) corresponds to the analytical expression

$$
< G^R(\mathbf{p} + \hbar\mathbf{q}'/2, \mathbf{p}' + \hbar\mathbf{q}/2; E_+) \, G^A(\mathbf{p}' - \hbar\mathbf{q}/2, \mathbf{p} - \hbar\mathbf{q}'/2; E) >
$$

$$
= \delta_{\mathbf{q}',\mathbf{q}} < G^R(\mathbf{p}_+, \mathbf{p}'_+; E_+) \, G^A(\mathbf{p}'_-, \mathbf{p}_-; E) > . \tag{8.3}
$$

The impurity-averaged density response function therefore has the form

$$
< \chi(\mathbf{q}', \mathbf{q}, \omega) > \; = \; \delta_{\mathbf{q}',\mathbf{q}} < \chi(\mathbf{q}, \mathbf{q}, \omega) > \; \equiv \; \chi(\mathbf{q}, \omega) \, \delta_{\mathbf{q}',\mathbf{q}} . \tag{8.4}
$$

Fourier transforming, we get the translational invariant expression

$$
\chi(\mathbf{x} - \mathbf{x}', \omega) \equiv \; <\chi(\mathbf{x}, \mathbf{x}', \omega)> \; = \; <\chi(\mathbf{x} - \mathbf{x}', 0, \omega)> . \tag{8.5}
$$

For the impurity averaged-density we therefore only get a response at the wave vector of the applied potential

$$
< \delta n(\mathbf{q}', \omega) > \; = \; < \chi(\mathbf{q}', \mathbf{q}, \omega) > V(\mathbf{q}, \omega) = \delta n(\mathbf{q}, \omega) \, \delta_{\mathbf{q}',\mathbf{q}} . \tag{8.6}
$$

The response at the imposed wave vector is

$$
\delta n(\mathbf{q}, \omega) = \chi(\mathbf{q}, \omega) \, V(\mathbf{q}, \omega) \tag{8.7}
$$

where the response function

$$
\chi(\mathbf{q}, \omega) = \chi^{RA}(\mathbf{q}, \omega) + \chi^{RR}(\mathbf{q}, \omega) + \chi^{AA}(\mathbf{q}, \omega) \tag{8.8}
$$

consists of the terms

$$\chi^{RA}(\mathbf{q}, \omega) = -\frac{i}{\pi} \frac{1}{V} \sum_{\mathbf{pp'}} \int_{-\infty}^{\infty} dE \; (f_0(E + \hbar\omega) - f_0(E))$$

$$< G^R(\mathbf{p}_+, \mathbf{p}'_+; E + \hbar\omega) \, G^A(\mathbf{p}'_-, \mathbf{p}_-; E) > \tag{8.9}$$

and

$$\chi^{RR}(\mathbf{q}, \omega) = \frac{-i}{\pi V} \sum_{\mathbf{pp'}} \int_{-\infty}^{\infty} dE \; f_0(E) < G^R(\mathbf{p}_+, \mathbf{p}'_+; E + \hbar\omega) \, G^R(\mathbf{p}'_-, \mathbf{p}_-; E) > \tag{8.10}$$

and

$$\chi^{AA}(\mathbf{q}, \omega) = \frac{i}{\pi} \frac{1}{V} \sum_{\mathbf{pp'}} \int_{-\infty}^{\infty} dE \; f_0(E) < G^A(\mathbf{p}_+, \mathbf{p}'_+; E) \, G^A(\mathbf{p}'_-, \mathbf{p}_-; E - \hbar\omega) > . \tag{8.11}$$

In the first term we note that the E-integration is constrained to values near the Fermi energy, $E \simeq \epsilon_F$. A term where we have products of two propagators of the same kind, say, two retarded propagators, suggests that such a term is of order $\hbar/p_F l$; however, due to the large range of E-integration for such a term it will be of the same order of magnitude as the first term, eq.(8.9). We show this in detail in section 8.3.1.

8.2 Four-Point Vertex

Having expressed the response function in terms of propagators, we can now make use of our previous diagrammatic experience in order to explore the average value. The typical structure of the impurity-averaged response function is the impurity average of a product of propagators; for example,

$$\frac{1}{V} \, \Phi_{\mathbf{pp'}}(E, \mathbf{q}, \omega) \equiv \; < G^R(\mathbf{p}_+, \mathbf{p}'_+; E + \hbar\omega) \, G^A(\mathbf{p}'_-, \mathbf{p}_-; E) >$$

$$= \left\langle \quad \right\rangle . \tag{8.12}$$

The lower momentum labeling on Φ is half the sum of the momenta on the left and right, respectively, and the momentum value in the parenthesis the difference between the upper and lower momentum labeling.

In order to perform the impurity average, we expand G^R and G^A in terms of the impurity potential, and make all possible impurity correlation connections.

There emerge two distinct classes of diagrams: a class where the upper (say R) and lower (say A) particle lines are not connected by impurity correlator lines, giving the product of the impurity-averaged propagators, and a class where they are connected, specified by the four-point vertex function, $\Gamma_{\mathbf{p},\mathbf{p}'}^{RA}(E,\mathbf{q},\omega)$,

$$(8.13)$$

Analytically the equation reads,[2]

$$\Phi_{\mathbf{p},\mathbf{p}'}(E,\mathbf{q},\omega) = G^R(\mathbf{p}_+,E_+)\,G^A(\mathbf{p}_-,E)\,[V\,\delta_{\mathbf{p},\mathbf{p}'}$$

$$+ \Gamma_{\mathbf{p}\mathbf{p}'}^{RA}(E,\mathbf{q},\omega)\,G^R(\mathbf{p}'_+,E_+)G^A(\mathbf{p}'_-,E)] \qquad (8.14)$$

where $\Gamma_{\mathbf{p},\mathbf{p}'}^{RA}(E,\mathbf{q},\omega)$ encompasses all the amputated skeleton diagrams where the upper and lower particle lines are connected by impurity correlators in all possible ways ($\hbar\mathbf{Q} \equiv \mathbf{p} + \mathbf{p}'$)

---

[2]Here V denotes the volume of the system, and no confusion with the notation for the applied potential should arise.

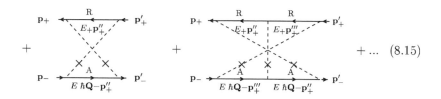

$$(8.15)$$

The exhibited diagrams belong to two important classes of diagrams, the ladder and maximally crossed diagrams, whose significance we shall discuss in detail later. We have not depicted multiple scattering diagrams. This, however, does not imply that we at present are assuming the Gaussian approximation.

As usual it is convenient to introduce irreducible sets of diagrams, here the two-line irreducible four-point vertex function $U_{\mathbf{p},\mathbf{p}'}^{RA}(E,\mathbf{q},\omega)$, consisting of all the skeleton diagrams that can not be cut in two by cutting only an upper and a lower particle line. In terms of the irreducible vertex we have the iterative equation

$$\Gamma_{\mathbf{p},\mathbf{p}'}^{RA}(E,\mathbf{q},\omega) = U_{\mathbf{p},\mathbf{p}'}^{RA}(E,\mathbf{q},\omega) + \frac{1}{V}\sum_{\mathbf{p}''} U_{\mathbf{p},\mathbf{p}''}^{RA}(E,\mathbf{q},\omega)$$

$$G^{R}(\mathbf{p}''_{+},E_{+})\,G^{A}(\mathbf{p}''_{-},E)\,\Gamma_{\mathbf{p}'',\mathbf{p}'}^{RA}(E,\mathbf{q},\omega) \qquad (8.16)$$

which simply expresses that any given diagram is either two-line irreducible, or two-line reducible once, twice, etc. (we occasionally suppress the retarded and advanced labels as they are specified by the context, and are of no relevance for the topological classification). Diagrammatically we have that the four-point vertex satisfies the iterative equation

$$(8.17)$$

the energy variable, E, is nonexchangeable between the particle lines, reflecting that the scattering is elastic, and are therefore suppressed.

The irreducible four-point function $U_{\mathbf{p},\mathbf{p}'}^{RA}(E,\mathbf{q},\omega)$ has the perturbative diagrammatic structure ($\hbar\mathbf{k} \equiv \mathbf{p} - \mathbf{p}'$, $\hbar\mathbf{Q} \equiv \mathbf{p} + \mathbf{p}'$)

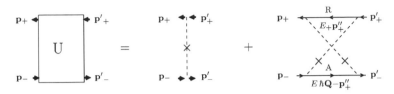

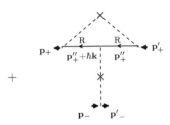

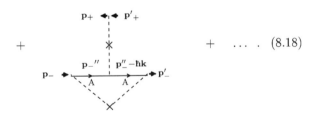

$$+ \quad \ldots \ . \quad (8.18)$$

If we append a full propagator line at the right or left entries of U we get topologically a skeleton self-energy diagram since the U-diagrams are two-line irreducible.

Exercise 8.1 *Draw all the skeleton diagrams for the irreducible four-point function U with three impurity correlators (recall exercise 3.2 on page 150).*

8.3 Impurity-Averaged Response Function

We shall now assess the physical content of the various contributions to the impurity-averaged density response function, eq.(8.8). Employing the developed diagrammatic representation of the involved correlation functions facilitates an easy assessment of the order of magnitude of the various contributions in the expansion parameter, $\hbar/\epsilon_F\tau$.

8.3.1 Regular Part of Response Function

The RR and AA terms in the response function will only give rise to regular terms, i.e., analytic terms exhibiting no poles in the complex ω-plane. In these terms we can neglect the contribution from diagrams where the upper and lower lines are connected. Consider, for example, the term

$$\Phi^{RR}_{\mathbf{pp}'}(E,\mathbf{q},\omega) \equiv V \langle G^R(\mathbf{p}_+,\mathbf{p}'_+, E+\hbar\omega)\, G^R(\mathbf{p}'_-,\mathbf{p}_-, E)$$

$$(8.19)$$

The diagrammatic expansion is identical to that of eq.(8.15). Concerning the ladder and maximally crossed diagrams we first note that for the case of a momentum independent impurity correlator, each loop momentum integration can be done separately, and gives a contribution at most of order $\hbar/p_F l$ relative to the first term in eq.(8.19). This follows since the propagators are peaked at the Fermi momentum with the width $\hbar/l$ ($\hbar\omega \ll \epsilon_F$, $q \ll k_F$), and ξ-integrating, $\xi_{\mathbf{p}} \equiv \epsilon_{\mathbf{p}} - \epsilon_F$, from minus to plus infinity, a product of retarded propagators gives zero

$$\int_{-\infty}^{\infty} d\xi\, G^R(E+\hbar\omega,\mathbf{p}_+)\, G^R(E,\mathbf{p}_-) = 0 \qquad (8.20)$$

as the poles are in the same half-plane. An impurity correlator which varies only on the scale of k_F (corresponding to a short ranged impurity potential, recall section 3.8) does not change the above estimate.

Vertex correction diagrams, such as the last two depicted in eq.(8.18), are also small by the factor $\hbar/p_F l$ as we again have two retarded propagators sharing a common internal momentum variable. We therefore only need to calculate the contribution to χ^{RR} from the first diagram

$$< G^R(\mathbf{p}_+, \mathbf{p}'_+, E_+)\, G^R(\mathbf{p}'_-, \mathbf{p}_-, E) > = \qquad \cdot \left(\delta_{\mathbf{p},\mathbf{p}'} + \mathcal{O}(\hbar/p_F l)\right) \quad (8.21)$$

where the second term signifies that (when integrated over $\mathbf{p}$ or $\mathbf{p}'$ as needed for the diagram with only one impurity correlator connecting the upper and lower particle lines) the terms with impurity lines connecting two retarded lines at most contribute to the response function a term of the relative order of magnitude $\hbar/p_F l$.

Expressing the product of two retarded propagators in terms of their difference, partial fractions,

$$G^R(E + \hbar\omega, \mathbf{p})\, G^R(E, \mathbf{p}) = \frac{G^R(E + \hbar\omega, \mathbf{p}) - G^R(E, \mathbf{p})}{-\hbar\omega + \Sigma^R(\mathbf{p}, E + \hbar\omega) - \Sigma^R(\mathbf{p}, E)} \quad (8.22)$$

we get, setting $\mathbf{q} = 0$,

$$\chi^{RR}(\mathbf{0}, \omega) = \frac{i}{\pi V} \sum_{\mathbf{p}} \int_{-\infty}^{\infty} dE\, f_0(E) \frac{1}{\hbar\omega - \Sigma^R(\mathbf{p}, E + \hbar\omega) + \Sigma^R(\mathbf{p}, E)}$$

$$\left(\frac{1}{E + \hbar\omega - \epsilon_{\mathbf{p}} - \Sigma^R(\mathbf{p}, E + \hbar\omega)} - \frac{1}{E - \epsilon_{\mathbf{p}} - \Sigma^R(\mathbf{p}, E)} \right) . \quad (8.23)$$

In the limit of small ω this expression has an explicit $1/\omega$-singular factor, and by Taylor expansion we get

$$\chi^{RR}(\mathbf{0}, \omega) = \frac{i}{\pi V} \sum_{\mathbf{p}} \int_{-\infty}^{\infty} dE\, f_0(E) \frac{\partial G^R(\mathbf{p}, E)}{\partial E} + \mathcal{O}(\hbar\omega/\epsilon_F)$$

$$= -\frac{i}{\pi V} \sum_{\mathbf{p}} \int_{-\infty}^{\infty} dE\, \frac{\partial f_0(E)}{\partial E}\, G^R(\mathbf{p}, E) + \mathcal{O}(\hbar\omega/\epsilon_F) . \quad (8.24)$$

Similarly we have for the AA-term as $\chi^{AA}(\mathbf{q}, \omega) = [\chi^{RR}(-\mathbf{q}, -\omega)]^*$, and we get for the regular terms the $\mathbf{q} = 0$ value

$$\chi^{RR}(\mathbf{0}, \omega) + \chi^{AA}(\mathbf{0}, \omega) = -\frac{i}{\pi V} \sum_{\mathbf{p}} \int_{-\infty}^{\infty} dE\, \frac{\partial f_0(E)}{\partial E} (G^R(\mathbf{p}, E) - G^A(\mathbf{p}, E))$$

$$= 2 N_0(\epsilon_F) \left(1 + \mathcal{O}((T/T_F)^2, (\hbar\omega/\epsilon_F), (\hbar/\epsilon_F \tau))\right) \quad (8.25)$$

essentially the density of states at the Fermi energy. We have used that since the temperature of a metal is far below the degeneracy temperature, $T \ll T_F$, the derivative of the Fermi function is peaked at the Fermi energy, $E \simeq \epsilon_F$, thereby restraining the momentum integration to the vicinity of the Fermi surface. In the sum of the RR and AA-term we can therefore perform the momentum integration before the integration over E. To the order of accuracy, $\hbar/p_F l$, we can neglect the momentum dependence in the self-energy, and perform the ξ-integration, $\xi \equiv E - \epsilon_F$, over the Lorentzian spectral function extending the lower integration limit from $-\epsilon_F$ to $-\infty$ as correction terms are of order $\mathcal{O}(\hbar/\epsilon_F \tau)^3$ [3]

$$
\begin{aligned}
\frac{1}{V} \sum_{\mathbf{p}} A(\mathbf{p}, E) &= \frac{\hbar}{V} \sum_{\mathbf{p}} \frac{2\Im m \Sigma^R(\mathbf{p}, E)}{(E - \epsilon_F - \xi_{\mathbf{p}} - \Re e \Sigma^R(\mathbf{p}, E))^2 + (\Im m \Sigma^R(\mathbf{p}, E))^2} \\
&\simeq N_0(\epsilon_F) \int_{-\infty}^{\infty} d\xi \frac{\hbar/\tau}{(E - \epsilon_F - \xi - \Re e \Sigma^R(\mathbf{p}_F, E))^2 + (\frac{\hbar}{2\tau})^2} \\
&= 2\pi N_0 \qquad\qquad\qquad\qquad\qquad\qquad (8.26)
\end{aligned}
$$

where $N_0 \equiv N_0(\epsilon_F)$ is the density of states (per spin) at the Fermi energy, and $E \simeq \epsilon_F$.

We therefore have for the density response function

$$
\chi(\mathbf{q}, \omega) = \chi^{RA}(\mathbf{q}, \omega) + 2N_0 + r(q/k_F, \hbar\omega/\epsilon_F) \qquad (8.27)
$$

where the regular function r consists of the higher derivative terms from the Taylor expansion in ω and q of the propagators in eq.(8.23). These terms have sufficient convergence to perform the momentum integration before the integration over E. Integrating products of retarded functions over ξ renders such terms, and thereby r, of relative order $\hbar/p_F l$, and they are therefore negligible in comparison with the density-of-states term we kept. The regular part of the response function, the RR and AA terms, is thus simply a constant contribution to the density response function, the density of states at the Fermi energy.

8.3.2 Nonanalytic Part of the Response Function

The nonanalytic part of the response function

$$
\begin{aligned}
\chi^{RA}(\mathbf{q}, \omega) &= -\frac{i}{\pi} \frac{1}{V} \sum_{\mathbf{p}\mathbf{p}'} \int_{-\infty}^{\infty} dE \; (f_0(E + \hbar\omega) - f_0(E_-)) \\
&\qquad < G^R(\mathbf{p}_+, \mathbf{p}'_+; E + \hbar\omega) \, G^A(\mathbf{p}'_-, \mathbf{p}_-; E) > \qquad (8.28)
\end{aligned}
$$

gives according to the above analysis the interesting contribution.

---

[3] The result is also immediate from eq.(3.74) on page 152, which states that to lowest order in $\hbar/\epsilon_F \tau$, the density of states of a Fermi gas is unaffected by a random potential. In view of the specific heat being proportional to the density of states at the Fermi energy this is not a surprising result for elastic scattering.

We are assuming that the Fermi gas is degenerate as is always appropriate for metals; i.e., all other energies are small compared to the Fermi energy, $\hbar\omega, kT \ll \epsilon_F$. Consequently the gate function, eq.(7.214), is a delta function, and fixes the energy variable at the Fermi energy, $E \simeq \epsilon_F$, so that

$$\chi^{RA}(\mathbf{q}, \omega) = \frac{i\hbar\omega}{\pi}\Phi(\mathbf{q}, \omega) + \mathcal{O}(\hbar\omega/\epsilon_F, q/k_F) \qquad (8.29)$$

where

$$\Phi(\mathbf{q}, \omega) \equiv \frac{1}{V^2}\sum_{\mathbf{pp'}}\Phi_{\mathbf{pp'}}(\mathbf{q}, \omega) \qquad (8.30)$$

and

$$\Phi_{\mathbf{pp'}}(\mathbf{q}, \omega) \equiv \Phi_{\mathbf{pp'}}(\epsilon_F, \mathbf{q}, \omega) . \qquad (8.31)$$

For the density response function we therefore finally obtain the expression, with corrections at most of order $\hbar\omega/\epsilon_F$, q/k_F,

$$\chi(\mathbf{q}, \omega) = \frac{i\hbar\omega}{\pi}\Phi(\mathbf{q}, \omega) + 2N_0 \qquad (8.32)$$

We have thus expressed the density response function in terms of the impurity average of the product of a retarded and an advanced propagator. This is to be expected, as this is precisely the quantity that describes the motion of a particle in random potential (recall the structure of the density matrix).

As a consequence of particle conservation, we note that the integral over space of the density is time independent, and consequently $\delta n(\mathbf{q} = \mathbf{0}, t) = 0$. Since changing the Hamiltonian by a time-dependent but spatially constant term does not change the number of particles, we according to eq.(8.7) have the constraint $\chi(\mathbf{q} = \mathbf{0}, \omega) = 0$. As a consequence of particle conservation (unitarity) we therefore obtain the leading small ω behavior

$$\Phi(\mathbf{q} = \mathbf{0}, \omega \to 0) \simeq \frac{2\pi N_0}{-i\hbar\omega} . \qquad (8.33)$$

8.4   Impurity-Averaged Current

We shall here calculate the impurity-averaged conductivity using the momentum representation for the current density, eq.(7.209) and eq.(7.211). For the impurity-averaged current density we only get a response at the wave vector of the driving field, and we have

$$< j_\alpha(\mathbf{q}', \omega) > = \sum_\beta < Q_{\alpha\beta}(\mathbf{q}', \mathbf{q}, \omega) > A_\beta(\mathbf{q}, \omega) = j_\alpha(\mathbf{q}, \omega)\delta_{\mathbf{q}',\mathbf{q}} \qquad (8.34)$$

since the impurity average recovers translation invariance

$$< Q_{\alpha\beta}(\mathbf{q}', \mathbf{q}, \omega) > = \delta_{\mathbf{q}',\mathbf{q}} < Q_{\alpha\beta}(\mathbf{q}, \mathbf{q}, \omega) > \equiv \delta_{\mathbf{q}',\mathbf{q}} Q_{\alpha\beta}(\mathbf{q}, \omega) \qquad (8.35)$$

the sum of incoming momenta equals the sum of outgoing momenta in any diagram representing the impurity average of a product of two propagators. For the impurity-averaged current density response we then have

$$j_\alpha(\mathbf{q}, \omega) = \sum_\beta Q_{\alpha\beta}(\mathbf{q}, \omega) A_\beta(\mathbf{q}, \omega) \tag{8.36}$$

where

$$Q_{\alpha\beta}(\mathbf{q}, \omega) = K_{\alpha\beta}(\mathbf{q}, \omega) - K_{\alpha\beta}(\mathbf{q}, 0) = K_{\alpha\beta}(\mathbf{q}, \omega) - \frac{ne^2}{m}\delta_{\alpha\beta} . \tag{8.37}$$

Here n denotes the equilibrium density of the electron gas, and the response function

$$< K_{\alpha\beta}(\mathbf{q}', \mathbf{q}, \omega) > = \delta_{\mathbf{q}', \mathbf{q}} < K_{\alpha\beta}(\mathbf{q}, \mathbf{q}, \omega) > \equiv K_{\alpha\beta}(\mathbf{q}, \omega) \, \delta_{\mathbf{q}', \mathbf{q}} \tag{8.38}$$

is specified by[4]

$$K_{\alpha\beta}(\mathbf{q}, \omega) = K_{\alpha\beta}^{RA}(\mathbf{q}, \omega) + K_{\alpha\beta}^{RR}(\mathbf{q}, \omega) + K_{\alpha\beta}^{AA}(\mathbf{q}, \omega) \tag{8.39}$$

where

$$K_{\alpha\beta}^{RA}(\mathbf{q}, \omega) = \frac{i}{\pi}\left(\frac{e}{m}\right)^2 \frac{1}{V} \sum_{\mathbf{pp'}} \int_{-\infty}^{\infty} dE \; (f_0(E) - f_0(E + \hbar\omega))$$

$$p_\alpha p'_\beta < G^R(\mathbf{p}_+, \mathbf{p}'_+; E + \hbar\omega)\, G^A(\mathbf{p}'_-, \mathbf{p}_-; E) > \tag{8.40}$$

and

$$K_{\alpha\beta}^{RR}(\mathbf{q}, \omega) = -\frac{i}{\pi}\left(\frac{e}{m}\right)^2 \frac{1}{V} \sum_{\mathbf{pp'}} p_\alpha p'_\beta \int_{-\infty}^{\infty} dE \, f_0(E)$$

$$< G^R(\mathbf{p}_+, \mathbf{p}'_+; E + \hbar\omega)\, G^R(\mathbf{p}'_-, \mathbf{p}_-; E) > \tag{8.41}$$

and

$$K_{\alpha\beta}^{AA}(\mathbf{q}, \omega) = \frac{i}{\pi}\left(\frac{e}{m}\right)^2 \frac{1}{V} \sum_{\mathbf{pp'}} p_\alpha p'_\beta \int_{-\infty}^{\infty} dE \, f_0(E)$$

$$< G^A(\mathbf{p}_+, \mathbf{p}'_+; E)\, G^A(\mathbf{p}'_-, \mathbf{p}_-; E - \hbar\omega) > . \tag{8.42}$$

For the impurity-averaged conductivity tensor

$$<\sigma_{\alpha\beta}(\mathbf{q}', \mathbf{q}, \omega)> = \sigma_{\alpha\beta}(\mathbf{q}, \omega) \, \delta_{\mathbf{q}', \mathbf{q}} \tag{8.43}$$

---

[4]We recall from section 7.8 that the three terms should be kept together under the momentum summation for reasons of convergence. For clarity of presentation, however, we write the three terms separately.

we accordingly have

$$\sigma_{\alpha\beta}(\mathbf{q},\omega) \;=\; \frac{K_{\alpha\beta}^{RA}(\mathbf{q},\omega)}{i\omega} \;+\; \frac{K_{\alpha\beta}^{RR}(\mathbf{q},\omega) + K_{\alpha\beta}^{AA}(\mathbf{q},\omega) - K_{\alpha\beta}(\mathbf{q},0)}{i\omega} \;. \tag{8.44}$$

The analysis of the various contributions to the conductivity is analogous to the one we just performed for the density response function. The K^{RR} and K^{AA} terms will only give rise to a regular term. The value of the regular term at zero wave vector and zero frequency, $\mathbf{q} = \mathbf{0}$, $\omega = 0$, is seen to cancel the diamagnetic term since by using the identity

$$\mathbf{p} = m(G^{R(A)}(\mathbf{p}, E))^{-2}\nabla_{\mathbf{p}}G^{R(A)}(\mathbf{p}, E) + \mathcal{O}(\hbar/p_F l) \tag{8.45}$$

we have

$$\int_{-\infty}^{\infty} dE\, f_0(E) \int \frac{d\mathbf{p}}{(2\pi\hbar)^3}\, p_\alpha\, p_\beta \left(G^R(\mathbf{p}, E)G^R(\mathbf{p}, E) - G^A(\mathbf{p}, E)G^A(\mathbf{p}, E)\right)$$

$$= \; m \int_{-\infty}^{\infty} dE\, f_0(E) \int \frac{d\mathbf{p}}{(2\pi\hbar)^3}\, p_\alpha\, \nabla_{p_\beta}\left(G^R(\mathbf{p}, E) - G^A(\mathbf{p}, E)\right)$$

$$= \; -m\,\delta_{\alpha\beta} \int_{-\infty}^{\infty} dE\, f_0(E) \int \frac{d\mathbf{p}}{(2\pi\hbar)^3}\, (G^R(\mathbf{p}, E) - G^A(\mathbf{p}, E))$$

$$= \; i\pi m n\,\delta_{\alpha\beta}\;. \tag{8.46}$$

In the last equality we have used that the electron density is specified in terms of the Fermi function

$$n \;=\; \frac{2}{V}\sum_{\mathbf{p}} f_0(\epsilon_{\mathbf{p}})\;. \tag{8.47}$$

To lowest order in the expansion parameter, $\hbar/p_F l$, we therefore get for the conductivity tensor

$$i\omega\,\sigma_{\alpha\beta}(\mathbf{q},\omega) \;=\; \frac{i}{\pi}\left(\frac{e}{m}\right)^2 \frac{1}{V}\sum_{\mathbf{p}\mathbf{p}'} p_\alpha\, p'_\beta \int_{-\infty}^{\infty} dE\,\Big\{\big(f_0(E) - f_0(E + \hbar\omega)\big)$$

$$< G^R(\mathbf{p}_+, \mathbf{p}'_+; E + \hbar\omega)\, G^A(\mathbf{p}'_-, \mathbf{p}_-; E) >$$

$$-\; \frac{\hbar\omega}{2}\frac{\partial f_0(E)}{\partial E} < G^R(\mathbf{p}_+, \mathbf{p}'_+; E)G^R(\mathbf{p}'_-, \mathbf{p}_-; E)$$

$$+\; G^A(\mathbf{p}_+, \mathbf{p}'_+; E)G^A(\mathbf{p}'_-, \mathbf{p}_-; E) > \Big\}\;. \tag{8.48}$$

The integrand is now peaked at the Fermi surface,[5] allowing us to perform the momentum integration before the E-integration, and the last term is seen to be

---

[5] The arbitrary energy scale E_F of chapter 3, thus turns out for the electron gas to be the Fermi energy.

of relative order $\hbar/p_F l$. We have left out the higher order $\omega \neq 0$-terms from the K^{RR} and K^{AA} contributions, as the integrand converges sufficiently fast in order to perform the ξ-integration before the E-integration, leaving such terms to be at most of relative order $\hbar/p_F l$ (analogous to the case for the regular contribution to the density response function). We therefore have for the impurity-averaged current density

$$j_\alpha(\mathbf{q}, \omega) = \sum_\beta \sigma_{\alpha\beta}(\mathbf{q}, \omega) E_\beta(\mathbf{q}, \omega) \tag{8.49}$$

where the conductivity tensor is given by $(p_F = m v_F)$

$$\sigma_{\alpha\beta}(\mathbf{q}, \omega) = \frac{e^2 v_F^2}{\pi} \int_{-\infty}^{\infty} dE \, \frac{f_0(E) - f_0(E + \hbar\omega)}{\omega}$$

$$\frac{1}{V} \sum_{\mathbf{pp'}} \hat{p}_\alpha \hat{p}'_\beta < G^R(\mathbf{p}_+, \mathbf{p}'_+; E + \hbar\omega) \, G^A(\mathbf{p}'_-, \mathbf{p}_-; E) > . \tag{8.50}$$

As to be expected, only electrons at the Fermi surface contribute to the longitudinal conductivity.

We have again obtained that a transport coefficient, here the conductivity, is specified in terms of the impurity average of the retarded and advanced propagator, the quantity which describes the motion of a particle in a random potential. We therefore have a complete diagrammatic description of the conductivity in terms of the four-point vertex. Alternatively, we can express the conductivity tensor in terms of the impurity-dressed current vertex, the three-point vector vertex Γ,

$$\sigma_{\alpha\beta}(\mathbf{q}, \omega) = \left(\frac{e}{m}\right)^2 \frac{p_F}{\pi} \int_{-\infty}^{\infty} dE \, \frac{f_0(E) - f_0(E + \hbar\omega)}{\omega}$$

$$\int \frac{d\mathbf{p}}{(2\pi\hbar)^3} \, \hat{p}_\alpha \, G^R(\mathbf{p}_+, E + \hbar\omega) \, G^A(\mathbf{p}_-, E) \, \Gamma_\beta(p_F, E, \mathbf{q}, \omega) . \tag{8.51}$$

Diagrammatically we encounter the conductivity diagrams (only ladder and maximally crossed diagrams are shown explicitly)[6]

_____

[6]For a degenerate Fermi gas the energy-variable integration disappears in favor of setting the energy variable equal to the Fermi energy, and each piece in a conductivity diagram reflects a number in the conductivity expression in accordance with the Feynman rules.

$$+ \quad \cdots \tag{8.52}$$

where for the current vertex in the momentum representation we have[7]

$$= \quad \frac{e}{m} \mathbf{p} \; . \tag{8.53}$$

For the impurity-dressed current vertex

$$\mathbf{\Gamma}_E(\mathbf{p}, \mathbf{q}, \omega) \qquad = \tag{8.54}$$

we have the diagrammatic expansion

$$+ \quad \cdots \tag{8.55}$$

---

[7]Except for the first diagram, the bubble diagram, where $\mathbf{p} \to p_F \hat{\mathbf{p}}$.

A conductivity diagram has two vertices, one where the electron is excited by the electric field, the excitation vertex, and one where the current is measured, the measuring vertex (here the excitation vertex is chosen to be a vector vertex just as the current vertex, as we chose to represent the electric field by a vector potential).

In the degenerate case, $\hbar\omega, kT \ll \epsilon_F$, we get

$$\sigma_{\alpha\beta}(\mathbf{q},\omega) = \frac{e^2}{m^2}\frac{\hbar p_F}{\pi}\int\frac{d\mathbf{p}}{(2\pi\hbar)^3}\,\hat{p}_\alpha\,G^R(\mathbf{p}_+,\epsilon_F^+)\,G^A(\mathbf{p}_-,\epsilon_F)\,\Gamma_\beta(\mathbf{p}_F,\epsilon_F,\mathbf{q},\omega). \quad (8.56)$$

8.5 Boltzmann Limit

We now embark on the calculation of the conductivity of a degenerate electron gas in a random potential. We shall in this section establish the content of the Boltzmann theory in terms of conductivity diagrams.

In order to orient ourselves we start by making the crudest of approximations

$$<G^R(\mathbf{p}_+,\mathbf{p}'_+;E_+)\,G^A(\mathbf{p}'_-,\mathbf{p}_-;E)> \;\rightarrow\; <G^R(\mathbf{p}_+,\mathbf{p}'_+;E_+)><G^A(\mathbf{p}'_-,\mathbf{p}_-;E)>$$

$$= \delta_{\mathbf{p},\mathbf{p}'}\,G^R(\mathbf{p}_+,E_+)G^A(\mathbf{p}_-,E) \quad (8.57)$$

corresponding to considering only the first diagram in eq.(8.52). The contribution to the conductivity from this so-called bubble diagram is

$$\sigma_{\alpha\beta}^D(\mathbf{q},\omega) = \frac{e^2\,v_F^2}{\pi}\int_{-\infty}^{\infty}dE\,\frac{f_0(E)-f_0(E+\hbar\omega)}{\omega}$$

$$\int\frac{d\mathbf{p}}{(2\pi\hbar)^3}\,\hat{p}_\alpha\hat{p}_\beta\,G^R(\mathbf{p}_+,E+\hbar\omega)\,G^A(\mathbf{p}_-,E) \quad (8.58)$$

where the impurity-averaged propagators were obtained in chapter 3, eq.(3.72). Since the electron gas is assumed degenerate, the combination of Fermi functions keeps the variable E of the order of the Fermi energy, $E \simeq \epsilon_F$. The momentum integral is then readily calculated to the standard order of accuracy by the residue method, $\xi_{\mathbf{p}} \equiv \epsilon_{\mathbf{p}} - \epsilon_F$,

$$\int\frac{d\mathbf{p}}{(2\pi\hbar)^3}\,\hat{p}_\alpha\hat{p}_\beta\,G^R(\mathbf{p},E+\hbar\omega)\,G^A(\mathbf{p},E)$$

$$= \int\frac{d\hat{\mathbf{p}}}{4\pi}\int_0^{\infty}d\epsilon_{\mathbf{p}}\,N_0(\epsilon_{\mathbf{p}})\,\hat{p}_\alpha\hat{p}_\beta\,G^R(\mathbf{p},E+\hbar\omega)\,G^A(\mathbf{p},E)$$

$$= \frac{N_0(\epsilon_F)}{3}\delta_{\alpha,\beta}\int_{-\infty}^{\infty}d\xi\,\frac{1}{E_+-\epsilon_F-\xi+i\hbar/2\tau}\cdot\frac{1}{E-\epsilon_F-\xi-i\hbar/2\tau}+\mathcal{O}(\hbar/p_F l)$$

$$= \frac{\pi m n \tau}{\hbar p_F^2}\frac{1}{1-i\omega\tau}\,\delta_{\alpha\beta}\;+\;\mathcal{O}(\hbar/p_F l)\,. \quad (8.59)$$

For the conductivity we get, to within corrections of order $\hbar/p_F l$, the Drude result

$$\sigma_{\alpha\beta}^D(\omega) = \sigma_0(\omega)\,\delta_{\alpha\beta} \qquad \sigma_0(\omega) = \frac{\sigma_0}{1 - i\omega\tau} \qquad \sigma_0 = \frac{ne^2\tau}{m} \qquad (8.60)$$

where $\tau \equiv \tau(\epsilon_F)$ is the momentum relaxation rate for an electron at the Fermi surface.

In order to interpret the Drude conductivity formula, let us Fourier transform back to real space, and from eq.(8.50) we obtain

$$j_\alpha(\mathbf{x},\omega) = \sum_\beta \int d\mathbf{x}' \,\sigma_{\alpha,\beta}(\mathbf{x} - \mathbf{x}',\omega) E_\beta(\mathbf{x}',\omega) \qquad (8.61)$$

where the impurity-averaged conductivity tensor, $\sigma_{\alpha\beta}(\mathbf{x}-\mathbf{x}',\omega) \equiv \,<\sigma_{\alpha\beta}(\mathbf{x},\mathbf{x}',\omega)>$, is given by

$$\sigma_{\alpha\beta}(\mathbf{x} - \mathbf{x}',\omega) = \frac{1}{\pi}\left(\frac{e\hbar}{m}\right)^2 \int_{-\infty}^{\infty} dE\, \frac{f_0(E) - f_0(E + \hbar\omega)}{\omega}$$

$$< G^R(\mathbf{x},\mathbf{x}';E + \hbar\omega)\,\overset{\leftrightarrow}{\nabla}_{x_\alpha}\overset{\leftrightarrow}{\nabla}_{x'_\beta}\,G^A(\mathbf{x}',\mathbf{x};E) > . \qquad (8.62)$$

Considering the dc case, we obtain upon inserting eq.(3.73) into eq.(8.62) for the case of the bubble diagram the conductivity

$$<\sigma_{\alpha\beta}(\mathbf{x},\mathbf{x}')> = \sigma_0\,\frac{3(x_\alpha - x'_\alpha)(x_\beta - x'_\beta)}{|\mathbf{x} - \mathbf{x}'|^4\, l}\,e^{-|\mathbf{x}-\mathbf{x}'|/l}\,. \qquad (8.63)$$

The Drude conductivity is local on the scale of the mean free path, and is effectively a delta function

$$<\sigma_{\alpha\beta}(\mathbf{x},\mathbf{x}')> = \sigma_0\,\delta_{\alpha\beta}\,\delta(\mathbf{x} - \mathbf{x}')\,. \qquad (8.64)$$

The bubble or Drude diagram only reproduces the Boltzmann result for the conductivity, eq.(5.112), for the case of isotropic scattering. The information in the Drude diagram, as regards the motion in the random potential, is

$$< G^R(\mathbf{x},\mathbf{x}';\epsilon_F) > \overset{\leftrightarrow}{\nabla}_{x_\alpha}\,\overset{\leftrightarrow}{\nabla}_{x'_\beta}< G^A(\mathbf{x}',\mathbf{x};\epsilon_F) > \,\propto\, \frac{e^{-|\mathbf{x}-\mathbf{x}'|/l}}{4\pi|\mathbf{x} - \mathbf{x}'|^2 l} \qquad (8.65)$$

the probability that a particle with energy ϵ_F has its first collision after traveling a distance $|\mathbf{x} - \mathbf{x}'|$. The effect on the conductivity of an angular dependence in the impurity-scattering cross section is not included in the bubble diagram. In order to include that we must study diagrams with explicit appearance of the impurity correlator. According to the analysis of section 3.6 we should to lowest order in the parameter $\hbar/\epsilon_F\tau$ neglect diagrams where impurity lines cross. We are then left with the ladder diagrams in eq.(8.52). In fact, each of the ladder diagrams are of the same order of magnitude as the Drude diagram, and they should therefore be considered together. To establish this we consider the ladder insertion

$$\zeta(\mathbf{q}, \omega) \equiv$$

$$= n_i \int \frac{d\mathbf{p}'}{(2\pi\hbar)^3} |V_{imp}(\mathbf{p} - \mathbf{p}')|^2 G^R(E + \hbar\omega, \mathbf{p}'_+) G^A(E, \mathbf{p}'_-)$$

$$= n_i \int \frac{d\hat{\mathbf{p}}'}{4\pi} |V_{imp}(p_F(\hat{\mathbf{p}} - \hat{\mathbf{p}}'))|^2 \int_0^\infty d\epsilon_{\mathbf{p}'} N_0(\epsilon_{\mathbf{p}'}) I(\epsilon_{\mathbf{p}'}, \hat{\mathbf{p}}', \mathbf{q}, E, \omega) . \quad (8.66)$$

where, $E \simeq \epsilon_F$, $p \simeq p_F$, $q \ll k_F$, $\omega \ll \epsilon_F/\hbar$, and

$$I(\epsilon_{\mathbf{p}'}, \hat{\mathbf{p}}', \mathbf{q}, E, \omega) \equiv G^R(E + \hbar\omega, \mathbf{p}' + \hbar\mathbf{q}) G^A(E, \mathbf{p}')$$

$$= \frac{1}{E + \hbar\omega - \epsilon_{\mathbf{p}'} - \frac{\hbar\sqrt{2m\epsilon_{\mathbf{p}'}}}{m}\hat{\mathbf{p}}' \cdot \mathbf{q} - \frac{\hbar^2 \mathbf{q}^2}{2m} + i\hbar/2\tau}$$

$$\frac{1}{E - \epsilon_{\mathbf{p}'} - i\hbar/2\tau} \quad (8.67)$$

and we have assumed a weak momentum dependence of the impurity potential. We can now perform the momentum integration as a contour integral. The integration contour along the real axis is pinched between the poles of the propagators, and the small value of the correlator (or equivalently $1/\tau$) is compensated by the consequent closeness of the poles to the real axis. We therefore expect the insertion to be of order unity. Shifting integration variable, $\xi_{\mathbf{p}'} = \epsilon_{\mathbf{p}'} - \epsilon_F$, and performing the integration, we get in three dimensions (to order $\mathcal{O}(\hbar/p_F l)$)

$$\zeta(\mathbf{q}, \omega) = \frac{i}{2ql} \ln\left(\frac{ql + \omega\tau + i}{-ql + \omega\tau + i}\right)$$

$$= 1 + i\omega\tau - \frac{1}{d}(ql)^2 . \quad (8.68)$$

In any dimensions we get the last equality in the diffusive regime, $ql, \omega\tau \ll 1$, where $l = v_F \tau$, and $\tau \equiv \tau(\epsilon_F)$ is the momentum relaxation rate at the Fermi surface in the Born approximation [8]

$$\frac{\hbar}{\tau(\epsilon_F)} = 2\pi n_i N_0 \int \frac{d\hat{\mathbf{p}}'_F}{4\pi} |V_{imp}(\mathbf{p}_F - \mathbf{p}'_F)|^2 . \quad (8.69)$$

---

[8]Including multiple scattering, we encounter the t-matrix and the exact scattering cross section as discussed in section 3.7.

The last expression in eq.(8.68) is immediately obtained by Taylor-expanding the propagator

$$
G^R(E + \hbar\omega, \mathbf{p}' + \hbar\mathbf{q}) = G^R(E, \mathbf{p}') - \left(\hbar\omega - \frac{\hbar\mathbf{q}\cdot\mathbf{p}'}{m}\right)[G^R(E, \mathbf{p}')]^2
$$

$$
+ \left(\hbar\omega - \frac{\hbar\mathbf{q}\cdot\mathbf{p}'}{m}\right)^2 [G^R(E, \mathbf{p}')]^3 + \dots \quad (8.70)
$$

and performing the integration term by term.

In terms of the three-point vector vertex, the ladder diagrams (including the bubble) are generated by the iterative equation[9]

$$(8.71)$$

Analytically we have that the three-point vector vertex in the ladder approximation, $\mathbf{\Gamma}^L$, satisfies the equation

$$
\mathbf{\Gamma}^L_E(\mathbf{p}, \mathbf{q}, \omega) = \mathbf{p} + n_i \int \frac{d\mathbf{p}'}{(2\pi\hbar)^3} \, |V_{imp}(\mathbf{p} - \mathbf{p}')|^2 G^R(\mathbf{p}'_+, E_+) G^A(\mathbf{p}'_-, E)\mathbf{\Gamma}^L_E(\mathbf{p}', \mathbf{q}, \omega).
$$
$$(8.72)$$

We shall only be interested in the normal skin effect, where the wavelength of the electric field is much larger than the mean free path, $ql \ll 1$.[10] We can therefore set the wave vector of the electric field $\mathbf{q}$ equal to zero in the propagators, and thereby in the vertex function as its scale of variation consequently is the Fermi wave vector $k_F = p_F/\hbar$.[11] We are therefore only interested in the spatially averaged current density, $\mathbf{j}(\omega) \equiv \mathbf{j}(\mathbf{q} = 0, \omega)$, or equivalently the conductance tensor (inverse resistance tensor)

$$
< G_{\alpha\beta}(\omega) > = L^{-2} \int d\mathbf{x} \int d\mathbf{x}' \, < \sigma_{\alpha\beta}(\mathbf{x}, \mathbf{x}', \omega) > \quad (8.73)
$$

---

[9]A general feature of the kinetic-equation approach, as compared to the linear-response approach, is that in the former one needs to consider far fewer diagrams. This apparent simplification can in actual calculations, however, soon be overshadowed by complications due to the fact that in the kinetic approach the diagrams represent nonequilibrium quantities.

[10]For an applied field of wavelength much shorter than the mean free path, $q \gg 1/l$, the corrections to the bare vertex can be neglected.

[11]In the calculation of the residual resistance, we expect that we do not need to discuss the Coulomb interaction between the electrons, since we have only weak density modulations. In a bulk metal the strong Coulomb interaction demands charge neutrality, and in a conductivity measurement we only encounter a weak spatially inhomogeneous electric field, i.e., $q \ll k_F$. The inevitable mass renormalization due to electron-electron interaction is a tiny effect in a metal.

where $< \sigma_{\alpha\beta} >$ is the impurity average of the conductivity tensor, and L is the length of the sample.

From the defining equation, eq.(8.72), we then obtain that $\Gamma^L_{\epsilon_F}(\mathbf{p}, \mathbf{q} = \mathbf{0}, \omega)$ is directed along $\mathbf{p}$, $\Gamma^L_{\epsilon_F}(\mathbf{p}, \mathbf{q} = \mathbf{0}, \omega) = \gamma_L(\omega)\,\mathbf{p}$, and we obtain for the ladder vertex equation, $E \simeq \epsilon_F$,

$$\gamma_L(\omega)\,\mathbf{p} = \mathbf{p} + n_i \int \frac{d\mathbf{p}'}{(2\pi\hbar)^3}\, |V_{imp}(\mathbf{p} - \mathbf{p}')|^2\, G^R(\mathbf{p}', E_+)\, G^A(\mathbf{p}', E)\, \gamma_L(\omega)\,\mathbf{p}' \; . \quad (8.74)$$

Solving the equation we obtain

$$\gamma_L(\omega) \;=\; \frac{1 - i\omega\tau}{1 - i\omega\tau - \frac{2\pi n_i N_0 \tau}{\hbar} \int \frac{d\hat{\mathbf{p}}'_F}{4\pi}\, |V_{imp}(\mathbf{p}_F - \mathbf{p}'_F)|^2\, \hat{\mathbf{p}}_F \cdot \hat{\mathbf{p}}'_F}$$

$$\;=\; \frac{1 - i\omega\tau}{1 - i\omega\tau_{tr}}\, \frac{\tau_{tr}}{\tau} \qquad\qquad\qquad (8.75)$$

where $\tau_{tr} \equiv \tau_{tr}(\epsilon_F)$ is the transport relaxation time in the Born approximation

$$\frac{\hbar}{\tau_{tr}(\epsilon_F)} \;=\; 2\pi n_i N_0 \int \frac{d\hat{\mathbf{p}}'_F}{4\pi}\, |V_{imp}(\mathbf{p}_F - \mathbf{p}'_F)|^2 (1 - \hat{\mathbf{p}}_F \cdot \hat{\mathbf{p}}'_F) \; . \quad (8.76)$$

If we include multiple scattering we have for the transport relaxation time

$$\frac{\hbar}{\tau_{tr}(\epsilon_F)} \;=\; 2\pi n_i N_0 \int \frac{d\hat{\mathbf{p}}'_F}{4\pi}\, |t^R_{\mathbf{p}_F \mathbf{p}'_F}(\epsilon_F)|^2 (1 - \hat{\mathbf{p}}_F \cdot \hat{\mathbf{p}}'_F) \; . \quad (8.77)$$

The contribution from the Drude and ladder diagrams gives, according to eq.(8.51), for the conductivity

$$\sigma(\omega) \;=\; \sigma_D(\omega)\, \gamma_L(\omega) \; . \qquad\qquad (8.78)$$

Inclusion of the vertex diagrams where the retarded and advanced particle lines are connected by the ladder, thus leads again to the Drude-type conductivity formula, except for the momentum relaxation time τ now being replaced by the transport time, and we obtain the Boltzmann conductivity (recall eq.(5.117))

$$\sigma(\omega) = \frac{\sigma_0}{1 - i\omega\tau_{tr}} \; , \qquad \sigma_0 = \frac{ne^2\tau_{tr}}{m} \; . \qquad (8.79)$$

The appearance of the transport time expresses the simple fact, that small angle scattering is not effective in degrading the current. For isotropic scattering the two relaxation times are identical, as each scattering direction is weighted equally. The transport time is the characteristic time a particle can travel before the direction of its velocity is randomized.

It is not a surprise that the ladder approximation reproduces the Boltzmann result. Performing the integration over the length of the momentum in the ladder vertex equation, eq.(8.74), we obtain the linearized Boltzmann equation, eq.(5.91), determining the linear nonequilibrium contribution to the distribution function.

The Boltzmann conductivity is local, and for the Boltzmann current density response to a spatially homogeneous electric field we have

$$\mathbf{j}(t) = \int\limits_{0}^{\infty} dt' \, \sigma(t') \, \mathbf{E}(t - t') \tag{8.80}$$

where

$$\sigma(t) = \int\limits_{-\infty}^{\infty} \frac{d\omega}{2\pi} \, e^{-i\omega t} \, \sigma(\omega) \, . \tag{8.81}$$

Considering the response to a sharp electric pulse

$$\mathbf{E}(t) = \mathbf{E}_0 \, \delta(t) \tag{8.82}$$

gives, according to eq.(8.79), for the decay of the current

$$\mathbf{j}(t) = \sigma(t) \, \mathbf{E}_0 = \frac{\sigma_0}{\tau_{tr}} \, \mathbf{E}_0 \, e^{-t/\tau_{tr}} \, . \tag{8.83}$$

After a time span τ_{tr} there is thus no measurable current.

The diagrams left out in the ladder approximation contains conductivity diagrams where impurity lines cross, and these diagrams are relatively smaller by the factor $\hbar/p_F l$. We therefore again encounter the Landau criterion for the validity of the Boltzmann description $p_F l \gg \hbar$. As already mentioned in footnote 14 on page 193 the Landau criterion is not sufficient in low-dimensional systems, $d \leq 2$. A two-dimensional metal at zero temperature is not a conductor, but for arbitrary small amount of disorder in fact an insulator![12] Instead of conducting behavior one encounters the phenomenon of localization as we shall discuss in the next chapter.

8.6   Response Function Relationships

The density and current response functions are related due to particle conservation. As the starting point for deriving the relationship we use the continuity equation, the expectation value of the operator equation, eq.(5.80),

$$\frac{\partial n(\mathbf{x}, t)}{\partial t} + \nabla \cdot \mathbf{j}(\mathbf{x}, t) = 0 \, . \tag{8.84}$$

Fourier-transforming, the continuity equation reads

$$\omega \, e \, \delta n(\mathbf{q}, \omega) - \mathbf{q} \cdot \mathbf{j}(\mathbf{q}, \omega) = 0 \tag{8.85}$$

where we now let $\mathbf{j}$ denote the charge current density. Assuming that the potential, $V = e\phi$, felt by the particles is electromagnetic, due to the charge e of the particles,

---

[12]In a three-dimensional metal at a critical amount of disorder a metal undergoes a metal-insulator transition, as discussed in section 9.3.4.

the density response to an electrostatic potential ϕ is

$$\delta n(\mathbf{q}, \omega) = -\chi(\mathbf{q}, \omega) e \phi(\mathbf{q}, \omega) . \tag{8.86}$$

As regards the impurity-averaged charge current density we have for the linear response expression

$$\mathbf{j}(\mathbf{q}, \omega) = \underline{\sigma}(\mathbf{q}, \omega) \mathbf{E}(\mathbf{q}, \omega) = -i \underline{\sigma}(\mathbf{q}, \omega) \mathbf{q} \, \phi(\mathbf{q}, \omega) \tag{8.87}$$

as $\mathbf{E} = -\nabla \phi$. Using the continuity equation, we then get the relation between the density response function and the conductivity tensor

$$e \omega \chi(\mathbf{q}, \omega) = \frac{i}{e} \mathbf{q} \cdot \underline{\sigma}(\mathbf{q}, \omega) \mathbf{q} . \tag{8.88}$$

In the isotropic case, $\sigma_{\alpha\beta}(\mathbf{q}, \omega) = \sigma(\mathbf{q}, \omega) \delta_{\alpha\beta}$, we have $\mathbf{q} \cdot \underline{\sigma}(\mathbf{q}, \omega)\mathbf{q} = \sigma(\mathbf{q}, \omega) q^2$ and thereby

$$\sigma(\mathbf{q}, \omega) = \frac{-i\omega e^2}{q^2} \chi(\mathbf{q}, \omega) . \tag{8.89}$$

The spatially averaged current density, $\mathbf{J}(\mathbf{q} = 0, \omega)$, is thus specified by the frequency-dependent conductivity according to

$$\sigma(\omega) \equiv \sigma(\mathbf{0}, \omega) = -i\omega e^2 \lim_{q \to 0} \frac{\chi(\mathbf{q}, \omega)}{q^2} . \tag{8.90}$$

It is instructive to consider the space-time density correlation function in the classical limit where we have for a single particle[13]

$$\tilde{\chi}_{cl}^{(n)}(\mathbf{x}, t; \mathbf{x}', t') = < \delta(\mathbf{x} - \mathbf{x}(t)) \, \delta(\mathbf{x}' - \mathbf{x}(t')) > \tag{8.91}$$

In a translational invariant system we may write ($V = L^d$ denoting the volume of the system)

$$\tilde{\chi}_{cl}^{(n)}(\mathbf{x}, t; \mathbf{x}', t') \equiv \tilde{\chi}_{cl}^{(n)}(\mathbf{x} - \mathbf{x}', t, t') = \frac{1}{V} \sum_{\mathbf{q}} e^{i\mathbf{q} \cdot (\mathbf{x} - \mathbf{x}')} \tilde{\chi}_{cl}^{(n)}(\mathbf{q}, t, t') \tag{8.92}$$

where

$$\tilde{\chi}_{cl}^{(n)}(\mathbf{q}, t, t') = \frac{1}{V} < e^{-i\mathbf{q} \cdot (\mathbf{x}(t) - \mathbf{x}(t'))} > . \tag{8.93}$$

For a steady state we can Fourier transform with respect to time and obtain

$$\tilde{\chi}_{cl}^{(n)}(\mathbf{q}, \omega) = \frac{1}{V} \int_{-\infty}^{\infty} d(t - t') \, e^{i\omega(t - t')} < e^{-i\mathbf{q} \cdot (\mathbf{x}(t) - \mathbf{x}(t'))} > \tag{8.94}$$

and thereby the relation

$$\omega^2 \tilde{\chi}_{cl}^{(n)}(\mathbf{q}, \omega) = \frac{1}{V} \int_{-\infty}^{\infty} d(t - t') \, e^{i\omega(t - t')} < \mathbf{q} \cdot \dot{\mathbf{x}}(t) \, \mathbf{q} \cdot \dot{\mathbf{x}}(t') e^{-i\mathbf{q} \cdot (\mathbf{x}(t) - \mathbf{x}(t'))} > \tag{8.95}$$

---

[13]The generalization of the following to an assembly of particles is straightforward as the total density operator is the sum of the single-particle density operators.

or in terms of the velocity correlation function $(\mathbf{v}(t) = \dot{\mathbf{x}}(t))$

$$\lim_{q \to 0} \frac{\omega^2}{q^2} \tilde{\chi}_{cl}^{(n)}(\mathbf{q}, \omega) = \frac{1}{Vd} \int_{-\infty}^{\infty} d(t - t') \, e^{i\omega(t-t')} < \mathbf{v}(t) \cdot \mathbf{v}(t') > . \tag{8.96}$$

8.7   Diffusion and Ladder Diagrams

Let us consider the diagrammatic equation for the impurity-averaged density matrix when we only include the ladder diagrams

$$\tag{8.97}$$

where

$$\tag{8.98}$$

is the sum of ladder diagrams.

For simplicity we assume a delta function impurity correlator

$$< V(\mathbf{x})V(\mathbf{x}') > = u^2 \, \delta(\mathbf{x} - \mathbf{x}') = n_i V_{imp}^2 = n_i |V_{imp}(\mathbf{p} = 0)|^2 \tag{8.99}$$

corresponding strictly to the potential $V_{imp}(\mathbf{x}) = V_{imp}\delta(\mathbf{x})$, but effectively to a range of the impurity potential much shorter than the mean free path (as discussed in section 3.8). In this case the matrix element $<\mathbf{p}|V_{imp}(\hat{\mathbf{x}})|\mathbf{p}'>$, and thereby the impurity correlator, is independent of the momentum transfer

$$= u^2 . \tag{8.100}$$

The impurity momentum relaxation time, τ, determining the position of the pole of the impurity-averaged propagator, is then, according to eq.(3.70), related to the impurity correlator strength by

$$\frac{\hbar}{\tau} \equiv \frac{\hbar}{\tau(\epsilon_F)} = 2\pi n_i N_0 \int \frac{d\hat{\mathbf{p}}'}{4\pi} |V_{imp}(\sqrt{2m\epsilon_F}(\hat{\mathbf{p}} - \hat{\mathbf{p}}'))|^2 = 2\pi N_0 u^2 \qquad (8.101)$$

where we have introduced the notation $N_0 \equiv N_0(\epsilon_F)$ for the density of states at the Fermi energy.

We shall be interested in the large-distance, and long-time-limit behavior of the particle motion, $|\mathbf{x} - \mathbf{x}'| \gg l$, $|t - t'| \gg \tau$. We can therefore make use of the fact that the impurity-averaged propagator has short spatial range, and we shall soon realize that in this limit the first diagram in eq.(8.97) gives a small contribution compared to the sum of ladder diagrams.

The ladder diagrams gives for the probability to find the particle at position $\mathbf{x}$ at time t the contribution

$$P_D(\mathbf{x}, t) = \int d\mathbf{r} d\mathbf{r}' d\tilde{\mathbf{x}} d\tilde{\mathbf{x}}' \int_{-\infty}^{\infty} dt_1 dt_1' dt_2 dt_2' \, G^R(\mathbf{x}, t; \mathbf{r}, t_1) G^A(\mathbf{r}, t_2'; \mathbf{x}, t)$$

$$D_{\mathbf{r}\mathbf{r}'}(t_1, t_1', t_2, t_2') \, G^R(\mathbf{r}', t_1'; \tilde{\mathbf{x}}, t') G^A(\tilde{\mathbf{x}}', t'; \mathbf{r}', t_2) \, \rho(\tilde{\mathbf{x}}, \tilde{\mathbf{x}}') \quad (8.102)$$

where $\rho(\tilde{\mathbf{x}}, \tilde{\mathbf{x}}') = \rho(\tilde{\mathbf{x}}, \tilde{\mathbf{x}}', t')$ is the density matrix at time t'. Fourier-transforming with respect to time we get

$$P_D(\mathbf{x}, \omega) = \hbar \int d\mathbf{r} d\mathbf{r}' d\tilde{\mathbf{x}} d\tilde{\mathbf{x}}' \int_{-\infty}^{\infty} \frac{dE}{2\pi} \, G^R_{E_+}(\mathbf{x}, \mathbf{r}) \, G^A_E(\mathbf{r}, \mathbf{x}) \, D_{\mathbf{r}\mathbf{r}'}(E, \omega)$$

$$G^R_{E_+}(\mathbf{r}', \tilde{\mathbf{x}}) \, G^A_E(\tilde{\mathbf{x}}', \mathbf{r}') \, \rho(\tilde{\mathbf{x}}, \tilde{\mathbf{x}}') \qquad (8.103)$$

where

$$D_{\mathbf{r}\mathbf{r}'}(E, \omega) \equiv$$

$$(8.104)$$

Introducing the spatial Fourier transform

$$D_{\mathbf{r}\mathbf{r}'}(E,\omega) \;=\; \frac{1}{V}\sum_{\mathbf{q}} e^{i\mathbf{q}\cdot(\mathbf{r}-\mathbf{r}')}\, D_E(\mathbf{q},\omega) \tag{8.105}$$

we have for $D_E(\mathbf{q},\omega)$ the diagrammatic expansion

Assuming the initial state of the particle is a spherical symmetric outgoing Gaussian wave packet with radial momentum of magnitude p_F, the main contribution to the E-integration in eq.(8.103) is from $E \simeq \epsilon_F = p_F^2/2m$, and we can set E equal to ϵ_F except in the propagators attached to the initial density matrix. The ladder diagrams constitute a geometric series and are readily summed to give the so-called Diffuson

$$D_E(\mathbf{q},\omega) \;=\; \frac{u^2}{1-\zeta(q,\omega)} \;. \tag{8.107}$$

We have already calculated the needed insertion in the previous section, and we obtain for $ql, \omega\tau \ll 1$, that $D_E(\mathbf{q},\omega)$ is independent of E, $E \simeq \epsilon_F$, (absorbing a factor of τu^{-2})

$$D(\mathbf{q},\omega) \;\equiv\; \tau u^{-2} D_E(\mathbf{q},\omega) \;=\; \frac{1}{-i\omega + D_0 q^2} \tag{8.108}$$

where $D_0 = v_F^2\tau/d$ is the diffusion constant in d dimensions. The function $D(\mathbf{q}, \omega)$ is seen to have a pole in the lower ω-half plane at $\omega = -iD_0q^2$, the so-called diffusion pole. We note that the exact infrared property of the four-point function due to particle conservation, eq.(8.33), is already captured by the ladder diagrams, and the form of the Diffuson or diffusion propagator is thus a consequence of particle conservation. Since $D_{\mathbf{r}\mathbf{r}'}$ is long ranged for small ω, the scale being set by $L_\omega = \sqrt{D_0/\omega}$,[14] we can in eq.(8.103) replace its argument $\mathbf{r}$ by $\mathbf{x}$ on account of the short range of the propagators, $D_{\mathbf{r}\mathbf{r}'} \to D_{\mathbf{x}\mathbf{r}'}$. We then perform the integration over $\mathbf{r}$ and obtain

$$\int d\mathbf{r}\, G_{\epsilon_F^+}^R(\mathbf{x}, \mathbf{r}) G_{\epsilon_F}^A(\mathbf{r}, \mathbf{x}) = \frac{2\pi N_0 \tau \hbar^{-1}}{1 - i\omega\tau} = \frac{u^{-2}}{1 - i\omega\tau} \tag{8.109}$$

to get for the probability distribution in the ladder approximation

$$P_D(\mathbf{x}, \omega) = \frac{\hbar u^{-2}}{1 - i\omega\tau} \int\limits_{-\infty}^{\infty} \frac{dE}{2\pi\hbar} \int d\mathbf{r}'\, D_{\mathbf{x}\mathbf{r}'}(\epsilon_F, \omega)$$

$$\int d\tilde{\mathbf{x}}\, d\tilde{\mathbf{x}}'\, G_{E_+}^R(\mathbf{r}', \tilde{\mathbf{x}})\, G_E^A(\tilde{\mathbf{x}}', \mathbf{r}')\, \rho(\tilde{\mathbf{x}}, \tilde{\mathbf{x}}') . \tag{8.110}$$

Assuming that the spatial dependence of the initial density matrix is smooth on the scale of the mean free path, we can substitute $\tilde{\mathbf{x}} \to \mathbf{r}'$ and $\tilde{\mathbf{x}}' \to \mathbf{r}'$ in $\rho(\tilde{\mathbf{x}}, \tilde{\mathbf{x}}')$, and perform the integration to obtain

$$\int d\tilde{\mathbf{x}} \int d\tilde{\mathbf{x}}' \int\limits_{-\infty}^{\infty} \frac{dE}{2\pi} G_{E_+}^R(\mathbf{r}', \tilde{\mathbf{x}})\, G_E^A(\tilde{\mathbf{x}}', \mathbf{r}') = \frac{\tau\hbar^{-1}}{1 - i\omega\tau} \tag{8.111}$$

where in the denominator the ω-term should be dropped in the long time limit. We then obtain for the behavior of the particle motion at long distances and long times

$$P_D(\mathbf{x}, \omega) = \int d\mathbf{r}'\, D(\mathbf{x}, \mathbf{r}', \omega)\, \rho(\mathbf{r}', \mathbf{r}') = \int d\mathbf{r}'\, D(\mathbf{x}, \mathbf{r}', \omega)\, P(\mathbf{r}') \tag{8.112}$$

where $P(\mathbf{r}') \equiv \rho(\mathbf{r}', \mathbf{r}')$ is the probability density at time t' to find the particle at position $\mathbf{r}'$, and $D(\mathbf{x}, \mathbf{r}', \omega)$ is the Fourier transform of $D(\mathbf{q}, \omega)$, eq.(8.108).

Assuming that the probability density at the initial time, $P(\mathbf{r}')$, is localized near some point $\mathbf{x}'$ (on a scale much larger than the mean free path l, but much smaller than L_ω) we get for the conditional probability

$$P_D(\mathbf{x}, \mathbf{x}', \omega) \equiv P_D(\mathbf{x}, \omega) = D(\mathbf{x}, \mathbf{x}', \omega) . \tag{8.113}$$

---

[14] Justifying the neglect of the first diagram for long distances and long time behavior, since for $\omega \ll 1/\tau$ we have $L_\omega \gg l$.

When Fourier transforming, we pick up from the ω-integration the contribution from the diffusion pole and get for the conditional probability

$$
\begin{aligned}
P_D(\mathbf{x}, t; \mathbf{x}', t') &= D(\mathbf{x}, t; \mathbf{x}', t') \\[2mm]
&= \int_{-\infty}^{\infty} \frac{d\mathbf{q}}{(2\pi)^d} \int_{-\infty}^{\infty} \frac{d\omega}{2\pi} \, e^{i\mathbf{q}\cdot(\mathbf{x}-\mathbf{x}')-i\omega(t-t')} \, \frac{1}{-i\omega + D_0 q^2} \\[2mm]
&= \left(\frac{1}{4\pi D_0(t-t')} \right)^{d/2} e^{-\frac{(\mathbf{x}-\mathbf{x}')^2}{4D_0(t-t')}} \quad .
\end{aligned}
\tag{8.114}
$$

We have thus established that the ladder diagrams describe diffusion, Brownian motion. We shall therefore often refer to the function $D(\mathbf{x}, t; \mathbf{x}', t')$ as the diffusion propagator, or Diffuson for short. Diffusion was discussed in section 5.6 starting from the Boltzmann theory, and we have now established what diffusion amounts to in terms of diagrams.

The position of the particle can in the diffusive approximation be considered a stochastic variable, and its random positions at specified times, $\mathbf{x}(t)$, a stochastic process, the so-called Wiener process. The relation, obtained from eq.(8.112) by Fourier transformation,

$$
P_D(\mathbf{x}, t) = \int d\mathbf{x}' \, D(\mathbf{x}, t; \mathbf{x}', t') \, P_D(\mathbf{x}', t')
\tag{8.115}
$$

is the signature of a Markovian process, here for the process in question, the diffusion process.[15] The Repeated use of the relation eq.(8.115) generates, in view of eq.(8.114), for the diffusion propagator the path integral expression

$$
D(\mathbf{x}, t; \mathbf{x}', t') = \int_{\mathbf{x}_{t'}=\mathbf{x}'}^{\mathbf{x}_t=\mathbf{x}} \mathcal{D}\mathbf{x}_{\bar{t}} \, e^{-S_{\mathcal{E}}[\mathbf{x}_{\bar{t}}]} = \int_{\mathbf{x}_{t'}=\mathbf{x}'}^{\mathbf{x}_t=\mathbf{x}} \mathcal{D}\mathbf{x}_{\bar{t}} \, e^{-\int_{t'}^{t} d\bar{t} \, L_{\mathcal{E}}(\dot{\mathbf{x}}_{\bar{t}})}
\tag{8.116}
$$

where the Euclidean action $S_{\mathcal{E}}[\mathbf{x}_{\bar{t}}]$ is specified by the Euclidean Lagrangian

$$
L_{\mathcal{E}}(\dot{\mathbf{x}}_t) = \frac{\dot{\mathbf{x}}_t^2}{4D_0}
\tag{8.117}
$$

and we obtain that the probability density of diffusive paths is given by

$$
P_D[\mathbf{x}_{\bar{t}}] \equiv e^{-S_{\mathcal{E}}[\mathbf{x}_{\bar{t}}]} = e^{-\int_{t'}^{t} d\bar{t} \, \frac{\dot{\mathbf{x}}_{\bar{t}}^2}{4D_0}} \, .
\tag{8.118}
$$

We note that the velocity entering the above Wiener measure is not the local velocity but the velocity averaged over Boltzmannian paths.

We have recovered the result of section 5.6, that the diffusion equation is equivalent to the imaginary-time Schrödinger equation for a particle of mass $\hbar/2D_0$.

---

[15]The Markovian principle that the *future* is independent of the *past* if we know the *present*, i.e., the causality principle of classical physics in the context of a stochastic dynamic system.

Equivalently, we obtain from eq.(8.108) by Fourier-transforming, and demanding that $D(\mathbf{x}, t < t'; \mathbf{x}', t') = 0$, that the diffusion propagator or Diffuson satisfies the equation

$$\left(\frac{\partial}{\partial t} - D_0 \Delta_{\mathbf{x}} \right) D(\mathbf{x}, t; \mathbf{x}', t') = \delta(\mathbf{x} - \mathbf{x}')\,\delta(t - t') . \tag{8.119}$$

By analogy with the equation for the Green's function of the Schrödinger equation we then again obtain the path integral solution eq.(8.116).

Exercise 8.2 *Show that for a diffusing particle we have the Gaussian property for the characteristic function*

$$< e^{i\mathbf{q}(\mathbf{x} - \mathbf{x}')} > \equiv \int d(\mathbf{x} - \mathbf{x}') P_D(\mathbf{x}, t; \mathbf{x}', t')\, e^{i\mathbf{q}\cdot(\mathbf{x} - \mathbf{x}')} = e^{-\frac{1}{2}q^2 <(\mathbf{x} - \mathbf{x}')^2>}$$

$$= e^{-D_0 q^2 (t - t')} . \tag{8.120}$$

We shall in chapter 11 need the average over diffusive paths of the following quantity

$$< e^{i\mathbf{q}\cdot(\mathbf{x}(t) - \mathbf{x}(t'))} >_D = \frac{\int \mathcal{D}\mathbf{x}_{\bar{t}}\, P_D[\mathbf{x}_{\bar{t}}]\, e^{i\mathbf{q}\cdot(\mathbf{x}(t) - \mathbf{x}(t'))}}{\int \mathcal{D}\mathbf{x}_{\bar{t}}\, P_D[\mathbf{x}_{\bar{t}}]} . \tag{8.121}$$

Introducing t and t' as intermediate times in the discretized expression for the path integral we obtain

$$< e^{i\mathbf{q}\cdot(\mathbf{x}(t) - \mathbf{x}(t'))} >_D = \int d(\mathbf{x} - \mathbf{x}') P_D(\mathbf{x}, t; \mathbf{x}', t')\, e^{i\mathbf{q}\cdot(\mathbf{x} - \mathbf{x}')} \tag{8.122}$$

and in view of eq.(8.120), we have

$$< e^{i\mathbf{q}\cdot(\mathbf{x}(t) - \mathbf{x}(t'))} >_D = e^{-D_0 q^2 |t - t'|} . \tag{8.123}$$

It is useful to connect the above analysis to the density correlation function in the classical limit, and consider the correlation function on the form established in eq.(8.93) in the diffusive limit

$$\tilde{\chi}_D^{(n)}(\mathbf{q}, t, t') = \frac{1}{V} < e^{i\mathbf{q}\cdot(\mathbf{x}(t) - \mathbf{x}(t'))} >_D = \frac{1}{V} e^{-D_0 q^2 |t - t'|} . \tag{8.124}$$

For the density fluctuations we therefore have in the diffusion approximation, $t > t'$,

$$\tilde{\chi}_D^n(\mathbf{x}, t, \mathbf{x}', t') = \frac{1}{V} \sum_{\mathbf{q}} e^{i\mathbf{q}\cdot(\mathbf{x} - \mathbf{x}')} e^{-D_0 q^2 (t - t')} = \left(\frac{1}{4\pi D_0 (t - t')} \right)^{d/2} e^{-\frac{(\mathbf{x} - \mathbf{x}')^2}{4 D_0 (t - t')}}$$

$$= D(\mathbf{x}, t; \mathbf{x}', t') . \tag{8.125}$$

8.8   Particle Conservation

The topological character of the definition of various irreducible sets of diagrams leads to certain interrelations between them, and expresses in diagrammatic terms the conservation laws obeyed by the system. In the following we shall explore the consequence of particle conservation, establishing a relation between the self-energy and the irreducible four-point vertex.[16]

For example, if we in the diagrams for the irreducible four-point vertex U look at the case with the external labeling

$$\text{(8.126)}$$

and append a propagator (with momentum $\mathbf{p}_+ - \mathbf{p}_1$) between the two $(\mathbf{p}_+ - \mathbf{p}_1)$-entries, we get a self-energy diagram as U was two-particle irreducible, and consequently the constructed diagram is one-particle irreducible. To see that we in this way precisely get all the self-energy diagrams we notice that if we in a self-energy diagram with a given number of impurity correlators plug out a propagator successively at all possible places, we precisely generate the U-diagrams of that order. For example, plugging out a propagator successively in the first two diagrams of eq.(3.59) on page 149 gives us the first four diagrams for U depicted in eq.(8.18) on page 326.

In the above consideration we have not been concerned with the R and A labeling as only the topology of the diagrams mattered. Defining

$$\Delta\Sigma_{\mathbf{p}}(E, \mathbf{q}, \omega) \equiv \left(\Sigma^R(E_+, \mathbf{p}_+) - \Sigma^A(E, \mathbf{p}_-) \right) \qquad (8.127)$$

where[17]

$$\Sigma^R(\mathbf{p}_+, E_+) = \sum_{\mathcal{D}_N} \frac{1}{V^N} \sum_{\{\mathbf{p}_i\}} U_{\mathbf{p}_N}^{(0)} .. U_{\mathbf{p}_1}^{(0)} \, G^R(E_+, \mathbf{p}_+ - \mathbf{p}_N) \, G^R(E_+, \mathbf{p}_+ + \mathcal{D}_N(\{\mathbf{p}_i\}))$$

$$.. G^R(E_+, \mathbf{p}_+ - \mathbf{p}_1) \qquad (8.128)$$

is the retarded self-energy in terms of skeleton diagrams. We have introduced the notation for the impurity correlator, $U_i^{(0)} \equiv U_{\mathbf{p}_i}^{(0)} \equiv n_i |V_{imp}(\mathbf{p}_i)|^2$. The sum is over all possible skeleton self-energy diagrams, $\mathcal{D}_N$, classified according to their number of impurity correlators N. A diagram of order N has $n = 2N - 1$ propagator lines with N independent internal momenta, but has a momentum labeling, $\mathcal{D}_N(\{\mathbf{p}_i\})$,

---

[16]We follow the presentation of reference [32].

[17]For notational simplicity, we assume a Gaussian impurity average. The conclusion, however, being independent of this assumption as the argument involves only a rearrangement of the propagators.

depending on the topology of the diagram. However, this is a point of no further nuisance for the argument to follow, as only the topology of a diagram matters. Similarly we have for the advanced self-energy

$$\Sigma^A(\mathbf{p}_+, E) = \sum_{\mathcal{D}_N} \frac{1}{V^N} \sum_{\{\mathbf{p}_i\}} U^{(0)}_{\mathbf{p}_N} .. U^{(0)}_{\mathbf{p}_1} \, G^A(E, \mathbf{p}_+ - \mathbf{p}_N) \, G^A(E, \mathbf{p}_+ + \mathcal{D}_N(\{\mathbf{p}_i\}))$$

$$.. \, G^A(E, \mathbf{p}_+ - \mathbf{p}_1) \, . \qquad (8.129)$$

Introducing

$$\Delta G_{\mathbf{p}}(E_+, \mathbf{q}, \omega) \equiv G^R(E_+, \mathbf{p}_+) - G^A(E, \mathbf{p}_-) \qquad (8.130)$$

we can rewrite

$$\Delta \Sigma_{\mathbf{p}}(E, \mathbf{q}, \omega) = \sum_{\mathcal{D}_N} \frac{1}{V^N} \sum_{\{\mathbf{p}_i\}} U^{(0)}_1 U^{(0)}_2 .. U^{(0)}_N \, (G^R_1 G^R_2 .. G^R_N - G^A_1 G^A_2 .. G^A_N)$$

$$(8.131)$$

on the form

$$\Delta \Sigma_{\mathbf{p}}(E, \mathbf{q}, \omega) = \sum_{\mathcal{D}_N} \frac{1}{V^N} \sum_{\{\mathbf{p}_i\}} U^{(0)}_1 U^{(0)}_2 .. U^{(0)}_N \, \{ G^R_1 G^R_2 .. G^R_{N-1} \Delta G_N$$

$$+ \; G^R_1 G^R_2 .. \Delta G_{N-1} G^A_N + .. + \Delta G_1 G^A_2 .. G^A_{N-1} \Delta G^A_N \} \quad (8.132)$$

alternating terms canceling each other. The terms in the curly bracket all correspond topologically to a self-energy diagram. Removing the ΔG-propagator corresponds to plugging out a propagator line in all possible ways, i.e., precisely generating the U-diagrams topologically. We therefore have the relation

$$\Delta \Sigma_{\mathbf{p}}(E, \mathbf{q}, \omega) = \frac{1}{V} \sum_{\mathbf{p}'} U^{RA}_{\mathbf{p}, \mathbf{p}'}(\mathbf{q}, \omega) \, \Delta G_{\mathbf{p}'}(E, \mathbf{q}, \omega) \qquad (8.133)$$

as the ΔG is precisely at the place separating the R-lines from the A-lines.

The only property we have used to establish the above relation between the irreducible four-point function and the self-energy is, besides the topological character of the Σ and U diagrams, the particular sequence of R and A labeling. The latter is just an expression of unitarity (that time evolution of a state is a unitary transformation), or equivalently, particle conservation.

Attaching a propagator line onto a U-diagram can be done in several ways, and we could equally well have established the identity

$$\Delta \Sigma_{\mathbf{p}}(E, \mathbf{q}, \omega) = \frac{1}{V} \sum_{\mathbf{p}'} \Delta G_{\mathbf{p}'}(E, \mathbf{q}, \omega) U^{RA}_{\mathbf{p}', \mathbf{p}}(\mathbf{q}, \omega) \, . \qquad (8.134)$$

8.9 Quantum Kinetic Equation

In section 8.4 we displayed the conductivity diagrams in terms of the three-point vector vertex function. We could of course equally well have discussed the conductivity diagrams in terms of the four-point vertex function. In section 8.5 we showed that the equation satisfied by the three-point vector vertex function in the non-crossing or ladder approximation is identical to the linearized Boltzmann equation. We shall in the following consider the four-point vertex function, and find the transport-like equation it satisfies. However, now we shall include quantum effects.[18] This will be crucial when we in chapter 9 study the motion of a particle in a random potential.

According to eq.(8.13) and eq.(8.17) we can express the impurity average of the propagator product, the four-point vertex function, in terms of the irreducible four-point vertex function $U_{\mathbf{p},\mathbf{p}'}$, and we have

$$\Phi_{\mathbf{p},\mathbf{p}'}(\mathbf{q},\omega) = G^R_{\mathbf{p}_+} G^A_{\mathbf{p}_-} V \delta_{\mathbf{p}\mathbf{p}'} + G^R_{\mathbf{p}_+} G^A_{\mathbf{p}_-} \frac{1}{V} \sum_{\mathbf{p}''} U^{RA}_{\mathbf{p},\mathbf{p}''}(\mathbf{q},\omega) \Phi_{\mathbf{p}'',\mathbf{p}'}(\mathbf{q},\omega) \quad (8.135)$$

where we have suppressed the energy variable, which in view of eq.(8.29) is fixed at the Fermi energy, $E \simeq \epsilon_F$. Diagrammatically we have

$$(8.136)$$

where the short $\mathbf{p}''$-lines are just labeling and do not represent internal propagators in accordance with eq.(8.135).

We express the product of the retarded and advanced Green's functions in terms of their difference

$$G^R_{\mathbf{p}_+} G^A_{\mathbf{p}_-} = \frac{G^A_{\mathbf{p}_-} - G^R_{\mathbf{p}_+}}{[G^R_{\mathbf{p}_+}]^{-1} - [G^A_{\mathbf{p}_-}]^{-1}} = -\frac{\hbar^{-1}\Delta G_{\mathbf{p}}(\mathbf{q},\omega)}{\omega - (\mathbf{p}\cdot\mathbf{q})/m - \Delta\Omega_{\mathbf{p}}(\mathbf{q},\omega)} \quad (8.137)$$

where

$$\Delta\Omega_{\mathbf{p}}(\mathbf{q},\omega) \equiv \frac{1}{\hbar}\Delta\Sigma_{\mathbf{p}}(E,\mathbf{q},\omega) \equiv \frac{1}{\hbar}\left(\Sigma^R_{\mathbf{p}_+}(E_+) - \Sigma^A_{\mathbf{p}_-}(E)\right) . \quad (8.138)$$

---

[18]In this section we follow the presentation of reference [33].

We suppress the dependence on the energy and the frequency as it runs with the R- and A-labels, and plays no role in the following argument.

Multiplication of eq.(8.135) by the denominator of eq.(8.137) leads to

$$\sum_{\mathbf{p''}} \left\{ (\omega - \mathbf{v_p} \cdot \mathbf{q} - \Delta\Omega_{\mathbf{p}}(\mathbf{q}, \omega)) \, \delta_{\mathbf{p''},\mathbf{p}} + \frac{1}{\hbar V} \Delta G_{\mathbf{p}}(\mathbf{q}, \omega) U_{\mathbf{p},\mathbf{p''}}^{RA}(\mathbf{q}, \omega) \right\} \Phi_{\mathbf{p''},\mathbf{p'}}(\mathbf{q}, \omega)$$

$$= \frac{-V}{\hbar} \Delta G_{\mathbf{p}}(\mathbf{q}, \omega) \delta_{\mathbf{pp'}}. \tag{8.139}$$

Making use of the unitarity condition, eq.(8.133), we can rewrite the equation as the transport-like equation with the presence of a source

$$(\omega - \mathbf{v_p} \cdot \mathbf{q}) \, \Phi_{\mathbf{p},\mathbf{p'}}(\mathbf{q}, \omega) - I[\Phi_{.,p'}(\mathbf{q}, \omega)] = \frac{-V}{\hbar} \Delta G_{\mathbf{p}}(\mathbf{q}, \omega) \, \delta_{\mathbf{pp'}} \tag{8.140}$$

where we have the collision integral[19]

$$I[\Phi_{.,p'}(\mathbf{q}, \omega)] = \frac{-1}{\hbar V} \sum_{\mathbf{p''}} U_{\mathbf{p},\mathbf{p''}}^{RA}(\mathbf{q}, \omega) \left\{ \Delta G_{\mathbf{p}}(\mathbf{q}, \omega) \Phi_{\mathbf{p''},\mathbf{p'}}(\mathbf{q}, \omega) - \Delta G_{\mathbf{p''}}(\mathbf{q}, \omega) \Phi_{\mathbf{p},\mathbf{p'}}(\mathbf{q}, \omega) \right\}. \tag{8.141}$$

Summing the equation over $\mathbf{p}$ and $\mathbf{p'}$ we obtain

$$\omega \, \Phi(\mathbf{q}, \omega) - q \, \Phi_j(\mathbf{q}, \omega) - \frac{1}{V^2} \sum_{\mathbf{p},\mathbf{p'}} \Delta\Omega_{\mathbf{p}}(\mathbf{q}, \omega) \Phi_{\mathbf{p},\mathbf{p'}}(\mathbf{q}, \omega)$$

$$= -\frac{1}{\hbar V} \sum_{\mathbf{p}} \Delta G_{\mathbf{p}}(\mathbf{q}, \omega) - \frac{1}{\hbar V^3} \sum_{\mathbf{p},\mathbf{p'},\mathbf{p''}} \Delta G_{\mathbf{p}}(\mathbf{q}, \omega) U_{\mathbf{p},\mathbf{p''}}^{RA}(\mathbf{q}, \omega) \Phi_{\mathbf{p''},\mathbf{p'}}(\mathbf{q}, \omega) \tag{8.142}$$

where

$$\Phi(\mathbf{q}, \omega) \equiv \frac{1}{V^2} \sum_{\mathbf{p},\mathbf{p'}} \Phi_{\mathbf{p},\mathbf{p'}}(\mathbf{q}, \omega) = \frac{1}{V} \sum_{\mathbf{p},\mathbf{p'}} < G^R(\mathbf{p}_+, \mathbf{p'}_+; E_+) G^A(\mathbf{p'}_-, \mathbf{p}_-; E) > \tag{8.143}$$

is the previously introduced zeroth moment, eq.(8.30), and we have introduced the first moment of $\Phi_{\mathbf{p},\mathbf{p'}}$ (similarly suppressing any dependence on the Fermi energy)

$$\Phi_j(\mathbf{q}, \omega) \equiv \frac{1}{V^2} \sum_{\mathbf{p},\mathbf{p'}} \left(\frac{\mathbf{p} \cdot \hat{\mathbf{q}}}{m} \right) \Phi_{\mathbf{p},\mathbf{p'}}(\mathbf{q}, \omega) . \tag{8.144}$$

Making use of the unitarity condition, the identity eq.(8.134),

$$\Delta\Omega_{\mathbf{p'}}(\mathbf{q}, \omega) = \frac{1}{\hbar V} \sum_{\mathbf{p}} \Delta G_{\mathbf{p}} U_{\mathbf{p},\mathbf{p'}}^{RA} . \tag{8.145}$$

---

[19] We note that $U_{\mathbf{p},\mathbf{p'}}^{RA}(\mathbf{q}, \omega)$ has the interpretation of a transition probability between momentum states $\mathbf{p'}$ and $\mathbf{p} + \mathbf{q}$ due to the random potential.

we get

$$\omega\, \Phi(\mathbf{q},\omega) - q\, \Phi_j(\mathbf{q},\omega) = -\frac{1}{\hbar V} \sum_{\mathbf{p}} \Delta G_{\mathbf{p}}(\mathbf{q},\omega) \; . \tag{8.146}$$

For $q \ll k_F$, $\hbar\omega \ll \epsilon_F$, the integrand on the right-hand side is proportional to the spectral weight, and as previously calculated, eq.(8.26), we get[20]

$$-\frac{1}{V} \sum_{\mathbf{p}} \Delta G_{\mathbf{p}}(0,0) = i \int \frac{d\mathbf{p}}{(2\pi\hbar)^d}\, A(\mathbf{p}, \epsilon_F)$$

$$= 2\pi i N_0(\epsilon_F)\,(1 + \mathcal{O}(\hbar/\epsilon_F \tau)) \; . \tag{8.147}$$

Using the peaked character of the spectral function we thus arrive at the equation

$$\omega\, \Phi(\mathbf{q},\omega) - q\, \Phi_j(\mathbf{q},\omega) = \frac{2\pi i N_0}{\hbar} \left(1 + \mathcal{O}\left(\frac{\hbar\omega}{\epsilon_F}, \frac{q}{k_F}\right)\right) \; . \tag{8.148}$$

It is not surprising that eq.(8.148) is equivalent to the continuity equation with a source term as the essential ingredient amounted to using the unitarity condition, the identity eq.(8.133). For $q = 0$ we can conclude that $\Phi(\mathbf{0},\omega)$ diverges as $1/\omega$ for small ω, in accordance with our previous observation, eq.(8.33).

Performing a summation over $\mathbf{p}'$ in eq.(8.139) we obtain

$$\left(\omega - \frac{\mathbf{p}\cdot\mathbf{q}}{m} - \Delta\Omega_{\mathbf{p}}(\mathbf{q},\omega)\right) \tilde{\Phi}_{\mathbf{p}}(\mathbf{q},\omega) = -\hbar^{-1}\Delta G_{\mathbf{p}}(\mathbf{q},\omega)\left[1 + \frac{1}{V}\sum_{\mathbf{p}''} U_{\mathbf{p},\mathbf{p}''}\, \tilde{\Phi}_{\mathbf{p}''}(\mathbf{q},\omega)\right] \tag{8.149}$$

where

$$\tilde{\Phi}_{\mathbf{p}}(\mathbf{q},\omega) \equiv \sum_{\mathbf{p}'} \Phi_{\mathbf{p},\mathbf{p}'}(\mathbf{q},\omega) \; . \tag{8.150}$$

The dependence in $\tilde{\Phi}_{\mathbf{p}}(\mathbf{q},\omega)$ on the length of $\mathbf{p}$ and $\mathbf{q}$ is determined by the peaked structure of the spectral weight, $\tilde{\Phi}_{\mathbf{p}}(\mathbf{q},\omega) \propto \Delta G_{\mathbf{p}}(\mathbf{q},\omega)$, and the dependence of $\tilde{\Phi}$ on q is therefore weak, of order q/k_F. We expect that the effect of disorder will be to smear out directional dependences, and make the angular $\hat{\mathbf{p}}$-dependence in the preceding function weak. In the spherical harmonic expansion

$$\tilde{\Phi}_{\mathbf{p}}(\mathbf{q},\omega) = \Delta G_{\mathbf{p}}(0,0)\big(c_0(q,\omega) + c_1(q,\omega)\,\hat{\mathbf{p}}\cdot\hat{\mathbf{q}} + \dots\big) \tag{8.151}$$

we therefore only need to take into account the first harmonic. Comparing with the Legendre expansion

$$\tilde{\Phi}_{\mathbf{p}}(\mathbf{q},\omega) = \sum_{l=0}^{\infty} P_l(\hat{\mathbf{p}}\cdot\hat{\mathbf{q}})\, \tilde{\Phi}_l(p,q,\omega) \tag{8.152}$$

---

[20]We have used the weak-disorder estimate, eq.(3.72), for the impurity-averaged propagators. The quantum interference processes, which we discuss in the next chapter, do not effect the single-particle propagators. The peaked character of the spectral weight is therefore set by the mean free path as in the weak-disorder case, the lowest approximation in $\hbar/p_F l$.

we get for the zeroth moment

$$\frac{1}{V} \sum_{\mathbf{p}} \tilde{\Phi}_{\mathbf{p}}(\mathbf{q}, \omega) = \frac{1}{V} \sum_{\mathbf{p}} \tilde{\Phi}_0(p, q, \omega) = -2\pi N_0 \, c_0(q, \omega) \qquad (8.153)$$

and for the first moment

$$\frac{1}{V} \sum_{\mathbf{p}} (\hat{\mathbf{p}} \cdot \hat{\mathbf{q}}) \, \tilde{\Phi}_{\mathbf{p}}(\mathbf{q}, \omega) = \int \frac{d\mathbf{p}}{(2\pi\hbar)^d} \, \tilde{\Phi}_1(p, q, \omega) = \frac{-2\pi N_0}{d} \, c_1(q, \omega) \,. \qquad (8.154)$$

The first-order spherical harmonic expansion is therefore the same as the one generated by the first moment projections of $\Phi_{\mathbf{p}'',\mathbf{p}'}(\mathbf{q}, \omega)$

$$\tilde{\Phi}_{\mathbf{p}}(\mathbf{q}, \omega) \simeq \frac{\Delta G_{\mathbf{p}}(0, 0)}{-2\pi i N_0 \, V} \sum_{\mathbf{p}',\mathbf{p}''} \left[1 + \frac{d}{p_F^2} (\mathbf{p} \cdot \hat{\mathbf{q}})(\mathbf{p}'' \cdot \hat{\mathbf{q}}) \right] \Phi_{\mathbf{p}'',\mathbf{p}'}(\mathbf{q}, \omega) \,. \qquad (8.155)$$

Multiplying eq.(8.149) by the factor $\mathbf{p} \cdot \hat{\mathbf{q}} \, \tau_{tr}(\mathbf{p})/im$ and summing over $\mathbf{p}$ we get

$$\sum_{\mathbf{p}} \left(\frac{\mathbf{p} \cdot \hat{\mathbf{q}} \, \tau_{tr}(\mathbf{p})}{im} \right) \left(\omega - \frac{\mathbf{p} \cdot \mathbf{q}}{m} - \frac{i}{\tau} \right) \tilde{\Phi}_{\mathbf{p}}(\mathbf{q}, \omega)$$

$$= -\hbar^{-1} \sum_{\mathbf{p}} \left(\frac{\mathbf{p} \cdot \hat{\mathbf{q}} \tau_{tr}(\mathbf{p})}{im} \right) \Delta G_{\mathbf{p}}(0, 0) \left[1 + \frac{1}{V} \sum_{\mathbf{p}''} U_{\mathbf{p},\mathbf{p}''} \, \tilde{\Phi}_{\mathbf{p}''}(\mathbf{q}, \omega) \right] (8.156)$$

and as the self-energy term only varies on the scale of the Fermi momentum and energy, it has been substituted by the constant i/τ.

Making use of the weak angular dependence we obtain by inserting the expansion eq.(8.155) ($\tau_{tr} \equiv \tau_{tr}(\epsilon_F)$)

$$[-i\omega\tau_{tr} + K(\mathbf{q}, \omega)] \Phi_j(\mathbf{q}, \omega) + iq D_0 \Phi(\mathbf{q}, \omega) = 0 \qquad (8.157)$$

where the function $K(\mathbf{q}, \omega)$ is specified by the irreducible four-point function

$$K(\mathbf{q}, \omega) = 1 + \frac{d\tau_{tr}}{\pi\hbar m n V^2} \sum_{\mathbf{p},\mathbf{p}'} (\mathbf{p} \cdot \hat{\mathbf{q}}) \Delta G_{\mathbf{p}} [U_{\mathbf{p},\mathbf{p}'}(\mathbf{q}, \omega) - U_0(\mathbf{p} - \mathbf{p}')] (\mathbf{p}' \cdot \hat{\mathbf{q}}) \Delta G_{\mathbf{p}'} \,.$$

$$(8.158)$$

A term containing the impurity correlator

$$U_0(\mathbf{p} - \mathbf{p}') = n_i \, |V_{imp}(\mathbf{p} - \mathbf{p}')|^2 \qquad (8.159)$$

has been added and subtracted, and in the former case been evaluated giving the constant 1. We have repeatedly used that

$$\frac{d}{-2\pi i N_0 p_F^2 V} \sum_{\mathbf{p}} (\mathbf{p} \cdot \hat{\mathbf{q}})^2 \, \Delta G_{\mathbf{p}} \, g(\mathbf{p}) \simeq g(p_F) \qquad (8.160)$$

for any function $g(\mathbf{p})$ slowly varying relative to $\Delta G_{\mathbf{p}}$.

Using the property of weak angular dependence, a closed set of equations for Φ and Φ_j has been obtained which we solve to get

$$\Phi(\mathbf{q},\omega) = \frac{2\pi\hbar^{-1}N_0}{-i\omega + \frac{D_0q^2}{-i\omega\tau_{tr}+K(\mathbf{q},\omega)}} \tag{8.161}$$

and

$$\Phi_j(\mathbf{q},\omega) = \frac{2\pi\hbar^{-1}N_0D_0\,q}{\omega(-i\omega\tau_{tr}+K(\mathbf{q},\omega))+iD_0q^2} \cdot \tag{8.162}$$

We note

$$\lim_{\omega\to 0}\lim_{\mathbf{q}\to 0}\Phi_j(\mathbf{q},\omega) = 0 \tag{8.163}$$

so that there is no singular behavior in $\Phi_j(\mathbf{0},\omega)$ in the small ω regime.

Using the relationship between the density response function and the four-point function, eq.(8.32), we get for the density response function for a Fermi gas in a random potential

$$\chi(\mathbf{q},\omega) = \frac{2N_0D_0q^2}{-i\omega(-i\omega\tau_{tr}+K(\mathbf{q},\omega))+D_0\,q^2} \cdot \tag{8.164}$$

Recalling eq.(8.90), we obtain for the conductivity

$$\sigma(\omega) = \frac{\sigma_0}{K(\mathbf{0},\omega)-i\omega\tau_{tr}} \cdot \tag{8.165}$$

If we approximate the irreducible four-point function by the impurity correlator. the first term in eq.(8.18), we have $K=1$, and we get the Boltzmann conductivity from eq.(8.165).

In the limit $\omega\tau_{tr} \ll K(\mathbf{q},\omega)$ (which we in the next chapter demonstrate is essentially no restriction in the regions of interest) we have[21]

$$\Phi(\mathbf{q},\omega) = \frac{2\pi\hbar^{-1}N_0}{-i\omega+\tilde{D}(\mathbf{q},\omega)\,q^2} \tag{8.166}$$

where we have introduced the diffusivity

$$\tilde{D}(\mathbf{q},\omega) \equiv \frac{D_0}{K(\mathbf{q},\omega)} \cdot \tag{8.167}$$

The zeroth moment function $\Phi(\mathbf{q},\omega)$ is specified by the four-point vertex function, and we have shown that it has a diffusive structure. This link between the four-point vertex function and the *diffusion*-pole is utilized in the next chapter to construct the self-consistent theory of localization.

---

[21] The form of the obtained expression is of course evident because of particle conservation. The matter of importance here is that we have obtained an expression, eq.(8.158), for the (inverse) diffusivity. In the next chapter we shall use the diagrammatic technique to analyze the diffusivity, and obtain the self-consistent theory of localization.

Using the relationship between the density response function and the four-point vertex function, eq.(8.32), we get for the density response function for a Fermi gas in a random potential

$$\chi(\mathbf{q},\omega) \;=\; \frac{2N_0\tilde{D}(\mathbf{q},\omega)q^2}{-i\omega + \tilde{D}(\mathbf{q},\omega)q^2} \;=\; \frac{\tilde{D}(\mathbf{q},\omega)q^2}{-i\omega + \tilde{D}(\mathbf{q},\omega)q^2}\, \chi(\mathbf{q},0) \qquad (8.168)$$

and as the diffusivity has a smooth q-dependence on the scale of the Fermi wave vector the density response function has a diffusion pole.

For an external potential smoothly varying on the scale of the mean free path, $ql \ll 1$, the function K has only a weak dependence on q, of order q/k_F as q occurs added to the large momentum of the propagator, and we can therefore set q equal to zero. We introduce $\tilde{D}(\omega) \equiv \tilde{D}(\mathbf{q} = \mathbf{0},\omega)$ and $K(\omega) \equiv K(\mathbf{q} = \mathbf{0},\omega)$.

According to eq.(8.168) and eq.(8.90) we have the Einstein relation

$$\sigma(\omega) \;=\; 2e^2 N_0\,\tilde{D}(\omega)\,. \qquad (8.169)$$

The low-frequency behavior of K distinguishes whether we have a conductor or an insulator. In the former case $K(0) = 1$, and in the latter case $K(\omega)$ diverges at low frequencies. We note, that the divergence in K can only come from the irreducible four-point vertex.

8.10 Time-Reversal Symmetry

For a time-reversal invariant situation the four-point vertex Γ satisfies the relation in the momentum variables

$$\Gamma_{\mathbf{p},\mathbf{p}'}(E,\mathbf{q},\omega) \;=\; \Gamma_{(\mathbf{p}-\mathbf{p}'+\mathbf{q})/2,(\mathbf{p}'-\mathbf{p}+\mathbf{q})/2}(E,\mathbf{p}+\mathbf{p}',\omega)\,. \qquad (8.170)$$

The energy variables on the upper and lower particle lines remain the same on both sides of the equation, and we suppress the dependence on the variable E in the following. As usual, the lower index momentum labeling on Γ is per definition half the sum of the momenta on the left and right, respectively, and the value in the parenthesis the difference between the upper and lower momentum labeling.

The validity of this equality is easily established diagrammatically. First we twist the four-point vertex diagrams by twisting for example the lower line, leaving us with the integration over the same product of numbers, and therefore with an identity. Furthermore, since we twist the full vertex, we generate all the diagrams of the full four-point vertex, but now with the same direction on the particle lines, as expressed by the second equality sign

$$= \quad \boxed{\Gamma} \quad = \quad \boxed{\Gamma} \qquad (8.171)$$

Last, we have applied time-reversal invariance to reverse the arrow direction on all the lower propagator lines. This interchange of direction accompanied by a sign change in the momenta on the lower line is valid for the time-reversal invariant case as is easily verified order by order in (skeleton) perturbation theory. Let us consider, say, a diagram with three impurity correlators attached to the lower line. The part of the diagram represented by

$$= \qquad (8.172)$$

is identical to the number represented by the second diagram where line directions and signs of momenta are inverted, provided we have for the impurity-averaged propagator

$$G^{A(R)}(-\mathbf{p}, E) = G^{A(R)}(\mathbf{p}, E) . \qquad (8.173)$$

For time-reversal invariant dynamics, we have according to eq.(2.207) for the propagators

$$G^{R(A)}(\mathbf{p}, \mathbf{p}', E) = G^{R(A)}(-\mathbf{p}', -\mathbf{p}, E) . \qquad (8.174)$$

Upon impurity averaging we regain the translation invariance property of the impurity averaged propagator, and thereby eq.(8.173) and eq.(8.172) have been demonstrated.

In the time-reversal invariant situation, we can identify any set of diagrams $\mathcal{D}$, with its twisted and one-line reversed set of diagrams

$$\mathcal{D}_{\mathbf{p},\mathbf{p}'}(\mathbf{q}, \omega) = \bar{\mathcal{D}}_{(\mathbf{p}-\mathbf{p}'+\mathbf{q})/2,(\mathbf{p}'-\mathbf{p}+\mathbf{q})/2}(\mathbf{p} + \mathbf{p}', \omega) . \qquad (8.175)$$

In contrast to the case of the full four-point vertex Γ, the topological structure of the diagrams is different as the second identity in eq.(8.171) is illegitimate for a subclass of diagrams. What is achieved by the twist and time reversing of one of the lines, however, is that one set of diagrams is related to a set of diagrams with a different topology, and a different momentum labeling.

As an important example of exploiting time-reversal invariance in such a manner, we consider the twisted diffusion diagrams ($\hbar\mathbf{Q} \equiv \mathbf{p} + \mathbf{p}'$)

$$
= D^{(\mathbf{p}+\mathbf{p}',\omega)}_{(\mathbf{p}-\mathbf{p}'+\mathbf{q})/2,(\mathbf{p}'-\mathbf{p}+\mathbf{q})/2} \,. \tag{8.176}
$$

The first equality sign is obtained by just twisting the ladder diagrams, and lastly we applied time-reversal invariance to change the arrow direction and signs of momenta on the lower propagator lines. For the case of time-reversal invariance we thus have the relation between the twisted ladder diagrams and the Diffuson:

$$C_{\mathbf{p},\mathbf{p}'}(\mathbf{q},\omega) \;=\; D^{(\mathbf{p}+\mathbf{p}',\omega)}_{(\mathbf{p}-\mathbf{p}'+\mathbf{q})/2,(\mathbf{p}'-\mathbf{p}+\mathbf{q})/2} \tag{8.177}$$

In section 8.7 we calculated the Diffuson for small momentum difference (between upper and lower line momentum input values) and we therefore obtain for the twisted ladder diagrams for small total momentum

$$C_{\mathbf{p},\mathbf{p}'}(\mathbf{q},\omega) \;=\; \frac{u^2/\tau}{-i\omega + D_0(\mathbf{p}+\mathbf{p}')^2\hbar^{-2}} \tag{8.178}$$

where the last equality is valid in the momentum regime, $|\mathbf{p}+\mathbf{p}'|\,l \ll \hbar$, and $\omega\tau \ll 1$, $q \ll k_F$.

In the case of time-reversal invariance, we can thus relate a singular behavior of one class of diagrams to a singularity in different variables in another set of diagrams. In the next chapter this property will be exploited to classify all four-point diagrams, and diagrammatically derive the self-consistent theory of localization.

Chapter 9

Localization

In this chapter the quantum mechanical motion of a particle in a random potential at zero temperature is addressed. After presenting the scaling theory of localization, and verifying its predictions in the weak-disorder regime, the self-consistent theory of localization is presented.

In a seminal paper of 1958, P. W. Anderson showed that a particle's motion in a sufficiently disordered three-dimensional system behaves quite differently from that predicted by classical physics according to the Boltzmann theory [34]. In fact, at zero temperature diffusion will be absent, as particle states are localized in space due to the random potential. A sufficiently disordered system therefore behaves as an insulator and not as a conductor! By changing the impurity concentration, a transition from metallic to insulating behavior occurs. This is called the Anderson metal-insulator transition. In this chapter we shall discuss the phenomenon of Anderson localization using the developed diagrammatic technique.

In a pure metal, the Bloch or plane wave eigenstates of the Hamiltonian are current carrying

$$<\hat{\mathbf{j}}>_{ext} = \int d\mathbf{x} <\mathbf{p}|\hat{\mathbf{j}}(\mathbf{x})|\mathbf{p}> = e\,\mathbf{v_p} \,. \tag{9.1}$$

In a sufficiently disordered system, a typical energy eigenstate has a finite extension, and does not carry any current

$$<\hat{\mathbf{j}}>_{loc} = \mathbf{0} \,. \tag{9.2}$$

The last statement is not easily made rigorous, and the phenomenon of localization is quite subtle. We shall return to the discussion of wave function localization in section 9.3.3.

Astonishing progress in the understanding of transport in disordered systems has taken place since the introduction of the scaling theory of localization [37]. A key ingredient in the subsequent development of the understanding of the transport properties of disordered systems was the intuition provided by diagrammatic perturbation theory. We shall exploit this in the present chapter, as well as in chapter 11 where we will discuss the weak localization effect. We start by considering the scaling theory.[1]

---

[1] The scaling theory of localization has its inspiration in the original work of Wegner [35] and

9.1   Scaling Theory of Localization

We shall consider a macroscopically homogeneous conductor, i.e., one with a spatially uniform impurity concentration, at zero temperature. The conductance of a d-dimensional hypercube of linear dimension L is according to eq.(8.73) proportional to the conductivity

$$G(L) = L^{d-2}\,\sigma(L)\,. \tag{9.3}$$

The central idea of the scaling theory of localization is that the conductance rather than the conductivity is the quantity of importance for determining the transport properties of a macroscopic sample. The conductance has dimension of $e^2/\hbar$, independent of the spatial dimension of the sample, and we introduce the dimensionless conductance of a hypercube

$$g(L) \equiv \frac{G(L)}{\frac{e^2}{\hbar}}\,. \tag{9.4}$$

The one-parameter scaling theory of localization is based on the assumption that the dimensionless conductance solely determines the conductivity behavior of a disordered system. Consider fitting n^d identical blocks of length L (i.e., having the same impurity concentration, and mean free path smaller than the size of the system, $l < L$) into a hypercube of linear dimension nL. The dc conductance of the hypercube $g(nL)$ is then related to the conductance of each block $g(L)$ by

$$g(nL) = f(n, g(L)) \tag{9.5}$$

This is the one-parameter scaling assumption, the conductance of each block solely determines the conductance of the larger block; there is no extra dependence on L or microscopic parameters such as l or λ_F.

For a continuous variation of the linear dimension of a system, the one-parameter scaling assumption results in the logarithmic derivative being solely a function of the dimensional conductance

$$\frac{d\ln g}{d\ln L} = \beta(g)\,. \tag{9.6}$$

This can be seen by differentiating eq.(9.5) to get

$$\frac{d\ln g(L)}{d\ln L} = \frac{L}{g}\frac{dg}{dL} = \frac{L}{g}\frac{dg(nL)}{dL}\Bigg|_{n=1} = \frac{1}{g}\frac{dg(nL)}{dn}\Bigg|_{n=1} = \frac{1}{g}\frac{df(n,g)}{dn}\Bigg|_{n=1} \equiv \beta(g(L))\,. \tag{9.7}$$

The physical significance of the scaling function, β, is as follows: If we start out with a block of size L, with a value of the conductance $g(L)$ for which $\beta(g)$ is positive, then the conductance according to eq.(9.6) will increase upon enlarging the system, and vice versa for $\beta(g)$ negative. The β-function thus specifies the transport properties at that degree of disorder for a system in the thermodynamic, infinite volume, limit.

Thouless [36].

In the limit of weak disorder, large conductance $g \gg 1$, we expect metallic conduction to prevail. The conductance is thus described by classical transport theory, i.e., Ohm's law prevails, $G(L) = L^{d-2} \sigma_0$, and the conductivity is independent of the linear size of the system, and we obtain the limiting behavior for the scaling function

$$\beta(g) = d - 2 , \qquad g \gg 1 \qquad (9.8)$$

the scaling function having an asymptotic limit depending only on the dimensionality of the system.

In the limit of strong disorder, small conductance $g \ll 1$, we expect with Anderson [34] that localization prevails, so that the conductance assumes the form $g(L) \propto e^{-L/\xi}$, where ξ is called the localization length, the length scale beyond which the resistance grows exponentially with length.[2] In the low conductance, so-called strong localization, regime we thus obtain for the scaling function, c being a constant,

$$\beta(g) = \ln g + c , \qquad g \ll 1 \qquad (9.9)$$

a logarithmic dependence in any dimension.

Since there is no intrinsic length scale to tell us otherwise, it is physically reasonable in this consideration to draw the scaling function as a monotonic non-singular function connecting the two asymptotes. We therefore obtain the behavior of the scaling function depicted in figure 9.1. This is precisely the picture expected in three and one dimensions. In three dimensions the unstable fix-point signals the metal-insulator transition predicted by Anderson. The transition occurs at a critical value of the disorder where the scaling function vanishes, $\beta(g_c) = 0$. If we start with a sample with conductance larger than the critical value, $g > g_c$, then upon increasing the size of the sample the conductance increases since the scaling function is positive. In the thermodynamic limit, the system becomes a metal with conductivity σ_0. Conversely, starting with a more disordered sample with conductance less than the critical value, $g < g_c$, upon increasing the size of the system, the conductance will flow to the insulating regime, since the scaling function is negative. In the thermodynamic limit the system will be an insulator with zero conductance. This is the localized state. In one dimension it can be shown exactly, that all states are exponentially localized for arbitrarily small amount of disorder [38], and the metallic state is absent, in accordance with the scaling function being negative. An astonishing prediction follows from the scaling theory in the two-dimensional case where the one-parameter scaling function is also negative. There is no true metallic state in two dimensions![3]

---

[2]This expectation we demonstrate to hold true in section 9.3.5. At this point we just argue that if the envelope function for a typical electronic wave function is exponentially localized (as demonstrated in section 9.3.3), the conductance will have the stated length dependence, where ξ is the localization length of a typical wave function in the random potential, as it is proportional to the probability for the electron to be at the edge of the sample.

[3]In this day and age, low-dimensional electron systems are routinely manufactured. For example, a two-dimensional electron gas can be created in the inversion layer of an MBE grown GaAs-AlGaAs heterostructure. Two-dimensional localization effects provide a useful tool for probing material characteristics, as we discuss in chapter 11.

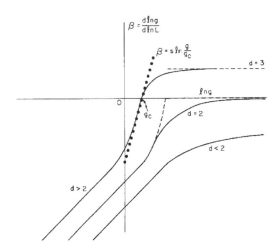

Figure 9.1 The scaling function as function of $\ln g$ (from E. Abrahams, P. W. Anderson, D. C. Licciardello, and T. V. Ramakrishnan, Phys. Rev. Lett. **42**, 673 (1979)).

The prediction of the scaling theory of the absence of a true metallic state in two dimensions was at variance with the previously conjectured theory of *minimal metallic conductivity*. The classical conductivity obtained from the Boltzmann theory has the form, in two and three dimensions $(d = 2, 3)$,[4]

$$\sigma_0 \;=\; \frac{e^2}{\hbar} \frac{k_F l}{d\pi^{d-1}} k_F^{d-2} \;. \tag{9.10}$$

According to Mott [39], the conductivity in three (and two) spatial dimensions should decrease as the disorder increases, until the mean free path becomes of the order of the Fermi wavelength of the electron, $l \sim \lambda_F$. The minimum metallic conductivity should thus occur for the amount of disorder for which $k_F l \sim 2\pi$, and in two dimensions have the universal value $e^2/\hbar$. Upon further increasing the disorder, the conductivity should discontinuously drop to zero.[5] This is in contrast to the scaling theory, which predicts the conductivity to be a continuous function of disorder. The metal-insulator transition thus resembles a second-order phase transition, in contrast to Mott's first-order conjecture (corresponding to a scaling function represented by the dashed line in figure 9.1).[6]

---

[4]In one dimension, the Boltzmann conductivity is $\sigma_0 = 2e^2 l/\pi\hbar$. However, the conclusion to be drawn from the scaling theory is that even the slightest amount of disorder invalidates the Boltzmann theory in one and two dimensions.

[5]In three dimensions in the thermodynamic limit the *conductance* drops to zero at the critical value according to the scaling theory.

[6]The impressive experimental support for the existence of a minimal metallic conductivity

The phenomenological scaling theory offers a comprehensive picture of the conductance of disordered systems, and predicts that all states in two dimensions are localized irrespective of the amount of disorder. To gain confidence in this surprising result, one should check the first correction to the metallic limit. We therefore calculate the first quantum correction to the scaling function and verify that it is indeed negative.

9.2 Coherent Backscattering

In diagrammatic terms, the quantum corrections to the classical conductivity are described by conductivity diagrams where impurity lines connecting the retarded and advanced lines cross. Such diagrams are nominally smaller, determined by the quantum parameter $\hbar/p_F l$, than the classical contribution. The subclass of diagrams, where the impurity lines cross a maximal number of times, is of special importance since their sum exhibits singular behavior (as we already noted in section 8.10). Such a type of diagram is illustrated below:

$$(9.11)$$

The maximally crossed diagrams describe the first quantum correction to the classical conductivity, the weak-localization effect, a subject we discuss in full detail in chapter 11.

In the frequency and wave vector region of interest each insertion in a maximally crossed diagram is of order 1, just as in the case of the ladder diagrams. Diagrams with maximally crossing impurity lines are therefore all of the same order of magnitude and must accordingly all be summed ($\hbar\mathbf{Q} \equiv \mathbf{p} + \mathbf{p}'$);

$$(9.12)$$

From the maximally crossed diagrams, we obtain analytically, by applying the Feynman rules, the correction to the conductivity of a degenerate Fermi gas,

in two dimensions is now believed to either reflect the cautiousness one must exercise when attempting to extrapolate measurements at finite temperature to zero temperature, or invoke a crucial importance of electron-electron interaction in dirty metals even at very low temperatures.

$\hbar\omega, kT \ll \epsilon_F,$[7]

$$\delta\sigma_{\alpha,\beta}(\mathbf{q},\omega) = \left(\frac{e}{m}\right)^2 \frac{\hbar}{\pi} \int \frac{d\mathbf{p}}{(2\pi\hbar)^d} \int \frac{d\mathbf{p}'}{(2\pi\hbar)^d}\, p_\alpha\, p'_\beta\, \tilde{C}_{\mathbf{p},\mathbf{p}'}(\epsilon_F, \mathbf{q}, \omega)\, G^R(\mathbf{p}_+, \epsilon_F + \hbar\omega)$$

$$G^R(\mathbf{p}'_+, \epsilon_F + \hbar\omega) G^A(\mathbf{p}'_-, \epsilon_F) G^A(\mathbf{p}_-, \epsilon_F)\,. \qquad (9.13)$$

To describe the sum of the maximally crossed diagrams, we have introduced the so-called Cooperon $\tilde{C}$,[8] corresponding to the diagrams ($\epsilon_F^+ \equiv \epsilon_F + \hbar\omega$)

$$\qquad\qquad\qquad\qquad\qquad\qquad (9.14)$$

---

[7]In fact we shall in this section assume zero temperature as we shall neglect any influence on the maximally crossed diagrams from inelastic scattering. Interaction effects will be the main topic of section 11.3.

[8]The nickname refers to the singularity in its momentum dependence being for zero total momentum, as is the case for the Cooper pairing correlations resulting in the superconductivity instability.

In the last equality we have twisted the A-line around in each of the diagrams, and by doing so, we of course do not change the numbers being multiplied together.

Let us consider the case where the random potential is delta-correlated[9]

$$<V(\mathbf{x})V(\mathbf{x}')> \ = u^2\,\delta(\mathbf{x}-\mathbf{x}')\,. \tag{9.15}$$

Since the impurity correlator in the momentum representation then is a constant, u^2, all internal momentum integrations become independent. As a consequence, the dependence of the Cooperon on the external momenta will only be in the combination $\mathbf{p}+\mathbf{p}'$, for which we introduce the notation $\hbar\mathbf{Q}\equiv\mathbf{p}+\mathbf{p}'$, as well as $\tilde{C}_\omega(\mathbf{p}+\mathbf{p}')\equiv\tilde{C}_{\mathbf{p},\mathbf{p}'}(\epsilon_F,\mathbf{0},\omega)\equiv\tilde{C}_\omega(\mathbf{Q})$, and we have

$$\tilde{C}_\omega(\mathbf{Q}) \equiv \ C \tag{9.16}$$

---

[9]As we already noted in section 8.5, the case of a short-range potential goes through as usual, the only change being the appearance of the transport time instead of the momentum relaxation time. For a discussion of the effects of anisotropy we refer to [40].

For convenience we have extracted a factor from the maximally crossed diagrams which we shortly demonstrate, eq.(9.23), is simply the constant u^2 in the relevant parameter regime. We shall therefore also refer to C as the Cooperon. Diagrammatically we obtain according to eq.(9.16)

$$
\boxed{C} \;=\; 1 \;+\; \times \!\!\overset{\text{R}}{\underset{\text{A}}{\boxed{\mathbf{p}''_+}}}\, \boxed{C} \;.
\qquad (9.17)
$$

$$\hbar\mathbf{Q}-\mathbf{p}''_+$$

Analytically the Cooperon satisfies the equation

$$
C_\omega(\mathbf{Q}) = 1 + u^2 \int \frac{d\mathbf{p}''}{(2\pi\hbar)^d}\, G^R(\mathbf{p}''_+, \epsilon_F + \hbar\omega) G^A(\mathbf{p}''_+ - \hbar\mathbf{Q}, \epsilon_F)\, C_\omega(\mathbf{Q}) \;.
\qquad (9.18)
$$

It is obvious that a change in the wave vector of the external field can be compensated by a shift in the momentum integration variable, leaving the Cooperon independent of any spatial inhomogeneity in the electric field which is smooth on the atomic scale.

The Cooperon equation is a simple geometric series which we immediately can sum[10]

$$
\begin{aligned}
C_\omega(\mathbf{Q}) &= \left(1 + \zeta(\mathbf{Q},\omega) + \zeta^2(\mathbf{Q},\omega) + \zeta^3(\mathbf{Q},\omega) + \dots \right) \\[4pt]
&= 1 + \zeta(\mathbf{Q},\omega)\, C_\omega(\mathbf{Q}) \\[6pt]
&= \frac{1}{1 - \zeta(\mathbf{Q},\omega)} \;.
\end{aligned}
\qquad (9.19)
$$

Diagrammatically we can express the result

$$
C_\omega(\mathbf{Q}) \;=\; \cfrac{1}{1 \;-\; \times \!\!\overset{\text{R}}{\underset{\text{A}}{\boxed{\begin{smallmatrix}\epsilon_F^+\mathbf{p}''_+\\[2pt]\epsilon_F\hbar\mathbf{Q}-\mathbf{p}''_+\end{smallmatrix}}}}} \;.
\qquad (9.20)
$$

We have previously calculated the insertion $\zeta(\mathbf{Q},\omega)$, eq.(8.68), and for the region of interest, $\omega\tau, Ql \ll 1$, we have

$$
\zeta(\mathbf{Q},\omega) \;=\; 1 + i\omega\tau - D_0\tau Q^2
\qquad (9.21)
$$

---

[10]This result we already derived in section 8.10, where we established the relation between the Diffuson and its twisted diagrams in the case of time-reversal invariance.

and for the Cooperon

$$C_\omega(\mathbf{Q}) = \frac{\frac{1}{\tau}}{-i\omega + D_0 Q^2} \cdot \tag{9.22}$$

The Cooperon exhibits singular infrared behavior.

In the singular region the prefactor in eq.(9.16) equals the constant u^2 as

$$= u^2 \zeta(\mathbf{Q}, \omega) \simeq u^2 \tag{9.23}$$

i.e., in the region of interest we thus have $\tilde{C} = u^2 C$. As far as regards the singular behavior we could equally well have defined the Cooperon by the set of diagrams

$$\tilde{C}_\omega(\mathbf{Q}) = \quad + \quad \dots \tag{9.24}$$

as adding a constant to a singular function does not change the singular behavior.

Changing in the conductivity expression, eq.(9.13), one of the integration variables, $\mathbf{p}' = -\mathbf{p} + \hbar\mathbf{Q}$, we get for the contribution of the maximally crossed diagrams

$$\delta\sigma_{\alpha\beta}(\mathbf{q}, \omega) = \left(\frac{e}{m}\right)^2 \frac{\hbar}{\pi} \int \frac{d\mathbf{p}}{(2\pi\hbar)^d} \int' \frac{d\mathbf{Q}}{(2\pi)^d} \, p_\alpha \left(-p_\beta + \hbar Q_\beta\right) \frac{u^2/\tau}{-i\omega + D_0 Q^2}$$

$$G_{\epsilon_F^+}^R(\mathbf{p}_+) \, G_{\epsilon_F^+}^R(-\mathbf{p}_+ + \hbar\mathbf{Q}) \, G_{\epsilon_F}^A(-\mathbf{p}_- + \hbar\mathbf{Q}) \, G_{\epsilon_F}^A(\mathbf{p}_-) \tag{9.25}$$

where the prime on the $\mathbf{Q}$-integration signifies that we only need to integrate over the region $Ql < 1$ from which the large contribution is obtained. Everywhere except in the Cooperon we can therefore neglect $\mathbf{Q}$ as $|\mathbf{p} - \hbar\mathbf{Q}| \sim p \sim p_F$. Assuming a smoothly varying external field on the atomic scale, $q \ll k_F$,[11] we can perform the momentum integration, and obtain to leading order in $\hbar / p_F l$

$$\int \frac{d\mathbf{p}}{(2\pi\hbar)^d} \; p_\alpha p_\beta \; G^R_{\epsilon_F}(\mathbf{p}_+) G^R_{\epsilon_F}(-\mathbf{p}_+) \; G^A_{\epsilon_F}(-\mathbf{p}_-) G^A_{\epsilon_F}(\mathbf{p}_-) \; = \; \frac{4\pi\tau^3 N_d(\epsilon_F) p_F^2}{\hbar^3 d} \, \delta_{\alpha\beta}$$

$$(9.26)$$

where we have also safely neglected the ω dependence in the propagators as for the region giving the large contribution, we have $\omega < 1/\tau \ll \epsilon_F/\hbar$.

At zero frequency we have for the first quantum correction to the conductivity of an electron gas

$$\delta\sigma(L) \; = \; -\frac{2e^2 D_0}{\pi\hbar} \int \!\!\!\!\!{}^{\prime} \frac{d\mathbf{Q}}{(2\pi)^d} \frac{1}{D_0 Q^2} \; . \tag{9.27}$$

In the one- and two-dimensional case the integral diverges for small Q, and we need to assess the lower cut-off.[12] In order to understand the lower cut-off we note that the maximally crossed diagrams lend themselves to a simple physical interpretation. The R-line in the Cooperon describes the amplitude for the scattering sequence of an electron (all momenta being near the Fermi surface as the contribution is otherwise small)

$$\mathbf{p}' \rightarrow \mathbf{p}_1 \rightarrow .. \rightarrow \mathbf{p}_N \rightarrow \mathbf{p} \simeq -\mathbf{p}' \tag{9.28}$$

whereas the A-line describes the complex conjugate amplitude for the opposite, i.e., time-reversed, scattering sequence

$$\mathbf{p}' \rightarrow -\mathbf{p}_N \rightarrow .. \rightarrow -\mathbf{p}_1 \rightarrow \mathbf{p} \simeq -\mathbf{p}' \tag{9.29}$$

i.e., the Cooperon describes a quantum interference process: the quantum interference between time-reversed scattering sequences. The physical process responsible for the quantum correction is thus coherent backscattering.[13] The random potential acts as sets of mirrors such that an electron in momentum state $\mathbf{p}$ ends up backscattered into momentum state $-\mathbf{p}$. The quantum correction to the conductivity is thus negative as the conductivity is a measure of the initial and final correlation of the velocities as reflected in the factor $\mathbf{p} \cdot \mathbf{p}'$ in the conductivity expression.

---

[11]In a conductor a spatially varying electric field will due to the mobile charges be screened (as we discuss further in section 10.5). In a metal, say, an applied electric field is smoothly varying on the atomic scale, $q \ll k_F$, and we can set q equal to zero as it appears in combination with large momenta, $p, p' \sim p_F$.

[12]Langer and Neal [41] were the first to study the maximally crossed diagrams, and noted that they give a divergent result at zero temperature. However, in their analysis they did not assess the lower cut-off correctly.

[13]The coherent backscattering effect was considered for light waves already in 1968 [42]. It is amusing that a quantitative handling of the phenomena had to await the study of the analogous effect in solid-state physics, and the diagrammatic treatment of electronic transport in metals a decade later.

The quantum interference process described by the above scattering sequences corresponds in real space to the quantum interference between the two alternatives for a particle to traverse a closed loop in opposite (time-reversed) directions.[14]

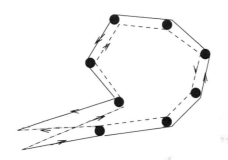

Figure 9.2 Coherent backscattering process.

We are considering the phenomenon of conductivity, where currents through connecting leads are taken in and out of a sample, say, at opposing faces of a hypercube. The maximal size of a loop allowed to contribute to the coherent backscattering process is thus the linear size of the system, as we assume that an electron reaching the end of the sample is irreversibly lost to the environment (leads and battery).[15] For a system of linear size L we then have for the quantum correction to the conductivity

$$\delta\sigma(L) = -\frac{2e^2 D_0}{\pi\hbar} \int_{1/L}^{1/l} \frac{d\mathbf{Q}}{(2\pi)^d} \frac{1}{D_0 Q^2} . \qquad (9.30)$$

Performing the integral in the two-dimensional case gives for the first quantum correction to the dimensionless conductance[16]

$$\delta g(L) = -\frac{1}{\pi^2} \ln \frac{L}{l} . \qquad (9.31)$$

We note that the first quantum correction to the conductivity indeed is negative, describing the precursor effect of localization. For the asymptotic scaling function we then obtain

$$\beta(g) = -\frac{1}{\pi^2 g} , \qquad g \gg 1 \qquad (9.32)$$

---

[14]This all important observation of the physical origin of the quantum correction to the conductivity (originally expressed in reference [43]) we shall take advantage of in chapter 11, where the real space treatment of weak localization is done in detail.

[15]An electron is assumed never to reenter from the leads phase coherently, and the Cooperon equation should be solved with the boundary condition that the Cooperon vanishes on the lead boundaries.

[16]The precise magnitudes of the cutoffs are irrelevant for the scaling function in the two-dimensional case, as a change can only produce the logarithm of a constant in the dimensionless conductance.

and the first quantum correction to the scaling function is thus seen to be negative in concordance with the scaling picture.

Exercise 9.1 *Show that in dimensions one and three, the first quantum correction to the dimensionless conductance is*

$$\delta g(L) \;=\; \begin{cases} -\frac{1}{\pi^2}\left(1 - \frac{l}{L}\right) & d = 1 \\[2ex] -\frac{1}{\pi^3}\left(\frac{L}{l} - 1\right) & d = 3 \end{cases} \tag{9.33}$$

and thereby for the scaling function to lowest order in $1/g$

$$\beta(g) \;=\; (d - 2) - \frac{a}{g} \tag{9.34}$$

where

$$a \;=\; \begin{cases} \frac{1}{\pi^2} & d = 1 \\[2ex] \frac{1}{\pi^3} & d = 3 \;. \end{cases} \tag{9.35}$$

We can introduce the length scale characterizing localization, the localization length, qualitatively as follows: for a sample much larger than the localization length, $L \gg \xi$, the sample is in the localized regime and we have $g(L) \simeq 0$. To estimate the localization length, we equate it to the length for which $g(\xi) \simeq g_0$, i.e., the length scale where the scale dependent part of the conductance is comparable to the Boltzmann conductance. The lowest-order perturbative estimate based on eq.(9.31) and eq.(9.33) gives in two and one dimensions the localization lengths $\xi^{(2)} \simeq l \exp \pi k_F l/2$ and $\xi^{(1)} \simeq l$, respectively.

The one-parameter scaling hypothesis has been shown to be valid for the average conductance in the above considered model [35]. Whether the one-parameter scaling picture for the disorder model studied is true for higher-order cumulants of the conductance, $< g^n >$, is a difficult question which seems to have been answered in the negative in reference [44]. However, a different question is whether deviations from one-parameter scaling are observable, in the sense that a sample has to be so close to the metal-insulator transition that real systems cannot be made homogeneous enough. Furthermore, electron-electron interaction can play a profound role in real materials invalidating the model studied, and leaving room for a metal-insulator transition in low-dimensional systems [45].

We can also calculate the zero-temperature frequency dependence of the first quantum correction to the conductivity for a sample of large size, $L \gg \sqrt{D_0/\omega} \equiv L_\omega$. From eq.(9.25) we have

$$\delta\sigma_{\alpha\beta}(\omega) \;=\; \delta\sigma(\omega)\, \delta_{\alpha\beta} \tag{9.36}$$

where

$$\delta\sigma(\omega) \;=\; -\frac{2e^2 D_0}{\hbar\pi} \int\limits_0^{1/l} \frac{d\mathbf{Q}}{(2\pi)^d} \, \frac{1}{-i\omega + D_0 Q^2} \; . \tag{9.37}$$

Calculating the integral, we get for the frequency dependence of the quantum correction to the conductivity in, say, two dimensions [46]

$$\frac{\delta\sigma(\omega)}{\sigma_0} = -\frac{1}{\pi k_F l} \ln\frac{1}{\omega\tau} \; . \tag{9.38}$$

We note that for the perturbation theory to remain valid the frequency can not be too small, $\omega\tau \simeq 1$.

The quantum correction to the conductivity in two dimensions is seen to be universal

$$\delta\sigma(\omega) \;=\; -\frac{1}{2\pi^2}\frac{e^2}{\hbar} \ln\frac{1}{\omega\tau} \; . \tag{9.39}$$

Let us calculate the first quantum correction to the current density response to a spatially homogeneous electric pulse, recall eq.(8.80),

$$\delta\mathbf{j}(t) \;=\; \delta\sigma(t)\,\mathbf{E}_0 \tag{9.40}$$

where

$$\delta\sigma(t) \;=\; -\frac{2e^2 D_0}{\hbar\pi} \int\limits_{-\infty}^{\infty} \frac{d\omega}{2\pi}\, e^{-i\omega t} \int\limits_{1/L}^{1/l} \frac{d\mathbf{Q}}{(2\pi)^d}\, \frac{1}{-i\omega + D_0 Q^2} \;=\; -\frac{2e^2 D_0}{\hbar\pi} \int\limits_{1/L}^{1/l} \frac{d\mathbf{Q}}{(2\pi)^d}\, e^{-iD_0 Q^2 t} \tag{9.41}$$

which in the two-dimensional case becomes

$$\delta\sigma(t) \;=\; \frac{e^2}{2\pi^2 \hbar t} \left(e^{-\frac{t}{2\tau}} - e^{-\frac{D_0 t}{L^2}} \right) \; . \tag{9.42}$$

After the short time τ the classical contribution, eq.(8.83), and the above quantum contribution in the direction of the force on the electron dies out, and an *echo* in the current due to coherent backscattering occurs

$$\mathbf{j}(t) \;=\; -\frac{e^2}{2\pi^2 \hbar t} e^{-t/\tau_D}\,\mathbf{E}_0 \; . \tag{9.43}$$

on the large time scale $\tau_D \equiv L^2/D_0$, the time it takes an electron to diffuse across the sample (for even larger times $t \gg \tau_D$ quantum corrections beyond the first dominates the current).

Exercise 9.2 *Show that in dimensions one and three, the frequency dependence of the first quantum correction to the conductivity is*

$$\frac{\delta\sigma(\omega)}{\sigma_0} = \begin{cases} -\frac{1+i}{2\sqrt{2}}\frac{1}{\sqrt{\omega\tau}} & d = 1 \\[2ex] (1-i)\frac{3\sqrt{3}}{2\sqrt{2}}\frac{\sqrt{\omega\tau}}{(k_F l)^2} & d = 3 \ . \end{cases} \tag{9.44}$$

In dimension d the quantum correction to the conductivity is of relative order $1/(k_F l)^{d-1}$*. In strictly one dimension the weak localization regime is thus absent; i.e., there is no regime where the first quantum correction is small compared to the Boltzmann result, we are always in the strong localization regime.*

From the formulas, eq.(7.143) and eq.(9.39), we find that in a quasi-two-dimensional system, where the thickness of the film is much smaller than the length scale introduced by the frequency of the time-dependent external field, $L_\omega \equiv (D_0/\omega)^{1/2}$, the quantum correction to the conductance exhibits the singular frequency behavior

$$\delta < G_{\alpha\beta}(\omega) > \ = \ -\frac{e^2}{2\pi^2\hbar}\delta_{\alpha\beta} \ln \frac{1}{\omega\tau} \ . \tag{9.45}$$

The quantum correction to the conductance is in the limit of a large two-dimensional system only finite because we consider a time-dependent external field, and the conductance increases with the frequency. This feature can be understood in terms of the coherent backscattering picture. In the presence of the time-dependent electric field the electron can at arbitrary times exchange a quantum of energy $\hbar\omega$ with the field, and the coherence between two otherwise coherent alternatives will be partially disrupted. The more ω increases, the more the coherence of the backscattering process is suppressed, and consequently the tendency to localization, as a result of which the conductivity increases.

The first quantum correction plays a role even at finite temperatures, and in chapter 11 we show that from an experimental point of view there are important quantum corrections to the Boltzmann conductivity even at weak disorder. We have realized, that if the time-reversal invariance for the electron dynamics can be broken, the coherence in the backscattering process is disrupted, and localization is suppressed. The interaction of an electron with its environment invariably breaks the coherence, and we discuss the effects of electron-phonon and electron-electron interaction in section 11.3. A more distinct probe for influencing localization is to apply a magnetic field which we discuss in section 11.4.

We have realized that the precursor effect of localization, weak localization, is due to coherent backscattering. The constructive interference between propagation along time-reversed loops, which increases the probability for a particle to return to its starting position. The phenomenon of localization can be understood qualitatively as follows: The main amplitude of the electronic wave function incipient on the first impurity in figure 9.2 is not scattered into the loop depicted, but continues in its forward direction. However, this part of the wave also encounters coherent backscattering along another closed loop feeding constructively back into the original loop, and thereby increasing the probability of return. This process repeats at any impurity, and the random potential acts as a mirror, making it impossible for a particle to diffuse away from its starting point. We now turn to a quantitative discussion of localization.

9.3 Self-consistent Theory of Localization

In the previous section we indeed demonstrated that the first quantum correction to the conductance is negative in concordance with the prediction of the scaling theory of localization. We shall now go beyond first-order perturbation theory in the quantum parameter λ_F/l, and construct the self-consistent theory of localization following reference [33]. The self-consistent theory[17] provides a good approximate description of Anderson localization, as comparison with numerical results testify, except possibly very close to the metal-insulator transition in three dimensions.[18] To probe the motion of the electrons we shall consider the density response, which according to eq.(8.168) is specified by the diffusivity. In order to establish the self-consistent theory of localization we shall utilize the diagrammatic structure of the skeleton perturbation expansion of the four-point function describing the motion of a particle in a random potential. We assume for simplicity the isotropic scattering model where $\tau_{tr} = \tau$.

9.3.1 Weak-Localization Regime

In the previous section it was shown that the first quantum correction to the conductivity is governed by the infra-red (small ω) behavior of the Cooperon. Let us therefore first investigate the contribution from the Cooperon to the diffusivity; i.e., we approximate the irreducible vertex U in eq.(8.158) by the Cooperon $\tilde{C}_\omega(\mathbf{p}+\mathbf{p}')$, and obtain in this approximation for the diffusivity (for $q \ll k_F$ we can set $\mathbf{q}$ equal to zero)

$$\frac{D_0}{\tilde{D}_C(\omega)} = 1 + \frac{d\tau}{\pi\hbar mnV^2} \sum_{\mathbf{p},\mathbf{p}'}{}' (\mathbf{p} \cdot \hat{\mathbf{q}}) \, \Delta G_\mathbf{p} \left[\frac{u^2/\tau}{-i\omega + D_0(\mathbf{p}+\mathbf{p}')^2\hbar^{-2}} \right] (\mathbf{p}' \cdot \hat{\mathbf{q}}) \, \Delta G_{\mathbf{p}'}$$

(9.46)

where the prime indicates that the summation is restricted to the singular region $|\mathbf{p} + \mathbf{p}'| < \hbar/l$. Changing momentum variable to the total momentum, $\hbar\mathbf{Q} \equiv$

---

[17]As all self-consistent theories, such as also the one we employ in the next chapter to describe the electron-electron interaction in a metal, it is uncontrolled, in the sense that no small parameter estimates the accuracy of the theory. However, we do not have any general tool to calculate properties of strongly interacting many-body systems (except in one-dimensional systems, where it is possible to obtain exact results).

[18]In the field theoretic formulation of the localization problem, the self-consistent theory is known to be equivalent to in the effective action to keep all vacuum diagrams up to two-particle irreducible level; see reference [47].

$\mathbf{p} + \mathbf{p}'$, we obtain

$$\frac{D_0}{\tilde{D}_C(\omega)} = 1 + \frac{d\tau}{\pi \hbar m n V^2} {\sum_{\mathbf{p}, \mathbf{Q}}}' (\mathbf{p} \cdot \hat{\mathbf{q}}) \, \Delta G_{\mathbf{p}} \left[\frac{u^2/\tau}{-i\omega + D_0 \mathbf{Q}^2} \right] (\hbar \mathbf{Q} - \mathbf{p}) \cdot \hat{\mathbf{q}} \, \Delta G_{\hbar \mathbf{Q} - \mathbf{p}} \cdot$$

$$(9.47)$$

The singular behavior of the Cooperon is in the small Q-limit, $Ql \ll 1$, so that Q can be set to zero in the spectral function which is peaked at k_F, leading to the simplification of the expression eq.(9.47)

$$\frac{D_0}{\tilde{D}_C(\omega)} = 1 - \frac{d\tau}{\pi \hbar m n V^2} {\sum_{\mathbf{p}, \mathbf{Q}}}' (\mathbf{p} \cdot \hat{\mathbf{q}})^2 (\Delta G_{\mathbf{p}})^2 \left[\frac{u^2/\tau}{-i\omega + D_0 Q^2} \right]. \qquad (9.48)$$

The momentum integral is readily evaluated, and the diffusivity is to lowest order in the quantum parameter given by

$$\frac{D_0}{\tilde{D}_C(\omega)} = 1 + \frac{1}{\pi \hbar N_0 V} {\sum_{\mathbf{Q}}}' \frac{1}{-i\omega + D_0 Q^2}$$

$$= 1 + \frac{1}{k_F l} \frac{d}{\pi} k_F^{2-d} \int_0^{1/l} dQ \, \frac{Q^{d-1}}{-i\omega/D_0 + Q^2}. \qquad (9.49)$$

In less than two dimensions, $d \leq 2$, the first quantum correction is seen to diverge

$$\frac{D_0}{\tilde{D}_C(\omega)} - 1 = \left(\frac{\hbar}{\pi p_F l} \right)^{d-1} \begin{cases} \frac{\pi}{\sqrt{2}} \frac{1}{\sqrt{\omega \tau}} & d = 1 \\[2ex] \ln \frac{1}{\omega \tau} & d = 2. \end{cases} \qquad (9.50)$$

In order for perturbation theory to be valid, the zero frequency limit can not be taken. If, on the other hand, ω is not too small, the second term in the expression eq.(9.49) is much less than one, and from the Einstein relation, eq.(8.169), the weak-localization expression for the conductivity is recovered

$$\frac{\delta \sigma(\omega)}{\sigma_0} = -\frac{\hbar}{p_F l} \frac{d}{\pi} k_F^{2-d} \int_0^{1/l} dQ \, \frac{Q^{d-1}}{-i\omega/D_0 + Q^2}. \qquad (9.51)$$

However, the zero frequency limit is precisely the one of interest as localization is signaled by the infrared divergence of the inverse diffusivity, $K(\omega)$. Based on the perturbative result, eq.(9.49), a natural guess for a self-consistent equation for the diffusivity is obtained by substituting on the right-hand side of eq.(9.49) the diffusivity instead of the diffusion constant

$$\frac{D_0}{\tilde{D}(\omega)} = 1 + \frac{1}{\pi \hbar N_0 V} {\sum_{\mathbf{Q}}}' \left[\frac{1}{-i\omega + \tilde{D}(\omega) Q^2} \right]. \qquad (9.52)$$

Obtaining this conjectured self-consistent equation can be based on a diagrammatic classification. In order to obtain the result, the key point to notice is that a quantity where exactly the desired denominator appears is known, viz. the four-point function $\Phi(\mathbf{q}, \omega)$ of eq.(8.166). If the irreducible four-point function U can be related to Φ, a self-consistent equation for $\tilde{D}$ is thus according to eq.(8.158) obtained.

9.3.2 Self-consistent Equation

We now demonstrate that the conjectured self-consistent equation for the diffusivity, eq.(9.52), can be justified diagrammatically by taking into account the most singular contribution to the irreducible vertex function $U_{\mathbf{p},\mathbf{p}'}$.

A four point vertex diagram can be classified according to whether it is *immediately left or right two-line reducible*, i.e., has an impurity correlator (or t-matrix if we include multiple scattering) line connecting the particle lines at the utmost right or left of the diagram. We define the auxiliary four-point vertex consisting of all *immediately left and right two-line irreducible* diagrams, i.e., the sum of all the diagrams which have no impurity line connecting the R and A line at the very right or left of the diagram. This vertex function is denoted $\gamma_{\mathbf{p},\mathbf{p}'}(\mathbf{q},\omega)$.

The four-point vertex consists of the *immediately left and right irreducible* diagrams γ, the Diffuson D, the *immediately left reducible* diagrams, and the *immediately right reducible* diagrams, and finally the both *immediately left and right reducible* diagrams

$$(9.53)$$

corresponding to the equation (the energy variables, $E + \hbar\omega$ on the upper retarded line and E on the lower, advanced, line, are suppressed, and in view of eq.(8.29) fixed at the Fermi energy, $E \simeq \epsilon_F$)

$$\Gamma_{\mathbf{p},\mathbf{p}'}(\mathbf{q},\omega) = \gamma_{\mathbf{p},\mathbf{p}'}(\mathbf{q},\omega) + D(\mathbf{q},\omega) + D(\mathbf{q},\omega)\sum_{\mathbf{p}''} R_{\mathbf{p}''}(\mathbf{q},\omega)\,\gamma_{\mathbf{p}'',\mathbf{p}'}(\mathbf{q},\omega)$$

$$+ \quad D(\mathbf{q},\omega) \sum_{\mathbf{p}''} \gamma_{\mathbf{p},\mathbf{p}''}(\mathbf{q},\omega) R_{\mathbf{p}''}(\mathbf{q},\omega)$$

$$+ \quad (D(\mathbf{q},\omega))^2 \sum_{\mathbf{p}'',\mathbf{p}'''} R_{\mathbf{p}'''}(\mathbf{q},\omega)\, \gamma_{\mathbf{p}''',\mathbf{p}''}(\mathbf{q},\omega) R_{\mathbf{p}''}(\mathbf{q},\omega) . \qquad (9.54)$$

The notation $R_{\mathbf{p}}(\mathbf{q},\omega) \equiv G^R_{\mathbf{p}_+} G^A_{\mathbf{p}_-}$ and $D(\mathbf{q},\omega) \equiv D_{\mathbf{p},\mathbf{p}'}(\mathbf{q},\omega)$ has been introduced. Furthermore, we have used the fact that since $E \simeq \epsilon_F$, $D_{\mathbf{p},\mathbf{p}'}(\mathbf{q},\omega)$ is slowly varying compared to the peaked function $R_{\mathbf{p}'}(\mathbf{q},\omega)$ in the variable $\mathbf{p}'$, and has been taken outside the summation. According to eq.(8.13) the function $\Phi(\mathbf{q},\omega)$ is related to the four-point vertex by

$$\Phi(\mathbf{q},\omega) \quad = \quad \frac{1}{V^2} \sum_{\mathbf{p},\mathbf{p}'} \Phi_{\mathbf{p},\mathbf{p}'}(\mathbf{q},\omega)$$

$$= \quad \frac{1}{V} \sum_{\mathbf{p}} R_{\mathbf{p}}(\mathbf{q},\omega) + \frac{1}{V^2} \sum_{\mathbf{p},\mathbf{p}'} R_{\mathbf{p}}(\mathbf{q},\omega)\, \Gamma_{\mathbf{p},\mathbf{p}'}(\mathbf{q},\omega)\, R_{\mathbf{p}'}(\mathbf{q},\omega) . \quad (9.55)$$

For the considered case of a delta impurity correlator ($U_0 = u^2$), the four-point vertex $\Gamma_{\mathbf{p}\mathbf{p}'}(\mathbf{q},\omega)$ is independent of $\mathbf{p}$ and $\mathbf{p}'$, and when the expression eq.(9.54) for Γ is inserted into eq.(9.55), we just have products over $\mathbf{q},\omega$-dependent functions[19]

$$\Phi(\mathbf{q},\omega) \quad = \quad R(\mathbf{q},\omega) + Z(\mathbf{q},\omega) + D(\mathbf{q},\omega)\, R^2(\mathbf{q},\omega) + 2D(\mathbf{q},\omega)\, R(\mathbf{q},\omega)\, Z(\mathbf{q},\omega)$$

$$+ \quad (D(\mathbf{q},\omega))^2\, R^2(\mathbf{q},\omega)\, Z(\mathbf{q},\omega) . \qquad (9.56)$$

The above equation can be rewritten as

$$\Phi = R\,(1 + DR) + Z\,(1 + DR)^2 \qquad (9.57)$$

where we have introduced the notation

$$R = R(\mathbf{q},\omega) \equiv \frac{1}{V} \sum_{\mathbf{p}} R_{\mathbf{p}} , \qquad Z = Z(\mathbf{q},\omega) \equiv \frac{1}{V^2} \sum_{\mathbf{p},\mathbf{p}'} R_{\mathbf{p}} \gamma_{\mathbf{p},\mathbf{p}'}(\mathbf{q},\omega) R_{\mathbf{p}'} .$$

$$(9.58)$$

According to the expression for the Diffuson, eq.(8.107),

$$D \equiv D(\mathbf{q},\omega) = \frac{U_0}{1 - U_0\, R} \qquad (9.59)$$

we have

$$1 + R\,D = 1 + \frac{R\,U_0}{1 - U_0\, R} = \frac{1}{1 - U_0\, R} = D\,U_0^{-1} \qquad (9.60)$$

and thereby

$$1 + R\,D = D\,U_0^{-1} . \qquad (9.61)$$

---

[19]For a potential with range small compared to the mean free path, the peaked character of $R_{\mathbf{p}}$ restricts momenta to the Fermi Surface, and upon performing the angular integration the discussion is equivalent.

Using this equality we can rewrite eq.(9.57) as

$$[D(\mathbf{q}, \omega)]^2 Z(\mathbf{q}, \omega) = U_0^2\, \Phi(\mathbf{q}, \omega) - D(\mathbf{q}, \omega) + U_0\,. \tag{9.62}$$

Since $\Gamma_{\mathbf{p},\mathbf{p}'}(\mathbf{q}, \omega)$ is the full vertex function we have according to eq.(8.171)

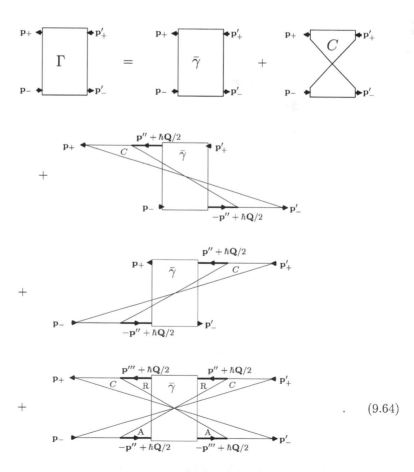

where the particle lines on the right-hand side run parallel. The diagrammatic expansion, however, is analogous to the one in eq.(9.53) as the topological classification made no reference to the directions of the particle lines. We now twist the diagrams in this expansion on the right-hand side of eq.(9.63) and obtain

We shall now invoke time-reversal symmetry. In that case the two Cooperons in the last diagram are seen to have the same argument $\mathbf{Q}, \omega$, where $\hbar\mathbf{Q} = \mathbf{p} + \mathbf{p}'$. The last four diagrams in eq.(9.64) are explicitly two-line irreducible, and are thus part of the irreducible vertex $U_{\mathbf{p},\mathbf{p}'}(\mathbf{q}, \omega)$. The very last diagram in eq.(9.64) contains two Cooperons with identical arguments, and are therefore more singular than the other diagrams. We therefore have for the dominant contribution to the irreducible four-point function

$$U_{\mathbf{p},\mathbf{p}'}(\mathbf{q},\omega) \;\rightarrow\; U_{\mathbf{p},\mathbf{p}'}^{\text{domin}}(\mathbf{q},\omega) \;=\; [C(\mathbf{p}+\mathbf{p}',\omega)]^2 Z(\mathbf{p}+\mathbf{p}',\omega)$$

$$\rightarrow\; [C(\mathbf{p}+\mathbf{p}',\omega)]^2 Z(\mathbf{p}+\mathbf{p}',\omega) + C(\mathbf{p}+\mathbf{p}',\omega) \;. \qquad (9.65)$$

The Cooperon added in the last line, ensures that we get the correct limiting behavior in the weak-disorder limit, the aim of the self-consistent approach being to interpolate between the weak- and strong-localization regimes.

In the time reversal invariant situation we then get, using the relation eq.(9.62), for the dominant contribution to the irreducible four-point function

$$U_{\mathbf{p},\mathbf{p}'}^{\text{domin}}(\mathbf{q},\omega) - U_0 \;=\; U_0^2\, \Phi(\mathbf{p}+\mathbf{p}',\omega) \;. \qquad (9.66)$$

Inserting the dominating contribution on the right-hand side of eq.(8.158) we obtain

$$\frac{D_0}{\tilde{D}(\mathbf{q},\omega)} \;=\; 1 + \frac{d\tau}{\pi\hbar m n V^2} \sum_{\mathbf{p},\mathbf{p}'} (\mathbf{p}\cdot\hat{\mathbf{q}})\Delta G_{\mathbf{p}}\, U_0\, \Phi(\mathbf{p}+\mathbf{p}',\omega)(\mathbf{p}'\cdot\hat{\mathbf{q}})\Delta G_{\mathbf{p}'} \qquad (9.67)$$

and performing the integration over one of the momenta we have

$$\frac{D_0}{\tilde{D}(\mathbf{q},\omega)} \;=\; 1 + \frac{1}{2(\pi N_0)^2 V} {\sum_{\mathbf{Q}}}'\, \Phi(\mathbf{Q},\omega) \;. \qquad (9.68)$$

Upon inserting the expression eq.(8.168) yields a self-consistent equation for the diffusivity

$$\frac{D_0}{\tilde{D}(\mathbf{q},\omega)} = 1 + \frac{\hbar k_F^{2-d}}{\pi m} \int_0^{1/l} dQ\, \frac{Q^{d-1}}{-i\omega + \tilde{D}(\mathbf{Q},\omega)Q^2} \;. \qquad (9.69)$$

The result, eq.(9.69), is independent of the small external momentum $\mathbf{q}$ and we can set $\mathbf{q} = 0$ in $\tilde{D}(\mathbf{q},\omega)$, and since Q is small, $Q \leq 1/l \ll k_F$, we can neglect the $\mathbf{Q}$-dependence in $\tilde{D}(\mathbf{Q},\omega)$ as well, and we obtain

$$\frac{D_0}{\tilde{D}(\omega)} = 1 + \frac{\hbar k_F^{2-d}}{\pi m} \int_0^{1/l} dQ \frac{Q^{d-1}}{-i\omega + \tilde{D}(\omega)Q^2} \;. \qquad (9.70)$$

This is precisely the self-consistent equation, conjectured for the diffusivity, eq.(9.52).

We note the different roles played by the two infrared divergences in the theory; the Diffuson, which we used to classify the diagrams of the four-point vertex, and the Cooperon, introduced by twisting the diagrams (the quantity representing the coherent backscattering process which drives the localization transition), and the crucial role of time reversal symmetry that relates the two.[20]

Writing the equation in terms of the inverse diffusivity $K(\omega)$, eq.(8.158), we find that the solution of the self-consistent equation has at small frequencies, $\omega \ll \omega_0^2 \tau$, the form

$$K(\omega) = 1 + \frac{i\omega_0^2 \tau}{\omega} \tag{9.71}$$

where ω_0 is determined by

$$1 = \frac{\hbar k_F^{2-d}}{\pi m} \int_0^{1/l} dQ \, \frac{Q^{d-1}}{\omega_0^2 \tau + D_0 Q^2} \,. \tag{9.72}$$

In two dimensions, for example, we find from eq.(9.72) for the frequency scale where localization becomes of importance:

$$\omega_0 = \frac{1}{\sqrt{2}\,\tau} \, e^{-\frac{\pi k_F l}{2}} \,. \tag{9.73}$$

The discussed solution, eq.(9.71), is infrared divergent, and corresponds to the insulating phase, and the diffusivity vanishes as $\omega \to 0$ according to

$$\tilde{D}(\omega) = -i\omega \, \frac{D_0}{\omega_0^2 \tau} \,. \tag{9.74}$$

9.3.3 Localization Length

In order to get a quantitative criterion for the localization length, we can consider the spectral correlation function

$$\mathcal{A}(\mathbf{x}, \mathbf{x}', E, E + \hbar\omega) \equiv \frac{\hbar}{\pi^2 N(E)} < \Im m G^R(\mathbf{x}, \mathbf{x}', E + \hbar\omega) \, \Im m G^R(\mathbf{x}', \mathbf{x}, E)>$$

$$= \frac{\hbar}{N(E)} < \sum_{\lambda, \lambda'} \psi_\lambda(\mathbf{x}) \, \psi_\lambda^*(\mathbf{x}') \psi_{\lambda'}(\mathbf{x}') \, \psi_{\lambda'}^*(\mathbf{x}) \, \delta(E + \hbar\omega - \epsilon_{\lambda'}) \, \delta(E - \epsilon_\lambda) > \tag{9.75}$$

where

$$N(E) = < \sum_\lambda \psi_\lambda(\mathbf{x}) \, \psi_\lambda^*(\mathbf{x}) \, \delta(E - \epsilon_\lambda) > \tag{9.76}$$

is the impurity-averaged density of states.

According to eq.(7.26) we have the relationship to the density response function

$$\chi(\mathbf{q}, \omega) = \int_{-\infty}^{\infty} dE \int_{-\infty}^{\infty} d\omega' \, \frac{f_0(E + \hbar\omega') - f_0(E)}{\omega' - \omega - i0} \, \frac{N(E)}{\hbar} \, \mathcal{A}(\mathbf{q}, E, E + \hbar\omega') \tag{9.77}$$

---

[20]It has therefore not been possible to extend the self-consistent theory to the case of broken time-reversal symmetry.

where $\mathcal{A}(\mathbf{q}, E, E + \hbar\omega)$ is the Fourier transform of $\mathcal{A}(\mathbf{x}, \mathbf{x}', E, E + \hbar\omega)$. Since the Fourier transform is real,

$$\mathcal{A}(\mathbf{q}, E, E + \hbar\omega) = \frac{\hbar}{2\pi^2 V N(E)} \, \Re e \left(\Phi(E, \mathbf{q}, \omega) - \Phi^{RR}(E, \mathbf{q}, \omega) \right) \qquad (9.78)$$

we have

$$\Im m \, \chi(\mathbf{q}, \omega) = \pi \int\limits_{-\infty}^{\infty} dE \, \frac{N(E)}{\hbar} \, (f_0(E + \hbar\omega) - f_0(E)) \, \mathcal{A}(\mathbf{q}, E, E + \hbar\omega) \, . \qquad (9.79)$$

In particular we note that at zero temperature

$$\mathcal{A}(\mathbf{q}, \epsilon_F, \epsilon_F + \hbar\omega) = -\frac{\Im m \, \chi(\mathbf{q}, \omega)}{\pi \omega N(\epsilon_F)} \, . \qquad (9.80)$$

Since

$$< G^R(\mathbf{x}, \mathbf{x}', E + \hbar\omega) \, G^A(\mathbf{x}', \mathbf{x}, E) > = \frac{1}{V} \sum_{\mathbf{q}} e^{i\mathbf{q}\cdot(\mathbf{x}-\mathbf{x}')} \, \Phi(E, \mathbf{q}, \omega) \qquad (9.81)$$

we obtain by using the spectral representation of the propagators, eq.(2.154), that

$$\Phi(E, \mathbf{q}, \omega) = \int\limits_{-\infty}^{\infty} dE' \int\limits_{-\infty}^{\infty} d\omega' \, \frac{N(E') \, \mathcal{A}(\mathbf{q}, E, E + \hbar\omega)}{(E + \hbar\omega - E' + i0)(E - E' - \hbar\omega' - i0)} \, . \qquad (9.82)$$

The term in the spectral correlation function, eq.(9.75), where $\lambda' = \lambda$ gives a delta function contribution, proportional to $\delta(\omega)$. In the case of extended wave functions the coefficient of the singular term vanishes in the thermodynamic limit, whereas from the region of energies where the states are localized we have a singular contribution.[21] States which have equal energy, thus also have spatially correlated wave functions. For the Fourier transform of the spectral correlation function we have

$$\mathcal{A}(\mathbf{q}, E, E + \hbar\omega) = A_E(\mathbf{q}) \, \delta(\omega) + A_E^r(\mathbf{q}, \omega) \qquad (9.83)$$

where

$$A_E(\mathbf{x}) = \int \frac{d\mathbf{q}}{(2\pi)^d} \, e^{i\mathbf{q}\cdot\mathbf{x}} \, A_E(\mathbf{q}) = \frac{\hbar}{N(E)} < \sum_{\lambda} |\psi_\lambda(\mathbf{x})|^2 \, |\psi_\lambda(\mathbf{0})|^2 \, \delta(E - \epsilon_\lambda) > \qquad (9.84)$$

and $A_E^r(\mathbf{q}, \omega)$ is a regular function vanishing at $\omega = 0$. We obtain from eq.(9.82) and eq.(9.83) the small ω behavior

$$\Phi(\mathbf{q}, \omega \to 0) \simeq \frac{2\pi N(E)}{-i\hbar\omega} \, A_E(\mathbf{q}) \, . \qquad (9.85)$$

---

[21]This is the localization criterion of Berezinskii and Gor'kov [48].

In the insulating phase

$$\xi = \xi(E) \equiv \lim_{\omega \to 0} \sqrt{\frac{D(\omega)}{-i\omega}} \equiv \lim_{\omega \to 0} \xi(\omega) \tag{9.86}$$

is a positive constant, and the self-consistent theory gives

$$\Phi(E, \mathbf{q}, \omega) = \frac{2\pi N_0}{-i\hbar\omega} \frac{1}{1 + \xi^2 q^2} \tag{9.87}$$

or

$$A_E(\mathbf{q}) = \frac{1}{1 + \xi^2 q^2} . \tag{9.88}$$

According to eq.(9.84) we therefore find that the wave functions in a random potential are exponentially localized. In three dimensions, for example, we have

$$A_E(\mathbf{x}) = \frac{2\pi^2}{\xi^2 x} e^{-x/\xi} \tag{9.89}$$

and we identify ξ as the wave function localization length.

In two dimensions we obtain from eq.(9.73) for the wave function localization length within the self-consistent theory[22]

$$\xi = \xi(\epsilon_F) = l\, e^{\frac{\pi k_F l}{2}} . \tag{9.90}$$

Exercise 9.3 *Show that the localization length for a particle with energy E is determined by the expression*

$$\xi^2(E) = \frac{1}{2dN(E)} \int d^d\mathbf{x}\, \mathbf{x}^2 < \sum_\lambda \delta(E - \epsilon_\lambda)\, |\psi_\lambda(\mathbf{x})|^2\, |\psi_\lambda(\mathbf{0})|^2 > . \tag{9.91}$$

Solution

The result follows from the sum-rule

$$\int_{-\infty}^{\infty} d\omega\, \mathcal{A}(\mathbf{q}, E, E + \hbar\omega) = 1 \tag{9.92}$$

and the fact that in the small $\mathbf{q}$-limit, $\mathbf{q} \to 0$, we have

$$\mathcal{A}(\mathbf{q}, E, E + \hbar\omega) = (1 - \xi^2(E)\mathbf{q}^2)\, \delta(\omega) + \mathbf{q}^2\, A_E^r(\omega) . \tag{9.93}$$

---

[22]We note that the perturbative estimate on page 370 gave the same order of magnitude indicating that the higher order terms in the scaling function are small (see also the footnote on page 382).

Let us consider the localization length within the self-consistent theory in more detail. Multiplying both sides of eq.(9.70) with $\tilde{D}(\omega)/D_0$ yields

$$\frac{\tilde{D}(\omega)}{D_0} = 1 - \frac{dk_F^{2-d}}{\pi k_F l} \int_0^{1/l} dQ \frac{Q^{d-1}}{-i\omega/\tilde{D}(\omega) + Q^2}$$

$$= 1 - \frac{d}{\pi} \left(\frac{1}{k_F l}\right)^{d-1} \left(\frac{\xi(\omega)}{l}\right)^{2-d} \int_0^{\frac{\xi(\omega)}{l}} dx \frac{x^{d-1}}{1 + x^2} . \tag{9.94}$$

Since the diffusivity vanishes in the insulating phase we get the equation to determine the localization length

$$1 = \frac{dk_F^{2-d}}{\pi k_F l} \int_0^{1/l} dQ \frac{Q^{d-1}}{-i\omega/\tilde{D}(\omega) + Q^2}$$

$$= 1 - \frac{d}{\pi} \left(\frac{1}{k_F l}\right)^{d-1} \left(\frac{\xi(\omega)}{l}\right)^{2-d} \int_0^{\frac{\xi(\omega)}{l}} dx \frac{x^{d-1}}{1 + x^2} . \tag{9.95}$$

In dimensions $0 < d < 2$ the integral converges for arbitrary $\xi = \xi(\omega = 0)$. In one dimension we get the localization length

$$\xi = 2.61\, l \tag{9.96}$$

i.e., the localization length is of the order of the mean free path, in fair agreement with the exact result $\xi = 4\, l$ [38].

In two dimensions we get

$$\xi = \xi(\epsilon_F) = l \exp\left\{\frac{1}{2}(\pi k_F l - 1)\right\} = \begin{cases} \exp\frac{\pi k_F l}{2} & k_F l \gg 1 \\ \sqrt{\pi k_F l} & k_F l \ll 1 . \end{cases} \tag{9.97}$$

In the weak-disorder limit, $k_F l \gg 1$, the localization is exponentially weak in two dimensions, and experimentally weak-localization effects can easily be probed in two-dimensional systems, the subject of chapter 11. We note the nonanalytic dependence of the localization length on the disorder parameter $1/k_F l$.

The localization length calculated from the self-consistent theory agrees up to a numerical factor with the lowest-order perturbative estimate of section 9.2. This indicates that the higher-order terms in the scaling function, $1/g^n, n = 2, 3, ..$ are small. In fact, Wegner has shown that the expansion starts out with a finite term of order $1/g^4$ [49].

9.3.4   Critical Exponents

In this section we will discuss the case where the spatial dimension is larger than two, $d > 2$, and the system can exhibit a metal-insulator transition. In dimensions larger than two, there always exists a solution of the self-consistent equation for which the diffusivity at zero frequency is nonzero, $\tilde{D}(0) \neq 0$, provided the disorder parameter $\lambda \equiv \hbar/\pi p_F l$ is smaller than a certain critical disorder value λ_c. Using the identity

$$\frac{x^{d-1}}{1+x^2} = x^{d-3}\left[1 - \frac{1}{1+x^2}\right]$$

the expression eq.(9.94) can be rewritten as

$$\frac{\sigma(\omega)}{\sigma_0} = 1 - \frac{d}{d-2}\frac{1}{\pi}\left(\frac{1}{k_F l}\right)^{d-1} + \frac{d}{\pi}\frac{(\frac{1}{k_F l})^{d-1}}{(\frac{\xi(\omega)}{l})^{d-2}}\int_0^{\frac{\xi(\omega)}{l}} dx\, \frac{x^{d-3}}{1+x^2}. \tag{9.98}$$

In the metallic phase, $\lambda < \lambda_c$, where $\xi(\omega \to 0) \to \infty$, the third term vanishes for all dimensions $d > 2$. In the limit of small ω we have

$$\frac{\sigma(0)}{\sigma_0} = 1 - \left(\frac{\lambda}{\lambda_c}\right)^{d-1} = C(\lambda/\lambda_c)\left(1 - \frac{\lambda}{\lambda_c}\right), \qquad \lambda \le \lambda_c \tag{9.99}$$

where $C(x) = \sum_{n=0}^{d-2} x^n$, and the critical value of disorder is identified as

$$\lambda_c = \left(\frac{d-2}{d\pi^{d-2}}\right)^{\frac{1}{d-1}}, \qquad d \ge 2. \tag{9.100}$$

The dc conductivity vanishes at the critical disorder value $\lambda = \lambda_c$, and the transition is approached according to

$$\sigma(0) \propto \lambda_c - \lambda \tag{9.101}$$

i.e., with a critical exponent s for the conductivity equal to 1

$$\sigma(0) \propto |\lambda_c - \lambda|^s, \qquad s = 1, \qquad d > 2. \tag{9.102}$$

In the insulating regime, $\lambda > \lambda_c$, the dc conductivity vanishes, and eq.(9.98) determines the localization length

$$1 = \left(\frac{\lambda}{\lambda_c}\right)^{d-1}\left[1 - (d-2)\left(\frac{\xi}{l}\right)^{2-d}\int_0^{\xi/l} dx\, \frac{x^{d-3}}{1+x^2}\right]. \tag{9.103}$$

For dimensions $2 < d < 4$, the integral in eq.(9.103) converges as the localization length diverges, yielding a certain constant $c(d)$ depending on the dimension d. We are only interested in the scaling behavior of the localization length as the transition is approached, and there one has

$$\xi = c(d)\, l\left|1 - \frac{\lambda}{\lambda_c}\right|^{\frac{-1}{d-2}}, \qquad c(d) = \left[2\pi\frac{d-1}{2-d}\sin\frac{\pi d}{2}\right]^{\frac{-1}{d-2}} \tag{9.104}$$

i.e., the critical exponent ν for the localization length is $1/(d-2)$

$$\xi \propto |\lambda_c - \lambda|^{-\nu}, \qquad \nu = \frac{1}{d-2}, \qquad 2 < d < 4. \tag{9.105}$$

For dimensions larger than four, $d > 4$, one finds

$$1 = \left(\frac{\lambda}{\lambda_c}\right)^{d-1} \left[1 - \frac{d-2}{d-4}\left(\frac{\xi}{l}\right)^{-2}\right] \qquad (9.106)$$

and thereby the scaling behavior

$$\xi \propto (\lambda - \lambda_c)^{-1/2}, \qquad \lambda > \lambda_c \qquad (9.107)$$

i.e., the critical exponent for the localization length is

$$\nu = \frac{1}{2}, \qquad d > 4. \qquad (9.108)$$

In dimensions $2 < d < 4$ the critical exponents satisfy the scaling relation

$$s = \nu(d-2), \qquad 2 < d < 4. \qquad (9.109)$$

which in three dimensions makes the two exponents identical, $\nu = s$.

The critical exponents were originally obtained by Wegner [35] employing the renormalization group in a field theoretic treatment of the disorder problem. Instead of studying the localization problem using diagrammatic perturbation theory, in this approach the impurity average is performed from the outset, and one is led to the field theoretic description of transport properties. For the impurity case one encounters the nonlinear σ model [35] [44]. The renormalization group technique can then be applied, thus making an interesting connection between quantum transport theory and the theory of phase transitions. The self-consistent theory of localization is the mean field approximation of the field theoretical model [47], and we have found that the critical dimension is $d = 2$, and the upper critical dimension is $d = 4$.

9.3.5   Scaling Behavior

The phenomenon of Anderson localization can be understood in terms of the random potential acting as a mirror backscattering the electronic wave function, thereby leading to a spatial localization of the particle.[23] To probe the spatial localization we investigate the length-dependent scaling behavior. In a finite system there is a finite probability for a particle to reach the sample ends. As discussed in section 9.2, this influence on the conductivity is represented by a lower cutoff $1/L$ on the Q-integration. The diffusivity at zero frequency for a sample of length L, $\tilde{D}(L)$ is thus given by

$$\frac{\tilde{D}(L)}{D_0} = 1 - \frac{dk_F^{2-d}}{\pi k_F l} \int_{1/L}^{1/l} dQ \, \frac{Q^{d-1}}{\xi^{-2} + Q^2}. \qquad (9.110)$$

---

[23]In two dimensions even the slightest amount of disorder leads to localization, and the localization phenomenon is thus quite different from a localization due to the particle being in a bound state in a potential.

In the insulating phase, where $\tilde{D}(\omega = 0) = 0$, we have from eq.(9.94)

$$1 = \frac{dk_F^{2-d}}{\pi k_F l} \int_0^{\xi/l} dx \frac{x^{d-1}}{1+x^2} . \tag{9.111}$$

Subtracting eq.(9.111) from eq.(9.110) leads to

$$\frac{\tilde{D}(L)}{D_0} = \frac{d}{\pi k_F l}(k_F \xi)^{2-d} \int_0^{\xi/L} dx \frac{x^{d-1}}{1+x^2} . \tag{9.112}$$

Let us compute the current density J flowing in a sample of finite length L under the influence of the applied electric field E in, say, the x-direction. The current density J is determined by requiring it to be equal and opposite to the diffusion current under open-circuit conditions. The current density at the end of the sample is given by

$$J = -e\tilde{D}(L) \frac{d(\delta n)}{dx}\bigg|_{x=L} \tag{9.113}$$

where $\delta n(x)$ is the electronic density change induced by the electric potential $U(x) = eE(L - x)$. The density change due to the external electric field is in linear response

$$\delta n(x) = \int_0^L dx' \, \chi(x - x') \, U(x'). \tag{9.114}$$

The density response function in eq.(9.114) is according to eq.(8.166) the Fourier transform of

$$\chi(\mathbf{q}, 0) = \frac{q^2}{\xi^{-2} + q^2} \, 2N_0 \tag{9.115}$$

and we obtain

$$\chi(x) = \int_{-\infty}^{+\infty} \frac{dq}{2\pi} e^{-iqx} \chi(\mathbf{q}, 0) = 2N_0 \left[\delta(x) - \frac{1}{2\xi} e^{-\frac{|x|}{\xi}} \right] . \tag{9.116}$$

For the current density we therefore have

$$J(L) = c_d e^2 E \xi^{2-d} \left(1 + \frac{L}{\xi} \right) \exp(-L/\xi) \int_0^{\xi/L} dx \frac{x^{d-1}}{1+x^2}, \tag{9.117}$$

where

$$c_d = \frac{2}{\pi} \frac{S_d}{(2\pi)^d} \tag{9.118}$$

and S_d is the surface area of the unit sphere in d dimensions $S_d = (2^{d-1}\pi^{d/2}\Gamma(d/2))^{-1}$, or $4\pi, 2\pi$ and 1 in three, two and one dimensions, respectively.

For the dimensionless conductance we then obtain

$$g(L/\xi) = c_d(1 + L/\xi) \, e^{-L/\xi} \, h_d(L/\xi) \tag{9.119}$$

where

$$h_d(y) = \int_0^1 dx \, \frac{x^{d-1}}{y^2 + x^2} \, . \tag{9.120}$$

In one and two dimensions we have[24]

$$h_1(y) = \frac{1}{y} \arctan \frac{1}{y} \, , \qquad h_2(y) = \frac{1}{2} \ln(1 + 1/y^2) \tag{9.121}$$

leading to

$$g(L) = \begin{cases} \frac{2\xi}{\pi^2 L}\left(1 + \frac{L}{\xi}\right) e^{-L/\xi} \arctan \xi/L & d = 1 \\[2mm] \frac{1}{2\pi^2}\left(1 + \frac{L}{\xi}\right)e^{-L/\xi} \ln\left(1 + \frac{\xi^2}{L^2}\right) & d = 2 \end{cases} \tag{9.122}$$

where ξ is the localization length given by eq.(9.104) for $\lambda < \lambda_c$. In the insulating regime, the conductance $g(L)$ decreases exponentially, and is negligible small when the length of the sample is larger than the localization length ξ.

The self-consistent theory thus gives for the scaling function in the strong-disorder regime (in all dimensions)

$$\beta(g) = \ln g \, , \qquad g \ll 1 \, . \tag{9.123}$$

This was precisely the input we used in our discussion of the scaling theory of localization.

---

[24]Note that for $d > 2, h_d(y) = \frac{1}{d-2} - y^2 h_{d-2}(y)$.

Chapter 10

Interactions in Metals

In this chapter we shall consider the interactions between the constituents of say a metal, i.e., electrons and ions. By adopting a mean field approach, the dynamics of the electrons can be obtained by perturbation theory from the properties of the noninteracting electrons, and the effective electron-electron interaction in good conductors is considered.[1] We shall not be interested in properties due to deviations from a spherical Fermi surface, and throughout we consider the isotropic model of a metal. The dynamics of the ions can for our purposes be treated in the harmonic approximation. In chapter 6 we introduced the formalism for describing a particle interacting linearly with oscillators without referring to the physical nature of the oscillators. As stipulated, this is a generic case which has wide applications. Indeed in this chapter, we shall give an account of the electron-phonon interaction, profiting from the results of chapter 6. In the case of phonons, the oscillators represent collective degrees of freedom, and we shall show how such an effective description comes about. We then calculate the collision rates due to electron-phonon and electron-electron interaction.

10.1 Isotropic Model of a Metal

A solid, such as a metal, is an assembly of nuclei and electrons. From a first principles point of view, the dynamics of such a system constitutes an unsolvable many-body problem as the number of involved particles is astronomical ($\sim 10^{23}$). We shall therefore be interested in an approximate description of the system, which is justifiable for the particular type of phenomena which is of our interest.

Electron diffraction or X-ray experiments reveal the grainy character of a metal. Charge is separated spatially into two pieces: the nuclei and the tightly bound electrons (core electrons which are concentrated in spatially well-localized regions), and the conduction electrons, which have their density spread throughout the solid. We are interested in the low-frequency dynamics, and can assume that the core electrons follow the motion of the nucleus adiabatically. The core electrons stay in

---

[1] We shall use the words *metal* and *good conductor*, as applies to a heavily doped semiconductor, synonymously.

their ground-state configuration, and we have the Hamiltonian for the conduction electron-ion system

$$\hat{H} = H(\{\hat{\mathbf{x}}_{n_e}\}_{n_e}, \{\hat{\mathbf{x}}_{m_i}\}_{m_i}, \{\hat{\mathbf{p}}_{n_e}\}_{n_e}, \{\hat{\mathbf{p}}_{m_i}\}_{m_i}) = \hat{H}_e + \hat{H}_{ions} + \hat{H}_{i-e} \qquad (10.1)$$

where the Hamiltonian for the conduction electrons is

$$\hat{H}_e = H(\{\hat{\mathbf{x}}_{n_e}\}_{n_e}, \{\hat{\mathbf{p}}_{n_e}\}_{n_e}) = \sum_{n_e=1}^{N_e} \frac{\hat{\mathbf{p}}_{n_e}^2}{2m} + \frac{1}{2} \sum_{n_e \neq m_e} V(\hat{\mathbf{x}}_{n_e} - \hat{\mathbf{x}}_{m_e}) \qquad (10.2)$$

and similarly for the Hamiltonian for the ions

$$\hat{H}_{ions} = H(\{\hat{\mathbf{x}}_{m_i}\}_{m_i}, \{\hat{\mathbf{p}}_{m_i}\}_{m_i}) = \sum_{m_i=1}^{N_e} \frac{\hat{\mathbf{p}}_{m_i}^2}{2M} + \frac{1}{2} \sum_{n_i \neq m_i} V_{Z^2}(\hat{\mathbf{x}}_{n_i} - \hat{\mathbf{x}}_{m_i}) \qquad (10.3)$$

and for the interaction we have

$$\hat{H}_{i-e} = \sum_{n_e, m_i} V_Z(\hat{\mathbf{x}}_{n_e} - \hat{\mathbf{x}}_{m_i}) . \qquad (10.4)$$

The interactions between the particles is the Coulomb interaction, for example we have for the ion-ion interaction

$$V_{Z^2}(\hat{\mathbf{x}}_{n_i} - \hat{\mathbf{x}}_{m_i}) = \frac{Z^2 e^2}{4\pi\epsilon_0 |\hat{\mathbf{x}}_{n_i} - \hat{\mathbf{x}}_{m_i}|} \qquad (10.5)$$

where Z accounts for the valence of the ions, $Z = 1, 2, \ldots$

The Hamiltonian is formally identical to the Hamiltonian for the plasma of nuclei and electrons, and we now invoke the circumstance that the system is in the solid phase, the state where the translation invariance of the ionic system is broken. In carefully grown metals and semiconductors, the nuclei occupy to a high degree of regularity points on a lattice, and we have a crystalline solid. As a starting point for describing a metal we shall therefore take the idealization of a perfect crystal: identical ions with equilibrium positions located at regular spacing, and an equal number of neutralizing electrons.

The dynamics of the system is determined by the masses of the constituents and their mutual interactions. However, due to the large disparity in the masses of the electrons and ions, $m/M \sim 10^{-4}$, the motion of the ions are slow compared to that of the electrons. Let us therefore first assume that the ions are fixed in space; i.e., their masses M are assumed infinite so that we can neglect their dynamics. The dynamic problem is thus reduced to that of N interacting electrons in the periodic background potential of the ions. Next, let us assume that we are only interested in charge oscillations of long (compared to the interatomic distance) wavelength. We can then represent the ions by a smeared-out background of positive charge, whose sole effect is to keep the system overall charge neutral, the jellium model.[2] We thus neglect the periodic potential inherent to a crystal, the

---

[2]For the time being the smeared-out background charge is assumed fixed. Later in this chapter we allow for the background charge to move.

breaking of the translational symmetry of the electronic system down to the crystal lattice symmetry, with its consequences of band structure of allowed energies, and Fermi surface structure. As for the phenomena we shall consider, the effect of the periodic potential can in principle easily be incorporated, and will only lead to renormalization of parameters, such as for example the electronic mass and matrix elements, and introduce the actual instead of the spherical Fermi surface averages. However, no qualitative different features will appear.[3] In the following we shall have this isotropic model of a metal in mind.

We have reduced the problem to that of N electrons with their mutual Coulomb interaction in the inert smeared-out ionic background

$$H(\{\hat{\mathbf{x}}_{n_e}\}_{n_e}, \{\hat{\mathbf{p}}_{n_e}\}_{n_e}) = \sum_{n_e=1}^{N_e} \frac{\hat{\mathbf{p}}_{n_e}^2}{2m} + \frac{1}{2} \sum_{n_e \neq m_e} V(\hat{\mathbf{x}}_{n_e} - \hat{\mathbf{x}}_{m_e}) + V_b(\hat{\mathbf{x}}_{n_e}) . \quad (10.6)$$

The interaction potential of an electron with the inert background is

$$V_b(\mathbf{x}_{n_e}) = -n_i \int d\mathbf{x} \frac{Ze^2}{4\pi\epsilon_0 |\mathbf{x} - \mathbf{x}_{n_e}|} \quad (10.7)$$

where n_i denotes the density of ions,[4] and thus an irrelevant constant, and V the Coulomb interaction between the electrons

$$V(\mathbf{x} - \mathbf{x}') = \frac{e^2}{4\pi\epsilon_0 |\mathbf{x} - \mathbf{x}'|} . \quad (10.8)$$

The problem at hand is still an unsolvable many-electron problem. In the following we shall be interested only in states with charge density variations of wavelength long compared to the interatomic distance. We therefore expect that the fluctuations in the potential felt by an electron is small, and the detailed information contained in the many-body Hamiltonian should therefore not be relevant. In view of this we resort to a mean-field description where any electron is assumed to move in an average field created by the charge of all the other electrons and the inert background. Our problem of N interacting electrons then reduces to N problems for noninteracting electrons in the mean field, $\overline{V}$,

$$H_{MF} = \sum_{n_e=1}^{N_e} \left(\frac{\hat{\mathbf{p}}_{n_e}^2}{2m} + \overline{V}(\hat{\mathbf{x}}_{n_e}, t) \right) \quad (10.9)$$

where the mean or effective field

$$\overline{V}(\hat{\mathbf{x}}_{n_e}, t) = \int d\mathbf{x} \frac{e^2 \, \delta n(\mathbf{x}, t)}{4\pi\epsilon_0 |\mathbf{x} - \hat{\mathbf{x}}_{n_e}|} \quad (10.10)$$

---

[3]We are thus not interested in effects involving interband transitions, or magnetoresistance of open Fermi surface materials, or other band structure specific effects. For band structure effects we refer to reference [25].

[4]No confusion with the notation for the impurity concentration should arise.

is the potential energy due to the average density of charges, the mean value of the electronic charge minus the inert background charge density, $\delta n = n_e - n_0$. The theory is self-consistent, i.e., the mean field depends on the induced charge distribution, which in turn depends on the mean field. This time dependent self-consistent approximation is referred to as the Hartree approximation, and we now turn to obtain a closed set of equations relating the two self-consistently coupled quantities.

10.2   Mean-Field Electron Dynamics

The mean field Hamiltonian is specified in terms of the mean field, which in turn is specified in terms of the average charge density. We therefore investigate the dynamics of the system in order to get an explicit expression for the mean field. For N fermions in a mean field we have the equation satisfied by the one-particle density matrix

$$\left(\frac{\partial}{\partial t} - \frac{i}{\hbar}\left(\frac{\hbar^2}{2m}(\Delta_{\mathbf{x}} - \Delta_{\mathbf{x}'}) - \overline{V}(\mathbf{x},t) + \overline{V}(\mathbf{x}',t)\right)\right)\rho(\mathbf{x},\mathbf{x}',t) = 0 \quad (10.11)$$

or in terms of the coordinates $\mathbf{r} = \mathbf{x} - \mathbf{x}'$, $\mathbf{R} = (\mathbf{x} + \mathbf{x}')/2$

$$\left(\frac{\partial}{\partial t} + \frac{\hbar}{im}\nabla_{\mathbf{R}}\cdot\nabla_{\mathbf{r}} + \frac{i}{\hbar}\left(\overline{V}(\mathbf{R}+\mathbf{r}/2,t) - \overline{V}(\mathbf{R}-\mathbf{r}/2,t)\right)\right)f(\mathbf{R},\mathbf{r},t) = 0 \quad (10.12)$$

where we have introduced the notation $f(\mathbf{R},\mathbf{r},t) \equiv \rho(\mathbf{R},\mathbf{r},t)$. Transforming to Wigner coordinates we get the quantum kinetic equation

$$\int d\mathbf{r}\int\frac{d\mathbf{p}'}{(2\pi\hbar)^3}\, e^{\frac{i}{\hbar}\mathbf{r}\cdot(\mathbf{p}'-\mathbf{p})}f(\mathbf{R},\mathbf{p}',t)\left(\overline{V}(\mathbf{R}+\mathbf{r}/2,t) + \overline{V}(\mathbf{R}-\mathbf{r}/2,t)\right)$$

$$+ \left(\frac{\partial}{\partial t} + \mathbf{v}_{\mathbf{p}}\cdot\frac{\partial}{\partial\mathbf{R}}\right)f(\mathbf{R},\mathbf{p},t) = 0 . \quad (10.13)$$

To probe the dynamics of the electrons we apply an external field, V_a, and the total mean field then consists of the applied and an induced field

$$\overline{V}(\mathbf{x},t) = V_a(\mathbf{x},t) + V_{\text{ind}}(\mathbf{x},t) \quad (10.14)$$

where the induced field according to eq.(10.10) is given by

$$V_{\text{ind}}(\mathbf{x},t) = \int d\mathbf{x}'\, V(\mathbf{x}-\mathbf{x}')\,\delta n(\mathbf{x}',t) . \quad (10.15)$$

The excess electronic charge (the factor of 2 accounts for the spin degeneracy)

$$\delta n(\mathbf{x},t) = 2\int\frac{d\mathbf{p}}{(2\pi\hbar)^3}\,\delta f(\mathbf{x},\mathbf{p},t) \quad (10.16)$$

is expressed in terms of the deviation from the equilibrium distribution, the Fermi function,

$$f(\mathbf{x}, \mathbf{p}, t) = f_0(\epsilon_{\mathbf{p}}) + \delta f(\mathbf{x}, \mathbf{p}, t) \tag{10.17}$$

as the equilibrium part is canceled by the background charge. We can thus rewrite the average potential in terms of the deviation of the Wigner function from the equilibrium distribution

$$\overline{V}(\mathbf{x}, t) = V_a(\mathbf{x}, t) + 2 \int d\mathbf{x}' \, V(\mathbf{x} - \mathbf{x}') \int \frac{d\mathbf{p}}{(2\pi\hbar)^3} \, \delta f(\mathbf{x}', \mathbf{p}, t) \tag{10.18}$$

and the Fourier transform of the mean field satisfies the equation

$$\overline{V}(\mathbf{q}, \omega) = V_a(\mathbf{q}, \omega) + 2V(\mathbf{q}) \int \frac{d\mathbf{p}}{(2\pi\hbar)^3} \, \delta f(\mathbf{q}, \mathbf{p}, \omega) \, . \tag{10.19}$$

In the absence of the external field, $V_a = 0$, there is in the equilibrium case no excess electronic charge, $\delta n = 0$, and the mean field vanishes, $\overline{V} = 0$.

We expect that charge density fluctuations in a metal are strongly suppressed due to the presence of highly mobile conduction electrons. We can therefore use the linear response expression, eq.(7.12), for the density deviation

$$\delta n(\mathbf{x}, t) = \int_{-\infty}^{\infty} dt' \int d\mathbf{x}' \, \chi(\mathbf{x}, t; \mathbf{x}', t') V_a(\mathbf{x}', t') \tag{10.20}$$

where χ is the density response function. Abundance of mobile charges is preferable, and the accuracy of the description is better the higher the density of electrons. In linear response we can take

$$V_a(\mathbf{x}, t) = V_a(\mathbf{q}, \omega) \, e^{i\mathbf{q}\cdot\mathbf{x} - i(\omega + i\epsilon)t} \tag{10.21}$$

where the convergence factor ensures that the applied field vanishes in the remote past where we assume the equilibrium distribution, the Fermi function,

$$\lim_{t \to -\infty} f(\mathbf{x}, \mathbf{p}, t) = f_0(\epsilon_{\mathbf{p}}) \, . \tag{10.22}$$

Because of the temporal and spatial homogeneity of the unperturbed system we have

$$\chi(\mathbf{x}, t; \mathbf{x}', t') = \chi(\mathbf{x} - \mathbf{x}', t - t') \tag{10.23}$$

and the Fourier transform of the linear density deviation has a nonzero value only at the applied frequency and wave vector

$$\delta n(\mathbf{q}', \omega') = \delta_{\mathbf{q}, \mathbf{q}'} \, 2\pi \, \delta(\omega - \omega') \, \delta n(\mathbf{q}, \omega) \tag{10.24}$$

where

$$\delta n(\mathbf{q}, \omega) = V_a(\mathbf{q}, \omega) \, \chi(\mathbf{q}, \omega + i\epsilon) \, . \tag{10.25}$$

We introduce the dielectric function

$$\epsilon(\mathbf{q}, \omega) \equiv \frac{V_a(\mathbf{q}, \omega)}{\overline{V}(\mathbf{q}, \omega)} \tag{10.26}$$

and note, according to eq.(10.19) and eq.(10.25), the relationship to the density response function

$$\frac{1}{\epsilon(\mathbf{q}, \omega)} = 1 + V(\mathbf{q}) \chi(\mathbf{q}, \omega) . \tag{10.27}$$

Inserting the linear response expression for the density, eq.(10.20), into the expression for the Hartree potential, we are linearizing in the external field, and we must do the same in the quantum kinetic equation for the Wigner function, eq.(10.13),

$$0 = \left(\frac{\partial}{\partial t} + \mathbf{v_p} \cdot \frac{\partial}{\partial \mathbf{R}} \right) \delta f(\mathbf{R}, \mathbf{p}, t)$$

$$+ \frac{i}{\hbar} \int d\mathbf{r} \int \frac{d\mathbf{p}'}{(2\pi\hbar)^3} \, e^{\frac{i}{\hbar} \mathbf{r} \cdot (\mathbf{p}' - \mathbf{p})} f_0(\epsilon_{\mathbf{p}'}) \left(\overline{V}(\mathbf{R} + \mathbf{r}/2, t) - \overline{V}(\mathbf{R} - \mathbf{r}/2, t) \right). \tag{10.28}$$

Together with eq.(10.19) we have simplified the two equations relating the mean field and the distribution function to a point where we can get a closed expression for the effective potential. Fourier-transforming eq.(10.28), we have ($\mathbf{p}_{\pm} = \mathbf{p} \pm \mathbf{q}/2$)

$$\hbar \left(\omega + i\epsilon - \mathbf{v_p} \cdot \mathbf{q} \right) \delta f(\mathbf{q}, \mathbf{p}, \omega) = \overline{V}(\mathbf{q}, \omega) \left(f_0(\epsilon_{\mathbf{p}_-}) - f_0(\epsilon_{\mathbf{p}_+}) \right) . \tag{10.29}$$

Inserting the solution

$$\delta f(\mathbf{q}, \mathbf{p}, \omega) = \frac{f_0(\epsilon_{\mathbf{p}_-}) - f_0(\epsilon_{\mathbf{p}_+})}{\hbar \omega + i\epsilon - \hbar \mathbf{v_p} \cdot \mathbf{q}} \, \overline{V}(\mathbf{q}, \omega)$$

$$= \frac{f_0(\epsilon_{\mathbf{p}_-}) - f_0(\epsilon_{\mathbf{p}_+})}{\hbar \omega + \epsilon_{\mathbf{p}_-} - \epsilon_{\mathbf{p}_+} + i\epsilon} \, \overline{V}(\mathbf{q}, \omega) \tag{10.30}$$

of the linearized quantum kinetic equation into eq.(10.19) we get the mean field expressed in terms of the applied field and the intrinsic (Coulomb) interaction

$$\overline{V}(\mathbf{q}, \omega) = V_a(\mathbf{q}, \omega) + 2V(\mathbf{q}) \int \frac{d\mathbf{p}}{(2\pi\hbar)^3} \frac{f_0(\epsilon_{\mathbf{p}_-}) - f_0(\epsilon_{\mathbf{p}_+})}{\hbar \omega + \epsilon_{\mathbf{p}_-} - \epsilon_{\mathbf{p}_+} + i\epsilon} \, \overline{V}(\mathbf{q}, \omega) . \tag{10.31}$$

Solving for the mean field we obtain for the dielectric function in the linearized mean field approximation

$$\epsilon_{LMF}(\mathbf{q}, \omega) = 1 - V(\mathbf{q}) \chi_0(\mathbf{q}, \omega) \tag{10.32}$$

where

$$\chi_0(\mathbf{q}, \omega) \equiv 2 \int \frac{d\mathbf{p}}{(2\pi\hbar)^3} \frac{f_0(\epsilon_{\mathbf{p}_-}) - f_0(\epsilon_{\mathbf{p}_+})}{\hbar \omega + \epsilon_{\mathbf{p}_-} - \epsilon_{\mathbf{p}_+} + i\epsilon} . \tag{10.33}$$

Inserting eq.(10.32) into eq.(10.27), we obtain for the density response function in the linearized mean field approximation

$$\chi_{LMF}(\mathbf{q}, \omega) = \frac{\chi_0(\mathbf{q}, \omega)}{1 - V(\mathbf{q})\,\chi_0(\mathbf{q}, \omega)} \;. \tag{10.34}$$

For later use we introduce the conductivity

$$\mathbf{j}(\mathbf{q}, \omega) = -\frac{i}{e}\mathbf{q}\,\sigma(\mathbf{q}, \omega)\,\overline{V}(\mathbf{q}, \omega) \tag{10.35}$$

and using the continuity equation we obtain the relation between the conductivity and the dielectric function

$$\epsilon(\mathbf{q}, \omega) = 1 + \frac{i\sigma(\mathbf{q}, \omega)}{\omega\,\epsilon_0} \;. \tag{10.36}$$

For a noninteracting Fermi gas, $V = 0$, the mean field consists solely of the applied field, $\overline{V} = V_a$, and eq.(10.48) gives for the density deviation

$$\delta n(\mathbf{q}, \omega) = \chi_0(\mathbf{q}, \omega)V_a(\mathbf{q}, \omega) \tag{10.37}$$

identifying χ_0 as the free Fermi gas density response function in the linearized Hartree approximation. The density response of an interacting Fermi gas is thus in the linearized Hartree approximation expressible in terms of the bare interaction, the intrinsic (Coulomb) interaction, and the density response function of a noninteracting Fermi gas.[5]

10.3   Mean-Field Diagrammatics

We would like to characterize the kind of induced-charge fluctuations included in the linearized mean field approximation for the effective potential. This can be achieved by establishing the diagrammatic interpretation of the approximation.

10.3.1   Noninteracting Fermi Gas Response

We recall the expression eq.(8.9) for noninteracting particles

$$\chi_0^{RA}(\mathbf{q}, \omega) = \frac{-i}{\pi V}\sum_{\mathbf{P}}\int_{-\infty}^{\infty} dE\;(f_0(E_+) - (f_0(E_-))$$

$$\frac{1}{E_+ - \epsilon_{\mathbf{p}_+} + i0} \cdot \frac{1}{E_- - \epsilon_{\mathbf{p}_-} - i0} \;. \tag{10.38}$$

---

[5]The linearized Hartree approximation is in the original literature referred to as the random phase approximation [50]. The quantum fluctuations in the density has a phase dependence which depends on the positions of the electrons. For a large number of electrons the phases become erratic, and the sum of such random phase terms tends to average to zero, leaving only the mean density to generate the field.

This expression is almost identical to χ_0. In fact, performing the integration by the method of residues we find that for energy exchanges with the external field which are small compared to the Fermi energy, $\hbar\omega \ll \epsilon_F$, the contribution from the poles of the Fermi functions are small. We thus only pick up the contribution from the propagator pole

$$\chi_0^{RA}(\mathbf{q},\omega) \simeq \int \frac{d\mathbf{p}}{(2\pi\hbar)^3} \frac{f_0(\epsilon_{\mathbf{p}_+}) - f_0(\epsilon_{\mathbf{p}_-})}{\epsilon_{\mathbf{p}_-} - \epsilon_{\mathbf{p}_+} + \hbar\omega + i0} = -\chi_0(\mathbf{q},\omega) . \qquad (10.39)$$

We therefore have a diagrammatic interpretation of the degenerate free Fermi gas response function

$$-\chi_0(\mathbf{q},\omega) = \qquad \qquad (10.40)$$

where we sum the product of a free retarded and advanced propagator over all momenta $\mathbf{p}$ in accordance with the usual diagrammatic rule, but only integrate over the variable E in the thermal layer around the Fermi surface according to the presence of the constraining gate function in eq.(10.38).

10.3.2   Effective Interaction

The diagrammatic interpretation of the degenerate free Fermi gas density response function allows us also to get a diagrammatic interpretation of the interacting Fermi gas density response function. We shall thereby be able to establish which type of induced density fluctuations we are including in the linearized mean field approximation. A metal is a degenerate Fermi system, and the occupation of the single-particle levels is thus described by a Fermi sea slightly spilled over the Fermi surface. The energy exchange between a field and the electrons can thus only take place through transitions in the thermal layer around the Fermi surface as described by the constraining gate function appearing in eq.(10.39) and eq.(10.40). A typical excitation by a potential takes an electron across the Fermi surface, thereby creating an intermediate state with an electron in a level above the Fermi surface occupied, and a level below the Fermi surface which is empty. But this is precisely the interpretation of the bubble diagram, eq.(10.40): the retarded line represents the amplitude for the propagating electron, and the advanced line can be interpreted as describing an electron moving backwards in time, which is equivalent to a particle with opposite charge moving forward in time, the absence of an electron, a so-called hole.

If we let the applied potential be created by a charged test particle, the mean field is the effective potential felt by an electron in the interacting electron gas. Since we are considering the thermodynamic limit, an electron can be considered a test particle, and we can assume that the applied field is generated by one of the electrons

$$V_a(\mathbf{q}, \omega) = \frac{e^2}{\epsilon_0 \mathbf{q}^2} \equiv V(\mathbf{q}) \tag{10.41}$$

and we have for the effective electron-electron interaction

$$\overline{V}(\mathbf{q}, \omega) = \frac{V(\mathbf{q})}{\epsilon(\mathbf{q}, \omega)} . \tag{10.42}$$

We can now give a simple interpretation of the corresponding effective field in the interacting electron gas as it has the power series expansion

$$
\begin{aligned}
\overline{V}(\mathbf{q}, \omega) &= \frac{V(\mathbf{q})}{1 - V(\mathbf{q})\chi_0(\mathbf{q}, \omega)} \\
&= V(\mathbf{q}) + V(\mathbf{q})\,\chi_0(\mathbf{q}, \omega)\,V(\mathbf{q}) \\
&+ V(\mathbf{q})\,\chi_0(\mathbf{q}, \omega)\,V(\mathbf{q})\,\chi_0(\mathbf{q}, \omega)\,V(\mathbf{q}) + ...
\end{aligned}
\tag{10.43}
$$

The linearized mean-field approximation for the effective field thus corresponds to summing all the bubble diagrams[6]

where the thin wiggly line denotes the intrinsic (Coulomb) interaction, and it is understood that the integration over the variable E is only over the thermal layer around the Fermi Surface. The effective interaction, represented diagrammatically by a thick wiggly line, is not only the bare (Coulomb) interaction, but includes how the other electrons, the induced charge fluctuations, influence the interaction. From the diagrammatic description we realize, that in the linearized mean field

---

[6]We have according to eq.(10.34) an equivalent diagrammatic expansion of the density response function.

approximation these fluctuations are determined by the dynamics of particle-hole excitations in the system: a particle-hole pair is created and annihilated, upon which another pair is created and annihilated, etc. In the linearized mean field approximation we thus exclude more complicated density configurations than those build from particle-hole density fluctuations.

The effective interaction between two electrons in a metal is thus not the bare Coulomb interaction as in vacuum because of the presence of all the other electrons. An electron can through its Coulomb interaction excite another electron out of the Fermi sea thereby giving rise to a nonuniform charge density, described by the propagation of particle-hole pair charge fluctuations through the system, which in turn will screen the interaction.[7] The density response function, or density fluctuation propagator, is also referred to as the polarization, describing the change in the bare propagator of the interaction

$$\overline{V}(\mathbf{q}, \omega) \;=\; \frac{1}{V^{-1}(\mathbf{q}) - \chi_0(\mathbf{q}, \omega)} \tag{10.45}$$

or diagrammatically

$$\tag{10.46}$$

The only quantum aspect playing a role in the mean field approximation is the quantum statistics of the electrons, that they obey Pauli's exclusion principle. The kinematics, however, is purely classical as we have seen, in fact simply free electron kinematics since the propagators in the polarization are the free propagators.

10.4   Plasma Oscillations

Let us obtain the dispersion relation for the density oscillations of the interacting Fermi gas. Inserting the distribution function, eq.(10.30), into the Fourier-transform of eq.(10.16)

$$\delta n(\mathbf{q}, \omega) = 2 \int \frac{d\mathbf{p}}{(2\pi\hbar)^3} \, \delta f(\mathbf{q}, \mathbf{p}, \omega) \tag{10.47}$$

we get for the density deviation

$$\delta n(\mathbf{q}, \omega) \;=\; \chi_0(\mathbf{q}, \omega) \, \overline{V}(\mathbf{q}, \omega) \;=\; \frac{\chi_0(\mathbf{q}, \omega)}{\epsilon(\mathbf{q}, \omega)} \, V_a(\mathbf{q}, \omega) \tag{10.48}$$

---

[7]The screening of the electron-electron interaction in a metal is analogous to how interactions gets renormalized in a quantum field theory such as Quantum Electrodynamics, where quantum fluctuations, pair creation of particles and antiparticles across the mass gap, screens the interaction. In the present case, however, we are dealing with thermal fluctuations only, and there is no gap in the excitation spectrum (as would be the case in semiconductors).

and by Fourier-transforming

$$\delta n(\mathbf{q}, t) = \int\limits_{-\infty}^{\infty} d\omega \, e^{-i\omega t} \, \frac{V_a(\mathbf{q}, \omega) \, \chi_0(\mathbf{q}, \omega)}{\epsilon(\mathbf{q}, \omega)} . \tag{10.49}$$

The zeros of the dielectric function, $\epsilon(\mathbf{q}, \omega = \omega(\mathbf{q})) = 0$, gives the poles in the integrand. They thus describe the possible electronic density oscillations the electron gas can be excited to in the charge-neutralizing inert background

$$\delta n(\mathbf{q}, t) \propto e^{-i\omega(\mathbf{q}) t} . \tag{10.50}$$

Whereas the poles of the propagator describe the possible single-particle excitations, the poles in the dielectric function describe collective excitations, the electronic density oscillations, characterized by the dispersion relation $\omega = \omega(\mathbf{q})$, the zeros of the dielectric function.[8]

For the long-wavelength charge oscillations, $\mathbf{q} \cdot \mathbf{p}_F \ll m\omega$, we can Taylor-expand the denominator of eq.(10.33) and obtain

$$\chi_0(\mathbf{q}, \omega) = \frac{nq^2}{m\omega^2} \left(1 + <\mathbf{v}^2> \frac{q^2}{\omega^2} \right) \tag{10.51}$$

where

$$<\mathbf{v}^2> = \int \frac{d\mathbf{p}}{(2\pi\hbar)^3} \frac{\mathbf{p}^2}{m^2} \, f_0(\epsilon_{\mathbf{p}}) \tag{10.52}$$

is the equilibrium velocity fluctuations in a free Fermi gas. For Coulomb interaction we get from eq.(10.32) the electronic plasma oscillation dispersion

$$\omega(q) = \sqrt{\omega_p^2 + <\mathbf{v}^2> q^2} \tag{10.53}$$

where

$$\omega_p \equiv \sqrt{\frac{ne^2}{m\epsilon_0}} \tag{10.54}$$

is called the plasma frequency.

The long-range nature of the Coulomb interaction results in a gap in the excitation spectrum of the charge density oscillations, $\omega(q = 0) \neq 0$. Even long-wavelength deviations from charge neutrality are unfavored. As anticipated, exciting plasma oscillations are increasingly energetically costly in the high density limit due to the density dependence of the energy gap. The gap in the plasmon excitation spectrum can be observed by sending a beam of electrons through a thin metal foil.

---

[8]A zero of the dielectric function in the upper ω-half-plane signals, according to eq.(10.50), an instability, a phase transition. Barring this situation, we have according to eq.(10.27) that the dielectric function is analytic in the upper ω-half-plane.

Exercise 10.1 *Show that the equilibrium velocity fluctuations in a free Fermi gas in the degenerate and classical limits are given respectively by*

$$
< \mathbf{v}^2 > \; = \; \begin{cases} \frac{3}{5} v_F^2 & T \ll T_F \\[2mm] \frac{3kT}{m} & T \gg T_F \; . \end{cases}
\tag{10.55}
$$

10.5 Static Screening

In this section we elaborate on the important phenomena of screening by considering the static limit at zero temperature. The static noninteracting Fermi gas response can at zero temperature be calculated, and we get for the susceptibility

$$
\chi_0(\mathbf{q}, 0) \; = \; -2N_0 \, \chi_3(q/2k_F)
\tag{10.56}
$$

where the function

$$
\chi_3(x) \; = \; \frac{1}{2} \; + \; \frac{1 - x^2}{4x} \ln \left| \frac{1 + x}{1 - x} \right| \; .
\tag{10.57}
$$

is seen to be nonanalytic at the value $x = 1$, where the derivative has a singularity. In fact, in the static case we can evaluate the free Fermi gas response function for wave vectors small compared to the Fermi wave vector directly (to order $\mathcal{O}((q/k_F)^2)$[9]

$$
\chi_0(\mathbf{q}, 0) = 2 \int \frac{d\mathbf{p}}{(2\pi\hbar)^3} \frac{\partial f_0(\epsilon_\mathbf{p})}{\partial \epsilon_\mathbf{p}} = -2 \int \frac{d\mathbf{p}}{(2\pi\hbar)^3} \left(\frac{\partial f_0(\epsilon_\mathbf{p})}{\partial \mu} \right)_T = - \left(\frac{\partial n}{\partial \mu} \right)_T .
\tag{10.58}
$$

For a noninteracting Fermi gas at temperatures low compared to the degeneracy temperature, $T \ll T_F$, we have that the change in density per unit chemical potential is equal to the density of states at the Fermi energy

$$
\frac{\partial n}{\partial \mu} \; = \; 2N_0 \; = \; \frac{mk_F}{\pi^2 \hbar^2} \; .
\tag{10.59}
$$

Let us consider the introduction of a foreign atom into an otherwise pure metal, an impurity atom. For example, introducing a divalent atom, such as magnesium, into a monovalent metal, such as copper. In comparison to the pure crystal, at the site of the impurity nucleus there is an extra plus charge, and in addition an extra roaming conduction electron. In this case the applied field is the Coulomb field of

---

[9]This linear response function at zero frequency is expressed in terms of a thermodynamic quantity. Furthermore, the zero wave vector limit is insensitive to details of the model and holds true even for a relativistic electron gas.

the extra unit of plus charge (choosing the site of the impurity nucleus as position reference)

$$V_a(\mathbf{x}) = -\frac{e^2}{4\pi\epsilon_0 |\mathbf{x}|} . \tag{10.60}$$

We then obtain, according to eq.(10.26), for the effective interaction energy felt by a conduction electron

$$\overline{V}(\mathbf{q}) = -\frac{e^2}{\epsilon_0} \frac{1}{q^2 + \kappa_s^2 \chi_3(q/2k_F)} \tag{10.61}$$

as the static dielectric function equals

$$\epsilon(\mathbf{q}, 0) = 1 - \frac{e^2}{\epsilon_0 q^2} \chi_0(\mathbf{q}, 0) = 1 + \frac{\kappa_s^2}{q^2} \chi_3(q/2k_F) \tag{10.62}$$

where

$$\kappa_s^2 = \frac{e^2}{\epsilon_0} \left(\frac{\partial n}{\partial \mu}\right)_T = \frac{2N_0 e^2}{\epsilon_0} = \frac{4}{\pi} \frac{k_F}{a_0} \tag{10.63}$$

and a_0 denotes the Bohr radius, $a_0 = 4\pi\epsilon_0 \hbar^2/me^2$.

The cohesion of a metal is due to metallic binding; i.e., the metallic bonding between the ions provided by the almost spatially homogeneous charge of the freely roaming conduction electrons outweighs the repulsion between the ions. However, the average kinetic energy of a conduction electron (of the order of the Fermi energy ϵ_F) can not be excessively less than the magnitude of the attractive average potential energy provided by the ionic lattice (of order $e^2 n^{1/3}/\epsilon_0 \sim e^2/\epsilon_0 a_0$). Otherwise it would be energetically favorable for the metal to dissociate. In a metal the screening length and the Fermi wavelength and the interatomic distance are thus all of the same order of magnitude, $k_F \sim a_0^{-1} \sim \kappa_s$.

The induced electron density distribution due to the introduction of the, say, positive charge Ze, is according to eq.(10.62) at zero temperature

$$\delta n(\mathbf{x}) = Z \int \frac{d^3 q}{(2\pi)^3} e^{i\mathbf{q}\cdot\mathbf{x}} \frac{\kappa_s^2 \chi_3(q/2k_F)}{q^2 + \kappa_s^2 \chi_3(q/2k_F)}$$

$$= \frac{Z\kappa_s^2}{(2\pi)^2 x} \int_{-\infty}^{\infty} dq \, q \sin qx \, \frac{\chi_3(q/2k_F)}{q^2 + \kappa_s^2 \chi_3(q/2k_F)} . \tag{10.64}$$

To calculate the integral we note that the integrand has two branch cuts due to the logarithm in the function χ_3. Upon shifting the integration contour into the upper complex q-plane, the long-distance behavior is determined by the contributions from the branch cuts, which we choose to stretch from $-i\infty$ to $i\infty$ crossing the real axis at $q = \pm 2k_F$, respectively. The function χ_3 has jumps $\pm i\pi$ across the branch cuts, giving for the induced electronic density at large distances, $x \gg \kappa_s^{-1}$,

$$\delta n(\mathbf{x}) \simeq \frac{Z}{\pi} \frac{\kappa_s^2/k_F^2}{(4 + \kappa_s^2/2k_F^2)^2} \frac{\cos 2k_F |\mathbf{x}|}{|\mathbf{x}|^3} \tag{10.65}$$

the so-called Friedel oscillations in the screening charge. The long-range Friedel oscillations have their origin in the nonanalyticity of the function χ_3, which in turn reflects the discontinuity in the zero-temperature Fermi distribution at the Fermi energy.[10] At finite temperatures the oscillations will be damped by the factor $\exp\{-2\pi mkT|\mathbf{x}|/\hbar^2 k_F\}$.

Analogously, the existence of a sharp Fermi surface[11] leads to long-range oscillations in the effective potential, $|\mathbf{x}| \gg \kappa_s^{-1}$,

$$\overline{V}(\mathbf{x}) \simeq -\frac{e^2}{4\pi\epsilon_0}\frac{\cos(2k_F|\mathbf{x}|)}{|\mathbf{x}|^3} . \tag{10.66}$$

If we are only interested in the potential averaged over the microscopic length scale of the Fermi wavelength, the above branch cut contribution averages to zero, and we are left with the contribution from the pole at $q \simeq \kappa_s$ in the inverse static dielectric function, giving for the effective static interaction in real space:

$$\overline{V}(\mathbf{x}) = -\frac{e^2}{4\pi\epsilon_0}\frac{e^{-\kappa_s|\mathbf{x}|}}{|\mathbf{x}|} . \tag{10.67}$$

A qualitative change has taken place, the long-range character of the Coulomb interaction has been cut off, and the mean field has a finite spatial range, κ_s^{-1}, the screening length. The extra plus charge of the impurity nucleus attracts electrons, and an electron interacts effectively with the extra plus charge and its surrounding cloud of excess electronic charge; the static plus charge has been screened. The effective potential is thus much weaker than the Coulomb potential from the impurity charge, and in a metal the range of the Coulomb potential from the extra charge on an impurity atom is cut off at atomic distance, the Thomas-Fermi screening length.[12] We note that indeed at large distances, $|\mathbf{x}| > \kappa_s^{-1}$, or small wave vectors, $|\mathbf{q}| \ll \kappa_s$, the effective potential is small, and we expect the linearized mean field theory to be valid at large distances for the spatially averaged mean field. Note, that at ultra-small distances $|\mathbf{x}| < \kappa_s^{-1}$, we get also the correct behavior for the mean field since we there must recover the bare Coulomb strength. We have thus obtained an expression for the spatially averaged mean field, which we expect to interpolate well except possibly at wave vectors of the order of the screening wave vector, $|\mathbf{q}| \sim \kappa_s$.

The nature of the Friedel oscillations can be illuminated by calculating the electronic density in the presence of the impurity in the mean field approximation

$$n(\mathbf{x}) = \sum_{|\mathbf{p}|<p_F} \psi_{\mathbf{p}}^*(\mathbf{x})\,\psi_{\mathbf{p}}(\mathbf{x}) . \tag{10.68}$$

---

[10]The Friedel oscillations give rise to the random sign of the exchange interaction between impurity spins in a spin-glass.

[11]The Coulomb interaction by itself does not smear out the sharp Fermi surface. This follows from the fact that the imaginary part of the self-energy due to electron-electron interaction vanishes near the Fermi surface as $(E - \epsilon_F)^2$, as we show in section 10.8.4, and therefore only a renormalization lowering the step in the momentum distribution function by the factor $|\partial\Re e\Sigma^R(E_F,\mathbf{p}_F)/\partial E|$ takes place, Luttinger's theorem [51].

[12]For substitutional impurities we are indeed justified in using a local model for the impurity potential as we did in our discussion in section 3.4 of the impurity-averaged propagator.

The exact eigenstates in the mean field are determined by the Lippmann-Schwinger equation, eq.(2.186),

$$\psi_{\mathbf{p}}(\mathbf{x}) = \frac{e^{\frac{i}{\hbar}\mathbf{p}\cdot\mathbf{x}}}{\sqrt{V}} + \int d\mathbf{x}'\, G_0^R(\mathbf{x},\mathbf{x}',\epsilon_{\mathbf{p}})\, \overline{V}(\mathbf{x}')\, \psi_{\mathbf{p}}(\mathbf{x}')\ . \tag{10.69}$$

In lowest-order perturbation theory, we assume that the mean field is weak, we obtain

$$\psi_{\mathbf{p}}(\mathbf{x}) = \frac{e^{\frac{i}{\hbar}\mathbf{p}\cdot\mathbf{x}}}{\sqrt{V}} + \int d\mathbf{x}'\, G_0^R(\mathbf{x},\mathbf{x}',\epsilon_{\mathbf{p}})\, \overline{V}(\mathbf{x}')\, \frac{e^{\frac{i}{\hbar}\mathbf{p}\cdot\mathbf{x}'}}{\sqrt{V}} \tag{10.70}$$

which gives for the induced density

$$\delta n(\mathbf{x}) = -\frac{m k_F^2}{4\pi^3\hbar^2}\int d\mathbf{x}'\, \overline{V}(\mathbf{x}')\, \frac{j_1(2k_F|\mathbf{x}-\mathbf{x}'|)}{|\mathbf{x}-\mathbf{x}'|^2} \tag{10.71}$$

where j_1 is the spherical Bessel function

$$j_1(x) = \frac{\sin x - x\cos x}{x^2}\ . \tag{10.72}$$

Poisson's equation then determines the mean field according to

$$\Delta_{\mathbf{x}}\overline{V}(\mathbf{x}) = -\frac{e^2\delta n(\mathbf{x})}{\epsilon_0} = \frac{m k_F^2}{4\pi^3\hbar^2}\int d\mathbf{x}'\, \overline{V}(\mathbf{x}')\frac{j_1(2k_F|\mathbf{x}-\mathbf{x}'|)}{|\mathbf{x}-\mathbf{x}'|^2}\ . \tag{10.73}$$

The slow third-order power law decay at large distances is then apparent in both the potential and charge density. The Friedel oscillations can thus be viewed as due to the diffraction of the electron wave function off the screened scatterer.

In the following we shall be interested in the interaction between volumes of charge (long-wavelength oscillations) much larger than atomic size, and we are therefore only interested in the density averaged over a distance much larger than the Fermi wavelength. The above calculated branch cut contribution, eq.(10.65), then averages to zero, and we are left with the contribution from the pole at $q \simeq \kappa_s$ in the inverse static dielectric function, giving rise to the local induced excess electronic density

$$\delta n(\mathbf{x}) = \frac{Z\kappa_s^2}{4\pi}\frac{e^{-\kappa_s|\mathbf{x}|}}{|\mathbf{x}|} = \frac{2N_0 Z e^2}{4\pi\epsilon_0}\frac{e^{-\kappa_s|\mathbf{x}|}}{|\mathbf{x}|} = -2N_0\overline{V}(\mathbf{x})\ . \tag{10.74}$$

The divergence in the above expression for the charge density is absent in an exact evaluation of the screening charge using eq.(10.64). We note that this approximation corresponds to a local relationship between the induced charge distribution and the effective field, as reflected by the relation

$$\int d\bar{\mathbf{x}}\,\chi(\mathbf{x},\bar{\mathbf{x}})\,\epsilon(\bar{\mathbf{x}},\mathbf{x}') = -2N_0\delta(\mathbf{x}-\mathbf{x}') = -\epsilon_0\kappa_s^2 e^{-2}\delta(\mathbf{x}-\mathbf{x}') \tag{10.75}$$

and is referred to as the complete screening model.

In the mean-field linear response description of the long-wavelength density fluctuations out of the Fermi sea, the electron kinetics is classical, in fact only free electron kinetics is involved, as we have noted in section 10.3.2. That this is the case can also be seen since the above result is nothing but the Thomas-Fermi result. In the Thomas-Fermi screening theory a quasi-classical wave packet description of the electron motion is adopted. The electron moves in the smooth mean field and can be ascribed the energy

$$\epsilon(\mathbf{p}, \mathbf{x}) = \epsilon_{\mathbf{p}} + \overline{V}(\mathbf{x}) \tag{10.76}$$

so that we have for the excess charge distribution

$$
\begin{aligned}
\delta n(\mathbf{x}) \;&=\; 2 \int \frac{d\mathbf{p}}{(2\pi\hbar)^3} \left[f_0(\epsilon_{\mathbf{p}} + \overline{V}(\mathbf{x})) - f_0(\epsilon_{\mathbf{p}}) \right] \\[2mm]
&\simeq\; -2\overline{V}(\mathbf{x}) \int \frac{d\mathbf{p}}{(2\pi\hbar)^3} \left(-\frac{\partial f_0(\epsilon_{\mathbf{p}})}{\partial \epsilon_{\mathbf{p}}} \right) \\[2mm]
&=\; -2N_0\, \overline{V}(\mathbf{x}) \tag{10.77}
\end{aligned}
$$

where we have assumed $|\overline{V}(\mathbf{x})| \ll \epsilon_F$. The Thomas-Fermi theory thus corresponds to neglecting the branch cut contributions through setting $\chi_3 = 1$. Determining the mean field self-consistently from the Poisson equation we again obtain the screened Coulomb potential

$$\overline{V}(\mathbf{x}) = -\frac{e^2}{4\pi\epsilon_0} \frac{e^{-|\mathbf{x}|/\lambda_{TF}}}{|\mathbf{x}|} \;. \tag{10.78}$$

where $\lambda_{TF} = (\epsilon_0/2N_0 e^2)^{1/2} = \kappa_s^{-1}$ is the Thomas-Fermi screening length. Since the average potential (Coulomb) energy is assumed much smaller than the average kinetic energy, $n^{-1/3} \ll a_0$, the screening length is seen to be larger than the average distance between the electrons.

The Thomas-Fermi result has a simple thermodynamic interpretation. The electrochemical potential is in equilibrium a constant throughout space, and equals the chemical potential far from the inserted charge. The Thomas-Fermi theory is simply the approximation where we add the local chemical potential and the work needed done due to the mean field in bringing an electron into the position in question

$$\mu(n + \delta n(\mathbf{x})) + \overline{V}(\mathbf{x}) = \mu(n) \;. \tag{10.79}$$

The result, eq.(10.77), then follows upon Taylor-expanding.

Exercise 10.2 *Show that the free Fermi gas response function at zero temperature in the one-dimensional case is specified by, $x = q/2k_F$,*

$$\chi_1(x) = \frac{1}{2x} \ln \left| \frac{1+x}{1-x} \right| \;. \tag{10.80}$$

The logarithmic singularity at wave vector $2k_F$ leads to an instability in the electronic charge density, a charge density wave appears. The instability is stabilized by alternating opposite shifts in ionic positions, the Peierls transition, whereby the unit cell is doubled in length and a gap appears in the electronic spectrum at the Fermi energy due to the doubling of the periodicity of the periodic potential. The phenomenon can take place in higher dimensions in materials where the Fermi surface has parallel pieces, a nested Fermi surface.

In a nondegenerate electron gas we can calculate the static susceptibility according to

$$\chi_0(\mathbf{q}, \omega = 0) \equiv 2 \int \frac{d\mathbf{p}}{(2\pi\hbar)^3} \frac{\partial f_0(\epsilon_{\mathbf{p}})}{\partial \epsilon_{\mathbf{p}}} \tag{10.81}$$

and since in the classical limit, $\partial f_0(\epsilon_{\mathbf{p}})/\partial \epsilon_{\mathbf{p}} = -f_0(\epsilon_{\mathbf{p}})/kT$, we get for the dielectric function

$$\epsilon(\mathbf{q}) = 1 + \frac{ne^2}{\epsilon_0 kT q^2} = 1 + \frac{\kappa_{DH}^2}{q^2} \tag{10.82}$$

where

$$\kappa_{DH} = \sqrt{\frac{ne^2}{\epsilon_0 kT}} \tag{10.83}$$

is the Debye-Hückel screening wave number, which of course is the thermodynamic result, eq.(10.58), in the classical limit $(\partial n/\partial \mu)_T = n/kT$. The implicit assumption of the Thomas-Fermi screening theory, that the average potential energy is smaller than the average kinetic energy, now corresponds to the criterion, $e^2 n^{1/3}/\epsilon_0 \ll kT$. The Debye-Hückel screening is the relevant one for the screening of ionized impurities in a nondegenerate semiconductor.

The classical theory is simply the Debye–Hückel theory of electrolytes, because in a classical Coulomb gas, the charge density profile is given by the Boltzmann distribution,

$$n(\mathbf{x}) = n e^{\overline{V}(\mathbf{x})/kT} \simeq n\left(1 + \frac{\overline{V}(\mathbf{x})}{kT} + ...\right). \tag{10.84}$$

The conditions at infinity, where we assume that there is no influence of the external charge, $\overline{V} = 0$, determines the prefactor. Linearizing, $|\overline{V}(\mathbf{x})| \ll kT$, we have together with the Poisson equation a closed set of equations.

10.6   Lattice Dynamics

We now take into account that the ionic mass is finite, allowing the ions to oscillate around their equilibrium positions. To each configuration of ions $(\mathbf{x}_1, \mathbf{x}_2, ..)$ we ascribe a potential energy $V(\mathbf{x}_1, \mathbf{x}_2, ..)$. The regular lattice configuration $(\mathbf{R}_1, \mathbf{R}_2, ..)$ corresponds to the minimum of the potential

$$\frac{\partial V(\mathbf{x}_1, \mathbf{R}_2, \mathbf{R}_3, ..)}{\partial \mathbf{x}_1}\bigg|_{\mathbf{x}_1 = \mathbf{R}_1} = \mathbf{0} \tag{10.85}$$

and similarly in the other variables. Assuming small displacements of the ions, we can use the approximation

$$V(\mathbf{x}_1, \mathbf{x}_2, ..) = \frac{1}{2} \sum_{i,j,\alpha,\beta} (\mathbf{x}_i - \mathbf{R}_i)_\alpha \, G_{\alpha,\beta}(i,j) \, (\mathbf{x}_j - \mathbf{R}_j)_\beta$$

$$\equiv \frac{1}{2} \sum_{k,k'} q_k \, G_{kk'} \, q_{k'} \tag{10.86}$$

where $q_k \equiv (\mathbf{x}_i - \mathbf{R}_i)_\alpha$ is simply a relabeling of the Cartesian coordinates, and we choose the equilibrium configuration energy as reference, $V(\mathbf{R}_1, \mathbf{R}_2, ..) = 0$, and

$$G_{\alpha,\beta}(i,j) \equiv \frac{\partial^2 V(\mathbf{R}_1, \mathbf{R}_2, .., \mathbf{x}_i, \mathbf{R}_{i+1}, .., \mathbf{x}_j, \mathbf{R}_{j+1}, ..)}{\partial x_i^\alpha \partial x_j^\beta}\bigg|_{\mathbf{x}_{i,j} = \mathbf{R}_{i,j}} \quad . \tag{10.87}$$

The matrix $G_{\alpha,\beta}(i,j)$ is real and symmetric, $G_{\alpha,\beta}(i,j) = G_{\beta,\alpha}(j,i)$, and can therefore be diagonalized by an orthogonal matrix $\underline{S}$. The diagonal elements of $\underline{S}\,\underline{G}\,\underline{S}^{-1}$ are the eigenfrequencies ω_{n_i} of the lattice vibrations, and since the matrix is positive (since the equilibrium configuration is stable) the frequencies are positive, $\omega_{n_i} > 0$. The problem of N interacting oscillators is thus equivalent to N different noninteracting oscillators, as the potential energy eq.(10.86) is diagonalized by introducing the normal coordinates

$$Q_k = \sum_{k'} S_{kk'} \, q_{k'} \quad . \tag{10.88}$$

Having diagonalized Hamilton's function for the ions, we have in the same token diagonalized the lattice Hamiltonian into separate harmonic oscillators

$$H(\{\hat{P}_{n_i}\}_{n_i}, \{\hat{Q}_{n_i}\}_{n_i}) = \sum_{n_i=1}^{3N_i} \frac{\hat{P}_{n_i}^2}{2M} + \frac{1}{2}M \sum_{n_i=1}^{3N_i} \omega_{n_i}^2 \hat{Q}_{n_i}^2 \tag{10.89}$$

where, since the transformation $\hat{P}_k = \sum_{k'} S_{kk'} \, \hat{p}_{k'}$ is orthogonal, $\hat{P}_{n_i}$ is the operator complementary to the normal mode operator $\hat{Q}_{n_i}$. A normal mode operator describes a collective displacement of all the ions. The excited states of a harmonic oscillator is described by referring to the number of quanta in the oscillator, and in the present context these quanta are called phonons.

The canonical commutation relations for the normal mode operators follows from the commutation relations for the ionic degrees of freedom, $\hat{q}_k \equiv (\hat{\mathbf{x}}_i - \mathbf{R}_i)_\alpha$,

$$[\hat{q}_k, \hat{p}_{k'}] = i\hbar \, \delta_{k,k'} \,, \qquad [\hat{Q}_{n_i}, \hat{P}_{n_i'}] = i\hbar \, \delta_{n_i, n_i'} \quad . \tag{10.90}$$

In the following we are only interested in the long-wavelength dynamics of the lattice vibrations, and a continuum description is sufficient. For simplicity we shall assume an isotropic lattice where the long-wavelength dynamics is characterized by a single parameter, which we now turn to determine in terms of the microscopic parameters of the model.

As the ions oscillate around their equilibrium positions we can have compressed and rarefied areas in space of the background charge. Since the mass of the electron is small compared to the ionic mass, $m/M \sim 10^{-4}$, the conduction electrons are highly mobile and will adiabatically follow the slow motion of the background. The interaction between the background charges will thus at small frequencies be equivalent to two external charges inserted into an *electron gas with a fixed neutralizing background charge*. According to the foregoing analysis the interaction between the background charges is screened. For the background-background interaction energy we have

$$
\begin{aligned}
V'_{b-b} &= \frac{1}{2} \int d\mathbf{x} \int d\mathbf{x}' \, (n_i + \delta n_b(\mathbf{x})) \overline{V}_{bb}(\mathbf{x} - \mathbf{x}')(n_i + \delta n_b(\mathbf{x}')) \\
&= \frac{1}{2} \int d\mathbf{x} \int d\mathbf{x}' \, \delta n_b(\mathbf{x}) \overline{V}_{bb}(\mathbf{x} - \mathbf{x}') \delta n_b(\mathbf{x}') \\
&+ \frac{1}{2} n_i^2 \int d\mathbf{x} \int d\mathbf{x}' \, \overline{V}_{bb}(\mathbf{x} - \mathbf{x}')
\end{aligned}
\tag{10.91}
$$

where we have used that the total number of ions is conserved

$$
\int d\mathbf{x} \, \delta n_b(\mathbf{x}) = 0 .
\tag{10.92}
$$

For the effective interaction energy $\overline{V}_{bb}$ between unit volumes of background charge we have the instantaneous screened interaction

$$
\overline{V}_{bb}(\mathbf{x}) = \frac{Z^2 e^2}{4\pi \epsilon_0 |\mathbf{x}|} e^{-\kappa_s |\mathbf{x}|} .
\tag{10.93}
$$

The last term in eq.(10.91) is a constant which does not influence the dynamics, and the interaction energy governing the background dynamics is

$$
V_{b-b} = \frac{1}{2} \int d\mathbf{x} \int d\mathbf{x}' \, \delta n_b(\mathbf{x}) \overline{V}_{bb}(\mathbf{x} - \mathbf{x}') \delta n_b(\mathbf{x}') .
\tag{10.94}
$$

For oscillations of the background with wavelengths long compared to the lattice spacing, the screened Coulomb interaction is effectively a delta function

$$
\overline{V}_{bb}(\mathbf{x} - \mathbf{x}') = \frac{Z^2}{2N_0} \delta(\mathbf{x} - \mathbf{x}')
\tag{10.95}
$$

giving for the background-background interaction energy

$$
V_{b-b} = \frac{Z^2}{4N_0} \int d\mathbf{x} \, \delta n_b(\mathbf{x}) \, \delta n_b(\mathbf{x}) .
\tag{10.96}
$$

For wavelengths long compared to the interatomic distance, we can use a continuum description of the oscillations in the background. In terms of the displacement field $\mathbf{u}(\mathbf{x}, t)$, describing the displacement of the background at position $\mathbf{x}$ at time t, we have for small displacements

$$\frac{\delta n_b(\mathbf{x}, t)}{n_i} = - \nabla \cdot \mathbf{u}(\mathbf{x}, t) . \tag{10.97}$$

Together with the kinetic energy of the background we then get the Lagrange functional valid for small displacements

$$L[\mathbf{u}] = \frac{M n_i}{2} \int d\mathbf{x} \left[\left(\frac{\partial \mathbf{u}(\mathbf{x}, t)}{\partial t} \right)^2 - c^2 \left(\nabla \cdot \mathbf{u}(\mathbf{x}, t) \right)^2 \right] \tag{10.98}$$

which is the Lagrange functional for an elastic medium with the sound velocity given by

$$c^2 = \frac{Zn}{2N_0 M} = \frac{Z}{3} \frac{m}{M} v_F^2 , \tag{10.99}$$

where $n = Z n_i$ is the equilibrium electron density. We note, that the longitudinal sound velocity is typically 100 times smaller than the Fermi velocity. The equation of motion following from Hamilton's principle is the equation

$$\frac{1}{c^2} \frac{\partial^2 \mathbf{u}(\mathbf{x}, t)}{\partial t^2} = \nabla(\nabla \cdot \mathbf{u}(\mathbf{x}, t)) \tag{10.100}$$

for the displacement field of the background. We have hereby reached the expected conclusion that a solid supports sound waves.

Introducing the momentum density of the ions

$$\Pi(\mathbf{x}, t) \equiv M n_i \frac{\partial \mathbf{u}(\mathbf{x}, t)}{\partial t} \tag{10.101}$$

we have the Hamilton functional

$$H_b = \int d\mathbf{x} \left[\frac{1}{2 M n_i} \left(\Pi(\mathbf{x}, t) \right)^2 + \frac{M n_i c^2}{2} \left(\nabla \cdot \mathbf{u}(\mathbf{x}, t) \right)^2 \right] \tag{10.102}$$

for an elastic medium.

10.6.1   Linear Chain

In order to familiarize ourselves with the emergence of the continuum limit, and the quantization of the background dynamics, we consider the one-dimensional case first, and consider a chain of N ions with masses M_a and lattice spacing a connected by springs with equal force constants k.

We measure the position of ions from their equilibrium positions, i.e., we introduce the displacements of the ions, $u_n(t) \equiv x_n(t) - x_n^0$, and their classical dynamics is determined by the coupled set of Newton's equations

$$M_a \ddot{u}_n(t) = k(u_{n+1}(t) - u_n(t)) - k(u_n(t) - u_{n-1}(t)) \qquad n = 1, 2, \ldots \tag{10.103}$$

Figure 10.1 Linear chain of masses connected by springs.

We are interested in bulk properties, so we loop the chain by imposing periodic boundary conditions; i.e., we neglect surface effects by identifying the (displacements of) ions labeled $N+1$ and 1, $u_{N+1}(t) = u_N(t)$. The equations of motion for the ions, eq.(10.103), are then generated by the Hamilton function

$$H_{LC}(\{u_n\}_{n=1,..,N}, \{p_n\}_{n=1,..,N}) = \frac{1}{2M_a} \sum_{n=1}^{N} p_n^2 + \frac{1}{2}k \sum_{n=1}^{N} (u_{n+1} - u_n)^2 \quad (10.104)$$

and the other set of Hamilton equations are $p_n = M\dot{u}_n$.

The dynamics generated by the linear chain Hamilton function with periodic boundary conditions sustains running waves

$$u_n(t) = u_0 \, e^{iqna} e^{-i\omega_q t} \quad (10.105)$$

with dispersion[13]

$$\omega_q = \sqrt{\frac{4k}{M_a}} \, \left| \sin \frac{qa}{2} \right|. \quad (10.106)$$

Since the wave vectors q and $q + 2\pi/a$ represent the same displacement configuration of the ions, we restrict the q-values to lie in the first Brillouin zone which, due to the periodic boundary condition, consists of the set of N discrete values

$$B_1 = \left\{ q \,|\, q = \frac{2\pi}{Na} p \,, \quad p = -\frac{N}{2} + 1, -\frac{N}{2} + 2, \, .. \,, \frac{N}{2} - 1, \frac{N}{2} \right\} \quad (10.107)$$

in the interval $]-\pi/a, \pi/a]$, and we have for definiteness assumed N even.

Let us now take the continuum limit. Adding more ions and springs and scaling the mass and spring constant, i.e., letting the lattice constant approach zero, $a \to 0$, and the number of ions infinity, $N \to \infty$, in such a manner that the product remains constant, $Na = L$, where L is the length of the chain. As we scale down the lattice spacing, the ionic mass per unit length, M_a/a, becomes the continuum mass density for the smeared-out background

$$\frac{M_a}{a} \to M n_i \quad (10.108)$$

where n_i is the density of ions in the one-dimensional crystal, each ion having the mass M. Scaling the length of the spring to half its size doubles its force constant as each "twist" of the spring stretches twice as much; i.e., the product

---

[13]If we had a unit cell with a basis we would get additional mode branches, for example, an optical branch due to the difference in ionic masses.

ka is constant. The dispersion relation, eq.(10.106), becomes linear (the Debye model) in the continuum limit

$$\omega_q = \lim_{a \to 0} qa \sqrt{\frac{k}{M_a}} \equiv cq \, . \tag{10.109}$$

In the continuum limit the sound waves are thus dispersionless, and the sound velocity is given by

$$c = \sqrt{\frac{ka}{Mn_i}} \tag{10.110}$$

the sound velocity of the continuum, the velocity with which density disturbances of all wave vectors travel in the chain in the continuum limit. Comparing with eq.(10.99) we have identified the continuum-limit spring constant in terms of our microscopic parameters

$$ka = \frac{n^2}{2N_0} = \frac{2}{3} n \epsilon_F \, . \tag{10.111}$$

To facilitate obtaining the continuum-limit Lagrange functional, we introduce the notation $u(x_n, t) \equiv u_n(t)$, $x_n = na$. For the kinetic energy we have in the continuum limit

$$\frac{1}{2M} \sum_{n=1}^{N} p_n^2 = \frac{M}{2a} \sum_{n=1}^{N} a \left(\frac{\partial u(x_n, t)}{\partial t} \right)^2 \to \frac{n_i M}{2} \int_0^L dx \left(\frac{\partial u(x, t)}{\partial t} \right)^2 \, . \tag{10.112}$$

Noting that

$$u_{n+1}(t) - u_n(t) \equiv u(x_{n+1}, t) - u(x_n, t) \simeq a \frac{\partial u(x_n, t)}{\partial x_n} \tag{10.113}$$

we get for the potential energy

$$\frac{k}{2} \sum_{n=1}^{N} (u_{n+1} - u_n)^2 = \frac{ka}{2} \sum_{n=1}^{N} a \left(\frac{\partial u(x_n, t)}{\partial x_n} \right)^2 \to \frac{Mn_i c^2}{2} \int_0^L dx \left(\frac{\partial u(x, t)}{\partial x} \right)^2 \tag{10.114}$$

and we have the continuum Lagrange functional

$$L[u] = \frac{n_i M}{2} \int_0^L dx \left(\left(\frac{\partial u(x, t)}{\partial t} \right)^2 - c^2 \left(\frac{\partial u(x, t)}{\partial x} \right)^2 \right) \, . \tag{10.115}$$

The continuum form of the equation of motion following from Hamilton's principle is the one-dimensional version of eq.(10.100)

$$\left(\frac{\partial^2}{\partial x^2} - \frac{1}{c^2} \frac{\partial^2}{\partial t^2} \right) u(x, t) = 0 \tag{10.116}$$

which alternatively can be obtained by taking the continuum limit of the discretized version of the equation of motion, eq.(10.103).

The quantum mechanics of the linear chain is specified by the canonical commutation relations for the position and momentum variables of the individual ions

$$[\hat{u}_n, \hat{p}_m] = i\hbar\, \delta_{n,m} \qquad [\hat{u}_n, \hat{u}_m] = 0 \qquad [\hat{p}_n, \hat{p}_m] = 0 \ . \tag{10.117}$$

Analogously to the introduction of the normal modes, we introduce the non-hermitian operators, $q \in B_1$, through the linear transformation

$$\hat{Q}_q \equiv \frac{1}{\sqrt{N}} \sum_{n=1}^{N} \hat{u}_n\, e^{-inaq} \ , \qquad \hat{P}_q \equiv \frac{1}{\sqrt{N}} \sum_{n=1}^{N} \hat{p}_n\, e^{iqna} \tag{10.118}$$

and on account of the orthogonality relations

$$\sum_{n=1}^{N} e^{i(q-q')na} = N\, \delta_{q,q'} \ , \qquad \sum_{q \in B_1} e^{iq(n-n')a} = N\, \delta_{n,n'} \tag{10.119}$$

we have

$$\hat{u}_n = \frac{1}{\sqrt{N}} \sum_{q \in B_1} \hat{Q}_q\, e^{inaq} \ , \qquad \hat{p}_n = \frac{1}{\sqrt{N}} \sum_{q \in B_1} \hat{P}_q\, e^{-iqna} \ . \tag{10.120}$$

The operators are seen to obey the commutation relations

$$[\hat{Q}_q, \hat{P}_{q'}] = i\hbar\, \delta_{q,q'} \ . \tag{10.121}$$

Just as in the classical case the linear transformation decouples the oscillators, and for the linear chain Hamiltonian corresponding to Hamilton's function eq.(10.104) we have

$$\hat{H}_{LC} = \sum_q \left(\frac{\hat{P}_q \hat{P}_{-q}}{2M} + k(1 - \cos qa)\hat{Q}_q \hat{Q}_{-q} \right) \ . \tag{10.122}$$

The linear chain Hamiltonian can be transformed into normal mode form of N-independent harmonic oscillators

$$\hat{H}_{LC} = \sum_q \hbar\omega_q \left(\hat{a}_q^\dagger \hat{a}_q + \frac{1}{2} \right) \tag{10.123}$$

by introducing normal mode operators, the annihilation operator

$$\hat{a}_q = \frac{i}{(2M\hbar\omega_q)^{1/2}} \left(\hat{P}_{-q} - iM\omega_q\hat{Q}_q \right) \tag{10.124}$$

and the creation operator

$$\hat{a}_q^\dagger = \frac{-i}{(2M\hbar\omega_q)^{1/2}} \left(\hat{P}_q + iM\omega_q\hat{Q}_{-q} \right) \tag{10.125}$$

which on account of eq.(10.121) satisfy the commutation relations

$$[\hat{a}_q, \hat{a}_{q'}^\dagger] = \delta_{q,q'} \qquad [\hat{a}_q^\dagger, \hat{a}_{q'}^\dagger] = 0 \qquad [\hat{a}_q, \hat{a}_{q'}] = 0 \ . \tag{10.126}$$

The introduced harmonic oscillators, labeled by the wave numbers, describe collective displacements of all the ions in the chain, and the quanta in these oscillations are referred to as phonons. The vibrations of the linear chain are thus created in discrete units.

In the continuum limit the quantized chain dynamics is specified by the lattice displacement field operator

$$\hat{u}(x) = \sqrt{\frac{\hbar}{2Mn_iL}} \sum_{q \in B_1} \frac{1}{\sqrt{\omega_q}} e^{iqx} (\hat{a}_q + \hat{a}_{-q}^{\dagger}) . \qquad (10.127)$$

10.6.2   Three-dimensional Case

In order to quantize the background dynamics in the three-dimensional case we could repeat the analysis of the one-dimensional case, and go back to the harmonically interacting ions, use canonical quantization, and then introduce normal mode operators. Instead we shall here quantize the normal mode operators directly. We therefore consider the Fourier expansion of a real field.[14] We impose boundary conditions on the finite system and get a countable set of Fourier coefficients:

$$\mathbf{u}(\mathbf{x}, t) = \frac{1}{V} \sum_{\mathbf{k} \neq 0} \mathbf{u}_{\mathbf{k}}(t) e^{i\mathbf{k} \cdot \mathbf{x}} \qquad (10.128)$$

and we discard the $\mathbf{k} = \mathbf{0}$ mode from the summation, since we do not wish to consider an overall translation of the solid. Assuming no shear or vorticity we have the conditions

$$\nabla \times \mathbf{u}(\mathbf{x}, t) = \mathbf{0} \qquad \mathbf{k} \times \mathbf{u}_{\mathbf{k}}(t) = \mathbf{0} \qquad \mathbf{k} \parallel \mathbf{u}_{\mathbf{k}}(t) \qquad (10.129)$$

and only longitudinal waves are sustained.[15] The equation of motion then following from eq.(10.100) is (since both the divergence and curl of $\Box\mathbf{u}(\mathbf{x}, t)$ are seen to vanish)

$$\Box\mathbf{u}(\mathbf{x}, t) = 0 , \qquad \ddot{\mathbf{u}}_{\mathbf{k}}(t) + c^2 k^2 \mathbf{u}_{\mathbf{k}}(t) = 0 \qquad (10.130)$$

specified by the D'Lambertian

$$\Box = \left(\Delta - \frac{1}{c^2} \frac{\partial^2}{\partial t^2} \right) \qquad (10.131)$$

with the general solution in terms of running waves

$$\mathbf{u}_{\mathbf{k}}(t) = \mathbf{x}_{\mathbf{k}} e^{-i\omega_{\mathbf{k}} t} + \mathbf{y}_{\mathbf{k}} e^{i\omega_{\mathbf{k}} t} , \qquad \omega_{\mathbf{k}} = c |\mathbf{k}| \equiv c k . \qquad (10.132)$$

We impose, say, periodic boundary conditions, so that

$$\mathbf{k} = \frac{2\pi}{L} (n_x, n_y, n_z) , \qquad n_{x,y,z} = \pm 1, \pm 2, ... \qquad (10.133)$$

[14]Here the displacement field, but the analysis is equivalent for the quantization of the electromagnetic field (however, in that case the field is transverse in the Coulomb gauge).

[15]In the isotropic model only longitudinal modes give rise to density changes.

where L is the linear dimension of the assumed cubic-shaped body.

Since $\mathbf{u}(\mathbf{x}, t)$ is real we have $\mathbf{u}_\mathbf{k}^*(t) = \mathbf{u}_{-\mathbf{k}}(t)$, and thereby $\mathbf{x}_\mathbf{k} = \mathbf{y}_{-\mathbf{k}}^*$ so that introducing

$$\mathbf{a}_\mathbf{k}(t) \equiv \mathbf{x}_\mathbf{k}\, e^{-i\omega_\mathbf{k} t}\,, \qquad \mathbf{a}_{-\mathbf{k}}^*(t) \equiv \mathbf{y}_\mathbf{k}\, e^{i\omega_\mathbf{k} t} \tag{10.134}$$

we have the normal mode expansion for a real field

$$\mathbf{u}(\mathbf{x}, t) = \frac{1}{V} \sum_{\mathbf{k} \neq 0} [\mathbf{a}_\mathbf{k}(t)\, e^{i\mathbf{k}\cdot\mathbf{x}} + \mathbf{a}_\mathbf{k}^*(t)\, e^{-i\mathbf{k}\cdot\mathbf{x}}] \tag{10.135}$$

or equivalently for the Fourier components

$$\mathbf{u}_\mathbf{k}(t) = \mathbf{a}_\mathbf{k}(t) + \mathbf{a}_{-\mathbf{k}}^*(t)\,. \tag{10.136}$$

Inserting the normal mode form, eq.(10.135), into the expression for the background Hamilton functional, eq.(10.102), and using

$$\int_V d\mathbf{x}\, e^{-i\mathbf{x}\cdot(\mathbf{k}-\mathbf{k}')} = V\delta_{\mathbf{k},\mathbf{k}'} \tag{10.137}$$

and identities like

$$(\mathbf{k} \cdot \mathbf{a}_\mathbf{k}(t))(\mathbf{k} \cdot \mathbf{a}_{-\mathbf{k}}(t)) = -k^2 a_\mathbf{k}(t)\, a_{-\mathbf{k}}(t)\,, \qquad \mathbf{a}_\mathbf{k}(t) \equiv a_\mathbf{k}(t)\, \hat{\mathbf{k}} \tag{10.138}$$

we obtain the normal mode form for the background Hamilton function

$$H_b = \frac{1}{2} \sum_\mathbf{k} \hbar\omega_\mathbf{k}\, [c_\mathbf{k}(t)c_\mathbf{k}^*(t) + c_\mathbf{k}^*(t)c_\mathbf{k}(t)] \tag{10.139}$$

where

$$c_\mathbf{k}(t) = \left(\frac{2M n_i\, \omega_\mathbf{k}}{\hbar V}\right)^{1/2} \frac{\mathbf{k} \cdot \mathbf{a}_\mathbf{k}(t)}{k}\,. \tag{10.140}$$

We can also introduce the real canonical variables

$$\mathbf{Q}_\mathbf{k}(t) = \sqrt{\frac{n_i}{V}}\, \Big(\mathbf{a}_\mathbf{k}(t) + \mathbf{a}_\mathbf{k}^*(t)\Big) \tag{10.141}$$

and

$$\mathbf{P}_\mathbf{k}(t) = M\dot{\mathbf{Q}}_\mathbf{k}(t) = -i\omega_\mathbf{k}\, M\, \sqrt{\frac{n_i}{V}}\, \Big(\mathbf{a}_\mathbf{k}(t) - \mathbf{a}_\mathbf{k}^*(t)\Big) \tag{10.142}$$

and we have the Hamilton function for the elastic background precisely of the form

$$H_b = \sum_\mathbf{k} \left(\frac{\mathbf{P}_\mathbf{k}^2(t)}{2M} + \frac{1}{2} M\omega_\mathbf{k}^2\, \mathbf{Q}_\mathbf{k}^2(t)\right) \tag{10.143}$$

as our model Hamiltonian of chapter 6.

We expect the classical description of the lattice dynamics to be invalid at low temperatures, where quantum effects become important. Analogously to the linear chain the quantum dynamics of the background is governed by the Hamiltonian

$$\hat{H}_b = \frac{1}{2} \sum_\mathbf{k} \hbar\omega_\mathbf{k}\, [\hat{c}_\mathbf{k}(t)\, \hat{c}_\mathbf{k}^\dagger(t) + \hat{c}_\mathbf{k}^\dagger(t)\, \hat{c}_\mathbf{k}(t)] \tag{10.144}$$

where the operators satisfy the equal-time commutation relations

$$[\hat{c}_{\mathbf{k}}, \hat{c}_{\mathbf{k'}}^{\dagger}] = \delta_{\mathbf{k},\mathbf{k'}} \qquad [\hat{c}_{\mathbf{k}}^{\dagger}, \hat{c}_{\mathbf{k'}}^{\dagger}] = 0 \qquad [\hat{c}_{\mathbf{k}}, \hat{c}_{\mathbf{k'}}] = 0 \tag{10.145}$$

or in the Schrödinger picture

$$\hat{H}_b = \frac{1}{2} \sum_{|\mathbf{k}| < k_D} \hbar \omega_{\mathbf{k}} \left(\hat{c}_{\mathbf{k}}^{\dagger} \hat{c}_{\mathbf{k}} + \frac{1}{2} \right) . \tag{10.146}$$

We have introduced the ultraviolet cutoff, k_D, since displacements with too short a wavelength, much shorter than the inverse lattice spacing, are unphysical because they do not reflect displacements of ions. The value of the cutoff is of the order of the inverse lattice spacing, and can be determined experimentally by considering the specific heat of the lattice, since in the classical, high-temperature limit the specific heat, by equipartition, is just a measure of the number of degrees of freedom, the number of ions in the lattice times the number of spatial dimensions.

In a real crystal there are in addition to the longitudinal phonons considered above two additional acoustic branches of transverse phonons (and possibly optical branches), all in general having complicated dispersion relations. In the following we shall consider the Debye model where all the branches of the phonon spectrum are replaced by three branches, and for simplicity we assume they have identical linear dispersion.

We introduce the number of oscillator modes per unit frequency per unit volume ($\omega_D \equiv c\, k_D$)

$$F(\omega) = \frac{3}{V} \sum_{\mathbf{k}} \delta(\omega - \omega_{\mathbf{k}}) \, \theta(\omega_D - \omega_{\mathbf{k}}) \tag{10.147}$$

where the factor of 3 is due to the equal contributions from the three identical branches. Noting that $3N$ is the number of ionic degrees of freedom in the three-dimensional case, equipartition demands

$$3N = V \int_0^{\infty} d\omega \, F(\omega) \tag{10.148}$$

and we have for the Debye frequency $\omega_D = (6\pi^2 n_i)^{1/3} = (2/Z)^{1/3} k_F c$, and for the density of oscillator states in the Debye model[16]

$$F(\omega) = \frac{9 n_i \, \omega^2}{\omega_D^3} \, \theta(\omega_D - \omega) . \tag{10.149}$$

The Debye energy $\hbar \omega_D \sim \epsilon_F c/v_F$ is typically 100 times smaller than the Fermi energy for a metal.

---

[16]We could of course just as well simply have counted the number of modes allowed by the cutoff and boundary condition. Since there are three modes for each $\Delta \mathbf{k}$-volume of size $(2\pi)^3/V$ this gives for the Debye wave vector $k_D^3 = 6\pi^2 n_i$.

In the Debye model we obtain for the specific heat[17]

$$
c_V = \frac{1}{V}\frac{\partial <\hat{H}_b>}{\partial T} = \frac{\partial}{\partial T}\left(\frac{1}{V}\sum_{\mathbf{k}}\theta(\omega_D - \omega_{\mathbf{k}})\,\hbar\omega_{\mathbf{k}}\left(n(\omega_{\mathbf{k}}) + \frac{1}{2}\right)\right)
$$

$$
= \frac{\partial}{\partial T}\left(\int_0^\infty d\omega\, F(\omega)\,\hbar\omega\, n(\omega)\right)
$$

$$
= 9n_i k\left(\frac{T}{\theta_D}\right)^3\int_0^{\Theta_D/T}dx\,\frac{x^4\,e^x}{(e^x - 1)^2} \tag{10.150}
$$

where we have introduced the Debye temperature, $\Theta_D \equiv \hbar\omega_D/k$.[18]

Exercise 10.3 *Verify the limiting behavior of the temperature dependence for the specific heat in the Debye model*

$$
c_V = \begin{cases} 3n_i k & T \gg \Theta_D \\[2mm] \frac{12\pi^4}{5}\,n_i k\left(\frac{T}{\Theta_D}\right)^3 & T \ll \Theta_D. \end{cases} \tag{10.151}
$$

The average number of phonons per unit volume, $N_{ph}(T)$, is proportional to the third power of the temperature at temperatures much lower than the Debye temperature (this is valid for a real crystal because at low enough temperatures only the long-wavelength acoustic oscillations, which have linear dispersion, are excited)

$$
N_{ph}(T) = \frac{1}{V}\sum_{\mathbf{k}}\theta(\omega_D - \omega_{\mathbf{k}})\,n(\omega_{\mathbf{k}}) = \int_0^\infty d\omega\, F(\omega)\,n(\omega)
$$

$$
= 9n_i\left(\frac{T}{\Theta_D}\right)^3\int_0^{\Theta_D/T}dx\,\frac{x^2}{e^x - 1}. \tag{10.152}
$$

---

[17]Incidentally, quantum mechanics was conceived in 1900 when Planck investigated the temperature behavior of black-body radiation, and found that it was not described by equipartition, $c_V = 3n_i k$, as demanded by classical physics, but could be explained by postulating the existence of the quantum of action, governing the discrete exchange of energy between matter and electromagnetic radiation. The presented analogous analysis for the specific heat of solids is due to Debye (1912).

[18]Here k denotes the Boltzmann constant and should not be confused with our notation for the length of a wave vector.

Exercise 10.4 *The energy in the crystal vibrations is defined up to a constant. Choosing the ground-state energy to be zero, we have for the average energy at temperature T*

$$U_{ph}(T) \; = \; \int_0^\infty d\omega \, F(\omega) \, \hbar\omega \, n(\omega) \; . \tag{10.153}$$

Express the vibrational energy in terms of the average number of phonons.

In the continuum limit the quantized lattice dynamics is specified by the displacement field operator

$$\hat{\mathbf{u}}(\mathbf{x}) \; = \; \sqrt{\frac{\hbar}{2Mn_iV}} \, \sum_{\mathbf{k}\neq 0} \frac{\mathbf{k}}{k\sqrt{\omega_{\mathbf{k}}}} \, \left(\hat{c}_{\mathbf{k}} \, e^{i\mathbf{k}\cdot\mathbf{x}} + \hat{c}_{\mathbf{k}}^\dagger \, e^{-i\mathbf{k}\cdot\mathbf{x}} \right) \; . \tag{10.154}$$

10.7 Electron-Phonon Interaction

A deformation of the background charge will create an effective potential felt by an electron at point $\mathbf{x}$ given by

$$\overline{V}_{e-b}(\mathbf{x}) \; = \; \int d\mathbf{x}' \, \overline{V}_{eb}(\mathbf{x} - \mathbf{x}') \, \delta n_b(\mathbf{x}') \tag{10.155}$$

where for the slow background variations the interaction will be screened according to

$$\overline{V}_{eb}(\mathbf{x}) = -\frac{Ze^2}{4\pi\epsilon_0|\mathbf{x}|} \, e^{-\kappa_s|\mathbf{x}|} \; . \tag{10.156}$$

In the mean-field description each electron is independent, the interaction energy of the electrons with the background charge fluctuations is the sum over the interaction energy of each electron, and the mean-field interaction energy of the electrons due to a deformation in the background density is

$$V_{e-b} = - \int d\mathbf{x} \, \overline{V}_{e-b}(\mathbf{x}) \, n_e(\mathbf{x}) = - \int d\mathbf{x} \! \int d\mathbf{x}' \, \delta n_b(\mathbf{x}) \, \overline{V}_{eb}(\mathbf{x} - \mathbf{x}') \, n_e(\mathbf{x}') \; . \tag{10.157}$$

The Hamiltonian for an electron interacting with the background is obtained by inserting the single-electron density operator

$$\hat{n}(\mathbf{x}) = \delta(\mathbf{x} - \hat{\mathbf{x}}_e) \tag{10.158}$$

and we have for the electron-background interaction energy operator

$$V_{e-b}(\hat{\mathbf{x}}_e) = \int d\mathbf{x} \, \overline{V}_{eb}(\hat{\mathbf{x}}_e - \mathbf{x}) \, \delta n_b(\mathbf{x}) \; . \tag{10.159}$$

For the long-wavelength oscillations of the background we can use the short-range jellium model of complete screening

$$\overline{V}_{eb}(\mathbf{x} - \mathbf{x}') = -\frac{Ze^2}{\epsilon_0\kappa_s^2} \, \delta(\mathbf{x} - \mathbf{x}') = -\frac{Z}{2N_0} \, \delta(\mathbf{x} - \mathbf{x}') \tag{10.160}$$

whereby the electron at position $\mathbf{x}$ acquires the potential energy, the deformation potential,

$$V_{e-b}(\mathbf{x}) = -\frac{Z}{2N_0}\,\delta n_b(\mathbf{x}) = \frac{n}{2N_0}\,\nabla_{\mathbf{x}} \cdot \mathbf{u}(\mathbf{x})\,. \qquad (10.161)$$

In order to understand the weak (compared to the Coulomb interaction) residual interaction between the electrons and the background charge, we note that the electrons move adiabatically along with the ions, and in doing so they move in the effective field V_{e-b}. The bulk of the electrons follows adiabatically the ions, since due to the Pauli principle they have no states to be scattered into, and simply acts to screen the Coulomb interaction between the ions and electrons. The small fraction of electrons in the thin shell around the Fermi surface of size the Debye energy, the maximal lattice vibration energy, will be able to be scattered into unoccupied states by the deformation potential. The interaction V_{e-b} provides the mechanism for the dissipative phenomena of damping of sound waves and scattering of electrons.

Quantizing the lattice dynamics we get for the electron-phonon interaction in the jellium model

$$\hat{V}_{e-ph}(\hat{\mathbf{x}}_e) \equiv \hat{V}_{e-b}(\hat{\mathbf{x}}_e) = \frac{n}{2N_0}\,\nabla_{\mathbf{x}_e} \cdot \hat{\mathbf{u}}(\hat{\mathbf{x}}_e)$$

$$= \frac{i}{2}\sqrt{\frac{\hbar}{N_0 V}}\sum_{|\mathbf{k}|\leq k_D}\sqrt{\omega_{\mathbf{k}}}\,[\hat{c}_{\mathbf{k}}\,e^{i\mathbf{k}\cdot\hat{\mathbf{x}}_e} - \hat{c}_{\mathbf{k}}^{\dagger}\,e^{-i\mathbf{k}\cdot\hat{\mathbf{x}}_e}]\,. \qquad (10.162)$$

The interaction between the lattice of ions and an electron is thus transmitted in discrete units, the quanta we called phonons.[19] Introducing the phonon field operator in the Schrödinger picture

$$\hat{\phi}(\mathbf{x}) \equiv c\sqrt{Mn_i}\,\nabla_{\mathbf{x}} \cdot \hat{\mathbf{u}}(\mathbf{x})$$

$$= i\sqrt{\frac{\hbar}{2V}}\sum_{|\mathbf{k}|\leq k_D}\sqrt{\omega_{\mathbf{k}}}\,[\hat{c}_{\mathbf{k}}\,e^{i\mathbf{k}\cdot\mathbf{x}} - \hat{c}_{\mathbf{k}}^{\dagger}\,e^{-i\mathbf{k}\cdot\mathbf{x}}] \qquad (10.163)$$

we have for the electron-phonon interaction in terms of the electron density operator

$$\hat{V}_{e-ph} = g\int d\mathbf{x}\,\hat{n}_e(\mathbf{x})\,\hat{\phi}(\mathbf{x}) \qquad (10.164)$$

where g is the electron-phonon coupling constant

$$g^2 = \frac{1}{2N_0} = \frac{4}{9}\,\frac{\epsilon_F^2}{Mn_i c^2} \qquad (10.165)$$

---

[19]Phonons refer to collective oscillations of the ions and their screening cloud of electrons, similarly as the effective electron-electron interaction describes the interaction between electrons and their screening clouds. Such objects are referred to as quasiparticles.

and the last equality is obtained by using the Bohm-Staver relation, eq.(10.99), between the sound velocity and the Fermi velocity.

If we do not invoke the Bohm-Staver relation we obtain from eq.(10.161)

$$\hat{V}_{e-ph}(\hat{\mathbf{x}}_e) \equiv \frac{n}{2N_0} \nabla_{\mathbf{x}_e} \cdot \hat{\mathbf{u}}(\hat{\mathbf{x}}_e)$$

$$= \frac{1}{\sqrt{V}} \sum_{|\mathbf{k}| \le k_D} [\alpha_\mathbf{k} \hat{c}_\mathbf{k} e^{i\mathbf{k}\cdot\hat{\mathbf{x}}_e} + \alpha_\mathbf{k}^* \hat{c}_\mathbf{k}^\dagger e^{-i\mathbf{k}\cdot\hat{\mathbf{x}}_e}] \qquad (10.166)$$

where

$$\alpha_\mathbf{k} = \frac{n}{2N_0} \frac{i}{c} \sqrt{\frac{\hbar\omega_\mathbf{k}}{2Mn_i}} . \qquad (10.167)$$

We therefore have

$$|\alpha_\mathbf{k}|^2 = \lambda \frac{\hbar\omega_\mathbf{k}}{2N_0} \qquad (10.168)$$

where we have introduced the dimensionless electron-phonon coupling constant

$$\lambda = \frac{Z}{3} \frac{\epsilon_F}{Mc^2} . \qquad (10.169)$$

We note that the dimensionless electron-phonon coupling constant equals one-half, $\lambda = 1/2 = N_0 g^2$, if the sound velocity obeys the Bohm-Staver relation.

The above expression, eq.(10.166), for the electron-phonon interaction only described the coupling to the longitudinal phonons. In a real crystal we have also coupling to transverse phonons (and possibly optical phonons), and for the interaction of an electron with the lattice we have in general (we have suppressed the cutoffs)

$$\hat{V}_{e-ph}(\hat{\mathbf{x}}_e) = \frac{1}{\sqrt{V}} \sum_{\mathbf{k},b} \hat{e}_b(\mathbf{k})(\alpha_b(\mathbf{k}) \hat{c}_\mathbf{k} e^{i\mathbf{k}\cdot\hat{\mathbf{x}}_e} + \alpha_b^*(\mathbf{k}) \hat{c}_\mathbf{k}^\dagger e^{-i\mathbf{k}\cdot\hat{\mathbf{x}}_e}) . \quad (10.170)$$

where $\hat{e}_b(\mathbf{k})$ is the polarization vector for phonon branch b, and $\alpha_b(\mathbf{k})$ the electron-phonon coupling.

We finally get the Hamiltonian describing a metal in the isotropic model

$$\hat{H} = \hat{H}_{el-gas} + \hat{H}_b + \hat{V}_{e-ph} \qquad (10.171)$$

where

$$\hat{H}_{el-gas} = \sum_{i=1}^{N} \frac{\hat{\mathbf{p}}_i^2}{2m} + \frac{1}{2} \sum_{i \ne j} \overline{V}(\hat{\mathbf{x}}_i - \hat{\mathbf{x}}_j) \qquad (10.172)$$

is the effective Hamiltonian for the electron gas, and the background Hamiltonian $\hat{H}_b$ and the electron-phonon interaction $\hat{V}_{e-b}$ has previously been specified. If external fields are present these must be added.

10.8 Boltzmann Theory

In this section we shall consider the kinetic equation for the electrons appropriate for the interactions present in a metal. The main interest will be on calculating the inelastic scattering rates of electrons.

10.8.1 Particle-Phonon Interaction

In chapter 6 we derived the Boltzmann equation for a single particle interacting with oscillators

$$\frac{\partial f}{\partial t} + \mathbf{F} \cdot \frac{\partial f}{\partial \mathbf{p}} + \mathbf{v_p} \cdot \frac{\partial f}{\partial \mathbf{x}} = I_{pt}[f] .$$ (10.173)

In order to use the results of chapter 6 for the electron-phonon interaction we simply substitute for the coupling constant α_q

$$|\alpha_q|^2 \to |\alpha_\mathbf{k}|^2 = \lambda \frac{\hbar \omega_\mathbf{k}}{2N_0}$$ (10.174)

where λ is the dimensionless electron-phonon coupling constant, and we obtain the collision integral

$$I_{\mathbf{p},t}[f] = -\int \frac{d\mathbf{p}'}{(2\pi\hbar)^3} \left\{ W^+(\mathbf{p}', \mathbf{p}) \left[f(\mathbf{p}, t)(1 + n_{\mathbf{p}-\mathbf{p}'}) - f(\mathbf{p}', t) n_{\mathbf{p}-\mathbf{p}'} \right] \right.$$

$$\left. + W^-(\mathbf{p}', \mathbf{p}) \left[f(\mathbf{p}, t) n_{\mathbf{p}'-\mathbf{p}} - f(\mathbf{p}', t)(1 + n_{\mathbf{p}'-\mathbf{p}}) \right] \right\} .$$ (10.175)

where for the case of electron-phonon interaction we have for the transition probability

$$W^\pm(\mathbf{p}', \mathbf{p}) = \frac{2\pi}{\hbar} |\alpha_{\mathbf{p}'-\mathbf{p}}|^2 \delta(\epsilon_{\mathbf{p}'} - \epsilon_\mathbf{p} \pm \hbar\omega_{\pm(\mathbf{p}-\mathbf{p}')}) .$$ (10.176)

The above kinetic equation is appropriate for an electron in a nondegenerate semiconductor.

The equilibrium distribution, $I[f_0] = 0$, is readily found to be the Maxwell-Boltzmann distribution

$$f_0(\epsilon_\mathbf{p}) = f_{MB}(\epsilon_\mathbf{p}) = \frac{1}{V} \left(\frac{2\pi\hbar^2}{mkT} \right)^{3/2} e^{-\epsilon_\mathbf{p}/kT} .$$ (10.177)

Let us study the simplest nonequilibrium situation, where the distribution only is out of equilibrium for a single momentum value

$$f_{\mathbf{p}'}(t) = f_{MB}(\epsilon_{\mathbf{p}'}) + \delta f_\mathbf{p}(t) \delta_{\mathbf{p},\mathbf{p}'}$$ (10.178)

and we assume no external fields. The Boltzmann equation then reduces to

$$\frac{\partial \delta f_\mathbf{p}(t)}{\partial t} = -\frac{\delta f_\mathbf{p}}{\tau_\mathbf{p}(T)}$$ (10.179)

whose solution describes the exponential relaxation to equilibrium

$$f_{\mathbf{p}}(t) = f_{MB}(\epsilon_{\mathbf{p}}) + \delta f_{\mathbf{p}}(t=0)\, e^{-t/\tau_{\mathbf{p}}(T)} \tag{10.180}$$

with the relaxation time given by

$$\frac{1}{\tau_{\mathbf{p}}(T)} = \int \frac{d\mathbf{p}'}{(2\pi\hbar)^3} \left(W^+(\mathbf{p}, \mathbf{p}')(1 + n(\mathbf{p} - \mathbf{p}')) + W^-(\mathbf{p}, \mathbf{p}')\, n(\mathbf{p}' - \mathbf{p}) \right)$$

$$= \frac{2\pi}{\hbar} \int \frac{d\mathbf{p}'}{(2\pi\hbar)^3} \, |\alpha_{\mathbf{p}-\mathbf{p}'}|^2 \left[\delta(\epsilon_{\mathbf{p}} - \epsilon_{\mathbf{p}'} - \hbar\omega_{\mathbf{p}-\mathbf{p}'})\, (n(\mathbf{p} - \mathbf{p}') + 1) \right.$$

$$+ \left. \delta(\epsilon_{\mathbf{p}'} - \epsilon_{\mathbf{p}} + \hbar\omega_{\mathbf{p}'-\mathbf{p}})\, n(\mathbf{p}' - \mathbf{p}) \right]$$

$$= \frac{2\pi}{\hbar} \int \frac{d\mathbf{p}'}{(2\pi\hbar)^3} \, |\alpha_{\mathbf{p}-\mathbf{p}'}|^2 \, \delta(\epsilon_{\mathbf{p}} - \epsilon_{\mathbf{p}'} - \hbar\omega_{\mathbf{p}-\mathbf{p}'})\, (2n(\omega_{\mathbf{p}-\mathbf{p}'}) + 1) \tag{10.181}$$

where in the last line we have used the fact that $n(\mathbf{p}) = n(\omega_{\mathbf{p}})$ is the equilibrium phonon distribution, and that the phonon dispersion satisfies $\omega_{-\mathbf{p}} = \omega_{\mathbf{p}}$ due to time-reversal symmetry. At zero temperature, the particle lowers its energy due to spontaneous emission until it reaches the ground state, $\mathbf{p} = \mathbf{0}$, for which the lifetime is infinite, $\tau_{\mathbf{p}=\mathbf{0}}(T=0) = \infty$.

We note that according to eq.(6.223), the relaxation time is given in terms of the imaginary part of the equilibrium self-energy

$$\frac{\hbar}{\tau_{\mathbf{p}}(T)} = i\left(\Sigma^R(\epsilon_{\mathbf{p}}, \mathbf{p}) - \Sigma^A(\epsilon_{\mathbf{p}}, \mathbf{p}) \right) = -2\,\Im m\, \Sigma^R(\epsilon_{\mathbf{p}}, \mathbf{p}). \tag{10.182}$$

10.8.2   Degenerate Fermi System

In a metal the conduction electrons constitute a degenerate Fermi gas, and we must include the quantum statistical effects of the presence of the other electrons. Since the electrons are fermions, they obey Pauli's exclusion principle, which only allows scattering into unoccupied states. The collision integral which ensures that the equilibrium distribution is the Fermi function is readily obtained by ascribing the probability $(1 - f)$ for the probability for the outgoing state to be vacant

$$I_{\mathbf{p}}^{(1)}[f] = -\int \frac{d\mathbf{p}'}{(2\pi\hbar)^3} \left\{ W^+(\mathbf{p}', \mathbf{p})[f_{\mathbf{p}}(1 - f_{\mathbf{p}'})(1 + n_{\mathbf{p}-\mathbf{p}'}) - f_{\mathbf{p}'}(1 - f_{\mathbf{p}})n_{\mathbf{p}-\mathbf{p}'}] \right.$$

$$+ \left. W^-(\mathbf{p}', \mathbf{p})[f_{\mathbf{p}}(1 - f_{\mathbf{p}'})n_{\mathbf{p}'-\mathbf{p}} - f_{\mathbf{p}'}(1 - f_{\mathbf{p}})(1 + n_{\mathbf{p}'-\mathbf{p}})] \right\}. \tag{10.183}$$

Since all the electron distribution functions depend on the same time, the one dictated by the left-hand side of the kinetic equation, we have suppressed this dependence. We can therefore memorize the four scattering terms by the same Boltzmann diagrams as in eq.(6.171), and the same Boltzmann dictionary except for the change in interpretation

$$\mathbf{p} \equiv 1 - f_{\mathbf{p}} \, . \tag{10.184}$$

10.8.3 Electron-Phonon Relaxation Time

Assuming the nonequilibrium situation

$$f_{\mathbf{p}'} = f_0(\epsilon_{\mathbf{p}'}) + \delta f_{\mathbf{p}} \, \delta_{\mathbf{p},\mathbf{p}'} \tag{10.185}$$

we again get the relaxation-time equation eq.(10.179), however, with the electronic relaxation time due to electron-phonon interaction given by

$$\frac{1}{\tau_{\mathbf{p}}(T)} = \frac{2\pi}{\hbar} \int \frac{d\mathbf{p}'}{(2\pi\hbar)^3} \, |\alpha_{\mathbf{p}-\mathbf{p}'}|^2 \, [\delta(\epsilon_{\mathbf{p}} - \epsilon_{\mathbf{p}'} - \hbar\omega_{\mathbf{p}-\mathbf{p}'})\{1 + n_{\mathbf{p}-\mathbf{p}'} - f_0(\epsilon_{\mathbf{p}'})\}$$

$$+ \ \delta(\epsilon_{\mathbf{p}} - \epsilon_{\mathbf{p}'} + \hbar\omega_{\mathbf{p}'-\mathbf{p}})\{n_{\mathbf{p}'-\mathbf{p}} + f_0(\epsilon_{\mathbf{p}'})\}] \tag{10.186}$$

where we have used the identities obeyed by the Bose and Fermi functions

$$f_0(x + y)[1 - f_0(x)] = [f_0(x) - f_0(x + y)]n(y) \tag{10.187}$$

and

$$f_0(-x) = 1 - f_0(x) \quad , \quad n(-y) = 1 + n(y) \quad . \tag{10.188}$$

To perform the momentum integration we introduce the momentum transfer $\mathbf{q} = \mathbf{p}' - \mathbf{p}$, and the angle $\cos\theta \equiv \hat{\mathbf{p}} \cdot \hat{\mathbf{p}}'$. Of interest are electrons near the Fermi surface $|\mathbf{p}| \simeq p_F$, and as $\hbar\omega_D \ll \epsilon_F$ the energy conservation restricts $\mathbf{p}'$ to be close to the Fermi surface. Since the momenta are pinned to the Fermi surface we have $d(\cos\theta) = -qdq/p_F^2$, and we can convert the angular integration into an integration over the magnitude of the momentum transfer

$$\int \frac{d\mathbf{p}'}{(2\pi\hbar)^3} \quad \rightarrow \quad \frac{N_0}{2p_F^2} \int_0^\infty d\epsilon_{\mathbf{p}'} \int_0^{q_m} dq \, q \tag{10.189}$$

where $q_m \equiv \min\{\hbar k_D, 2p_F\}$. Introducing the dimensionless coupling-weighted phonon density of states,

$$\alpha^2 F(\omega) = \frac{N_0}{2\hbar p_F^2} \int_0^{q_m} dq \, q \, |\alpha_{\mathbf{q}}|^2 \, \delta(\omega - \omega_{\mathbf{q}}) \tag{10.190}$$

we obtain, $|\epsilon_{\mathbf{p}} - \epsilon_F| \ll \epsilon_F$,

$$\frac{1}{\tau(\epsilon_{\mathbf{p}}, T)} = 2\pi \int_0^\infty d\omega \, \alpha^2 F(\omega)[1 + 2n(\omega) - f_0(\epsilon_{\mathbf{p}} - \hbar\omega) + f_0(\epsilon_{\mathbf{p}} + \hbar\omega)] \, . \tag{10.191}$$

In the Debye model (assuming $k_D > 2k_F$) we have

$$\alpha^2 F(\omega) = \lambda \frac{(\hbar\omega)^2}{(2p_F c)^2} \theta(2k_F c - \omega).$$ (10.192)

Exercise 10.5 *Show that the Fermi surface average of the weighted phonon density of states $\alpha^2 F$ can be specified in terms of the phonon spectral function*

$$\alpha^2 F(\omega) = \frac{\lambda}{\pi\hbar} \int d\hat{\mathbf{k}}' \, \Im m \, D^R(\mathbf{k} - \mathbf{k}', \omega)$$ (10.193)

where $D^R(\mathbf{k} - \mathbf{k}', \omega)$ is the Fourier transform of the retarded free phonon propagator

$$D^R(\mathbf{x}, t; \mathbf{x}', t') = -i\theta(t - t') < [\hat{\phi}(\mathbf{x}, t), \hat{\phi}(\mathbf{x}', t')] >$$ (10.194)

specified in terms of the phonon field operator in the Heisenberg picture

$$\hat{\phi}(\mathbf{x}, t) \equiv c\sqrt{Mn_i} \, \nabla_{\mathbf{x}} \cdot \hat{\mathbf{u}}(\mathbf{x}, t)$$

$$= \sqrt{\frac{\hbar}{2V}} \sum_{|\mathbf{k}| \le k_D} \sqrt{\omega_{\mathbf{k}}} \, [\hat{c}_{\mathbf{k}}(t) e^{i\mathbf{k}\cdot\mathbf{x}} - \hat{c}_{\mathbf{k}}^\dagger(t) e^{-i\mathbf{k}\cdot\mathbf{x}}]$$ (10.195)

and the average is with respect to the thermal equilibrium state of the phonons.

For an electron at the Fermi surface, $\epsilon_{\mathbf{p}} = \epsilon_F$, we note the identity

$$\frac{2}{\sinh \frac{\hbar\omega}{kT}} = [1 + 2n(\omega) - f^0(\epsilon_F - \hbar\omega) + f^0(\epsilon_F + \hbar\omega)]$$ (10.196)

and get for the relaxation time of an electron at the Fermi surface ($k_D > 2k_F$)

$$\frac{1}{\tau_{e-ph}(T)} \equiv \frac{1}{\tau_{e-ph}(\epsilon_F, T)} = \pi\lambda \frac{(kT)^3}{\hbar(p_F c)^2} \int_0^{x_m} dx \, \frac{x^2}{\sinh x}$$ (10.197)

where $x_m \equiv 2p_F c/kT$. At temperatures low compared to the Debye temperature, the integral can be expressed in terms of Riemann's zeta function, $\zeta(3) \simeq 1.2$, and at high temperatures the integral is easily evaluated:

$$\frac{1}{\tau_{e-ph}(T)} \equiv \frac{1}{\tau(\epsilon_F, T)} = \begin{cases} \frac{7\pi\zeta(3)}{2} \lambda \frac{(kT)^3}{\hbar(p_F c)^2} & kT \ll 2p_F c \\[2ex] 2\pi\lambda \frac{kT}{\hbar} & kT \gg 2p_F c. \end{cases}$$ (10.198)

If we use the kinetic formula for the resistivity we get for high temperatures a linear temperature dependence due to electron-phonon scattering

$$\rho = \frac{m}{ne^2 \tau_{e-ph}(T)} = 2\pi\lambda\, kT\, \frac{m}{ne^2\hbar}. \tag{10.199}$$

The linear temperature dependence is simply due to the fact that at high temperatures the average of the squared displacement of an ion is proportional to the temperature (equipartition). The size of the scattering object is therefore proportional to the temperature and so is the scattering rate.

Using the same formula at low temperatures leads to an incorrect result since, as we know from the impurity case, small angle scattering is ineffective in degrading the current. At temperatures much lower than the Debye temperature, $T \ll T_D$, the scattering is essentially elastic, and we can use the result obtained for elastic scattering that the efficiency of small angle scattering in degrading the current is specified by the factor $(1 - \cos\theta)$, where θ is the angle between the initial and scattered momentum of the electron. Since the typical phonon energy is given by the temperature, $\hbar\omega_{\mathbf{q}} \sim kT$, and we have for the momentum transfer $q = 2k_F \sin\frac{\theta}{2}$, we obtain

$$1 - \cos\theta = 2\sin^2\frac{\theta}{2} = \frac{q^2}{2k_F^2} \sim T^2 \frac{k^2}{\hbar^2 c^2} \tag{10.200}$$

and the correct low-temperature power law dependence of the resistivity is T^5, the Bloch-Grüneisen behavior.

The electron-phonon relaxation rate τ_{e-ph}^{-1} describes the decay of an electron in an energy (and momentum) state due to electron-phonon interaction. In the realm of classical kinetics it is not a quantity that in general can easily be extracted by measurement, as all energy transfers are weighted equally.[20] However, we shall in section 11.3.1 show that the electron-phonon relaxation rate to an excellent degree of accuracy is identical to the phase-breaking rate determining the temperature dependence of magnetoresistance in the weak-localization regime. The phase breaking rate is a physical quantity characteristic of quantum kinetics, and as discussed in chapter 11 describes the phase coherence of the electron propagation, i.e., its wave function.

At zero temperature the relaxation rate is easily calculated because the Bose function vanishes and the Fermi function has a step function behavior, $f^0(\epsilon_{\mathbf{p}} \pm \hbar\omega) = \theta(\epsilon_{\mathbf{p}} \mp \hbar\omega - \epsilon_F)$,

$$\frac{1}{\tau_{e-ph}(\epsilon_{\mathbf{p}}, T = 0)} = 2\pi \int_0^\infty d\omega\, \alpha^2 F(\omega) \left[\theta(\epsilon_{\mathbf{p}} - \epsilon_F - \hbar\omega) + \theta(\epsilon_F - \epsilon_{\mathbf{p}} - \hbar\omega) \right]$$

$$= \begin{cases} \dfrac{\pi}{6}\lambda \dfrac{|\epsilon_{\mathbf{p}} - \epsilon_F|^3}{\hbar(p_F c)^2} & |\epsilon_{\mathbf{p}} - \epsilon_F| \leq 2p_F c \\[3mm] \dfrac{4\pi\lambda}{3\hbar} p_F c & |\epsilon_{\mathbf{p}} - \epsilon_F| \geq 2p_F c. \end{cases} \tag{10.201}$$

---

[20]Of course in certain parameter regimes it is possible, as the preceding example of high temperatures of the resistivity demonstrates.

At zero temperature the lifetime of an electron on the Fermi surface is infinite because no scattering can take place according to Pauli's exclusion principle since all states with lower energy are occupied, and excitation of an electron by the environment can happen only at finite temperatures.

We have assumed that the lattice is in thermal equilibrium; i.e., no heating of the ions occurs. Such a situation can be realized if the sample is in good thermal contact with its surroundings.

10.8.4   Electron-Electron Scattering

We are now quite confident in writing down kinetic equations, and whenever we have interaction between objects we can write down the equation governing their kinetics. Explicit time dependence, inertia, and possible external fields immediately gives the left-hand side of the kinetic equation, and once we know the scattering mechanism we can write down the collision integral describing stochastically the transitions between states. As an example we consider fermions with two-particle interaction, for which we have the collision integral[21]

$$
I_{\mathbf{p}}^{(1)}[f] = -\int_{\mathbf{p}_1 \mathbf{p}_2 \mathbf{p}_2'} W(\mathbf{p}_2, \mathbf{p}_1; \mathbf{p}_2', \mathbf{p})[f(\mathbf{p})f(\mathbf{p}_2')(1 - f(\mathbf{p}_1))(1 - f(\mathbf{p}_2))]
$$

$$
+ \int_{\mathbf{p}_1' \mathbf{p}_2 \mathbf{p}_2'} W(\mathbf{p}_2, \mathbf{p}; \mathbf{p}_2', \mathbf{p}_1')[f(\mathbf{p}_1')f(\mathbf{p}_2')(1 - f(\mathbf{p}))(1 - f(\mathbf{p}_2))] \quad (10.202)
$$

where we have introduced the notation

$$
\int_{\mathbf{p}_1' \mathbf{p}_2 \mathbf{p}_2'} \equiv \int \frac{d\mathbf{p}_1'}{(2\pi\hbar)^3} \int \frac{d\mathbf{p}_2}{(2\pi\hbar)^3} \int \frac{\mathbf{p}_2'}{(2\pi\hbar)^3} \quad . \qquad (10.203)
$$

The collision integral is easily memorized by drawing the corresponding gain and loss Boltzmann diagrams

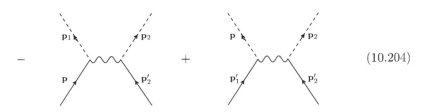

$$(10.204)$$

In general for electron-electron collisions in a crystal we only have the constraint that the total energy, spin, and crystal momentum (modulo a reciprocal lattice vector) is conserved in the scattering process, and that the transition probability is antisymmetric with respect to interchange of the fermions. Let us in the following

---

[21] Here we do not take into account the possible effect of spin on the scattering properties.

neglect the effect of the periodic potential of the crystal, and consider two-particle collisions where the interaction is simply potential scattering, and the potential depends only on the relative distance between the particles. We then have for the transition probability

$$W(\mathbf{p}_2, \mathbf{p}_1; \mathbf{p}'_2, \mathbf{p}'_1) = W(\mathbf{p}_1 - \mathbf{p}'_1)\, \delta(\mathbf{p}_1 + \mathbf{p}_2 - (\mathbf{p}'_1 + \mathbf{p}'_2))\, \delta(\epsilon_{\mathbf{p}_1} + \epsilon_{\mathbf{p}_2} - (\epsilon_{\mathbf{p}'_1} + \epsilon_{\mathbf{p}'_2})).$$
(10.205)

At low temperatures the lifetime of a particle in a state $\mathbf{p}'_1$ above the Fermi surface is determined by the process where the electron scatters an electron across the Fermi surface from state $\mathbf{p}'_2$ to state $\mathbf{p}_2$. In the final state we thus have two electrons in states $(\mathbf{p}_1, \mathbf{p}_2)$ above the Fermi surface and a hole below. The decay rate of the electron due to this process can be estimated from the phase space available for the scattering process according to Fermi's golden rule

$$\frac{1}{\tau_{e-e}(\mathbf{p}'_1)} \propto W \int d\mathbf{p}_1 d\mathbf{p}'_2\, \delta(\epsilon_{\mathbf{p}_1} + \epsilon_{\mathbf{p}'_1 + \mathbf{p}'_2 - \mathbf{p}_1} - \epsilon_{\mathbf{p}'_1} - \epsilon_{\mathbf{p}'_2})$$
(10.206)

where momentum conservation is used to eliminate one of the momentum integrations, and we have assumed no strong momentum dependence of the interaction in accordance with our knowledge of the effective electron-electron interaction (since its scale is the Fermi energy, and the phase space restriction sets a scale which due to the exclusion principle is the temperature). The angle between $\mathbf{p}_1$ and $\mathbf{p}'_1 + \mathbf{p}'_2$ is specified by energy conservation, and performing the integration over this angle is immediately done due to the presence of the delta function; we obtain

$$\frac{1}{\tau_{e-e}(\mathbf{p}'_1)} \propto W \int dp_1 dp'_2$$
(10.207)

and we just need to assess the integration limits. The upper limit on p_2 is p_F since the state is below the Fermi surface. If we assume that the initial state is close to the Fermi surface $p'_1 - p_F \ll p_F$, then for the envisaged decay process the other momentum states are also close to the Fermi surface due to the exclusion principle. We thus have the typical configuration of the four momenta subject to the momentum conservation constraint as depicted in figure 10.2. Projecting the vectors onto the conserved momentum in the scattering process, $\mathbf{p}'_1 + \mathbf{p}'_2 = \mathbf{p}_1 + \mathbf{p}_2$, gives for the magnitudes the relationship $p_1 \simeq p'_1 + p'_2 - p_2$. Since $p_2 > p_F$, we have the relationships $0 < p_1 - p_F < (p'_1 - p_F) + (p'_2 - p_F)$ and $p_F - p'_1 < p'_2 - p_F < 0$. For the phase space integral we therefore get

$$\frac{1}{\tau_{e-e}(\mathbf{p}'_1)} \propto W \int dp_1 dp'_2 \propto (p'_1 - p_F)^2.$$
(10.208)

Close to the Fermi surface we have for the energy of a state measured from the Fermi energy $\xi_{\mathbf{p}} = v_F(p - p_F)$, and from dimensional analysis we therefore get the following result for the relaxation time due to electron-electron interaction:

$$\frac{1}{\tau_{e-e}(\xi)} \propto \frac{\xi^2}{\hbar \epsilon_F}$$
(10.209)

the proportionality factor expected to be of order 1.

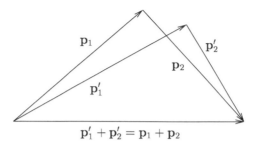

Figure 10.2 Momentum conservation constraint.

At finite temperature we have for the considered decay process:

$$\frac{1}{\tau_{e-e}(\mathbf{p}'_1)} \propto \int d\mathbf{p}_1 d\mathbf{p}'_2 \, \delta(\xi_{\mathbf{p}_1} + \xi_{\mathbf{p}'_1+\mathbf{p}'_2-\mathbf{p}_1} - (\xi_{\mathbf{p}'_1} + \xi_{\mathbf{p}'_2})) f_0(\xi_{\mathbf{p}'_2})\,(1 - f_0(\xi_{\mathbf{p}_1}))(1 - f_0(\xi_{\mathbf{p}_2}))$$

$$(10.210)$$

and for states near the Fermi surface, $|\xi_{\mathbf{p}}| \ll \epsilon_F$, we obtain a quadratic temperature dependence of the relaxation rate[22]

$$\frac{1}{\tau_{e-e}(T)} \propto \frac{(kT)^2}{\hbar \epsilon_F}\,. \qquad (10.211)$$

The characteristic quadratic temperature dependence of the electron-electron scattering rate is thus simply a consequence of energy and momentum conservation in the two-particle collision and the phase space restrictions due to the exclusion principle.[23]

In order to quantify the above results we repeat the analysis of the preceding section, but now for the electron-electron interaction, and we obtain for the energy relaxation due to electron-electron interaction ($\mathbf{p}_2 = \mathbf{p} + \mathbf{p}'_2 - \mathbf{p}_1$)

$$\frac{1}{\tau_{e-e}(\mathbf{p})} = \int_{\mathbf{p}_1 \mathbf{p}'_2} W(\mathbf{p}_1 - \mathbf{p}'_1) \, \delta(\xi_{\mathbf{p}_1} + \xi_{\mathbf{p}+\mathbf{p}'_2-\mathbf{p}_1} - \xi_{\mathbf{p}} - \xi_{\mathbf{p}'_2})$$

$$\left(f_{\mathbf{p}'_2}\,(1 - f_{\mathbf{p}_1})(1 - f_{\mathbf{p}_2}) \; + \; f_{\mathbf{p}_2}\,(1 - f_{\mathbf{p}'_2})\,f_{\mathbf{p}} \right) \qquad (10.212)$$

where we have introduced the notation for the Fermi function

$$f_{\mathbf{p}} = f_0(\xi_{\mathbf{p}})\,. \qquad (10.213)$$

---

[22]The long lifetime of excitations near the Fermi surface due to the exclusion principle is the basis of Landau's phenomenological Fermi liquid theory of strongly interacting fermions. In this theory the excited states at low temperatures of the Fermi liquid are described in terms of single-particle states labeled by the quantum numbers momentum, spin, charge.

[23]The corresponding T^2 contribution to the resistance is negligible in metals, and consideration of two-particle scattering is mainly of interest in quantum liquids such as ^{3}He.

The transition probability is specified by the screened Coulomb interaction

$$W(\mathbf{p}) = \left| \frac{V(\mathbf{p})}{\epsilon(\mathbf{p})} \right|^2 = \left| \frac{\frac{e^2}{\epsilon_0}}{\hbar^{-2}\mathbf{p}^2 + \kappa_s^2} \right|^2. \tag{10.214}$$

Since the temperature is much lower than the degeneracy temperature, the magnitude of the four momenta are as previously noted approximately equal, and at most a distance $|\epsilon_{\mathbf{p}} - \epsilon_F|$ from the Fermi surface. Introduce the angle between $\mathbf{p}_1'$ and $\mathbf{p}_2'$, and the angle between the plane spanned by $\mathbf{p}_1'$ and $\mathbf{p}_2'$, and the plane spanned by $\mathbf{p}_1$ and $\mathbf{p}_2$, and polar angles with respect to the polar axis along the direction of $\mathbf{p}_1 + \mathbf{p}_2$. The integrations can then be performed, and for the electron-electron interaction collision rate we obtain for an electron at the Fermi surface the temperature dependence

$$\frac{1}{\tau_{e-e}(T)} = \begin{cases} \frac{\pi^2 e^2}{32\epsilon_0 v_F^2 \kappa_s \hbar^3} (kT)^2 & \kappa_s \ll k_F \\[2ex] \frac{\pi^3}{16} \frac{(kT)^2}{\hbar \epsilon_F} & \kappa_s \gg k_F . \end{cases} \tag{10.215}$$

For an electron in energy state ξ we get at zero temperature the relaxation rate

$$\frac{1}{\tau_{e-e}(\xi)} = \begin{cases} \frac{\pi^2}{64} \frac{\kappa_s}{k_F} \frac{\xi^2}{\hbar \epsilon_F} & \kappa_s \ll k_F \\[2ex] \frac{\pi}{16} \frac{\xi^2}{\hbar \epsilon_F} & \kappa_s \gg k_F . \end{cases} \tag{10.216}$$

For the case where the screening wave vector is much larger than the Fermi wave vector, $\kappa_s \gg k_F$, the interaction is pointlike and the collision rate is independent of the electronic charge.

10.8.5 Electron-Electron Interaction in a Dirty Metal

In a dirty metal the effective electron-electron interaction changes character, because instead of moving ballistically the impurity scattering leads to diffusive motion of the electrons.

Let us assume that the screening length is much smaller than the Fermi wave vector; i.e., the range of the screened Coulomb potential is much larger than the spacing between the electrons. The exchange correction to the energy ϵ_λ due to electron-electron interaction is then much larger than the direct or Hartree term, and we have for the exchange self-energy

$$\Sigma_\lambda^{ex} = -\sum_{\lambda' occ.} \int d\mathbf{x} \int d\mathbf{x}' \, \overline{V}(\mathbf{x} - \mathbf{x}') \, \psi_\lambda^*(\mathbf{x}) \, \psi_{\lambda'}^*(\mathbf{x}') \, \psi_\lambda(\mathbf{x}') \, \psi_{\lambda'}(\mathbf{x}) \tag{10.217}$$

where the summation is over all occupied states λ', i.e., all the states below the Fermi level since we assume zero temperature. We are interested in the mean energy shift averaged over all states with energy ξ (measured from the Fermi energy)

$$\Sigma^{ex}(\xi) = \frac{1}{N_0 V} \sum_\lambda < \delta(\xi - \xi_\lambda) \Sigma_\lambda^{ex}) > \tag{10.218}$$

for which we obtain the expression, say $\xi > 0$,

$$\Sigma^{ex}(\xi) = -\frac{1}{N_0 V} \int\limits_{-\infty}^{0} d\xi' \int dx \int dx'\, \overline{V}(\mathbf{x} - \mathbf{x}')$$

$$< \sum_{\lambda,\lambda'}{}' \delta(\xi - \xi_\lambda)\, \delta(\xi' - \xi_{\lambda'})\, \psi_\lambda^*(\mathbf{x})\, \psi_{\lambda'}^*(\mathbf{x}')\, \psi_\lambda(\mathbf{x}')\, \psi_{\lambda'}(\mathbf{x}) > \qquad (10.219)$$

where the prime on the summation indicates it is only over states λ' occupied and states λ unoccupied. The impurity-averaged quantity is the spectral correlation function we encountered earlier, eq.(9.75), except for the restrictions on the summations. However, these are irrelevant as the main contribution comes from $\xi' \simeq \xi$, and we obtain, according to eq.(9.80), in the weak-disorder limit, $1/k_F l \ll 1$,

$$\Sigma^{ex}(\xi) = -\frac{1}{2\pi} \int\limits_{\xi/\hbar}^{\infty} d\omega \int \frac{d\mathbf{q}}{(2\pi)^d}\, \overline{V}(\mathbf{q})\, \frac{D_0\, q^2}{\omega^2 + (D_0\, q^2)^2}\,. \qquad (10.220)$$

In the above model of a static interaction the average change in energy is purely real. The result obtained can be used to calculate the change in density of states. To lowest order in the electron-electron interaction we have for the change in density of states due to the electron-electron interaction

$$\delta N(\xi) \equiv \; <N(\xi)> -N_0(\xi) = -N_0(\xi)\, \frac{\partial \Sigma^{ex}(\xi)}{\partial \xi}$$

$$= \frac{N_0}{2\pi\hbar} \int \frac{d\mathbf{q}}{(2\pi)^d}\, \overline{V}(\mathbf{q})\, \frac{D_0\, q^2}{\left(\frac{\xi}{\hbar}\right)^2 + (D_0\, q^2)^2} \qquad (10.221)$$

as according to eq.(3.74) on page 152 the change in the density of states due to disorder is negligible in the weak-disorder limit.

Exercise 10.6 *Verify that if $\overline{V}$ is a short-range potential, the change in the density of states near the Fermi surface due to electron-electron interaction is in the weak-disorder limit*

$$\frac{\delta N_3(\xi)}{N_3(0)} = \frac{\overline{V}(\mathbf{q} = 0)}{4\sqrt{2}\pi^2}\, \frac{\sqrt{|\xi|}}{(\hbar D_0)^{3/2}} \qquad (10.222)$$

in three dimensions, and in two dimensions

$$\frac{\delta N_2(\xi)}{N_2(0)} = \frac{\overline{V}(\mathbf{q} = 0)}{(2\pi)^2 \hbar D_0}\, \ln \frac{|\xi|\tau}{\hbar}\,. \qquad (10.223)$$

The singularity in the density of states is due to the spatial correlation of the exact impurity wave functions of almost equal energy, as described by the singular behavior of the spectral correlation function. The singularity in the density of states gives rise to the zero-bias anomaly, a dip in the conductivity of a tunnel junction at low voltages [52].

Quite generally the propagator in the energy representation satisfies in the presence of disorder and electron-electron interaction the equation

$$G^R_{\lambda\lambda'}(E) \;=\; G^{(0)R}_{\lambda\lambda'}(E) \;+\; \sum_{\lambda_1\lambda'_1} G^{(0)R}_{\lambda\lambda_1}(E)\, \Sigma^R_{\lambda_1\lambda'_1}(E)\, G^R_{\lambda'_1\lambda'}(E) \qquad (10.224)$$

where the propagator in the absence of electron-electron interaction $G^{(0)R}_{\lambda\lambda'}(E) = G^{(0)R}_\lambda(E)\,\delta_{\lambda\lambda'}$ is diagonal, and specified in terms of the exact impurity eigenstates (here in the momentum representation)

$$G^{R(A)}_0(\mathbf{p},\mathbf{p}',E) \;=\; \sum_\lambda \frac{\psi_\lambda(\mathbf{p})\,\psi^*_\lambda(\mathbf{p}')}{E - \epsilon_\lambda\,(\overset{+}{\underset{-}{}})\,i0} \;\equiv\; \sum_\lambda \psi_\lambda(\mathbf{p})\,\psi^*_\lambda(\mathbf{p}')\,G^{(0)R(A)}_\lambda(E)\,.$$

$$(10.225)$$

Since energy eigenstates are only spatially correlated if they have the same energy, as observed in section 9.3.3, only the diagonal terms $\Sigma^R_\lambda(E) \equiv \Sigma^R_{\lambda\lambda}(E)$ contribute in eq.(10.224), and we obtain that the propagator is approximately diagonal and specified by

$$G^R_\lambda(E) \;=\; \frac{1}{E - \epsilon_\lambda - \Sigma^R_\lambda(E)}\,. \qquad (10.226)$$

The imaginary part of the self-energy describes the decay of an exact impurity eigenstate due to electron-electron interaction. When calculating the inelastic decay rate, we should only count processes starting with the same energy, and on the average in the random potential we are therefore interested in the quantity

$$\Sigma^R_{E'}(E) \;=\; \frac{1}{N_0 V} \sum_\lambda \;<\delta(E'-\xi_\lambda)\,\Sigma^R_\lambda(E)>\,. \qquad (10.227)$$

To lowest order in the electron-electron interaction we can set E equal to E' because their difference is the real part of the self-energy, and we get for the inelastic collision rate

$$\frac{1}{\tau_{e-e}(E,T)} \;=\; -2\,\Im m\,\Sigma^R_E(E) \;=\; i\left(\Sigma^R_E(E) - \Sigma^A_E(E)\right)$$

$$=\; -\frac{1}{2\pi\hbar N_0 V}\sum_\lambda <\!\left(\Sigma^R_\lambda(E) - \Sigma^A_\lambda(E)\right)\!\left(G^{(0)R}_\lambda(E) - G^{(0)A}_\lambda(E)\right)\!> \quad (10.228)$$

where we have expressed the delta-function in eq.(10.227) in terms of the spectral function. We thus have to impurity average a product of a self-energy and a propagator. In the weak-disorder limit $k_F l \gg 1$, the diagrams contributing the collision rate are specified in terms of the Diffuson and the effective electron-electron interaction as depicted in figure 10.3.

The effective electron-electron interaction itself will also be changed due to the presence of impurities. To understand this change, we note that a fluctuation in the density of electrons creates an electric potential, which in turn is felt by an electron. Describing the density fluctuations in the Gaussian approximation they

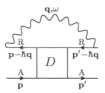

Figure 10.3 Typical diagram encountered for dirty limit scattering rate.

are characterized by the density correlation function, which is specified by the density response function through the fluctuation-dissipation theorem:

$$\tilde{\chi}^{(n)}(\mathbf{q}, \omega) \;=\; \hbar \coth \frac{\hbar \omega}{2kT} \; \Im m \, \chi(\mathbf{q}, \omega) \; . \qquad (10.229)$$

The relation between the density and potential fluctuations, analogous to the relation between the electric field and current fluctuations (see eq.(7.197) in section 7.6), is specified by the potential fluctuation correlator according to

$$e^2 < \phi(\mathbf{q}, \omega)\phi(-\mathbf{q}, -\omega) > \;=\; \hbar \coth \frac{\hbar \omega}{2kT} \; \frac{\Im m \, \chi(\mathbf{q}, \omega)}{|\chi(\mathbf{q}, \omega)|^2} \; . \qquad (10.230)$$

Since we have a diagrammatic interpretation of the density response function or equivalently the effective interaction, we can assess the quantitative change on the effective electron-electron interaction due to impurity scattering. In the diagrammatic expansion of the effective electron-electron interaction, eq.(10.44), we must impurity average all the bubble diagrams. To lowest order in the disorder parameter $1/k_F l$, we should insert the impurity ladder into the bubble diagram; i.e., we encounter the diagrams of the type

$$\qquad (10.231)$$

The impurity-averaged bubble diagram was evaluated in chapter 8, and according to eq.(8.168) we thus have in the diffusive limit

$$\chi(\mathbf{q}, \omega) \;=\; \frac{2N_0 D_0 \, q^2}{-i\omega + D_0 \, q^2} \qquad (10.232)$$

and thereby (in the three-dimensional case) for the dielectric function, $ql, \omega\tau \ll 1$,

$$\epsilon(\mathbf{q}, \omega) \;=\; 1 + \frac{e^2}{\epsilon_0 \, q^2} \frac{2N_0 D_0 \, q^2}{-i\omega + D_0 \, q^2} \;=\; 1 + \frac{D_0 \, \kappa_s^2}{-i\omega + D_0 \, q^2} \; . \qquad (10.233)$$

We then obtain for the inelastic collision rate or energy relaxation rate in terms of the Diffuson and the dynamically screened electron-electron interaction in the presence of impurities, specified by the expression in eq.(10.230), $\overline{V}^R(\mathbf{q}, \omega) = V(\mathbf{q})/\epsilon(\mathbf{q}, \omega)$,

$$\frac{1}{\tau_{e-e}(E,T)} = -\frac{1}{2\hbar V^2}\Im\left(\int\frac{d\mathbf{q}}{(2\pi)^3}\int\frac{d\omega}{2\pi}\, D(\mathbf{q},\omega)(\overline{V}^R(\mathbf{q},\omega) - \overline{V}^A(\mathbf{q},\omega))\,u^4\right.$$

$$\sum_{\mathbf{pp'}} G^R(E-\hbar\omega,\mathbf{p}-\hbar\mathbf{q})\,G^A(E,\mathbf{p'})G^R(E-\hbar\omega,\mathbf{p'}-\hbar\mathbf{q})\,G^A(E,\mathbf{p})$$

$$\left.\left(\tanh\frac{E-\hbar\omega}{2kT}+\coth\frac{\hbar\omega}{2kT}\right)\right) \tag{10.234}$$

In order to respect Pauli's exclusion principle we have, as in section 10.8.2, performed the substitution $\coth\frac{\hbar\omega}{2kT} \to \coth\frac{\hbar\omega}{2kT} + \tanh\frac{E-\hbar\omega}{2kT}$.[24] The momentum integrals over the impurity-averaged propagators are immediately performed and we obtain

$$\frac{1}{\tau_{e-e}(E,T)} = \frac{1}{\hbar}\int\frac{d\mathbf{q}}{(2\pi)^3}\int\frac{d\omega}{2\pi}\,\Im\overline{V}^R(\mathbf{q},\omega)\,\Re eD(\mathbf{q},\omega)\left(\tanh\frac{E-\hbar\omega}{2kT}+\coth\frac{\hbar\omega}{2kT}\right) \tag{10.235}$$

from which we can calculate the collision rate.

We could also calculate the collision rate or energy relaxation rate in the dirty limit by solving the Boltzmann equation with the two-particle interaction modified by the impurity scattering

$$\frac{\partial f(\epsilon)}{\partial t} = 2\pi\int_{-\infty}^{\infty}d\omega\int_{-\infty}^{\infty}\frac{d\epsilon'}{2\pi\hbar}\,P(\omega)\,R(\epsilon,\epsilon',\omega) \tag{10.236}$$

where

$$R(\epsilon,\epsilon',\omega) = f(\epsilon)\,f(\epsilon'-\omega)\,(1-f(\epsilon-\omega))\,(1-f(\epsilon'))$$

$$- f(\epsilon-\omega)\,f(\epsilon')\,(1-f(\epsilon))\,(1-f(\epsilon'-\omega)) \tag{10.237}$$

and

$$P(\omega) = \frac{2N_0\tau^2}{\pi\hbar}\int\frac{d\mathbf{q}}{(2\pi)^3}\left(\frac{V(\mathbf{q})}{|\epsilon(\mathbf{q},\omega)|}\frac{(D_0\,q)^2}{\omega^2+(D_0\,q)^2}\right)^2 \tag{10.238}$$

is analogous to $\alpha^2 F$ for the electron-phonon case. In the three-dimensional case we have, $\omega\tau < 1$,

$$P(\omega) = \frac{\omega^{-1/2}}{8\sqrt{2}\pi^2\hbar N_0 D_0^{3/2}}. \tag{10.239}$$

---

[24]We note that in accordance with the exclusion principle the lifetime of an electron on the Fermi surface, $E = 0$, at zero temperature is then infinite.

We therefore get for an electron on the Fermi surface in a dirty metal the electron-electron collision rate at temperatures $kT < \hbar/\tau$[25]

$$\frac{1}{\tau_{e-e}(T)} = \int\limits_{0}^{\infty} d\omega \; P(\omega) \frac{2\omega}{\sinh\frac{\hbar\omega}{kT}} = c \; \frac{\tau^{1/2}}{k_F l} \frac{(kT)^{3/2}}{\sqrt{\hbar\epsilon_F \tau}} \tag{10.240}$$

where c is a constant of order unity ($\zeta(3/2) \simeq 2.612$)

$$c = \frac{3\sqrt{3\pi}}{16} \zeta(3/2)(\sqrt{8} - 1) \; . \tag{10.241}$$

For an electron in energy state ξ, $\xi < \hbar/\tau$, we get analogously in the dirty limit for the electron-electron collision rate at zero temperature

$$\frac{1}{\tau_{e-e}(\xi)} = \frac{\sqrt{6}}{4} \frac{\tau^{1/2}}{\hbar^{3/2}(k_F l)^2} \xi^{3/2} \; . \tag{10.242}$$

The scattering rate due to electron-electron interaction is thus enhanced in a dirty metal compared to the clean case [53] [54], diffusion enhanced electron-electron interaction. The interpretation of this enhancement can be given in terms of the previous phase space argument determining the relaxation time, and the breaking of translational invariance due to the presence of disorder. The violation of momentum conservation in the virtual scattering processes due to impurities gives more phase space for final states. Alternatively, viewing the collision in real space, due to the motion being diffusive instead of ballistic the electrons spend more time close together where the interaction is strong, or, wave functions of equal energy in a random potential are spatially correlated thereby leading to an enhanced electron-electron interaction. The scattering process now includes quantum interference between the elastic and inelastic processes as signified by the collision rate $\hbar/\tau_{e-e}$ being depending on $\hbar$.

We note that the expression for the energy relaxation rate in two dimensions diverges in the infrared for a dirty metal in the above lowest-order perturbative calculation. However, this is not alarming since we do not expect the relaxation rate to be the relevant measurable quantity, as in this quantity scattering at all energies is weighted equally. We shall return to this problem in section 11.3.2 where we calculate the phase relaxation rate of the electronic wave function in a dirty two-dimensional metallic film, a quantity which does not diverge due to collisions with small energy transfer. We shall there need the expression for the effective electron-electron interaction at low energies and momenta in a dirty metal for which we have, according to eq.(10.243) and eq.(10.232),

$$<\phi(\mathbf{q},\omega)\phi(-\mathbf{q},-\omega)> = \frac{2kT}{e^2(2N_0)^2} \frac{2N_0}{D_0 q^2} = \frac{2kT}{\sigma_0 q^2} \; . \tag{10.243}$$

---

[25]From the region of large ω and q we get the clean limit rate, eq.(10.215), which dominates at temperatures $kT \gg \hbar/\tau$.

Chapter 11

Weak Localization

We start this chapter by discussing the weak-localization contribution to the conductivity in the position representation, before turning to discuss the effects of interactions on the weak-localization effect, the destruction of the phase coherence of the wave function due to electron-phonon and electron-electron interaction. Then anomalous magnetoresistance and the weak-localization Aharonov-Bohm effect are presented. The weak antilocalization effect due to spin-orbit scattering is investigated as well as the effect of spin-flip scattering. Finally we discuss mesoscopic fluctuations.

The theory of weak localization dates back to the seminal work on the scaling theory of localization [37],[1] and developed rapidly into a comprehensive understanding of the quantum corrections to the Boltzmann conductivity. Based on the insight provided by the diagrammatic technique, the first quantum correction, the weak-localization effect, was soon realized to be the result of a simple type of quantum mechanical interference (as already noted in section 9.2), and the resulting physical insight eventually led to a quantitative understanding of mesoscopic phenomena in disordered conductors. We shall use the quantum interference picture in parallel with the quantitative diagrammatic technique, to discuss the weak-localization phenomenon.

11.1 Quantum Correction to Conductivity

In section 8.5 we derived the Boltzmann expression for the classical conductivity as the weak-disorder limiting case where the quantum mechanical wave nature of the motion of an electron is neglected. In terms of diagrams this corresponded to neglecting conductivity diagrams where impurity correlators cross, because such contributions are smaller by the factor λ_F/l, and thus constitute quantum corrections to the classical conductivity.

A special class of diagrams where impurity correlators crossed a maximal number of times was in section 9.2 seen, in the time-reversal invariant situation, to

---

[1]Presented in chapter 9.

exhibit singular behavior although the diagrams nominally are of order $\hbar/p_F l$.[2]

$$
\text{(11.1)}
$$

In fact, in section 9.3 we studied in the self-consistent theory, how this singularity drives the Anderson metal-insulator transition. We shall therefore consider the explicitly time-dependent situation where the frequency ω of the external field is not equal to zero, in order to cut off the singular behavior. In this case (and others to be studied shortly) the first quantum correction to the conductivity in the parameter λ_F/l is a small correction to the Boltzmann conductivity (recall eq.(9.38)), and we speak of the weak-localization effect.

In the discussion of interaction effects and magnetoresistance it will be convenient to use the spatial representation for the conductivity. The free-electron model and a delta-correlated random potential, eq.(8.99), will be used for convenience.

In the position representation the impurity-averaged current density

$$
j_\alpha(\mathbf{x},\omega) \equiv\ <j_\alpha(\mathbf{x},\omega)>\ =\ \sum_\beta \int d\mathbf{x}'\ <\sigma_{\alpha\beta}(\mathbf{x},\mathbf{x}',\omega)> E_\beta(\mathbf{x}',\omega) \tag{11.2}
$$

is, besides regular corrections of order $\mathcal{O}(\hbar/p_F l)$, specified by the conductivity tensor, eq.(8.62),

$$
\sigma_{\alpha\beta}(\mathbf{x}-\mathbf{x}',\omega) \equiv\ <\sigma_{\alpha\beta}(\mathbf{x},\mathbf{x}',\omega)>\ =\ \frac{1}{\pi}\left(\frac{e\hbar}{m}\right)^2 \int_{-\infty}^{\infty} dE\ \frac{f_0(E)-f_0(E+\hbar\omega)}{\omega}
$$

$$
<G^R(\mathbf{x},\mathbf{x}';E+\hbar\omega)\ \overset{\leftrightarrow}{\nabla}_{x_\alpha}\overset{\leftrightarrow}{\nabla}_{x'_\beta}\ G^A(\mathbf{x}',\mathbf{x};E)> \ . \tag{11.3}
$$

The contribution to the conductivity from the maximally crossed diagrams is conveniently exhibited in twisted form where they become ladder-type diagrams

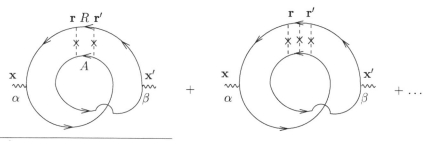

---

[2]In addition to these maximally crossed diagrams, there are additional diagrams of the same order of magnitude (also coming from the regular terms). However, they give contributions to the conductivity which are insensitive to low magnetic fields and temperatures in comparison to the contribution from the maximally crossed diagrams.

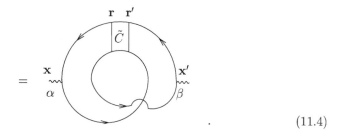

$$
= \qquad\qquad . \tag{11.4}
$$

The sum of the maximally crossed diagrams, the Cooperon $\tilde{C}_\omega(\mathbf{r}, \mathbf{r}'; E)$, is in the position representation specified by the diagrams

$$
\cdots \;+\; \cdots \;+\; \ldots \;=\; \mathbf{r}\;\boxed{\tilde{C}}\;\mathbf{r}' \;. \tag{11.5}
$$

The analytical expression for the quantum correction to the conductivity is therefore ($E_+ \equiv E + \hbar\omega$)

$$
\delta\sigma_{\alpha\beta}(\mathbf{x} - \mathbf{x}', \omega) \;=\; \frac{1}{\pi}\left(\frac{e\hbar}{m}\right)^2 \int d\mathbf{r} \int d\mathbf{r}' \int_{-\infty}^{\infty} dE\; \frac{f_0(E) - f_0(E + \hbar\omega)}{\omega}\; \tilde{C}_\omega(\mathbf{r}, \mathbf{r}'; E)
$$

$$
G_{E_+}^R(\mathbf{x} - \mathbf{r}) G_{E_+}^R(\mathbf{r}' - \mathbf{x}') \overset{\leftrightarrow}{\nabla}_{x_\alpha} \overset{\leftrightarrow}{\nabla}_{x'_\beta} G_E^A(\mathbf{x}' - \mathbf{r}) G_E^A(\mathbf{r}' - \mathbf{x}) \;. \tag{11.6}
$$

The impurity-averaged propagator decays, according to eq.(3.73), exponentially as a function of its spatial variable with the scale set by the impurity mean free path. The spatial scale of variation of the sum of the maximally crossed diagrams is typically much larger. For the present case where we neglect effects of inelastic interactions, we recall from eq.(9.22) that the spatial range of the Cooperon is $L_\omega \equiv \sqrt{D_0/\omega}$, which for $\omega\tau \ll 1$ is much larger than the mean free path, since $D_0 = v_F l/d$ is the diffusion constant in d dimensions.[3] The impurity-averaged propagators attached to the maximally crossed diagrams will therefore require the starting and end points of $\tilde{C}_\omega(\mathbf{r}, \mathbf{r}', E)$ to be within the distance of a mean free path, in order for a nonvanishing contribution to the integral. On the scale of variation of the Cooperon this amounts to setting its arguments equal, and we can

---

[3]For samples of size larger than the mean free path, $L > l$, the diffusion process is effectively three-dimensional, so that one should use the value $d = 3$ in the expression for the diffusion constant. In strictly two-dimensional systems, such as for the electron gas in the inversion layer in a heterostructure at low temperatures, the value $d = 2$ should be used.

therefore substitute $\mathbf{r} \to \mathbf{x}, \mathbf{r}' \to \mathbf{x}$, and obtain

$$\delta\sigma_{\alpha\beta}(\mathbf{x} - \mathbf{x}', \omega) = \frac{1}{\pi}\left(\frac{e\hbar}{m}\right)^2 \int_{-\infty}^{\infty} dE \frac{f_0(E_-) - f_0(E_+)}{\omega} \tilde{C}_\omega(\mathbf{x}, \mathbf{x}; E) \int d\mathbf{r} \int d\mathbf{r}'$$

$$G_{E_+}^R(\mathbf{x} - \mathbf{r}) G_{E_+}^R(\mathbf{r}' - \mathbf{x}') \overset{\leftrightarrow}{\nabla}_{x_\alpha} \overset{\leftrightarrow}{\nabla}_{x'_\beta} G_{E_-}^A(\mathbf{x}' - \mathbf{r}) G_{E_-}^A(\mathbf{r}' - \mathbf{x}) . \quad (11.7)$$

The gate combination of the Fermi functions renders for the degenerate case, $\hbar\omega, kT \ll \epsilon_F$, the energy variable in the thermal layer around the Fermi surface, and we have for the first quantum correction to the conductivity of a degenerate electron gas

$$\delta\sigma_{\alpha\beta}(\mathbf{x} - \mathbf{x}', \omega) = \frac{\hbar}{\pi}\left(\frac{e}{m}\right)^2 \tilde{C}_\omega(\mathbf{x}, \mathbf{x}; \epsilon_F) \, \Phi_{\alpha,\beta}(\mathbf{x} - \mathbf{x}') \quad (11.8)$$

where

$$\Phi_{\alpha,\beta}(\mathbf{x} - \mathbf{x}') \equiv \int d\mathbf{r} \int d\mathbf{r}' G_{\epsilon_F}^R(\mathbf{x} - \mathbf{r}) G_{\epsilon_F}^R(\mathbf{r}' - \mathbf{x}') \overset{\leftrightarrow}{\nabla}_{x_\alpha} \overset{\leftrightarrow}{\nabla}_{x'_\beta} G_{\epsilon_F}^A(\mathbf{x}' - \mathbf{r}) G_{\epsilon_F}^A(\mathbf{r}' - \mathbf{x}). \quad (11.9)$$

Clearly this function is local with the scale of the mean free path, and in appendix D it is shown that to lowest order in $\hbar/p_F l$ we have

$$\Phi_{\alpha,\beta}(\mathbf{x} - \mathbf{x}') = -\frac{(2\pi N_0 \tau)^2}{2\hbar^2} \frac{(x_\alpha - x'_\alpha)(x_\beta - x'_\beta)}{|\mathbf{x} - \mathbf{x}'|^4} e^{-|\mathbf{x} - \mathbf{x}'|/l} \cos^2 k_F |\mathbf{x} - \mathbf{x}'| . \quad (11.10)$$

Since the function $\Phi_{\alpha,\beta}(\mathbf{x} - \mathbf{x}')$ decays on the scale of the mean free path, and appears in connection with the Cooperon, which is a smooth function on this scale, it acts effectively as a delta function

$$\Phi_{\alpha,\beta}(\mathbf{x} - \mathbf{x}') = -\frac{(2\pi N_0 \tau)^2 l}{3\hbar^2} \delta_{\alpha\beta} \, \delta(\mathbf{x} - \mathbf{x}') . \quad (11.11)$$

We therefore obtain that the first quantum correction, the weak-localization contribution, to the conductivity is local

$$\delta\sigma_{\alpha\beta}(\mathbf{x} - \mathbf{x}', \omega) = \delta\sigma(\mathbf{x}, \omega) \, \delta_{\alpha\beta} \, \delta(\mathbf{x} - \mathbf{x}') \quad (11.12)$$

and specified by[4]

$$\delta\sigma(\mathbf{x}, \omega) = -\frac{2e^2 D_0 \tau}{\pi\hbar} C_\omega(\mathbf{x}, \mathbf{x}) . \quad (11.13)$$

As we already noted in section 9.2 the Cooperon is independent of the energy of the electron (here the Fermi energy since only electrons at the Fermi surface contribute to the conductivity) $C_\omega(\mathbf{x}, \mathbf{x}') \equiv C_\omega(\mathbf{x}, \mathbf{x}', \epsilon_F)$, and we have introduced $C_\omega(\mathbf{x}, \mathbf{x}') \equiv u^{-2} \tilde{C}_\omega(\mathbf{x}, \mathbf{x}')$.

---

[4]We could also have evaluated the conductivity, eq.(11.6), directly by Fourier-transforming the propagators, and recalling eq.(9.26).

The quantum correction to the conductance of a disordered degenerate electron gas is

$$\delta G_{\alpha\beta}(\omega) \equiv \; <\delta G_{\alpha\beta}(\omega)> \; = \; L^{-2} \int d\mathbf{x} \int d\mathbf{x}' <\delta\sigma_{\alpha\beta}(\mathbf{x},\mathbf{x}',\omega)>$$

$$= \; -\frac{2e^2 D_0 \tau}{\pi\hbar} L^{-2}\delta_{\alpha\beta} \int d\mathbf{x} \, C_\omega(\mathbf{x},\mathbf{x}) \; . \tag{11.14}$$

11.2   Quantum Interference and the Cooperon

In this section, we shall elucidate in more detail than in section 9.2 the physical process in real space described by the maximally crossed diagrams, and in addition consider the influence of external fields. The weak-localization effect can be understood in terms of a simple kind of quantum mechanical interference. By following the scattering sequences appearing in the diagrammatic representation of the Cooperon contribution to the conductivity, see eq.(11.1), we realize that the quantum correction to the conductivity consists of products of the form "amplitude for scattering sequence of an electron off impurities in real space *times* the complex conjugate of the amplitude for the opposite scattering sequence." The quantum correction to the conductivity is thus the result of quantum mechanical interference between amplitudes for an electron traversing a loop in opposite directions. To lowest order in λ_F/l we only need to include the stationary, i.e., classical, paths determined by the electron bumping into impurities, as illustrated in figure 11.1.

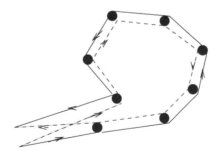

Figure 11.1 Trajectories involved in the weak-localization quantum interference process.

The solid line, say, in figure 11.1 corresponds to the propagation of the electron represented by the retarded propagator, and the broken line to the propagation represented by the advanced propagator, the complex conjugate of the amplitude for scattering off impurities in the opposite sequence. The starting and end points refer to the points $\mathbf{x}$ and $\mathbf{x}'$ in eq.(11.6), respectively.[5] According to the formula, eq.(11.13), for the quantum correction to the conductivity, we only need to consider scattering sequences which start and end at the same point on the scale of the mean free path, as demanded by the impurity-averaged propagators attached to the maximally crossed diagrams in eq.(11.6).

In the time-reversal invariant situation, the contribution to the return probability from the maximally crossed diagrams equals the contribution from the ladder diagrams (as already noted in section 8.10), and the return probability including the weak-localization contribution is thus twice the classical result[6]

$$P_{cl+wl}(\mathbf{x}, t; \mathbf{x}, t') \;=\; 2\, P_{cl}(\mathbf{x}, t; \mathbf{x}, t') \;=\; 2 \left(\frac{1}{4\pi D_0(t - t')} \right)^{d/2} \tag{11.15}$$

where the last expression is valid in the diffusive limit. To see how this comes about in the interference picture, let us consider the return probability in general. The quantity of interest is therefore the amplitude K for an electron to arrive at a given space point $\mathbf{x}$ at time $t/2$ when initially it started at the same space point at time $-t/2$. According to Feynman, this amplitude is given by the path integral expression

$$K\left(\mathbf{x}, t/2; \mathbf{x}, -t/2\right) \;=\; \int_{\mathbf{x}_{-t/2}=\mathbf{x}}^{\mathbf{x}_{t/2}=\mathbf{x}} \mathcal{D}\mathbf{x}_t\, e^{\frac{i}{\hbar}S[\mathbf{x}_t]} \;\equiv\; \sum_c A_c \tag{11.16}$$

where the path integral includes all paths which start and end at the same point. For the return probability we have

$$P = |K|^2 = \left| \sum_c A_c \right|^2 \;=\; \sum_c |A_c|^2 \;+\; \sum_{c \neq c'} A_c A_{c'}^* \tag{11.17}$$

where A_c is the amplitude for the path c. In the sum over paths we only need to include to order λ_F/l the stationary, i.e., classical, paths determined by the electron bumping into impurities. The sum of the absolute squares is then the classical contribution to the return probability, and the other terms are quantum interference terms. In the event that the particle only experiences the impurity potential, we have for the amplitude for the particle to traverse the path c,

$$A_c \;=\; e^{\frac{i}{\hbar} \int_{-\frac{t}{2}}^{\frac{t}{2}} d\bar{t}\, \{\frac{1}{2}m\dot{\mathbf{x}}_c^2(\bar{t}) - V(\mathbf{x}_c(\bar{t}))\}} \;. \tag{11.18}$$

Due to the impurity potential, the amplitude has a random phase. A first conjecture would be to expect that upon impurity averaging, the interference terms

---

[5]The angle between initial and final velocities are exaggerated in figure 11.1 since we recall that in order for the Cooperon to give a large contribution their angle must be less than $1/k_F l$.

[6]The fact that impurity lines cross, do not per se make a diagram of order $1/k_F l$ relative to a noncrossed diagram. In case of the conductivity diagrams this is indeed the case for the maximally crossed diagrams because the circumstances needed for a large contribution set a constrain on the correlation of the initial and final velocity, $\mathbf{p}' \simeq -\mathbf{p} + \hbar\mathbf{Q}$ (recall also when estimating self-energy diagrams in section 3.6 the importance of the incoming and outgoing momenta being equal). However, in the quantity of interest here the position is fixed.

in general average to an insignificant small value, and we would be left with the classical contribution to the conductivity. However, there are certain interference terms which are resilient to the impurity average. It is clear that impurity averaging can not destroy the interference between time-reversed trajectories since we have for the amplitude for traversing the time-reversed trajectory, $\mathbf{x}_{\bar{c}}(t) = \mathbf{x}_c(-t)$,

$$
A_{\bar{c}} = e^{\frac{i}{\hbar}\int_{-\frac{t}{2}}^{\frac{t}{2}}d\bar{t}\,\{\frac{1}{2}m\dot{\mathbf{x}}_{\bar{c}}^2(\bar{t}) - V(\mathbf{x}_{\bar{c}}(\bar{t}))\}} = e^{\frac{i}{\hbar}\int_{-\frac{t}{2}}^{\frac{t}{2}}d\bar{t}\,\{\frac{1}{2}m[-\dot{\mathbf{x}}_c(-\bar{t})]^2 - V(\mathbf{x}_c(-\bar{t}))\}} = A_c(11.19)
$$

In this time-reversal invariant situation the amplitudes for traversing a closed loop in opposite directions are identical, $A_{\bar{c}} = A_c$, and the corresponding interference term contribution to the return probability is independent of the disorder, $A_c A_{\bar{c}}^* = 1$! The two amplitudes for the time-reversed electronic trajectories which return to the starting point thus interfere constructively in case of time-reversal invariance. In correspondence to this enhanced localization, there is a decrease in conductivity which can be calculated according to eq.(11.13).

The foregoing discussion based on the physical understanding of the weak localization effect will now be substantiated by deriving the equation satisfied by the Cooperon. The Cooperon $C_\omega(\mathbf{x}, \mathbf{x}')$ is generated by the iterative equation

$$(11.20)$$

where we have introduced the diagrammatic notation[7]

$$\equiv \delta(\mathbf{x} - \mathbf{x}') \quad . \tag{11.21}$$

The Cooperon equation, eq.(11.20), is most easily obtained by adding the term

$$= u^2\,\delta(\mathbf{x} - \mathbf{x}') \tag{11.22}$$

---

[7]No confusion with our previous diagrammatic notation for the density matrix should arise.

to the infinite sum of terms represented by the function $\tilde{C}$, eq.(11.5). Alternatively, one can proceed as in section 9.2, now exploiting the local character of the propagators. In any event, we have in the singular region $\tilde{C} \simeq u^2 C$.

The Cooperon equation in the spatial representation is

$$C_\omega(\mathbf{x}, \mathbf{x}') = \delta(\mathbf{x} - \mathbf{x}') + \int d\mathbf{x}'' \, \tilde{J}^C_\omega(\mathbf{x}, \mathbf{x}'') \, C_\omega(\mathbf{x}'', \mathbf{x}') \qquad (11.23)$$

where according to the Feynman rules the insertion is given by

$$\tilde{J}^C_\omega(\mathbf{x}, \mathbf{x}') = u^2 \, G^R_{\epsilon_F + \hbar\omega}(\mathbf{x}, \mathbf{x}') \, G^A_{\epsilon_F}(\mathbf{x}, \mathbf{x}') \,. \qquad (11.24)$$

The Cooperon is slowly varying on the scale of the mean free path, the spatial range of the function $\tilde{J}^C_\omega(\mathbf{x}, \mathbf{x}')$, and a low-order Taylor-expansion of the Cooperon on the right-hand side of eq.(11.23) is therefore sufficient. Upon partial integration, the integral equation then becomes, for a second-order Taylor expansion, a differential equation for the Cooperon

$$\left\{ [1 - \int d\mathbf{r} \, \tilde{J}^C_\omega(\mathbf{r})] - \frac{1}{2d} \left(\int d\mathbf{r} \, r^2 \tilde{J}^C_\omega(\mathbf{r}) \right) \nabla^2_\mathbf{x} \right\} C_\omega(\mathbf{x}, \mathbf{x}') = \delta(\mathbf{x} - \mathbf{x}') \qquad (11.25)$$

where $\tilde{J}^C_\omega(\mathbf{r})$ by definition equals $\tilde{J}^C_\omega(\mathbf{x}, \mathbf{x}')$ for $\mathbf{r} = \mathbf{x} - \mathbf{x}'$. The integrals needed can now be performed by substituting the expression for the impurity-averaged propagator eq.(3.73); assuming $\epsilon_F \tau \gg \hbar$ and $\hbar\omega \ll \epsilon_F$, we obtain (recall eq.(8.109))

$$\int d\mathbf{r} \, \tilde{J}^C_\omega(\mathbf{r}) = \frac{1}{1 - i\omega\tau} \,, \qquad \frac{1}{2d} \int d\mathbf{r} \, r^2 \tilde{J}^C_\omega(\mathbf{r}) = \frac{D_0 \tau}{(1 - i\omega\tau)^3} \,. \qquad (11.26)$$

Hence, for small frequencies, $\omega\tau \ll 1$, we obtain the following equation determining the Cooperon

$$\left\{ -i\omega - D_0 \nabla^2_\mathbf{x} \right\} C_\omega(\mathbf{x}, \mathbf{x}') = \frac{1}{\tau} \delta(\mathbf{x} - \mathbf{x}') \,. \qquad (11.27)$$

This equation is of course simply the position representation of the equation for the Cooperon already derived in the momentum representation, eq.(9.22). Indeed we recover, that the Cooperon only varies on the large length scale $L_\omega = (D_0/\omega)^{1/2}$. The typical size of an interference loop is much larger than the mean free path, and we only need the large-scale behavior of the Boltzmannian paths of figure 9.2, the smooth diffusive loops of figure 11.2.

The fact that in the time-reversal invariant case we have obtained that the Cooperon satisfies the same diffusion-type equation as the Diffuson is not surprising. The Diffuson is, according to eq.(8.98), determined by a similar diagrammatic integral equation as the Cooperon, however, with the important difference that one of the particle lines, say the advanced one, is reversed. The Diffuson will therefore be determined by the same integral equation as the Cooperon, except for $\tilde{J}^C_\omega$ now being substituted by the diffusion insertion $\tilde{J}^D_\omega$ given by

$$\tilde{J}^D_\omega(\mathbf{x}, \mathbf{x}') = u^2 \, G^R_{\epsilon_F + \hbar\omega}(\mathbf{x}, \mathbf{x}') \, G^A_{\epsilon_F}(\mathbf{x}', \mathbf{x}) \,. \qquad (11.28)$$

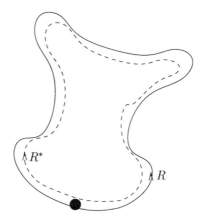

Figure 11.2 Diffusive loops.

In a time-reversal invariant situation the two insertions are equal, $\tilde{J}^C_\omega = \tilde{J}^D_\omega$, and we recover that the Diffuson and the Cooperon satisfy the same equation (and we have hereby rederived the result, eq.(11.15)).

In the time-reversal invariant situation the amplitudes for traversing a closed loop in opposite directions are identical, and in such a coherent situation one must trace the complete interference pattern of wave reflection in a random medium, and one encounters the phenomenon of localization discussed in chapter 9. However, we also realize that the interference effect is sensitive to the breaking of time-reversal invariance. By breaking the coherence between the amplitudes for traversing time-reversed loops the tendency to localization of an electron can be suppressed.[8] In moderately disordered conductors we can therefore arrange for conditions so that the tendency to localization of the electronic wave function has only a weak though measurable influence on the conductivity. The first quantum correction then gives the dominating contribution in the parameter λ_F/l, and we speak of the so-called weak-localization regime. The destruction of phase coherence is due to the interaction of the electron with its environment, such as electron-electron interaction, electron-phonon interaction, interaction with magnetic impurities, or interaction with an external magnetic field. From an experimental point of view the breaking of coherence between time-reversed trajectories by an external magnetic field is of special importance, and we start by discussing this case.

11.2.1 Quantum Interference in a Magnetic Field

The influence of a magnetic field on the quantum interference process described by the Cooperon is readily established in view of the already presented formulas.

---

[8]By disturbance, the coherence can be disrupted, and the tendency to localization be suppressed, thereby decreasing the resistance. Normally, disturbances *increase* the resistance.

In the weak magnetic field limit, $l^2 < l_B^2$, where $l_B = (\hbar/2eB)^{1/2}$ is the magnetic length, or equivalently $\omega_c\tau < \hbar/\epsilon_F\tau$, the bending of a classical trajectory with energy ϵ_F can be neglected on the scale of the mean free path. Classical magnetoresistance effects are then negligible, because they, according to section 5.4.2, are of importance only when $\omega_c\tau \geq 1$. The amplitude for propagation along a straight-line classical path determined by the impurities is then only changed due to the presence of the magnetic field by the additional phase picked up along the straight line, the line integral along the path of the vector potential $\mathbf{A}$ describing the magnetic field. In the presence of such a weak static magnetic field the propagator is thus changed according to[9]

$$G_E^R(\mathbf{x},\mathbf{x}') \rightarrow G_E^R(\mathbf{x},\mathbf{x}') \exp\left\{ \frac{ie}{\hbar} \int_{\mathbf{x}'}^{\mathbf{x}} d\bar{\mathbf{x}} \cdot \mathbf{A}(\bar{\mathbf{x}}) \right\} . \tag{11.29}$$

The resulting change in the Cooperon insertion is then

$$\tilde{J}_\omega^C(\mathbf{x},\mathbf{x}') \quad \rightarrow \quad \tilde{J}_\omega^C(\mathbf{x},\mathbf{x}') \exp\left\{ \frac{2ie}{\hbar} \int_{\mathbf{x}'}^{\mathbf{x}} d\bar{\mathbf{x}} \cdot \mathbf{A}(\bar{\mathbf{x}}) \right\}$$

$$= \quad \tilde{J}_\omega^C(\mathbf{x},\mathbf{x}') \exp\left\{ \frac{2ie}{\hbar}(\mathbf{x}-\mathbf{x}') \cdot \mathbf{A}(\mathbf{x}) \right\} . \tag{11.30}$$

The factor of 2 reflects that in weak-localization interference terms between time-reversed trajectories, the additional phases due to the magnetic field add.

Repeating the Taylor-expansion leading to eq.(11.25), we now obtain in the Cooperon equation additional terms due to the presence of the magnetic field

$$\left\{ -i\omega - D_0 \left(\nabla_{\mathbf{x}} - \frac{2ie}{\hbar}\mathbf{A}(\mathbf{x}) \right)^2 \right\} C_\omega(\mathbf{x},\mathbf{x}') \;=\; \frac{1}{\tau} \delta(\mathbf{x}-\mathbf{x}') . \tag{11.31}$$

Introducing the Fourier transform

$$C_{t,t'}(\mathbf{x},\mathbf{x}') \;=\; \int \frac{d\omega}{2\pi} \, e^{-i\omega(t-t')} \, C_\omega(\mathbf{x},\mathbf{x}') \tag{11.32}$$

we obtain in the space-time representation the Cooperon equation

$$\left\{ \frac{\partial}{\partial t} - D_0 \left(\nabla_{\mathbf{x}} - \frac{2ie}{\hbar}\mathbf{A}(\mathbf{x}) \right)^2 \right\} C_{t,t'}(\mathbf{x},\mathbf{x}') \;=\; \frac{1}{\tau} \delta(\mathbf{x}-\mathbf{x}') \, \delta(t-t') . \tag{11.33}$$

We note that this equation is formally identical to the imaginary-time Green's function equation for a particle of mass $\hbar/2D_0$ and charge $2e$ moving in the magnetic field described by the vector potential $\mathbf{A}$. The solution of this equation can be expressed as the path integral (recall eq.(8.119) on page 347, and exercise 2.2 on page 96)

$$C_{t,t'}(\mathbf{x},\mathbf{x}') \;=\; \frac{1}{\tau} \int_{\mathbf{x}_{t'}=\mathbf{x}'}^{\mathbf{x}_t=\mathbf{x}} \mathcal{D}\mathbf{x}_t \; e^{-\int_{t'}^{t} d\bar{t} \left(\frac{\dot{\mathbf{x}}_{\bar{t}}^2}{4D_0} + \frac{2ie}{\hbar}\dot{\mathbf{x}}_{\bar{t}}\cdot\mathbf{A}(\mathbf{x}_{\bar{t}}) \right)} . \tag{11.34}$$

---

[9]The exact propagator in a homogeneous magnetic field is calculated in appendix A.

11.2.2 Quantum Interference in a Time-Dependent Field

Let us now obtain the equation satisfied by the Cooperon when the particle interacts with an environment as described by the Lagrangian L_1. The total Lagrangian
is then $L = L_0 + L_1$, where[10]

$$L_0(\mathbf{x}, \dot{\mathbf{x}}) = \frac{1}{2}m\dot{\mathbf{x}}^2 - V(\mathbf{x}) \tag{11.35}$$

describes the particle in the impurity potential. We first present a derivation of
the Cooperon equation based on the interference picture of the weak-localization
effect, before presenting the diagrammatic derivation.[11]

The conditional probability density for an electron to arrive at position $\mathbf{x}$ at
time t given it was at position $\mathbf{x}'$ at time t' is given by the absolute square of the
propagator

$$P(\mathbf{x}, t; \mathbf{x}', t') = |K(\mathbf{x}, t; \mathbf{x}', t')|^2 . \tag{11.36}$$

In the quasi-classical limit, which is the one of interest, $\lambda_F \ll l$, we can in the path
integral expression for the propagator replace the path integral by the sum over
classical paths

$$K(\mathbf{x}, t; \mathbf{x}, t') = \int_{\mathbf{x}_{t'}=\mathbf{x}}^{\mathbf{x}_t=\mathbf{x}} \mathcal{D}\mathbf{x}_t \, e^{\frac{i}{\hbar}S[\mathbf{x}_t]} \simeq \sum_{\mathbf{x}_t^{cl}} A[\mathbf{x}_t^{cl}] \, e^{\frac{i}{\hbar}S[\mathbf{x}_t^{cl}]} \tag{11.37}$$

where the prefactor takes into account the Gaussian fluctuations around the classical path (see appendix A). We assume that we may neglect the influence of L_1
on the motion of the electrons, and the classical paths are determined by L_0, i.e.,
by the large kinetic energy and the strong impurity scattering. The paths in the
summation are therefore solutions of the classical equation of motion

$$m\ddot{\mathbf{x}}_t^{cl} = -\nabla V(\mathbf{x}_t^{cl}) . \tag{11.38}$$

The quantum interference contribution to the return probability in time span t
from the time-reversed loops is in the quasi-classical limit

$$P\left(\mathbf{x}, \frac{t}{2}; \mathbf{x}, -\frac{t}{2}\right) = \sum_{\mathbf{x}_t^{cl}} |A[\mathbf{x}_t^{cl}]|^2 \, e^{\frac{i}{\hbar}(S[\mathbf{x}_t^{cl}] - S[\mathbf{x}_{-t}^{cl}])} \tag{11.39}$$

where $\mathbf{x}_{-t/2}^{cl} = \mathbf{x} = \mathbf{x}_{t/2}^{cl}$. We are interested in the return probability for an electron
constrained to move on the Fermi surface, i.e., its energy is equal to the Fermi
energy ϵ_F. For the weak-localization quantum interference contribution to the
return probability we therefore obtain

$$C(t) = \frac{1}{N_0} \sum_{\mathbf{x}_t^{cl}} \left| A[\mathbf{x}_t^{cl}] \right|^2 e^{i\varphi[\mathbf{x}_t^{cl}]} \, \delta(\epsilon[\mathbf{x}_t^{cl}] - \epsilon_F) \tag{11.40}$$

---

[10] A possible dynamics of the environment plays no role for the present discussion, and its
Lagrangian is suppressed.

[11] Essentially we follow the presentations of reference [55] and [56].

where the sum is over classical trajectories of duration t that start and end at the same point, and

$$\epsilon[\mathbf{x}_t^{cl}] = \frac{1}{2}m\,[\dot{\mathbf{x}}_t^{cl}]^2 + V(\mathbf{x}_t^{cl}) \qquad (11.41)$$

is the energy of the electron on a classical trajectory. The normalization factor follows from the fact that the density of classical paths in the quasi-classical limit equals the density of states.[12] We have introduced the phase difference between a pair of time-reversed paths

$$\varphi[\mathbf{x}_t^{cl}] = \frac{1}{\hbar}\left(S[\mathbf{x}_t^{cl}] - S[\mathbf{x}_{-t}^{cl}]\right) . \qquad (11.42)$$

As noted previously in section 11.2.1, a substantial cancellation occurs in the phase difference since L_0 is an even function of the velocity and the quenched disorder potential is independent of time. Hence, the phase difference is a small quantity given by

$$\varphi[\mathbf{x}_t^{cl}] = \frac{1}{\hbar}\int_{-t/2}^{t/2} d\bar{t}\,\{L_1(\mathbf{x}_{\bar{t}}^{cl}, \dot{\mathbf{x}}_{\bar{t}}^{cl}, \bar{t}) - L_1(\mathbf{x}_{-\bar{t}}^{cl}, -\dot{\mathbf{x}}_{-\bar{t}}^{cl}, \bar{t})\}$$

$$= \frac{1}{\hbar}\int_{-t/2}^{t/2} d\bar{t}\,\{L_1(\mathbf{x}_{\bar{t}}^{cl}, \dot{\mathbf{x}}_{\bar{t}}^{cl}, \bar{t}) - L_1(\mathbf{x}_{\bar{t}}^{cl}, -\dot{\mathbf{x}}_{\bar{t}}^{cl}, -\bar{t})\}$$

$$\equiv \int_{-t/2}^{t/2} d\bar{t}\,\tilde{\varphi}(\mathbf{x}_t^{cl}) \qquad (11.43)$$

where in the last term in the second equality we have replaced the integration variable $\bar{t}$ by $-\bar{t}$. We recognize that L_1 though small, plays an important role here since it destroys the phase coherence between the time-reversed trajectories.

We must now average the quantum interference term with respect to the impurity potential. Since the dependence on the impurity potential in eq.(11.40) is only implicit through its determination of the classical paths, averaging with respect to the random impurity potential is identical to averaging with respect to the probability functional for the classical paths in the random potential. In view of the expression appearing in eq.(11.40), we thus encounter the probability to find a classical path $\mathbf{x}_t$ of duration t which start and end at the same point, and for which the particle has the energy ϵ_F

$$P_t[\mathbf{x}_t] = \frac{1}{N_0}<\sum_{\mathbf{x}_t^{cl}}\left|A[\mathbf{x}_t^{cl}]\right|^2 \delta(\epsilon[\mathbf{x}_t^{cl}] - \epsilon_F)\,\delta[\mathbf{x}_t^{cl} - \mathbf{x}_t]>_{\text{imp}} = <C(t)>_{\text{imp}}^{\varphi=0} .$$

$$(11.44)$$

---

[12]The Bohr-Sommerfeld quantization rule.

The second delta function is, as indicated, in the functional sense, allowing only the classical path in question to contribute to the path integral. The classical probability of return in time t of a particle with energy ϵ_F is given by

$$P_R^{(cl)}(t) = \int_{\mathbf{x}_{-t/2}=\mathbf{x}}^{\mathbf{x}_{t/2}=\mathbf{x}} \mathcal{D}\mathbf{x}_t \, P_t[\mathbf{x}_t] \, . \tag{11.45}$$

We therefore get, according to eq.(11.40), for the impurity average of the weak-localization quantum interference term, the Cooperon,

$$C_{\frac{t}{2},-\frac{t}{2}}(\mathbf{x},\mathbf{x}) = <C(t)>_{\mathrm{imp}} = \int_{\mathbf{x}_{-t/2}=\mathbf{x}}^{\mathbf{x}_{t/2}=\mathbf{x}} \mathcal{D}\mathbf{x}_t \, P_t[\mathbf{x}_t] \, e^{i\varphi[\mathbf{x}_t^{cl}]} \, . \tag{11.46}$$

In many situations of interest, an adequate expression for the probability density of classical paths in a random potential, $P_t[\mathbf{x}_t]$, is obtained by considering the classical paths as realizations of Brownian motion[13]; i.e., the classical motion is assumed a diffusion process, and the probability distribution of paths is given by eq.(8.118). Performing the impurity average gives in the diffusive limit for the weak-localization interference term[14]

$$C_{\frac{t}{2},\frac{-t}{2}}(\mathbf{x},\mathbf{x}) = \int_{\mathbf{x}_{-t/2}=\mathbf{x}}^{\mathbf{x}_{t/2}=\mathbf{x}} \mathcal{D}\mathbf{x}_t \, e^{-\int_{-t/2}^{t/2} dt \, (\frac{\dot{\mathbf{x}}_t^2}{4D_0} - i\bar{\varphi}(\mathbf{x}_t^{cl}))} \tag{11.47}$$

where D_0 is the diffusion constant for a particle with energy ϵ_F, $D_0 = v_F^2 \tau/d$.

Let us now obtain the equation satisfied by the Cooperon in the presence of a time-dependent electromagnetic field. In that case we have for the interaction the Lagrangian

$$L_1(\mathbf{x}_t, \dot{\mathbf{x}}_t, t) = e\dot{\mathbf{x}}_t \cdot \mathbf{A}(\mathbf{x}_t, t) - e\phi(\mathbf{x}_t, t) \, . \tag{11.48}$$

Since the coherence between time-reversed trajectories is partially upset, it is convenient to introduce arbitrary initial and final times, and we have for the phase difference between a pair of time-reversed paths

$$\varphi[\mathbf{x}_t^{cl}] = \frac{1}{\hbar}\{S[\mathbf{x}_t^{cl}] - S[\mathbf{x}_{t_i+t_f-t}^{cl}]\}$$

$$= \int_{t_i}^{t_f} dt \left(L_1(\mathbf{x}_t^{cl}, \dot{\mathbf{x}}_t^{cl}, t) - L_1(\mathbf{x}_{t_i+t_f-t}^{cl}, \dot{\mathbf{x}}_{t_i+t_f-t}^{cl}, t) \right) \tag{11.49}$$

---

[13] An exception to this is discussed in section 11.3.1.

[14] In case the classical motion in the random potential is adequately described as the diffusion process, we immediately recover the result eq.(11.15) for the return probability.

as the contributions to the phase difference from L_0 cancels, and we are left with

$$\varphi[\mathbf{x}_t^{cl}] = \frac{e}{\hbar} \int_{t_i}^{t_f} dt \Big\{ \dot{\mathbf{x}}_{cl}(t) \cdot \mathbf{A}(\mathbf{x}_{cl}(t), t) + \phi(\mathbf{x}_{cl}(t_i + t_f - t), t_i + t_f - t) - \phi(\mathbf{x}_{cl}(t), t)$$

$$- \dot{\mathbf{x}}_{cl}(t_i + t_f - t) \cdot \mathbf{A}(\mathbf{x}_{cl}(t_i + t_f - t), t_i + t_f - t) \Big\} . \tag{11.50}$$

Introducing the shift in the time variable

$$t' \equiv t - T , \qquad T \equiv \frac{1}{2}(t_f + t_i) \tag{11.51}$$

we get

$$\varphi[\mathbf{x}_t^{cl}] = \frac{e}{\hbar} \int_{\frac{t_i - t_f}{2}}^{\frac{t_f - t_i}{2}} dt' \Big\{ \dot{\mathbf{x}}_{cl}(t' + T) \cdot \mathbf{A}(\mathbf{x}_{cl}(t' + T), t' + T)$$

$$- \dot{\mathbf{x}}_{cl}(T - t') \cdot \mathbf{A}(\mathbf{x}_{cl}(T - t'), T - t')$$

$$- \phi(\mathbf{x}_{cl}(t' + T), t' + t) + \phi(\mathbf{x}_{cl}(T - t'), T - t') \Big\} . \tag{11.52}$$

The electromagnetic field is assumed to have a negligible effect on determining the classical paths, and we can shift the time argument specifying the position on the path to be symmetric about the moment in time T, and thereby rewrite the phase difference, $t \equiv t_f - t_i$,

$$\varphi[\mathbf{x}_t^{cl}] = \frac{e}{\hbar} \int_{-\frac{t}{2}}^{\frac{t}{2}} d\bar{t} \Big\{ \dot{\mathbf{x}}_{\bar{t}}^{cl} \cdot \mathbf{A}_T(\mathbf{x}_{\bar{t}}^{cl}, \bar{t}) - \phi(\mathbf{x}_{\bar{t}}^{cl}, \bar{t}) \Big\} \tag{11.53}$$

where

$$\phi_T(\mathbf{x}, t) = \phi(\mathbf{x}, T + t) - \phi(\mathbf{x}, T - t) \tag{11.54}$$

and

$$\mathbf{A}_T(\mathbf{x}, t) = \mathbf{A}(\mathbf{x}, T + t) + \mathbf{A}(\mathbf{x}, T - t) . \tag{11.55}$$

An electric field can be represented solely by a scalar potential, and we immediately conclude that only if the field is different on time-reversed trajectories can it lead to destruction of phase coherence. In particular, an electric field constant in time does not affect the phase coherence, and thereby does not influence the weak-localization effect.

The differential equation corresponding to the path integral, eq.(11.47), therefore gives for the Cooperon equation for the case of a time-dependent electromagnetic field

$$\left\{ \frac{\partial}{\partial t} + \frac{e}{\hbar} \phi_T(\mathbf{x}_t, t) - D_0 \left(\nabla_{\mathbf{x}} - \frac{ie}{\hbar} \mathbf{A}_T(\mathbf{x}, t) \right)^2 \right\} C_{t,t'}^T(\mathbf{x}, \mathbf{x}') = \delta(\mathbf{x} - \mathbf{x}') \, \delta(t - t') .$$

$$\tag{11.56}$$

When the sample is exposed to a time-independent magnetic field, we recover the static Cooperon equation, eq.(11.31).

Above we have presented the derivation of the Cooperon equation in the presence of a time-dependent electromagnetic field based on the quantum interference picture of the weak-localization effect. We now provide the derivation of this result based on the diagrammatic representation of the Cooperon following the presentation of reference [58].

We seek the equation satisfied by the Cooperon in the presence of the vector potential $\mathbf{A}(\mathbf{x}, t) = \mathbf{A}e^{i\mathbf{q}\cdot\mathbf{x} - i\omega t}$, and should insert a current vertex into the propagators in the Cooperon in all possible ways. When the field is inserted into a propagator in a maximally crossed diagram, the subsequent propagators will all have their energy and momentum boosted by the wave vector and frequency of the vector potential according to the current vertex formula, eq.(8.53). The Cooperon in the presence of the field therefore satisfies the equation, $E_\pm = E \pm \hbar\omega/2$,

$$C_{\mathbf{Q}}(E_+, E_-) = C_{\mathbf{Q}}^{(0)}(E_+ - E_-)\left(1 + \hat{\zeta}(\mathbf{Q}, E_+, E_-, \mathbf{q})C_{\mathbf{Q}}(E_+, E_-)\right) \quad (11.57)$$

where

$$C_{\mathbf{Q}}^{(0)}(E_+ - E_-) = C_{\mathbf{Q}}(\omega) = \frac{1/\tau}{-i(E_+ - E_-)/\hbar + D_0 Q^2} \quad (11.58)$$

is the Cooperon in the absence of the vector potential. The operator $\hat{\zeta}$ describes the effect of inserting the electromagnetic field in all possible ways in the retarded and advanced lines of the insertion $\zeta(\mathbf{Q}, \omega)$ of the Cooperon. We only need to account for the electromagnetic field to second order. There will therefore be five contributions corresponding to the five ways the current vertex can be inserted into the insertion consisting of the box of a retarded and advanced propagator, and $\hat{\zeta}$ is therefore the differential operator

$$\hat{\zeta}(\mathbf{Q}, E_+, E_-, \mathbf{q}) = \sum_{j=1,2}\sum_{k=0}^{j} \zeta_{jk}\, e^{j\mathbf{q}\cdot\frac{\partial}{\partial\mathbf{Q}} + k\hbar\omega\cdot\frac{\partial}{\partial E_+} + (j-k)\hbar\omega\cdot\frac{\partial}{\partial E_-}} \quad (11.59)$$

where ζ_{jk} describes the effect of inserting the current vertex for $\mathbf{q} = 0$ and $\omega = 0$. The index j describes how many times the vector potential is inserted, and k whether it is inserted into the retarded or advanced propagator. However, we noted in section 8.4, eq.(8.53), that inserting the zero wave vector and frequency current vertex amounted to differentiating the propagator

$$G^{R(A)}(\mathbf{p}, E) \to G^{R(A)}(\mathbf{p}, E)\,\mathbf{v}_{\mathbf{p}}\cdot e\mathbf{A}\,G^{R(A)}(\mathbf{p}, E) = e\mathbf{A}\cdot\frac{\partial}{\partial\mathbf{p}}G^{R(A)}(\mathbf{p}, E)\,. \quad (11.60)$$

The differentiation appears under the momentum integral in the insertion $\zeta(\mathbf{Q}, \omega)$, and we can therefore shift to a differentiation with respect to $\mathbf{Q}$. Calculating for the five possible insertions we get for the two first-order terms

$$\zeta_{1k} = e\mathbf{A}\cdot\frac{\partial\zeta(\mathbf{Q}, \omega)}{\partial\mathbf{Q}} = -2eD_0\,\tau\,\mathbf{A}\cdot\mathbf{Q} \quad (11.61)$$

and for the three second-order terms

$$\zeta_{20} \;=\; \frac{1}{2}\,\zeta_{21} \;=\; \zeta_{22} \;=\; \frac{e^2}{2}\,A_\alpha A_\beta \frac{\partial^2 \zeta(\mathbf{Q},\omega)}{\partial Q_\alpha Q_\beta} \;=\; -e^2 D_0 \tau \mathbf{A}^2 \; . \tag{11.62}$$

The equation satisfied by the Cooperon

$$\{[C_{\mathbf{Q}}^{(0)}(E_+,E_-)]^{-1} - \hat{\zeta}(\mathbf{Q},E_+,E_-,\mathbf{q})\}\, C_{\mathbf{Q}}(E_+,E_-) \;=\; 1 \tag{11.63}$$

thus becomes in this case

$$\left\{ -\frac{i}{\hbar}(E_+ - E_-) + D_0 \left(\mathbf{Q} - e\mathbf{A} e^{\mathbf{q}\cdot\frac{\partial}{\partial \mathbf{Q}}} \left(e^{\hbar\omega\cdot\frac{\partial}{\partial E_+}} + e^{\hbar\omega\cdot\frac{\partial}{\partial E_-}} \right) \right)^2 \right\} C_{\mathbf{Q}}(E_+,E_-) \;=\; \frac{1}{\tau}. \tag{11.64}$$

Fourier transforming we then get

$$\left\{ 2\frac{\partial}{\partial t} - D_0 \left(\nabla_{\mathbf{x}} - \frac{ie}{\hbar}\mathbf{A}_T(\mathbf{x},t) \right)^2 \right\} C_{t,t'}^{T,T'}(\mathbf{x},\mathbf{x}') \;=\; \frac{1}{\tau}\,\delta(\mathbf{x}-\mathbf{x}')\,\delta(t-t')\,\delta(T-T') \tag{11.65}$$

where

$$\mathbf{A}_T(\mathbf{x},t) \;=\; \mathbf{A}(\mathbf{x},T+t/2) + \mathbf{A}(\mathbf{x},T-t/2) \; . \tag{11.66}$$

Because of the time dependence of the external field, the Cooperon has the time labeling

$$\tag{11.67}$$

where we have introduced the notation

$$T = \frac{1}{2}(t_1 + t_1') \quad , \quad T' = \frac{1}{2}(t_2 + t_2') \quad , \quad t = t_1 - t_1' \quad , \quad t' = t_2 - t_2' \tag{11.68}$$

and

$$C(\mathbf{x},\mathbf{x}';t_1,t_1',t_2,t_2') \;\equiv\; C_{t,t'}^{T,T'}(\mathbf{x},\mathbf{x}') \; . \tag{11.69}$$

Since there is no differentiation with respect to the variable T, it is only a parameter in the Cooperon equation, we have

$$C_{t,t'}^{T,T'}(\mathbf{x},\mathbf{x}) \;=\; C_{t,t'}^{T}(\mathbf{x},\mathbf{x}')\,\delta(T-T') \tag{11.70}$$

where $C_{t,t'}^T(\mathbf{x}, \mathbf{x}')$ satisfies the equation

$$\left\{ 2\frac{\partial}{\partial t} + ie\phi_T(\mathbf{x}, t) - D_0\left(\nabla_{\mathbf{x}} - \frac{ie}{\hbar}\mathbf{A}_T(\mathbf{x}, t)\right)^2 \right\} C_{tt'}^T(\mathbf{x}, \mathbf{x}') = \frac{1}{\tau}\delta(\mathbf{x} - \mathbf{x}')\,\delta(t - t') .$$

(11.71)

Here we have also inserted the effect of a time-dependent scalar potential on the Cooperon, where $\phi_T(\mathbf{x}, t)$ is given by

$$\phi_T(\mathbf{x}, t) = \phi(\mathbf{x}, T + t/2) - \phi(\mathbf{x}, T - t/2) .$$

(11.72)

Instead of performing the analogous analysis as above, we note that this result follows from the gauge covariance property of the Cooperon.

We now derive the conductivity formula applying to the case in question. In the case of an external electromagnetic field influencing the Cooperon we consider the quantum correction to the kinetic propagator, the matrix element of eq.(6.200),

$$\delta f(\mathbf{x}_1, t_1, \mathbf{x}_1', t_1') = \sum \qquad\qquad\qquad (11.73)$$

where the sum is over all maximally crossed diagrams. For the quantum correction to the current we then have

$$\delta\mathbf{j}(\mathbf{x}, t) = \frac{e\hbar}{2im}\left(\frac{\partial}{\partial \mathbf{x}} - \frac{\partial}{\partial \mathbf{x}'}\right)\delta\rho(\mathbf{x}, t, \mathbf{x}', t)\Big|_{\mathbf{x}'=\mathbf{x}} .$$

(11.74)

The structure of the general maximally crossed diagram with n impurity correlators is

$$\delta\rho = \sum_{j=0}^{2n+1} (G^R)^j \rho (G^A)^{2n-j} .$$

(11.75)

If the equilibrium kinetic propagator ρ occurs in the above diagram at a place different from the ones indicated by circles, the contribution vanishes to the order of accuracy. In that case, viz., we encounter the product of two retarded or two advanced propagators sharing the same momentum integration variable, and since the impurity correlator effectively decouples the momentum integrations such terms are smaller by the factor $\hbar/\epsilon_F\tau$.

Displaying a maximally crossed kinetic propagator diagram on twisted form we have (we use the notation $1 \equiv (\mathbf{x}_1, t_1)$ etc.) the diagram in figure 11.3.

Because of the four different places where the kinetic propagator can occur we explicitly keep the four outermost impurity correlators, and obtain for the quantum

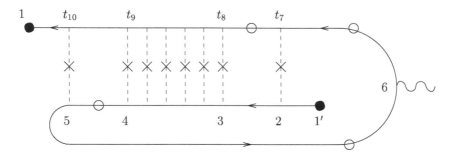

Figure 11.3 Twisted maximally crossed kinetic propagator diagram.

correction to the kinetic propagator

$$\delta\rho(\mathbf{x}_1, t_1, \mathbf{x}_1', t_1') = \frac{eu^8}{2im} \int d\Gamma \; G^R(\mathbf{x}_1, t_1; \mathbf{x}_5, t_{10}) \, G^R(\mathbf{x}_5, t_{10}; \mathbf{x}_4, t_9) \, G^0(\mathbf{x}_3, t_8; \mathbf{x}_2, t_7)$$

$$\mathbf{A}(\mathbf{x}_6, t_6) \cdot G^0(\mathbf{x}_2, t_7; \mathbf{x}_6, t_6) \overset{\leftrightarrow}{\nabla}_{\mathbf{x}_6} G^0(\mathbf{x}_6, t_6; \mathbf{x}_5, t_5)$$

$$G^0(\mathbf{x}_5, t_5; \mathbf{x}_4, t_4) \, C(\mathbf{x}_4, t_9, t_6; \mathbf{x}_3, t_8, t_3) \, G^o(\mathbf{x}_5, t_5; \mathbf{x}_4, t_4)$$

$$G^A(\mathbf{x}_3, t_3; \mathbf{x}_2, t_2) \, G^A(\mathbf{x}_2, t_2; \mathbf{x}_1', t_1') \tag{11.76}$$

where the propagators with a circle indicate where the kinetic propagator can appear (i.e., we have a sum of four terms, and the kinetic propagator is always sandwiched in between retarded propagators to the left and advanced propagators to the right), and we have introduced the abbreviation

$$d\Gamma = dt_7 dt_8 dt_9 dt_{10} \prod_{i=1}^{6} d\mathbf{x}_i dt_i \; . \tag{11.77}$$

Since the propagators carry the large momentum p_F, we can take for the explicitly appearing linear response vector potential

$$\mathbf{A}(t) = \mathbf{A}_{\omega_1} \, e^{-i\omega_1 t} \; . \tag{11.78}$$

The eight exhibited propagators in eq.(11.76) can be taken to be the equilibrium ones, and by Fourier transforming the propagators, and performing the integration over the momenta, we obtain for the quantum correction to the current density, $E_1^\pm = E_1 \pm \hbar\omega_1/2$,

$$\delta\mathbf{j}(\mathbf{x}, \omega_2) = \frac{4e^2 D_0 \tau}{i\pi} \mathbf{A}(\omega_1) \int\limits_{-\infty}^{\infty} \frac{dE_1}{2\pi\hbar} \left(f_0(E_1^-) - f_0(E_1^+) \right) \int dt_1 dt_1' dt_2 dt_2' \, \delta(t_2' - t_1')$$

$$e^{\frac{i}{\hbar}(E_1^- t_1' - E_1^+ t_2 - \hbar\omega_2 t_1)} \, C(\mathbf{x}, \mathbf{x}; t_1, t_1', t_2, t_2') \tag{11.79}$$

or equivalently

$$\delta \mathbf{j}(\mathbf{x}, \omega_2) = \frac{4e^2 D_0 \tau}{i\pi} \mathbf{A}(\omega_1) \int\limits_{-\infty}^{\infty} \frac{dE_1}{2\pi\hbar} \left(f_0(E_1^-) - f_0(E_1^+) \right)$$

$$\int\limits_{-\infty}^{\infty} dt \int\limits_{-\infty}^{\infty} dT \; C_{t,-t}^T(\mathbf{x}, \mathbf{x}) \; e^{iT(\omega_1 - \omega_2) + \frac{t}{2}(\omega_1 + \omega_2)} \; . \tag{11.80}$$

For the quantum correction to the conductivity in the presence of a time-dependent electromagnetic field

$$\delta \mathbf{j}(\mathbf{x}, \omega_2) = \delta\sigma(\mathbf{x}, \omega_2, \omega_1) \, \mathbf{E}(\omega_1) \tag{11.81}$$

we therefore obtain[15]

$$\delta\sigma(\mathbf{x}, \omega_2, \omega_1) = -\frac{4e^2 D_0 \tau}{\pi\omega} \int\limits_{-\infty}^{\infty} \frac{dE_1}{2\pi\hbar} \left(f_0(E_1^-) - f_0(E_1^+) \right)$$

$$\int\limits_{-\infty}^{\infty} dt \int\limits_{-\infty}^{\infty} dT \; e^{iT(\omega_2 - \omega_1) + \frac{i}{2} t(\omega_1 + \omega_2)} C_{t,-t}^T(\mathbf{x}, \mathbf{x}) \; . \tag{11.82}$$

In the degenerate case we have

$$\delta\sigma(\mathbf{x}, \omega_2, \omega_1) = -\frac{4e^2 D_0 \tau}{\pi\hbar} \int\limits_{-\infty}^{\infty} dt \int\limits_{-\infty}^{\infty} dT \; C_{t,-t}^T(\mathbf{x}, \mathbf{x}) \; e^{iT(\omega_1 - \omega_2) + \frac{t}{2}(\omega_1 + \omega_2)} \; . \tag{11.83}$$

In a time-independent magnetic field we recover the expression eq.(11.13) for the quantum correction to the conductivity.

11.3 Phase Breaking in Weak Localization

The phase coherence between the amplitudes for pairs of time-reversed trajectories is interrupted when the environment of the electron, besides the dominating random potential, is taken into account. At nonzero temperatures, energy exchange due to the interaction with the environment will partially upset the coherence between time-reversed paths involved in the weak-localization phenomenon. The constructive interference is then partially destroyed.

Quantitatively the effect on weak localization by inelastic interactions with energy transfers ΔE of the order of the temperature, $\Delta E \sim kT$, strongly inelastic processes, can be understood by the observation that the single-particle Green's

---

[15]For an electron gas in thermal equilibrium f_0 is the Fermi function, but in principle we could at this stage have any distribution not violating Pauli's exclusion principle. However, that would then necessitate a discussion of energy relaxation processes tending to drive the system toward the equilibrium distribution.

function will be additionally damped due to interactions. If in addition to disorder we have an interaction, say with phonons, the self-energy will in lowest order in the interaction be changed according to

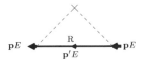

 $\qquad(11.84)$

and we will get an additional contribution to the imaginary part of the self-energy

$$\Im m \Sigma^R \; = \; -\frac{\hbar}{2\tau} - \frac{\hbar}{2\tau_{in}} \; . \qquad (11.85)$$

Upon redoing the calculation leading to eq.(8.68), we obtain in the limit $\tau_{in} \gg \tau$

$$\zeta(\mathbf{Q}, \omega) = 1 - \frac{\tau}{\tau_{in}} + i\omega\tau + D_0\tau Q^2 \; . \qquad (11.86)$$

This will in turn lead to the change in the Cooperon equation, $\omega \to \omega + i/\tau_{in}$, and we get the real space Cooperon equation[16]

$$\left\{ -i\omega - D_0\nabla_\mathbf{x}^2 + \frac{1}{\tau_{in}} \right\} C_\omega(\mathbf{x}, \mathbf{x}') \; = \; \frac{1}{\tau}\,\delta(\mathbf{x} - \mathbf{x}') \; . \qquad (11.87)$$

The effect on weak localization of electron-electron interaction and electron-phonon interaction have been studied in detail experimentally [59] [60], and can phenomenologically be accounted for adequately by introducing a temperature-dependent phase-breaking rate $1/\tau_\varphi$ in the Cooperon equation, describing the temporal exponential decay $C(t) \to C(t)\exp\{-t/\tau_\varphi\}$ of phase coherence. In many cases the inelastic scattering rate, $1/\tau_{in}$, is identical to the phase-breaking rate, $1/\tau_\varphi$. This is for example the case for electron-phonon interaction, as we shortly demonstrate. However, one should keep in mind that the inelastic scattering rate is defined as the damping of an energy state for the case where abinitio all scattering processes are weighted equally, irrespective of the amount of energy transfer. In a clean metal we observed in section 10.8 that the energy relaxation rate due to electron-phonon or electron-electron interaction is determined by energy transfers

---

[16]In the Cooperon, contributions from diagrams where besides impurity correlator lines also interaction lines connecting the retarded and advanced particle line appear should be included for consistency. However, for strongly inelastic processes these contributions are small.

of the order of the temperature as a consequence of the exclusion principle (at temperatures below the Debye temperature). In section 10.8.5 we noted that in a three-dimensional sample the energy relaxation rate in a dirty metal is larger than in a clean metal due to a strong enhancement of the electron-electron interaction with small energy transfer. When calculating the weak localization phase-breaking rate we must therefore pay special attention to the low-energy electron-electron interaction. In a thin film or in the two-dimensional case the energy relaxation rate even diverged in perturbation theory, due to the abundance of collisions with small energy transfer. However, the physically measurable phase-breaking rate do of course not suffer such a divergence since the phase change caused by an inelastic collision is given by the energy transfer times the remaining time to elapse on the trajectory. Collisions with energy transfer of the order of (the phase-breaking rate) $\hbar\omega \sim \hbar/\tau_\varphi$ or less are therefore inefficient in destroying the phase coherence between the amplitudes for traversing typical time-reversed trajectories of duration the phase coherence time τ_φ.[17] In terms of diagrams this is reflected by the fact that interaction lines can connect the upper and lower particle lines in the Cooperon, whereas there is no such process for the diagrammatic representation of the inelastic scattering rate (recall figure 10.3). This distinction is of importance in the case of a thin metallic film, the quasi two-dimensional case, where there is an abundance of scatterings with small energy transfer due to diffusion-enhanced electron-electron interaction.

In the time-reversal invariant situation, the Cooperon is equal to the classical probability that an electron at the Fermi level in time t returns to its starting point. If coherence is disrupted by interactions, the constructive interference is partially destroyed. This destruction of phase coherence results in the decay in time of coherence, described by the factor $\exp\{-t/\tau_\varphi\}$ in the expression for the Cooperon, the probability not to suffer a phase breaking collision, described by the phase-breaking rate $1/\tau_\varphi$. In view of the quantum interference picture of the weak localization effect, we shall also refer to τ_φ as the wave function phase relaxation time.

A comprehensive understanding of the phase coherence length in weak localization, the length scale $L_\varphi \equiv \sqrt{D_0\tau_\varphi}$ over which the electron diffuses quantum mechanically coherently, has been established, and this has given valuable information about inelastic scattering processes. The phase coherence length L_φ is at low temperatures much larger than the impurity mean free path l, explaining the slow spatial variation of the Cooperon on the scale of the mean free path, which we have repeatedly exploited.

11.3.1 Electron-Phonon Interaction

In this section we calculate the phase-breaking rate due to electron-phonon interaction using the simple interference picture described in the previous section. We

---

[17] A similar situation is the difference between the transport and momentum relaxation time. The transport relaxation time is the one appearing in the conductivity, reflecting that small angle scattering is ineffective in degrading the current.

start from the one-electron Lagrangian, which is given by[18]

$$L(\mathbf{x}, \dot{\mathbf{x}}) = \frac{1}{2}m\dot{\mathbf{x}}^2 - V(\mathbf{x}) - e\phi(\mathbf{x}, t) \tag{11.88}$$

where V is the impurity potential, and the deformation potential is specified in terms of the lattice displacement field, eq.(10.161),

$$e\phi(\mathbf{x}, t) = \frac{n}{2N_0} \nabla_{\mathbf{x}} \cdot \mathbf{u}(\mathbf{x}, t) . \tag{11.89}$$

It is important to note that the impurities move in phase with the distorted lattice; hence the impurity potential has the form

$$V(\mathbf{x}) = \sum_i V_{\text{imp}}(\mathbf{x} - (\mathbf{R}_i + \mathbf{u}(\mathbf{x}, t)) \tag{11.90}$$

where $\mathbf{R}_i$ is the equilibrium position of the i'th ion. The impurity scattering is thus only elastic in the frame of reference that locally moves along with the lattice. We therefore shift to this moving frame of reference by changing the electronic coordinate according to $\mathbf{x} \to \mathbf{x}+\mathbf{u}$. The impurity scattering then becomes static on account of generating additional terms of interaction. Expanding the Lagrangian eq.(11.88) in terms of the displacement, and neglecting terms of relative order m/M, such as the term $m\dot{\mathbf{u}} \cdot \mathbf{v}/2$, the transformed Lagrangian can be written as $L = L_0 + L_1$, where L_0 is given in eq.(11.35), and[19]

$$L_1(\mathbf{x}_t, \dot{\mathbf{x}}_t) = m\dot{\mathbf{x}}_t \cdot (\dot{\mathbf{x}}_t \cdot \nabla) \mathbf{u}(\mathbf{x}_t, t) - \frac{1}{3}\dot{\mathbf{x}}_t^2 \nabla \cdot \mathbf{u}(\mathbf{x}_t, t) . \tag{11.91}$$

In the last line we have used the relation $n/2N_0 = mv_F^2/3$, and the fact that the magnitude of the velocity is conserved in elastic scattering. We therefore obtain for the phase difference[20]

$$\varphi[\mathbf{x}_t^{cl}] = \frac{1}{\hbar} \int\limits_{-t/2}^{t/2} dt \left\{ \nabla_\beta u_\alpha(\mathbf{x}_t^{cl}, t) - \nabla_\beta u_\alpha(\mathbf{x}_t^{cl}, -t) \right\} \left[\dot{x}_t^\alpha \dot{x}_t^\beta - \frac{1}{3} \delta_{\alpha\beta} \dot{\mathbf{x}}_t^2 \right] \tag{11.92}$$

where summation over repeated Cartesian indices is implied, and we have chosen the classical paths to satisfy the boundary condition, $\mathbf{x}_{-t/2}^{cl} = \mathbf{0} = \mathbf{x}_{t/2}^{cl}$.

We must now average the quantum interference term as given in eq.(11.40) with respect to the lattice vibrations, and with respect to the random positions of the impurities. Since the Lagrangian for the lattice vibrations is a quadratic form

---

[18]We follow reference [56].

[19]This result can also be obtained without introducing the moving frame of reference. By simply Taylor-expanding eq.(11.90) and using Newton's equation we obtain a Lagrangian which differs from the one in eq.(11.91) by only a total time derivative, and therefore generates the same dynamics.

[20]In neglecting the Jacobian of the nonlinear transformation to the moving frame, we neglect the influence of the lattice motion on the paths.

in the displacement $\mathbf{u}$, and the phase difference $\varphi[\mathbf{x}_t^{cl}]$ is linear in the displacement, the phonon average can by Wick's theorem be computed according to (see exercise 6.125 on page 255)[21]

$$< e^{i\varphi[\mathbf{x}_t^{cl}]} >_{ph} = e^{-\frac{1}{2}<\varphi[\mathbf{x}_t^{cl}]^2>_{ph}} . \tag{11.93}$$

For the argument of the exponential we obtain ($\mathbf{v}_t \equiv \dot{\mathbf{x}}_t^{cl}$)

$$< \varphi[\mathbf{x}_t^{cl}]^2 >_{ph} = \frac{m^2}{\hbar^2} \int\limits_{-t/2}^{t/2} dt_1 \int\limits_{-t/2}^{t/2} dt_2 \left[\sum_{\pm} (\pm) D_{\alpha\beta\gamma\delta}(\mathbf{x}_{t_1}^{cl} - \mathbf{x}_{t_2}^{cl}, t_1 \mp t_2) \right]$$

$$\left[v_{t_1}^\alpha v_{t_1}^\beta - \frac{1}{3} \delta_{\alpha\beta} \mathbf{v}_{t_1}^2 \right] \left[v_{t_2}^\gamma v_{t_2}^\delta - \frac{1}{3} \delta_{\gamma\delta} \mathbf{v}_{t_2}^2 \right] \tag{11.94}$$

where the phonon correlator

$$D_{\alpha\beta\gamma\delta}(\mathbf{x}, t) = < \nabla_\beta u_\alpha(\mathbf{x}, t) \nabla_\delta u_\gamma(\mathbf{0}, 0) > \tag{11.95}$$

is an even function of the time difference t.

Concerning the average with respect to impurity positions, we will resort to an approximation which, since the exponential function is a convex function, can be expressed as the inequality

$$< C(t) >_{imp} \geq < C_I(t) >_{imp}^{(0)} < C(t) >_{imp}^{\varphi=0} e^{-\frac{1}{2}<<\varphi[\mathbf{x}_t^{cl}]^2>_{ph}>_{imp}} \tag{11.96}$$

where we have introduced the notation for the impurity average

$$<< (\varphi[\mathbf{x}_t^{cl}])^2 >_{ph}>_{imp} = \frac{\int\limits_{\mathbf{x}_{-t/2}=\mathbf{x}}^{\mathbf{x}_{t/2}=\mathbf{x}} \mathcal{D}\mathbf{x}_t \, P_t[\mathbf{x}_t] \, < (\varphi[\mathbf{x}_t^{cl}])^2 >_{ph}}{\int\limits_{\mathbf{x}_{-t/2}=\mathbf{x}}^{\mathbf{x}_{t/2}=\mathbf{x}} \mathcal{D}\mathbf{x}_t \, P_t[\mathbf{x}_t]} . \tag{11.97}$$

The phase difference eq.(11.92) depends on the local velocity of the electron, which is a meaningless quantity in Brownian motion.[22] It is therefore necessary when considering phase breaking due to electron-phonon interaction to consider the time-reversed paths involved in the weak-localization quantum interference process as realizations of Boltzmannian motion. At a given time, a Boltzmannian path is completely specified by its position and by the direction of its velocity as discussed in section 5.5. We are dealing with the Markovian process described by the Boltzmann propagator $F(\mathbf{v}, \mathbf{x}, t; \mathbf{v}', \mathbf{x}', t')$, where we now use the velocity as variable instead of the momentum as used in section 5.5. On account of the Markovian property, the four-point correlation function required in eq.(11.97) (the start

---

[21]We have suppressed the hat on $\mathbf{u}$ indicating that the displacement is an operator with respect to the lattice degrees of freedom (or we have envisaged treating the lattice vibrations in the path integral formulation).

[22]The velocity entering in the Wiener measure, eq.(8.118), is not the local velocity, but an average of the velocity on a Boltzmannian path; recall section 8.7.

and (identical) end point and two intermediate points according to eq.(11.94)) may be expressed as a product of three conditional probabilities of the type eq.(5.175), and we obtain

$$
<<\varphi[\mathbf{x}_t^{cl}]^2>_{\text{ph}}>_{\text{imp}} \quad = \quad \frac{4m^2}{\hbar^2} \int\limits_{-t/2}^{t/2} dt_1 \int\limits_{-t/2}^{t/2} dt_2 \int d\mathbf{x}_1 \int d\mathbf{x}_2 \int \frac{d\hat{\mathbf{v}}_1 d\hat{\mathbf{v}}_2}{(4\pi)^2}
$$

$$
\overline{F}(0, \frac{t}{2}; \mathbf{x}_1, \mathbf{v}_1, t_1) F(\mathbf{x}_1, \mathbf{v}_1, t_1; \mathbf{x}_2, \mathbf{v}_2, t_2) \, \overline{F}(\mathbf{x}_2, \mathbf{v}_2, t_2; 0, -\frac{t}{2})
$$

$$
\left[\sum_{\pm} (\pm) D_{\alpha\beta\gamma\delta}(\mathbf{x}_{t_1}^{cl} - \mathbf{x}_{t_2}^{cl}, t_1 \mp t_2) \right] \left[v_{t_1}^{\alpha} v_{t_1}^{\beta} - \frac{1}{3} \delta_{\alpha\beta} \mathbf{v}_{t_1}^2 \right] \left[v_{t_2}^{\gamma} v_{t_2}^{\delta} - \frac{1}{3} \delta_{\gamma\delta} \mathbf{v}_{t_2}^2 \right] .
$$

$$(11.98)$$

We use the notation that an angular average of the Boltzmann propagator F with respect to one of its velocities is indicated by a bar. For example, we have for the return probability

$$
< C(t) >_{\text{imp}}^{(\varphi=0)} \quad = \quad \overline{\overline{F}}(\mathbf{x}, t; \mathbf{x}', 0) \equiv \int \frac{d\hat{\mathbf{v}}'}{4\pi} \, \overline{F}(\mathbf{x}, t; \mathbf{v}', \mathbf{x}', t') . \qquad (11.99)
$$

The space-dependent quantities may be expressed by Fourier integrals according to eq.(5.177). Since the Boltzmann propagator is retarded, $F(\mathbf{v}, \mathbf{x}, t; \mathbf{v}', \mathbf{x}', t')$ vanishes for t earlier than t', we can expand the upper t_1-integration to infinity and the lower t_2-integration to minus infinity. Only thermally excited phonons contribute to the destruction of phase coherence, and we conclude that $D_{\alpha\beta\gamma\delta}(\mathbf{x}_{t_1}^{cl} - \mathbf{x}_{t_2}^{cl}, t_1 \mp t_2)$ is essentially zero for $|t_1 \pm t_2| \geq \hbar/kT$.[23] We can therefore extend the domain of integration to infinity with respect to $|t_1 \pm t_2|$ provided that $|t| \gg \hbar/kT$, and obtain in the convex approximation

$$
<C(t)>_{\text{imp}} = <C(t)>_{\text{imp}}^{(\varphi=0)} \exp \Big\{ - \frac{2m^2}{\hbar^2 <C(t)>_{\text{imp}}^{(\varphi=0)}} \int \frac{d\mathbf{k}d\mathbf{k}'d\omega d\omega'}{(2\pi)^8} \int \frac{d\hat{\mathbf{v}}_1 d\hat{\mathbf{v}}_2}{(4\pi)^2}
$$

$$
\overline{F}(\mathbf{v}_1; \mathbf{k}, \omega) F(\mathbf{v}_1, \mathbf{v}_2, \mathbf{k} + \mathbf{k}', \omega + \omega') D_{\alpha\beta\gamma\delta}(-\mathbf{k}', -\omega')
$$

$$
\left[\overline{F}(\mathbf{v}_2, \mathbf{k}, \omega) e^{-i\omega t} - \overline{F}(\mathbf{v}_2, \mathbf{k}, \omega + 2\omega') e^{-i(\omega+\omega')t} \right]
$$

$$
\left[v_{t_1}^{\alpha} v_{t_1}^{\beta} - \frac{1}{3} \delta_{\alpha\beta} \mathbf{v}_{t_1}^2 \right] \left[v_{t_2}^{\gamma} v_{t_2}^{\delta} - \frac{1}{3} \delta_{\gamma\delta} \mathbf{v}_{t_2}^2 \right] \Big\} \qquad (11.100)
$$

---

[23] As we noted below eq.(6.56) on page 242, there is no explicit reference to the lattice dynamics in the reduced density matrix propagator, a point also elaborated in reference [61] where it is shown that the zero point fluctuations of the lattice can not disrupt the weak-localization coherence.

We expect that the argument of the exponential above increases linearly in t for large times. Since the classical return probability in three dimensions has the time dependence $<C(t)>_{\text{imp}}^{(\varphi=0)} \propto t^{-3/2}$ (recall the form of the diffusion propagator), the integral above should not decrease faster than $t^{-1/2}$. Such a slow decrease is obtained from the $(\mathbf{k}, \omega)$-integration only from the combination $\overline{F}(\mathbf{v}_1; \mathbf{k}, \omega)\,\overline{F}(\mathbf{v}_2; \mathbf{k}, \omega)$, which according to eq.(5.181) features an infrared singular behavior $(-i\omega + D_0 k^2)^{-2}$ for small k and ω. In fact, it is just this combination which leads to a time-dependence proportional to $t^{-1/2}$, and compared to that, all other contributions may be neglected. For the important region of integration we thus have $\omega \ll \omega'$, since ω' is determined by the phonon correlator which gives the large contribution to the integral for the typical value $\hbar\omega' \simeq kT$. We are therefore allowed to approximate $F(\mathbf{v}_1, \mathbf{v}_2; \mathbf{k}+\mathbf{k}', \omega+\omega')$ by $F(\mathbf{v}_1, \mathbf{v}_2; \mathbf{k}', \omega')$. In addition, the same arguments show that the second term in the square bracket may be omitted. We thus obtain

$$<C(t)>_{\text{imp}} = <C(t)>_{\text{imp}}^{(\varphi=0)} e^{-t/\tau_\varphi} \tag{11.101}$$

where the phase-breaking rate due to electron-phonon interaction is given by

$$\frac{1}{\tau_\varphi} = \frac{2m^2}{\hbar^2} \int \frac{d\mathbf{k}'d\omega'}{(2\pi)^4} \int \frac{d\hat{\mathbf{v}}_1 d\hat{\mathbf{v}}_2}{(4\pi)^2} F(\mathbf{v}_1, \mathbf{v}_2; \mathbf{k}', \omega') \, D^{\alpha\beta\gamma\delta}(\mathbf{k}', \omega') \left[v_1^\alpha v_1^\beta - \frac{1}{3}\delta_{\alpha\beta} \mathbf{v}_1^2 \right]$$

$$\left[v_2^\alpha v_2^\beta - \frac{1}{3}\delta_{\alpha\beta} \mathbf{v}_2^2 \right] . \tag{11.102}$$

For simplicity we consider the Debye model where the lattice vibrations are specified by the density n_i and the mass M of the ions, and by the longitudinal c_l and the transverse c_t sound velocities.[24] We assume the phonons to have three-dimensional character. In case of longitudinal vibrations, we have the normal mode expansion of the displacement field

$$\mathbf{u}(\mathbf{r}, t) = \frac{i}{\sqrt{N}} \sum_{\mathbf{k}} \hat{\mathbf{k}} Q_{\mathbf{k}}(t) e^{i\mathbf{k}\cdot\mathbf{r}} \tag{11.103}$$

where N is the number of ions in the normalization volume. For the phonon average we have

$$< Q_{\mathbf{k}}(t) Q_{\mathbf{k}'}(t') > = \delta_{\mathbf{k},-\mathbf{k}'} \frac{\hbar}{2M\omega_{\mathbf{k}}} H(\omega_{\mathbf{k}}) \cos\omega_{\mathbf{k}}(t - t') \tag{11.104}$$

where $\omega_{\mathbf{k}} = c_l k$, provided that k is less than the cutoff wave vector k_D, and we obtain for the Fourier transform of the longitudinal phonon correlator

$$D_L^{\alpha\beta\gamma\delta}(\mathbf{k}, \omega)] = \frac{1}{2} k^\alpha k^\beta k^\gamma k^\delta H(\omega_{\mathbf{k}}) [\delta(\omega - \omega_{\mathbf{k}}) + \delta(\omega + \omega_{\mathbf{k}})] . \tag{11.105}$$

---

[24]The jellium model does not allow inclusion of Umklapp processes in the electron-phonon scattering.

Strictly speaking, we encounter in the above derivation $H(\omega) = 2n(\omega) + 1$, where n is the Bose distribution function. However, the present single electron theory does not take into account that the Pauli principle forbids scattering of an electron into occupied states. Referring to section 10.8.3, we recall that Pauli's exclusion principle can be incorporated by the replacement (f_0 is the Fermi function)

$$\frac{1}{2}H(\omega) \quad \rightarrow \quad 2n(\omega) + 1 - f_0(\epsilon_F - \hbar\omega) + f_0(\epsilon_F + \hbar\omega)$$

$$= \quad \coth\frac{\hbar\omega}{2kT} - \tanh\frac{\hbar\omega}{2kT} = \frac{2}{\sinh\frac{\hbar\omega}{kT}} \ . \tag{11.106}$$

Upon inserting eq.(11.105) in the expression eq.(11.102) for the phase-breaking rate, we encounter the directional average of expressions of the type

$$\hat{k}_\alpha \hat{k}_\beta [v_\alpha v_\beta - \delta_{\alpha\beta}\frac{\mathbf{v}^2}{3}] = k^{-2}[(\mathbf{k} \cdot \mathbf{v})^2 - k^2\frac{\mathbf{v}^2}{3}] \ . \tag{11.107}$$

Altogether the angular averages appear in the combination

$$\Phi_L(kl) = \frac{18}{\pi v_F^3 k^3}\left\{ I(k,\omega)\left[\int\frac{d\hat{\mathbf{v}}}{4\pi}\frac{(\mathbf{k} \cdot \mathbf{v})^2 - k^2\frac{\mathbf{v}^2}{3}}{-i\omega + i\mathbf{v} \cdot \mathbf{k} + 1\tau}\right]^2 + \int\frac{d\hat{\mathbf{v}}}{4\pi}\frac{[(\mathbf{k} \cdot \mathbf{v})^2 - k^2\frac{\mathbf{v}^2}{3}]^2}{-i\omega + i\mathbf{v} \cdot \mathbf{k} + 1/\tau}\right\}$$

$$= \frac{2}{\pi}\left(\frac{kl\arctan kl}{kl - \arctan kl} - \frac{3}{kl}\right) \tag{11.108}$$

where the result in the last line is obtained since $\omega = c_l k \ll v_F k$. For the phase-breaking rate due to longitudinal phonons we thus obtain

$$\frac{1}{\tau_{\varphi,l}} = \frac{\pi\hbar^2}{6mMc_l}\int_0^{k_D}dk \ k^2 \ \Phi_L(kl) \frac{1}{\sinh \hbar c_l k/kT} \ . \tag{11.109}$$

We note the limiting behaviors

$$\frac{1}{\tau_{\varphi,l}} = \begin{cases} \frac{7\pi\zeta(3)}{12}\frac{(kT)^3}{\hbar n Mc_l^4} & \hbar c_l/l \ll kT \ll \hbar c_l k_D \\ \\ \frac{\pi^4}{30} l \frac{(kT)^4}{\hbar n Mc_l^5} & kT \ll \hbar c_l/l \ . \end{cases} \tag{11.110}$$

The expression eq.(11.108) for the function Φ_L demonstrates in a direct way the important compensation that takes place in the case of longitudinal phonons between the two mechanisms contained in L_1. First, the term $(\mathbf{k} \cdot \mathbf{v})^2$ corresponds to $m\mathbf{v} \cdot (\mathbf{v} \cdot \nabla)\mathbf{u}$ and represents the coupling of the electrons to the vibrating impurities. Second, the term $-k^2\mathbf{v}^2/3$ is connected with $-m\mathbf{v}^2\nabla \cdot \mathbf{u}/3$, and originates from the interaction of the electrons with the lattice vibrations. Without this compensation, each of the mechanisms would appear to be enhanced in an impure metal, and would lead to an enhanced phase-breaking rate proportional to $(kT)^2/(nMc_l^3 l)$.

For the case of transverse vibrations, we note that $D_T^{\alpha\beta\gamma\delta}$ is of similar form as eq.(11.105) where, however, $\hat{k}_\alpha \hat{k}_\gamma$ has to be replaced by $(\delta_{\alpha\gamma} - \hat{k}_\alpha \hat{k}_\gamma)$ and an additional factor of 2 which accounts for the multiplicity of transverse modes. We then obtain a phase-breaking rate due to interaction with transverse phonons, $\tau_{\varphi,t}$, which is similar to the expression in eq.(11.109) with c_l and ϕ_L replaced by c_t and

$$\Phi_T(kl) = \frac{3}{\pi} \frac{2k^3 l^3 + 3kl - 3(k^2 l^2 + 1) \arctan kl}{k^4 l^4} \tag{11.111}$$

respectively. In particular, we obtain the limiting behaviors for the phase-breaking rate due to transverse phonons

$$\frac{1}{\tau_{\varphi,t}} = \begin{cases} \frac{\pi^2}{2} \frac{(kT)^2}{mMc_t^3 l} & \hbar c_t/l \ll kT \ll \hbar c_t k_D \\[2ex] \frac{\pi^4}{20} l \frac{(kT)^4}{\hbar^2 mMc_t^5} & kT \ll \hbar c_t/l \ . \end{cases} \tag{11.112}$$

We note that in the high temperature region, $\hbar c_t/l \ll kT \ll \hbar c_t k_D$, the transverse contribution is negligible in comparison with the longitudinal one if $c_t \simeq c_l$. But the transverse rate dominates in the case where the transverse sound velocity is much smaller than the longitudinal one. Such a situation may quite well be realized in some amorphous metals; then, it is possible to observe a phase-breaking rate of the form $\tau_\varphi^{-1} \propto T^2/l$ at higher, but not too high, temperatures.[25] The predictions of the theory are in good agreement with magnetoresistance measurements and carefully conducted experiments of the temperature dependence of the resistance [57].

The physical meaning of the second term in eq.(11.100) is as follows: It is appreciable only if the lattice deformation stays approximately constant during the time the electron spends on its path and leads, in this case, to a cancellation of the first term. Equivalently, we have that electron-phonon interactions with small energy transfers do not lead to destruction of phase coherence. The effect of this term is thus effectively to introduce a lower cutoff in the integral of eq.(11.109) at wave vector $k_0 = 1/c_l \tau_{\varphi,l}$. However, there are no realistic models of phonon spectra where this effect is of importance. We therefore have the relationships $\omega' \simeq kT/\hbar \gg \omega \simeq 1/\tau_\varphi$. It is therefore no surprise that the calculated phase-breaking rates are identical to the inelastic collision rates in a dirty metal [62]. When considering phase breaking due to electron-electron interaction, which we now turn to, the small energy transfer interactions are of importance.

11.3.2 Electron-Electron Interaction

In this section we consider the temperature dependence of the phase-breaking rate due to electron-electron interaction.[26] As already discussed at the beginning of

---

[25]A quadratic temperature dependence of the phase-breaking rate is often observed experimentally.

[26]We follow reference [58].

this section special attention to electron-electron interaction with small energy transfer must be exercised due to the diffusion enhancement. In diagrammatic terms we therefore need to take into account diagrams where the electron-electron interaction connects also the upper and lower particle lines in the Cooperon.

In section 10.8.5 we noted that the effective electron-electron interaction at low energies can be represented by a fluctuating field. The correlation function was in a dirty metal given by the expression in eq.(10.243). We can therefore obtain the effect on the Cooperon of the quasi-elastic electron-electron interaction by averaging the Cooperon with respect to a time-dependent electromagnetic field using the proper correlator. We therefore consider the equation for the Cooperon in the presence of an electromagnetic field, eq.(11.71),

$$\left\{ 2\frac{\partial}{\partial t} - D_0 \left(\nabla_{\mathbf{x}} - \frac{ie}{\hbar} \mathbf{A}_T(\mathbf{x}, t) \right)^2 + \frac{1}{\tau_T^{e-e}} \right\} C_{t,t'}^T(\mathbf{x}, \mathbf{x}') = \frac{1}{\tau} \delta(\mathbf{x} - \mathbf{x}')\, \delta(t - t')$$

(11.113)

where we have chosen a gauge in which the scalar potential vanishes, and $1/\tau_T^{e-e}$ is the energy relaxation rate due to high-energy processes, i.e., processes with energy transfers $\sim kT$.[27]

To account for the electron-electron interaction with small energy transfers, we must perform the Gaussian average of the Cooperon with respect to the fluctuating field. This is facilitated by writing the solution of the Cooperon equation as the path integral

$$C_{t,t'}^T(\mathbf{R}, \mathbf{R}') = \frac{1}{2\tau} \int_{\mathbf{x}_{t'}=\mathbf{x}'}^{\mathbf{x}_t=\mathbf{x}} \mathcal{D}\mathbf{x}_t \, e^{-S[\mathbf{x}_t]}$$

(11.114)

where the Euclidean action consists of two terms

$$S = S_0 + S_{\mathbf{A}}$$

(11.115)

where

$$S_0[\mathbf{x}_t] = \int_t^{t'} dt_1 \left(\frac{\dot{\mathbf{x}}_{t_1}^2}{4D_0} + \frac{1}{\tau_T^{e-e}} \right)$$

(11.116)

and

$$S_{\mathbf{A}}[\mathbf{x}_t] = \frac{ie}{\hbar} \int_t^{t'} dt_1 \, \dot{\mathbf{x}}_{t_1} \cdot \mathbf{A}_T(\mathbf{x}_{t_1}, t_1) \,.$$

(11.117)

In terms of diagrams, the Gaussian average corresponds to connecting the external field lines pairwise in all possible ways by the correlator of the field fluctuations, thereby producing the effect of the low-energy electron-electron interaction. Since

---

[27] As will become clear in the following, the separation in high- and low-energy transfers takes place at energies of the order of the temperature. However, in the following we shall not need to specify the separation explicitly.

the fluctuating vector potential appears linearly in the exponential Cooperon expression, the Gaussian average with respect to the fluctuating field is readily done

$$C_{t,t'}^{T}(\mathbf{R}, \mathbf{R}') = \frac{1}{2\tau} \int_{\mathbf{r}_{t'}'=\mathbf{R}'}^{\mathbf{r}_t=\mathbf{R}} \mathcal{D}\mathbf{r}_t \; e^{-(S_0[\mathbf{x}_t] + <S_\mathbf{A}>[\mathbf{x}_t])} \tag{11.118}$$

where the averaged action $<S_A>$ is expressed in terms of the correlator of the vector potential

$$<S_\mathbf{A}>[\mathbf{x}_t] = \frac{e^2}{2\hbar^2} \int_{t'}^{t} dt_1 \int_{t'}^{t} dt_2 \; \dot{x}_\mu(t_1) \, \dot{x}_\nu(t_2) < A_\mu^T(\mathbf{x}_{t_2}, t_1) A_\nu^T(\mathbf{x}_{t_2}, t_2) > . \tag{11.119}$$

If we recall the definition of $\mathbf{A}_T(\mathbf{x}_t, t)$, eq.(11.66), we have

$$< A_\mu^T(\mathbf{x}_{t_2}, t_1) A_\nu^T(\mathbf{x}_{t_2}, t_2)> = 2 \int \frac{d\mathbf{q}}{(2\pi)^d} \int \frac{d\omega}{2\pi} e^{i\mathbf{q}\cdot(\mathbf{x}_{t_1}-\mathbf{x}_{t_2})} < A_\mu A_\nu >_{\mathbf{q}\omega}$$

$$\left(\cos\omega \frac{t_1 + t_2}{2} + \cos\omega \frac{t_1 - t_2}{2} \right) \tag{11.120}$$

where we have introduced the notation

$$< A_\mu A_\nu >_{\mathbf{q}\omega} \equiv \; < A_\mu(\mathbf{q}, \omega) A_\nu(-\mathbf{q}, -\omega) > . \tag{11.121}$$

The electric field fluctuations could equally well have been represented by a scalar potential

$$< A_\mu(\mathbf{q}, \omega) A_\nu(-\mathbf{q}, -\omega) > = \frac{1}{\omega^2} < E_\mu(\mathbf{q}, \omega) E_\nu(-\mathbf{q}, -\omega) >$$

$$= \frac{q_\mu q_\nu}{\omega^2} < \phi(\mathbf{q}, \omega)\phi(-\mathbf{q}, -\omega) > . \tag{11.122}$$

In section 10.8.5 we showed that the electron-electron interaction with small energy transfers, $\hbar\omega \ll kT$, is determined by the temperature and conductivity of the sample[28]

$$< A_\mu A_\nu >_{\mathbf{q}\omega} = \frac{2kT}{\omega^2\sigma_0} \frac{q_\mu q_\nu}{q^2} . \tag{11.123}$$

Upon partial integration we notice the identity (the boundary terms are seen to vanish as $\mathbf{x}_{-t} = \mathbf{x}_t$)

$$\int_{t'}^{t} dt_1 \int_{t'}^{t} dt_2 \; q_\mu q_\nu \, \dot{x}_\mu(t_1) \, \dot{x}_\nu(t_2) e^{i\mathbf{q}\cdot(\mathbf{x}_{t_1}-\mathbf{x}_{t_2})} \left(\cos\frac{\omega(t_1+t_2)}{2} + \cos\frac{\omega(t_1-t_2)}{2} \right)$$

$$= - \int_{t'}^{t} dt_1 \int_{t'}^{t} dt_2 \; e^{i\mathbf{q}\cdot(\mathbf{x}_{t_1}-\mathbf{x}_{t_2})} \frac{\omega^2}{4} \left(\cos\frac{\omega(t_1+t_2)}{2} - \cos\frac{\omega(t_1-t_2)}{2} \right) \tag{11.124}$$

---

[28]Since the time label T now has disappeared, no confusion should arise in the following where T denotes the temperature.

and obtain

$$<S_A>[\mathbf{x}_t] = -\frac{e^2 kT}{2\sigma_0}\int'\frac{d\mathbf{q}}{(2\pi)^d}\int\frac{d\omega}{2\pi}\int_{t'}^{t}dt_1\int_{t'}^{t}dt_2\frac{e^{i\mathbf{q}\cdot(\mathbf{x}_{t_1}-\mathbf{x}_{t_2})}}{q^2}\left(\cos\frac{\omega(t_1+t_2)}{2} - \cos\frac{\omega(t_1-t_2)}{2}\right).$$

(11.125)

Performing the integration over ω and t_2, the expression for the Cooperon becomes

$$C^T_{t,-t}(\mathbf{x},\mathbf{x}') = \frac{1}{2\tau}\int_{\mathbf{x}_{-t}=\mathbf{x}'}^{\mathbf{x}_t=\mathbf{x}}\mathcal{D}\mathbf{x}_t\, e^{-\int_{-t}^{t}dt_1\left\{\frac{\dot{\mathbf{x}}_{t_1}}{4D_0}+\frac{1}{\tau_T^{e-e}}+\frac{2e^2kT}{\sigma_0}\int'\frac{d\mathbf{q}}{(2\pi)^2}\,q^{-2}\left(1-\cos(\mathbf{q}\cdot(\mathbf{x}_{t_1}-\mathbf{x}_{-t_1}))\right)\right\}}.$$

(11.126)

The singular term is regularized by remembering that in eq.(11.125) the ω-integration actually should have been terminated at the, in the present context, large frequency $kT/\hbar$. The factor $\exp\{i\mathbf{q}\cdot(\mathbf{x}_{t_1}-\mathbf{x}_{t_2})\}$ does therefore not reduce strictly to 1 for the first term in the parenthesis in eq.(11.125) as $|\mathbf{x}_{t_1}-\mathbf{x}_{t_2}| \geq (D_0\hbar/kT)^{1/2}$, and this oscillating phase factor provides the convergence of the integral. We should therefore cut off the $\mathbf{q}$-integral at the wave vector satisfying $q = (kT/\hbar D_0)^{1/2} \equiv L_T^{-1}$, as indicated by the prime on the $\mathbf{q}$-integration in the two previous equations.

Introducing new variables

$$\mathbf{R}_t = \frac{\mathbf{x}_t + \mathbf{x}_{-t}}{\sqrt{2}} \quad , \quad \mathbf{r}_t = \frac{\mathbf{x}_t - \mathbf{x}_{-t}}{\sqrt{2}}$$

(11.127)

the path integral separates in two parts[29]

$$C_{t,-t}(\mathbf{R},\mathbf{R}) = \frac{1}{2\sqrt{2}\tau}\int_{-\infty}^{\infty}d\mathbf{R}_0\int_{\mathbf{R}_{t=0}=\mathbf{R}_0}^{\mathbf{R}_t=\sqrt{2}\mathbf{R}}\mathcal{D}\mathbf{R}_t\, e^{-\int_0^t dt'\frac{\dot{\mathbf{R}}_{t'}^2}{2D_0}}$$

$$\int_{\mathbf{r}_0=0}^{\mathbf{r}_t=0}\mathcal{D}\mathbf{r}_t\, e^{-\int_0^t dt'\left\{\frac{\dot{\mathbf{r}}_{t'}^2}{4D_0}+\frac{2}{\tau_T^{e-e}}+\frac{2e^2kT}{\sigma_0}\int'\frac{d\mathbf{q}}{(2\pi)^2}\,q^{-2}\left(1-\cos(\sqrt{2}\mathbf{q}\cdot\mathbf{r}_{t'})\right)\right\}}.$$

(11.128)

The path integral with respect to $\mathbf{R}_t$ gives the probability that a particle started at position $\mathbf{R}_0$ at time $t=0$ by diffusion reaches the point $\sqrt{2}\,\mathbf{R}$ (recall eq.(8.118)). Integrating this probability over all possible starting points is identical to integrating over all final points and by normalization gives unity. We are thus left with the expression for the Cooperon

$$C_{t,-t} = \frac{1}{2\sqrt{2}\tau}\int_{\rho_0=0}^{\rho_t=0}\mathcal{D}\mathbf{r}_{\bar{t}}\, e^{-\int_0^t d\bar{t}\left(\frac{\dot{\mathbf{r}}_{\bar{t}}^2}{4D_0}+V(\mathbf{r}_t)\right)}$$

(11.129)

where we have introduced the notation

$$V(\mathbf{r}) = \frac{2}{\tau_T^{e-e}} + \frac{2e^2kT}{\sigma_0}\int'\frac{d\mathbf{q}}{(2\pi)^d}\,q^{-2}\left(1-\cos(\sqrt{2}\,\mathbf{q}\cdot\mathbf{r})\right).$$

(11.130)

---

[29]This is immediately obtained using the standard discretized representation of a path integral (see eq.(1.40) of chapter 1).

As expected from translational invariance, the Cooperon is independent of position.

We have thus reduced the problem of calculating the quantum correction to the conductivity,

$$\delta\sigma(\omega) = -\frac{4e^2 D_0 \tau}{\pi\hbar} \int_{-\infty}^{\infty} dt \, e^{i\omega t} \, C_{t,-t}(\mathbf{r},\mathbf{r}) \qquad (11.131)$$

in the presence of electron-electron interaction, to solving the imaginary time Schrödinger problem

$$\{\partial_t - D_0 \triangle_{\mathbf{r}} + V(\mathbf{r})\}C_{t,t'}(\mathbf{r},\mathbf{r}') = \frac{1}{2\sqrt{2}\,\tau} \delta(\mathbf{r}-\mathbf{r}')\,\delta(t-t') \, . \qquad (11.132)$$

In the three-dimensional case the first term in the integrand of eq.(11.130) gives rise to a temperature dependence of the form $T^{3/2}$. This is the same form as the one we found in section 10.8.5 for the inelastic scattering rate due to electron-electron interaction in a dirty metal. This term can thus be joined with the first term of eq.(11.130). We note, that the description of the low-energy behavior thus joins up smoothly with the description of the high-energy behavior, as it should.

We thus have for the potential in the three-dimensional case

$$V_3(\mathbf{r}) = \frac{2}{\tau_T^{e-e}} + \tilde{V}_3(\mathbf{r}) \qquad (11.133)$$

where

$$\tilde{V}_3(\mathbf{r}) = \frac{-e^2 kT}{\sqrt{2}\pi\hbar^2\sigma_0} \left\{ \begin{array}{ll} \frac{1}{r} & r \gg L_T \\[2mm] \frac{2\sqrt{2}}{\pi} L_T^{-1} & r \ll L_T. \end{array} \right. \qquad (11.134)$$

Fourier-transforming eq.(11.132) with respect to time and taking the static limit we obtain

$$\{-D_0 \triangle_{\mathbf{r}} + V_3(\mathbf{r})\}C_{\omega=0}(\mathbf{r},\mathbf{r}') = \frac{1}{2\sqrt{2}\tau} \delta(\mathbf{r}-\mathbf{r}') \, . \qquad (11.135)$$

Solving this equation to first order in the potential $\tilde{V}_3$ gives

$$C_1(0,0,\omega=0) = -\frac{e^2 kT}{4\pi\hbar^2 \tau D_0^2 \sigma_0} \left\{ \frac{L_\epsilon}{\pi L_T} \left(e^{-2\sqrt{2}\frac{L_\epsilon}{L_T}} - 1 \right) + Ei\left(-2\sqrt{2}\frac{L_\epsilon}{L_T} \right) \right\} \qquad (11.136)$$

where Ei is the exponential integral[30] and

$$L_\epsilon = \sqrt{D_0 \tau_T^{e-e}} \, . \qquad (11.137)$$

In accordance with the calculation of the inelastic lifetime in section 10.8.5 we have

$$\frac{L_T}{L_\epsilon} \sim \frac{(\frac{kT\tau}{\hbar})^{1/4}}{k_F l} \, . \qquad (11.138)$$

---

[30] $Ei(x) = \int_{-\infty}^{x} dt \, \frac{e^t}{t}$ for $x < 0$.

We can therefore expand the expression in eq.(11.136), and obtain for the quantum correction to the conductivity

$$\delta\sigma = \frac{e^2}{2\pi^2\hbar L_\epsilon}\left(1 + \frac{4\pi e^2 kTL_\epsilon}{\hbar^2 D_0\sigma_0}\ln\frac{L_T}{L_\epsilon}\right) \tag{11.139}$$

where the second term is the correction due to collisions with small energy transfer, proportional to $T^{1/4}\ln T$.

In the two-dimensional case we obtain from eq.(11.130) for the potential

$$V_2(\mathbf{r}) = \frac{2}{\tau_T^{e-e}} + \frac{e^2 kT}{\pi\hbar^2\sigma_0}\int_0^{L_T^{-1}}dq\,\frac{1 - J_0(\sqrt{2}\,qr)}{q} \tag{11.140}$$

where J_0 denotes the Bessel function. We observe the limiting behavior of the potential

$$V_2(\mathbf{r}) = \frac{2}{\tau_T^{e-e}} + \frac{e^2 kT}{\pi\hbar^2\sigma_0}\begin{cases} \frac{1}{4}\left(\frac{r}{L_T}\right)^2 & r \ll L_T \\[2ex] \left(1 - J_0(\frac{\sqrt{2}r}{L_T})\right)\ln\frac{\sqrt{2}r}{L_T} - \mathcal{C} + \ln 2 & r \gg L_T\,. \end{cases} \tag{11.141}$$

where $\mathcal{C}$ is the Euler constant.

We then get the following equation for the Cooperon in the region of large values of r

$$\left\{-D_0\triangle_\mathbf{r} + \frac{2}{\tau_T^{e-e}} + \frac{e^2 kT}{\pi\hbar^2\sigma_0}\ln\frac{\sqrt{2}\,r}{L_T}\right\}C_{\omega=0}(\mathbf{r},\mathbf{r}') = \frac{1}{2\sqrt{2}\tau}\delta(\mathbf{r} - \mathbf{r}')\,. \tag{11.142}$$

The electron typically diffuses coherently the distance $\sqrt{D_0\tau_T^{e-e}}$. According to section 10.8.5 we have for the relaxation in two dimensions for processes with large energy transfers

$$\sqrt{D_0\tau_T^{e-e}} \sim \sqrt{\frac{D_0^2 N_2(0)\hbar^2}{kT}} \sim (k_F l)^{1/2}L_T \tag{11.143}$$

where $N_2(0)$ denotes the density of states at the Fermi energy in two dimensions. The electron thus diffuses coherently far into the region where the potential is logarithmic, and the slow change of the potential allows the substitution

$$\frac{2}{\tau_T^{e-e}} + \frac{e^2 kT}{\pi\hbar^2\sigma_0}\ln\frac{\sqrt{2}\,r}{L_T} \rightarrow \frac{e^2 kT}{\pi\hbar^2\sigma_0}\ln\frac{\sqrt{2\tau_T^{e-e}D_0}}{L_T}\,. \tag{11.144}$$

Inserting into eq.(11.142), we can read off the phase-breaking rate due to electron-electron interaction in a dirty conductor in two dimensions[31]

$$\frac{1}{\tau_\varphi} = \frac{kT}{4\pi\hbar^2 D_0 N_2(0)}\ln 2\pi\hbar D_0 N_2(0)\,. \tag{11.145}$$

[31] Many experiments are performed on thin metallic films. For such a quasi-two-dimensional case we can express the result for the phase breaking due to electron-electron interaction in a film of thickness a as $\frac{1}{\tau_\varphi} = \frac{e^2 kT}{2\pi a\sigma_0\hbar^2}\ln\frac{\pi a\sigma_0\hbar}{e^2}$.

The phase-breaking rate due to diffusion-enhanced electron-electron interaction depends thus in two dimensions linearly on the temperature at low temperatures, $kT < \hbar/\tau$.

The above result for the phase-breaking rate can be understood as a consequence of the phase-breaking rate setting the lower energy cutoff, $\hbar/\tau_\varphi$, for the efficiency of inelastic scattering events in destroying phase coherence. To show this we note that the path integral expression for the Cooperon, eq.(11.118), is the weighted average with respect to diffusive paths. Since this weight is convex, we have

$$C_t \;\geq\; C_t^{(0)} \; e^{-<<(\varphi[\mathbf{x}_t^{cl}])^2 >_{\text{ee}}>_{\text{imp}}} \tag{11.146}$$

where the second bracket signifies the average with respect to diffusive paths of the phase difference between the two interfering alternatives, eq.(11.53),

$$<< (\varphi[\mathbf{x}_t^{cl}])^2 >_{\text{ee}}>_{\text{imp}} \;=\; \frac{\displaystyle \int_{\mathbf{x}_{-t/2}=\mathbf{x}}^{\mathbf{x}_{t/2}=\mathbf{x}} \mathcal{D}\mathbf{x}_t \, P_t[\mathbf{x}_t] \; < (\varphi[\mathbf{x}_t^{cl}])^2 >_{\text{ee}}}{\displaystyle \int_{\mathbf{x}_{-t/2}=\mathbf{x}}^{\mathbf{x}_{t/2}=\mathbf{x}} \mathcal{D}\mathbf{x}_t \, P_t[\mathbf{x}_t]} \tag{11.147}$$

and $C_t^{(0)}$ is the return probability in the absence of the fluctuating field, i.e., the denominator in the above equation. The first bracket signifies the Gaussian average over the fluctuating field, i.e., the low-energy electron-electron interaction,

$$< (\varphi[\mathbf{x}_t^{cl}])^2 >_{\text{ee}} \;=\; \frac{e^2}{\hbar^2} \int_{-t/2}^{t/2} dt_1 \int_{-t/2}^{t/2} dt_2 \; \Big(< \phi(\mathbf{x}_{t_1}^{cl} - \mathbf{x}_{t_2}^{cl}, t_1 - t_2)\, \phi(\mathbf{0},0) >_{\text{ee}}$$

$$-\; < \phi(\mathbf{x}_{t_1}^{cl} - \mathbf{x}_{t_2}^{cl}, t_1 + t_2)\, \phi(\mathbf{0},0) >_{\text{ee}} \Big) \tag{11.148}$$

where we now choose to let the scalar potential represent the fluctuating field. Fourier-transforming we encounter

$$<< \phi(\mathbf{x}_{t_1}^{cl} - \mathbf{x}_{t_2}^{cl}, t_1 - t_2)\, \phi(\mathbf{0},0) >_{\text{ee}}>_{\text{imp}} \;=\; 2\int \frac{d\mathbf{q}}{(2\pi)^d}\int \frac{d\omega}{2\pi} < e^{i\mathbf{q}\cdot(\mathbf{x}_{t_1}^{cl}-\mathbf{x}_{t_2}^{cl})}>_{\text{imp}} < \phi\,\phi >_{\mathbf{q}\omega}$$

$$(\cos\omega(t_1 + t_2) - \cos\omega(t_1 - t_2)) \tag{11.149}$$

where the correlator for the fluctuating potential is specified in eq.(10.243). For a diffusion process we have according to eq.(8.123)[32]

$$< e^{i\mathbf{q}\cdot(\mathbf{x}_{t_1}^{cl}-\mathbf{x}_{t_2}^{cl})} >_{\text{imp}} \;=\; e^{i\mathbf{q}\cdot<(\mathbf{x}_{t_1}^{cl}-\mathbf{x}_{t_2}^{cl})>_{\text{imp}}} \;=\; e^{-D_0 q^2|t_1-t_2|} \tag{11.150}$$

and we get

$$<< (\varphi[\mathbf{x}_t^{cl}])^2 >_{\text{ee}}>_{\text{imp}} = \frac{2e^2 kT}{\pi\sigma_0} \int_{-t/2}^{t/2}\!\!dt_1 \int_{-t/2}^{t/2}\!\!dt_2 \int \frac{d\mathbf{q}}{(2\pi)^d}\int \frac{d\omega}{2\pi}\, e^{-\frac{1}{2}D_0 q^2|t_1-t_2|-i\omega(t_1-t_2)} \tag{11.151}$$

---

[32]The last equality is an approximation due to the constraint, $\mathbf{x}_{-t/2} = \mathbf{x}_{t/2}$, however, for large times a very good one.

where the ω-integration is limited to the region $1/\tau_\varphi \leq |\omega| \leq kT/\hbar$. The averaged phase difference is seen to increase linearly in time:

$$\frac{1}{2} << (\varphi[\mathbf{x}_t^{cl}])^2 >_{\text{ee}}>_{\text{imp}} = \frac{t}{\tau_\varphi} \tag{11.152}$$

at a rate in accordance with the previous result for the phase-breaking rate, eq.(11.145).

The lack of effectiveness in destroying phase coherence by interactions with small energy transfers is reflected in the compensation at small frequencies between the two cosine terms appearing in the expression for the phase difference, eq.(11.149). In the case of diffusion-enhanced electron-electron interaction this compensation is crucial as there is an abundance of scattering events with small energy transfer, whereas the compensation was immaterial for electron-phonon interaction where the typical energy transfer is determined by the temperature.

Whereas the phase-breaking rate for electron-phonon interaction is model dependent, i.e., material dependent, we note the interesting feature that the phase-breaking rate for diffusion-enhanced electron-electron interaction is universal. In two dimensions we can rewrite

$$\frac{1}{\tau_\varphi} = \frac{e^2 \sigma_0 kT}{2\pi \hbar^2} \ln \frac{k_F l}{2} . \tag{11.153}$$

Phase-breaking rates in accordance with eq.(11.145) have been extracted from numerous magnetoresistance measurements; confer for example references [59] and [60]. We note that at sufficiently low temperatures the electron-electron interaction dominates the phase-breaking rate in comparison with the electron-phonon interaction.

11.3.3 Temperature Dependence of Resistance

At finite temperature, the quantum correction to the dc conductance of a two-dimensional system is given by

$$\delta G(T) = -\frac{2D_0 e^2}{\hbar \pi} \int_0^{1/l} \frac{dQ Q}{2\pi} \frac{1}{D_0 Q^2 + 1/\tau_\varphi} \tag{11.154}$$

The phase-breaking rate we have just shown has a power law dependence on the temperature

$$\frac{\hbar}{\tau_\varphi} = K \, T^p . \tag{11.155}$$

For the temperature dependence of the quantum correction to the conductance in the two-dimensional case we then obtain

$$\delta G(T) = -\frac{e^2}{2\pi^2 \hbar} \int_0^{1/l^2} dx \frac{1}{x + \frac{1}{D_0 \tau_\varphi}} = -\frac{e^2}{2\pi^2 \hbar} \ln \left(1 + \frac{\tau_\varphi}{2\tau} \right)$$

$$\simeq -\frac{e^2}{2\pi^2 \hbar} \ln \frac{D_0 \tau_\varphi}{l^2} \tag{11.156}$$

where we assume that the distance the electron diffuses before having inelastic scattering, the phase coherence length, $L_\varphi = \sqrt{D_0 \tau_\varphi}$, satisfies

$$l \ll L_\varphi \ll L \tag{11.157}$$

where L is the size of the sample.

For the first quantum correction to the resistance we then have

$$\delta R(T) = -R_{cl}^2 \, \delta G(T) \tag{11.158}$$

and obtain thereby the result for the temperature dependence of the resistance

$$\delta R(T) = -R_{cl}^2 \, \frac{e^2}{2\pi^2\hbar} \, p \, \ln \frac{T}{T_0} \tag{11.159}$$

where

$$T_0 = \left(\frac{D_0 \hbar}{K l^2} \right)^{1/p} . \tag{11.160}$$

Experimentally, the logarithmic temperature dependence was observed originally in reference [63]. In the relevant temperature regime of the experiment ($T \leq 1K$) the temperature T_0 is large, $T < T_0$, and it was found that in accordance with eq.(11.159), the resistance of a thin metallic film as a function of temperature increases with decreasing temperature. This effect of the temperature is due to the suppression of localization, thus diminishing the resistance value with increasing temperature. As the temperature increases, interference from large loops, $L \gg L_\varphi$, is destroyed by interactions. The coherent backscattering that inhibits the motion of the particle, is thus increasingly suppressed as the temperature increases, leading to the nonmetallic behavior of the conduction in thin metal films at low temperatures.

Varying the temperature, however, is not the proper diagnostic tool to uniquely reveal localization effects as diffusion-enhanced electron-electron interaction gives rise to similar temperature dependence of the resistance [52].

Exercise 11.1 *Find the weak-localization temperature dependence of the resistance in the three-dimensional case.*

11.4 Anomalous Magnetoresistance

From an experimental point of view, the disruption of coherence between time-reversed trajectories by an externally controlled magnetic field is the tool by which to study the weak-localization effect. Magnetoresistance measurements in the weak-localization regime has considerably enhanced the available information regarding inelastic scattering times (and spin-flip and spin-orbit scattering times as discussed in section 11.6). The weak-localization effect thus plays an important diagnostic role in materials science.

The influence of a magnetic field on the Cooperon was established in section 11.2.1, and we have the Cooperon equation

$$\left\{-i\omega - D_0\{\nabla_\mathbf{x} - \frac{2ie}{\hbar}\mathbf{A}(\mathbf{x})\}^2 + 1/\tau_\varphi\right\} C_\omega(\mathbf{x},\mathbf{x}') \; = \; \frac{1}{\tau}\,\delta(\mathbf{x}-\mathbf{x}') \,. \qquad (11.161)$$

We can now safely study the dc conductivity, i.e., assume that the external electric field is static, so that its frequency is equal to zero, $\omega = 0$, as the Cooperon in an external magnetic field is no longer infrared divergent. The Cooperon is formally identical to the imaginary-time Schrödinger Green's function for a fictitious particle with mass equal to $\hbar/2D_0$ and charge $2e$ moving in a magnetic field (see exercise 2.2 on page 96). To solve the Cooperon equation for the magnetic field case, we can thus refer to the equivalent quantum mechanical problem of a particle in an external homogeneous magnetic field [13]. Considering the case of a homogeneous magnetic field,[33] and choosing the z-direction along the magnetic field and representing the vector potential in the Landau gauge, $\mathbf{A} = B\left(-y,0,0\right)$, the corresponding Hamiltonian is

$$\hat{H} \; = \; \frac{D_0}{\hbar}(\hat{p}_x + 2eB\hat{y})^2 \; + \; \frac{D_0}{\hbar}\left(\hat{p}_y^2 \,+\, \hat{p}_z^2\right) \,. \qquad (11.162)$$

The problem separates

$$\psi(x,y) \; = \; e^{\frac{i}{\hbar}p_x x}\, e^{\frac{i}{\hbar}p_z z}\, \chi(y) \qquad (11.163)$$

where the function χ satisfies the equation

$$-\frac{\hbar D_0}{2}\frac{d^2\chi(y)}{dy^2} \; + \; \frac{1}{2}\frac{\hbar}{2D_0}\,\tilde{\omega}_c^2\left(y - \frac{p_x}{2eB}\right)^2 \chi(y) \; = \; \tilde{E}\,\chi(y) \qquad (11.164)$$

the shifted harmonic oscillator problem where $\tilde{\omega}_c$ is the *cyclotron* frequency for the fictitious particle, $\tilde{\omega}_c \equiv 4D_0|e|B/\hbar$, so that the energy spectrum is $E = \tilde{E} + \hbar D_0 Q_z^2 = \hbar\tilde{\omega}_c(n+1/2) + \hbar D_0 Q_z^2$, $n = 0,1,2,...$; $Q_z = p_z/\hbar = 2\pi n_z/L_z$, $n_z = 0,\pm1,\pm2,...$. In the *particle in a magnetic field*-analog, n is the orbital quantum number and p_x the quantum number describing the position of the cyclotron orbit, and describes here the possible locations of closed loops. The Cooperon in the presence of a homogeneous magnetic field of strength B thus has the spectral representation

$$C_0(\mathbf{x},\mathbf{x}') \; = \; \sum_{Q_z}{}' \sum_{n=0}^{n_{max}} \int\frac{dp_x}{2\pi\hbar}\frac{\psi_{n,p_x,Q_z}(\mathbf{x})\,\psi^*_{n,p_x,Q_z}(\mathbf{x}')}{4D_0|e|B\tau\hbar^{-1}(n+1/2) + D_0\tau Q_z^2 + \tau/\tau_\varphi} \qquad (11.165)$$

where the ψ_{n,p_x,Q_z} are the Landau wave functions

$$\psi_{n,p_x,Q_z}(\mathbf{x}) \; = \; \frac{1}{\sqrt{L_z}}\, e^{\frac{i}{\hbar}p_x x}\, e^{iQ_z z}\, \chi_n(y - p_x/2eB) \,. \qquad (11.166)$$

---

[33]The case of an inhomogeneous magnetic field is treated in reference [64].

and $\chi_n(y)$ is the harmonic oscillator wave function. In accordance with the derivation of the Cooperon equation, we can only describe variations on length scales larger than the mean free path. The sum over the *orbital*-quantum number n should therefore terminate when $D_0\tau|e|Bn_{max} \sim \hbar$, i.e., at values of the order of $n_{max} \simeq l_B^2/l^2$, where $l_B \equiv (\hbar/|e|B)^{1/2}$ is the magnetic length.

To calculate the Cooperon for equal spatial values, $C_0(\mathbf{x},\mathbf{x})$, we actually do not need all the information contained in eq.(11.165), since by normalization of the wave functions in the completeness relation we have

$$\int_{-\infty}^{\infty} \frac{dp_x}{2\pi\hbar} \chi_n^* \left(y - \frac{p_x}{2eB} \right) \chi_n \left(y - \frac{p_x}{2eB} \right) = -\frac{2eB}{2\pi\hbar} \int_{-\infty}^{\infty} dy\, |\chi_n(y)|^2 = -\frac{2eB}{2\pi\hbar}$$

(11.167)

and thereby

$$C_0(\mathbf{x},\mathbf{x}) = -\frac{2eB}{2\pi\hbar L_z} \sideset{}{'}\sum_{Q_z} \sum_{n=0}^{n_{max}} \frac{1}{4D_0|e|B\tau\hbar^{-1}(n+1/2) + D_0\tau Q_z^2 + \tau/\tau_\varphi}.$$

(11.168)

11.4.1 Magnetoresistance in Thin Films

We now consider the magnetoresistance of a film of thickness a, choosing the direction of the magnetic field perpendicular to the film.[34] Provided the thickness of the film is smaller than the phase coherence length, $a \ll L_\varphi$ (the thin film, or quasi-two-dimensional criterion), or the usually much weaker restriction that it is smaller than the magnetic length, $a \ll l_B$, only the smallest value of $Q_z = 2\pi n_z/a$, $n_z = 0, \pm 1, \pm 2, ...$ contributes to the sum. Since the smallest value is $Q_z = 0$, we obtain, according to eq.(11.13), for the quantum correction to the conductivity

$$\delta\sigma(B) = \frac{e^3 B D_0 \tau}{\pi^2 \hbar^2 a} \sum_{n=0}^{n_{max}} \frac{1}{4D_0|e|B\tau\hbar^{-1}(n+1/2) + \tau/\tau_\varphi}.$$

(11.169)

Employing the property of the di-gamma-function ψ (see for example reference [65])

$$\psi(x + n) = \psi(x) + \sum_{n=0}^{n-1} \frac{1}{x+n}$$

(11.170)

we get for the magnetoconductance

$$\delta G_{\alpha\beta}(B) = \frac{e^2}{4\pi^2\hbar} \tilde{f}_2(4D_0|e|B\hbar^{-1}\tau_\varphi)\, \delta_{\alpha\beta}$$

(11.171)

where

$$\tilde{f}_2(x) = \psi\left(\frac{1}{2} + \frac{1}{x} \right) + \psi\left(\frac{3}{2} + n_{max} + \frac{1}{x} \right).$$

(11.172)

---

[34] The strictly two-dimensional case can also be realized experimentally, for example by using the two-dimensional electron gas accumulating in the inversion layer in a MOSFET or heterostructure.

The magnetoconductance of a thin film is now obtained by subtracting the zero field conductance. In the limit $B \to 0$, the sum can be estimated to become

$$\sum_{n=0}^{n_{max}} \frac{1}{4D_0|e|B\tau\hbar^{-1}(n+1/2) + \tau/\tau_\varphi} \quad \to \quad \ln(n_{max}4D_0|e|B\hbar^{-1}\tau_\varphi) \,. \quad (11.173)$$

Using the property of the di-gamma-function

$$\lim_{n\to\infty} \psi\left(\frac{3}{2}+n+\frac{1}{x}\right) \simeq \ln n \quad (11.174)$$

we finally arrive at the low-field magnetoconductance of a thin film

$$\Delta G_{\alpha\beta}(B) \equiv \delta G_{\alpha\beta}(B) - \delta G_{\alpha\beta}(B\to 0) = \frac{e^2}{2\pi^2\hbar}\, f_2(B/B_\varphi)\, \delta_{\alpha\beta} \quad (11.175)$$

where

$$f_2(x) = \ln x + \psi\left(\frac{1}{2}+\frac{1}{x}\right) \quad (11.176)$$

and $B_\varphi = \hbar/4D_0|e|\tau_\varphi$, the (temperature-dependent) characteristic scale of the magnetic field for the weak-localization effect, is determined by the inelastic scattering. This scale is indeed small compared to the scale for classical magnetoresistance effects $B_{cl} \sim m/|e|\tau$, as $B_\varphi \sim B_{cl}\,\hbar/\epsilon_F\tau_\varphi$.[35] The weak-localization magnetoconductance is seen to be sensitive to very small magnetic fields, namely when the magnetic length becomes comparable to the phase coherence length, $l_B \sim L_\varphi$, or equivalently, $\omega_c\tau \sim \hbar/\epsilon_F\tau_\varphi$. Since the impurity mean free time, τ, can be much smaller than the phase coherence time τ_φ, the above description can be valid over a wide magnetic field range where classical magnetoconductance effects are absent. Classical magnetoconductance effects are governed by the orbit bending scale, $\omega_c\tau \sim 1$, whereas the weak-localization quantum effect sets in when a loop of typical area L_φ^2 encloses a flux quantum.[36] We note the limiting behavior of the function

$$f_2(x) = \begin{cases} \frac{x^2}{24} & \text{for} \quad x \ll 1 \\[2mm] \ln x & \text{for} \quad x \gg 1 \,. \end{cases} \quad (11.177)$$

The magnetoconductance is positive, and seen to have a quadratic upturn at low fields, and saturates beyond the characteristic field in a universal fashion, i.e.,

---

[35]In terms of the mass of the electron we have for the mass of the fictitious particle $\hbar/2D_0 \sim m\hbar/\epsilon_F\tau$, and the low magnetic field sensitivity can be viewed as due to the smallness of the fictitious mass in the problem.

[36]Beyond the low-field limit, $\omega_c\tau < \hbar/\epsilon_F\tau$, the expression for the magnetoconductance can not be given in closed form, and its derivation is more involved, since we must account for the orbit bending due to the magnetic field, the Lorentz force [66]. When the impurity mean free time τ becomes comparable to the phase coherence time τ_φ, we are no longer in the diffusive regime, and a Boltzmannian description must be introduced [67].

independent of sample parameters.[37] The magnetoresistance is therefore negative, $\Delta R = -\Delta G / G_{cl}^2$, which is a distinct sign that the effect is not classical, since we are considering a macroscopic system.[38]

The negative anomalous magnetoresistance can be understood qualitatively from the simple interference picture of the weak-localization effect. The presence of the magnetic field breaks the time-reversal invariance, and upsets the otherwise identical values of the phase factors in the amplitudes for traversing the time-reversed weak-localization loops. The quantum interference term for a loop c is due to the presence of the magnetic field changed according to

$$A_c A_{\bar{c}}^* \;\rightarrow\; |A_c^{(B=0)}|^2 \exp\left\{\frac{2ie}{\hbar} \oint_c d\bar{\mathbf{x}} \cdot \mathbf{A}(\bar{\mathbf{x}})\right\} \;=\; |A_c^{(B=0)}|^2 \, e^{\frac{2ie}{\hbar}\Phi_c} \,. \qquad (11.178)$$

where Φ_c is the flux enclosed by the loop c. The weak-localization interference term acquires a random phase depending on the loop size, and the strength of the magnetic field, decreasing the probability of return, and thereby increasing the conductivity. The negative contribution from each loop in the impurity field to the conductance is modulated in accordance with the phase shift prescription for amplitudes by the oscillatory factor, giving the expression

$$<G(B)> - <G(O)> = \frac{e^2}{\hbar} < \sum_c |A_c^{(B=0)}|^2 \left\{1 - \cos(2\pi\Phi_c/\Phi_0)\right\} e^{-t_c/\tau_\varphi} > \,.$$
$$(11.179)$$

The summation is over all classical loops in the impurity field returning to within a distance of the mean free path to a given point, and t_c is the duration for traversing the loop c, and Φ_0 is the flux quantum $\Phi_0 = 2\pi\hbar/2|e|$. The sum should be performed weighted with the probability for the realization of the loop in question, as expressed by the brackets. The weight of loops that are longer than the phase coherence length is suppressed, as their coherence are destroyed by inelastic scattering. In weak magnetic fields, only the longest loops are influenced by the phase shift due to the magnetic field. It is evident from eq.(11.179) that the low field magnetoconductance is positive and quadratic in the field.[39] The continuing monotonic behavior as a function of the magnetic field until saturation is simply a geometric property of diffusion, viz. that small diffusive loops are prolific. Instead of verifying this statement, let us turn the argument around and use our physical understanding of the weak localization effect to learn about the distribution of the areas of diffusive loops in two dimensions. Rewriting eq.(11.168) we have in two

---

[37]Experimental observations of the low field magnetoresistance of thin metallic films are in remarkable good agreement with the theory. The weak-localization effect is thus of importance for extracting information about inelastic scattering strengths, which is otherwise hard to come at. For reviews of the experimental results, see references [59] and [60].

[38]The classical magnetoresistance of a macroscopic sample calculated on the basis of the Boltzmann equation is positive.

[39]The minimum value of the magnetoresistance is *exactly* for zero magnetic field, and the weak localization effect is thus one of the few effects that can be used as a reference for zero magnetic field.

dimensions

$$C_0(\mathbf{x}, \mathbf{x}) = \int_0^\infty dt\, e^{-t/\tau_\varphi} \frac{B}{\tau\Phi_0} \sum_{n=0}^{n_{max}} e^{-\frac{4\pi BD_0}{\Phi_0}(n+1/2)t} . \qquad (11.180)$$

For times $t > \tau$ we can let the summation run over all natural numbers and we can sum the geometric series to obtain

$$C_0(\mathbf{x}, \mathbf{x}) = \int_0^\infty dt\, \frac{e^{-t/\tau_\varphi}}{4\pi\tau D_0 t} \frac{2\pi BD_0 t}{\Phi_0} \frac{1}{\sinh \frac{2\pi BD_0 t}{\Phi_0}} . \qquad (11.181)$$

The factors independent of the magnetic field are the return probability and the dephasing factor. Representing the factors depending on the field strength, which describes the influence of the magnetic field on the quantum interference process, by its cosine transform

$$\frac{2\pi BD_0 t}{\Phi_0} \frac{1}{\sinh \frac{2\pi BD_0 t}{\Phi_0}} = \int_{-\infty}^\infty dS\, \cos \frac{SB}{\Phi_0}\, f_t(S) \qquad (11.182)$$

and inverting gives

$$f_t(S) = \frac{1}{4D_0 t} \frac{1}{\cosh^2\left(\frac{S}{4D_0 t}\right)} . \qquad (11.183)$$

We can therefore write for the weak localization contribution to the conductance

$$\delta G(B) = -\frac{2e^2 D_0 \tau}{\pi\hbar} \int_0^\infty dt\, \frac{e^{-t/\tau_\varphi}}{4\pi\tau D_0 t} \int_0^\infty dS\, f_t(S) \cos\left(\frac{BS}{\Phi_0}\right) \qquad (11.184)$$

and we note that $f_t(S)$ is normalized, and has the interpretation of the probability for a diffusive loop of duration t to enclose the area S.

For the average size of a diffusive loop of duration t we have

$$<S>_t \equiv \int_0^\infty dS\, S f_t(S) = 4D_0 t \ln 2 \qquad (11.185)$$

i.e., the typical size of a diffusive loop of duration t is proportional to $D_0 t$.

For the fluctuations we have

$$<S^2>_t \equiv \int_{-\infty}^\infty dS\, S^2 f_t(S) = 8\pi^2 (D_0 t)^2 \qquad (11.186)$$

and we can write

$$f_t(S) = \frac{\pi}{\sqrt{2 <S^2>_t}} \frac{1}{\cosh^2 \frac{\pi S}{\sqrt{2<S^2>_t}}} . \qquad (11.187)$$

The probability distribution for diffusive loops is thus a steadily decreasing function of the area.

11.4.2 Three-Dimensional Sample

In the three-dimensional case we perform the Q_z-integration in eq.(11.165) and obtain

$$\delta\sigma(B) = -\frac{e^2}{\pi^3 l_B} \sum_{n=0}^{n_{max}} \frac{1}{\sqrt{n + \frac{1}{2} + \delta}} \tan^{-1}\left(\frac{l_B}{2l\sqrt{n + \frac{1}{2} + \delta}}\right) \tag{11.188}$$

where we have introduced

$$\delta = \frac{l_B^2}{4L_\varphi^2} = \frac{\hbar}{4D_0|e|B\tau_\varphi}. \tag{11.189}$$

In the low magnetic field limit, $\omega_c \tau \ll \hbar/\epsilon_F \tau$, the sum will approximately be given by the integral

$$\delta\sigma(B) = -\frac{e^2}{\pi^3 \hbar l_B} \int_0^{n_{max}+1} dx \frac{1}{\sqrt{x+\delta}} \tan^{-1}\left(\frac{l_B}{2l\sqrt{x+\delta}}\right) \tag{11.190}$$

so that

$$\Delta\sigma(B) \equiv \delta\sigma(B) - \delta\sigma(B=0)$$

$$- \frac{e^2}{\pi^3 \hbar l_B} \left(\int_0^{n_{max}+1} dx \frac{\tan^{-1}\frac{l_B}{2l\sqrt{x+\delta}}}{\sqrt{x+\delta}} - \sum_{n=0}^{n_{max}} \frac{\tan^{-1}(\frac{l_B}{2l\sqrt{n+\frac{1}{2}+\delta}})}{\sqrt{n+\frac{1}{2}+\delta}} \right). \tag{11.191}$$

We should not forget about the upper cutoff. However, the subtraction renders the above expression well-defined. In the low magnetic field limit, $l_B \ll l$, where $n_{max} \gg 1$, we obtain the result

$$\Delta\sigma(B) = \frac{e^2}{\pi^3 \hbar l_B} f_3(4|e|D_0 B\tau_\varphi) \tag{11.192}$$

where

$$f_3(x) = \sum_{n=0}^{\infty} \left\{ 2\left(\sqrt{n+1+\frac{1}{x}} - \sqrt{n+\frac{1}{x}}\right) - \frac{1}{\sqrt{n+\frac{1}{2}+\frac{1}{x}}} \right\}. \tag{11.193}$$

To find the asymptotic limit of the function f_3 for small values, $x \ll 1$, we expand the above expression in $n + 1/2 + 1/x$ and obtain

$$f_3(x) = \frac{1}{32} \sum_{n=0}^{\infty} \left(n+1+\frac{1}{x}\right)^{-5/2}. \tag{11.194}$$

This Riemann zeta-function has the limits (see for instance reference [65] p. 1073)

$$f_3(x) = \begin{cases} \frac{1}{48} x^{3/2} & x \ll 1 \\ 0.605 & x \gg 1. \end{cases} \tag{11.195}$$

where in the limit of large x, the function f_3 is seen to approach a constant value which has been calculated numerically [68].

We note that the weak-localization magnetoresistance in three dimensions is independent of the relative directions of the magnetic field and the current; i.e., the longitudinal and transverse magnetoresistance are identical. Furthermore, we notice that at large magnetic fields the magnetoresistance tends to a universal constant (independent of sample parameters). The low-field square-root dependence of the magnetoconductance cleared up a long-standing mystery in the field of magnetotransport in doped semiconductors.

11.5   Aharonov-Bohm Effect

The most striking manifestation of the quantum interference involved in the weak-localization effect is obtained by confining the motion of the electrons to a thin cylindrical shell (or a ring), which is penetrated by a magnetic flux directed along the axis of the cylinder. In that case, all closed loops enclose the same flux $\Phi = B\pi R^2$, where R is the radius of the cylinder, and the conductivity should oscillate as a function of the flux Φ threading the cylinder with the period Φ/Φ_0. This is an example of the Aharonov-Bohm effect in solid state physics.[40]

We now turn to the quantitative description of the weak-localization Aharonov-Bohm effect, and must therefore solve the Cooperon equation in cylindrical coordinates

$$\left\{ -D_0 \left(\left(\frac{1}{R} \frac{\partial}{\partial \varphi} - \frac{2ie}{\hbar} A_\varphi \right)^2 + \frac{\partial^2}{\partial z^2} \right) + \frac{1}{\tau_\varphi} \right\} C_0(z, z', \varphi, \varphi') = \frac{1}{\tau R} \delta(z-z') \delta(\varphi-\varphi') .$$

(11.196)

We have chosen a gauge where only the azimuthal component of the vector potential is nonzero, and we assume that the thickness a of the cylindrical shell satisfies the criterion $B^2 L_\varphi^2 a^2 \ll \Phi_0^2$. We can then assume that the vector potential is constant within the cylindrical shell, and we obtain the solution

$$C_0(z, z', \varphi, \varphi') = \frac{1}{2\pi R L_z D_0 \tau} \sum_{n, Q_z}{}' \frac{e^{iQ_z(z-z')+in(\varphi-\varphi')}}{Q_z^2 + (\frac{n}{R} - \frac{\Phi}{R\Phi_0})^2 + L_\varphi^{-2}}$$

(11.197)

where the restriction to the singular regime requires $|n|, \Phi/\Phi_0 < R/l$. Assuming $l \ll R$, we can extend the rapidly converging n-summation to $\pm\infty$.

In the case of a ring, the quasi-one-dimensional case, $L_z \ll L_\varphi$, we only have to keep the $Q_z = 0$-term in the Q_z-sum. The summation over n is readily done by the method of residues

$$\sum_{n=-\infty}^{\infty} \frac{1}{(n - \frac{\Phi}{\Phi_0})^2 + \left(\frac{R}{L_\varphi}\right)^2} = \sum_i res(g, z_i)$$

(11.198)

---

[40]This amazing manifestation of the quantum mechanical superposition principle at the macroscopic level was suggested in reference [69]. Here we follow the presentation of reference [70].

where

$$g(z) \equiv \pi \frac{\cos \pi z}{\sin \pi z} f(z) \quad , \quad f(z) \equiv \frac{1}{(z - \frac{\Phi}{R\Phi_0})^2 + \left(\frac{R}{L_\varphi}\right)^2} \tag{11.199}$$

and we obtain for the flux dependence of the conductance of a ring

$$\delta G(\Phi) = -\frac{e^2}{\pi \hbar} L_\varphi \frac{\sinh \frac{2\pi R}{L_\varphi}}{\cosh \frac{2\pi R}{L_\varphi} - \cos \frac{2\pi \Phi}{\Phi_0}} \cdot \tag{11.200}$$

In the limit $R \ll L_\varphi$ (which can be achieved for an arbitrary large radius by lowering the temperature) we get the result

$$\delta G(\Phi) = -\frac{2e^2}{\hbar} R \frac{1}{1 + \frac{1}{2}\left(\frac{2\pi R}{L_\varphi}\right)^2 - \cos \frac{2\pi \Phi}{\Phi_0}} \cdot \tag{11.201}$$

We note that the modulation amplitude can be made arbitrarily large.

In the case of a cylinder, the quasi-two-dimensional case, we have for the flux dependence of the conductance

$$\delta G(\Phi) = -\frac{2e^2}{\pi \hbar} \frac{1}{2\pi R} \int_{-1/l}^{1/l} dQ_z \sum_{n=-\infty}^{\infty} \frac{1}{Q_z^2 + \frac{1}{R^2}(n - \frac{\Phi}{\Phi_0})^2 + L_\varphi^{-2}} \cdot \tag{11.202}$$

We improve the convergence by adding and subtracting the zero-flux result, and can then extend the Q_z-integration to $\pm\infty$, and obtain (see for example the table of integrals [65] p. 978)

$$\Delta G(\Phi) \equiv \delta G(\Phi) - \delta G(\Phi = 0)$$

$$= -\frac{e^2}{\pi^2 \hbar} \left\{ \ln \frac{L_\varphi}{l} - 2 \sum_{n=1}^{\infty} K_0 \left(\frac{2\pi R n}{L_\varphi}\right) \left(1 - \cos \frac{2\pi n \Phi}{\Phi_0}\right) \right\} \tag{11.203}$$

where K_0 is the modified Bessel function, the McDonald function. The flux dependence tends to zero exponentially for $R \gg L_\varphi$, and in the opposite limit, $R \ll L_\varphi$ (achieved at low temperatures) the amplitude of the flux modulation will tend to infinity since

$$K_0(x) = \begin{cases} \sqrt{\frac{\pi}{2x}} \, e^{-x} & \text{for } x \gg 1 \\ -\ln x & \text{for } x \ll 1. \end{cases} \tag{11.204}$$

For a review of the experimental confirmation of the weak-localization Aharonov-Bohm effect, we refer to reference [71].

11.6 Magnetic-Impurity and Spin-Orbit Scattering

In this section we take into account spin-orbit scattering off the impurities and allow for the impurities to accommodate magnetic scattering.[41] On general physical grounds we deduce that magnetic-impurity scattering destroys quantum interference since the electronic path is partially kept track of, due to the flipped impurity spins revealing the visit of an electron, and quantum interference between distinguishable trajectories is excluded. Spin-orbit scattering, in contrast to magnetic-impurity scattering, does not violate time-reversal symmetry, and one could be inclined to assert that it should have no influence on the weak-localization quantum interference. However, the coupling between the orbital and spin degrees of freedom of an electron provided by the spin-orbit scattering leads to highly nontrivial effects.

To accommodate the additional scattering mechanisms, the impurity potential matrix element has to be changed from merely a constant (in the momentum representation) to the scattering amplitude including spin-flip and spin-orbit scattering

$$f_{\alpha\alpha'}(\mathbf{p}, \mathbf{p}') \; = \; V\,\delta_{\alpha\alpha'} \; + \; V_S\,\mathbf{S}_i\cdot\vec{\sigma}_{\alpha\alpha'} \; - \; iV_{so}\,(\mathbf{p}\times\mathbf{p}')\cdot\vec{\sigma}_{\alpha\alpha'} \qquad (11.205)$$

where $\mathbf{S}_i$ and $\vec{\sigma}$ denote the impurity and electron spin, respectively. Performing the standard positional impurity averaging, we obtain in the Born approximation the impurity correlator

$$U_{\alpha\alpha';\beta\beta'}(\mathbf{p},\mathbf{p}') \; = \; n_i\Big(|V|^2\,\delta_{\alpha\alpha'}\,\delta_{\beta\beta'} \; + \; |V_S|^2\,\mathbf{S}_i^2\,\vec{\sigma}_{\alpha\alpha'}\cdot\vec{\sigma}_{\beta\beta'}$$

$$- \; |V_{so}|^2\,(\mathbf{p}\times\mathbf{p}')\cdot\vec{\sigma}_{\alpha\alpha'}\,(\mathbf{p}'\times\mathbf{p})\cdot\vec{\sigma}_{\beta\beta'}\Big)$$

$$\equiv \qquad \begin{array}{c}\mathbf{p}\,\alpha \;\longleftarrow\!\top\!\longleftarrow\, \mathbf{p}'\,\alpha' \\[4pt] \ast \\[4pt] \mathbf{p}'\,\beta\,\longleftarrow\!\perp\!\longleftarrow\,\mathbf{p}\,\beta'\end{array} \qquad . \qquad (11.206)$$

Diagrammatically we have the Cooperon equation

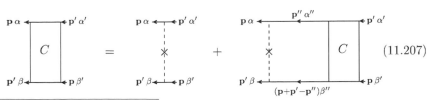

$$\qquad\qquad\qquad\qquad\qquad\qquad\qquad\qquad\qquad\qquad\qquad (11.207)$$

---

[41]We follow the presentation in reference [72]. Magnetoresistance in many-valley semiconductors is discussed in reference [73].

Due to the orbital structure of the spin-orbit scattering the Cooperon equation will in general be a complicated integral equation in the angular variables. However, in the limit, $\tau_{so}, \tau_s \gg \tau$, where an electron changes momentum many times before its spin is flipped the integral kernel, the correlator eq.(11.206), can be approximated by its average over momentum directions, the Fermi surface average. We thus get the following equation for the Cooperon:

$$C_{\alpha\alpha';\beta\beta'}(\mathbf{p}, \mathbf{p}') = \overline{U}_{\alpha\alpha';\beta\beta'} + \overline{U}_{\alpha\alpha'';\beta\beta''} \, \tilde{\zeta}(\mathbf{Q}, \omega) \, C_{\alpha''\alpha';\beta''\beta'}(\mathbf{Q}, \omega) \qquad (11.208)$$

where the Fermi surface averaged impurity correlator has the form:

$$\overline{U}_{\alpha\alpha';\beta\beta'} = u^2 \, \delta_{\alpha\alpha'} \, \delta_{\beta\beta'} + \frac{1}{3}(u_s^2 - u_{so}^2)\vec{\sigma}_{\alpha\alpha'} \cdot \vec{\sigma}_{\beta\beta'} \qquad (11.209)$$

with

$$u_s^2 = n_i |V_s|^2 < \mathbf{S}_i^2 > \qquad (11.210)$$

and

$$u_{so}^2 = n_i |V_{so}|^2 \, \overline{(\mathbf{p} \times \mathbf{p}')} \,. \qquad (11.211)$$

The bracket denotes an average over the assumed random directions of the impurity spins, and the bar denotes the Fermi surface average. The insertion is, $\hbar\mathbf{Q} \equiv \mathbf{p} + \mathbf{p}'$,

$$\tilde{\zeta}(\mathbf{Q}, \omega) = 2\pi N_0 \tau' \left(1 + i\omega\tau' - D_0\tau'\mathbf{Q}^2\right) \qquad (11.212)$$

where τ' is the total elastic scattering time

$$\frac{1}{\tau'} = \frac{1}{\tau} + \frac{1}{\tau_s} + \frac{1}{\tau_{so}} \,, \quad \frac{\hbar}{\tau_s} = 2\pi N_0 u_s^2 \,, \quad \frac{\hbar}{\tau_{so}} = 2\pi N_0 u_{so}^2 \,. \qquad (11.213)$$

The form of the insertion $\tilde{\zeta}$ is a simple consequence of the averaged propagator for the present case being diagonal in the spin degrees of freedom and given by

$$G_{\alpha\beta}^{R(A)}(E, \mathbf{p}) = \frac{1}{E - \epsilon_{\mathbf{p}} \, (\pm) \, \frac{i\hbar}{2\tau'}} \, \delta_{\alpha\beta} \,. \qquad (11.214)$$

We have assumed three-dimensionality with respect to spin-scattering properties so that[42]

$$\frac{1}{\tau_s^x} = \frac{1}{\tau_s^y} = \frac{1}{\tau_s^z} = \frac{1}{3}\frac{1}{\tau_s} \qquad (11.215)$$

and

$$\frac{1}{\tau_{so}^x} = \frac{1}{\tau_{so}^y} = \frac{1}{\tau_{so}^z} = \frac{1}{3}\frac{1}{\tau_{so}} \qquad (11.216)$$

where

$$\frac{\hbar}{\tau_s} = 2\pi N_0 u_s^2 \qquad (11.217)$$

---

[42]In a system where the motion of the electrons are strictly two-dimensional, the spin-orbit scattering is seen to be absent for the considered scattering mechanism, since a coupling $\hat{z} \cdot \vec{\sigma}$ can not flip the spin along the z-direction.

and

$$\frac{\hbar}{\tau_{so}} = 2\pi N_0 u_{so}^2 \; . \tag{11.218}$$

The spin structure of the Cooperon, determined by the spin structure of the impurity correlator, is given by

$$C_{\alpha\alpha';\beta\beta'} = A(\mathbf{Q},\omega)\,\delta_{\alpha\alpha'}\,\delta_{\beta\beta'} + B(\mathbf{Q},\omega)\,\vec{\sigma}_{\alpha\alpha'}\cdot\vec{\sigma}_{\beta\beta'} \; . \tag{11.219}$$

Using the identity

$$(\vec{\sigma}_{\alpha\alpha''}\cdot\vec{\sigma}_{\beta\beta''})(\vec{\sigma}_{\alpha''\alpha'}\cdot\vec{\sigma}_{\beta''\beta'}) = 3\,\delta_{\alpha\alpha'}\,\delta_{\beta\beta'} - 2\,\vec{\sigma}_{\alpha\alpha'}\cdot\vec{\sigma}_{\beta\beta'} \tag{11.220}$$

it is straightforward, in the limit $\tau \ll \tau_{so}, \tau_s$, to obtain

$$A(\mathbf{Q},\omega) = \frac{1}{2\tau}\left(\frac{\frac{3}{2}}{-i\omega + D_0\mathbf{Q}^2 + \frac{2}{3\tau_s} + \frac{4}{3\tau_{so}}} + \frac{\frac{1}{2}}{-i\omega + D_0\mathbf{Q}^2 + \frac{2}{\tau_s}}\right) \tag{11.221}$$

and

$$B(\mathbf{Q},\omega) = \frac{1}{\frac{2}{3}\tau}\left(\frac{1}{-i\omega + D_0\mathbf{Q}^2 + \frac{2}{3\tau_s} + \frac{4}{3\tau_{so}}} - \frac{1}{-i\omega + D_0\mathbf{Q}^2 + \frac{2}{\tau_s}}\right) \; . \tag{11.222}$$

We thus have the following conductivity correction

$$\delta\sigma(\omega) = -\frac{e^2 D_0}{\pi\hbar L^d}\sum_{\mathbf{Q}}{}' C_{\alpha\beta;\beta\alpha}(\mathbf{Q},\omega) = -\frac{e^2 D_0}{\pi\hbar L^d}\sum_{\mathbf{Q}}{}' 2\big(A(\mathbf{Q},\omega) + 3\,B(\mathbf{Q},\omega)\big)$$

$$= -\frac{2e^2 D_0}{\pi\hbar L^d}\sum_{\mathbf{Q}}{}'\left(\frac{\frac{3}{2}}{-i\omega + D_0\mathbf{Q}^2 + \frac{2}{3\tau_s} + \frac{4}{3\tau_{so}}} - \frac{\frac{1}{2}}{-i\omega + D_0\mathbf{Q}^2 + \frac{2}{\tau_s}}\right). \tag{11.223}$$

We notice that the magnetic impurity scattering in accordance with our expectation suppresses the weak-localization effect. In the absence of magnetic scattering, we observe that the spin-orbit scattering will not eliminate the singularity in the conductivity, but instead change the sign of the quantum correction (and reduce the strength by a factor of 2)! This effect is referred to as weak antilocalization, and has been observed in many substances for which impurities give rise to strong spin-orbit scattering. The scaling function will thus for the case of spin-orbit scattering cross the axis as in the three-dimensional case, before it at large conductance approaches zero, i.e., exhibit nonmonotonic behavior.[43] We have assumed that the sample length L is much greater than the spin-orbit length scale, $L_{so} \equiv \sqrt{D_0\tau_{so}}$, the length scale for randomizing the spin direction due to spin-orbit scattering. The scaling function thus depends on the symmetry of the scattering potential.

---

[43]Such a nontrivial scaling behavior seems to be confirmed by a four-loop renormalization group calculation [74]. Due to the importance of high-order terms in the perturbative expansion of the scaling function (here the importance of the finite $1/g^5$-term) a self-consistent theory of localization for systems with spin-orbit scattering has not been achieved.

If we decompose the spin states in singlet,

$$C_s(\mathbf{Q}, \omega) = A(\mathbf{Q}, \omega) - 3B(\mathbf{Q}, \omega) = \frac{1}{-i\omega + D_0\mathbf{Q}^2 + \frac{2}{\tau_s}} \tag{11.224}$$

and triplet,

$$C_t(\mathbf{Q}, \omega) = A(\mathbf{Q}, \omega) + B(\mathbf{Q}, \omega) = \frac{1}{-i\omega + D_0\mathbf{Q}^2 + \frac{2}{3\tau_s} + \frac{4}{3\tau_{so}}} \tag{11.225}$$

we recognize that the singlet part of the Cooperon is unaffected by spin-orbit scattering.

The weak antilocalization effect; i.e., the appearance of the minus sign is a consequence of the interference being between time-reversed scattering sequences. On the time-reversed trajectory the electron experiences the spin rotations in opposite sequence and opposite directions

$$\psi_c = \underline{\underline{R}}\,\psi_i\,, \qquad \psi_{\bar{c}} = \underline{\underline{R}}^{-1}\,\psi_i \tag{11.226}$$

where

$$\underline{\underline{R}}(\phi, \theta, \psi) = \begin{pmatrix} e^{\frac{i}{2}(\phi+\psi)}\cos\frac{\theta}{2} & i\sin\frac{\theta}{2}e^{\frac{i}{2}(\psi-\phi)} \\ i\sin\frac{\theta}{2}e^{-\frac{i}{2}(\psi-\phi)} & e^{-\frac{i}{2}(\phi+\psi)}\cos\frac{\theta}{2} \end{pmatrix} \tag{11.227}$$

is the unitary rotation operator in spin space parametrized by the Euler angles (recall exercise 1.15 on page 54). For the following example of initial spin state represented by the spinor

$$\psi_i = \begin{pmatrix} 1 \\ 0 \end{pmatrix} \tag{11.228}$$

we have for the final spin states for the two interfering alternatives

$$\psi_c = \begin{pmatrix} R_{11} \\ R_{12} \end{pmatrix}, \qquad \psi_{\bar{c}} = \begin{pmatrix} R_{11}^* \\ R_{21}^* \end{pmatrix} = \begin{pmatrix} R_{11}^* \\ -R_{12} \end{pmatrix} \tag{11.229}$$

where we notice the appearance of a minus sign. For the interference term we then get

$$\psi_c^*\psi_{\bar{c}} + \psi_{\bar{c}}^*\psi_c = R_{11}^*R_{11}^* + R_{12}^*R_{21}^* + R_{11}R_{11} + R_{21}R_{12} = 2\,\Re e(R_{11}^2|R_{12}|^2)$$

$$= 2\,\Re e\left(\cos^2\frac{\theta}{2}\,e^{i(\phi+\psi)} - \sin^2\frac{\theta}{2}\right). \tag{11.230}$$

The first term in the parentheses averages to zero due to the random phase factor, whereas the second term on the average produces the factor $-1/2$ in eq.(11.223).

The magnetoresistance calculation in the presence of spin-flip and spin-orbit scattering is parallel to the one already performed in section 11.4, and without repeating it we notice that it corresponds to the substitution $D_0Q^2 \to 4D_0|e|B\tau\hbar^{-1}$, and we obtain for the magnetoconductivity in the two-dimensional case

$$\Delta\sigma(B) = -\frac{e^2}{2\pi^2\hbar}\left(\psi\left(\frac{1}{2} + \frac{B_1}{B}\right) - \frac{3}{2}\psi\left(\frac{1}{2} + \frac{B_2}{B}\right) + \frac{1}{2}\psi\left(\frac{1}{2} + \frac{B_3}{B}\right)\right) \tag{11.231}$$

where

$$B_1 = B_\tau + B_{so} + B_s \ , \ \ B_2 = \frac{4}{3}B_{so} + \frac{2}{3}B_s + B_\varphi \ , \ \ B_3 = 2B_s + B_\varphi \qquad (11.232)$$

and we have introduced the characteristic elastic magnetic field scales

$$B_\tau = \frac{\hbar}{4D_0|e|\tau} \ , \ \ B_{so} = \frac{\hbar}{4D_0|e|\tau_{so}} \ , \ \ B_s = \frac{\hbar}{4D_0|e|\tau_s} \ . \qquad (11.233)$$

A sample with spin-orbit scattering will thus show a positive magnetoresistance, and the extremal value at zero magnetic field will be a minimum.[44]

11.7   Mesoscopic Fluctuations

In the following we shall show that when the size of a sample becomes comparable to the phase coherence length, $L \sim L_\varphi$, the individuality of the sample will be manifest in its transport properties. Such a sample is said to be mesoscopic. Characteristically the conductance will exhibit sample-specific, noiselike but reproduceable, aperiodic oscillations as a function of, say, magnetic field or chemical potential (i.e., density of electrons). The sample behavior is thus no longer characterized by its average characteristics, such as the average conductance, i.e., the average impurity concentration. The statistical assumption of phase-incoherent and therefore independent subsystems, allowing for such an average description, is no longer valid when the transport takes place quantum mechanically coherently throughout the whole sample. As a consequence, a mesoscopic sample does not possess the property of being self-averaging; i.e., the relative fluctuations in the conductance do not vanish in a central limit fashion inversely proportional to the volume in the large-volume limit. To describe the fluctuations from the average value we need to study the higher moments of the conductance distribution such as the variance $\Delta G_{\alpha\beta,\gamma\delta}$. We shall first study the fluctuations in the conductance at zero temperature, and consider the variance

$$\Delta G_{\alpha\beta,\gamma\delta} \ = \ < (G_{\alpha\beta}- < G_{\alpha\beta} >)(G_{\gamma\delta}- < G_{\gamma\delta} >) > \ . \qquad (11.234)$$

For the conductance fluctuations we have the expression

$$< G_{\alpha\beta} G_{\gamma\delta} > \ = \ (L^{-2})^2 \int\!d\mathbf{x}_2\!\int\!d\mathbf{x}'_2\!\int\!d\mathbf{x}_1\!\int\!d\mathbf{x}'_1 \ < \sigma_{\alpha\beta}(\mathbf{x}_2,\mathbf{x}'_2)\, \sigma_{\gamma\delta}(\mathbf{x}_1,\mathbf{x}'_1)> \ . \qquad (11.235)$$

The diagrams for the variance of the conductance fluctuations can still be managed within the standard impurity diagram technique in the weak disorder limit, $\epsilon_F\tau \gg \hbar$, and a typical conductance fluctuation diagram is depicted in figure 11.4. (here the box denotes the Diffuson).[45]

---

[44]An elegant verification of the weak antilocalization effect was demonstrated in experiments by Bergmann [59], where a thin film of magnesium where spin-orbit scattering is absent, were covered with an increasing amount of a submonolayer of gold, thereby increasing the spin-orbit scattering.

[45]The diagram is in the position representation, and the momentum labels should presently be ignored, but will be explained shortly.

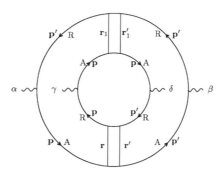

Figure 11.4 Conductance fluctuation diagram.

The construction of the conductance fluctuation diagrams follows from impurity averaging two conductivity diagrams: Draw two conductivity bubble diagrams, where the propagators include the impurity scattering. Treating the impurity scattering perturbatively, we get impurity vertices that we, upon impurity averaging as usual have to pair in all possible ways. Since we subtract the squared average conductance in forming the variance, ΔG, the diagrams for the variance consist only of diagrams where the two conductance loops are connected by impurity lines. As already noted in the discussion of weak localization, the dominant contributions to such loop-type diagrams are from the infrared and long-wavelength divergence of the Cooperon, and here additionally from the Diffuson.

To calculate the contribution to the variance from the Diffuson diagram depicted in figure 11.4, we write the corresponding expression down in the spatial representation in accordance with the usual Feynman rules for conductivity diagrams. Let us consider a hypercube of size L. If we assume that the sample size is bigger than the impurity mean free path, $L > l$, the spatial extension of the integration over the external, excitation and measuring, vertices can be extended to infinity, since the propagators have the spatial extension of the mean free path. We can therefore introduce the Fourier transform for the propagators since no reference to the finiteness of the system is necessary for such local quantities. Furthermore, since the spatial extension of the Diffuson is long range compared to the mean free path, we can set the spatial labels of the Diffusons equal to each other, i.e., $r_1 = r$ and $r'_1 = r'$. All the spatial integrations, except the ones determined by the Diffuson, can then be performed, leading to the momentum labels for the propagators as depicted in figure 11.4. Let us study the fluctuations in the dc conductance, so that the frequency, ω, of the external field is zero. The energy labels have for visual clarity been deleted from figure 11.4, since we only have elastic scattering and therefore one label, say ϵ, for the outer ring and one for the inner, ϵ'. According to the Feynman rules, we obtain for the Diffuson diagram the following analytical expression:

$$< G_{\alpha\beta}G_{\gamma\delta} >_D \;\; = \;\; L^{-4} \left(\frac{e^2\hbar^2 u^2}{4\pi m^2} \right)^2 \int_{-\infty}^{\infty} d\epsilon \, \frac{\partial f(\epsilon)}{\partial \epsilon} \int_{-\infty}^{\infty} d\epsilon' \, \frac{\partial f(\epsilon')}{\partial \epsilon'} \int \frac{d\mathbf{p}}{(2\pi\hbar)^3} \int \frac{d\mathbf{p}'}{(2\pi\hbar)^3}$$

$$G_\epsilon^R(\mathbf{p}')G_\epsilon^A(\mathbf{p}')G_{\epsilon'}^A(\mathbf{p}')G_{\epsilon'}^R(\mathbf{p}')G_\epsilon^R(\mathbf{p})G_{\epsilon'}^A(\mathbf{p})G_\epsilon^A(\mathbf{p})G_{\epsilon'}^R(\mathbf{p})$$

$$p_\alpha \, p_\gamma \, p_\delta' \, p_\beta' \int d\mathbf{r} \int d\mathbf{r}' \, |D(\mathbf{r},\mathbf{r}',\epsilon-\epsilon')|^2 \; . \tag{11.236}$$

In order to obtain the above expression we have noted that

$$D(\mathbf{r},\mathbf{r}',\epsilon'-\epsilon) \;\; = \;\; [D(\mathbf{r},\mathbf{r}',\epsilon-\epsilon')]^* \tag{11.237}$$

which follows from the relationship between the retarded and advanced propagators. At zero temperature, the Fermi functions set the energy variables in the propagators in the conductance loops to the Fermi energy, and the Diffuson frequency to zero. At zero temperature we therefore get for the considered Diffuson diagram the following analytical expression, $D(\mathbf{r},\mathbf{r}') \equiv D(\mathbf{r},\mathbf{r}',0)$,

$$< G_{\alpha,\beta}G_{\gamma,\delta} >_D \;\; = \;\; L^{-4} \left(\frac{e^2\hbar^2 u^2}{4\pi m^2} \right)^2 \int \frac{d\mathbf{p}}{(2\pi\hbar)^3} \int \frac{d\mathbf{p}'}{(2\pi\hbar)^3} \, p_\alpha \, p_\gamma \, p_\delta' \, p_\beta'$$

$$[G_{\epsilon_F}^R(\mathbf{p})G_{\epsilon_F}^A(\mathbf{p})G_{\epsilon_F}^R(\mathbf{p}')G_{\epsilon_F}^A(\mathbf{p}')]^2 \int d\mathbf{r} \int d\mathbf{r}' \, |D(\mathbf{r},\mathbf{r}')|^2 \; . \tag{11.238}$$

It is important to note that the same Diffuson appears twice. This is the leading singularity we need to keep track of. If we try to construct variance diagrams containing, say, three Diffusons, we will observe that they cannot carry the same wave vector, and will give a contribution smaller by the factor $\hbar/\epsilon_F\tau$. The momentum integrations at the current vertices can easily be performed by the residue method (recall eq.(9.26))

$$j_{\alpha\gamma} \;\; = \;\; \int \frac{d\mathbf{p}}{(2\pi\hbar)^3} \, p_\alpha \, p_\gamma \, [G_{\epsilon_F}^R(\mathbf{p})G_{\epsilon_F}^A(\mathbf{p})]^2 \;\; = \;\; \frac{4\pi}{3} \frac{p_F^2 N_0}{\hbar^3} \tau^3 \, \delta_{\alpha\gamma} \tag{11.239}$$

and we obtain for the considered Diffuson diagram the expression

$$< G_{\alpha\beta}G_{\gamma\delta} >_D \;\; = \;\; L^{-4} \left(\frac{e^2 D_0 \tau}{2\pi\hbar} \right)^2 \delta_{\alpha\gamma} \, \delta_{\delta\beta} \int d\mathbf{r} \int d\mathbf{r}' \, |D(\mathbf{r},\mathbf{r}')|^2 \; . \tag{11.240}$$

To calculate the Diffuson integrals we need to address the finite size of the sample and its attachment to the current leads, since the Diffuson has no inherent length scale cutoff. At the surface where the sample is attached to the leads, the Diffuson vanishes

$$D(\mathbf{r},\mathbf{r}') = 0 \qquad \mathbf{r} \;\; or \;\; \mathbf{r}' \;\; \text{on lead surfaces} \tag{11.241}$$

in accordance with the assumption, that once an electron reaches the lead it never returns to the disordered region phase coherently. On the other surfaces the current vanishes; i.e., the normal derivative of the Diffuson must vanish (recall eq.(5.201) and eq.(5.202))

$$\frac{\partial D(\mathbf{r}, \mathbf{r}')}{\partial \mathbf{n}} = 0 \qquad \mathbf{r} \ or \ \mathbf{r}' \ \text{on non-lead surfaces with surface normal } \mathbf{n} \ .$$

(11.242)

We assume that the leads have the same size as the sample surface.[46] Therefore by solving the diffusion equation for the Diffuson, with the above mixed (Dirichlet-von Neumann) boundary condition, we obtain the expression

$$\int d\mathbf{r} \int d\mathbf{r}' \, |D(\mathbf{r}, \mathbf{r}')|^2 = \left(\sum_n \frac{1/\tau}{D_0 \, q_n^2} \right)^2$$

(11.243)

where $n \equiv (n_x, n_y, n_z)$ is the eigenvalue index in the three-dimensional case

$$q_{n_\alpha} = \frac{\pi}{L} n_\alpha \qquad n_\alpha = n_x, n_y, n_z$$

(11.244)

where

$$n_x = 1, 2, ..., \qquad n_{y,z} = 0, 1, 2, ...$$

(11.245)

and we have assumed that the current leads are along the x-axis. Less than three dimensions corresponds to neglecting the n_y and n_z's. We therefore obtain from the considered Diffuson diagram the contribution to the conductance fluctuations[47]

$$< G_{\alpha\beta} G_{\gamma\delta} >_D = \left(\frac{e^2}{4\pi\hbar} \right)^2 c_d \, \delta_{\alpha,\gamma} \, \delta_{\delta,\beta}$$

(11.246)

where the constant c_d depends on the sample dimension. The summation in eq.(11.243) should, in accordance with the validity of the diffusion regime, be restricted to values satisfying $n_x^2 + n_y^2 + n_z^2 \leq N$, where N is of the order of $(L/l)^2$. However, the sum converges rapidly and the constants c_d are seen to be of order unity. The dimensionality criterion is essentially the same as in the theory of weak localization, as we shall show in the discussion below of the physical origin of the fluctuation effects. The important thing to notice is that the long-range nature of the Diffuson provides the L^4 factor that makes the variance, average of the squared conductance, independent of sample size (recall eq.(7.143)). The diagram depicted in figure 11.4 is only one of the two possible pairings of the current vertices, and we obtain an additional contribution from the diagram where, say, current vertices γ and δ are interchanged.

In addition to the contribution from the diagram in figure 11.4 there is the other possible singular Diffuson contribution to the variance from the diagram depicted in figure 11.5.

---

[46]This "thick lead" assumption is not of importance. Because of the relationship between the fluctuations in the density of states and the time scale for diffusing out of the sample, the result will be the same for any kind of lead attachment [75].

[47]Because of these inherent mesoscopic fluctuations, we realize that the conductance discussed in the scaling theory of localization is the average conductance.

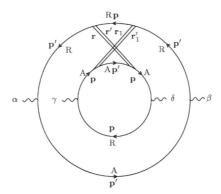

Figure 11.5 The other possible conductance fluctuation diagram.

This diagram contributes the same amount as the one in figure 11.4, but with a different pairing of the current vertices. We note that the diagram in figure 11.5 allows for only one assignment of current vertices.[48]

Reversing the direction in one of the loops gives rise to similar diagrams, but now with the Cooperon appearing instead of the Diffuson. Because the boundary conditions on the Cooperon are the same as for the Diffuson, in the absence of a magnetic field, the Cooperon contributes an equal amount. For the total contribution to the variance of the conductance, we therefore have (allowing for the spin degree of freedom of the electron would quadruple the value) at zero temperature

$$\Delta G_{\alpha\beta,\gamma\delta} = \left(\frac{e^2}{2\pi\hbar}\right)^2 c_d \left(\delta_{\alpha\gamma}\,\delta_{\delta\beta} + \delta_{\alpha\delta}\,\delta_{\gamma\beta} + \delta_{\alpha,\beta}\,\delta_{\gamma,\delta}\right). \qquad (11.247)$$

The variance of the conductance at zero temperature, and for the chosen geometry of a hypercube, is seen to be independent of size and dimension of the sample and degree of disorder, and the conductance fluctuations appear in the metallic regime described above to be universal.[49]

Since the average classical conductance, according to eq.(8.73), is proportional to L^{d-2}, Ohm's law, we find that the relative variance, $\Delta G < G >^{-2}$, is proportional to L^{4-2d}. This result should be contrasted with the behavior L^{-2d} of thermodynamic fluctuations, compared to which the quantum interference induced mesoscopic fluctuations are huge, reflecting the absence of self-averaging.

The dominating role of the lowest eigenvalue in eq.(11.243) indicates that mesoscopic fluctuations, studied in situations with less invasive probes than the current leads necessary for studying conductance fluctuations, can be enhanced compared

---

[48]The contribution from the diagram in figure 11.4 can, through the Einstein relation, be ascribed to fluctuations in the diffusion constant, whereas the diagram in figure 11.5 gives the contribution from the fluctuations in the density of states, the two types of fluctuations being independent [76].

[49]However, for a noncubic sample, the variance will be geometry dependent, [77], [78].

to the universal value. In the case of the conductance fluctuations, the necessary connection of the disordered region to the leads, which cut off the singularity in the Diffuson by the lowest eigenvalue, $n_x = 1$, reflecting the fact that due to the physical boundary conditions at the interface between sample and leads, the maximal time for quantum interference processes to occur uninterrupted is the time it takes the electron to diffuse across the sample, L^2/D_0. When considering other ways of observing mesoscopic fluctuations, the way of observation will in turn introduce the destruction of phase coherence necessary for rendering the fluctuations finite.

In order to understand the origin of the conductance fluctuations, we note that, just as the conductance essentially is given by the probability for diffusing between points in a sample, the variance is likewise the product of two such probabilities. When we perform the impurity average, certain of the quantum interference terms will not be averaged away, since certain pairs of paths are coherent. This is similar to the case of coherence involved in the weak-localization effect, but in the present case of the variance of quite a different nature. For example, the quantum interference process described by the diagram in figure 11.4 is depicted in figure 11.6, where the solid line corresponds to the outer conductance loop, and the dashed line to the inner conductance loop. The wavy portion of the lines corresponds to the long-range diffusion process.

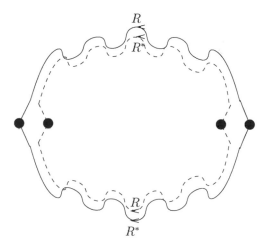

Figure 11.6 Statistical correlation described by the diagram in figure 11.4.

When one takes the impurity average of the variance, the quantum interference terms can pair up for each diffusive path in the random potential, but now they correspond to amplitudes for propagation in different samples. The diagrams for the variance, therefore, do not describe any physical quantum interference process, since we are not describing a probability but a product of probabilities. The variance gives the statistical correlation between amplitudes in different samples. The interference process corresponding to the diagram in figure 11.5 is likewise depicted in figure 11.7.

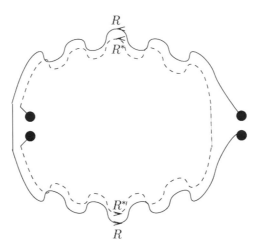

Figure 11.7 Statistical correlation described by the diagram in figure 11.5.

When a specific mesoscopic sample is considered, no impurity average is effectively performed as in the macroscopic case. The quantum interference terms in the conductance, which for a macroscopic sample averages to zero if we neglect the weak-localization effect, are therefore responsible for the mesoscopic fluctuations. In the weak-disorder regime the conductivity (or equivalently the diffusivity by Einstein's relation) is specified by the probability for the particle to propagate between points in space. According to eq.(11.17)

$$P = P_{cl} + 2 \sum_{c,c'} \sqrt{|A_c A_{c'}|} \cos(\phi_c - \phi_{c'}) \qquad (11.248)$$

as

$$A_c = |A_c| e^{i\phi_c} \quad , \quad \phi_c = \frac{1}{\hbar} S[\mathbf{x}_c(t)] \qquad (11.249)$$

where $|A_c|$ specifies the probability for the classical path c, and its phase is specified by the action. When the points in space in questions are farther apart than the mean free path, the ensemble average of the quantum interference term in the probability vanishes. The weak localization can be neglected because for random phases we have $< \cos(\phi_c - \phi_{c'}) >_{imp} = 0$. However, for the mean square of the probability, we encounter $< \cos^2(\phi_c - \phi_{c'}) >_{imp} = 1/2$, and obtain

$$< P^2 >_{imp} = < P >_{imp}^2 + 2 \sum_{c,c'} |A_c| |A_{c'}| . \qquad (11.250)$$

Because of quantum interference there is thus a difference between $< P^2 >_{imp}$ and $< P >_{imp}^2$ resulting in mesoscopic fluctuations. Since the effect is determined by the phases of paths, it is nonlocal.

The result in eq.(11.247) is valid in the metallic regime, where the average conductance is larger than $e^2/\hbar$. To go beyond the metallic regime would necessitate introducing the quantum corrections to diffusion, the first of which is the weak-localization type, which diagrammatically corresponds to inserting Cooperons in between Diffusons. Such an analysis is necessary for a study of the fluctuations in the strongly disordered regime, as performed in reference [44].

The Diffuson and Cooperon in the conductance fluctuation diagrams do not describe diffusion and return probability, respectively, in a given sample, but quantum-statistical correlations between motion in different samples, i.e., different impurity configurations, as each conductance loop in the figures 11.4 and 11.5 corresponds to different samples. In order to stress this important distinction, we shall in the following mark with a tilde the Diffusons and Cooperons appearing in fluctuation diagrams.

We now assess the effects of finite temperature on the conductance fluctuations. Besides the explicit temperature dependence due to the Fermi functions appearing in eq.(11.236), the ladder diagrams will be modified by interaction effects. The presence of the Fermi functions corresponds to an energy average over the thermal layer near the Fermi surface, and through the energy dependence of the Diffuson and Cooperon introduces the temperature-dependent length scale $L_T = \sqrt{D_0\hbar/kT}$. Since the loops in the fluctuation diagrams correspond to different conductivity measurements, i.e., different samples, interaction lines (due to for example electron-phonon or electron-electron interaction) are not allowed to connect the loops in a fluctuation diagram. The diffusion pole of the Diffuson appearing in a fluctuation diagram is therefore not immune to interaction effects. This was only the case when the Diffuson describes diffusion within a sample, since then the diffusion pole is a consequence of particle conservation and therefore unaffected by interaction effects. The consequence is that, just as in the case for the Cooperon, inelastic scattering will lead to a cutoff given by the phase-breaking rate $1/\tau_\varphi$. In short, the temperature effects will therefore ensure that up to the length scale of the order of the phase-coherence length, the conductance fluctuations are determined by the zero-temperature expression, and beyond this scale the conductance of such phase-incoherent volumes add as in the classical case.[50] A sample is therefore said to be mesoscopic when its size is in between the microscopic scale, set by the mean free path, and the macroscopic scale, set by the phase-coherence length, $l < L < L_\varphi$. A sample is therefore only self-averaging with respect to the impurity scattering for samples of size larger than the phase-coherence length.[51] A sample will therefore only exhibit the weak-localization effect when its size is much larger than the phase-coherence length but much smaller than the localization length $L_\varphi < L < \xi$.

An important way to reveal the conductance fluctuations experimentally is to measure the magnetoresistance of a mesoscopic sample. To study the fluctuation effects in magnetic fields, we must study the dependence of the variance on the

---

[50] For example for a wire we have $g(L) = g(L_\varphi)\, L/L_\varphi$.

[51] The conductance entering the scaling theory of localization is thus assumed averaged over phase-incoherent volumes.

magnetic fields $\Delta G_{\alpha\beta}(\mathbf{B}_+, \mathbf{B}_-)$, where $\mathbf{B}_+$ is the sum and $\mathbf{B}_-$ the difference in the magnetic fields influencing the outer and inner loops. Since the conductance loops can correspond to samples placed in different field strengths, the diffusion pole appearing in a fluctuation diagram will not be immune to the presence of magnetic fields, as in the case when the Diffuson describes diffusion within a given sample, since particle conservation is, of course, unaffected by the presence of a magnetic field. According to the low-field prescription for inclusion of magnetic fields, eq.(11.29), we get for the Diffuson

$$D_0 \left\{ \left(-i\nabla_{\mathbf{x}} - \frac{e}{\hbar}\mathbf{A}_-(\mathbf{x}) \right)^2 + 1/\tau_\varphi \right\} \tilde{D}(\mathbf{x}, \mathbf{x}') = \frac{1}{\tau}\delta(\mathbf{x} - \mathbf{x}') \qquad (11.251)$$

where $\mathbf{A}_-$ is the vector potential corresponding to the difference in magnetic fields, $\mathbf{B}_- = \nabla_{\mathbf{x}} \times \mathbf{A}_-$, and we have introduced the phase-breaking rate in view of the above consideration. In the case of the Diffuson, the magnetic field induced phases subtract, accounting for the appearance of the difference of the vector potentials $\mathbf{A}_-$. For the case of the Cooperon, the two phases add, and we obtain

$$D_0 \left\{ \left(-i\nabla_{\mathbf{x}} - \frac{e}{\hbar}\mathbf{A}_+(\mathbf{x}) \right)^2 + 1/\tau_\varphi \right\} \tilde{C}(\mathbf{x}, \mathbf{x}') = \frac{1}{\tau}\delta(\mathbf{x} - \mathbf{x}') \qquad (11.252)$$

where $\mathbf{A}_+$ is the vector potential corresponding to the sum of the fields, $\mathbf{B}_+ = \nabla \times \mathbf{A}_+$.

The *magneto-fingerprint* of a given sample, i.e., the dependence of its conductance on an external magnetic will show an erratic pattern with a given peak to valley ratio and a correlation field strength B_c. This, however, is not immediately the information we obtain by calculating the variance

$$\Delta G_{\alpha\beta,\gamma\delta}(\mathbf{B}_+, \mathbf{B}_-) = < [G_{\alpha\beta}(\mathbf{B}_1) - < G_{\alpha\beta}(\mathbf{B}_1) >][G_{\gamma\delta}(\mathbf{B}_2) - < G_{\gamma\delta}(\mathbf{B}_2) >] >$$
$$(11.253)$$

where $\mathbf{B}_1$ is the field in, say, the inner loop, $\mathbf{B}_1 = (\mathbf{B}_+ + \mathbf{B}_-)/2$, and $\mathbf{B}_2$ is the field in the outer loop, $\mathbf{B}_2 = (\mathbf{B}_+ - \mathbf{B}_-)/2$. In the variance, the magnetic fields are fixed in the two samples, and we are averaging over different impurity configurations, thus describing a situation in which the actual impurity configuration is changed, a hardly controllable endeavor from an experimental point of view. However, if the magnetoconductance of a given sample, $G(B)$, varies randomly with magnetic field, the two types of averages – the one with respect to magnetic field and the one with respect to impurity configuration – are equivalent, and the characteristics of the magneto-fingerprint can be extracted from the correlation function in eq.(11.253). The physical reason for the validity of such an *ergodic* hypothesis [79], [80], that changing magnetic field is equivalent to changing impurity configuration is, that since the electronic motion in the sample is quantum mechanically coherent the wave function pattern is sensitive to the position of all the impurities in the sample, just as the presence of the magnetic field is felt throughout the sample by the electron.[52] The extreme sensitivity to impurity configuration is also witnessed by

---

[52]The validity of the ergodic hypothesis has been substantiated in reference [81].

the fact that changing the position of one impurity by an atomic distance, $1/k_F$, is equivalent to shifting all the impurities by arbitrary amounts, i.e., to create a completely different sample [82], [83].

The ergodic hypothesis can be elucidated by the following consideration. In the mean square of the probability for propagating between two points in space we encounter the correlation function

$$< \Big(\cos(\phi_c(B_1) - \phi_{c'}(B_1)) \Big) \Big(\cos(\phi_c(B_2) - \phi_{c'}(B_2)) \Big) >_{\mathrm{imp}} \qquad (11.254)$$

where $(\phi_c(B) - \phi_{c'}(B))$ depends on the phases picked up due to the magnetic field, i.e., the flux through the area enclosed by the trajectories c and c'. When the magnetic field B_1 changes its value to B_2 (where the correlation function equals one half), the phase factor changes by 2π times the flux through the area enclosed by the trajectories c and c' in units of the flux quantum. This change, however, is equivalent to what happens when changing to a different impurity configuration for fixed magnetic field, i.e., the quantity we calculate.[53]

In order to calculate the variance in eq.(11.253) we must solve eq.(11.251) and eq.(11.252) with the appropriate mixed boundary value conditions in the presence of magnetic fields, and insert the solutions into contributions like that in eq.(11.240). However, determination of the characteristic correlations of the aperiodic magnetoconductance fluctuations can be done by inspection of eq.(11.251) and eq.(11.252). The correlation field B_c is determined by the sample-to-sample change in the magnetic field, i.e., $\mathbf{B}_-$. According to eq.(11.251) and eq.(11.252), this field is determined either by the sample size, through the gradient term, or the phase coherence length. When the phase-coherence length is longer than the sample size, the correlation field is therefore of order of the flux quantum divided by the sample area, $B_c \sim \phi_0/L^2$, where ϕ_0 is the normal flux quantum $\phi_0 = 2\pi\hbar/|e|$, since the typical diffusion loops, like those depicted in figures 11.6 and 11.7 enclose an area of the order of the sample, L^2. We note that in magnetic fields exceeding $\max\{\phi_0/L^2, \phi_0/L_\varphi^2\}$, the Cooperon no longer contributes to the field dependence of the conductance fluctuations, because its dependence on magnetic field is suppressed according to the weak-localization analysis.[54]

We note that the weak-localization and mesoscopic fluctuation phenomena are a general feature of wave propagation in a random media, be the wave nature classical, such as sound and light,[55] or of quantum origin such as for the motion of electrons. The weak-localization effect was in fact originally envisaged for the multiple scattering of electromagnetic waves [42]. [56] The coherent backscattering effect has been studied experimentally for light waves (for a review on classical wave

---

[53]Another way of revealing the mesoscopic fluctuations is to change the Fermi energy (i.e., the density of conduction electron as is feasible in an inversion layer). The typical energy scale E_c for these fluctuations is analogously determined by the typical time τ_{trav} it takes an electron to traverse the sample according to $E_c \sim \hbar/\tau_{\mathrm{trav}}$. In the diffusive regime we have $\tau_{\mathrm{trav}} \sim L^2/D_0$.

[54]For an account of the experimental discovery of conductance fluctuations, see reference [84].

[55]Here we refer to conditions described by Maxwell's equations.

[56]It is telling that it took the application of Feynman diagrams in the context of electronic motion in disordered conductors to understand the properties of classical waves in random media.

propagation in random media, see reference [85]). For the wealth of interesting weak-localization and mesoscopic fluctuation effects, we refer to the references cited in for example the references [59], [60], [71], [84], [86], [87], [88], [89].

Appendix A

Path Integrals and Propagators

In classical mechanics only the classical paths are of physical relevance; however, stating the quantum law of motion involved all paths. The way in which the various alternative paths contribute to the expression for the propagator was realized by Dirac [90], who noted that the conditional amplitude for an infinitesimal time step is related to Lagrange's function, L, according to

$$<\mathbf{x}, t + \Delta t|\mathbf{x}', t> \propto e^{\frac{i}{\hbar}\Delta t\, L(\mathbf{x},(\mathbf{x}-\mathbf{x}')/\Delta t)} \tag{A.1}$$

however, with L expressed in terms of the coordinates at time t and $t + \Delta t$. This gem of Dirac's was turned into brilliance by Feynman, and provided the intuitive approach to quantum mechanics as described in section 1.1. We shall here obtain the path integral expression for Dirac's transformation function $<\mathbf{x}, t|\mathbf{x}', t'>$.

Propagating in small steps by inserting complete sets at intermediate times we have for the propagator

$$<\mathbf{x}, t|\mathbf{x}', t'> = \int d\mathbf{x}_1 \int d\mathbf{x}_2 .. \int d\mathbf{x}_N \ <\mathbf{x}, t|\mathbf{x}_N, t_N><\mathbf{x}_N, t_N|\mathbf{x}_{N-1}, t_{N-1}>$$

$$<\mathbf{x}_{N-1}, t_{N-1}|\mathbf{x}_{N-2}, t_{N-2}> .. <\mathbf{x}_1, t_1|\mathbf{x}', t'> \ . \tag{A.2}$$

We are consequently interested in the transformation function for infinitesimal times, and from eq.(1.113) we obtain

$$<\mathbf{x}_n, t_n|\mathbf{x}_{n-1}, t_{n-1}> = <\mathbf{x}_n|e^{-\frac{i}{\hbar}\Delta t \hat{H}(t_n)}|\mathbf{x}_{n-1}>$$

$$= \delta(\mathbf{x}_n - \mathbf{x}_{n-1}) + \frac{\Delta t}{i\hbar} <\mathbf{x}_n|\hat{H}(t_n)|\mathbf{x}_{n-1}> + \mathcal{O}(\Delta t^2) \ \ (A.3)$$

where $\Delta t = t_n - t_{n-1} = (t - t')/(N + 1))$, as we have inserted N intermediate resolutions of the identity.

In the following we shall consider a particle of mass m in a potential V for which we have the Hamiltonian

$$\hat{H}(t) = \frac{\hat{\mathbf{p}}^2}{2m} + V(\hat{\mathbf{x}}, t) \ . \tag{A.4}$$

Inserting a complete set of momentum states, we get

$$<\mathbf{x}_n|H(\hat{\mathbf{x}}, \hat{\mathbf{p}}, t_n)|\mathbf{x}_{n-1}> \quad = \quad <\mathbf{x}_n|H(\mathbf{x}_n, \hat{\mathbf{p}}, t_n)|\mathbf{x}_{n-1}>$$

$$= \quad \int \frac{d\mathbf{p}_n}{(2\pi\hbar)^d} \, e^{\frac{i}{\hbar}\mathbf{p}_n \cdot (\mathbf{x}_n - \mathbf{x}_{n-1})} \, H(\mathbf{x}_n, \mathbf{p}_n, t_n) \quad (A.5)$$

where we encounter Hamilton's function on phase space

$$H(\mathbf{x}_n, \mathbf{p}_n, t_n) \quad = \quad \frac{\mathbf{p}_n^2}{2m} + V(\mathbf{x}_n, t_n) \, . \quad (A.6)$$

Inserting into eq.(A.3), we get

$$<\mathbf{x}_n, t_n|\mathbf{x}_{n-1}, t_{n-1}> \quad = \quad \int \frac{d\mathbf{p}_n}{(2\pi\hbar)^d} \, e^{\frac{i}{\hbar}\mathbf{p}_n \cdot (\mathbf{x}_n - \mathbf{x}_{n-1})} \left(1 + \frac{\Delta t}{i\hbar} H(\mathbf{x}_n, \mathbf{p}_n, t_n) + \mathcal{O}(\Delta t^2) \right)$$

$$= \quad \int \frac{d\mathbf{p}_n}{(2\pi\hbar)^d} \, e^{\frac{i}{\hbar}[\mathbf{p}_n \cdot (\mathbf{x}_n - \mathbf{x}_{n-1}) - \Delta t H(\mathbf{x}_n, \mathbf{p}_n, t_n)]} + \mathcal{O}(\Delta t^2) \, . \quad (A.7)$$

Inserting additional internal times, we approach the limit $\Delta t \to 0$, or equivalently $N \to \infty$, obtaining for the transformation function

$$<\mathbf{x}, t|\mathbf{x}', t'> \quad = \quad \lim_{N \to \infty} \int \prod_{n=1}^{N} d\mathbf{x}_n \prod_{n=1}^{N+1} \frac{d\mathbf{p}_n}{(2\pi\hbar)^d} \, e^{\frac{i}{\hbar}[\mathbf{p}_n \cdot (\mathbf{x}_n - \mathbf{x}_{n-1}) - \Delta t H(\mathbf{x}_n, \mathbf{p}_n, t_n)]}$$

$$\equiv \quad \int \frac{\mathcal{D}\mathbf{x}_{\bar{t}} \mathcal{D}\mathbf{p}_{\bar{t}}}{(2\pi\hbar)^d} \, e^{\frac{i}{\hbar} \int_{t'}^{t} d\bar{t} \, [\mathbf{p}_{\bar{t}} \cdot \dot{\mathbf{x}}_{\bar{t}} - H(\mathbf{x}_{\bar{t}}, \mathbf{p}_{\bar{t}}, \bar{t})]} \quad (A.8)$$

where $\mathbf{x}_0 \equiv \mathbf{x}'$, and $\mathbf{x}_{N+1} \equiv \mathbf{x}$. In the last equation we have just written the limit of the sum as a path integral, and the measure has been identified by the explicit limiting procedure.

The Hamilton function is quadratic in the momentum, and we have Gaussian integrals which can be performed

$$\int_{-\infty}^{\infty} \frac{d\mathbf{p}_n}{(2\pi\hbar)^d} \, e^{\frac{i}{\hbar}\Delta t(\mathbf{p}_n \cdot \frac{\mathbf{x}_n - \mathbf{x}_{n-1}}{\Delta t} - \frac{\mathbf{p}_n^2}{2m})} \quad = \quad \left(\frac{m}{2\pi i\hbar\Delta t} \right)^{d/2} e^{\frac{i}{2\hbar}m\left(\frac{\mathbf{x}_n - \mathbf{x}_{n-1}}{\Delta t}\right)^2 \Delta t} \quad (A.9)$$

and we thus get for the transformation function

$$<\mathbf{x}, t|\mathbf{x}', t'> = \lim_{N \to \infty} \frac{1}{\left(\frac{m}{2\pi i\hbar\Delta t}\right)^{-d/2}} \int \prod_{n=1}^{N} \frac{d\mathbf{x}_n}{\left(\frac{m}{2\pi i\hbar\Delta t}\right)^{-d/2}} e^{\frac{i}{\hbar}\Delta t \sum_{n=1}^{N+1} \left[\frac{m(\mathbf{x}_n - \mathbf{x}_{n-1})^2}{2\Delta t} - V(\mathbf{x}_n, t_n) \right]}$$

$$\equiv \quad \int_{\mathbf{x}_{t'}=\mathbf{x}'}^{\mathbf{x}_t=\mathbf{x}} \mathcal{D}\mathbf{x}_{\bar{t}} \, e^{\frac{i}{\hbar} \int_{t'}^{t} d\bar{t} \, L(\mathbf{x}_{\bar{t}}, \dot{\mathbf{x}}_{\bar{t}}, \bar{t})} \quad (A.10)$$

where L in the continuum limit is seen to be Lagrange's function

$$L(\mathbf{x}_t, \dot{\mathbf{x}}_t, t) = \frac{1}{2}m\dot{\mathbf{x}}_t^2 - V(\mathbf{x}_t, t) = \dot{\mathbf{x}}_t \cdot \mathbf{p}_t - H(\mathbf{x}_t, \mathbf{p}_t, t) \qquad (A.11)$$

related to Hamilton's function through a Legendre transformation. The integration measure has here been obtained for the case where we take the piecewise linear approximation for a path.[1]

We note that the path integral formalism is useful for obtaining the expressions for the propagators in spatially homogeneous fields, since we are then dealing with a quadratic Lagrange function which can be easily integrated. So let us consider a quadratic Lagrange function[2]

$$L(\mathbf{x}_t, \dot{\mathbf{x}}_t, t) = a(t)\dot{\mathbf{x}}_t^2 + b(t)\dot{\mathbf{x}}_t \cdot \mathbf{x}_t + c(t)\mathbf{x}_t^2 + e(t)\mathbf{x}_t + f(t). \qquad (A.12)$$

The action for the classical path starting at $\mathbf{x}'$ at time t', and reaching $\mathbf{x}$ at time t is

$$S_{cl}(\mathbf{x}, t; \mathbf{x}', t') \equiv S_{\mathbf{x}, t; \mathbf{x}', t'}[\mathbf{x}_{cl}(t)] = \int_{t'}^{t} d\bar{t}\, L(\mathbf{x}_{\bar{t}}^{cl}, \dot{\mathbf{x}}_{\bar{t}}^{cl}, \bar{t}) \qquad (A.13)$$

where $\mathbf{x}_{cl}(t)$ is the solution of Newton's equation of motion

$$0 = \left.\frac{\delta S}{\delta \mathbf{x}_t}\right|_{\mathbf{x}(t)=\mathbf{x}_{cl}(t)} = 2a(t)\ddot{\mathbf{x}}_{cl}(t) + 2\dot{a}(t)\dot{\mathbf{x}}_{cl}(t) - \dot{b}(t)\mathbf{x}_{cl}(t) + 2c(t)\mathbf{x}_{cl}(t) + e(t) \quad (A.14)$$

with the boundary conditions $\mathbf{x}_{cl}(t') = \mathbf{x}'$ and $\mathbf{x}_{cl}(t) = \mathbf{x}$.

We specify an arbitrary path according to its deviation from the classical path, $\mathbf{x}(t) = \mathbf{x}_{cl}(t) + \delta\mathbf{x}_t$. Due to the stationarity of the classical path there are no linear terms when we express the action in terms of $\delta\mathbf{x}_t$, and the quadratic terms are easily picked out:

$$S[\mathbf{x}(t)] = S_{cl}(\mathbf{x}, t; \mathbf{x}', t') + \int_{t'}^{t} d\bar{t}\, [a(\bar{t})\delta\dot{\mathbf{x}}_{\bar{t}}^2 + b(\bar{t})\delta\dot{\mathbf{x}}_{\bar{t}} \cdot \delta\mathbf{x}_{\bar{t}} + c(\bar{t})\delta\mathbf{x}_{\bar{t}}^2]. \qquad (A.15)$$

In the path integral on discretized form for the propagator we shift each internal integration over $\mathbf{x}_n$ to integration over $\delta\mathbf{x}_n$, $\mathbf{x}_n = \mathbf{x}_n^{cl} + \delta\mathbf{x}_n$, immediately giving the form

$$K(\mathbf{x}, t; \mathbf{x}', t') = A(t, t')\, e^{\frac{i}{\hbar}S_{cl}(\mathbf{x}, t; \mathbf{x}', t')} \qquad (A.16)$$

where the prefactor

$$A(t, t') = \int_{\delta\mathbf{x}_{t'}=\mathbf{0}}^{\delta\mathbf{x}_t=\mathbf{0}} \mathcal{D}(\delta\mathbf{x}_{\bar{t}})\, e^{\frac{i}{\hbar}\int_{t'}^{t} d\bar{t}\,[a(\bar{t})\delta\dot{\mathbf{x}}_{\bar{t}}^2 + b(\bar{t})\delta\dot{\mathbf{x}}_{\bar{t}}\delta\mathbf{x}_{\bar{t}} + c(\bar{t})\delta\mathbf{x}_{\bar{t}}^2]} \qquad (A.17)$$

is independent of the initial and final points $\mathbf{x}$ and $\mathbf{x}'$, since the limits for the paths in the path integral are fixed by having no deviations at the start and end points

---

[1] Other measures can be used, such as expanding the paths on a complete set of functions, so that the sum over all paths becomes the integral over all the expansion coefficients.

[2] Taking the most general quadratic form is of course straightforward.

$\delta\mathbf{x}_{t'} = \mathbf{0} = \delta\mathbf{x}_t$. The prefactor can usually be found by simple considerations. For example, for a free particle the prefactor is simply asserted by the initial condition for the propagator as noted in section 1.1.

In general the quadratic expansion around the classical path only gives the quasi-classical propagator

$$S_{\mathbf{x},t;\mathbf{x}',t'}[\mathbf{x}(t)] = S_{cl}(\mathbf{x},t;\mathbf{x}',t') + \frac{1}{2!} \int_{t'}^{t}d\bar{t}\int_{t'}^{t}d\bar{t}' \; \delta\mathbf{x}_{\alpha}(\bar{t}) \frac{\delta^2 S_{\mathbf{x},t;\mathbf{x}',t'}[\mathbf{x}_{cl}(t)]}{\delta\mathbf{x}_{\alpha}(\bar{t})\,\delta\mathbf{x}_{\beta}(\bar{t}')} \; \delta\mathbf{x}_{\beta}(\bar{t}') + ...$$

$$(A.18)$$

In the stationary phase approximation, the path integral to be performed is Gaussian, and we obtain the result in eq.(1.16).

Exercise A.1 *Show, that for a particle of mass m the retarded propagator in a constant force field, $\mathbf{F}$, is given by*

$$G_{\mathbf{F}}^{R}(\mathbf{x},t;\mathbf{x}',t') = G_{0}^{R}(\mathbf{x},t;\mathbf{x}',t')\, e^{\frac{i}{2\hbar}(t-t')(\mathbf{F}\cdot(\mathbf{x}+\mathbf{x}') - \frac{1}{12m}\mathbf{F}^2(t-t')^2)} \; . \qquad (A.19)$$

Solution

The classical action, $S_{cl}(\mathbf{x},t;\mathbf{x}',t')$, is obtained by solving Newton's equation

$$m\ddot{\mathbf{x}}_t = \mathbf{F} \; , \qquad \mathbf{x}_{t'} = \mathbf{x}' \, , \; \mathbf{x}_t = \mathbf{x} \qquad (A.20)$$

where the boundary conditions determine the constants

$$\mathbf{x}_t = \frac{\mathbf{F}}{2m}t^2 + \mathbf{a}\,t + \mathbf{b} \qquad (A.21)$$

to be

$$\mathbf{a} = \frac{\mathbf{x}-\mathbf{x}'}{t-t'} - \frac{\mathbf{F}}{2m}(t+t') \, , \qquad \mathbf{b} = \frac{\mathbf{x}+\mathbf{x}'}{2} - \frac{\mathbf{x}-\mathbf{x}'}{t-t'}\frac{t+t'}{2} + \frac{\mathbf{F}}{2m}tt' \; . \qquad (A.22)$$

The Lagrange function takes for the classical path the form

$$L(\dot{\mathbf{x}}_t,\mathbf{x}_t) = \frac{1}{2}m\dot{\mathbf{x}}_t^2 + \mathbf{F}\cdot\mathbf{x}_t = \frac{\mathbf{F}^2}{m}t^2 + 2\mathbf{a}\cdot\mathbf{F}t + \frac{1}{2}m\mathbf{a}^2 + \mathbf{b}\cdot\mathbf{F} \qquad (A.23)$$

and the corresponding classical action becomes

$$S_{cl}(\mathbf{x},t;\mathbf{x}',t') = \frac{\mathbf{F}^2}{3m}(t^3 - t'^3) + \mathbf{a}\cdot\mathbf{F}(t^2 - t'^2) + \left(\mathbf{b}\cdot\mathbf{F} + \frac{1}{2}m\mathbf{a}^2\right)(t-t') \; . \qquad (A.24)$$

Inserting the constants gives the stated exponent, and since the prefactor does not depend on the force according to eq.(A.17) (the coefficient to the linear term in the position is absent) it is the same as for the free particle propagator, and we obtain the stated result for the propagator in the presence of a constant force.

Exercise A.2 *Show that for a particle of charge e and mass m the propagator in a magnetic field, assumed time-independent and spatially homogeneous* $\mathbf{B} = B\hat{\mathbf{z}}$, *is given, in the symmetric gauge* $\mathbf{A}(\mathbf{x}) = \frac{1}{2}\mathbf{B} \times \mathbf{x}$, *by*

$$
G_{\mathbf{B}}^R(\mathbf{x}, t; \mathbf{x}', t') = -i\theta(t - t') \left(\frac{m}{2\pi i \hbar(t - t')} \right)^{3/2} \frac{\frac{\omega_c(t-t')}{2}}{\sin \frac{\omega_c(t-t')}{2}}
$$

$$
e^{\frac{i}{\hbar}\left(\frac{m}{2} \frac{(z-z')^2}{t-t'} - \frac{m\omega_c}{2}(xy' - yx') + \frac{m\omega_c}{4} \cot\left(\frac{\omega_c(t-t')}{2} \right)[(x-x')^2 + (y-y')^2] \right)}
$$

$$
= -i\theta(t - t') \left(\frac{m}{2\pi i \hbar(t - t')} \right)^{3/2} \frac{\frac{\omega_c(t-t')}{2}}{\sin \frac{\omega_c(t-t')}{2}}
$$

$$
e^{\frac{ei}{2\hbar}\left(\frac{(\mathbf{B}\cdot(\mathbf{x}-\mathbf{x}'))^2}{B\omega_c(t-t')} - \mathbf{B}\cdot(\mathbf{x}\times\mathbf{x}') + \frac{1}{2B}(\mathbf{B}\times(\mathbf{x}-\mathbf{x}'))^2 \cot \frac{\omega_c(t-t')}{2} \right)} \tag{A.25}
$$

where $\omega_c \equiv eB/m$ *is the Larmor or cyclotron frequency.*

Show, that the same expression for the momentum amplitude is reached as in the zero field time-of-flight analysis of section 1.1.2.

Solution

The solution of the classical equation of motion, $m\ddot{\mathbf{x}} = e\dot{\mathbf{x}} \times \mathbf{B}$, *has constant velocity along the magnetic field, and executes a circular motion in the perpendicular plane*

$$
\begin{pmatrix} x_t \\ y_t \end{pmatrix} = R_L \begin{pmatrix} \cos(\omega_c t + \phi) \\ -\sin(\omega_c t + \phi) \end{pmatrix} + \begin{pmatrix} c_1 \\ c_2 \end{pmatrix} \tag{A.26}
$$

where R_L *is the Larmor radius, related to the Larmor frequency through* $v_\perp = R_L |\omega_c|$, *where* $v_\perp$ *is the constant length of the velocity vector perpendicular to the magnetic field direction, and* $\mathbf{c} = (c_1, c_2)$ *is the center of the circular motion in the* $x - y$-plane.

In the symmetric gauge we obtain from the Lagrange function

$$
L = \frac{1}{2}m\dot{\mathbf{x}}_t^2 + e\dot{\mathbf{x}}_t \cdot \mathbf{A}(\mathbf{x}_t) \tag{A.27}
$$

for the classical action the expression

$$
S_{cl}(\mathbf{x}, t; \mathbf{x}', t') = \frac{m}{2}\frac{(z - z')^2}{t - t'} - \frac{m\omega_c}{2}(xy' - yx') + \frac{m\omega_c R_L^2}{2}\sin(\omega_c(t - t'))
$$

$$
= \frac{m}{2}\frac{(z - z')^2}{t - t'} - \frac{m\omega_c}{2}(xy' - yx')
$$

$$
+ \frac{m\omega_c}{4}\cot\left(\frac{\omega_c(t - t')}{2} \right)[(x - x')^2 + (y - y')^2] \tag{A.28}
$$

where the last rewriting makes the correctness of the zero field limit obvious.

The prefactor in the propagator we can, according to eq.(2.6), obtain from the result of exercise 1.23 on page 76 by analytical continuation of the partition function for a particle in a magnetic field

$$A(t - t') = -\frac{i}{V} Z_B \left(-i\frac{\hbar}{kT} \to (t - t') \right) \tag{A.29}$$

thereby obtaining the stated result.

We note the relationship

$$G_{\mathbf{B}}^A(\mathbf{x}, t; \mathbf{x}', t') = [G_{\mathbf{B}}^R(\mathbf{x}', t'; \mathbf{x}, t)]^* . \tag{A.30}$$

Let us now consider a time-of-flight experiment in the presence of a magnetic field. Suppose that we at a given time, say $t = 0$, ascertain that the particle is in a definite region of space, say, by administering fast opening and closing of some shutter arrangement. We can then ascribe the particle a wave function

$$\psi(\mathbf{x}, t = 0) = \psi_i(\mathbf{x}) \tag{A.31}$$

which is nonvanishing only in the region of size L_s near the shutter (the size being determined by the functioning of the shutter), i.e., for $\mathbf{x} \simeq \mathbf{0}$, as we choose our reference frame to have its origin in that region. Suppose the particle after a time span t is determined to be in a volume element $\Delta \mathbf{x}$ situated at position $\mathbf{x}$. We shall then say that the particle at time $t = 0$ had the momentum in the region

$$\Delta \mathbf{p} \equiv \left(\frac{m}{t}\right)^3 \left| \frac{\frac{\omega_c t}{2}}{\sin\frac{\omega_c t}{2}} \right|^2 \Delta \mathbf{x} \tag{A.32}$$

around the momentum value $\mathbf{p} \equiv (\mathbf{p}_\perp, p_z)$, where

$$p_z \equiv m\frac{z}{t} \tag{A.33}$$

and

$$\mathbf{p}_\perp \equiv \frac{mR_L\omega_c}{|\mathbf{x}_\perp|} \begin{pmatrix} \cos\frac{\omega_c t}{2} & -\sin\frac{\omega_c t}{2} \\ \sin\frac{\omega_c t}{2} & \cos\frac{\omega_c t}{2} \end{pmatrix} \begin{pmatrix} x \\ y \end{pmatrix} \equiv \underline{\underline{R}}^{-1}(t)\,\mathbf{x}_\perp . \tag{A.34}$$

This is the connection according to the classical equation of motion, thereby rendering $|\mathbf{x}_\perp|/R_L = 2|\sin\frac{\omega_c t}{2}|$.

Inverting the last equation we have

$$\begin{pmatrix} x \\ y \end{pmatrix} = \frac{2\sin\frac{\omega_c t}{2}}{m\omega_c} \begin{pmatrix} \cos\frac{\omega_c t}{2} & \sin\frac{\omega_c t}{2} \\ -\sin\frac{\omega_c t}{2} & \cos\frac{\omega_c t}{2} \end{pmatrix} \begin{pmatrix} p_x \\ p_y \end{pmatrix} \tag{A.35}$$

explaining the appearance of the Jacobian in eq.(A.32).

For the probability that the particle at time $t = 0$ has the momentum in question we have

$$P_p(\mathbf{p}, t = 0)\Delta\mathbf{p} \;=\; P_x(\mathbf{x}_\perp = \underline{R}(t)\,\mathbf{p}_\perp, z = p_z t/m, t)\,\Delta\mathbf{x}$$

$$= \left| \int d\mathbf{x}'\, G_{\mathbf{B}}^R(\mathbf{x}, t; \mathbf{x}', 0)\, \psi_i(\mathbf{x}') \right|^2 \Delta\mathbf{x}$$

$$= \left(\frac{m}{2\pi\hbar t} \right)^3 \Delta\mathbf{x}\, \left| \frac{\frac{\omega_c t}{2}}{\sin\frac{\omega_c t}{2}} \right|^2 \left| \int d\mathbf{x}'\, \psi_i(\mathbf{x}')\, e^{-\frac{i}{\hbar} z'\left(p_z - \frac{mz'}{2t} \right)} \right.$$

$$\left. e^{-\frac{i}{\hbar} x'\left(p_x - \frac{m\omega_c \cot\frac{\omega_c t}{2}}{4} x' \right)}\, e^{-\frac{i}{\hbar} y'\left(p_y - \frac{m\omega_c \cot\frac{\omega_c t}{2}}{4} y' \right)} \right|^2$$

$$= \frac{\Delta\mathbf{p}}{(2\pi\hbar)^3}\, \left| \int d\mathbf{x}'\, \psi_i(\mathbf{x}')\, e^{-\frac{i}{\hbar} z'\left(p_z - \frac{mz'}{2t} \right)} \right.$$

$$\left. e^{-\frac{i}{\hbar} x'\left(p_x - \frac{m\omega_c \cot\frac{\omega_c t}{2}}{4} x' \right)}\, e^{-\frac{i}{\hbar} y'\left(p_y - \frac{m\omega_c \cot\frac{\omega_c t}{2}}{4} y' \right)} \right|^2 \tag{A.36}$$

where we have used the result of the previous exercise, and we have utilized eq. (A.34) and eq. (A.33), and in the last equality eq. (A.32).

If we for given $\mathbf{x}_\perp$ choose the strength of the magnetic field and the shutter size to satisfy $L_s m\omega_c \cot\frac{\omega_c t}{2} \ll |p_x|, |p_y|$, and $mL_s/t \ll p_z$, we can neglect the quadratic terms in x', y', and z', and we obtain for the momentum amplitude function the same result as in section 1.1.2:

$$\psi(\mathbf{p}, t = 0) \;=\; \frac{1}{(2\pi\hbar)^{3/2}} \int d\mathbf{x}'\, e^{-\frac{i}{\hbar} \mathbf{x}' \cdot \mathbf{p}}\, \psi_i(\mathbf{x}'). \tag{A.37}$$

We end this appendix by relating the consistency condition for histories to path integral relations. Let us first consider the matrix element, $t' < t_1 < t$,

$$<\mathbf{x}, t| \hat{P}_\Delta(\mathbf{x}_1, t_1) |\mathbf{x}', t'> \;=\; \int_{\Delta\mathbf{x}_1} d\bar{\mathbf{x}}\; <\mathbf{x}, t|\bar{\mathbf{x}}, t_1><\bar{\mathbf{x}}, t_1|\mathbf{x}', t'> \tag{A.38}$$

of the projector

$$\hat{P}_\Delta(\mathbf{x}_1, t_1) \;=\; \int_{\Delta\mathbf{x}_1} d\bar{\mathbf{x}}\, |\bar{\mathbf{x}}, t_1><\bar{\mathbf{x}}, t_1| . \tag{A.39}$$

Writing the two propagators in eq.(A.38) in terms of their path integral expressions, we can rewrite the matrix element as a single but restricted path integral by choosing t_1 as an intermediate time

$$<\mathbf{x}, t|\hat{P}_\Delta(\mathbf{x}_1, t_1)|\mathbf{x}', t'> \quad = \quad \sum_{\mathbf{x}_t}^{(\mathbf{x}_1, t_1)_\Delta} e^{\frac{i}{\hbar} S[\mathbf{x}_t]} \tag{A.40}$$

where the sum is only over paths (all starting at $(\mathbf{x}', t')$ and ending at $(\mathbf{x}, t)$) that at time t_1 passes through the volume $\Delta \mathbf{x}$ around position $\mathbf{x}_1$. Such a bundle of paths is identical to the (coarse-grained) history

$$h_1 : \hat{P}_\Delta(\mathbf{x}_1, t_1)\, \hat{P}(\mathbf{x}, t) \tag{A.41}$$

with the initial state specified by $\hat{P}(\mathbf{x}', t')$, and the associated probability for the history

$$p(h_1) \quad = \quad Tr(\hat{P}_\Delta(\mathbf{x}_1, t_1)\, \hat{P}(\mathbf{x}', t')\, \hat{P}_\Delta(\mathbf{x}_1, t_1)\, \hat{P}(\mathbf{x}, t))$$

$$= \quad \left| \int_{\Delta\mathbf{x}_1} d\bar{\mathbf{x}} \ <\mathbf{x}, t|\bar{\mathbf{x}}, t_1><\bar{\mathbf{x}}, t_1|\mathbf{x}', t'> \right|^2 \tag{A.42}$$

is specified by the absolute square of the matrix element in eq.(A.40). The corresponding proposition is *the particle at position* $\mathbf{x}'$ *at time* t' *was subsequently at time* t_1 *in the volume* $\Delta\mathbf{x}$ *around position* $\mathbf{x}_1$ *and eventually at time* t *at position* $\mathbf{x}$.

Similarly we consider the history h_2 specified by a small volume around $\mathbf{x}_2$ at time t_1 (nonintersecting the volume around $\mathbf{x}_1$ in order for the two histories to be different). The additivity requirement for the consistency of the two histories is then in terms of the path integrals specified by

$$\left| \sum_{\mathbf{x}_t}^{(\mathbf{x}_1, t_1)_\Delta} e^{\frac{i}{\hbar} S[\mathbf{x}_t]} + \sum_{\mathbf{x}_t}^{(\mathbf{x}_2, t_1)_\Delta} e^{\frac{i}{\hbar} S[\mathbf{x}_t]} \right|^2$$

$$= \quad \left| \sum_{\mathbf{x}_t}^{(\mathbf{x}_1, t_1)_\Delta} e^{\frac{i}{\hbar} S[\mathbf{x}_t]} \right|^2 + \left| \sum_{\mathbf{x}_t}^{(\mathbf{x}_2, t_1)_\Delta} e^{\frac{i}{\hbar} S[\mathbf{x}_t]} \right|^2 . \tag{A.43}$$

The consistency condition is thus the requirement that the quantum interference terms between the two alternatives vanish.

Appendix B

Dirac's Delta Function

A sequence of functions, f_n, which becomes progressively peaked at a certain value, say at zero, so that for any smooth function ϕ we have

$$\lim_{n \to \infty} \int_{-\infty}^{\infty} dx \, f_n(x)\phi(x) = \phi(0) \qquad (B.1)$$

is called a Dirac sequence.

A Dirac sequence f_n can be made from any normalized function f

$$\int_{-\infty}^{\infty} dx \, f(x) = 1 \qquad (B.2)$$

by contracting a normalized function and making it progressively peaked

$$f_n(x) = nf(nx) \, . \qquad (B.3)$$

The sequence will then have the weak convergence property, eq.(B.1), for any smooth function ϕ, since for sufficiently large n we have that the integral is completely dominated by the "peakedness" of f_n at the origin, or by change of variable we can run the test function to its value at zero

$$\begin{aligned}
\lim_{n \to \infty} \int_{-\infty}^{\infty} dx \, \phi(x) f_n(x) &= \lim_{n \to \infty} \int_{-\infty}^{\infty} dx \, \phi(\frac{x}{n}) f(x) \\
&= \phi(0) \int_{-\infty}^{\infty} dx \, f(x) = \phi(0) \, .
\end{aligned} \qquad (B.4)$$

We shall cavalierly take the limiting procedure before the integration, and use for the limiting function the notation

$$\lim_{n \to \infty} f_n(x) = \delta(x) \qquad (B.5)$$

defining Dirac's delta function as a *function* having the property

$$\int_{-\infty}^{\infty} dx \, \phi(x) \, \delta(x) = \phi(0) \, . \qquad (B.6)$$

As a first example we can compress the gate function

$$f(x) = \frac{1}{2} \theta(1 - |x|) \qquad (B.7)$$

where θ is the step function

$$\theta(x) \equiv \begin{cases} 1 & \text{for} \quad x > 0 \\ 0 & \text{for} \quad x < 0 \end{cases} \tag{B.8}$$

which will then become a consecutively narrower and higher gate.

We can also have an oscillatory suppression by choosing the normalized function

$$f(x) = \frac{\sin x}{\pi x} \tag{B.9}$$

and thereby the Dirac sequence

$$f_n(x) = \frac{\sin nx}{\pi x} \tag{B.10}$$

since the integral in eq.(B.4) will again be dominated by the peak at zero, and away from zero the rapid oscillations of the function renders no contribution to the integral, and we have the representation of the δ-function

$$\delta(x) = \lim_{n \to \infty} \frac{\sin nx}{\pi x} \,. \tag{B.11}$$

For continuous label one speaks of Dirac families. For example from the normalized function

$$f(x) = \frac{e^{-x^2}}{\sqrt{\pi}} \tag{B.12}$$

we get the Dirac family

$$f_t(x) = \frac{1}{\sqrt{t}} f\left(\frac{x}{\sqrt{t}}\right) \tag{B.13}$$

and obtain the expression for the δ-function

$$\delta(x) = \lim_{t \to 0} f_t(x) = \lim_{t \to 0} \frac{e^{-(x/\sqrt{t})^2}}{\sqrt{t\pi}} \,. \tag{B.14}$$

Another useful representation of the δ-function is

$$\delta(x) = \frac{1}{\pi} \lim_{a \to 0} \frac{a}{x^2 + a^2} \,. \tag{B.15}$$

For the oscillatory function

$$f(x) = \frac{e^{ix^2}}{\sqrt{i\pi}} \tag{B.16}$$

we get

$$\delta(x) = \lim_{t \to 0} f_t(x) = \lim_{t \to 0} \frac{e^{i(x/\sqrt{t})^2}}{\sqrt{i\pi t}} \,. \tag{B.17}$$

We will also encounter functions whose derivatives are delta functions, such as the derivative of the step function, $\theta(x)$, which can be obtained from the Dirac family of smoothed out step functions. For example choosing

$$\theta_a(x) \;=\; \frac{1}{2}\left(1 \,+\, \tanh\frac{x}{a}\right) \tag{B.18}$$

we have

$$\theta(x) \;=\; \lim_{a\to 0}\theta_a(x)\,. \tag{B.19}$$

The derivative of the Fermi function

$$f_0(\epsilon) \equiv \frac{1}{e^{(\epsilon-\epsilon_F)/kT}+1} = \frac{1}{2}\left(1 - \tanh\left(\frac{\epsilon-\epsilon_F}{kT}\right)\right) = (1 - \theta_{kT}(\epsilon-\epsilon_F)) \tag{B.20}$$

will therefore at zero temperature become a delta function

$$\lim_{T\to 0}\frac{\partial f_0}{\partial \epsilon} \;=\; -\,\delta(\epsilon-\epsilon_F)\,. \tag{B.21}$$

A useful formula for splitting integrals into real and imaginary parts is

$$\frac{1}{E\pm i\epsilon} \;=\; \mp i\pi\delta(E) \,+\, \mathcal{P}\frac{1}{E} \tag{B.22}$$

where $\mathcal{P}$ denotes taking the Cauchy principal value of an integral where the integrand appears

$$\mathcal{P}\int_{-\infty}^{\infty}\!dE\,\frac{1}{E}\,f(E) \;\equiv\; \lim_{r\to 0}\left(\int_{-\infty}^{-r}\!dE\,\frac{1}{E}\,f(E) \,+\, \int_{r}^{\infty}\!dE\,\frac{1}{E}\,f(E)\right)\,. \tag{B.23}$$

As usual the equality is validated by evaluating the effect on a test function.[1] We can bypass the singularity at $E = 0$ by deforming the contour along the real axis with a small semicircle as no singularity are crossed. The presence of the infinitesimal imaginary part then becomes irrelevant and we can set ϵ to zero

$$\int_{-\infty}^{\infty}\!dE\,\frac{f(E)}{E+i\epsilon} \;=\; \int_{-\infty}^{-r}\!dE\,\frac{f(E)}{E} \,+\, \int_{r}^{\infty}\!dE\,\frac{f(E)}{E} \,+\, \int_{\pi}^{0}\!d\phi\; ire^{i\phi}\frac{f(re^{i\phi})}{re^{i\phi}}$$

$$=\; \mathcal{P}\int_{-\infty}^{\infty}\!dE\,\frac{f(E)}{E} \,-\, i\pi f(0)\,. \tag{B.24}$$

The contribution from the small half circle of radius r we calculate by parametrizing $E = re^{i\phi}$, which, as we let the radius of the half circle approach zero, precisely gives the δ-function contribution, and the rest of the integral is the principal value.

---

[1]Alternatively we note that

$$\frac{1}{E\pm i\epsilon} \;=\; \mp i\,\frac{\epsilon}{E^2+\epsilon^2} \,+\, \frac{E}{E^2+\epsilon^2}$$

and the first function on the right side acts according to eq.(B.15) as a delta function. The second function on the right side behaves as $1/E$ for $E \gg \epsilon$, and vanishes for $E \to 0$, and integrating it together with smooth functions of E gives the principal value.

The classical limit formula

$$\theta(-t) \lim_{\hbar \to 0} \frac{1}{\hbar} e^{-\frac{i}{\hbar}(E+i\epsilon)t} = \delta(t) \frac{i}{E + i\epsilon} \tag{B.25}$$

is proved by doing a shift of variables, $x \equiv t/\hbar$

$$\lim_{\hbar \to 0} \frac{1}{\hbar} \int_{-\infty}^{0} dt\, \phi(t) e^{-\frac{i}{\hbar}t(E+i\epsilon)} = \lim_{\hbar \to 0} \int_{-\infty}^{0} dx\, \phi(x\hbar) e^{-ix(E+i\epsilon)} = \frac{\phi(0)}{-i(E + i\epsilon)} \tag{B.26}$$

so that we have

$$\theta(-t) \lim_{\hbar \to 0} \frac{1}{\hbar} \cos\left\{\frac{1}{\hbar}Et\right\} = \theta(-t) \lim_{\hbar \to 0} \frac{1}{\hbar} \Re e\, e^{-\frac{i}{\hbar}t(E+i\epsilon)} = \pi\, \delta(t)\, \delta(E) . \tag{B.27}$$

A useful formula is

$$\delta(f(E)) = \sum_i \frac{1}{|f'(E_i)|}\, \delta(E - E_i) \tag{B.28}$$

where the E_i's are the zeros of the function f, $f(E_i) = 0$, and f' denotes the derivative of f.

A familiar example from electrostatics of a δ-function is the charge distribution of a point charge which is the singular limit of an, say spherical, extended charge distribution

$$\rho(\mathbf{x}) = \begin{cases} \rho_0 & |\mathbf{x}| \le a \\ 0 & |\mathbf{x}| > a . \end{cases} \tag{B.29}$$

Letting the size of the spherical charge distribution, a, shrink to zero and the charge density, ρ_0, approach infinity in such a way that $3\pi a^3 \rho_0/4 \to q$ (we assume for definiteness three dimensions), we get for the charge distribution of a point particle with charge q the δ-distribution

$$\rho(\mathbf{x}) = q\, \delta(\mathbf{x}) . \tag{B.30}$$

The electrostatic potential of a point particle

$$\phi(\mathbf{x}) = \frac{q}{4\pi\epsilon_0 |\mathbf{x}|} \tag{B.31}$$

satisfies the Poisson equation

$$\triangle \phi(\mathbf{x}) = -\rho(\mathbf{x})/\epsilon_0 = -\frac{q}{\epsilon_0} \delta(\mathbf{x}) \tag{B.32}$$

in accordance with the formula

$$\triangle \frac{1}{|\mathbf{x}|} = -4\pi\, \delta(\mathbf{x}) . \tag{B.33}$$

Appendix C

Antilinear Operators

An operator $\hat{A}$ on a vector space is said to be antilinear if [1]

$$\hat{A}\left(c_1|\psi_1> \,+ c_2|\psi_2>\right) \,=\, c_1^* \,\hat{A}\,|\psi_1> \,+ c_2^* \,\hat{A}\,|\psi_2> \,. \tag{C.1}$$

The operator $\hat{A}^\dagger$ which for a given antilinear operator $\hat{A}$ satisfies

$$<\psi_1|\hat{A}^\dagger|\psi_2> \;=\; <\psi_2|\hat{A}|\psi_1> \tag{C.2}$$

for arbitrary vectors $|\psi_1>$ and $|\psi_2>$, defines the hermitian conjugate operator to the antilinear operator $\hat{A}$ (note the difference relative to the definition for linear operators, eq.(1.177)). The hermitian conjugate of an antilinear operator is seen to be antilinear. When dealing with antilinear operators it is especially convenient to introduce the following notation: $|A\psi> \equiv \hat{A}|\psi>$,[2] and we can rewrite the defining equation for hermitian conjugation of an antilinear operator as

$$<A^\dagger\psi_2|\psi_1> \;=\; <A\psi_1|\psi_2> \,. \tag{C.3}$$

An antilinear operator $\hat{A}$ is called hermitian if $A^\dagger = \hat{A}$. The product $\hat{A}_1\,\hat{A}_2$ of two antilinear operators $\hat{A}_1$ and $\hat{A}_2$ is a linear operator having the adjoint $(\hat{A}_1\,\hat{A}_2)^\dagger = \hat{A}_2^\dagger\,\hat{A}_1^\dagger$.

An operator $\hat{T}$ is called an antiunitary operator if besides being antilinear, it satisfies the relation $\hat{T}^\dagger \,=\, \hat{T}^{-1}$; i.e.,

$$\hat{T}\,\hat{T}^\dagger \,=\, \hat{I} \,=\, \hat{T}^\dagger\,\hat{T} \tag{C.4}$$

or equivalently

$$<T\psi_f|T\psi_i> \;=\; <\psi_f|\psi_i>^* \tag{C.5}$$

---

[1]In this appendix $\hat{A}$ denotes an antilinear operator, but it should lead to no confusion with our notation in chapter 1 where a physical quantity was represented by an operator $\hat{A}$ which of course was linear (and hermitian).

[2]That Dirac's bra-ket notation thus in this context has to be amended is no surprise since it was constructed to deal exclusively with linear operators. No meaningful status is given to the sequence of symbols $<\psi_1|\hat{A}$ for an antilinear operator $\hat{A}$. In the matrix element $<\psi_2|\hat{A}|\psi_1>$ the antilinear operator always operates to the right on the vector.

for all vectors $|\psi_f>$ and $|\psi_i>$.

The product of a linear operator $\hat{L}$ and an antilinear operator $\hat{A}$ is antilinear and satisfies $(\hat{L}\,\hat{A})^\dagger = \hat{A}^\dagger\,\hat{L}^\dagger$, in particular c-numbers $(c \equiv c\hat{I})$ and antilinear operators can not be interchanged, instead $\hat{A}\,c = c^*\,\hat{A}$.

For any linear operator $\hat{L}$ we have the relation

$$<\psi|\hat{L}|\phi> = <T\phi|\hat{T}\,\hat{L}^\dagger\hat{T}^{-1}|T\psi> \qquad (C.6)$$

provided that $\hat{T}$ is an antiunitary operator. This is proved by the following consideration. Introducing the notation $|\phi_f> \equiv \hat{L}^\dagger\,|\psi_f>$, we have $<\psi_f|\,\hat{L} = <\phi_f|$, and we obtain

$$<\psi_f|\hat{L}|\psi_i> = <\phi_f|\psi_i> = <T\psi_i|T\phi_f> = <T\psi_i|\hat{T}|\phi_f>$$

$$= <T\psi_i|\hat{T}\,\hat{L}^\dagger|\psi_f> = <T\psi_i|\hat{T}\,\hat{L}^\dagger\,\hat{T}^{-1}|T\psi_f> \qquad (C.7)$$

where the second equality sign follows from $\hat{T}$ being antiunitary.

An important example of an antiunitary operator is the complex conjugation operator which we define as follows. First we choose a basis in the state space, say $\{|a>\}_a$, and we then define on an arbitrary state the operator $\hat{K}_{(a)}$

$$\hat{K}_{(a)}\,|\psi> = \sum_a <\psi|a>^*\,|a> . \qquad (C.8)$$

The definition of the complex conjugation operator $\hat{K}_{(a)}$ implies the following phase choice

$$\hat{K}_{(a)}\,|a> = |a> . \qquad (C.9)$$

The operator $\hat{K}_{(a)}$ is clearly antilinear and satisfies $\hat{K}_{(a)}^2 = \hat{I}$.

The matrix elements of $\hat{K}_{(a)}$ in the a-representation

$$<a|\hat{K}_{(a)}\,|a'> = <a|a'> = \delta_{a,a'} \qquad (C.10)$$

is a real and symmetric matrix, and from

$$<a|\hat{K}_{(a)}|a'> = <a'|\hat{K}_{(a)}|a> = <a|\hat{K}_{(a)}^\dagger|a'> \qquad (C.11)$$

and using the fact that if two antilinear operators have the same matrix elements in one representation they are identical,[3] we find that $\hat{K}_{(a)} = \hat{K}_{(a)}^\dagger$, i.e., $\hat{K}_{(a)}$ is hermitian and antiunitary.

For the complex conjugation operator defined with respect to another basis we have

$$\hat{K}_{(b)}\,|b> = |b> \qquad (C.12)$$

---

[3]Note that a linear and an antilinear operator can have the same matrix elements in one representation without being identical operators.

whereas

$$\hat{K}_{(a)} |b> \ = \ \sum_a <a|b>^* \ |a> \ .$$

(C.13)

From

$$<b|\hat{K}_{(a)}|b'> \ = \ \sum_a <b|a> <a|b'>^*$$

(C.14)

and

$$<b|\hat{K}_{(b)}|b'> \ = \ \delta_{b,b'}$$

(C.15)

we conclude that only if the transformation function, $<a|b>$, is real are the two complex conjugation operators, defined with respect to different basis's, identical. Even a phase transformation leaves the complex conjugation operators different.

Appendix D

Calculation of Φ

In this appendix we calculate the integral in eq.(11.9)

$$\Phi_{\alpha,\beta}(\mathbf{x} - \mathbf{x}') \equiv \int d\mathbf{r} \int d\mathbf{r}' \ G_{\epsilon_F}^R(\mathbf{x} - \mathbf{r}) G_{\epsilon_F}^R(\mathbf{r}' - \mathbf{x}') \overset{\leftrightarrow}{\nabla}_{x_\alpha} \overset{\leftrightarrow}{\nabla}_{x'_\beta} \ G_{\epsilon_F}^A(\mathbf{x}' - \mathbf{r}) G_{\epsilon_F}^A(\mathbf{r}' - \mathbf{x})$$

$$\equiv \frac{2}{4\hbar^2} \left(F(\mathbf{x} - \mathbf{x}') F_{\alpha,\beta}(\mathbf{x} - \mathbf{x}') - F_\alpha(\mathbf{x} - \mathbf{x}') F_\beta(\mathbf{x} - \mathbf{x}') \right) \tag{D.1}$$

to the desired degree of accuracy.

Introducing the Fourier transform of the propagators we have

$$F(\mathbf{x} - \mathbf{x}') \equiv \int \frac{d\mathbf{p}}{(2\pi\hbar)^3} \ e^{\frac{i}{\hbar}\mathbf{p}\cdot(\mathbf{x}-\mathbf{x}')} \ G_{\epsilon_F}^R(\mathbf{p}) G_{\epsilon_F}^A(\mathbf{p}) \tag{D.2}$$

$$F_\alpha(\mathbf{x} - \mathbf{x}') \equiv \int \frac{d\mathbf{p}'}{(2\pi\hbar)^3} \ e^{\frac{i}{\hbar}\mathbf{p}'\cdot(\mathbf{x}-\mathbf{x}')} \ p_\alpha \ G_{\epsilon_F}^R(\mathbf{p}') G_{\epsilon_F}^A(\mathbf{p}') \tag{D.3}$$

$$F_{\alpha,\beta}(\mathbf{x} - \mathbf{x}') \equiv \int \frac{d\mathbf{p}'}{(2\pi\hbar)^3} \ e^{\frac{i}{\hbar}\mathbf{p}'\cdot(\mathbf{x}-\mathbf{x}')} \ p_\alpha \ p_\beta \ G_{\epsilon_F}^R(\mathbf{p}') G_{\epsilon_F}^A(\mathbf{p}') \ . \tag{D.4}$$

We then observe the effect of angular averaging to obtain

$$F(\mathbf{a}) \equiv \int \frac{d\hat{\mathbf{p}}}{4\pi} \ e^{\frac{i}{\hbar}\mathbf{p}\cdot\mathbf{a}} \ = \ \frac{\hbar \sin \frac{pa}{\hbar}}{pa} \tag{D.5}$$

and

$$F_\alpha(\mathbf{a}) \equiv \int \frac{d\hat{\mathbf{p}}}{4\pi} \ p_\alpha \ e^{\frac{i}{\hbar}\mathbf{p}\cdot\mathbf{a}} = -i\hbar \frac{\partial}{\partial a_\alpha} \int \frac{d\hat{\mathbf{p}}}{4\pi} \ e^{\frac{i}{\hbar}\mathbf{p}\cdot\mathbf{a}}$$

$$= \ \frac{-i\hbar^2}{pa^2} \left(\frac{p}{\hbar} a_\alpha \cos \frac{pa}{\hbar} - \frac{a_\alpha}{a} \sin \frac{pa}{\hbar} \right) \tag{D.6}$$

and

$$F_{\alpha,\beta}(\mathbf{a}) \equiv \int \frac{d\hat{\mathbf{p}}}{4\pi}\, p_\alpha p_\beta\, e^{\frac{i}{\hbar}\mathbf{p}'\cdot\mathbf{a}} = (-i\hbar)^2 \frac{\partial^2}{\partial a_\beta \partial a_\alpha} \int \frac{d\hat{\mathbf{p}}}{4\pi}\, e^{\frac{i}{\hbar}\mathbf{p}'\cdot\mathbf{a}}$$

$$= (-i\hbar)\frac{\partial}{\partial a_\beta}\left(\frac{-i\hbar^2}{pa^2}\left(\frac{p}{\hbar}a_\alpha \cos\frac{pa}{\hbar} - \frac{a_\alpha}{a}\sin\frac{pa}{\hbar}\right)\right)$$

$$= \frac{-\hbar^3}{p}\left\{ \delta_{\alpha,\beta}\left(\frac{p}{\hbar a^2}\cos\frac{pa}{\hbar} - \frac{\sin\frac{pa}{\hbar}}{a^3}\right)\right.$$

$$\left. + \quad a_\alpha a_\beta \left(\frac{3\sin\frac{pa}{\hbar}}{a^5} - \frac{3p\cos\frac{pa}{\hbar}}{\hbar a^4} - \left(\frac{p}{\hbar}\right)^2\frac{\sin\frac{pa}{\hbar}}{a^3}\right)\right\} \ . \tag{D.7}$$

We can perform the integration over the length of the momentum using the residue theorem. Introducing $k \equiv p/\hbar = k_F\sqrt{1 + \xi/\epsilon_F}$, we get to order $\hbar/\epsilon_F\tau$

$$\int\limits_{-\epsilon_F}^{\infty} d\xi\, N_0(\xi) G_{\epsilon_F}^R(\mathbf{p}) G_{\epsilon_F}^A(\mathbf{p}) \left(\begin{array}{c} \cos k|\mathbf{x}-\mathbf{x}'| \\ \sin k|\mathbf{x}-\mathbf{x}'| \end{array}\right)$$

$$= \int\limits_{-\infty}^{\infty} d\xi\, N_0(\xi) \frac{1}{\xi - \frac{i\hbar}{2\tau}}\frac{1}{\xi + \frac{i\hbar}{2\tau}}\left(\begin{array}{c} \cos k|\mathbf{x}-\mathbf{x}'| \\ \sin k|\mathbf{x}-\mathbf{x}'| \end{array}\right)$$

$$= \frac{2\pi N_0\tau}{\hbar} e^{-\frac{|\mathbf{x}-\mathbf{x}'|}{2l}}\left(\begin{array}{c} \cos k_F|\mathbf{x}-\mathbf{x}'| \\ \sin k_F|\mathbf{x}-\mathbf{x}'| \end{array}\right) \tag{D.8}$$

and thereby

$$\int\frac{d\mathbf{p}}{(2\pi\hbar)^3}\, G_{\epsilon_F}^R(\mathbf{p})\, G_{\epsilon_F}^A(\mathbf{p}) = \frac{2\pi N_0\tau}{\hbar}\frac{\sin k_F|\mathbf{x}-\mathbf{x}'|}{k_F|\mathbf{x}-\mathbf{x}'|} e^{-\frac{|\mathbf{x}-\mathbf{x}'|}{2l}} \tag{D.9}$$

and

$$\int\frac{d\mathbf{p}}{(2\pi\hbar)^3}\, p_\alpha\, G_{\epsilon_F}^R(\mathbf{p})\, G_{\epsilon_F}^A(\mathbf{p}) = -2\pi i N_0\tau \frac{(\mathbf{x}-\mathbf{x}')_\alpha}{|\mathbf{x}-\mathbf{x}'|^2} e^{-\frac{|\mathbf{x}-\mathbf{x}'|}{2l}}$$

$$\left(\cos k_F|\mathbf{x}-\mathbf{x}'| - \frac{\sin k_F|\mathbf{x}-\mathbf{x}'|}{k_F|\mathbf{x}-\mathbf{x}'|}\right) \tag{D.10}$$

and

$$\int\frac{d\mathbf{p}}{(2\pi\hbar)^3}\, p_\alpha p_\beta\, G_{\epsilon_F}^R(\mathbf{p})\, G_{\epsilon_F}^A(\mathbf{p}) = -2\pi\hbar N_0\tau e^{-\frac{|\mathbf{x}-\mathbf{x}'|}{2l}}$$

$$\left\{\cos k_F|\mathbf{x}-\mathbf{x}'|\left(\frac{\delta_{\alpha\beta}}{|\mathbf{x}-\mathbf{x}'|^2} - \frac{3(\mathbf{x}-\mathbf{x}')_\alpha(\mathbf{x}-\mathbf{x}')_\beta}{|\mathbf{x}-\mathbf{x}'|^4}\right)\right.$$

$$- \sin k_F|\mathbf{x}-\mathbf{x}'|\left(\frac{\delta_{\alpha\beta}}{k_F|\mathbf{x}-\mathbf{x}'|^3} - (\mathbf{x}-\mathbf{x}')_\alpha(\mathbf{x}-\mathbf{x}')_\beta\right.$$

$$\left.\left.\left(\frac{3}{k_F|\mathbf{x}-\mathbf{x}'|^5} + \frac{k_F}{|\mathbf{x}-\mathbf{x}'|^3}\right)\right)\right\} . \tag{D.11}$$

We then obtain

$$\Phi_{\alpha,\beta}(\mathbf{x}-\mathbf{x}') = -\frac{(2\pi N_0\tau)^2}{2\hbar^2}\frac{e^{-|\mathbf{x}-\mathbf{x}'|/l}}{|\mathbf{x}-\mathbf{x}'|^2}\left\{\delta_{\alpha\beta}\left(\frac{\cos k_F|\mathbf{x}-\mathbf{x}'|}{k_F|\mathbf{x}-\mathbf{x}'|} - \frac{\sin k_F|\mathbf{x}-\mathbf{x}'|}{k_F^2|\mathbf{x}-\mathbf{x}'|^2}\right)\right.$$

$$+ \frac{(\mathbf{x}-\mathbf{x}')_\alpha(\mathbf{x}-\mathbf{x}')_\beta}{|\mathbf{x}-\mathbf{x}'|^2}\left(\cos^2 k_F|\mathbf{x}-\mathbf{x}'| - \sin k_F|\mathbf{x}-\mathbf{x}'|\right.$$

$$+ \frac{\sin^2 k_F|\mathbf{x}-\mathbf{x}'| + 3\sin k_F|\mathbf{x}-\mathbf{x}'|}{k_F^2|\mathbf{x}-\mathbf{x}'|^2}$$

$$\left.\left. - \frac{\sin 2k_F|\mathbf{x}-\mathbf{x}'| + 3\cos k_F|\mathbf{x}-\mathbf{x}'|}{k_F|\mathbf{x}-\mathbf{x}'|}\right)\right\} . \tag{D.12}$$

Since $\Phi_{\alpha,\beta}(\mathbf{x}-\mathbf{x}')$ is local, scale l, and appears in connection with the Cooperon, which is a smooth function on the scale of the mean free path, it is effectively a delta function. To determine the strength we note

$$\int d\mathbf{x}\,\Phi_{\alpha,\beta}(\mathbf{x}) = 4\pi\int_0^\infty dx\,x^2\,\Phi_{\alpha,\beta}(x)$$

$$= -\frac{(2\pi N_0\tau)^2}{6\hbar^2 k_F}\delta_{\alpha\beta}\left\{\frac{\frac{1}{k_F l} + 2k_F l}{\left(\frac{1}{k_F l}\right)^2 + 4} - \frac{1}{\left(\frac{1}{k_F l}\right)^2 + 1}\right.$$

$$\left. - \frac{1}{4k_F l}\ln\left(1 + \frac{4}{\left(\frac{1}{k_F l}\right)^2}\right)\right\} \tag{D.13}$$

To lowest order in $\hbar/p_F l$ only the $\cos^2$-term contributes and we have the stated result in eq.(11.10)

$$\Phi_{\alpha,\beta}(\mathbf{x}-\mathbf{x}') = -\frac{(2\pi N_0\tau)^2}{2\hbar^2}\frac{(\mathbf{x}-\mathbf{x}')_\alpha(\mathbf{x}-\mathbf{x}')_\beta}{|\mathbf{x}-\mathbf{x}'|^4}e^{-|\mathbf{x}-\mathbf{x}'|/l}\cos^2 k_F|\mathbf{x}-\mathbf{x}'| . \tag{D.14}$$

Appendix E

Static Correlation Function

In section 7.3.3 we derived that the longitudinal part of the static paramagnetic current correlation function equals the diamagnetic term

$$K_{\alpha\beta}(\mathbf{x}, \mathbf{x}', 0) = \frac{e^2 \rho_0(\mathbf{x}, \mathbf{x})}{m} \delta(\mathbf{x} - \mathbf{x}') \delta_{\alpha\beta} \tag{E.1}$$

using the fundamental property of gauge invariance. Here we first establish this relation using the equation of motion, and then relate the property to the dissipative property of a system.

Assuming the vector potential to be rotation free, the identity is proved by showing that for any well-behaved function Λ we have

$$-\frac{m}{\hbar^2} \nabla_{x_\alpha} \Lambda(\mathbf{x}) = \sum_\beta \int d\mathbf{x}' \left[\nabla_{x'_\beta} \Lambda(\mathbf{x}') \right] \int_{-\infty}^{\infty} dE_1 \int_{-\infty}^{\infty} dE_2$$

$$\frac{f_0(E_2) - f_0(E_1)}{E_1 - E_2} A(\mathbf{x}, \mathbf{x}'; E_1) \overset{\leftrightarrow}{\nabla}_{x_\alpha} \overset{\leftrightarrow}{\nabla}_{x'_\beta} A(\mathbf{x}', \mathbf{x}; E_2) . \tag{E.2}$$

In order to establish eq.(E.2) we do a partial integration, assuming $\Lambda(\mathbf{x} = \pm\infty) = 0$ or alternatively that K has a finite range so that the boundary term vanishes, and obtain according to eq.(7.96)[1]

$$\sum_\beta \int d\mathbf{x}' \, K_{\alpha\beta}(\mathbf{x}, \mathbf{x}', 0) \nabla_{x'_\beta} \Lambda(\mathbf{x}', \omega) = \int d\mathbf{x}' \Lambda(\mathbf{x}') \int_{-\infty}^{\infty} dE_1 \int_{-\infty}^{\infty} dE_2 \frac{f_0(E_2) - f_0(E_1)}{E_1 - E_2}$$

$$\left(\triangle_{\mathbf{x}'} A(\mathbf{x}, \mathbf{x}'; E_1) \right] \overset{\leftrightarrow}{\nabla}_{x_\alpha} A(\mathbf{x}', \mathbf{x}; E_2)$$

$$- A(\mathbf{x}, \mathbf{x}'; E_1) \overset{\leftrightarrow}{\nabla}_{x_\alpha} \triangle_{\mathbf{x}'} A(\mathbf{x}', \mathbf{x}; E_2) \right) . \tag{E.3}$$

---

[1] Since $(f_0(E_2) - f_0(E_1)) \delta(E_1 - E_2) = 0$, the imaginary infinitesimal in the denominator of eq.(7.96) can be dropped.

The equations of motion for the spectral weight (in the presence of a potential V)

$$\left\{ E + \frac{\hbar^2}{2m} \frac{\partial^2}{\partial \mathbf{x}^2} - V(\mathbf{x}) \right\} A(\mathbf{x}, \mathbf{x}'; E) = 0 \qquad (E.4)$$

and

$$\left\{ E + \frac{\hbar^2}{2m} \frac{\partial^2}{\partial \mathbf{x}'^2} - V(\mathbf{x}') \right\} A(\mathbf{x}, \mathbf{x}'; E) = 0 \qquad (E.5)$$

follows from eq.(2.154), and the Schrödinger equation.

Using the completeness property of the spectral function, eq.(2.158), and the relationship between the density matrix and the combination of the Fermi function and the spectral weight, eq.(7.49), we obtain that eq.(E.3) equals

$$\int d\mathbf{x}' \, \Lambda(\mathbf{x}')[\delta(\mathbf{x} - \mathbf{x}') \overset{\leftrightarrow}{\nabla}_{x_\alpha} \rho_0(\mathbf{x}', \mathbf{x}) - \rho_0(\mathbf{x}, \mathbf{x}') \overset{\leftrightarrow}{\nabla}_{x'_\alpha} \delta(\mathbf{x}' - \mathbf{x})] \qquad (E.6)$$

which we use to get the relation

$$\sum_\beta \int d\mathbf{x}' \, [\nabla_{x'_\beta} \Lambda(\mathbf{x}')] \int_{-\infty}^\infty dE_1 \int_{-\infty}^\infty dE_2 \, \frac{f_0(E_2) - f_0(E_1)}{E_1 - E_2} A(\mathbf{x}, \mathbf{x}'; E_1) \overset{\leftrightarrow}{\nabla}_{x_\alpha} \overset{\leftrightarrow}{\nabla}_{x'_\beta} A(\mathbf{x}', \mathbf{x}; E_2)$$

$$= -\frac{m}{\hbar^2} \rho_0(\mathbf{x}, \mathbf{x}) \nabla_{x_\alpha} \Lambda(\mathbf{x}) \qquad (E.7)$$

and thereby eq.(E.1).

We shall now exploit causality and a systems dissipative character to demonstrate that the static current correlation function equals the diamagnetic term. The conductivity tensor is analytic in the upper half plane, and using Cauchy's theorem for the function $\omega \sigma_{\alpha\beta}(\mathbf{x}, \mathbf{x}', \omega)$ we have

$$i\omega \sigma_{\alpha\beta}(\mathbf{x}, \mathbf{x}', \omega) = \mathcal{P} \int_{-\infty}^\infty \frac{d\omega'}{\pi} \frac{\omega' \sigma_{\alpha\beta}(\mathbf{x}, \mathbf{x}', \omega')}{\omega' - \omega} - \frac{e^2 \rho_0(\mathbf{x}, \mathbf{x})}{m} \delta_{\alpha\beta} \delta(\mathbf{x} - \mathbf{x}') \quad (E.8)$$

where we have used that at sufficiently high frequency any system becomes free; i.e., we have

$$\lim_{\omega \to \infty} \sigma_{\alpha\beta}(\mathbf{x}, \mathbf{x}', \omega) = -\frac{e^2 \rho_0(\mathbf{x}, \mathbf{x})}{i\omega m} \delta_{\alpha\beta} \delta(\mathbf{x} - \mathbf{x}') . \qquad (E.9)$$

For a dissipative system $\Re e \, \sigma_{\alpha\beta}(\mathbf{x}, \mathbf{x}', \omega \to 0)$ does not diverge, and we obtain the desired identity eq.(E.1)

$$\pi K_{\alpha\beta}(\mathbf{x}, \mathbf{x}', \omega = 0) = \int_{-\infty}^\infty d\omega \, \Re e \, \sigma_{\alpha\beta}(\mathbf{x}, \mathbf{x}', \omega) = \frac{\pi e^2 \rho_0(\mathbf{x}, \mathbf{x})}{m} \delta_{\alpha\beta} \delta(\mathbf{x} - \mathbf{x}') . \quad (E.10)$$

Bibliography

[1] The original papers of Feynman are reprinted in (ed.) J. Schwinger, *Selected Papers on Quantum Electrodynamics*, Dover Publications, New York, 1958.

[2] J. H. Van Vleck, Proc. Natl. Acad. Sci. USA **14**, 178 (1928).

[3] M. C. Gutzwiller, *Chaos in Classical and Quantum Mechanics*, Springer-Verlag, 1990.

[4] M. V. Berry and K. E. Mount, Rep. Prog. Phys. **35**, 315 (1972).

[5] R. P. Feynman and A. R. Hibbs, *Path Integrals and Quantum Mechanics*, McGraw-Hill, 1965.

[6] G. Baym, *Lectures on Quantum Mechanics*, Addison-Wesley, 1990.

[7] J. Schwinger, *Quantum Kinematics and Dynamics*, Addison-Wesley, 1991.

[8] A. Aspect, P. Grangier, and G. Roger, Phys. Rev. Lett. **47**, 460 (1981), and A. Aspect, J. Dalibard, and G. Roger, Phys. Rev. Lett. **49**, 1804 (1982).

[9] P. A. M. Dirac, *The Principles of Quantum Mechanics*, Oxford University Press, 1930 (fourth edition 1958).

[10] J. von Neumann, *Mathematical Foundation of Quantum Mechanics*, Princeton University Press, 1955.

[11] H. Weyl, *The Theory of Groups and Quantum Mechanics*, E. P. Dutton and Company, 1931.

[12] K. Gottfried, *Quantum Mechanics*: Vol. I, W. A. Benjamin, 1966.

[13] L. D. Landau and E. M. Lifshitz, *Quantum Mechanics*, Pergamon Press, 1977.

[14] J. S. Bell, Physics World (August 1990), pp. 33-40.

[15] R. B. Griffiths, J. Stat. Phys. **36**, 219 (1984). Phys. Rev. Lett. **70**, 2201 (1993).

[16] R. Omnès, *The Interpretation of Quantum Mechanics*, Princeton University Press, 1994.

[17] N. F. Mott, Proc. Roy. Soc. London **126**, 79 (1929). Reprinted in J. A. Wheeler and W. H. Zurek (eds.), *Quantum Theory and Measurement*, Princeton University Press, 1983.

[18] M. Gell-Mann and J. B. Hartle, in *Complexity, Entropy, and the Physics of Information*, W. H. Zurek (ed.), Addison-Wesley, Redwood City, Calif. (1990).

[19] S. F. Edwards, Philos. Mag. **3**, 1020 (1958).

[20] A. A. Abrikosov, L. P. Gor'kov, and I. E. Dzyaloshinski, *Quantum Field Theoretical Methods in Statistical Physics*. Pergamon, New York, 1965.

[21] G. Iche and P. Nozières, Physica **A 91**, 485 (1978).

[22] L. W. Boltzmann, Ber. Wien. Akad. **66**, 275 (1872), and *Vorlesungen über Gastheorie* (Barth, Leipzig, 1896). English translation: *Lectures on Gas Theory* (University of California Press, Berkeley, 1964).

[23] E. Wigner, Phys. Rev. **40**, 749 (1932).

[24] N. W. Ashcroft and N. D. Mermin, *Solid State Physics*, Holt, Rinehart and Winston, 1976.

[25] A. A. Abrikosov, *Fundamentals of the Theory of Metals*, North-Holland, 1988.

[26] K. Möhring and U. Smilansky, Nucl. Phys. **A338**, 227 (1980).

[27] A. O. Caldeira and A. J. Leggett, Ann. Phys. **149**, 374 (1983); **153**, 445 (1984) (E).

[28] J. Schwinger, J. Math. Phys. **2**, 407 (1961).

[29] R. Mills, *Propagators for Many-Particle Systems*, Gordon and Breach, New York, 1969.

[30] I. B. Levinson, Zh. Eksp. Teor. Fiz. **57**, 660 (1970) [Sov. Phys. JETP **30**, 362 (1970)].

[31] S. W. Lovesey, *Theory of Neutron Scattering from Condensed Matter*, Clarendon Press, Oxford, 1984.

[32] S. V. Maleev and B. P. Toperverg, Zh. Eksp. Teor. Fiz. **69**, 1440 (1975) [Sov. Phys. JETP **42**, 734 (1976)].

[33] D. Vollhardt and P. Wölfle, *Self-Consistent Theory of Anderson Localization*, in *Electronic Phase Transitions*, W. Hanke and Yu. V. Kopaev (eds.), Elsevier Science Publishers B.V., 1992.

[34] P. W. Anderson, Phys. Rev. **102**, 1008 (1958).

[35] F. Wegner, Z. Phys. **B25**, 327 (1976).

[36] D. J. Thouless in *Ill Condensed Matter*, R. Balian, R. Maynard, and G. Toulouse (eds.), Les Houches, Session XXXI, North-Holland, 1987.

[37] E. Abrahams, P. W. Anderson, D. C. Licciardello, and T. V. Ramakrishnan, Phys. Rev. Lett. **42**, 673 (1979). Reprinted in P. W. Anderson, *Basic Notions of Condensed Matter Physics*, Benjamin-Cummings, 1984.

[38] N. F. Mott and W. D. Twose, Adv. Phys. **10**, 107 (1961). R. Landauer, Philos. Mag. **21**, 863 (1970). V. L. Berezinskii, Zh. Eksp. Teor. Fiz. **65**, 1251 (1973) [Sov. Phys. JETP **38**, 620 (1974)]. A. A. Abrikosov and I. A. Ryzhkin, Adv. Phys. **27**, 147 (1978).

[39] N. F. Mott, in *Electronics and structural properties of amorphous semiconductors*, P. G. Le Comber and J. Mort (eds.), Acaademic Press, London, 1973.

[40] P. Wölfle and R. N. Bhatt, Phys. Rev. **B30**, 3542 (1984). D. Rainer and G. Bergmann, Phys. Rev. **B32**, 3522 (1985). S.-R. Eric Yang and J. Rammer, Phys. Rev. **B53**, 9568 (1996).

[41] J.S. Langer and T. Neal, Phys. Rev. Lett. **16**, 984 (1966).

[42] K. M. Watson, J. Math. Phys. **10**, 688 (1969).

[43] A. I. Larkin and D. E. Khmel'nitskii, Usp. Fyz. Nauk **136**, 536 (1982) [Sov. Phys. Usp. **25**, 185 (1982)]. D. E. Khmelnitskii, Physica B **126**, 235 (1984).

[44] B. L. Al'tshuler, V. E. Kravtsov, and I. V. Lerner in *Mesoscopic Phenomena in Solids*, B. L. Al'tshuler, P. A. Lee, and R. A. Webb (eds.), Elsevier Science Publishers B.V., North-Holland, 1991.

[45] For a review on interaction effects, see for example, D. Belitz and T. R. Kirkpatrick, Rev. Mod. Phys. **66**, 261 (1994).

[46] L. P. Gor'kov, A. I. Larkin, and D. E. Khmel'nitskii, Pis'ma Zh. Eksp. Teor. Fiz. **30**, 1251 (1979) [Sov. Phys. JETP Lett. **30**, 228 (1979)].

[47] M. V. Sadovskii, Sov. Sci. Rev. A. Phys. Vol. 7, 1 (1986).

[48] V. L. Berezinskii and L. P. Gor'kov, Zh. Eksp. Teor. Fiz. **77**, 2498 (1979) [Sov. Phys. JETP **50**, 1209 (1979)].

[49] F. Wegner, Nucl. Phys. **B316**, 663 (1989).

[50] D. Pines, *Elementary Excitations in Solids*, W. A. Benjamin, 1964. D. Pines and P. Nozières, *The theory of quantum liquids*, Vol. I, W. A. Benjamin, 1966.

[51] J. M. Luttinger, Phys. Rev. **119**, 1153 (1960).

[52] B. L. Al'tshuler and A. G. Aronov in *Electron-Electron Interactions in Disordered Systems*, A. L. Efros and M. Pollak (eds.), Elsevier Science Publishers B.V., North-Holland, 1985.

[53] A. Schmid, Z. Physik **271**, 251 (1974).

[54] B. L. Al'tshuler and A. G. Aronov, Zh. Eksp. Teor. Fiz. **75**, 1610 (1978) [Sov. Phys. JETP **48**, 812 (1978)], Pis'ma Zh. Eksp. Teor. Fiz. **30**, 514 (1979) [Sov. Phys. JETP Lett. **30**, 482 (1979)].

[55] S. Chakravarty and A. Schmid, Phys. Rep. **140**, 193 (1986).

[56] J. Rammer and A. Schmid, Contributed paper to the International Conference on Localization, Interaction and Transport Phenomena, Braunschweig, Abstracts p. 155, 1984. J. Rammer and A. Schmid, Phys. Rev. **B34**, 1352 (1986).

[57] K. S. Il'in, N. G. Ptitsina, A.V. Sergeev, G. N. Gol'tsman, E. M. Gershenzon, B. S. Karasik, E. V. Pechen and S. I. Krasnosvobodtsev, Physical Review **B 57**, 15623 (1998).

[58] B. L. Al'tshuler, A. G. Aronov, and D. E. Khmel'nitskii, J. Phys. **C15**, 7367 (1982).

[59] G. Bergmann, Phys. Rep. **107**, 1 (1984).

[60] B. L. Al'tshuler, A. G. Aronov, M. E. Gershenson, and Yu. V. Sharvin, Sov. Sci. Rev. A. Phys. Vol. 9, 223 (1987), I. M. Khalatnikov (ed.).

[61] J. Rammer, A. L. Shelankov, and A. Schmid, Phys. Rev. Lett. **60**, 1985 (1988) (C).

[62] A. Schmid, Z. Phys. **259**, 421 (1973).

[63] G. J. Dolan and D. D. Osheroff, Phys. Rev. Lett. **43**, 721 (1979).

[64] J. Rammer and A. L. Shelankov, Phys. Rev. **B36**, 3135 (1987).

[65] I. S. Gradstheyn and I. M. Ryzhik, *Table of Integrals, Series and Products*. Academic Press (1980).

[66] A. Kawabate, J. Phys. Soc. Japan **53**, 3540 (1984).

[67] H.-P. Wittmann and A. Schmid, J. Low Temp. Phys. **69**, 131 (1987).

[68] A. Kawabate, Solid State Commun. **34**, 431 (1980), J. Phys. Soc. Japan **49**, 628 (1980).

[69] B. L. Al'tshuler, A. G. Aronov, and B. Z. Spivak, Pis'ma Zh. Eksp. Teor. Fiz. **33**, 101 (1981) [Sov. Phys. JETP Lett. **33**, 94 (1981)].

[70] J. Rammer, Ph. D. thesis, University of Copenhagen, 1985. Published in part.

[71] A. G. Aronov and Yu. V. Sharvin, Rev. Mod. Phys. **59**, 755 (1987).

[72] S. Hikami, A. I. Larkin, and N. Nagaoka, Progr. Theor. Phys. **63**, 707 (1980).

[73] B. L. Al'tshuler, A. G. Aronov, A. I. Larkin, and D. E. Khmel'nitskii, Zh. Eksp. Teor. Fiz. **81**, 768 (1981) [Sov. Phys. JETP **54**, 411 (1981)].

[74] F. Wegner, Nucl. Phys. **B 316**, 663 (1989).

[75] R. A. Serota, S. Feng, C. Kane, and P. A. Lee, Phys. Rev. **B36**, 5031 (1987).

[76] B. L. Al'tshuler and B. I. Shklovskii, Zh. Eksp. Teor. Fiz. **91**, 220 (1986) [Sov. Phys. JETP **64**, 127 (1986)].

[77] B. L. Al'tshuler and D. E. Khmel'nitskii, Pis'ma Zh. Eksp. Teor. Fiz. **42**, 291 (1985) [Sov. Phys. JETP Lett. **42**, 359 (1985)].

[78] P. A. Lee, A. D. Stone, and H. Fukuyama, Phys. Rev. **B35**, 1039 (1987).

[79] B. L. Al'tshuler, Pis'ma Zh. Eksp. Teor. Fiz. **41**, 530 (1985) [Sov. Phys. JETP Lett. **41**, 648 (1985)].

[80] P. A. Lee and A. D. Stone, Phys. Rev. Lett. **55**, 1622 (1985).

[81] B. L. Al'tshuler, V. E. Kravtsov, and I. V. Lerner, Pis'ma Zh. Eksp. Teor. Fiz. **43**, 342 (1986) [Sov. Phys. JETP Lett. **43**, 441 (1986)].

[82] B. L. Al'tshuler and B. Z. Spivak, Pis'ma Zh. Eksp. Teor. Fiz. **42**, 363 (1985) [Sov. Phys. JETP Lett. **42**, 447 (1985)].

[83] S. Feng, P. A. Lee, and A. D. Stone, Phys. Rev. Lett. **56**, 1960 (1986); **56**, 2772 (E).

[84] S. Washburn and A. Webb, Adv. Phys. **35**, 375 (1986).

[85] P. Shen (ed.), *Scattering and Localization of Classical Waves in Random Media*, World Scientific, 1989.

[86] P. A. Lee and T. V. Ramakrishnan, Rev. Mod. Phys. **57**, 287 (1985).

[87] J. Rammer, Rev. Mod. Phys. **63**, 781 (1991).

[88] B. L. Al'tshuler, A. G. Aronov, D. E. Khmel'nitskii, and A. I. Larkin, *Coherent Effects in Disordered Conductors*, in I. M. Lifshitz (ed.), *Quantum Theory of Solids*, MIR, Moscow (1982).

[89] B. L. Al'tshuler, P. A. Lee and R. A. Webb (eds.), *Mesoscopic Phenomena in Solids*, Elsevier Science Publishers B.V., North-Holland, 1991.

[90] P. A. M. Dirac, Physik. Zeits. Sowjetunion **3**, 64 (1933). Reprinted in [1].

Index

Printed and bound by CPI Group (UK) Ltd, Croydon, CR0 4YY

17/10/2024

01775683-0016